STATISTICS AND PROBABILITY:
ESSAYS IN HONOR OF C. R. RAO

Statistics and Probability: Essays in Honor of C. R. Rao

Edited by

G. KALLIANPUR
P. R. KRISHNAIAH
J. K. GHOSH

1982

NORTH-HOLLAND PUBLISHING COMPANY—AMSTERDAM · NEW YORK · OXFORD

ISBN 0 444 86130 0

Published by:
NORTH-HOLLAND PUBLISHING COMPANY
AMSTERDAM · NEW YORK · OXFORD

Sole distributors for the U.S.A. and Canada:
Elsevier Science Publishing Company, Inc.
52 Vanderbilt Avenue
New York, N.Y. 10017

Library of Congress Cataloging in Publication Data

Main entry under title:

Statistics and probability.

"Publications of C.R. Rao": p.
1. Mathematical statistics--Addresses, essays, lectures. 2. Probabilities--Addresses, essays, lectures. 3. Rao, C. Radhakrishna (Calyampudi Radhakrishna), 1920- . I. Rao, C. Radhakrishna (Calyampudi Radhakrishna), 1920- . II. Kallianpur, G. III. Krishnaiah, Paruchuri R. IV. Ghosh, J. K.
QA276.16.S843 519.5 82-2169
ISBN 0-444-86130-0 AACR2

PRINTED IN THE NETHERLANDS

Dedicated to Professor C. R. Rao
on the occasion of his 60th birthday

C. R. Rao

PREFACE

In presenting this volume of essays in Statistics and Probability we wish to honor a distinguished scientist and one of the most eminent statisticians of our time.

The impact of C. R. Rao's work in the development of the modern theory of Statistics is so well recognized that it does not require comment in this Preface. However, scholars outside India might not be fully aware of the important role he has played in developing Statistics in India and more particularly, in the growth of the Indian Statistical Institute, and in promoting statistical education and training in South East Asian countries. After an initial period of study at Calcutta University in 1943 during which he received a Master's degree, Rao joined the Indian Statistical Institute which had been founded a decade or so earlier by P. C. Mahalanobis. The Institute had already made its mark in Statistical research, especially in such areas as Multivariate Analysis and the Design of Experiments. In the years that followed the Institute has produced, under Rao's direction, an entire generation of students and younger colleagues, many of whom have become eminent Statisticians and Probabilists in their own right.

C. R. Rao's career has not been confined to the field of Statistics in the narrow sense. After the Indian Statistical Institute assumed the functions of a University in the late fifties, its academic activities expanded enormously and many of the country's talented young scientists, economists, geologists, anthropologists, geneticists, and mathematicians were attracted to the Institute by Professor Rao's wide range of scientific interests. These interests are, to a large extent, represented by the essays in the volume, ranging from topics in Pure Mathematics, Probability Theory and Statistical Inference to those with a distinctly applied flavor. Even so, for reasons of space, it has not been possible to include several of the important areas of applications to which Rao himself has made significant contributions. Thus we regret that the volume contains no articles on Anthropology, Econometrics and Psychometrics (to name a few of these areas).

C. R. Rao was born on September 10, 1920 in Hadagali, Karnataka State, India. He has been up to now author or co-author of nine books and over 200 research publications, a list of which is given at the end of this volume. This volume is dedicated to C. R. Rao with admiration and affection on the occasion of his sixtieth birthday in appreciation of his many pioneering contributions to the field of statistics, which have found place in standard books on statistics.

We wish to express our appreciation to S. K. Mitra, K. R. Parthasarathy, B. Ramachandran and B. V. Rao, our colleagues on the Editorial Committee, for spending considerable amount of time and effort in preparation of this volume. We are grateful to R. F. Anderson, G. J. Babu, A. K. Basu, D. Basu, Sanjoy Bose, P. Bhimasankaram, Ratan DasGupta, S. DasGupta, R. H. Farrell, A. Hedayat, Subir

Ghosh, P. Jaganathan, V. M. Joshi, J. Jurečkova, S. Kotz, D. Majumdar, K. V. Mardia, M. G. Nadkarni, J. N. K. Rao, M. M. Rao, Y. R. Sarma, D. N. Shanbhag, Bikas K. Sinha, Bimal K. Sinha, A. C. Tamhane and S. J. Wolfe for reviewing the papers in this volume. Thanks are also due to the North-Holland Publishing Company for their excellent cooperation.

G. Kallianpur
P. R. Krishnaiah
J. K. Ghosh

CONTENTS

*Deceased 1981.

G. Kallianpur, P.R. Krishnaiah, J.K. Ghosh, eds., *Statistics and Probability: Essays in Honor of C.R. Rao*

COCHRAN'S THEOREM, RANK ADDITIVITY AND TRIPOTENT MATRICES

T.W. ANDERSON
Stanford University, Stanford, CA, U.S.A.

George P. H. STYAN
McGill University, Montreal, Que., Canada

Let $A_1,\dots,A_k$ be symmetric matrices and $A=\Sigma A_i$. A matrix version of Cochran's theorem is that (a) $A_i^2=A_i$, $i=1,\dots,k$, and (b) $A_iA_j=0$ $\forall i\neq j$, are necessary and sufficient conditions for (d) $\Sigma\,\mathrm{rank}(A_i)=\mathrm{rank}(A)$ whenever (c) $A=I$. This paper reviews extensions of the theorem and its statistical interpretations in the literature, presents various proofs of the above theorem, and obtains some generalizations. In particular, (c) above is replaced by $A^2=A$ and the condition of symmetry is deleted. The relations with (e) $\mathrm{rank}(A_i^2)=\mathrm{rank}(A_i)$, $i=1,\dots,k$, are explored. Another theorem covers the case of matrices not necessarily square. A is "tripotent" if $A^3=A$. Then $A_i^3=A_i$, $i=1,\dots,k$, and (b) are necessary and sufficient conditions for $A^3=A$, (d), and one further condition such as $A_iA=A_i^2$, $i=1,\dots,k$. Variations and statistical applications are treated. Tripotent is replaced by r-potent ($A^r=A$) for $r>3$.

1. Introduction

Let x be a $p\times 1$ random vector distributed according to a multivariate normal distribution with mean vector $\mathbf{0}$ and covariance matrix I_p. We will denote this by $x\sim\mathrm{N}(\mathbf{0}, I_p)$. Let $q_1,\dots,q_k$ be quadratic forms in x with ranks $r_1,\dots,r_k$, respectively, and suppose that $\Sigma q_i=x'x$. Then what has become well-known as Cochran's theorem is Theorem II of Cochran (1934, p. 179): *A necessary and sufficient condition that $q_1,\dots,q_k$ be independently distributed as χ^2 is that $\Sigma r_i=p$.*

Rao (1973, Section 3b.4) gives this result with $x\sim\mathrm{N}(\mu, I)$ as the Fisher–Cochran theorem. Fisher (1925) showed that if the quadratic form q in x is distributed as χ_h^2, then $x'x-q$ is distributed as χ_{p-h}^2 independently of q, cf. James (1952).

Our purpose in this paper is partly expository; we review various extensions of Cochran's theorem in a bibliographic and historical perspective, with special emphasis on matrix-theoretic analogues. While we present over 30 references, we note that Scarowsky (1973) has a rather complete discussion and bibliography on the distribution of quadratic forms in normal random variables. See also the bibliography by Anderson et al. (1972), where 90 research papers published through 1966 are listed under subject-matter code 2.5 (distribution of quadratic and bilinear forms in normal variables).

Our first section is devoted to a survey of results summarized in Theorems 1.1 and 1.2. The proofs are given in Section 2. In the following section the extensions from idempotent to tripotent matrices are given and proved. All matrices considered in this paper will be real.

To formulate our first matrix-theoretic extension of Cochran's theorem we let $\boldsymbol{A}_1, \ldots, \boldsymbol{A}_k$ be $p \times p$ symmetric matrices and write $\boldsymbol{A} = \Sigma \boldsymbol{A}_i$. Consider the following statements:

(a) $$\boldsymbol{A}_i^2 = \boldsymbol{A}_i, \quad i = 1, \ldots, k,$$

(b) $$\boldsymbol{A}_i \boldsymbol{A}_j = \boldsymbol{0} \quad \text{for all } i \neq j,$$

(c) $$\boldsymbol{A} = \boldsymbol{I},$$

(d) $$\sum \operatorname{rank}(\boldsymbol{A}_i) = \operatorname{rank}(\boldsymbol{A}).$$

Then the matrix-theoretic analogue of Cochran's theorem is:

$$(\mathrm{a}), (\mathrm{b}), (\mathrm{c}) \rightarrow (\mathrm{d}), \tag{1.1}$$

$$(\mathrm{c}), (\mathrm{d}) \rightarrow (\mathrm{a}), (\mathrm{b}). \tag{1.2}$$

The reason that these two versions of Cochran's theorem are equivalent follows from the following two well-known results:

Lemma 1.1. *Let $\boldsymbol{x} \sim \mathsf{N}(\boldsymbol{\mu}, \boldsymbol{\Sigma})$, with $\boldsymbol{\Sigma}$ positive definite, and let $\boldsymbol{A}$ be non-random and symmetric. Then $\boldsymbol{x}'\boldsymbol{A}\boldsymbol{x} \sim \chi_f^2(\delta^2)$, a non-central χ^2 distribution with f degrees of freedom and non-centrality parameter δ^2, if and only if $\boldsymbol{A}\boldsymbol{\Sigma}\boldsymbol{A} = \boldsymbol{A}$, and then $f = \operatorname{tr} \boldsymbol{A}\boldsymbol{\Sigma} = \operatorname{rank}(\boldsymbol{A})$ and $\delta^2 = \boldsymbol{\mu}'\boldsymbol{A}\boldsymbol{\mu}$.*

We write $\operatorname{tr} \boldsymbol{A}$ for the trace of $\boldsymbol{A}$ and note that when $\boldsymbol{A} = \boldsymbol{A}^2$ then $\operatorname{tr} \boldsymbol{A} = \operatorname{rank}(\boldsymbol{A})$; this result holds even when $\boldsymbol{A}$ is not symmetric [cf., e.g. Rao (1973, p. 28)].

When $\boldsymbol{\Sigma} = \boldsymbol{I}$, the condition in Lemma 1.1 reduces to $\boldsymbol{A}^2 = \boldsymbol{A}$, and this was first given by Craig (1943) with $\boldsymbol{\mu} = \boldsymbol{0}$ and then by Carpenter (1950) with $\boldsymbol{\mu}$ possibly non-zero. [Thus (a) is equivalent to $q_i = \boldsymbol{x}'\boldsymbol{A}_i\boldsymbol{x}$ having a χ^2-distribution with number of degrees of freedom equal to $\operatorname{rank}(\boldsymbol{A}_i)$.] Sakamoto (1944, Th. II, p. 5) gave the more general version with $\boldsymbol{\Sigma}$ positive definite and $\boldsymbol{\mu} = \boldsymbol{0}$. Cochran (1934, Corollary 1, p. 179) took $\boldsymbol{x} \sim \mathsf{N}(\boldsymbol{0}, \boldsymbol{I})$ and gave Lemma 1.1 with the condition that all the non-zero eigenvalues of $\boldsymbol{A}$ be equal to 1 instead of the condition $\boldsymbol{A}^2 = \boldsymbol{A}$.

Lemma 1.2. *Let $\boldsymbol{x}$ and $\boldsymbol{A}$ be defined as in Lemma 1.1 and let $\boldsymbol{B}$ be non-random and symmetric. Then $\boldsymbol{x}'\boldsymbol{A}\boldsymbol{x}$ and $\boldsymbol{x}'\boldsymbol{B}\boldsymbol{x}$ are independently distributed if and only if $\boldsymbol{A}\boldsymbol{\Sigma}\boldsymbol{B} = \boldsymbol{0}$.*

When $\boldsymbol{\Sigma} = \boldsymbol{I}$ the condition in Lemma 1.2 reduces to $\boldsymbol{A}\boldsymbol{B} = \boldsymbol{0}$, and this was first given by Craig (1943) with $\boldsymbol{\mu} = \boldsymbol{0}$ and then by Carpenter (1950) with $\boldsymbol{\mu}$ possibly non-zero. Again Sakamoto (1944, Th. I, p. 5) gave the more general version with $\boldsymbol{\Sigma}$ positive definite and $\boldsymbol{\mu} = \boldsymbol{0}$. Their proofs, however, turned out to be incorrect and the first correct proof of Lemma 1.2 (with $\boldsymbol{\mu} = \boldsymbol{0}$) seems to be by Ogawa (1948; 1949, cf. p. 85). Cochran (1934, Theorem III, p. 181) let $\boldsymbol{x} \sim \mathsf{N}(\boldsymbol{0}, \boldsymbol{I})$ and gave the condition in Lemma 1.2 as

$$|\boldsymbol{I} - \mathrm{i}s\boldsymbol{A}| \cdot |\boldsymbol{I} - \mathrm{i}t\boldsymbol{B}| = |\boldsymbol{I} - \mathrm{i}s\boldsymbol{A} - \mathrm{i}t\boldsymbol{B}| \tag{1.3}$$

for all real s and t, where $\mathrm{i} = \sqrt{-1}$ and $|\cdot|$ denotes determinant. Ogasawara and

Takahashi (1951, Lemma 1) gave a short proof that (1.3) implies $\boldsymbol{AB}=\mathbf{0}$ when the symmetric matrices $\boldsymbol{A}$ and $\boldsymbol{B}$ are not necessarily positive semi-definite.

Cochran's theorem was first extended to $\boldsymbol{x}\sim \mathrm{N}(\boldsymbol{\mu}, \boldsymbol{I}_p)$ by Madow (1940) and then to $\boldsymbol{x}\sim \mathrm{N}(\mathbf{0}, \boldsymbol{\Sigma})$, $\boldsymbol{\Sigma}$ positive definite, by Ogawa (1946, 1947), who also relaxed the condition (c) to $\boldsymbol{A}^2=\boldsymbol{A}$. Ogasawara and Takahashi (1951) extended Cochran's theorem to $\boldsymbol{x}\sim \mathrm{N}(\boldsymbol{\mu}, \boldsymbol{\Sigma})$, $\boldsymbol{\Sigma}$ positive definite, and to $\boldsymbol{x}\sim \mathrm{N}(\mathbf{0}, \boldsymbol{\Sigma})$, with $\boldsymbol{\Sigma}$ possibly singular. Extensions to $\boldsymbol{x}\sim \mathrm{N}(\boldsymbol{\mu}, \boldsymbol{\Sigma})$, with $\boldsymbol{\Sigma}$ possibly singular, have been given by Styan (1970, Theorem 6) and Tan (1977, Theorem 4.2); Ogasawara and Takahashi (1951) extended Lemmas 1.1 and 1.2 to $\boldsymbol{x}\sim \mathrm{N}(\boldsymbol{\mu}, \boldsymbol{\Sigma})$, with $\boldsymbol{\Sigma}$ possibly singular.

James (1952) appears to be the first to notice that (1.1) could be extended to

$$\text{(a)},\text{(c)} \rightarrow \text{(b)},\text{(d)},$$
$$\text{(b)},\text{(c)} \rightarrow \text{(a)},\text{(d)},$$

while

$$\text{(a)},\text{(b)} \rightarrow \boldsymbol{A}^2=\boldsymbol{A} \text{ and (d)}$$

follows at once from the definition of the χ^2-distribution.

Chipman and Rao (1964) and Khatri (1968) extended the matrix analogue of Cochran's theorem to square matrices which are not necessarily symmetric:

Theorem 1.1. *Let $\boldsymbol{A}_1,\ldots,\boldsymbol{A}_k$ be square matrices, not necessarily symmetric, and let $\boldsymbol{A}=\Sigma\boldsymbol{A}_i$. Consider the following statements*:

(a) $$\boldsymbol{A}_i^2=\boldsymbol{A}_i, \quad i=1,\ldots,k,$$

(b) $$\boldsymbol{A}_i\boldsymbol{A}_j=\mathbf{0} \quad \textit{for all } i\neq j,$$

(c) $$\boldsymbol{A}^2=\boldsymbol{A},$$

(d) $$\sum \operatorname{rank}(\boldsymbol{A}_i)=\operatorname{rank}(\boldsymbol{A}),$$

(e) $$\operatorname{rank}(\boldsymbol{A}_i^2)=\operatorname{rank}(\boldsymbol{A}_i), \quad i=1,\ldots,k.$$

Then

$$\text{(a)},\text{(b)} \rightarrow \text{(c)},\text{(d)},\text{(e)}, \tag{1.4}$$
$$\text{(a)},\text{(c)} \rightarrow \text{(b)},\text{(d)},\text{(e)}, \tag{1.5}$$
$$\text{(b)},\text{(c)},\text{(e)} \rightarrow \text{(a)},\text{(d)}, \tag{1.6}$$
$$\text{(c)},\text{(d)} \rightarrow \text{(a)},\text{(b)},\text{(e)}. \tag{1.7}$$

As Rao and Mitra (1971, p. 112) point out, the extra condition (e) in (1.6) is required (to cover possible asymmetry); for if $k=2$ and

$$\boldsymbol{A}_1=\begin{pmatrix} 0 & 1 \\ 0 & 0 \end{pmatrix}, \qquad \boldsymbol{A}_2=\begin{pmatrix} 0 & -1 \\ 0 & 0 \end{pmatrix},$$

then (b), (c) hold, but (a) and (d) do not. Banerjee and Nagase (1976) replace the extra condition (e) in (1.6) by

(f) $$\operatorname{rank}(A_i)=\operatorname{tr}A_i, \quad i=1,\ldots,k,$$

and prove that

$$\text{(b)},\text{(c)},\text{(f)}\rightarrow\text{(a)},\text{(d)}; \tag{1.8}$$

however, the condition (b) is now no longer required on the left of (1.8) since

$$\text{(c)},\text{(f)}\rightarrow\text{(a)},\text{(b)},\text{(d)}$$

follows from

$$\operatorname{rank}(A)=\operatorname{tr}A=\operatorname{tr}\sum A_i=\sum\operatorname{tr}A_i=\sum\operatorname{rank}(A_i)$$

and (1.7).

In Section 2 we present several proofs of Theorem 1.1.

Marsaglia and Styan (1974) extended Theorem 1.1 by considering an arbitrary sum of matrices, which may now be rectangular. The analogue of Theorem 1.1 is

Theorem 1.2. *Let $A_1,\ldots,A_k$ be $p\times q$ matrices, and let $A=\Sigma A_i$. Consider the following statements:*

(a′) $$A_iA^-A_i=A_i, \quad i=1,\ldots,k,$$

(b′) $$A_iA^-A_j=\mathbf{0} \quad \textit{for all } i\neq j,$$

(c′) $$\operatorname{rank}(A_iA^-A_i)=\operatorname{rank}(A_i), \quad i=1,\ldots,k,$$

(d′) $$\sum\operatorname{rank}(A_i)=\operatorname{rank}(A),$$

where A^- is some g-inverse of A. Then

$$\text{(a}')\rightarrow\text{(b}'),\text{(c}'),\text{(d}'), \tag{1.9}$$

$$\text{(b}'),\text{(c}')\rightarrow\text{(a}'),\text{(d}'), \tag{1.10}$$

$$\text{(d}')\rightarrow\text{(a}'),\text{(b}'),\text{(c}'). \tag{1.11}$$

If (a′) *or if* (b′) *and* (c′) *hold for some g-inverse A^-, then* (a′), (b′) *and* (c′) *hold for every g-inverse A^-.*

In Theorem 1.2 we define a g-inverse of A as any solution A^- to $AA^-A=A$, cf. Rao (1962), Rao and Mitra (1971).

The condition (c′) in Theorem 1.2 plays the role of condition (e) in Theorem 1.1.

Marsaglia and Styan (1974, Th. 13) proved (1.11), while Hartwig (1981) has established (1.9). The proposition (1.10), however, appears to be new and is proved in Section 2, where we also present several different proofs of (1.7). In Section 3 we extend Theorem 1.1 to tripotent matrices, following the work by Luther (1965), Tan (1975, 1976) and Khatri (1977). In Section 4 we discuss the applications of these algebraic theorems to statistics.

2. Some proofs

2.1. Proofs of Theorem 1.1

To prove (1.7) in Theorem 1.1 we begin by reducing condition (c) to a sum being $\boldsymbol{I}$, as in the earlier version of Cochran's theorem; then (1.7) reduces to (1.2). We may do this since if $\boldsymbol{A}$ is $p \times p$, not necessarily symmetric, then, as we shall show,

$$\boldsymbol{A}^2 = \boldsymbol{A} \leftrightarrow \operatorname{rank}(\boldsymbol{I}-\boldsymbol{A}) = p - \operatorname{rank}(\boldsymbol{A}). \tag{2.1}$$

(Note Fisher's 1925 result goes both ways, cf. Section 1, paragraph 2.) To prove (2.1) let $\boldsymbol{A}^2 = \boldsymbol{A}$; then $(\boldsymbol{I}-\boldsymbol{A})^2 = \boldsymbol{I}-\boldsymbol{A}$ and so

$$\operatorname{rank}(\boldsymbol{I}-\boldsymbol{A}) = \operatorname{tr}(\boldsymbol{I}-\boldsymbol{A}) = p - \operatorname{tr}\boldsymbol{A} = p - \operatorname{rank}(\boldsymbol{A}).$$

To go the other way we follow Krafft (1978, pp. 407–408) by noting that

$$\mathfrak{N}(\boldsymbol{A}) \subset \mathcal{C}(\boldsymbol{I}-\boldsymbol{A}), \tag{2.2}$$

where $\mathfrak{N}(\boldsymbol{A}) = \{\boldsymbol{x}\colon \boldsymbol{A}\boldsymbol{x} = \boldsymbol{0}\}$ is the null space of $\boldsymbol{A}$ and $\mathcal{C}(\boldsymbol{I}-\boldsymbol{A}) = \{(\boldsymbol{I}-\boldsymbol{A})\boldsymbol{x}\}$ is the column space of $\boldsymbol{I}-\boldsymbol{A}$. [If $\boldsymbol{x} \in \mathfrak{N}(\boldsymbol{A})$, then $\boldsymbol{A}\boldsymbol{x} = \boldsymbol{0}$ and $(\boldsymbol{I}-\boldsymbol{A})\boldsymbol{x} = \boldsymbol{x} \in \mathcal{C}(\boldsymbol{I}-\boldsymbol{A})$.] If $\operatorname{rank}(\boldsymbol{I}-\boldsymbol{A}) = p - \operatorname{rank}(\boldsymbol{A})$, then equality must hold in (2.2) and so $\boldsymbol{A}^2 = \boldsymbol{A}$.

We are grateful to a reviewer for suggesting an alternative proof of (2.1). Consider the equations

$$\begin{aligned}\begin{pmatrix}\boldsymbol{I} & \boldsymbol{A}\\ \boldsymbol{A} & \boldsymbol{A}\end{pmatrix} &= \begin{pmatrix}\boldsymbol{I} & \boldsymbol{I}\\ \boldsymbol{0} & \boldsymbol{I}\end{pmatrix}\begin{pmatrix}\boldsymbol{I}-\boldsymbol{A} & \boldsymbol{0}\\ \boldsymbol{0} & \boldsymbol{A}\end{pmatrix}\begin{pmatrix}\boldsymbol{I} & \boldsymbol{0}\\ \boldsymbol{I} & \boldsymbol{I}\end{pmatrix}\\ &= \begin{pmatrix}\boldsymbol{I} & \boldsymbol{0}\\ \boldsymbol{A} & \boldsymbol{I}\end{pmatrix}\begin{pmatrix}\boldsymbol{I} & \boldsymbol{0}\\ \boldsymbol{0} & \boldsymbol{A}-\boldsymbol{A}^2\end{pmatrix}\begin{pmatrix}\boldsymbol{I} & \boldsymbol{A}\\ \boldsymbol{0} & \boldsymbol{I}\end{pmatrix}.\end{aligned}$$

Then

$$\operatorname{rank}(\boldsymbol{A}-\boldsymbol{A}^2) + p = \operatorname{rank}(\boldsymbol{I}-\boldsymbol{A}) + \operatorname{rank}(\boldsymbol{A}) \tag{2.2a}$$

from which (2.1) follows at once.

The equation (2.2a) represents "rank additivity on the Schur complement", cf. Marsaglia and Styan (1974, p. 291). Let

$$\boldsymbol{M} = \begin{pmatrix}\boldsymbol{E} & \boldsymbol{F}\\ \boldsymbol{G} & \boldsymbol{H}\end{pmatrix}$$

and suppose that

$$\operatorname{rank}(\boldsymbol{E},\boldsymbol{F}) = \operatorname{rank}\begin{pmatrix}\boldsymbol{E}\\ \boldsymbol{G}\end{pmatrix} = \operatorname{rank}(\boldsymbol{E}),$$

$$\operatorname{rank}(\boldsymbol{G},\boldsymbol{H}) = \operatorname{rank}\begin{pmatrix}\boldsymbol{F}\\ \boldsymbol{H}\end{pmatrix} = \operatorname{rank}(\boldsymbol{H}).$$

Then

$$(\boldsymbol{M}/\boldsymbol{E}) = \boldsymbol{H} - \boldsymbol{G}\boldsymbol{E}^{-}\boldsymbol{F}$$

is the Schur complement of $\boldsymbol{E}$ in $\boldsymbol{M}$ and this is invariant over choices of $\boldsymbol{E}^{-}$. Similarly

$$(\boldsymbol{M}/\boldsymbol{H}) = \boldsymbol{E} - \boldsymbol{F}\boldsymbol{H}^{-}\boldsymbol{G}$$

is the Schur complement of $\boldsymbol{H}$ in $\boldsymbol{M}$ and this is invariant over choices of $\boldsymbol{H}^{-}$. It follows that

$$\begin{aligned}\operatorname{rank}(\boldsymbol{M})&=\operatorname{rank}(\boldsymbol{M}/\boldsymbol{E})+\operatorname{rank}(\boldsymbol{E})\\&=\operatorname{rank}(\boldsymbol{M}/\boldsymbol{H})+\operatorname{rank}(\boldsymbol{H}).\end{aligned}\tag{2.2b}$$

Setting

$$\boldsymbol{M}=\begin{pmatrix}\boldsymbol{I} & \boldsymbol{A}\\ \boldsymbol{A} & \boldsymbol{A}\end{pmatrix}$$

yields (2.2a) directly.

We now write $\boldsymbol{A}_0=\boldsymbol{I}-\boldsymbol{A}$, and in view of (2.1) we replace (c) by $\Sigma_{i=0}^{k}\boldsymbol{A}_i=\boldsymbol{I}$, and (d) by $\Sigma_{i=0}^{k}\operatorname{rank}(\boldsymbol{A}_i)=p$.

The proof of (1.7) by Cochran (1934, p. 180), cf. also Anderson (1958, p. 164) and Rao (1973, Section 3b.4), requires that $\boldsymbol{A}_1,\dots,\boldsymbol{A}_k$ be symmetric. In this event we may write

$$\boldsymbol{A}_i=\boldsymbol{P}_i\boldsymbol{P}_i'-\boldsymbol{Q}_i\boldsymbol{Q}_i',\quad i=0,1,\dots,k,\tag{2.3}$$

where $\boldsymbol{P}_i$ is $p\times p_i$, $\boldsymbol{Q}_i$ is $p\times q_i$, and $\boldsymbol{A}_i$ has p_i positive and q_i negative eigenvalues, cf., e.g. Anderson (1958, p. 346). In (2.3) we assume that $\boldsymbol{P}_i$ has rank p_i, $\boldsymbol{Q}_i$ has rank q_i and $p_i+q_i=r_i$, the rank of $\boldsymbol{A}_i$. Hence

$$\begin{aligned}\boldsymbol{I}_p&=\sum_{i=0}^{k}\boldsymbol{A}_i=\sum_{i=0}^{k}\boldsymbol{P}_i\boldsymbol{P}_i'-\sum_{i=0}^{k}\boldsymbol{Q}_i\boldsymbol{Q}_i'\\&=(\boldsymbol{P}_0,\dots,\boldsymbol{P}_k,\boldsymbol{Q}_0,\dots,\boldsymbol{Q}_k)\begin{pmatrix}\boldsymbol{I}_{p-q} & \boldsymbol{0}\\ \boldsymbol{0} & -\boldsymbol{I}_q\end{pmatrix}\begin{pmatrix}\boldsymbol{P}_0'\\ \vdots\\ \boldsymbol{P}_k'\\ \boldsymbol{Q}_0'\\ \vdots\\ \boldsymbol{Q}_k'\end{pmatrix}\\&=\boldsymbol{PJP}',\end{aligned}\tag{2.4}$$

say, where $q=\Sigma_{i=0}^{k}q_i$, since from (d) now $p=\Sigma_{i=0}^{k}r_i=\Sigma_{i=0}^{k}(p_i+q_i)=(\Sigma_{i=0}^{k}p_i)+q$. But (2.4) is positive definite and $\boldsymbol{P}$ is non-singular; hence $q=0$ and $\boldsymbol{J}=\boldsymbol{I}_p$. Thus $q_0=\cdots=q_k=0$ and (2.4) reduces to

$$\boldsymbol{I}_p=(\boldsymbol{P}_0,\dots,\boldsymbol{P}_k)\begin{pmatrix}\boldsymbol{P}_0'\\ \vdots\\ \boldsymbol{P}_k'\end{pmatrix}=\boldsymbol{PP}',$$

and so $\boldsymbol{P}=(\boldsymbol{P}_0,\dots,\boldsymbol{P}_k)$ is an orthogonal matrix. Hence $\boldsymbol{A}_i^2=\boldsymbol{P}_i\boldsymbol{P}_i'\boldsymbol{P}_i\boldsymbol{P}_i'=\boldsymbol{P}_i\boldsymbol{P}_i'=\boldsymbol{A}_i$ since $\boldsymbol{P}_i'\boldsymbol{P}_i=\boldsymbol{I}_{r_i}$, and $\boldsymbol{A}_i\boldsymbol{A}_j=\boldsymbol{P}_i\boldsymbol{P}_i'\boldsymbol{P}_j\boldsymbol{P}_j'=\boldsymbol{0}$ for all $i\neq j$ since then $\boldsymbol{P}_i'\boldsymbol{P}_j=\boldsymbol{0}$.

We now present four other proofs of (1.7); these four proofs do not require that $\boldsymbol{A}_0,\dots,\boldsymbol{A}_k$ be symmetric.

Following Craig (1938, p. 49), cf. also Aitken (1950, Section 6) and Rao and Mitra (1971, pp. 111–112), we may prove (1.7) using a rank-subadditivity argument. From (2.1) with $\boldsymbol{A}_k$ replacing $\boldsymbol{A}$ we have

$$\begin{aligned} p-\operatorname{rank}(\boldsymbol{A}_k) &\leqslant \operatorname{rank}(\boldsymbol{I}_p-\boldsymbol{A}_k) \\ &= \operatorname{rank}(\boldsymbol{A}_0+\cdots+\boldsymbol{A}_{k-1}) \\ &\leqslant \operatorname{rank}(\boldsymbol{A}_0)+\cdots+\operatorname{rank}(\boldsymbol{A}_{k-1}) \\ &= p-\operatorname{rank}(\boldsymbol{A}_k) \end{aligned} \tag{2.5}$$

when (d) holds. This inequality string, therefore, collapses, and $\operatorname{rank}(\boldsymbol{I}_p-\boldsymbol{A}_k)=p-\operatorname{rank}(\boldsymbol{A}_k)$, which implies $\boldsymbol{A}_k^2=\boldsymbol{A}_k$ by (2.1); repeating the argument with $\boldsymbol{A}_{k-1}, \boldsymbol{A}_{k-2}, \ldots$ yields (a). To see that this implies (b) we follow Rao and Mitra (1971, p. 112) by noting that the argument used in (2.5) implies that

$$(\boldsymbol{A}_i+\boldsymbol{A}_j)^2=\boldsymbol{A}_i+\boldsymbol{A}_j$$

and so

$$\boldsymbol{A}_i\boldsymbol{A}_j+\boldsymbol{A}_j\boldsymbol{A}_i=\boldsymbol{0}.$$

Premultiplying by $\boldsymbol{A}_i$ yields

$$\boldsymbol{A}_i\boldsymbol{A}_j+\boldsymbol{A}_i\boldsymbol{A}_j\boldsymbol{A}_i=\boldsymbol{0}, \tag{2.6}$$

while postmultiplying (2.6) by $\boldsymbol{A}_i$ yields

$$2\boldsymbol{A}_i\boldsymbol{A}_j\boldsymbol{A}_i=\boldsymbol{0}=\boldsymbol{A}_i\boldsymbol{A}_j\boldsymbol{A}_i.$$

Substituting into (2.6) yields (b).

Our next proof of (1.7) follows Chipman and Rao (1964, p. 4), cf. also Styan (1970, p. 571). We write

$$\boldsymbol{A}_i=\boldsymbol{B}_i\boldsymbol{C}_i',$$

where $\boldsymbol{B}_i$ and $\boldsymbol{C}_i$ are $p\times r_i$ of rank r_i. Then

$$\begin{aligned} \boldsymbol{I}_p=\sum\boldsymbol{A}_i &= \sum\boldsymbol{B}_i\boldsymbol{C}_i' \\ &= (\boldsymbol{B}_0,\ldots,\boldsymbol{B}_k)\begin{pmatrix}\boldsymbol{C}_0' \\ \\ \boldsymbol{C}_k'\end{pmatrix} \\ &= \boldsymbol{B}\boldsymbol{C}', \end{aligned}$$

say. By (d) $\boldsymbol{B}$ and $\boldsymbol{C}$ are both non-singular and so $\boldsymbol{C}'=\boldsymbol{B}^{-1}$ and

$$\boldsymbol{C}'\boldsymbol{B}=\boldsymbol{I}_p=\begin{pmatrix}\boldsymbol{C}_0'\boldsymbol{B}_0,\ldots,\boldsymbol{C}_0'\boldsymbol{B}_k \\ \vdots \qquad\qquad \vdots \\ \boldsymbol{C}_k'\boldsymbol{B}_0,\ldots,\boldsymbol{C}_k'\boldsymbol{B}_k\end{pmatrix},$$

which implies that

$$A_i^2 = B_i C_i' B_i C_i' = B_i C_i' = A_i,$$
$$A_i A_j = B_i C_i' B_j C_j' = \mathbf{0} \quad \text{for all } i \neq j.$$

Hence (1.7) is established.

Our fourth proof of (1.7) follows Loynes (1966), cf. also Searle (1971, p. 63). A rank-subadditivity argument is used similar to that used in (2.5):

$$\begin{aligned} p - \operatorname{rank}(A_k) &\leq \operatorname{rank}(I_p - A_k) \\ &\leq \operatorname{rank}(A_0, A_1, \ldots, A_{k-1}, I_p - A_k) \\ &= \operatorname{rank}(A_0, \ldots, A_{k-1}, I - A_0 - \cdots - A_{k-1} - A_k) \\ &= \operatorname{rank}(A_0, \ldots, A_{k-1}) \\ &\leq \operatorname{rank}(A_0) + \cdots + \operatorname{rank}(A_{k-1}) \\ &= p - \operatorname{rank}(A_k). \end{aligned}$$

Our fifth and final proof of (1.7) follows a suggestion made by a reviewer. Let the $pk \times p$ matrix

$$K = (I, I, \ldots, I)'$$

and let

$$D = \begin{pmatrix} A_1 & & \\ & \ddots & \\ & & A_k \end{pmatrix}.$$

Then

$$A = \sum_{i=1}^{k} A_i = K'DK$$

and

$$\begin{aligned} &(\mathrm{c}) \leftrightarrow K'DK = K'DKK'DK, \qquad (2.6\mathrm{a}) \\ &(\mathrm{d}) \leftrightarrow \operatorname{rank}(D) = \operatorname{rank}(K'DK) \\ &\qquad = \operatorname{rank}(K'D) = \operatorname{rank}(DK) \end{aligned}$$

using Sylvester's Law of Nullity. Applying the rank cancellation rules Lemmas 2.1 and 2.2 below to (2.6a) yields

$$D = DKK'D$$

or

$$\begin{pmatrix} A_1 & & \\ & \ddots & \\ & & A_k \end{pmatrix} = \begin{pmatrix} A_1 \\ \vdots \\ A_k \end{pmatrix} (A_1, \ldots, A_k),$$

which is (a) and (b). See also Marsaglia and Styan (1974, p. 285) and Srivastava and Khatri (1979, p. 14).

The rest of Theorem 1.1 is easily proved. To prove (1.4) we see that (a), (b)$\rightarrow$

$$\boldsymbol{A}^2 = \left(\sum \boldsymbol{A}_i\right)^2 = \sum \boldsymbol{A}_i^2 + \sum_{i \neq j} \boldsymbol{A}_i \boldsymbol{A}_j = \sum \boldsymbol{A}_i = \boldsymbol{A},$$

while

$$\sum \operatorname{rank}(\boldsymbol{A}_i) = \sum \operatorname{tr} \boldsymbol{A}_i = \operatorname{tr} \sum \boldsymbol{A}_i = \operatorname{tr} \boldsymbol{A} = \operatorname{rank}(\boldsymbol{A}), \tag{2.7}$$

and so (1.4) is established.

To prove (1.5) we see that (a), (c)$\rightarrow$(d) from (2.7) and so (1.5) follows from (1.7).

To prove (1.6) we see that (b), (c)$\rightarrow$

$$\boldsymbol{A}^2 = \left(\sum \boldsymbol{A}_i\right)^2 = \sum \boldsymbol{A}_i^2 + \sum_{i \neq j} \boldsymbol{A}_i \boldsymbol{A}_j = \sum \boldsymbol{A}_i^2 = \sum \boldsymbol{A}_i = \boldsymbol{A};$$

multiplying through by $\boldsymbol{A}_i$ yields

$$\boldsymbol{A}_i^3 = \boldsymbol{A}_i^2 \tag{2.8}$$

using (b). To see that (2.8)$\rightarrow$(a) we use the rank cancellation rule Lemma 2.1 below, cf. (2.13) in Marsaglia and Styan (1974, p. 271); this rule will also be useful later on.

Lemma 2.1 (Right-hand rank cancellation rule). *If*

$$\boldsymbol{LAX} = \boldsymbol{MAX} \text{ and } \operatorname{rank}(\boldsymbol{AX}) = \operatorname{rank}(\boldsymbol{A}) \tag{2.9}$$

for some conformable matrices $\boldsymbol{A}$, $\boldsymbol{L}$, $\boldsymbol{M}$ *and* $\boldsymbol{X}$, *then*

$$\boldsymbol{LA} = \boldsymbol{MA}. \tag{2.10}$$

Thus (2.8)$\rightarrow$(a) by replacing $\boldsymbol{L}$, $\boldsymbol{A}$ and $\boldsymbol{X}$ in (2.9) by $\boldsymbol{A}_i$ and $\boldsymbol{M}$ by $\boldsymbol{I}$. Then (2.9) becomes (2.8) and (e), while (2.10) becomes (a). [We note that the two matrices $\boldsymbol{A}_1$ and $\boldsymbol{A}_2$ displayed right after Theorem 1.1 satisfy (2.8) but not (e).] Then (d) follows from (1.4) or (1.5).

Proof of Lemma 2.1. Let $\boldsymbol{A} = \boldsymbol{BC}'$, where $\boldsymbol{B}$ and $\boldsymbol{C}$ have r columns and $r = \operatorname{rank}(\boldsymbol{A}) = \operatorname{rank}(\boldsymbol{B}) = \operatorname{rank}(\boldsymbol{C})$. Then $\operatorname{rank}(\boldsymbol{AX}) = \operatorname{rank}(\boldsymbol{BC}'\boldsymbol{X}) = \operatorname{rank}(\boldsymbol{C}'\boldsymbol{X}) = \operatorname{rank}(\boldsymbol{A})$, and so $\boldsymbol{C}'\boldsymbol{X}$ has full row rank. Thus $\boldsymbol{LAX} = \boldsymbol{MAX}$ equals $\boldsymbol{LBC}'\boldsymbol{X} = \boldsymbol{MBC}'\boldsymbol{X} \rightarrow \boldsymbol{LB} = \boldsymbol{MB} \rightarrow \boldsymbol{LBC}' = \boldsymbol{MBC}'$, which is (2.10).

Transposing the matrices in Lemma 2.1 yields:

Lemma 2.2 (Left-hand rank cancellation rule). *If*

$$\boldsymbol{LAX} = \boldsymbol{LAY} \text{ and } \operatorname{rank}(\boldsymbol{LA}) = \operatorname{rank}(\boldsymbol{A})$$

for some conformable matrices $\boldsymbol{A}$, $\boldsymbol{L}$, $\boldsymbol{X}$ *and* $\boldsymbol{Y}$, *then*

$$\boldsymbol{AX} = \boldsymbol{AY}.$$

2.2. *Proof of Theorem 1.2*

Premultiplying (a′) by A^- yields (a) of Theorem 1.1 with A_i replaced by A^-A_i. Moreover, condition (c) of Theorem 1.1 now always holds since $A^-A=\sum A^-A_i$ is always idempotent. Hence (1.5) implies that $A^-A_iA^-A_j=0$ for all $i\neq j$. Premultiplying by A_i and using (a′) yields (b′). Furthermore (1.5) implies that

$$\sum \operatorname{rank}(A^-A_i)=\operatorname{rank}\sum A^-A_i=\operatorname{rank}(A),$$

which reduces to (d′) since (a′) implies

$$\operatorname{rank}(A^-A_i)=\operatorname{rank}(A_i), \tag{2.11}$$

which follows from

$$\operatorname{rank}(A_i)=\operatorname{rank}(A_iA^-A_i)\leqslant\operatorname{rank}(A^-A_i)\leqslant\operatorname{rank}(A_i).$$

Thus (1.9) is established.

We now prove (1.11). Condition (d′) implies

$$\sum\operatorname{rank}(A_i)=\operatorname{rank}(A)=\operatorname{rank}(A^-A)=\operatorname{rank}\left(\sum A^-A_i\right)\leqslant\sum\operatorname{rank}(A^-A_i)$$
$$\leqslant\sum\operatorname{rank}(A_i) \tag{2.12}$$

and so (d) of Theorem 1.1 holds with A_i replaced by A^-A_i. Since (c) now always holds we obtain in lieu of (a)

$$A^-A_iA^-A_i=A^-A_i \tag{2.13}$$

by (1.7). But (2.12) implies (2.11) and so we may cancel the front A^- on both sides of (2.13) using Lemma 2.2 to yield (a′). The rest of (1.11) follows from (1.9).

To prove (1.10) we use the same technique which we used above to yield

$$A^-A_iA^-A_iA^-A_i=A^-A_iA^-A_i, \tag{2.14}$$

which is (2.8) with A_i replaced by A^-A_i. The rank condition (c′) and Lemma 2.1 allow us to cancel the A^-A_i on the right of both sides of (2.14) to yield

$$A^-A_iA^-A_i=A^-A_i. \tag{2.15}$$

Using (2.11) and Lemma 2.2 allows us to cancel the leading A^- on both sides of (2.15) and this yields (a′). The rest of (1.10) follows from (1.9) and the proof is complete.

We are grateful to a reviewer for suggesting an alternative proof of Theorem 1.2. To prove (1.9) we suppose that (a′) holds. Then A_iA^- is idempotent with rank equal to $\operatorname{rank}(A_i)$ and so

$$\sum_{i=1}^{k}\operatorname{rank}(A_i)=\sum_{i=1}^{k}\operatorname{rank}(A_iA^-)=\sum_{i=1}^{k}\operatorname{tr}A_iA^-=\operatorname{tr}\sum_{i=1}^{k}A_iA^-$$
$$=\operatorname{tr}AA^-=\operatorname{rank}(A);$$

thus (a′)⇒(d′). Now let K be defined as in (2.6a) and let

$$L=(I,I,\ldots,I)'.$$

The matrices $\boldsymbol{K}$ and $\boldsymbol{L}$ differ only in that the k identity matrices in $\boldsymbol{L}$ are $q\times q$ while those in $\boldsymbol{K}$ are $p\times p$. Then

$$\boldsymbol{A}=\boldsymbol{K}'\boldsymbol{D}\boldsymbol{L}$$

and

$$\begin{aligned}(\mathrm{d}')\leftrightarrow \operatorname{rank}(\boldsymbol{D})&=\operatorname{rank}(\boldsymbol{K}'\boldsymbol{D}\boldsymbol{L})\\ &=\operatorname{rank}(\boldsymbol{K}'\boldsymbol{D})=\operatorname{rank}(\boldsymbol{D}\boldsymbol{L}).\end{aligned}$$

Thus

$$\boldsymbol{K}'\boldsymbol{D}\boldsymbol{L}(\boldsymbol{K}'\boldsymbol{D}\boldsymbol{L})^{-}\boldsymbol{K}'\boldsymbol{D}\boldsymbol{L}=\boldsymbol{K}'\boldsymbol{D}\boldsymbol{L}$$

implies

$$\boldsymbol{D}\boldsymbol{L}(\boldsymbol{K}'\boldsymbol{D}\boldsymbol{L})^{-}\boldsymbol{K}'\boldsymbol{D}=\boldsymbol{D}$$

using the rank cancellation rules Lemmas 2.1 and 2.2. Hence

$$\boldsymbol{D}\boldsymbol{L}\boldsymbol{A}^{-}\boldsymbol{K}'\boldsymbol{D}=\boldsymbol{D}$$

or

$$\begin{pmatrix}\boldsymbol{A}_1\\ \vdots\\ \boldsymbol{A}_k\end{pmatrix}\boldsymbol{A}^{-}(\boldsymbol{A}_1,\ldots,\boldsymbol{A}_k)=\begin{pmatrix}\boldsymbol{A}_1 & & \\ & \ddots & \\ & & \boldsymbol{A}_k\end{pmatrix},$$

cf. Marsaglia and Styan (1974, p. 285). Therefore (d′)⇒(a′), (b′) and we have proved both (1.9) and (1.11). To prove (1.10) we consider the system of equations

$$\boldsymbol{A}_i\boldsymbol{A}^{-}\boldsymbol{A}\boldsymbol{A}^{-}\boldsymbol{A}=\boldsymbol{A}_i\boldsymbol{A}^{-}\boldsymbol{A},\quad i=1,\ldots,k,$$

which always hold. Given (b′), this system reduces to

$$\boldsymbol{A}_i\boldsymbol{A}^{-}\boldsymbol{A}_i\boldsymbol{A}^{-}\boldsymbol{A}_i=\boldsymbol{A}_i\boldsymbol{A}^{-}\boldsymbol{A}_i,\quad i=1,\ldots,k.$$

When (c′) holds we may apply the rank cancellation rule Lemma 2.1 and obtain

$$\boldsymbol{A}_i\boldsymbol{A}^{-}\boldsymbol{A}_i=\boldsymbol{A}_i,\quad i=1,\ldots,k,$$

and so (b′), (c′)→(a′). Since (a′)→(d′) from (1.9), the proof is complete.

We may extend Theorem 1.1 to tripotent matrices using Theorem 1.2. We do this in the next section.

3. Tripotent matrices

A square matrix $\boldsymbol{A}$ is said to be *tripotent* whenever $\boldsymbol{A}^3=\boldsymbol{A}$. Tripotent matrices have been studied by Luther (1965), Tan (1975, 1976) and Khatri (1977). These authors considered extending Theorem 1.1 to $\boldsymbol{A}_1,\ldots,\boldsymbol{A}_k$ tripotent. This is of interest in statistics since if $\boldsymbol{x}\sim \mathsf{N}(\boldsymbol{0},\boldsymbol{I})$ and $\boldsymbol{A}$ is symmetric nonrandom, then $\boldsymbol{x}'\boldsymbol{A}\boldsymbol{x}$ is distributed as the difference of two independently distributed χ^2-variates if and only if $\boldsymbol{A}^3=\boldsymbol{A}$, cf. Graybill (1969, p. 352) and Tan (1975, Theorem 3.5).

Consider, therefore, the following statements:

(a″) $$A_i^3 = A_i, \quad i=1,\ldots,k,$$

(b″) $$A_iA_j = \mathbf{0} \quad \text{for all } i \neq j,$$

(c″) $$A^3 = A,$$

(d″) $$\sum \operatorname{rank}(A_i) = \operatorname{rank}(A).$$

Then it is easy to see that (a″), (b″)→(c″). To see that (d″) is also implied we note that when $A^3 = A$ then, cf. Graybill (1969, Theorem 12.4.4) and Rao and Mitra (1971, Lemma 5.6.1),

$$\operatorname{rank}(A) = \operatorname{tr} A^2, \tag{3.1}$$

since A^2 is now idempotent, and has rank equal to $\operatorname{tr} A^2 = \operatorname{rank}(A^2) \geqslant \operatorname{rank}(A^3) = \operatorname{rank}(A) \geqslant \operatorname{rank}(A^2)$. Thus

$$\sum \operatorname{rank}(A_i) = \sum \operatorname{tr} A_i^2 = \operatorname{tr} \sum A_i^2 = \operatorname{tr} A^2 = \operatorname{rank}(A)$$

when (a″), (b″), (c″) hold. Notice that we have not supposed that $A_1, \ldots, A_k$ are symmetric; the equality (3.1) holds even when A is not symmetric.

As Khatri (1977, p. 88) has pointed out, (c″) and (d″) need not imply (a″) and (b″) even if the A_i's are symmetric; e.g. if

$$A_1 = \tfrac{1}{3}\begin{pmatrix} 4 & 2 \\ 2 & 1 \end{pmatrix}, \qquad A_2 = -\tfrac{1}{3}\begin{pmatrix} 1 & 2 \\ 2 & 4 \end{pmatrix}, \qquad A = \begin{pmatrix} 1 & 0 \\ 0 & -1 \end{pmatrix},$$

then $\operatorname{rank}(A_1) + \operatorname{rank}(A_2) = 1 + 1 = 2 = \operatorname{rank}(A)$ and $A^3 = A$, but $A_1^3 \neq A_1$, $A_2^3 \neq A_2$, $A_1A_2 \neq \mathbf{0}$. It is, therefore, of interest to see what extra condition could be added to (c″) and (d″) so as to imply (a″) and (b″). Khatri (1977, Lemma 10) uses

$$\operatorname{rank}(A) = \sum \left\{ \operatorname{rank}\left[A_i(A^2 + A)\right] + \operatorname{rank}\left[A_i(A^2 - A)\right] \right\}, \tag{3.2}$$

which is rather complicated. We may simplify (3.2) in various ways. To do this we first note that

$$A^3 = A \leftrightarrow A = A^-, \tag{3.3}$$

cf. Graybill (1969, Theorem 12.4.1), Rao and Mitra (1971, Lemma 5.6.2). Thus Theorem 1.2 implies that (c″), (d″) are equivalent to

$$A_iAA_i = A_i, \quad i=1,\ldots,k, \tag{3.4}$$

and

$$A_iAA_j = \mathbf{0} \quad \text{for all } i \neq j. \tag{3.5}$$

Summing (3.5) over all $j \neq i$ and adding to (3.4) yields

$$A_iA^2 = A_i, \quad i=1,\ldots,k. \tag{3.6}$$

Hence under (c″) and (d″) the condition (3.2) is equivalent to

$$\operatorname{rank}(A) = \sum \{ \operatorname{rank}[A_i(I + A)] + \operatorname{rank}[A_i(I - A)] \}, \tag{3.7}$$

which is a little simpler than (3.2). But (3.7) implies

$$\operatorname{rank}(A) \geqslant \sum \operatorname{rank}(2A_i) = \sum \operatorname{rank}(A_i) \geqslant \operatorname{rank}(A). \tag{3.8}$$

Thus equality holds throughout (3.8) and so (d″) holds and (3.7) implies

$$\operatorname{rank}(A_i) = \operatorname{rank}[A_i(I+A)] + \operatorname{rank}[A_i(I-A)], \quad i=1,\dots,k. \tag{3.9}$$

Summing (3.9) and using (d″) yields (3.7).

Some motivation for the condition (3.9), and hence also for the equivalent conditions (3.2) and (3.7), may be obtained from the following characterization of a tripotent matrix, extending Lemma 5.6.6 in Rao and Mitra (1971, p. 114):

Lemma 3.1. *Let* A *be a square matrix, not necessarily symmetric. Then* $A^3 = A$ *if and only if*

$$\operatorname{rank}(A) = \operatorname{rank}(A+A^2) + \operatorname{rank}(A-A^2). \tag{3.10}$$

Proof. We use Theorem 1.2 with $A_1 = A+A^2$, $A_2 = A-A^2$, and $A_1+A_2 = 2A$. If $A^3 = A$ then $\frac{1}{2}A = (2A)^-$ and (3.10) follows from (1.9) since

$$\begin{aligned}(A+A^2)\tfrac{1}{2}A(A+A^2) &= \tfrac{1}{2}A^3 + A^4 + \tfrac{1}{2}A^5 \\ &= A + A^2,\end{aligned}$$

and similarly

$$(A-A^2)\tfrac{1}{2}A(A-A^2) = A - A^2.$$

To go the other way we use (1.11). Then (3.10) implies

$$\mathbf{0} = (A+A^2)\tfrac{1}{2}A^-(A-A^2) = \tfrac{1}{2}A - \tfrac{1}{2}A^3$$

and so $A^3 = A$ and the proof is complete.

We are grateful to a reviewer for suggesting an alternative proof of Lemma 3.1. Let

$$M = \begin{pmatrix} A & A+A^2 \\ A+A^2 & 2A+2A^2 \end{pmatrix}.$$

Then applying (2.2b) on the two Schur complements

$$(M/A) = A - A^3$$

and

$$\big(M/(2A+2A^2)\big) = \tfrac{1}{2}(A - A^2)$$

yields

$$\operatorname{rank}(A-A^3) + \operatorname{rank}(A) = \operatorname{rank}(A-A^2) + \operatorname{rank}(A+A^2)$$

from which Lemma 3.1 follows at once.

Lemma 3.1 suggests using the condition

$$A_iA=A_i^2, \quad i=1,\ldots,k, \tag{3.11}$$

instead of (3.9), or (3.7) or (3.2). We obtain:

Theorem 3.1. *Let* $A_1,\ldots,A_k$ *be square matrices, not necessarily symmetric, and let* $A=\Sigma A_i$. *Then*

(a″) $$A_i^3=A_i, \quad i=1,\ldots,k,$$

and

(b″) $$A_iA_j=0 \quad \textit{for all } i\neq j,$$

hold if and only if

(c″) $$A^3=A,$$

(d″) $$\sum \operatorname{rank}(A_i)=\operatorname{rank}(A),$$

and

(e″) $$A_iA=A_i^2, \quad i=1,\ldots,k.$$

The condition (e″) *may be replaced by* (3.9), *by* (3.7), *by* (3.2), *by*

(e1) $$A_i^2A=A_i, \quad i=1,\ldots,k,$$

or by

(e2) $$A_iA=AA_i, \quad i=1,\ldots,k.$$

Proof. We have already shown that (a″), (b″) imply (c″), (d″) and hence also (e″), (3.9), (3.7), (3.2), (e1) and (e2). To go the other way let (c″) and (d″) hold. Then (3.4) and (3.5) are true. Substituting (e″) yields (a″) and $A_i^2A_j=0$ for all $i\neq j$; premultiplying by A_i yields (b″).

We have shown that when (c″), (d″) hold, then (3.9), (3.7) and (3.2) are equivalent. To see that (a″), (b″) are implied when any one of these three extra conditions holds we use Theorem 1.2 with the A_i's replaced by the $A_i(I+A)$ and the $A_i(I-A)$ in (3.7), which equation shows them to be rank-additive (the sum is $2A$). Then (1.11) implies that

$$A_i(I+A)(\tfrac{1}{2}A)A_i(I-A)=0, \tag{3.12}$$

using $\frac{1}{2}A=(2A)^-$. Substituting (3.4) and (3.6) into (3.12) yields

$$(A_i+A_i^2)(I-A)=0.$$

Postmultiplying by A_j $(j\neq i)$ and using (3.5) gives

$$A_iA_j=-A_i^2A_j. \tag{3.13}$$

However, (1.11) also implies, cf. (3.12),

$$A_i(I-A)(\tfrac{1}{2}A)A_i(I+A)=0,$$

which leads to

$$A_iA_j = A_i^2A_j. \tag{3.14}$$

Adding (3.13) and (3.14) yields (b″), and substituting (b″) into (3.4) gives (a″).

Now let (c″), (d″), (e1) hold. Then (3.4), (3.5) hold and premultiplying (3.5) by A_i yields (b″). Then (3.4) implies (a″).

Finally, we let (c″), (d″), (e2) hold. Then substitution of (e2) into (3.4) yields (e1), and the proof is complete.

We are grateful to a reviewer for drawing our attention to the following result:

Corollary 3.1. *Let $A_1, \ldots, A_k$ be square matrices, not necessarily symmetric, and let $A = \Sigma A_i$. Then* (3.7)→(c″), (d″), *and the other conditions in Theorem* 3.1.

Proof. From (3.8) we see that (3.7)→(d″). Moreover (3.9) implies, using (1.11) of Theorem 1.2, that

$$(A_i + A_iA)\tfrac{1}{2}A_i^-(A_i + A_iA) = A_i + A_iA, \quad i = 1, \ldots, k,$$

$$(A_i + A_iA)\tfrac{1}{2}A_i^-(A_i - A_iA) = \mathbf{0}, \quad i = 1, \ldots, k.$$

Adding these two sets of equations yields

$$A_iAA_i^-A_i = A_iA, \quad i = 1, \ldots, k,$$

while subtracting them gives

$$A_iAA_i^-A_iA = A_i, \quad i = 1, \ldots, k.$$

Thus

$$A_i = A_iA^2, \quad i = 1, \ldots, k,$$

and summing over $i = 1, \ldots, k$ yields (c″).

Khatri (1977, Lemma 10) proved that part of Theorem 3.1 with (e″) replaced by (3.2). He also claimed that (b″), (c″)→(a″), (d″), (e″). But this is not so for the same reason that this does not hold in Theorem 1.1; again if we let

$$A_1 = \begin{pmatrix} 0 & 1 \\ 0 & 0 \end{pmatrix}, \quad A_2 = \begin{pmatrix} 0 & -1 \\ 0 & 0 \end{pmatrix}, \tag{3.15}$$

then (b″) and (c″) hold, but (a″) and (d″) do not. If, however, we add the condition

(e) $$\operatorname{rank}(A_i^2) = \operatorname{rank}(A_i)$$

to (b″), (c″) as was done in Theorem 1.1, then (a″), (d″) do follow. From (b″), (c″) we have

$$\sum A_i^3 = \sum A_i$$

and so, cf. (2.8),

$$A_i^4 = A_i^2,$$

which implies $A_i^3 = A_i$ using (e) and Lemma 2.1.

The commutativity condition in Theorem 3.1:

(e2) $$A_iA=AA_i, \quad i=1,\ldots,k,$$

has been used before in a generalization of Cochran's Theorem. Recall the statements

(b″) $$A_iA_j=0 \quad \text{for all } i\neq j,$$

(d″) $$\sum \operatorname{rank}(A_i)=\operatorname{rank}(A),$$

(e) $$\operatorname{rank}(A_i^2)=\operatorname{rank}(A_i), \quad i=1,\ldots,k.$$

Then Marsaglia (1967, Theorem 3) and Marsaglia and Styan (1974, Theorem 15) proved that

$$\begin{aligned}&(b''),(d'')\leftrightarrow(d''),(e2),\\&(b''),(e)\rightarrow(d''),\end{aligned} \tag{3.16}$$

while Luther (1965, Theorem 1, p. 684) and Taussky (1966, Theorem 2), assuming the A_i's to be symmetric, proved that

$$(b'')\leftrightarrow(d''),(e2).$$

The condition (e) in (3.16) cannot be dropped in view of the example (3.15); when the A_i's are symmetric, however, (e) is automatically satisfied.

Luther (1965, Theorem 3, p. 689) and Tan (1976, Theorem 2.2) have given versions of Theorem 3.1 when the A_i's are symmetric. We obtain:

Theorem 3.2. *Let $A_1,\ldots,A_k$ be symmetric matrices and let $A=\sum A_i$. Then*

(a″) $$A_i^3=A_i, \quad i=1,\ldots,k,$$

and

(b″) $$A_iA_j=0 \quad \textit{for all } i\neq j$$

hold if and only if

(c″) $$A^3=A,$$

(d″) $$\sum_{i=1}^{k} \operatorname{rank}(A_i)=\operatorname{rank}(A),$$

and

(es) $$\operatorname{tr}AA_i\geqslant \operatorname{tr}A_i^2, \quad i=1,\ldots,k-1.$$

The condition (es) *may be replaced by*

(es1) $$\operatorname{tr}A^2\geqslant \sum_{i=1}^{k}\operatorname{tr}A_i^2$$

or by

(es2) $$\operatorname{rank}(A_i)\geqslant \operatorname{tr}A_i^2, \quad i=1,\ldots,k-1.$$

The condition (es2) was used by Luther (1965, Theorem 3, p. 689), who also showed that the condition (es) may be replaced by

$$\operatorname{tr} \boldsymbol{A}_i \boldsymbol{A}_j \geqslant 0 \quad \text{for all } i \neq j; \tag{3.17}$$

summing (3.17) over all $i \neq j$ yields (es1). Luther also considered the condition

$$\boldsymbol{A}_i^3 = \boldsymbol{A}_i, \quad i=1,\ldots,k-1, \tag{3.18}$$

and proved that (3.18), (c″), (d″)→(a″), (b″), while Khatri (1977, Lemma 10 and Note 9) showed that (a″), (c″), (d″)→(b″). But (a″) clearly implies (3.18), which implies (es2) with equality in view of (3.1). Tan (1976, Theorem 2.2) gives a condition which seems to be intended as (es1) with equality (Tan has an extra $\not\Sigma$ (in his notation) inside the trace on both sides of his condition).

Our proof of Theorem 3.2 uses the following result, cf. Graybill (1969, p. 235).

Lemma 3.2. *Let* $\boldsymbol{A}$ *be a square matrix. Then*

$$\operatorname{tr} \boldsymbol{A}'\boldsymbol{A} \geqslant \operatorname{tr} \boldsymbol{A}^2$$

with equality if and only if $\boldsymbol{A}$ *is symmetric.*

Proof. The result follows at once from

$$\operatorname{tr}(\boldsymbol{A}-\boldsymbol{A}')'(\boldsymbol{A}-\boldsymbol{A}') = 2(\operatorname{tr} \boldsymbol{A}'\boldsymbol{A} - \operatorname{tr} \boldsymbol{A}^2) \geqslant \boldsymbol{0}.$$

Proof of Theorem 3.2. That (a″), (b″) imply (c″), (d″), (es), (es1), (es2) follows from Theorem 3.1. To go the other way, let (c″), (d″) hold. Then (3.4) and (3.6) hold, and so

$$\operatorname{tr} \boldsymbol{A}_i \boldsymbol{A}^2 \boldsymbol{A}_i = \operatorname{tr}(\boldsymbol{A}\boldsymbol{A}_i)'\boldsymbol{A}\boldsymbol{A}_i \geqslant \operatorname{tr}(\boldsymbol{A}\boldsymbol{A}_i)^2 = \operatorname{tr} \boldsymbol{A}\boldsymbol{A}_i\boldsymbol{A}\boldsymbol{A}_i \tag{3.19}$$

becomes

$$\operatorname{tr} \boldsymbol{A}_i^2 \geqslant \operatorname{tr} \boldsymbol{A}\boldsymbol{A}_i. \tag{3.20}$$

The condition (es) implies equality in (3.20), and hence in (3.19) and so by Lemma 3.2 $\boldsymbol{A}\boldsymbol{A}_i = (\boldsymbol{A}\boldsymbol{A}_i)' = \boldsymbol{A}_i\boldsymbol{A}$, which is condition (e2) of Theorem 3.1, but only for $i=1,\ldots,k-1$. Substitution in (3.6) yields

$$\boldsymbol{A}\boldsymbol{A}_i\boldsymbol{A}_j = \boldsymbol{0}, \quad i=1,\ldots,k-1 \text{ and } j \neq i. \tag{3.21}$$

But (3.4) implies that $\operatorname{rank}(\boldsymbol{A}\boldsymbol{A}_i) = \operatorname{rank}(\boldsymbol{A}_i)$ and so by Lemma 2.2 we may cancel the $\boldsymbol{A}$ in (3.21) to get

$$\boldsymbol{A}_i\boldsymbol{A}_j = \boldsymbol{0}, \quad i=1,\ldots,k-1 \text{ and } j \neq i.$$

Thus

$$\boldsymbol{A}_i\boldsymbol{A}_k = \boldsymbol{0} = \boldsymbol{A}_k\boldsymbol{A}_i, \quad i=1,\ldots,k-1,$$

upon transposition and so (b″) holds. Substitution in (3.4) yields (a″).

Now suppose (c″), (d″), (es1) hold. Then so does (3.20) which we may sum to yield

$$\sum_{i=1}^{k} \operatorname{tr} \boldsymbol{A}_i^2 \geqslant \operatorname{tr} \boldsymbol{A}^2.$$

But (es1) indicates that this inequality goes the other way and so we must have equality, which in turn implies equality in (3.19) and that $\boldsymbol{A}\boldsymbol{A}_i$ is symmetric for all $i=1,\ldots,k$. Thus (a″), (b″) are implied as before.

Finally let (c″), (d″), (es2) hold. Then from (3.4) $\boldsymbol{A}\boldsymbol{A}_i$ is idempotent and so

$$\operatorname{tr} \boldsymbol{A}\boldsymbol{A}_i = \operatorname{rank}(\boldsymbol{A}\boldsymbol{A}_i) = \operatorname{rank}(\boldsymbol{A}_i) \geqslant \operatorname{tr} \boldsymbol{A}_i^2, \quad i=1,\ldots,k-1,$$

is condition (es) and our proof is complete.

We conclude this section with an extension of Theorem 3.1 to r-potent matrices. We define a square matrix $\boldsymbol{A}$ to be r-potent whenever $\boldsymbol{A}^r = \boldsymbol{A}$ for some positive integer $r \geqslant 2$. As Tan (1975, Lemma 1) has pointed out, the nonzero eigenvalues of an r-potent matrix are the $(r-1)$th roots of unity. Since a symmetric matrix has only real eigenvalues, a symmetric r-potent matrix must be tripotent; see also Tan (1976, p. 608). When r is even then the matrix must be idempotent.

We obtain:

Theorem 3.3. *Let* $\boldsymbol{A}_1,\ldots,\boldsymbol{A}_k$ *be square matrices, not necessarily symmetric, and let* $\boldsymbol{A} = \Sigma \boldsymbol{A}_i$. *Let* r *be a fixed positive integer* $\geqslant 2$. *Then*

(a) $$\boldsymbol{A}_i^r = \boldsymbol{A}_i, \quad i=1,\ldots,k,$$

and

(b) $$\boldsymbol{A}_i \boldsymbol{A}_j = \mathbf{0} \quad \textit{for all } i \neq j$$

hold if and only if

(c) $$\boldsymbol{A}^r = \boldsymbol{A},$$

(d) $$\sum_{i=1}^{k} \operatorname{rank}(\boldsymbol{A}_i) = \operatorname{rank}(\boldsymbol{A}),$$

and

$(e)_r$ $$\boldsymbol{A}_i \boldsymbol{A}^{r-2} = \boldsymbol{A}_i^{r-1}, \quad i=1,\ldots,k.$$

The condition $(e)_r$ *may be replaced by*

$(e1)_r$ $$\boldsymbol{A}_i^2 \boldsymbol{A}^{r-2} = \boldsymbol{A}_i, \quad i=1,\ldots,k,$$

or by

$(e2)_r$ $$\boldsymbol{A}_i \boldsymbol{A}^{r-2} = \boldsymbol{A}^{r-2} \boldsymbol{A}_i, \quad i=1,\ldots,k.$$

Tan (1975, Theorem 2.1) suggested that (c), (d)→(a), (b) but, cf. Khatri (1976), seems to have realized that this is not true (Tan, 1976). When $r=3$ the conditions $(e)_r$, $(e1)_r$, $(e2)_r$ become the conditions (e), (e1), (e2), respectively, of Theorem 3.1.

When $r=2$ the conditions $(e)_r$, $(e2)_r$ are automatically satisfied and Theorem 3.3 becomes part of Theorem 1.1. The condition $(e1)_r$, however, when $r=2$ becomes $A_i^2=A_i$, or (a), and so $(e1)_r$ may be too strong an extra condition to require that (c), (d)$\rightarrow$(a), (b) in Theorem 3.3. Under (c), (d), however, the condition $(e1)_r$ is equivalent to the commutativity condition

$$A_i(A_iA^{r-2})=(A_iA^{r-2})A_i, \quad i=1,\ldots,k, \tag{3.22}$$

which is in the same spirit as the condition $(e2)_r$. When $r=2$ the condition (3.22) is automatically satisfied.

To prove that the conditions (3.22) and $(e1)_r$ are equivalent when (c), (d) hold we note first that $A^{r-2}=A^-$ when A is r-potent. Then, cf. (3.4)–(3.6), we see that (c), (d) are equivalent to

$$A_iA^{r-2}A_i=A_i, \quad i=1,\ldots,k, \tag{3.23}$$

$$A_iA^{r-2}A_j=\mathbf{0} \quad \text{for all } i\neq j, \tag{3.24}$$

$$A_iA^{r-1}=A_i, \quad i=1,\ldots,k, \tag{3.25}$$

and (3.23) shows that (3.22)$\leftrightarrow(e1)_r$.

Proof of Theorem 3.3. Let (a), (b) hold. Then so does (c), and

$$\left(A_i^{r-1}\right)^2=A_i^{r+r-2}=A_i^{r-1}$$

is idempotent and so

$$\operatorname{rank}(A_i)=\operatorname{tr}A_i^{r-1},$$

cf. (3.1). Hence (a), (b), (c) imply

$$\sum\operatorname{rank}(A_i)=\sum\operatorname{tr}A_i^{r-1}=\operatorname{tr}\sum A_i^{r-1}=\operatorname{tr}A^{r-1}=\operatorname{rank}(A),$$

which is (d). Then (b) implies $(e)_r$ and $(e2)_r$ and turns $(e1)_r$ into (a). To go the other way, let (c), (d), $(e)_r$ hold. Then so do (3.23), (3.24). Substitution of $(e)_r$ into (3.23) yields (a), while substitution of $(e)_r$ into (3.24) yields $A_i^{r-1}A_j=\mathbf{0}=A_iA_j$ upon premultiplication by A_i and substituting (a).

Now let (c), (d), $(e1)_r$ hold. Then (3.23), (3.24) hold and postmultiplying $(e1)_r$ by A_j $(j\neq i)$ yields (b) by substituting (3.24). Then (a) follows from (3.23) by use of (b). Finally, suppose that (c), (d), $(e2)_r$ hold. Premultiplying $(e2)_r$ by A_i and substituting (3.23) yields $(e1)_r$ and so our proof is complete.

Tan (1975, Theorem 2.1) also suggested that (b), (c) and

$$\operatorname{rank}(A_i^{r-1})=\operatorname{rank}(A_i^{2(r-1)}), \quad i=1,\ldots,k, \tag{3.26}$$

imply (a), (d), but withdrew this, cf. Tan (1976, p. 608). It is straightforward, however, to see that (b), (c) imply

$$\sum A_i^r=\sum A_i$$

and hence

$$A_i^{r+1} = A_i^2. \tag{3.27}$$

The extra condition

$$\operatorname{rank}(A_i^2) = \operatorname{rank}(A_i),$$

cf. (e), applied to (3.27) then yields (a) in view of the rank cancellation rule Lemma 2.1. The extra condition (3.26) is, however, not sufficient (unless $r=2$), as is seen from the counter-example provided by (3.15).

4. Statistical applications

The analysis of variance involves the decomposition of a sum of squares of observations into quadratic forms. In classical cases these quadratic forms are independently distributed according to χ^2-distributions. Then ratios of them are proportional to F-statistics. Cochran's theorem provides an algebraic method of verifying the necessary properties of the quadratic forms to justify the F-tests.

As indicated in Lemma 1.1, when x has the distribution $\mathrm{N}(\mathbf{0}, I)$, then $A^2 = A$ implies $x'Ax$ has the χ^2-distribution with degrees of freedom equal to the number of unit eigenvalues of A, the other eigenvalues being 0. Lemma 1.2 states that $AB = \mathbf{0}$ implies independence of $x'Ax$ and $x'Bx$ because the joint characteristic function when $x \sim \mathrm{N}(\mathbf{0}, I)$ is, cf. (1.3),

$$\mathcal{E} e^{isx'Ax/2 + itx'Bx/2} = |I - isA - itB|^{-1/2} = |I - isA|^{-1/2} |I - itB|^{-1/2}.$$

As an example, consider the one-way analysis of variance. Let $y_{i\alpha}$ be normally distributed according to $\mathrm{N}(\mu_i, \sigma^2)$, $i = 1, \ldots, m$, $\alpha = 1, \ldots, n$, and suppose the mn variables are independent. Under the typical null hypothesis H: $\mu_1 = \cdots = \mu_m = \mu$, say, the exponent of the normal distribution is $\Sigma_{i=1}^m \Sigma_{\alpha=1}^n (y_{i\alpha} - \mu)^2$. Let

$$q_1 = n \sum_{i=1}^m (\bar{y}_i - \bar{y})^2 = n \sum_{i=1}^m \bar{y}_i^2 - mn\bar{y}^2,$$

$$q_2 = \sum_{i=1}^m \sum_{\alpha=1}^n (y_{i\alpha} - \bar{y}_i)^2 = \sum_{i=1}^m \sum_{\alpha=1}^n y_{i\alpha}^2 - n \sum_{i=1}^m \bar{y}_i^2,$$

$$q_3 = mn\bar{y}^2,$$

where $\bar{y}_i = \Sigma_{\alpha=1}^n y_{i\alpha}/n$ and $\bar{y} = \Sigma_{i=1}^m \bar{y}_i/m$. Let $y^{(i)} = (y_{i1}, \ldots, y_{in})'$, $y = (y^{(1)'}, \ldots, y^{(m)'})'$,

$$A_1 = \frac{1}{n}\left(I_m - \frac{1}{m}\varepsilon_m \varepsilon_m'\right) \otimes \varepsilon_n \varepsilon_n',$$

$$A_2 = I_m \otimes \left(I_n - \frac{1}{n}\varepsilon_n \varepsilon_n'\right),$$

$$A_3 = \frac{1}{mn}\varepsilon_m \varepsilon_m' \otimes \varepsilon_n \varepsilon_n',$$

where $\boldsymbol{\varepsilon}_n=(1,\ldots,1)'$ of n components and $\boldsymbol{\varepsilon}_m=(1,\ldots,1)'$ of m components. Then $q_i=\boldsymbol{y}'\boldsymbol{A}_i\boldsymbol{y}$. We easily verify that $\Sigma\boldsymbol{A}_i=\boldsymbol{I}_{mn}$, $\operatorname{rank}(\boldsymbol{A}_1)=m-1$, $\operatorname{rank}(\boldsymbol{A}_2)=m(n-1)$, and $\operatorname{rank}(\boldsymbol{A}_3)=1$. Then (a) and (b) hold. (Of course, in the simple example above the conditions could be verified directly.) By Lemmas 1.1 and 1.2 the quadratic forms are independently distributed as χ^2's, the last being non-central.

The multivariate analogue of the χ^2-distribution is the Wishart distribution. If $\boldsymbol{Y}_1,\ldots,\boldsymbol{Y}_N$ are independently distributed, each according to $\mathsf{N}(\mathbf{0},\boldsymbol{\Sigma})$, then the distribution of $\boldsymbol{S}=\Sigma_{\alpha=1}^N\boldsymbol{Y}_\alpha\boldsymbol{Y}_\alpha'$ is known as the Wishart distribution. [Cf. e.g., Chapter 7 of Anderson (1958).] If $q_1,\ldots,q_k$ have independent χ^2-distributions when the dimensionality of $\boldsymbol{Y}_\alpha$ is 1, then

$$\boldsymbol{Q}_1=\sum_{\alpha,\beta=1}^{N}a_{\alpha\beta}^{(1)}\boldsymbol{Y}_\alpha\boldsymbol{Y}_\beta',\ \ldots,\boldsymbol{Q}_k=\sum_{\alpha,\beta=1}^{N}a_{\alpha\beta}^{(k)}\boldsymbol{Y}_\alpha\boldsymbol{Y}_\beta'$$

have independent Wishart distributions; here $\boldsymbol{A}_i=(a_{\alpha\beta}^{(i)})$, $i=1,\ldots,k$. Cochran's theorem is correspondingly useful in multivariate analysis of variance.

It should be noted that when $\boldsymbol{A}_1,\ldots,\boldsymbol{A}_k$ are symmetric several proofs show that there exists an orthogonal matrix that simultaneously diagonalizes $\boldsymbol{A}_1,\ldots,\boldsymbol{A}_k$, the resulting diagonal matrices have 0's and 1's as diagonal elements, and the 1's in the transformed $\boldsymbol{A}_i$ correspond to 0's in the transformed $\boldsymbol{A}_j$, $j\neq i$. Cf. (2.3)–(2.4).

If $\boldsymbol{A}^3=\boldsymbol{A}$, the eigenvalues of $\boldsymbol{A}$ are 1, -1, and 0. Hence $\boldsymbol{x}'\boldsymbol{A}\boldsymbol{x}$ for $\boldsymbol{x}\sim\mathsf{N}(\mathbf{0},\boldsymbol{I})$ has the distribution of $\chi_1^2-\chi_2^2$, where χ_1^2 and χ_2^2 are independent, the number of degrees of freedom of χ_1^2 is the number of eigenvalues equal to 1 and the number of degrees of freedom of χ_2^2 is the number of eigenvalues equal to -1.

Components of variance are often estimated as differences of quadratic forms. Let $y_{i\alpha}=\mu+u_i+v_{i\alpha}$, $\alpha=1,\ldots,n$, $i=1,\ldots,m$, where μ is an unobservable constant and the unobservable u_i's and $v_{i\alpha}$'s are independently normally distributed with means 0 and variances $\mathscr{E}u_i^2=\sigma_u^2$ and $\mathscr{E}v_{i\alpha}^2=\sigma_v^2$. Then for q_1 and q_2 as defined above

$$\mathscr{E}q_1=(m-1)(n\sigma_u^2+\sigma_v^2),$$

$$\mathscr{E}q_2=(mn-m)\sigma_v^2.$$

Thus $q_1/(m-1)-q_2/(mn-m)$ is an unbiased estimator of $n\sigma_u^2$. Other differences of quadratic forms arise in other designs.

Press (1966) has given the distribution of an arbitrary quadratic form, which is a linear combination of χ^2's with possibly negative coefficients. Let $T=\alpha\chi_1^2-\beta\chi_2^2$, where χ_1^2 and χ_2^2 are independently distributed as χ^2-variables with m and n degrees of freedom, respectively, and $\alpha>0$, $\beta>0$. The density of T is

$$\left[K/\Gamma(\tfrac12 m)\right]t^{(1/2)(m+n)-1}e^{-t/2\alpha}\psi\left[\tfrac12 n,\tfrac12(m+n);t(\alpha+\beta)/2\alpha\beta\right],\quad t\geqslant 0,$$

$$\left[K/\Gamma(\tfrac12 n)\right](-t)^{(1/2)(m+n)-1}e^{t/2\beta}\psi\left[\tfrac12 m,\tfrac12(m+n);-t(\alpha+\beta)/2\alpha\beta\right],\quad t\leqslant 0,$$

where $K^{-1}=2^{(m+n)/2}\alpha^{m/2}\beta^{n/2}$ and

$$\psi(a,b,x)=\frac{\Gamma(1-b)}{\Gamma(1+a-b)}\,{}_1F_1(a,b;x)+\frac{\Gamma(b-1)}{\Gamma(a)}x^{1-b}\,{}_1F_1(1+a-b,2-b;x),$$

and ${}_1F_1(a, b; x)$ is the confluent hypergeometric function. Robinson (1965) gave a similar result for $\alpha=\beta=1$. In the special case of equal degrees of freedom ($n=m$) Pearson et al. (1932) gave the density of $Z=\chi_1^2-\chi_2^2$ as

$$\frac{z^{p-(1/2)}K_{n-(1/2)}(2z)}{2\sqrt{\pi}\,\Gamma(n)},$$

where $K_r(x)$ is the Bessel function of second order and imaginary argument.

In Theorem 3.2 (a″) indicates that $q_i=\boldsymbol{x}'\boldsymbol{A}_i\boldsymbol{x}$ is distributed as the difference of two χ^2-variables if $\boldsymbol{x}\sim \mathrm{N}(\boldsymbol{0}, \boldsymbol{I})$ and (b″) states that q_i and q_j are independent. Then (c″) and (d″) and either (es), (es1), or (es2) are conditions implying (a″) and (b″). In most cases (c″) is easily verified and (d″) is as in Section 1. Each of (es), (es1), and (es2) require computation of $\operatorname{tr}\boldsymbol{A}_i^2$, $i=1,\ldots,k-1$, and (es1) needs also $\operatorname{tr}\boldsymbol{A}_k^2$. Of the left-hand sides, $\operatorname{tr}\boldsymbol{A}^2$ may be easiest to compute.

Acknowledgements

The authors are grateful to an anonymous reviewer and to Akimichi Takemura for several helpful comments. See Takemura (1980) for further developments. This research was supported in part by the Office of Naval Research under Contract Number N00014-75-C-0442 (NR-042-034) and by the Natural Sciences and Engineering Research Council Canada, Grant Number A7274.

References

Aitken, A.C. (1950). On the statistical independence of quadratic forms in normal variates. *Biometrika* **37**, 93–96.

Anderson, T.W. (1958). *An Introduction to Multivariate Statistical Analysis*. Wiley, New York.

Anderson, T.W., Das Gupta, S. and Styan, G.P.H. (1972). *A Bibliography of Multivariate Statistical Analysis*. Oliver & Boyd, Edinburgh. (Reprinted 1977, Krieger, Huntington, NY.)

Banerjee, K.S. and Nagase, G. (1976). A note on the generalization of Cochran's theorem. *Comm. Statist. A–Theory Methods* **5**, 837–842.

Carpenter, O. (1950). Note on the extension of Craig's theorem to non-central variates. *Ann. Math. Statist.* **21**, 455–457.

Chipman, J.S. and Rao, M.M. (1964). Projections, generalized inverses, and quadratic forms. *J. Math. Anal. Appl.* **9**, 1–11.

Cochran, W.G. (1934). The distribution of quadratic forms in a normal system, with applications to the analysis of covariance. *Proc. Cambridge Philos. Soc.* **30**, 178–191.

Craig, A.T. (1938). On the independence of certain estimates of variance. *Ann. Math. Statist.* **9**, 48–55.

Craig, A.T. (1943). Note on the independence of certain quadratic forms. *Ann. Math. Statist.* **14**, 195–197.

Fisher, R.A. (1925). Applications of "Student's" distribution. *Metron* **5**, 90–104.

Graybill, F.A. (1969). *Introduction to Matrices with Applications in Statistics*. Wadsworth, Belmont, CA.

Graybill, F.A. and Marsaglia, G. (1957). Idempotent matrices and quadratic forms in the general linear hypothesis. *Ann. Math. Statist.* **28**, 678–686.

Hartwig, R.E. (1980). A note on rank-additivity. *Linear and Multilinear Algebra*, in press.

James, G.S. (1952). Notes on a theorem of Cochran. *Proc. Cambridge Philos. Soc.* **48**, 443–446.

Khatri, C.G. (1968). Some results for the singular normal multivariate regression models. *Sankhyā Ser. A* **30**, 267–280.

Khatri, C.G. (1976). Review of Tan (1975). *Math. Reviews* **51** (2159).

Khatri, C.G. (1977). Quadratic forms and extension of Cochran's theorem to normal vector variables. In: P.R. Krishnaiah, Ed., *Multivariate Analysis — IV*. North-Holland, Amsterdam, pp. 79–94.

Krafft, O. (1978). *Lineare Statistische Modelle und Optimale Versuchspläne*. Vandenhoeck & Ruprecht, Göttingen.

Loynes, R.M. (1966). On idempotent matrices. *Ann. Math. Statist.* **37**, 295–296.

Luther, N.Y. (1965). Decomposition of symmetric matrices and distributions of quadratic forms. *Ann. Math. Statist.* **36**, 683–690.

Madow, W.G. (1940). The distribution of quadratic forms in non-central normal random variables. *Ann. Math. Statist.* **11**, 100–103.

Marsaglia, G. (1967). Bounds on the rank of the sum of matrices. *Trans. Fourth Prague Conf. Information Theory, Statist. Decision Functions, Random Processes* (Prague, Aug. 31 – Sept. 11, 1965). Czech. Acad. Sci., 455–462.

Marsaglia, G. and Styan, G.P.H. (1974). Equalities and inequalities for ranks of matrices. *Linear and Multilinear Algebra* **2**, 269–292.

Ogasawara, T. and Takahashi, M. (1951). Independence of quadratic quantities in a normal system. *J. Sci. Hiroshima Univ. Ser. A* **15**, 1–9.

Ogawa, J. (1946). On the independence of statistics of quadratic forms (in Japanese). *Res. Mem. Inst. Statist. Math. Tokyo* **2**, 98–111.

Ogawa, J. (1947). On the independence of statistics of quadratic forms (translation of Ogawa, 1946). *Res. Mem. Inst. Statist. Math. Tokyo* **3**, 137–151.

Ogawa, J. (1948). On the independence of statistics between linear forms, quadratic forms and bilinear forms from normal distributions (in Japanese). *Res. Mem. Inst. Statist. Math. Tokyo* **4**, 1–40.

Ogawa, J. (1949). On the independence of bilinear and quadratic forms of a random sample from a normal population (translation of Ogawa, 1948). *Ann. Inst. Statist. Math. Tokyo* **1**, 83–108.

Pearson, K., Stouffer, S.A. and David, F.N. (1932). Further applications in statistics of the $T_m(x)$ Bessel function. *Biometrika* **24**, 293–350.

Press, S.J. (1966). Linear combinations of non-central chi-square variates. *Ann. Math. Statist.* **37**, 480–487.

Rao, C.R. (1962). A note on a generalized inverse of a matrix with applications to problems in mathematical statistics. *J. Roy. Statist. Soc. Ser. B* **24**, 152–158.

Rao, C.R. (1973). *Linear Statistical Inference and Its Applications*, 2nd ed. Wiley, New York.

Rao, C.R. and Mitra, S.K. (1971). *Generalized Inverse of Matrices and its Applications*. Wiley, New York.

Robinson, J. (1965). The distribution of a general quadratic form in normal variates. *Austral. J. Statist.* **7**, 110–114.

Sakamoto, H. (1944). On the independence of (two) statistics (Lecture at the annual Mathematics–Physics Meeting, July 19, 1943) (in Japanese). *Res. Mem. Inst. Statist. Math. Tokyo* **1** (9), 1–25.

Scarowsky, I. (1973). Quadratic forms in normal variables. Thesis, Dept. Math., McGill Univ.

Searle, S.R. (1971). *Linear Models*. Wiley, New York.

Srivastava, M.S. and Khatri, C.G. (1979). *An Introduction to Multivariate Statistics*. North-Holland, New York.

Styan, G.P.H. (1970). Notes on the distribution of quadratic forms in singular normal variables. *Biometrika* **57**, 567–572.

Takemura, A. (1980). On generalizations of Cochran's theorem and projection matrices, Technical Report No. 44, Office of Naval Research Contract N00014-75-C-0442, Department of Statistics, Stanford University, Stanford, CA.

Tan, W.Y. (1975). Some matrix results and extensions of Cochran's theorem. *SIAM J. Appl. Math.* **28**, 547–554.

Tan, W.Y. (1976). Errata: Some matrix results and extensions of Cochran's theory. *SIAM J. Appl. Math.* **30**, 608–610.

Tan, W.Y. (1977). On the distribution of quadratic forms in normal random variables. *Canad. J. Statist.* **5**, 241–250.

Taussky, O. (1966). Remarks on a matrix theorem arising in statistics. *Monatsh. Math.* **70**, 461–464.

G. Kallianpur, P.R. Krishnaiah, J.K. Ghosh, eds., *Statistics and Probability: Essays in Honor of C.R. Rao*
© North-Holland Publishing Company (1982) 25–37

INVARIANT SUBSPACES OF VECTOR VALUED FUNCTION SPACES ON BOHR GROUPS AND MULTIVARIATE PREDICTION THEORY

Somesh Chandra BAGCHI
Indian Statistical Institute, Calcutta, India

1. Introduction

A Bohr group is a compact abelian group, B, whose dual is a subgroup, Γ, of $\mathbb{R}$, where Γ is not isomorphic to the group of integers $\mathbb{Z}$. There is, then, a natural action of $\mathbb{R}$ on B through a one-to-one continuous homomorphism of $\mathbb{R}$ onto a dense subgroup of B. If γ is an element of Γ, the corresponding character on B is denoted by χ_γ. Let $\mathbf{H}$ be a Hilbert space. $L^2(B;\mathbf{H})$ is the Hilbert space of weakly measurable, $\mathbf{H}$-valued functions on B having square integrable norms for the Haar measure. A closed subspace, M, of $L^2(B;\mathbf{H})$ is called invariant if $\chi_\gamma \cdot M \subseteq M$ for every positive γ. M is called simply invariant if the inclusion is strict for all positive γ, and doubly invariant if $\chi_\gamma \cdot M = M$ for all γ.

The theory of invariant subspaces on the circle group is well known through the book of Helson [4]. For the Bohr groups, invariant subspaces were first studied by Helson and Lowdenslager in ref. [7]. They considered the scalar situation, i.e. invariant subspaces of $L^2(B)$ and proved that there is a one-to-one correspondence between simply invariant subspaces (which are normalised in some sense) on the one hand, and cocycles for the $\mathbb{R}$-action on B on the other. Through deep studies into the structure of cocycles in the subsequent work of Helson and others, a nearly complete picture of the invariant subspaces has now emerged. For a comprehensive account, see the lecture notes of ref. [6].

The first part of this paper is devoted to extending the Helson–Lowdenslager correspondence to invariant subspaces of $L^2(B;\mathbf{H})$ under the simplifying assumption that Γ is countable. The tool for this extension is the correspondence of cocycles and certain systems of imprimitivity which is available in the theory of group representations. Only the cocycles now take values in the group of unitary operators on an abstract Hilbert space (see ref. [10]). Our object is to restore the Helson–Lowdenslager picture by considering a larger class of cocycles. These cocycles take partial isometries as values. We develop the necessary theory of these cocycles in Section 3. In Section 6 we relate them to invariant subspaces.

In the second part of the paper we consider a multivariate stationary stochastic process with time Γ. We assume that the process has an absolutely continuous spectral representation and obtains conditions for the process to be purely non-deterministic (Theorem 10.1). This makes use of our description of invariant subspaces. Our result generalises the work of Helson and Lowdenslager in the univariate case (see ref. [8], I) and is also related to the corresponding result for multivariate processes with time $\mathbb{Z}$ (see ref. [8], II).

PART I

2. Preliminaries

By a pair (G_0, G) we will mean that G_0 and G are locally compact, second-countable abelian groups with a given continuous homomorphism $\phi: G_0 \to G$ such that ϕ is 1–1 and $\phi(G_0)$ is dense in G. The most important pair for us will be $(\mathbb{R}, B)$ where B is a Bohr group.

Let $\mathbf{H}$ be a separable Hilbert space. We denote by, $L^2(G;\mathbf{H})$, the Hilbert space of weakly measurable $\mathbf{H}$-valued functions on G with the inner product,

$$(F, F') = \int_G (F(x), F'(x))\,\mathrm{d}\sigma(x),$$

where σ is the Haar measure on G.

A *range function*, J on G, is a function on G such that for each x, $J(x)$ is a closed subspace of a given Hilbert space, $\mathbf{H}$. We will assume that J is measurable in the sense that the corresponding projection valued map is weakly measurable. Given a range function, J, we define a subspace,

$$JL_2(G;\mathbf{H}) = \{F \in L_2(G;\mathbf{H}) : F(x) \in J(x) \quad \text{a.e.}\}. \tag{1}$$

Let P stand for the canonical spectral measure on $L_2(G;\mathbf{H})$. That is, for a Borel set $E \subseteq G$, $P(E)$ is the multiplication operator by the characteristic function of E. The following fact is easy to prove.

Proposition 2.1. *The closed subspaces M of $L_2(G;\mathbf{H})$ which are invariant under every $P(E)$ are the subspaces of the form $JL_2(G;\mathbf{H})$. Further, M determines J uniquely almost everywhere.*

3. Partial isometry valued cocycles

For general results on cocycles and systems of imprimitivity we refer the reader to the book of Varadarajan [13]. Let (G_0, G) be a pair. $\mathbf{U(H)}$ is our notation for the set of unitary operators on a Hilbert space $\mathbf{H}$. $\mathbf{P(H)}$ stands for the set of partial isometries on $\mathbf{H}$. If $T \in \mathbf{P(H)}$ we write $\mathbf{Ker}(T)$, $\mathbf{In}(T)$ and $\mathbf{R}(T)$ for the kernel, initial space and the range of T respectively.

Definition 3.1. By a *cocycle* we mean a measurable operator function, A, on $G_0 \times G$ having values in **P(H)** such that,

$$A(g_1+g_2, x)=A(g_1, x)A(g_2, x+\phi(g_1)), \tag{2}$$

for a.e. (g_1, g_2, x) (with respect to the product measure $\lambda\times\lambda\times\sigma$ on $G_0\times G_0\times G$).

Two cocycles are identified if they agree except on a $\lambda\times\sigma$-null set. Before we can work with cocycles it is necessary to obtain a result about the range of the cocycle.

Lemma 3.2. *Let A be a cocycle. There exists a measurable range function J on G such that*

(i) $\mathbf{R}\, A(g, x)=J(x)$,

(ii) $\mathbf{In}\, A(g, x)=J(x+\phi(g))$,

for a.e. (g, x).

Proof. By definition, (2) is satisfied for all (g_1, g_2, x) belonging to a subset, C, of $G_0\times G_0\times G$ having full measure. Consider the transformation,

$$T:(g_1, g_2, x)\rightarrow(g_1+g_2, -g_2, x),$$

which is a measure preserving homomorphism of $G_0\times G_0\times G$ onto itself. Let C_0 be the intersection of the sets T^nC as n takes on all integer values. Then C_0 is of full measure and $TC_0=C_0$. If $(g_1, g_2, x)\in C_0$, then $(g_1+g_2, -g_2, x)\in C_0$ and from (2),

$$A(g_1, x)=A(g_1+g_2, x)A(-g_2, x+\phi(g_1)+\phi(g_2)). \tag{3}$$

From (2) and (3),

$$\mathbf{R}\, A(g_1+g_2, x)=\mathbf{R}\, A(g_2, x) \quad \text{for all } (g_1, g_2, x)\in C_0. \tag{4}$$

Since C_0 is of full measure, by Fubini's theorem, there exist a null set $N\subseteq G$ such that if $x\in G-N$, then $(g_1, g_2, x)\in C_0$ for a.e. (g_1, g_2). That is, for $x\in G-N$, the x-section C_0^x is of full measure in $G_0\times G_0$. Fixing x, consider a $g_1=g_1^x$ such that $(C_0^x)^{g_1}$ is of full measure in G_0. We have for each $x\in G-N$,

$$\mathbf{R}\, A(g_1^x+g_2, x)=\mathbf{R}\, A(g_1^x, x), \quad (\text{a.e. } g_2).$$

By using (4) this means that for $x\in G-N$

$$\mathbf{R}\, A(g, x)=\mathbf{R}\, A(g_1^x, x) \quad (\text{a.e. } g). \tag{5}$$

Define a range function J on G by setting $J(x)=\mathbf{R}\, A(g_1^x, x)$ if $x\in G-N$ and arbitrarily elsewhere. A standard argument using the relation (5) shows that J is measurable. We already have:

$$J(x)=\mathbf{R}\, A(g, x), \quad \text{for a.e. } x, \quad \text{for a.e. } g.$$

With both sides being measurable, assertion (i) follows.

To prove point (ii), consider a second transformation $(g_1, g_2, x)\rightarrow(-g_1, g_1+g_2, x+\phi(g))$ which is again a measure preserving homomorphism of $G_0\times G_0\times G$ onto itself. Let $C_1\subseteq C_0$ be a set of full measure which is invariant under this transformation. Fix $(g_1, g_2, x)\in C_1$. Then $(-g_1, g_1+g_2, x+\phi(g_1))\in C_1$. Since $C_1\subseteq$

C_0 we have:

$$A(g_1+g_2, x)=A(g_1, x)A(g_2, x+\phi(g_1)) \tag{6}$$

and

$$A(g_2, x+\phi(g_1))=A(-g_1, x+\phi(g_1))A(g_1+g_2, x). \tag{7}$$

From (6) and (7),

$$\mathbf{In}\, A(g_1+g_2, x)=\mathbf{In}\, A(g_2, x+\phi(g_1)), \tag{8}$$

for $(g_1, g_2, x)\in C_1$. However, $C_1\subseteq C_0$ and hence from (4)

$$\mathbf{R}\, A(g_1+g_2, x)=\mathbf{R}\, A(g_1, x). \tag{9}$$

By combining (8), (9) and (6) one has:

$$\mathbf{R}\, A(g_2, x+\phi(g_1))=\mathbf{In}\, A(g_1, x) \quad \text{for all } (g_1, g_2, x)\in C_1. \tag{10}$$

Now, by Fubini's theorem, choose a subset $D_1\subseteq G_0\times G$ having full measure such that for each $(g_1, x)\in D_1, (g_1, g_2, x)\in C_1$ for a.e. g_2. Again let D_2 be a subset of $G_0\times G$ having full measure such that if $(g_1, x)\in D_2$, then $x+\phi(g_1)$ is in $G-N$. In writing $D=D_1\cap D_2$, one concludes from (10) and the first half of the lemma that,

$$\mathbf{In}\, A(g_1, x)=J(x+\phi(g_1)) \quad \text{for all } (g_1, x)\in D.$$

In Lemma 2.2, the range function J is uniquely determined by A and will be called the *range of A*. The following result is a consequence of the ergodicity of the Haar measure σ on G for G_0-action (see ref. [3, ch. VIII, sec. 7]).

Corollary 3.3. *If J is the range function of a cocycle, A, then J has constant dimension a.e.*

Definition 3.4. Let A and A' be cocycles having range J and J', respectively, A' is said to be cohomologous to A if there exists a measurable function $\rho: G\to \mathbf{P}(\mathbf{H})$ such that:

(i) for a.e. $x\in G$, $\mathbf{In}\,\rho(x)=J(x)$ and
(ii) for a.e. $(g, x)\in G_0\times G$, $A'(g, x)=\rho(x)A(g, x)\rho(x+\phi(g))^*$.

If (i) and (ii) hold, it is obvious that $J'(x)=\rho(x)[J(x)]$ a.e. It is easy to see that 'A is cohomologous to A'' defines an equivalence relation. Consider now the cocycle,

$$A(g, x)=I_{\mathbf{H}_1}, \quad (g, x)\in G_0\times G,$$

when $I_{\mathbf{H}_1}$ is the projection on a non-zero closed subspace $\mathbf{H}_1\subseteq\mathbf{H}$. Cocycles which are cohomologous to such a cocycle for some $\mathbf{H}_1$ are called *coboundaries*. A coboundary is, thus, of the form,

$$A(g, x)=\rho(x)\rho^*(x+\phi(g)), \quad \text{a.e. } (g, x),$$

where $\rho: G\to\mathbf{P}(\mathbf{H})$ is a measurable function such that $\mathbf{In}\,\rho(x)$ is constant.

4. Systems of imprimitivity of the type (U, P^J)

Let A be a cocycle (with respect to G_0, G and $\mathbf{P}(\mathbf{H})$) with range J. P^J will stand for the restriction of the canonical spectral measure on G to $JL_2(G;\mathbf{H})$. Using the cocycle identity it can be shown that (see Varadarajan [13, Lemma 9.6]) there exists a representation U^A of G_0 such that for a.e. $g \in G_0$,

$$\left(U_g^A F\right)(x) = A(g, x) F(x + \phi(g)), \tag{11}$$

for each $F \in L_2(G; \mathbf{H})$, a.e. x. Now for every Borel set E of G we have:

$$\left(U_g^A\right)^{-1} P^J(E) U_g^A = P^J(E + \phi(g)), \quad \text{a.e. } g. \tag{12}$$

Using the continuity of the representation U_g^A, this identity holds for all $g \in G_0$.

If a representation U of G_0 and a spectral measure on G over the same Hilbert space satisfy (12) then (U, P) is called a *system of imprimitivity*. We have shown that a cocycle, A, gives rise to a system of imprimitivity. Conversely, we have:

Theorem 4.1. *Let (U, P^J) be a system of imprimitivity on (G_0, G) acting in $JL^2(G; \mathbf{H})$. Then there exists a unique cocycle A having range J such that $U = U^A$.*

Proof. We use the fact that the G_0-action on G is ergodic to conclude that P^J is homogeneous (see Lemma 9.10 of ref. [13]). Thus P^J is equivalent to the canonical spectral measure $P^{(n)}$ on $L^2(G; \mathbf{H}_n)$ where $\mathbf{H}_n$ is an n-dimensional $(1 \leqslant n \leqslant \infty)$ Hilbert space.

Accordingly, let

$$S: L^2(G; \mathbf{H}_n) \to JL^2(G; \mathbf{H}),$$

such that $SP^{(n)} = P^J S$. This implies that there exist a measurable function s, where $s(x)$ is an isomorphism of H_n onto $J(x)$, such that for all $F \in L^2(G; \mathbf{H}_n)$,

$$SF(x) = s(x) F(x), \quad (\text{a.e. } x).$$

Now $(S^* US, P^{(n)})$ is a system of imprimitivity acting in $L^2(G; \mathbf{H}_n)$. By Theorem 9.11 of ref. [13], there exists a cocycle A' (with unitary operator values) such that $S^* US = U^{A'}$. Defining,

$$A(g, x) = s(x) A'(g, x) s(x + \phi(g))^*,$$

we see that A is a cocycle and $U = U^A$. The uniqueness of A is a consequence of (11).

Finally, looking at the maps in the proof above and applying Theorem 9.7 of ref. [13] we obtain the following result.

Theorem 4.2. *(U^A, P^J) and $(U^{A'}, P^{J'})$ are equivalent if and only if A and A' are cohomologous. In fact, if,*

$$A'(g, x) = \rho(x) A(g, x) \rho(x + \phi(g))^*, \quad (a.e.\ (g, x)),$$

as in Definition 2.4, then $TU_g^A T^{-1} = U_g^{A'}$ for all $g \in G_0$, where for $F \in JL^2(G; \mathbf{H})$

$$(TF)(x) = \rho(x) F(x), \quad (a.e.\ x). \tag{13}$$

5. Invariant subspaces on Bohr groups

A Bohr group is a compact abelian group, B, whose discrete dual Γ is a subgroup of the additive group of real numbers. These groups are characterized by the property that their duals carry an Archimedean order. We will assume that Γ is countable and is not isomorphic to the group of integers. Then B is a second countable topological group. For $\gamma\in\Gamma$, χ_γ will stand for the corresponding character on B. σ denotes the normalized Haar measure on B.

A closed subspace $M\subseteq L^2(B;\mathbf{H})$ is called *invariant* if for each $\gamma>0$ $(\gamma\in\Gamma)$, $\chi_\gamma\cdot M\subseteq M$. M is called *doubly invariant* if $\chi_\gamma\cdot M=M$ for all $\gamma>0$ (and hence for all $\gamma\in\Gamma$) and *simply invariant* otherwise.

The double invariant subspaces are simple to describe. The operator of multiplication by χ_γ on the space $L^2(B;\mathbf{H})$ has the spectral representation,

$$\int_B \chi_\gamma(x)\,\mathrm{d}P(x),$$

where P is the canonical spectral measure. Therefore, by Stone's theorem, a doubly invariant subspace is invariant under $P(E)$ for every Borel subset $E\subseteq B$. By Proposition 2.1 we have:

Theorem 5.1. *For every measurable range function J on B whose values are closed subspaces of* $\mathbf{H}$, *$JL^2(B;\mathbf{H})$ is a doubly invariant subspace. Conversely, every doubly invariant subspace of $L^2(B;\mathbf{H})$ is of the form $JL^2(B;\mathbf{H})$ where J is a measurable range function determined almost everywhere.*

Let M be an invariant subspace of $L^2(B;\mathbf{H})$. If M is simply invariant, it can be easily seen that, $\chi_\gamma\cdot M$ is a strictly decreasing family. In this case, we define for $s\in\mathbb{R}$:

$$M_s=\bigcap_{\gamma<s}\chi_\gamma\cdot M.$$

The family $\{M_s, s\in\mathbb{R}\}$, is called the *characteristic family* of M and has the following properties:

$$\chi_\gamma M_s=M_{s+\gamma},\quad \text{for each } \gamma\in\Gamma,\quad s\in\mathbb{R}. \tag{14}$$

$$\{M_s, s\in\mathbb{R}\} \tag{15}$$

is a strictly decreasing left continuous family of invariant subspaces.

We call an invariant subspace $M\subseteq L_2(B;\mathbf{H})$ *exact* if:

$$M_\infty=\bigcap_{s\in\mathbb{R}} M_s=\{0\}.$$

It is easy to see that if M is exact, then M is a simply invariant subspace. On the other hand, if $M_{-\infty}$ is the closure of the subspace $U_{s\in\mathbb{R}}M_s$, $M_{-\infty}$ is a doubly invariant subspace. Representing $M_{-\infty}=JL_2(B;\mathbf{H})$ we call J *the range* of M. We will also write M_- for the smallest closed subspace containing M_s for every $s>0$.

Then $M_0 \supseteq M \supseteq M_-$ and simple examples would show that either or both of the inclusions can be strict. Call M left continuous if $M = M_0$. We note that every closed subspace M' such that $M_0 \supseteq M' \supseteq M_-$ is an invariant subspace and will have the same characteristic family as M.

Function theory on Bohr groups has been studied in great detail in Helson and Lowdenslager [8]. A function $F \in L^2(B; \mathbf{H})$ has the Fourier series expansion:

$$F = \sum_{\gamma \in \Gamma} a_\gamma \chi_\gamma, \quad (a_\gamma \in \mathbf{H}).$$

We call F *analytic* if the negative Fourier coefficients vanish, i.e. $a_\gamma = 0$ for $\gamma < 0$. The space consisting of all analytic functions is denoted by $H^2(B; \mathbf{H})$. If $\mathbf{H}_1 \subseteq \mathbf{H}$; $H^2(B; \mathbf{H}_1)$ consists of analytic functions whose Fourier coefficients $a_\gamma \in \mathbf{H}_1$. It is easy to see that $H^2(B; \mathbf{H}_1)$ is an exact invariant subspace.

6. Invariant spaces and cocycles

Let B be a Bohr group with dual Γ. Then we get a pair $(\mathbb{R}, B)$ where for $s \in \mathbb{R}$, $\phi(s) = \phi_s$ is the character on Γ:

$$\phi_s(\gamma) = \exp(is\gamma) \quad (\gamma \in \Gamma).$$

Let M be an exact left-continuous invariant subspace of $L^2(B; \mathbf{H})$. The characteristic family $\{M_s, s \in \mathbb{R}\}$ gives a spectral measure. Putting,

$$U_t = -\int_{\mathbb{R}} e^{its} \mathrm{d} M_s, \tag{16}$$

where we use M_s to denote the projection onto the subspace $M_s \subseteq JL^2(B; \mathbf{H})$, we have a representation of $\mathbb{R}$ over $JL^2(B; \mathbf{H})$. Using the fact that $\chi_\gamma . M_s = M_{s+\gamma}$ for each $\gamma \in \Gamma$, $s \in \mathbb{R}$, we have:

$$U_t \cdot \chi_\gamma = e^{it\gamma} \chi_\gamma \cdot U_t, \quad (t \in \mathbb{R}, \gamma \in \Gamma).$$

Since the spectral measure for the representation $\gamma \to \chi_\gamma$ of Γ on $JL^2(B; \mathbf{H})$ is the canonical spectral measure P^J, we have

$$U_t^{-1} P^J(E) U_t = P^J(E + \phi(t)),$$

where $t \in \mathbb{R}$ and $E \subseteq B$ is a Borel subset. That is, (U, P^J) is a system of imprimitivity. By Theorem 4.1, there exists a cocycle A having range J such that $U = U^A$.

Conversely, given a cocycle, A, having range J, we have a system of imprimitivity (U, P^J) for the pair $(\mathbb{R}, B)$. Expressing,

$$U_t = -\int_{\mathbb{R}} e^{its} \mathrm{d} M_s,$$

where $\{M_s, s \in \mathbb{R}\}$ is a left-continuous resolution of identity, it is easily shown that M_0 is an exact left-continuous simply invariant subspace having range J. We have proved the:

Theorem 6.1. *There exists a one-to-one correspondence between exact left-continuous invariant subspaces M of $L^2(B;\mathbf{H})$ and cocycles A for the pair $(\mathbb{R}, B)$. In this correspondence the range J of A is the range of the subspace M. If (U^A, P^J) is the system of imprimitivity corresponding to A, we have,*

$$U_t^A = -\int_{\mathbb{R}} \mathrm{e}^{its}\,\mathrm{d}M_s,$$

where $M = M_0$.

The following corollary will now follow from Corollary 2.3.

Corollary 6.2. *If M is an exact, left-continuous invariant subspace having range J, then J has constant dimension almost everywhere.*

Let ρ be a measurable partial isometry valued map on B. We denote T_ρ the operator induced by ρ on $L^2(B;\mathbf{H})$ as in (13). Theorem 3.2 and the proof of Theorem 5.1 yield.

Theorem 6.2. *Let A and A' be cohomologous cocycles:*

$$A'(t,x) = \rho(t,x)A(t,x)\rho(x+\phi_t)^* \quad (a.e.\,(t,x)).$$

If M and M' are the corresponding invariant subspaces, then:

$$M' = T_\rho(M).$$

Corollary 6.3. *The cocycle corresponding to the invariant subspace M is a coboundary if and only if M is of the form $T_\rho H^2(B;\mathbf{H}_1)$ where $\mathbf{H}_1 \subset \mathbf{H}$ is a closed subspace and ρ is a measurable partial isometry valued function such that* $\mathbf{In}\,\rho(s) = \mathbf{H}_1$ *a.e. x.*

The invariant subspaces described by the corollary above are the obvious ones as long as we insist on left-continuity. That there are others, would now follow from the existence of cocycles which are not coboundaries. Such cocycles were exhibited by Helson, Gamelin and Yale for the one-dimensional case (see refs. [3, 5, 9, 14]). These examples can be used to construct cocycles in higher dimensions which are not coboundaries by considering suitable direct sums. At this point one can ask if every cocycle in our situation is such a direct sum. For invariant subspaces the question is if every invariant subspace is a direct sum of one-dimensional invariant subspaces. For the special case of $B = T^2$ (T is the circle group), "irreducible" cocycles of all dimensions n ($n = 2, 3, \ldots, \infty$) have been exhibited in ref. [2] (see also ref. [10, footnote, p. 150]).

7. A theorem of Helson

In this section we proceed to obtain a criterion to decide when a function $F \in L^2(B;\mathbf{H})$ belongs to an exact invariant subspace M in terms of the associated cocycle A. This result was proved by Helson in ref. [5] (see also Muhly [10, Prop. 5.4]). Since the proof in our situation needs only the obvious modification over that of Helson's, we merely state the result. We need a definition first.

Definition 7.1. Let μ be the Cauchy measure on the line with density $d\mu(t)=1/\pi(1+t^2)dt$. By $H^2(\mathbb{R},\mu)$ we mean the subspace consisting of all $f\in L_2(\mathbb{R},\mu)$ for which,

$$h(x)=\frac{1}{\sqrt{2\pi}}\int_{\mathbb{R}} f(t)(1-it)^{-1}e^{itx}dt,$$

vanishes for a.e. $x<0$. Let $\mathbf{H}$ be a Hilbert space. Then $H^2(\mathbb{R},\mu;\mathbf{H})$ will denote the subspace of measurable functions, F, having μ-square summable norms such that for each $\xi\in\mathbf{H}$, the scalar function $t\to(F(t),\xi)$ is in $H^2(\mathbb{R},\mu)$.

Theorem 7.2. *Let M be an exact and left-continuous invariant subspace of $L^2(B;\mathbf{H})$ having range J. Let A be the $(\mathbb{R},B)$ cocycle corresponding to M. A function, F, in $JL^2(B;\mathbf{H})$ belongs to M if and only if, for a.e. x, the function $t\to A(t,x)F(x+\phi(t))$ belongs to $H^2(\mathbb{R},\mu;\mathbf{H})$.*

PART II

Applications to multivariate stationary stochastic processes

8. The problem

We shall be concerned in this part with a class of q-variate stochastic processes Y with time Γ, $Y=\{Y_\gamma:\gamma\in\Gamma\}$ where each Y_γ is a q-tuple of elements in a Hilbert space, $\mathbf{H}$. q will be a positive integer and the process will be weakly stationary in the sense that, writing $Y_\gamma=(Y_\gamma^1,\ldots,Y_\gamma^q)$, the inner product $(Y_\gamma^i,Y_{\gamma'}^j)$ will depend only on $\gamma-\gamma'$, once i and j, $1\leq i,j\leq q$, are fixed. We shall assume that in the spectral representation of Y, we have a spectral density (with respect to the Haar measure on B) and then we have the following general construction for Y.

Let W be a positive semi-definite matrix valued matrix function W on B, $W(x)=[W_{ij}(x)]_{i\leq i,j\leq q}$ where each $W_{ij}\in L^1(B)$. By $L^2(B,W)$ we will mean the space of $1\times q$-matrix valued measurable functions F on B for which $(F,F)_W<\infty$ where

$$(F,F')_W=\int_B F(x)W(x)F'^*(x)d\sigma(x),$$

$L^2(B;W)$ is a Hilbert space with this inner product $(\cdot,\cdot)_W$ (see Rosenberg [12]). We shall use the notation $\|F\|_W=(F,F)_W$. Now let I_j denote the $1\times q$-matrix having all entries equal to zero, except the jth which is equal to 1. As before, χ_γ is the character on B corresponding to $\gamma\in\Gamma$. We put

$$Y_\gamma=(\chi_\gamma I_1,\ldots,\chi_\gamma I_q),\quad \gamma\in\Gamma.$$

It is clear that Y is a stationary stochastic process in the above sense.

For each $t \in R$ we define $\mathbf{H}_t$, the space of the process up to time t as the smallest subspace of $L^2(B, W)$ containing the vectors $Y_\gamma^j, \gamma \leqslant t, 1 \leqslant j \leqslant q$. The space $\mathbf{H}_{-\infty} = \bigcap_{t > -\infty} \mathbf{H}_t$ is called the remote past and $\mathbf{H}_\infty = \bigvee_{t > -\infty} \mathbf{H}_t$ is called the space of the process. The process will be called purely deterministic if $\mathbf{H}_{-\infty} = \mathbf{H}_\infty$ and purely non-deterministic if $\mathbf{H}_{-\infty} = \{0\}$. Our problem in this chapter is to decide, in terms of W, when the process is purely non-deterministic.

We now provide the connection of this problem with the first part of the paper. Given W, let Q be the function: $Q(x) = W^{1/2}(x)$, the positive semidefinite square root ($x \in B$). Now if $F \in L^2(B, W)$, the function $Q \cdot F$ defined by,

$$Q \cdot F(x) = F(x) Q(x), \quad x \in B,$$

belongs to $L^2(B; \mathbb{C}^q)$ (elements of $\mathbb{C}^q$ being written as $1 \times q$-matrices). It is easy to see that $\|Q \cdot F\| = \|F\|_W$ where $\|\cdot\|$ stands for the norm in $L^2(B; \mathbb{C}^q)$. Now let M denote the image of $\mathbf{H}_0$ under the mapping $F \to Q \cdot F$. The following proposition is self-evident.

Proposition 9.1. *With the above notation*:

(i) *M is doubly invariant if and only if Y is purely deterministic.*

(ii) *M is an exact invariant subspace if and only if Y is purely non-deterministic.*

9. Some facts about continuous time-parameter stationary stochastic processes

On $\mathbb{R}$, we will mostly use the Cauchy measure μ, $\mathrm{d}\mu(t) = 1/\pi(1+t^2)\mathrm{d}t$. Let $W_1(t) = [f_{ij}(t)]_{1 \leqslant i, j \leqslant q}$ be a μ-almost everywhere positive semidefinite matrix function with $f_{ij} \in L^1(\mathbb{R}, \mu)$. As on B, we define the space $L^2(\mathbb{R}, W_1)$ with respect to μ. We consider the canonical stationary stochastic process,

$$X = \{X_s = (X_s^1, \ldots, X_s^q), s \in \mathbb{R})\},$$

where $X_s^j(t) = \mathrm{e}^{\mathrm{i}st} I_j$ (a.e. t). As before, we consider the spaces $\mathbf{H}_r(X)$ generated by the vectors X_s^j, $s \leqslant r$ and $1 \leqslant j \leqslant q$ and $\mathbf{H}_{-\infty}(X) = \bigcap_{r \in \mathbb{R}} \mathbf{H}_r(X)$.

Definition 9.1. Let $\mathbf{H}$ be a separable Hilbert space. A measurable range function J on $\mathbb{R}$ is said to be an analytic range function if there exists a sequence of functions $F_i \in H^2(R, \mu; \mathbf{H})$ (see Section 6), $i \geqslant 1$, such that for a.e. $s \in \mathbb{R}$, $J(s)$ is the closed subspace generated by $\{F_i(s); i \geqslant 1\}$.

For any positive semi-definite matrix W, $\det W$ will stand for the determinant of W regarded as a transformation on its range, where it is non-singular. The following theorem gives the result we want.

Theorem 9.2. *The following are equivalent*:

(a) *The canonical process X in $L^2(\mathbb{R}, W_1)$ is purely deterministic, that is $\mathbf{H}_{-\infty}(X) = 0$.*

(b) *$W_1 = A_1 A_1^*$ where A_1 is a $q \times q$ matrix valued function with entries belonging to $H_2(\mathbb{R}, \mu)$.*

(c) *$\int_R \log \det W_1(t) \, \mathrm{d}\mu(t) > -\infty$ and W_1 has analytic range.*

Proof. We will use the map,

$$\tau : \mathrm{e}^{\mathrm{i}\theta} \to t = \tan\tfrac{1}{2}(\theta - \pi),$$

of the circle T (parametrised by $[0, 2\pi)$) onto $\mathbb{R} \cup \{\infty\}$. This map transforms the normalised Lebesgue measure $(1/2\pi)\mathrm{d}\theta$ into the Cauchy measure μ on $\mathbb{R}$, and the usual Hardy space H^2 on T to the space $H^2(\mathbb{R}, \mu)$. Let $\tilde{W}_1(\theta) = W_1(\tau(\theta))$. Then $\tilde{W}_1$ is a weight function on T and we have the canonical process $\{\tilde{X}_n, n \in \mathbb{Z}\}$ with respect to $\tilde{W}_1$. It is known that $\tilde{X}$ is purely non-deterministic if and only if X is purely non-deterministic (see Rozanov [12, p. 112]). Our theorem is now a restatement for $\mathbb{R}$ of Theorem 13 in ref. [8] for T.

We make a simple remark here. Suppose the cannonical process X in $L^2(\mathbb{R}, W_1)$ has trivial remote past. Let $F \neq 0$ be fixed in $L^2(\mathbb{R}, W_1)$ and let $\mathrm{d}(F, \mathbf{H}_r(X))$ stand for the metric distance of F from $\mathbf{H}_r(X)$ in $L^2(\mathbb{R}, W_1)$. Then, from the way $\mathbf{H}_{-\infty}$ is defined, $\lim_{r \to -\infty} \mathrm{d}(F, \mathbf{H}_r(X)) = \| F \|_{W_1}$. Suppose now $F_r \in \mathbf{H}_r(X)$ is arbitrarily chosen for each $r \in \mathbb{R}$. Since $\| F - F_r \|_{W_1} \geqslant \mathrm{d}(F, \mathbf{H}_r(X))$, it follows that:

$$\liminf_{r \to -\infty} \| F - F_r \|_{W_1} \geqslant \| F \|_{W_1} > 0. \tag{17}$$

10. The main theorem

We use the terminology and notation explained in Sections 7 and 8.

Theorem 10.1. *Let W be a $q \times q$ positive semi-definite matrix valued weight function on B. The canonical process X in $L^2(B, W)$ is purely non-deterministic if and only if:*

(i) *For a.e. x, $W(x + \phi_t)$ has an analytic range function on $\mathbb{R}$.*

(ii) *For a.e. x, $\int_{\mathbb{R}} \log\det W(x + \phi_t)\,\mathrm{d}\mu(t) > -\infty$.*

Proof. Let $Q, Q(x) = W^{1/2}(x)$ be the pointwise positive semi-definite square root of W. Assume that Y is purely non-deterministic. Then by proposition $M = Q \cdot \mathbf{H}_0(Y)$ is an exact invariant subspace. If M is not left-continuous consider $M_0 \supseteq M$, the left-hand limit at 0, which is a left continuous exact invariant subspace. Let A be the corresponding cocycle. Using Theorem 7.2 we have, for each F in M_0, the functions $t \to F(x + \phi_t)A(t, x)$ belonging to $H^2(\mathbb{R}, \mu; \mathbb{C}^q)$ for a.e. $x \in B$ (the element $A(t, x) \in \mathbf{P}(\mathbb{C}^q)$ acts by matrix multiplication from the right). Now the constant function $I_j \in \mathbf{H}_0(Y)$, $I_j \cdot Q \in M$ for each j, $1 \leqslant j \leqslant q$. Thus,

$$t \to I_j Q(x + \phi_t) A(t, x),$$

defines a function in $H^2(\mathbb{R}, \mu; C^q)$ for each j, a.e. x. This means that the entries of the matrix function $Q(x + \phi_t)A(t, x)$ are in $H^2(\mathbb{R}, \mu)$.

Recall that M and A must have the same range function, say J. Since $M = Q \cdot \mathbf{H}_0(Y)$, it is easily seen that $J(x) = \mathbf{R}Q(x)$ a.e. Then $A(t, x)A^*(t, x)$ is the identity transformation on $J(x+\phi_t)=\mathbf{R}Q(x+\phi_t)$ a.e. (t, x). We have the factorisation

$$\begin{aligned}W(x+\phi_t)&=Q(x+\phi_t)Q(x+\phi_t)\\&=Q(x+\phi_t)A(t,x)A^*(t,x)Q(x+\phi_t)\\&=Q(x+\phi_t)A(t,x)[Q(x+\phi_t)A(t,x)]^*,\end{aligned}$$

for a.e. x for a.e. t. By Theorem 9.2 the necessity of (i) and (ii) follows.

Conversely, suppose W satisfies conditions (i) and (ii) of the theorem. We shall show that $\mathbf{H}_{-\infty}(Y)=\{0\}$. Let $F\in\mathbf{H}_{-\infty}(Y)$, that is, $F\in\mathbf{H}_t(Y)$ for all $t\in\mathbb{R}$. By definition $\mathbf{H}_t(Y)$ is the closure, $L^2(B,W)$, of all trigonometric polynomials having their spectrum contained in $[-t,\infty)$. For each n, choose a trigonometric polynomial P_n whose spectrum is contained in $[n,\infty)$ such that:

$$\|F-P_n\|_W<\frac{1}{n}\quad\text{and}\quad P_n\in\mathbf{H}_{-n}.$$

Now

$$\begin{aligned}\|F-P_n\|_W&=\int_B(F(x)-P_n(x))W(x)(F(x)-P_n(x))^*\,\mathrm{d}\sigma(x)\\&=\int_B\mathrm{d}\sigma(x)\int_{\mathbb{R}}(F(x+\phi_t)-P_n(x+\phi_t))W(x+\phi_t)(F(x+\phi_t)\\&\qquad-P_n(x+\phi_t))^*\,\mathrm{d}\mu(t)\\&=\int_B H_n(x)\,\mathrm{d}\sigma(x),\end{aligned}$$

where $H_n(x)$ is the inside integral in the previous step. Since $F\in L^2(B,W)$, Fubini's theorem would imply that $F(x+\phi_t)$ as a function of t would belong to $L^2(\mathbb{R},W(x+\phi_t))$ for a.e. x. Fix such an $x\in B$ for which conditions (i) and (ii) of the theorem are satisfied. Then by Theorem 9.2, the canonical process X in $L^2(\mathbb{R},W(x+\phi_t))$ has no remote past. Further, the function $t\to P_n(x+e_t)$ is a trigonometric polynomial on $\mathbb{R}$ having its spectrum contained in $[n,\infty)$ and so belonging to $\mathbf{H}_{-n}(X)$. Therefore, from the relation (17) in Section 9, if $t\to F(x+\phi_t)$ is a non-zero function, then $\liminf_{n\to\infty}H_n(x)>0$. But then we have from the Fatou–Lebesgue lemma

$$0=\lim_{n\to\infty}\int_B H_n(x)\,\mathrm{d}\sigma(x)\geqslant\int_B\liminf_{n\to\infty}H_n(x)\,\mathrm{d}\sigma(x).$$

This proves that for a.e. x. $F(x+\phi_t)=0$ a.e. t and hence $F=0$. Thus $\mathbf{H}_{-\infty}(Y)=\{0\}$.

Acknowledgements

The material presented here is taken from the author's doctoral thesis [1]. The author thankfully acknowledges help and suggestions received from his thesis supervisor Professor M.G. Nadkarni.

References

[1] Bagchi, S.C. (1973). Invariant subspaces of vector-valued function spaces on Bohr groups. Ph.d. Thesis. Indian Statistical Institute, Calcutta.

[2] Bagchi, S.C., Mathew, J. and Nadkarni, M.G. (1974). On systems of imprimitivity on locally compact abelian groups with dense actions. *Acta Math.* **133**, 287–304.

[3] Gamelin, T.W. (1969). *Uniform Algebras*. Prentice-Hall, NJ.

[4] Helson, H. (1964). *Lectures on Invariant Subspaces*. Academic Press, New York.

[5] Helson, H. (1965). Compact groups with ordered duals. *Proc. London Math. Soc.* **14A**, 144–156.

[6] Helson, H. (1975). Analyticity on compact groups with ordered duals. *Algebra in Analysis*, J.H. Williamson, Ed. Academic Press, New York.

[7] Helson, H. and Lowdenslager, D. (1960). Invariant subspaces. *Proc. Internat. Symp. Linear Spaces*, Jerusalem. 251–262.

[8] Helson, H. and Lowdenslager, D. (1958). Prediction theory and Fourier series in several variables I. *Acta Math.* **99**, 165–202; II, *Acta Math.* **106**, 175–213.

[9] Helson, H. and Kahane, J.P. (1972). Compact groups with ordered duals III. *J. London Math. Soc.* **4**, 573–575.

[10] Muhly, P.S. (1972). A structure theory for isometric representations of a class of semigroups. *J. Reine Angew. Math.* **255**, 135–154.

[11] Rosenberg, M. (1964). The square-integrability of matrix-valued functions with respect to a non-negative Hermitian measure. *Duke Math. J.* **31**, 291–298.

[12] Rozanov, Yu.A. (1967). *Stationary Stochastic Processes*. English Translation. Holden-Day

[13] Varadarajan, V.S. (1970). *Geometry of Quantum Theory*, Vol. 2. Van Nostrand Reinhold Co.

[14] Yale, K. (1971). Invariant subspaces and projective representations. *Pacific J. Math.* **36**, 557–565.

G. Kallianpur, P.R. Krishnaiah, J.K. Ghosh, eds., *Statistics and Probability: Essays in Honor of C.R. Rao*

A NOTE ON THE EFFECTIVE VARIANCE OF RANDOMLY STOPPED MEANS*

R.R. BAHADUR
The University of Chicago, Chicago, IL, U.S.A.

1. Introduction

Let $X_1, X_2, \ldots$ be a sequence of i.i.d. real valued random variables. It is assumed that the m.g.f. $\phi(t)=E[\exp(tX_1)]$ is finite in a neighborhood of $t=0$, and that

$$E(X_1)=0, \qquad \mathrm{Var}(X_1)=\sigma^2>0. \tag{1}$$

Let $N_1, N_2, \ldots$ be a sequence of random variables such that each N_n takes values in the set of positive integers. No particular relation between the random variables N_n and X_i is assumed, except of course that they be defined on the same probability space. It is assumed, however, that N_n is asymptotically constant in the following sense: A sequence $m_1, m_2, \ldots$ exists of positive integers m_n with $\lim_{n\to\infty} m_n=+\infty$ such that

$$N_n/m_n \to 1 \text{ in probability} \tag{2}$$

as $n\to\infty$. For each n, let

$$T_n=\left(\sum_{i=1}^{N_n} X_i\right)/N_n. \tag{3}$$

It then follows from (1) and (2) by the Doeblin–Anscombe theorem [see Chow and Teicher (1978)] that T_n is asymptotically normally distributed with asymptotic mean 0 and asymptotic variance σ^2/m_n as $n\to\infty$.

Choose $\varepsilon>0$ and consider the large deviation probability $P(|T_n|\geqslant\varepsilon)$. It is often convenient to study this probability in terms of a quantity called the effective standard deviation of T_n, which is denoted by $\tau_n(\varepsilon)$, and defined by

$$P(|T_n|\geqslant\varepsilon)=P\left(|N(0,1)|\geqslant\frac{\varepsilon}{\tau_n(\varepsilon)}\right), \tag{4}$$

$0\leqslant\tau_n\leqslant\infty$. The question considered in this note is whether σ^2/m_n is an adequate estimate of τ_n^2 in the following sense: with $r_n(\varepsilon)$ defined by

$$\tau_n^2(\varepsilon)=\frac{\sigma^2}{m_n}[1+r_n(\varepsilon)], \tag{5}$$

*Support for this research was provided in part by the National Science Foundation Grant No. MCS76-81435.

$-1 \leqslant r_n \leqslant +\infty$, we have

$$\lim_{\varepsilon \to 0} \overline{\lim_{n \to \infty}} |r_n(\varepsilon)| = 0. \tag{6}$$

If this is the case, we will say that the asymptotic effective variance of T_n equals its asymptotic variance. The validity of (6) is equivalent (cf. Section 2) to the statement that, for each sufficiently small ε, $-(\varepsilon^2/2\sigma^2)m_n$ is a precise asymptotic estimate of $\log P(|T_n| \geqslant \varepsilon)$.

The following construction shows that in the present context equality of the two variances always requires some additional condition. Suppose that $P(N_n = n) = 1/n^2$, $P(N_n = 2n) = 1 - (1/n^2)$. Then (2) holds with $m_n \equiv 2n$; indeed, $N_n = 2n$ for all sufficiently large n with probability one. Hence the asymptotic variance of T_n is $\sigma^2/2n$. However, if each N_n is independent of the sequence $\{X_i\}$ then (5) and (6) hold with $m_n \equiv n$, so the asymptotic effective variance of T_n is σ^2/n.

Returning to the general case, suppose that the given sequences $\{N_n\}$ and $\{m_n\}$ satisfy not only (2) but the following stronger condition: For any given $\delta > 0$ there exists a ρ with $0 < \rho < 1$ such that

$$P\left(\left|\frac{N_n}{m_n} - 1\right| > \delta\right) < \rho^{m_n} \tag{7}$$

for all sufficiently large n. It is shown in the following section that then (6) does hold for r_n defined by (5).

The present considerations have the following application to probabilities of large deviations of maximum likelihood and related estimates based on observation of a Markov chain. Consider a finite state space, and suppose that the one-step transition probabilities are sufficiently smooth positive functions of a parameter θ taking values in an open set in a finite dimensional space. Let $g(\theta)$ be a sufficiently smooth, real valued function of θ. For each n, let U_n be the maximum likelihood estimate of g based on the first n steps of the chain. Suppose that a particular θ obtains. It can be shown that there exists an affine function of the observed transition counts in the first n steps, say $T_{n,\theta}$, such that the difference $U_n - T_{n,\theta}$ is asymptotically negligible in a sufficiently strong sense. It is known [see Billingsley (1961)] that the observed transition counts may be represented as randomly stopped sums of i.i.d. zero–one variables. It is also known that here the stopping variables are asymptotically constant in the sense of the preceding paragraph. It follows hence, by an elaboration of the argument of the following section, that the asymptotic effective variance of $T_{n,\theta}$ equals its asymptotic variance, say $\sigma^2(\theta)/n$. The quantity $\sigma^2(\theta)$ may be found from formulae given in Billingsley (1961) or Kemeny and Snell (1960). Since $U_n - T_{n,\theta}$ is negligible, it follows that the asymptotic effective variance of U_n is $\sigma^2(\theta)/n$. It might be added that this conclusion, together with known lower bounds for probabilities of large deviations of any consistent estimate of g, implies that U_n is an efficient estimate in the sense of asymptotic effective variances.

2. Proof. For each positive integer j let $\bar{X}_j = (\Sigma_{i=1}^{j} X_i)/j$. Let $f(\varepsilon) = \inf\{-t\varepsilon + \log \phi(t) : t \geqslant 0\}$. Then, for each $\varepsilon > 0$,

$$P(\bar{X}_j \geqslant \varepsilon) \leqslant \exp[-j \cdot f(\varepsilon)] \tag{8}$$

for each j by Bernstein's inequality, and

$$\lim_{j\to\infty}\left\{j^{-1}\log P\left(\bar{X}_j \geqslant \varepsilon\right)\right\} = -f(\varepsilon) \tag{9}$$

by Chernoff's theorem [see, for example, Bahadur (1971)]. Since $\phi(t)<\infty$ for sufficiently small $|t|$, it follows easily from (1) that

$$f(\varepsilon)=\frac{\varepsilon^2}{2\sigma^2}+O(\varepsilon^3) \tag{10}$$

as $\varepsilon\downarrow 0$. For each positive integer j let p_j be the probability that X_1, $X_1+X_2,\ldots,$ and $X_1+\cdots+X_j$ are all positive. Then

$$\lim_{j\to\infty}\left\{j^{1/2}\cdot p_j\right\}=d, \tag{11}$$

where d is a positive constant [see Chung (1968), p. 272)].

Choose and fix δ, $0<\delta<1$, and let A_n be the event that $|N_n/m_n-1|>\delta$. Let $\rho=\rho(\delta)$ and $n_1=n_1(\delta)$ be such that $0<\rho<1$ and (7) holds for $n>n_1$. Choose $\varepsilon>0$, and let B_n be the event that, with T_n defined by (3), $T_n\geqslant\varepsilon$. Let C_n be the event that $\bar{X}_j\geqslant\varepsilon$ for at least one j such that $m_n(1-\delta)\leqslant j\leqslant m_n(1+\delta)$. Then $B_n\subset A_n\cup C_n$. It follows from (8) that $P(C_n)\leqslant(2m_n\delta+1)\exp[-m_n(1-\delta)f(\varepsilon)]$. It follows from (10) that there exists $\varepsilon_1=\varepsilon_1(\delta)>0$ such that $\rho<\exp[-(1-\delta)f(\varepsilon)]$ for $0<\varepsilon<\varepsilon_1$. It follows hence from (7) that, for $0<\varepsilon<\varepsilon_1$ and $n>n_1$, $P(B_n)\leqslant P(A_n)+P(C_n)\leqslant(2m_n\delta+2)\exp[-m_n(1-\delta)f(\varepsilon)]$. Hence, by the definition of B_n,

$$\overline{\lim_{n\to\infty}}\left\{m_n^{-1}\log P(T_n\geqslant\varepsilon)\right\}\leqslant-(1-\delta)f(\varepsilon) \tag{12}$$

for $0<\varepsilon<\varepsilon_1$.

Next we bound $P(T_n\geqslant\varepsilon)$ from below. Let $k_n=k_n(\delta)$ be the greatest integer less than $m_n(1-\delta)$, let $\lambda=(1+2\delta)/(1-\delta)$, and let D_n be the event that $\bar{X}_{k_n}\geqslant\lambda\varepsilon$. Let A_n and B_n be defined as in the preceding paragraph, and let F_n be the event that, for every i with $1\leqslant i\leqslant(2m_n\delta+1)$ the sum $X_{k_n+1}+\cdots+X_{k_n+i}$ is positive. Since $k_n/m_n\to(1-\delta)$,

$$\frac{k_n\lambda}{k_n+2m_n\delta+1}>1, \tag{13}$$

for all sufficiently large n, say for $n>n_2(\delta)$. It is readily seen from (13) that $(D_n\cap F_n)\subset(A_n\cup B_n)$. Hence

$$P(B_n)\geqslant P(D_n\cap F_n)-P(A_n) \tag{14}$$

for $n>n_2$. It follows from the definition of F_n by (11) that

$$\lim_{n\to\infty}\left\{m_n^{-1}\log P(F_n)\right\}=0. \tag{15}$$

Since $P(D_n \cap F_n) = P(D_n) \cdot P(F_n)$, we have

$$\lim_{n\to\infty} \{m_n^{-1} \log P(D_n \cap F_n)\} = -(1-\delta) f(\lambda\varepsilon) \tag{16}$$

by (9) and (15). It follows from (7), (10) and (16) that there exists $\varepsilon_2 = \varepsilon_2(\delta) > 0$ such that, if $0 < \varepsilon < \varepsilon_2$, the ratio $P(A_n)/P(D_n \cap F_n)$ is well defined for all sufficiently large n and converges to zero. Hence

$$\underline{\lim}_{n\to\infty} \{m_n^{-1} \log P(T_n \geqslant \varepsilon)\} \geqslant -(1-\delta) f(\lambda\varepsilon) \tag{17}$$

for $0 < \varepsilon < \varepsilon_2$, by (14) and (16).

Let $g_1, g_2, \ldots$ be a sequence of extended real valued functions of $\varepsilon > 0$. In the following, $\underline{\lim}\,\underline{\lim}\{g_n(\varepsilon)\}$ and $\overline{\lim}\,\overline{\lim}\{g_n(\varepsilon)\}$ denote the iterated inferior limit and superior limit of $g_n(\varepsilon)$, first as $n \to \infty$ and then as $\varepsilon \to 0$. With h a real valued constant, we write $\{g_n(\varepsilon)\} \doteq h$ if and only if $\underline{\lim}\,\underline{\lim}\{g_n(\varepsilon)\} = \overline{\lim}\,\overline{\lim}\{g_n(\varepsilon)\} = h$.

By dividing (12) and (17) by ε^2 and letting $\varepsilon \to 0$, it follows from (10) that

$$\overline{\lim}\;\overline{\lim}\left\{\left(m_n\varepsilon^2\right)^{-1} \log P(T_n \geqslant \varepsilon)\right\} \leqslant -\frac{(1-\delta)}{2\sigma^2} \tag{18}$$

and

$$\underline{\lim}\;\underline{\lim}\left\{\left(m_n\varepsilon^2\right)^{-1} \log P(T_n \geqslant \varepsilon)\right\} \geqslant -\frac{(1-\delta)\lambda^2}{2\sigma^2}. \tag{19}$$

Since δ is arbitrary, and $\lambda = (1+2\delta)/(1-\delta)$, it follows from (18) and (19) that the left-hand sides in (18) and (19) are both equal to $-1/2\sigma^2$. Thus

$$\left\{\left(m_n\varepsilon^2\right)^{-1} \log P(T_n \geqslant \varepsilon)\right\} \doteq -\frac{1}{2\sigma^2}. \tag{20}$$

It follows from symmetry that (20) continues to hold with $P(T_n \geqslant \varepsilon)$ replaced by $P(T_n \leqslant -\varepsilon)$. Since $P(T_n \geqslant \varepsilon) \leqslant P(|T_n| \geqslant \varepsilon) \leqslant 2\max\{P(T_n \leqslant -\varepsilon), P(T_n \geqslant \varepsilon)\}$, it now follows that we also have

$$\left\{\left(m_n\varepsilon^2\right)^{-1} \log P(|T_n| \geqslant \varepsilon)\right\} \doteq -\frac{1}{2\sigma^2}. \tag{21}$$

We will complete the proof by showing that (21) is always equivalent to the desired conclusion (6).

Choose $\varepsilon > 0$ and consider the following conditions: (i) $0 < P(|T_n| \geqslant \varepsilon) < 1$ for all sufficiently large n, and $P(|T_n| \geqslant \varepsilon) \to 0$, and (ii) $0 < \tau_n(\varepsilon) < \infty$ for all sufficiently large n, and $\tau_n(\varepsilon) \to 0$. It follows from (4), by elementary properties of the $N(0,1)$-distribution, that (i) and (ii) are equivalent, and that if they are satisfied,

$$\lim_{n\to\infty} \{\tau_n^2(\varepsilon) \cdot \log P(|T_n| \geqslant \varepsilon)\} = -\tfrac{1}{2}\varepsilon^2. \tag{22}$$

If (21) holds then (i), and therefore (22), hold for each sufficiently small ε; hence

$$\left\{\frac{m_n \cdot \tau_n^2(\varepsilon)}{\sigma^2}\right\} \doteq 1 \tag{23}$$

by (21) and (22). On the other hand, if (23) holds, then (ii), and therefore (22), hold for each sufficiently small $\varepsilon>0$; hence (21) holds, by (22) and (23). Thus (21) and (23) are equivalent. It is clear from (5) that (23) is equivalent to $\{r_n(\varepsilon)\} \doteq 0$. It is readily seen that $\{r_n(\varepsilon)\} \doteq 0$ is equivalent to (6). This completes the proof.

References

Bahadur, R.R. (1971). *Some Limit Theorems in Statistics*. SIAM, Philadelphia, PA.
Billingsley, P. (1961). Statistical methods in Markov chains. *Ann. Math. Statist.* **32**, 12–40.
Chow, Y.S. and Teicher, H. (1978). *Probability Theory*. Springer-Verlag, New York.
Chung, K.L. (1968). *A Course in Probability Theory*. Harcourt, Brace, and World, Inc., New York.
Kemeny, J.G. and Snell, J.L. (1960). *Finite Markov Chains*. Van Nostrand Rheinhold Company, New York.

G. Kallianpur, P.R. Krishnaiah, J.K. Ghosh, eds., *Statistics and Probability*: *Essays in Honor of C.R. Rao*

SOME ESTIMATION PROBLEMS FOR RANDOM FIELDS*

A.V. BALAKRISHNAN
UCLA, Los Angeles, CA, U.S.A.

1. Introduction

Recently there has been much interest in estimation and inference problems for random fields under the signal-plus-noise model: See [1–4], particularly with new application areas such as statistical geodesy [1]. But much of this work is of heuristic or approximate nature, and in this paper we obtain precise formulas for likelihood ratios and Kalman filtering by exploiting Hilbert-space techniques. The theory of weak random variables suffices for linear smoothing problems, where in particular we obtain the continuous-parameter generalization of the results in [5]. The likelihood-ratio and non-linear estimation theory involves the use of the authors's "non-linear white-noise" theory [6].

Section 2 deals with the linear smoothing problems using the estimation theory for weak random variables developed in [7]. We develop an exact formula for the likelihood ratio for random fields and show in particular how a Krein-factorization is possible for rectangular as well as non-rectangular domains in R^2, extending previous results in [8]. In Section 3, we derive a Kalman filtering formalism to obtain the Krein factorization (under appropriate sufficient conditions) for some non-rectangular domains in R^2. The "state space" for the filter has to be non-finite dimensional, it turns out. In Section 5, we develop a Cramer–Rao bound for estimation error based on our likelihood-ratio formula.

2. Linear estimation (smoothing) theory

We begin with linear estimation (smoothing) theory in the generality of Hilbert-space valued weak random variables. See [7] for the background theory. As long as only linear estimation is involved, one can use equivalently the "generalized" processes of Gelfand–Vilenkin [9]. Since we are, however, interested in the non-linear case, we prefer to use the weak random variable theory to provide a unified framework.

Let A denote a closed linear operator with domain dense in $\mathcal{H}$ such that zero is in its resolvent set; that, in other words, A^{-1} is defined and (can be extended to be) bounded. Let us use $\mathfrak{L}$ to denote A^{-1}. Let as consider the "equation":

$$AX = N$$

where N is white noise in $\mathcal{H}$, as defining the "process" X, since it has a unique

*Research supported in part under grant no. 78-3550, Applied Mathematics Division, AFOSR, United States Air Force.

solution for each "sample" N in $\mathcal{H}$. Then X is a (weak) Gaussian random variable in $\mathcal{H}$ and can be equivalently defined by

$$X=\mathcal{L}N.$$

In particular, the co-variance operator

$$E[XX^*]=\mathcal{L}\mathcal{L}^*.$$

Let us now define the "observation" or "observed process" as

$$v=CX+N_0$$

where C is a linear bounded operator mapping $\mathcal{H}$ into another separable Hilbert space $\mathcal{H}_0$ and N_0 is white Gaussian noise in $\mathcal{H}_0$ independent of the white noise N. Then

$$E[vv^*]=C\mathcal{L}\mathcal{L}^*C^*+I.$$

Let us use S to denote the process CX. The conditional expectations

$$\hat{S}=E[S|v], \qquad \hat{X}=E[X|v]$$

are both well defined (see [7]). We have

$$\hat{S}=C\hat{X} \quad \text{and} \quad \hat{X}=Lv$$

where L is determined from:

$$E[Xv^*]=LE[vv^*]$$

or

$$\mathcal{L}\mathcal{L}^*C^*=L(I+C\mathcal{L}\mathcal{L}^*C^*).$$

Hence

$$L=\mathcal{L}\mathcal{L}^*C^*(I+C\mathcal{L}\mathcal{L}^*C^*)^{-1}.$$

It is more convenient to express L in a slightly different way. Thus let

$$L=M\mathcal{L}^*C^*.$$

Then we have that

$$M=\mathcal{L}(I+\mathcal{L}^*C^*C\mathcal{L})^{-1}.$$

Moreover, since zero is in the resolvent set of A (and of A^*),

$$\mathcal{L}(I+\mathcal{L}^*C^*C\mathcal{L})^{-1}\mathcal{L}^*=(C^*C+A^*A)^{-1},$$

and hence, we have

$$L=(C^*C+A^*A)^{-1}C^*$$

and

$$\hat{X}=(C^*C+A^*A)^{-1}C^*V.$$

The corresponding error covariance matrix is readily seen to be:

$$E\left[(X-\hat{X})(X-\hat{X})^*\right]=(C^*C+A^*A)^{-1}.$$

Next

$$A^*A\hat{X}+C^*C\hat{X}=C^*v$$

which we can also rewrite as:

$$A^*A\hat{X}=C^*(v-C\hat{X})$$
$$=C^*(v-\hat{S}). \tag{2.1}$$

Let

$$\nu=v-\hat{S}.$$

Then ν is *not* white noise in $\mathcal{H}_0$. For,

$$E[\nu\nu^*]=(I-CL)(I+C\mathfrak{L}\mathfrak{L}^*C^*)(I-CL)^*$$
$$=\left(I-C(C^*C+A^*A)^{-1}C^*\right)$$

in using the well-known identity

$$\left(I-T(I+T^*T)^{-1}T^*\right)(I+TT^*)=I$$

for any bounded linear operator T mapping $\mathcal{H}$ into $\mathcal{H}_0$.

Let us see how this applies to random fields. Let $t\in\mathfrak{D}\subset R^n$ and let

$$v(t)=S(t)+N_0(t)$$

where $N_0(\cdot)$ is white noise in $L_2\mathfrak{D}$ and let $S(t)$ be a Gaussian field with covariance

$$R(t_1,t_2)=E[S(t_1)S(t_2)].$$

Let R denote the covariance operator $Rf=g$; $g(t)=\int_{\mathfrak{D}}R(t,s)f(s)\,\mathrm{d}s$ mapping $L_2(D)$ into itself. Assume that

$$\int_D E\left[S(t)^2\right]\mathrm{d}t<\infty$$

so that R is nuclear. Then R can be factored (even if not uniquely)

$$R=\mathfrak{L}\mathfrak{L}^*$$

where $\mathfrak{L}$ is Hilbert–Schmidt, mapping $\mathfrak{L}_2(D)$ into $\mathfrak{L}_2(D)$ and

$$\mathfrak{L}+\mathfrak{L}^*$$

is nuclear. Assume that R is one-to-one; that in other words, zero is not eigenvalue of R. Let

$$A=\mathfrak{L}^{-1}$$

so that A is closed, with domain dense in $\mathfrak{L}_2(\mathfrak{D})$. Then we can apply the previous theory to obtain

$$\hat{S}(t)=E\left[S(t)\,|\,v(s),s\in\mathfrak{D}\right].$$

Thus

$$\hat{S}=(I+A^*A)^{-1}v.$$

Example 1. In some applications we may be given $S(t)$ as the solution of a partial differential equation: e.g. (of importance in geophysical applications):

$$\nabla^2 S(t) = N(t),$$

where ∇^2 is the Laplacian and $\mathfrak{D} \subset R^3$, or R^2, and $S(t)$ satisfies homogeneous boundary conditions:

$$S(t) \text{ on } \Gamma$$

where Γ is the boundary of $\mathfrak{D}$. Here ∇^2 is clearly defined on the space of infinitely smooth functions with compact support in $\mathfrak{D}$ and A can be taken as its smallest closure. Then A has a bounded inverse $\mathfrak{L}$ and $\mathfrak{L}$ is Hilbert–Schmidt if $\mathfrak{D}$ is bounded. Moreover A and $\mathfrak{L}$ are self-adjoint. If we take C to be the identity so that we have:

$$v = S + N,$$

then

$$\hat{S} = (I + A^* A)^{-1} v.$$

If the eigenfunctions of ∇^2 are known (as for instance if $\mathfrak{D}$ is a rectangle), then we can of course express $\hat{S}$ as

$$\hat{S} = \sum_1^\infty \frac{[v, \phi_n]\phi_n}{1 + \lambda_n^2}$$

where ϕ_n are the eigenfunctions:

$$\nabla^2 \phi_n = \lambda_n \phi_n.$$

In this particular case (being self-adjoint) ϕ_n are also the eigenfunctions of the covariance operator $\mathfrak{L}^* \mathfrak{L}$. For an actual calculation for a rectangle in R^2, see [4], where the special nature of the eigenfunctions in this case leads to some computational advantage over other possible techniques. We note that our result (2.1) generalizes the discrete-time case treated by Novikov [5].

3. Likelihood-ratios

We begin with a general result (for a more detailed treatment of which we refer the reader to [6]). Let W be a separable Hilbert space and let N denote white noise in W, with characteristic functional

$$E[\exp i[N, h]] = \exp -\tfrac{1}{2}[h, h], \quad h \in W,$$

and let us denote the corresponding weak distribution (or finitely additive measure) by μ_G. Let S denote an independent Gaussian process with characteristic functional

$$E[\exp i[S, h]] = \exp \tfrac{1}{2}[Rh, h]$$

where we assume that R is nuclear so that in particular the corresponding finitely-additive measure μ_S is actually countably additive. Let

$$v=S+N$$

and because of the asserted independence,

$$[\exp i[v,h]]=\exp -\tfrac{1}{2}[(I+R)h,h].$$

Let $\phi(\cdot)$ denote the $R-N$ derivative of μ_V with respect to μ_G, so that for any cylinder set, C, with (finite-dimensional) Borel base,

$$\mu_V(C)=\lim \int_C \phi(P_n h)\,\mathrm{d}\mu_G,$$

where $\{P_n\}$ is any sequence of finite-dimensional projections converging strongly to the identity. In particular:

$$\phi(v)=\int_{\mathcal{H}} \exp -\tfrac{1}{2}\{[h,h]-2[v,h]\}\,\mathrm{d}\mu_S, \quad v\in W. \tag{3.1}$$

Since $(I+R)^{-1}$ is linear bounded and non-negative definite, it can be factorized (in many ways) as:

$$(I+R)^{-1}=KK^*.$$

Let

$$L=I-K.$$

Then it is readily shown that L is Hilbert–Schmidt and that $(L+L^*)$ is nuclear. Using this, we can express (3.1) as:

$$\phi(v)=\exp -\tfrac{1}{2}\{[Lv,Lv]-2[Lv,v]+\mathrm{Tr}(L+L^*)\}, \tag{3.2}$$

which is then the likelihood-ratio of "signal-plus-noise" to "noise alone". Let

$$\hat{S}=Lv.$$

Then

$$v-Lv=\nu$$

is white noise, and we can readily verify also that the error-covariance matrix

$$E\big[(S-\hat{S})(S-\hat{S})^*\big]=L+L^*$$

and being nuclear, we also have that

$$\mathrm{Tr}\,E\big[(S-\hat{S})(S-\hat{S})^*\big]=\mathrm{Tr}(L+L^*).$$

Note that in the case of Example 1, using the factorization:

$$Lf=\sum_1^\infty \frac{1}{\sqrt{1+\lambda_n^2}}[f,\phi_n]\phi_n,$$

the likelihood ratio can be expressed:

$$\exp -\tfrac{1}{2}\left\{\sum_1^\infty \frac{[V,\phi_n]^2}{1+\lambda_n^2} - 2\sum_1^\infty \frac{[V,\phi_n]}{\sqrt{1+\lambda_n^2}} + 2\sum_1^\infty \frac{1}{\sqrt{1+\lambda_n^2}}\right\}. \tag{3.3}$$

Unfortunately this version is generally not useful in parameter-estimation.

Of particular importance, is the case where W is an L_2-space over a separable Hilbert space:

$$W = L_2[(0,T);\mathcal{H}].$$

Then a particular factorization of $(I+R)^{-1}$ becomes important. Thus according to the Krein Factorization (see [7]):

$$(I+R)^{-1} = (I-L)(I-L^*) \tag{3.4}$$

where L is Volterra as well, so that we have the representation:

$$Lf = g; \qquad g(t) = \int_0^t L(t,s) f(s)\,\mathrm{d}s, \quad 0 \leqslant t \leqslant T$$

where $L(t,s)$ is Hilbert–Schmidt and

$$\int_0^T \mathrm{d}t \int_0^t \| L(t,s) \|_{H\cdot S}^2 \,\mathrm{d}s = \| L \|_{H\cdot S}^2.$$

Moreover now,

$$\hat{S}(t) = \int_0^t L(t,s) v(s)\,\mathrm{d}s, \quad 0 \leqslant t \leqslant T,$$

has the interpretation that

$$\hat{S}(t) = E[S(t) \mid v(s), s \leqslant t],$$

and the "innovation"

$$\nu(t) = v(t) - \hat{S}(t), \quad 0 < t < T$$

is white noise in W. Also:

$$E\big[(S(t)-\hat{S}(t))(S(t)-\hat{S}(t))^*\big] = \operatorname{Tr} L(t,t),$$

$$\operatorname{Tr}(L+L^*) = \int_0^T \operatorname{Tr} L(t,t)\,\mathrm{d}t.$$

Let us see how this theory applies to random fields. First let $\mathcal{D}$ be a bounded rectangle in R^2:

$$\mathcal{D} = [(t_1,t_2) \mid 0 < t_1 < T_1; 0 < t_2 \leqslant T_2].$$

Then we exploit the fact we can write,

$$L_2(\mathcal{D}) = L_2[(0,T_1); L_2(0,T_2)]$$

or, in our notation

$$L_2(D) = W = L_2[(0,T_1);\mathcal{H}]$$

where

$$\mathcal{H}=L_2(0,T_2).$$

Let

$$v(t_1,t_2)=S(t_1,t_2)+N(t_1,t_2), \quad (t_1,t_2)\in\mathcal{D}, \tag{3.5}$$

where $N(t_1,t_2)$ is white noise in $L_2(\mathcal{D})$ and the "signal" $S(t_1,t_2)$ is independent of the noise and

$$\int_0^{T_1}\int_0^{T_2}E\left(S(t_1,t_2)^2\right)\mathrm{d}t_1\,\mathrm{d}t_2<\infty$$

so that the corresponding measure μ_S is countably additive on (the Basel sets of) W. For each $t_1, 0<t_1<T_1$,

$$S(t_1,t_2), \quad 0<t_2<T_2$$

is an element of $L_2[0,T_2]$, and similarly for $N(t_1,t_2)$, so that we can rewrite (3.5) as:

$$v(t_1,\cdot)=S(t_1,\cdot)+N(t_1,\cdot), \quad 0<t_1<T_l \tag{3.6}$$

and we can now apply the Krein-factorization theory. We have

$$Rf=g; \qquad g(t)=\int_0^{T_1}R(t,s)f(s)\,\mathrm{d}s, \quad f(\cdot)\in W$$

and for h in $L_2(0,T_2)$:

$$R(t,s)h=w; \qquad w(t_2)=\int_0^{T_2}R(t,s;t_2,s_2)h(s_2)\,\mathrm{d}s_2, \quad 0<t_2<T_2.$$

Also,

$$(I+R)^{-1}=(I-L)(I-L)^*,$$

$$Lf=g; \qquad g(t)=\int_0^{t}L(t,s)f(s)\,\mathrm{d}s, \quad f(\cdot)\in W,$$

$$L(t,s)h=w; \qquad w(t_2)=\int_0^{T_2}L(t,s;\,t_2,s_2)L(s_2)\,\mathrm{d}s_2, \quad h(\cdot)\in L_2(0,T_2),$$

and

$$\nu(t_1,t_2)=v(t_1,t_2)-\hat{S}(t_1,t_2)$$

is white noise in $L_2(\mathcal{D})$ with

$$\hat{S}(t_1,t_2)=\int_0^{t_1}\mathrm{d}s\int_0^{T_2}L(t_1,s_1;\,t_2,s_2)V(s_1,s_2)\,\mathrm{d}s_2,$$

$$\operatorname{Tr}L(t,t)=\int_0^{T_2}\operatorname{Tr}L(t,t;\,t_2,t_2)\,\mathrm{d}t_2.$$

Note that

$$\hat{S}(t_1,t_2)=E\left[S(t_1,t_2)\,|\,V(s_1,s_2),\,0\leqslant s_1\leqslant t_1,\,0\leqslant s_2\leqslant T_2\right].$$

To apply this result to regions other than rectangular we have to choose an appropriate coordinate system. Thus let us consider the case where $\mathcal{D}$ is the unit circle in R^2:

$$\mathcal{D}=\left[(t_1,t_2)\,|\,t_1^2+t_2^2<1\right].$$

Choosing polar-coordinates (r,θ) we note that $L_2(\mathcal{D})$ is an L_2-space over a Hilbert space:

$$L_2(\mathcal{D})=L_2((0,2\pi);\mathcal{H}),$$

where $\mathcal{H}$ is now the Lebesgue measurable functions $f(\cdot)$ on $[0,1]$ such that

$$\int_0^1 f(r)^2 r\,\mathrm{d}r<\infty$$

with inner-product:

$$[f,g]=\int_0^1 f(r)\overline{g(r)}\,r\,\mathrm{d}r.$$

Hence we can apply the Krein factorization again to obtain

$$\hat{S}(r\cos\theta,r\sin\theta)=\int_0^\theta \mathrm{d}\sigma\int_0^1 L(\theta;\sigma;r;\gamma)V(\gamma\cos\sigma,\gamma\sin\sigma)\gamma\,\mathrm{d}\gamma,\quad 0\leqslant 0\leqslant 2\pi$$

where

$$(I+R)^{-1}=(I-L)(I-L^*)$$

where L is defined by:

$$Lf=g;\qquad g(\theta)=\int_0^\theta L(\theta,\sigma)f(\sigma)\,\mathrm{d}\sigma,\quad 0<\theta<2\pi;$$

$$f(\sigma)\in\mathcal{H},\quad f(\sigma)\sim f(\sigma,\gamma),$$

$$L(\theta,\sigma)f(\sigma)\sim\int_0^1 L(\theta;\sigma;r;\gamma)f(\sigma,\gamma)\gamma\,\mathrm{d}\gamma.$$

Note that,

$$\nu(t_1,t_2)=v(t_2,t_1)-\hat{S}(t_2,t_1),$$

defines white noise in $L_2(\mathcal{D})$.

4. Kalman filtering

We have seen that the likelihood functional involves a covariance factorization (Krein–Gohberg or other). However, such a factorization is difficult to carry out when the covariance contains unknown parameters (as in maximum-likelihood estimation). An alternate procedure then is to use the Kalman filtering technique (which has other advantages as well). Unfortunately the latter technique does not seem to be possible always for random fields. Here we discuss a class of problems where it is possible, for both rectangular and circular domains.

Once again we begin with a general Hilbert-Space theory and specialize it later. Let $\mathcal{H}$, $\mathcal{H}_0$ denote separable Hilbert Spaces and let $\mathfrak{N}$ denote the Hilbert Space of Hilbert–Schmidt Operators on $\mathcal{H}$ into $\mathcal{H}_0$. Let

$$W = L_2[(0,T);\mathcal{H}],$$

$$W_0 = L_2[(0,T);\mathcal{H}_0],$$

$$W_N = L_2[(0,T);\mathfrak{N}],$$

where $0<T<\infty$. Let $L(\cdot)$ W_N be given. Define the operator L on H into W_0 by:

$$Lh = L(\cdot)h.$$

Let $N(\cdot)$ denote white noise in W and let

$$S(t) = \int_0^t L(t-\sigma)N(\sigma)\,d\sigma, \quad \text{a.e.} 0<t<T. \tag{4.1}$$

Then

$$\int_0^T E[\|S(t)\|^2]\,dt = \int_0^T dt \int_0^t \|L(\sigma)\|^2_{H.S.}\,d\sigma$$
$$\leqslant T\|L\|^2 < \infty$$

so that $S(\cdot)$ has a nuclear covariance. Our first step is to obtain a "state-space" representation for $S(\cdot)$. For simplicity, we shall assume that $L(\cdot)$ is continuous. For this purpose let us define the shift-semigroup $T(t)$ over W_0 by:

$$T(t)f = g; \qquad g(s) = f(t+s), \quad 0<s<T-t,$$
$$= 0, \qquad s>T-t.$$

Let A denote its generator. Note that for x in W_0,

$$L^*x = \int_0^T L(s)^* Tx(s)\,ds.$$

Let $x(t)$ denote the "generalized" solution (see [7]) of the differential equation:

$$\frac{dx(t)}{dt} = Ax(t) + LN(t), \quad 0<t<T \text{ a.e.}, \tag{4.2}$$
$$x(0) = 0$$

where $x(t) \in W_0$ for each t. Let C denote the operator with domain equal to the class of continuous functions in W_0 and range in $\mathcal{H}_0$ and for $f \in \mathfrak{D}(C)$:

$$Cf = f(0).$$

Then we have the representation:

$$S(t) = Cx(t), \quad 0<t<T \tag{4.3}$$

using the fact that

$$x(t) = \int_0^t T(t-\sigma)LN(\sigma)\,d\sigma.$$

Using this representation we have that:

Theorem 4.1. *Let* $v(t)=S(t)+N_0(t)$, $0<t<T$ *where* $N_0(\cdot)$ *is white noise in* W_0. *Let*

$$\hat{S}(t)=E[S(t)\,|\,v(s),\, s\leqslant t],\quad 0\leqslant t\leqslant T.$$

Then $\hat{S}(t)$ *has the (Kalman) representation*:

$$\hat{S}(t)=C\hat{x}(t),$$

$$\dot{\hat{x}}(t)=A\hat{x}(t)+(CP(t))^*[v(t)-C\hat{x}(t)],$$

$$\hat{x}(0)=0,$$

where $P(t)$ *satisfies the Riccati equation*:

$$[\dot{P}(t)x,x]=[P(t)x,A^*x]+[Ax,P(t)x]+[L^*x,L^*x]-[CP(t)x,CP(t)x], \tag{4.4}$$

$$P(0)=0;\quad \textit{for } x\in\mathcal{D}(A^*).$$

For each t, $P(t)$ *is linear bounded, mapping* W_N *into itself, and* $P(t)$ *maps* $\mathcal{D}(A^*)$ *into* $\mathcal{D}(C)$.

Proof. See [10].

Let us see how to use this result when the domain $\mathcal{D}$ of the random field is a circle in R^2. Suppose that the signal $S(\cdot)$ has the representation:

$$S(r\cos\theta, r\sin\theta)=\int_0^\theta d\sigma\int_0^1 F(\theta-\sigma;r,\gamma)N(\gamma\cos\sigma,\gamma\sin\sigma)\gamma\,d\gamma \tag{4.5}$$

where $N(\cdot,\cdot)$ is white noise in $L_2(\mathcal{D})$ and

$$\int_0^{2\pi}\int_0^1 |F(\sigma;r;\gamma)|^2\gamma\,d\gamma\,d\sigma<\infty$$

and $F(\cdot,\cdot,\cdot)$ is continuous in all its variables. Then let $F(\theta)$ represent the operator:

$$F(\theta)f=g;\qquad g(r)=\int_0^1 F(\theta;r;\gamma)\gamma f(\gamma)\,d\gamma$$

mapping

$$L_2((0,2\pi);\mathcal{H})$$

where $\mathcal{H}$ is the L_2-space over (0.1) defined in Section 3. Let

$$v(t_1,t_2)=S(t_1,t_2)+N_0(t_1,t_2),\quad (t_1,t_2)\in\mathcal{D}$$

where $N_0(\cdot,\cdot)$ is again white noise in $L_2(\mathcal{D})$, and is independent of $N(\cdot,\cdot)$. Let

$$\hat{S}(r\cos\theta, r\sin\theta)=E[S(r\cos\theta, r\sin\theta)\,|\,v(\gamma\cos\delta,\gamma\sin\sigma),\quad 0\leqslant\sigma\leqslant\theta;\,0\leqslant\gamma\leqslant1].$$

Then writing

$$\tilde{S}(\theta, r) = S(r\cos\theta, r\sin\theta),$$
$$\tilde{N}(\theta, r) = N(r\cos\theta, r\sin\theta),$$

we see that we have the representation

$$\tilde{S}(\theta, \cdot) = \int_0^\theta F(\theta-\sigma)N(\sigma, \cdot)\,\mathrm{d}\sigma, \quad 0<\theta<2\pi,$$

and hence

$$\tilde{S}(\theta, \cdot) \qquad \text{is the function},$$
$$\hat{S}(r\cos\theta, r\sin\theta), \quad 0\leqslant r\leqslant 1,$$

and we have the representation from the theorem, with the notation as therein.

$$\hat{\tilde{S}}(\theta) = C\hat{x}(\theta), \quad 0<\theta<2\pi,$$
$$\frac{\mathrm{d}\hat{x}(\theta)}{\mathrm{d}\theta} = A\hat{x}(\theta) + (CP(\theta))^*(\tilde{v}(\theta, \cdot) - C\hat{x}(\theta))$$

with $P(\theta)$ satisfying (4.4). Note that "updating" involves a "radial" sweep. Note also that in the Krein factorization

$$\operatorname{Tr}(L+L^*) = \int_0^{2\pi} \operatorname{Tr} C(CP(\theta))^*\,\mathrm{d}\theta.$$

The question of when the signal process satisfies the condition (4.5), and whether it can be deduced from the covariance would appear to be difficult. On the other hand, such a representation does hold if $\tilde{S}(\theta, r)$ satisfies a partial differential equation of the form:

$$\frac{\partial^2 \tilde{S}}{\partial\theta^2} + D(r)\tilde{S} = \tilde{N}(r, \theta)$$

where $D(r)$ is a differential operator in r, $0<r<1$ and

$$[D(r)f, f] \geqslant c[f, f], \quad c>0,$$

where [,] denotes inner product in $\mathcal{H}$. In this case, we can clearly obtain a semigroup representation (in θ) directly and the usual theory (with C bounded) goes through.

For a treatment of the case where $\mathcal{D}$ is a rectangle see [8].

5. Non-linear estimation: C–R error bounds

Finally let us consider the non-linear estimation problem based on maximizing the likelihood functional. Let Θ denote the (finite-dimensional) unknown parameter vector and let

$$v(t) = S(\Theta_0; t) + N_0(t), \quad 0<t<T$$

where we have explicitly indicated the parametric dependence, Θ_0 being the "true" parameter value, and we take the general Hilbert-space formulation as in Section 3. Let $R(\Theta)$ denote the signal-covariance operator for each and let the Krein-factorization be:

$$(I+R(\Theta))^{-1}=(I+L(\Theta))(I-L(\Theta))^*$$

so that the negative log-likelihood functional is

$$(\Theta)=\tfrac{1}{2}\{[L(\Theta)v, L(\Theta)v]-2[L(\Theta)v, v]+\operatorname{Tr}(L(\Theta)+L(\Theta)^*)\}.$$

The gradient with respect to Θ is

$$\left\{\left[\frac{\partial}{\partial\alpha_i}L(\Theta)v, L(\Theta)v-v\right]+\tfrac{1}{2}\frac{\partial}{\partial\alpha_i}\operatorname{Tr}(L(\Theta)+L(\Theta)^*)\right\}$$

where $\Theta=\{\alpha_i\}$. Also, at the "true-value" $\Theta=\Theta_0$, the expected value is zero since

$$v-L(\Theta_0)v=\nu$$

is white noise and since we can write

$$L(\Theta_0)v=M(\Theta_0)\nu; \qquad (I-L(\Theta))^{-1}=(I+M(\Theta))$$

where $M(\Theta)$ is also Volterra,

$$E\left[\frac{\partial}{\partial\alpha_i}L(\Theta_0)v, L(\Theta_0)v-\nu\right]=-E\left(\left[\frac{\partial}{\partial\alpha_i}L(\Theta_0)(\nu+M(\Theta_0)\nu), \nu\right]\right)$$

$$=-\tfrac{1}{2}\frac{\partial}{\partial\alpha_i}\operatorname{Tr}(L(\Theta_0)+L(\Theta_0)^*),$$

the trace of the product $L(\Theta_0)M(\Theta_0)$ being zero, since it is nuclear and Volterra. To calculate the C–R bound we need to take second partial derivatives and since

$$E\left[\frac{\partial^2}{\partial\alpha_i\partial\alpha_j}L(\Theta_0)v, v\right]=\tfrac{1}{2}\frac{\partial^2}{\partial\alpha_i\partial\alpha_j}\operatorname{Tr}(L(\Theta_0)+L(\Theta_0)^*),$$

it follows that

$$E\left[\frac{\partial^2}{\partial\alpha_i\partial\alpha_j}\mathcal{V}(\Theta)\right]_{|\Theta=\Theta_0}$$

is given by

$$\operatorname{Tr}\left(\frac{\partial}{\partial\alpha_i}L(\Theta_0)\right)(I+R(\Theta_0))\left(\frac{\partial}{\partial\alpha_j}L(\Theta_0)\right)^*.$$

References

[1] Nash, R.S. and Jordan, S.K. (1978). Statistical geodesy—An engineering perspective. *Proc. IEEE* **66** (5).

[2] Woods, J.W. and Radewan, C.H. (1977). Kalman filtering in two dimensions. *Trans. Information Theory* **23** (1).

[3] Larrimore, W.E. (1977). Statistical inference on stationary random fields. *Proc. IEEE* **65** (6).
[4] Rose, S.C. (1980). Two-dimensional smoothing of a vector Laplacian field with application to geodesy. Thesis, UCLA Engineering.
[5] Novikov, A.A. (1978). Recurrent interpolation of partially observed random fields with discrete parameter. *Acad. Nauk USSR* **2**.
[6] Balakrishnan, A.V. (1980). Non-linear white noise theory. *Multivariate Anal.* **5**.
[7] Balakrishnan, A.V. (1980). Applied Functional Analysis. 2nd ed. Springer-Verlag, Berlin.
[8] Balakrishnan, A.V. (1980). Likelihood ratios and kalman filtering for random fields. In: *Proceedings of the IFIP W.C. on Stochastic Differential Systems, Visegrad, Hungary, 1980.* Springer-Verlag Lecture Notes in Information and Control, Springer, Berlin.
[9] Celfand, I.M. and Vilenkin, N.Ya. (1964). *Generalized Functions.* Vol. 4. Academic Press, New York.
[10] Balakrishnan, A.V. (1977). Stochastic filtering and control of linear systems. In: *Control Theory of Systems Governed by Partial Differential Equations.* Academic Press, New York.
[11] Balakrishnan, A.V. (1980). On a class of Riccati equations in a Hilbert space. *Appl. Math. Optim.*

G. Kallianpur, P.R. Krishnaiah, J.K. Ghosh, eds., *Statistics and Probability: Essays in Honor of C.R. Rao*

CONDITIONALITY VERSUS SIMILARITY IN THE ANALYSIS OF 2×2 TABLES*

G.A. BARNARD
University of Waterloo, Waterloo, Ontario, Canada

In testing hypotheses against specified alternatives, three concepts should be distinguished: The P value attained by the data, the α level for rejection, and the long-run risk of type I error. The α level should vary not only with the plausibility of the hypothesis tested, but also with the sensitivity of the experiment, if the long-run risk of type II error is to be minimised. When α varies, the long-run risk of type I error is the *average* of the α values used.

1. Conditionality versus similarity in the analysis of 2×2 tables

One of the hazards that one is exposed to when contributing to a festschrift for C.R. Rao is that his range of research is so wide and extensive and so full of wise remarks, that one may easily devote a paper to making a point which he himself has already made. In the hope of perhaps avoiding this, I propose to take up a question he put to me relatively recently. We were discussing 2×2 tables, in particular some of my own early work (Barnard, 1945). I had then argued that in a 2×2 comparative trial of Treatment I (probability of success p_1) versus Treatment II (probability of success p_2) with 3 experimental units having each treatment, the "significance level" for testing $p_1=p_2=p$ against $p_1>p_2$, with the result (Table 1)

Table 1

	Success	Failure	Total
Treatment I	3	0	3
Treatment II	0	3	3
	3	3	6

should be

$$\max_{0\leqslant p\leqslant 1}\left\{\text{Probability of getting } \frac{3\mid 0}{0\mid 3},\ \text{giving } p_1=p_2=p\right\}=\tfrac{1}{64},$$

rather than the value given by Fisher's "exact test":

$$\frac{3!3!3!3!}{6!}\times\frac{1}{3!0!0!3!}=\frac{1}{20}.$$

Rao asked me how I would explain why I would now argue for $\frac{1}{20}$ and not $\frac{1}{64}$. My brief reply was, that the $\frac{1}{64}$ figure combined two improbabilities: (1) the improbability of getting a total of 3 out of 6, and (2) the improbability, given the total of 3 out

*Research supported by the National Science and Engineering Research Council of Canada.

of 6, of all 3 being with Treatment I. The first improbability should not be counted against the hypothesis $p_1 = p_2 = p$, since it could not be readily removed by supposing $p_1 > p_2$, assuming, as would be reasonable in most contexts, that we do not take the extreme hypothesis $p_1 = 0$, $p_2 = 0$ seriously.

The problem of how one should statistically analyse 2×2 tables is especially important in connection with clinical trials of medical treatments, which have been discussed recently in some important papers (Geisser 1980, Peto et al. 1976, 1977). The issues involved are many and complex. Which of the very many models giving rise to 2×2 tables is appropriate to a given case, and whether it should be regarded as one involving hypothesis testing, or estimation, or something still more complex must be considered. In this note we confine ourselves to one of the simplest models, which already calls for the exercise of judgement in ways which will not appeal to those anxious to reduce the practice of statistics to the mechanical application of mathematical rules.

The main point to be made concerns the concept of "risk of type I error". There are three closely related concepts: (i) The P-value for given data x_0 in relation to a tested hypothesis H_0. This is the probability, on H_0, of getting a result as extreme as, or more extreme than, x_0. (ii) The significance level, or α value, at which H_0 would be rejected. This is the value such that H_0 is rejected iff $P \leqslant \alpha$. (iii) The long-run risk of type I error. This is the upper bound to the frequency, in a long series of hypothesis tests, of wrong rejections of the hypotheses tested. Fifty or sixty years ago, when the theory of hypothesis testing was being developed, tabulating and computational difficulties came in the way of clear distinctions between these three concepts. But now it is easy to compute exact P values — in the case of 2×2 tables, on hand-held calculators — and so the differences between the three concepts should be more readily seen.

The statistician analysing results of this kind should evaluate the P value. It is not for him, nor for the individual experimenter, alone to impose an α value on other persons. Different people will have different opinions as the plausibility of H_0, and these differences should be reflected in the α levels they use. This point is well made in the paper by Peto et al., loc. cit. But there is a further reason, discussed below, why α levels should vary with the data, if we are to achieve our aim of minimising the long-run risk of type II error. We prefer the word "risk" to "frequency" in these phrases, because the "frequencies" involved relate only to two mutually exclusive hypothetical worst cases — viz. those when H_0 is always true, and those when H_1 is always true.

Any real long run will not consist of repetitions of the same experiment testing the same hypothesis. Hypotheses must differ, and experiments are unlikely to be identical. If, as suggested, the α level should vary from case to case, the long-run risk of type I error will be

$$\bar{\alpha} = \sum_i \alpha_i / M \tag{1}$$

where α_i, $i = 1, 2, \ldots, M$ is the α level for the ith experiment in a long run of M independent experiments. An easy extension of the weak law of large numbers justifies the equation of frequencies to mean probabilities here.

Thus the P value will be determined by the data, by H_0, and the test being used. The α value will be chosen in the light of the plausibility of H_0, and of the considerations indicated below. The long run risk of type I error will be the *average* of the α values used in the long run considered.

2. Varying α levels are necessary to minimise type II error

In their classical work, Neyman and Pearson discussed the problem of testing a simple hypothesis H_0, saying that observable x has probability function $\phi_0(x)$, against a simple alternative H_1, saying that x has probability function $\phi_1(x)$. They showed that in order to minimise β, the "probability of type II error", subject to an upper bound on α, the "probability of type I error", we should reject H_0 iff the likelihood ratio

$$L(x)=\phi_1(x)/\phi_0(x) \tag{2}$$

exceeds a critical value $k(\alpha)$, chosen so that

$$\Pr(L(x)>k(\alpha)|H_0)=\alpha. \tag{3}$$

(We ignore trivial complications arising from discreteness of distributions.) In discussing the practical implications of this result, it has hitherto been assumed that in the long runs considered, α will be held constant. The use of fixed significance levels, $\alpha=0.05$ or 0.01, was introduced by Fisher for largely accidental reasons connected with Pearson's copyright in the tables of χ^2. Although it has many advantages, especially the dangerously seductive one of saving us the effort of thinking, for the reasons indicated above and below we now ought to abandon it. The recommendation of P values in elementary texts such as Freedman et al. (1978) is welcome, in that it paves the way for variations of α levels. That this must be done in order to minimise the long-run risk of type II error follows from a simple extension of the Neyman–Pearson lemma which the present author thought he was first to notice, until he came across Pitman's essay (1965), where a general proof of the extension is given. We give here a simple proof which assumes, for simplicity, that the distributions involved are continuous and nowhere null:

Lemma. *If H_{0i}, tested against H_{1i} has critical L-value $k_i(\alpha_i)$, and β-value β_i $(i=1,2)$, then unless*

$$k_1(\alpha_1)=k_2(\alpha_2)$$

we can find k such that

$$\Pr(L_1(x_1)>k|H_{01})+\Pr(L_2(x_2)>k|H_{02})=\alpha_1+\alpha_2$$

while

$$\Pr(L_1(x_1)\leqslant k|H_{11})+\Pr(L_2(x_2)\leqslant k|H_{12})<\beta_1+\beta_2.$$

That is, unless we keep the critical value of the likelihood ratio constant across experiments, we can lower the long-run $\bar{\beta}$ value while keeping the value fixed.

Proof.

$$\beta=\int_{L<k} L(x)\phi_0(x)\,\mathrm{d}x, \qquad \alpha=\int_{L>k} \phi_0(x)\,\mathrm{d}x.$$

Changing k by an infinitesimal $\mathrm{d}k$, changes α by infinitesimal $\mathrm{d}\alpha$ and β by infinitesimal $k\,\mathrm{d}\alpha$. Thus

$$\frac{\partial\beta}{\partial\alpha}=-k.$$

For two experiments, $\partial\beta_1/\partial\alpha_1=-k_1$, $\partial\beta_2/\partial\alpha_2=-k_2$; so if we impose the condition $\alpha_1+\alpha_2=$constant, $\mathrm{d}\alpha_1=-d\alpha_2$,

$$\begin{aligned} d(\beta_1+\beta_2)&=-k_1\mathrm{d}\alpha_1-k_2\mathrm{d}\alpha_2\\ &=(k_2-k_1)\mathrm{d}\alpha_1, \end{aligned}$$

and by choice of sign of $\mathrm{d}\alpha_1$ this can be made negative unless $k_1=k_2$, i.e. $\beta_1+\beta_2$ can be reduced, with $\alpha_1+\alpha_2=$constant, unless $k_1=k_2$. In practice we rarely test simple hypotheses against simple alternatives. If the hypotheses involve a nuisance parameter θ, we can reduce them to simple hypotheses by guessing "priors" $w_0(\theta), w_1(\theta)$ for θ, and then taking the likelihood ratio L to be

$$L(x)=\int\phi_1(x,\theta)w_1(\theta)\,d\theta\Big/\int\phi_0(x,\theta)w_0(\theta)\,d\theta.$$

Provided our w's approximately reflect the long-run frequencies of the various values of θ, using this L as the likelihood ratio will approximately optimise the performance of our tests.

The general conclusion is the common sense one, that, to minimise the risk of type II error we should use low α values in sensitive experiments, when H_1 and H_0 are easily distinguished, and we should raise our α values when the experiment is less sensitive, other things — in particular the plausibility of H_0 and H_1 — being equal.

2.1. *Critique of "similarity" and "unbiasedness"*

It often happens that the sensitivity of the experiment becomes known only as the experiment is performed. For example, in Eddington's vivid description of the eclipse expeditions of May 29, 1919 it appears that the errors at Principe were larger than those at Sobral due to the weather (Eddington 1921). An imaginary example is discussed by Cox and Hinkley (1974, pp. 95–96) in which x is a measurement of a quantity μ, subject to normally distributed error. In addition to x a number c is observed. If $c=1$ the error variance is 1, while if $c=2$ the error variance is 10^6. For testing $\mu=0$ versus $\mu=\mu_A>0$, keeping the critical likelihood ratio constant leads to a much higher value of α when $c=2$ than when $c=1$. Provided the marginal distribution of c is known, it is easy to determine α_1 and α_2 and the associated critical value of $L(x)$ to satisfy

$$\alpha_1\Pr(c=1)+\alpha_2\Pr(c=2)=\bar{\alpha},$$

for any specified $\bar{\alpha}$. Curiously, Cox and Hinkley seem to take for granted that the α

level should be the same for $c=1$ as for $c=2$, and suggest that the example indicates a conflict between the principle of conditionality, which requires conditioning on the value of c, and the requirement of maximum power. The conflict really is between the principle of conditionality and the mistaken principle of constancy of α. And we may note that, in this case, there is no difficulty in fixing $\bar{\alpha}$, in advance of the experiment, at any value that may be desired. The conflict perceived by Cox and Hinkley seems to derive from their notion that the α level should in some way 'measure' the extent to which the data are discrepant from the hypothesis H_0; but in so far as such a "measurement" can be made, it is the function of the P value, not the α value.

The idea of keeping the α level fixed, no matter what may appear concerning the sensitivity of the experiment, is carried to extreme in Neyman's concept of a "similar" test. When there is a nuisance parameter it often has a considerable influence on the sensitivity of the experiment and to require, as one does in requiring similarity, that the α level should be invariant with respect to changes in the nuisance parameter is therefore unreasonable. In the Behrens–Fisher problem Linnik and the present writer have shown in different ways that the requirement of similarity is inconsistent with other more compelling requirements. And in the case of the 2×2 table, we may formulate the problem as that of testing the hypothesis that the odds ratio,

$$\lambda=(p_1q_2/p_2q_1),$$

is equal to 1, against the alternative that it has some value, such as 3, greater than 1. Then the geometric mean of the odds,

$$\theta=\sqrt{(p_1p_2/q_1q_2)},$$

appears as a nuisance parameter. As Lehmann (1959, pp. 134–147) showed, requiring similarity of a test in this case is equivalent to requiring unbiasedness, and either requirement leads to the absurdity that we are required with frequency α, to reject the hypothesis $\lambda=1$ on the basis of the wholly uninformative result

$$\begin{matrix} 0 & 3 \\ 0 & 3. \end{matrix}$$

Just as unbiasedness of an estimate is a property with superficial attractiveness, which on further analysis can lead to absurdity, so also can the requirement of unbiasedness or similarity of a test. Indeed, the situation in the latter case is worse. If an estimate having otherwise desirable properties happens to be unbiased, this is a bonus; but if a test is similar, it means that its α level is unresponsive to changes in the sensitivity of the experiment, and this, as we have seen, will usually be disadvantageous.

2.2. *Application to the 2×2 table*

We illustrate the application of these ideas to the 2×2 table by reference to the case cited at the beginning, where we have two samples of 3 each. In this case the results can be fully specified by the total number r of successes, and the number a of successes with Treatment I. For the table given above we have $(a,r)=(3,3)$. Sixteen results altogether are possible:

(0,0)				1			
(0,1),	(1,1)			1,	λ		
(0,2),	(1,2),	(2,2)		1,	3λ,	λ^2	
(0,3),	(1,3),	(2,3),	(3,3)	1,	9λ,	$9\lambda^2$,	λ^3
	(1,4),	(2,4),	(3,4)		1,	3λ,	λ^2
		(2,5),	(3,5)			1,	λ
			(3,6)				1

Numbers proportional to the conditional probabilities, given r, are tabulated to the right. The actual conditional probabilities are obtained by dividing these numbers by their row sum. The marginal distribution of r is tabulated in Table 2,

Table 2

Probabilities of values of r, as functions of $\lambda=(p_1q_2/p_2q_1)$ and $\theta=\sqrt{(p_1p_2/q_1q_2)}$*

	r:	0	1	2	3	4	5	6
$\theta=1$	$\lambda=1$	0.0156	0.0938	0.2344	0.3125	0.2344	0.0938	0.0156
	5	0.0097	0.0784	0.2395	0.3448	0.2395	0.0784	0.0097
$\theta=4$	$\lambda=1$	0.00006	0.0015	0.0154	0.0819	0.2458	0.3932	0.2621
	5	0.00005	0.0015	0.0185	0.1063	0.2952	0.3864	0.1920

from which it can be seen that very little information about λ comes from r, compared with that which comes from the conditional distribution of a, given r. At the same time, the sensitivity of the data concerning λ is very much affected by the value of r. And the information about λ contained in r cannot be separated from the nuisance parameter θ. For all these reasons, it appears that we should use the conditional distribution, given the observed r, as the basis of our test.

If we do this, it can be seen that the sensitivity is null when $r=0$, is equivalent to a single trial, with probability $\lambda/(1+\lambda)$ when $r=1$, and increases to rather more than that of three independent trials with this probability when $r=3$. It then follows from our reasoning that our α levels should be relatively low when $r=3$, and should rise as r moves away from this value. However the sensitivity is so low when $r=1$ that it may be doubted whether λ would ever in practice be so large as to justify any rejection of $\lambda=1$ on this basis; and with $r=0$ there can clearly be no decision. Similar remarks apply to $r=5$ and $r=6$. With $r=2$ the P value for the most extreme result is 0.2, and this might be regarded as sufficient grounds for rejection of an implausible H_0.

Thus if we imagine someone willing to allow his long-run risk of type I error to rise up 0.05, such a person could set his α level when $r=3$ at 0.05, regardless of the plausibility of the hypothesis tested, while for $r=2$ or 4 he could set it at 0.2 for the less plausible hypotheses, provided these made up no more than 1 in 4 of the hypotheses considered and as long as he never rejected the hypothesis tested on the basis of the uninformative results $r=0,1,5$ or 6, he could (referring to Table 2) raise the proportion of 1 in 4 to 1 in 2.73, still without raising his long-run risk of type I error above 0.05.

Similar calculations for more realistically-sized 2×2 tables can easily be performed on modern hand-held calculators.

References

Barnard, G.A. (1945). A new test for 2×2 tables. *Nature* **156**, 177.

Cox, D.R. and Hinkley, D.V. (1974). *Theoretical Statistics*. Chapman and Hall, London.

Eddington, A.A. (1921). *Space, Time and Gravitation*. Chapter VI. Cambridge University Press, Cambridge.

Freedman, D., Pisani, R. and Purves, R. (1978). *Statistics*. Norton, New York.

Geisser, S. (1980). Randomisation, Stratification and other Stuff. Technical Report No. 374, School of Statistics, University of Minnesota, MN.

Lehmann, E.L. (1959). *Testing Statistical Hypotheses*. Wiley, New York.

Peto, R., Pike, M.C., Armitage, P., Breslow, N.E., Cox, D.R., Howard, S.V., Mantel, N., McPherson, K., Peto, J. and Smith, P.G. (1977). Design and Analysis of Randomised Clinical Trials requiring prolonged observation of each patient. *Brit. J. Cancer* **34**, 586–612; **35**, 2–39.

G. Kallianpur, P.R. Krishnaiah, J.K. Ghosh, eds., *Statistics and Probability: Essays in Honor of C.R. Rao*

HYPERBOLIC LIKELIHOOD

O. BARNDORFF-NIELSEN
Aarhus University, Denmark

Probability functions and likelihood functions which are nearly log-concave can often be well approximated by a hyperbolic probability or likelihood function. Various procedures for determining the parameters of an approximating hyperbolic function are considered, the discussion being cast in terms of likelihood.

1. Introduction

Probability functions or likelihood functions which are (nearly) log-concave can often be well approximated by a hyperbolic probability or likelihood function. [For definition, properties and various applications of the hyperbolic distributions, see, e.g. Bagnold and Barndorff-Nielsen (1980), Barndorff-Nielsen (1978, 1979) and Blæsild (1978).] The procedure for determining the parameters of the hyperbolic function in order to obtain an appropriate approximation must be chosen with regard to both the envisaged uses of the approximation and the practicality of the requisite calculations. The present note discusses some procedures for fitting one- and two-dimensional likelihood functions by a hyperbolic likelihood function. The interest in this lies primarily in those cases where, relative to the relevant reference set, the family of potentially observable likelihood functions, of which the actual likelihood function is a member, constitutes a translation family, either exactly or to a good approximation. For then the approximating hyperbolic likelihood function yields not only a summary description of the observed likelihood function but also an approximation to the distribution of the maximum likelihood estimator and approximate confidence intervals. Although the following discussion is phrased in terms of likelihood functions it will be apparent that the fitting procedures concerned may also be used to obtain approximations to probability functions.

2. Hyperbolic approximations to one-dimensional likelihood functions

Consider a likelihood function $L(\omega)$ of a one-dimensional parameter, ω, and suppose that the log-likelihood function, $l(\omega)=\ln L(\omega)$, is concave or nearly so. In many such cases, $L(\omega)$ can be well approximated by a hyperbolic likelihood function, i.e. the likelihood function corresponding to a single observation from a hyperbolic distribution with ω as the translation parameter. Denoting the hyperbolic likelihood function by $\tilde{L}(\omega)$ and its logarithm by $\tilde{l}(\omega)$ one has

$$\tilde{l}(\omega)=-\alpha\sqrt{\delta^2+(\tilde{x}-\omega)^2}+\beta(\tilde{x}-\omega), \tag{2.1}$$

and the problem now is to choose the pseudo-observation $\tilde{x}$ and the remaining 3 parameters δ, α and β of the hyperbolic distribution so that $\tilde{l}(\omega)$ approximates $l(\omega)$ closely, except of course for an additive constant. Of the manifold possible methods of doing this, here we will discuss a few which are simple to execute and which employ geometrical characteristics of the given likelihood function $L(\omega)$, such as the maximum likelihood estimate $\hat{\omega}$ and the observed information $\hat{j}$, which are of independent interest and are likely to be calculated in any case.

Let $\hat{l}^{(i)}$ denote the ith derivative of l evaluated at the maximum point of l, and let $\hat{\tilde{l}}^{(i)}$ have the analogous meaning. Also, for convenience, let $\kappa=\sqrt{\alpha^2-\beta^2}$, $\zeta=\delta\kappa$, and $\pi=\beta/\kappa$. The parameters ζ and π are invariant under location and scale transformations of the hyperbolic distribution and in terms of these parameters the probability (density) function of the hyperbolic distribution, corresponding to (2.1), may be written

$$\left\{2\delta\sqrt{1+\pi^2}\,K_1(\zeta)\right\}^{-1}\exp\left\{-\zeta\left(\sqrt{1+\pi^2}\sqrt{1+\left(\frac{x-\omega}{\delta}\right)^2}+\pi\left(\frac{x-\omega}{\delta}\right)\right)\right\}. \quad (2.2)$$

The derivative $\tilde{l}^{(i)}(\omega)$, for the first few values of i, are

$$\tilde{l}^{(1)}(\omega)=\alpha(\tilde{x}-\omega)/\sqrt{\delta^2+(\tilde{x}-\omega)^2}-\beta,$$

$$\tilde{l}^{(2)}(\omega)=-\alpha\delta^2/\{\delta^2+(\tilde{x}-\omega)^2\}^{3/2},$$

$$\tilde{l}^{(3)}(\omega)=-3\alpha\delta^2(\tilde{x}-\omega)/\{\delta^2+(\tilde{x}-\omega)^2\}^{5/2},$$

$$\tilde{l}^{(4)}(\omega)=-3\alpha\delta^2\{4(\tilde{x}-\omega)^2-\delta^2\}/\{\delta^2+(\tilde{x}-\omega)^2\}^{7/2}, \quad (2.3)$$

from which it follows that $\tilde{l}(\omega)$ takes its maximum at

$$\hat{\tilde{\omega}}=\tilde{x}-\delta\pi$$

and that

$$\hat{\tilde{l}}^{(2)}=-\delta^{-2}\zeta/\{1+\pi^2\},$$

$$\hat{\tilde{l}}^{(3)}=-3\delta^{-3}\zeta\pi/\{1+\pi^2\}^2,$$

$$\hat{\tilde{l}}^{(4)}=-3\delta^{-4}\zeta\{4\pi^2-1\}/\{1+\pi^2\}^3.$$

An immediate possibility for determining the approximating hyperbolic likelihood function is to solve the four equations

$$\hat{\tilde{\omega}}=\hat{\omega}, \qquad \hat{\tilde{l}}^{(i)}=\hat{l}^{(i)}, \quad i=2,3,4, \quad (2.4)$$

with respect to $\tilde{x}$, δ, α and β. It is convenient to start by transforming the equations slightly by introducing the quantities $\hat{j}=-\hat{l}^{(2)}$, i.e. the observed information, and

$$s=-\hat{l}^{(3)}/\hat{j}^{3/2} \qquad k=-\hat{l}^{(4)}/\hat{j}^2,$$

with analogous meanings for $\hat{\tilde{j}}$, $\tilde{s}$ and $\tilde{k}$. Note that s and k are a kind of local measure of skewness and kurtosis, respectively. Simple calculations show that

$$\tilde{s}=3\zeta^{-1/2}\pi/\sqrt{1+\pi^2}, \qquad \tilde{k}=3\zeta^{-1}\{4\pi^2-1\}/\{1+\pi^2\}$$

and one notes that

$$5\tilde{s}^2-3\tilde{k}=9\zeta^{-1}.$$

Thus, setting

$$t=5s^2-3k,$$

the system of equations (2.4) is solved by

$$\tilde{x}=\hat{\omega}+3s/\left(t\sqrt{\hat{j}}\right), \qquad \delta=3\sqrt{t-s^2}/\left(t\sqrt{\hat{j}}\right),$$

$$\alpha=3\sqrt{t\hat{j}}/(t-s^2), \qquad \beta=3\sqrt{\hat{j}}s/(t-s^2). \tag{2.5}$$

The fitting of a hyperbolic likelihood function determined by equations (2.5), or equivalently (2.4), hinges entirely on local properties of the observed log-likelihood function near the maximum likelihood estimate $\hat{\omega}$, and in many cases a better approximation can be achieved by employing some non-local features of $l(\omega)$.

A natural idea here is to retain the first two equations of (2.4) and to supplement these by requiring that the two log-likelihood functions, when normed to have maximum value 0, should coincide not only at 0 on the log-likelihood scale but also at c units down this scale, for some suitable, chosen value of c. Let $\bar{l}(\omega)$ and $\bar{\tilde{l}}(\omega)$ denote the two normed log-likelihood functions, i.e.

$$\bar{l}(\omega)=l(\omega)-l(\hat{\omega}), \qquad \bar{\tilde{l}}(\omega)=-\alpha\sqrt{\delta^2+(\tilde{x}-\omega)^2}+\beta(\tilde{x}-\omega)+\zeta$$

and let $\underline{\omega}$ and $\overline{\omega}$, where $\underline{\omega}<\overline{\omega}$, be the two roots of the equation $\bar{l}(\omega)=-c$ (it is assumed that this equation has exactly two roots). Then instead of the system (2.4) we now have the four equations

$$\hat{\tilde{\omega}}=\hat{\omega}, \qquad \hat{\tilde{j}}=\hat{j}, \qquad \bar{\tilde{l}}(\underline{\omega})=-c, \qquad \bar{\tilde{l}}(\overline{\omega})=-c, \tag{2.6}$$

which, after some algebraic calculations, turn out to have the solution

$$\begin{aligned}
\tilde{x}&=\hat{\omega}+(c/\hat{j})v/u^2,\\
\delta&=(c/\hat{j})\sqrt{u^2-v^2}/u^2,\\
\alpha&=cu/(u^2-v^2),\\
\beta&=cv/(u^2-v^2),
\end{aligned} \tag{2.7}$$

where

$$u=\sqrt{(\overline{\omega}-\underline{\omega})^2/4-2c/\hat{j}},$$
$$v=\hat{\omega}-(\underline{\omega}+\overline{\omega})/2.$$

The two procedures for fitting a hyperbolic likelihood function to an observed likelihood function L discussed above were derived through viewing the fitting problem not as one of directly fitting to L but as a question of obtaining a good approximation to the logarithm of the observed likelihood function, i.e. l. Alternatively, one may construct approximations to the score function $l^{(1)}$, which carries the same statistical information as the log-likelihood function l. A well-known advantage of working with the score function, rather than the log-likelihood function, is that the question of how well L can be approximated by a normal likelihood function is reduced to that of the approximate linearity of $l^{(1)}$ in some neighbourhood of $\hat{\omega}$.

Suppose that $-l^{(1)}$ has a sigmoid shape. One may then try to approximate $-l^{(1)}$ by some sigmoid curve of a simpler functional form, choosing the free variables of the latter so that, for instance, the two curves intersect at the maximum likelihood point $\hat{\omega}$ and have the same first, second and third derivative at the point, $\hat{\hat{\omega}}$, say, where the slope of $-l^{(1)}$ is maximal. With the approximating curve as minus a hyperbolic score function this latter alternative is expressed by the equations

$$\tilde{l}^{(1)}(\hat{\omega})=0, \qquad \tilde{l}^{(2)}(\hat{\hat{\omega}})=l^{(2)}(\hat{\hat{\omega}}),$$

$$\tilde{l}^{(3)}(\hat{\hat{\omega}})=0, \qquad \tilde{l}^{(4)}(\hat{\hat{\omega}})=l^{(4)}(\hat{\hat{\omega}}), \tag{2.8}$$

where $\hat{\omega}$ and $\hat{\hat{\omega}}$ are determined from $l^{(1)}(\hat{\omega})=0$ and $l^{(3)}(\hat{\hat{\omega}})=0$, respectively. The solution to (2.8) is given by

$$\tilde{x}=\hat{\hat{\omega}}, \qquad \delta=\sqrt{3\hat{\hat{j}}/l^{(\mathrm{iv})}(\hat{\hat{\omega}})}\,,$$

$$\alpha=\hat{\hat{j}}\,\delta, \qquad \beta=\alpha\pi/\sqrt{1+\pi^2}\,, \tag{2.9}$$

where

$$\hat{\hat{j}}=-l^{(2)}(\hat{\hat{\omega}}), \qquad \pi=(\hat{\hat{\omega}}-\hat{\omega})/\delta.$$

It should be noted that in case ω is a translation parameter the quantities $\hat{j}$, s, k, t, u, v, $\hat{\hat{\omega}}-\hat{\omega}$, $\hat{\hat{j}}$ and $l^{(4)}(\hat{\hat{\omega}})$ — and hence also δ, α and β in (2.5), (2.7) and (2.9) — are ancillary statistics (provided, of course, that $l(\omega)$ is based on a sample $x_1,\ldots, x_n$ with $n \geqslant 4$).

The above considerations may be illustrated by the case of a sample $x_1,\ldots,x_n$ from a hyperbolic location parameter family. Thus suppose $x_1,\ldots,x_n$ follow a distribution

$$\frac{\sqrt{\alpha_0^2-\beta_0^2}}{2\,\delta\alpha_0 K_1\left(\delta_0\sqrt{\alpha_0^2-\beta_0^2}\right)}\exp\left\{-\alpha_0\sqrt{\delta_0^2+(x-\omega)^2}+\beta_0(x-\omega)\right\} \tag{2.10}$$

where δ_0, α_0 and β_0 are known. The probability function $p(\hat{\omega};\omega|c)$ of the conditional distribution of the maximum likelihood estimator, $\hat{\omega}$, given the configuration $c=(c_1,\ldots,c_n)=(x_1-\hat{\omega},\ldots,x_n-\hat{\omega})$ is of the form

$$p(\hat{\omega};\omega|c)=a(c)e^{l(\hat{\omega}(x)-\hat{\omega}+\omega)},$$

where $x=(x_1,\ldots,x_n)$,

$$l(\omega)=\sum_{i=1}^{n}\left\{-\alpha_0\sqrt{\delta_0^2+(x_i-\omega)^2}+\beta_0(x_i-\omega)\right\}, \tag{2.11}$$

and $a(c)$ is the norming constant. It does not seem possible to find a simple expression for $a(c)$, but if a satisfactory approximation to $l(\omega)$ by a hyperbolic likelihood function $\tilde{l}(\omega)$ can be found then this yields readily an approximation to $a(c)$ and hence to the distribution $p(\hat{\omega}; \omega|c)$.

The log-likelihood function (2.1) is, obviously, asymptotically linear for $\omega \to \pm\infty$ and it might be thought that in these particular circumstances a better approximation than those suggested above would be achieved by choosing $\tilde{l}(\omega)$ to have the same asymptotic slopes as $l(\omega)$ while retaining the conditions $\hat{\tilde{\omega}} = \hat{\omega}$ and $\hat{\tilde{j}} = \hat{j}$. However, generally, this approach leads to approximations which are, on the whole, less satisfactory.

Example. Let $\alpha_0 = 1$, $\beta_0 = 0.5$ and $\delta_0 = 1$. Then the graph of the probability function (2.10) with $\omega = 0$ is as shown in Fig. 1.

A simulated sample of five observations from the hyperbolic distribution of Fig. 1 was constructed. The values obtained were $x_1 = -0.26$, $x_2 = 3.92$, $x_3 = 1.16$, $x_4 = -0.41$, $x_5 = -0.60$. The log-likelihood function, l, for this sample is presented in Fig. 2 together with the approximating normal log-likelihood function, which fits poorly, especially in the lower tail.

Approximating l by the hyperbolic log-likelihood function $\tilde{l}$ which has the same maximum point as l and the same curvature at that point and which coincides with l at the level $l(\hat{\omega}) - c$, i.e. using the approximation determined by (2.7), and choosing, somewhat arbitrarily, c equal to 8, one obtains the degree of fit shown in Fig. 3. This will be satisfactory for most purposes. The parameter values of the approximating hyperbolic distribution are $\tilde{x} = -0.29$, $\delta = 1.11$, $\alpha = 3.69$ and $\beta = 1.14$, and the information in the sample $x_1, \ldots, x_5$ concerning the unknown parameter ω is thus equivalent, approximately, to the information on ω that would be provided by a single observation $\tilde{x} = -0.29$ from a hyperbolic distribution with ω as translation parameter and with known $\delta = 1.11$, $\alpha = 3.69$ and $\beta = 1.14$. The solutions, in the present case, to the equation $l(\omega) = l(\hat{\omega}) - c$ are $\underline{\omega} = -4.77$ and $\overline{\omega} = 1.98$, and the

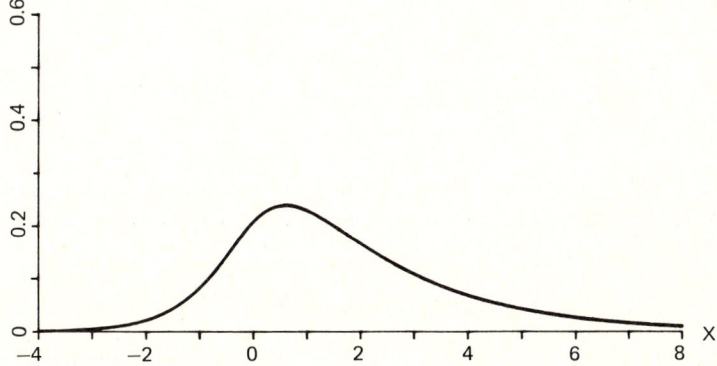

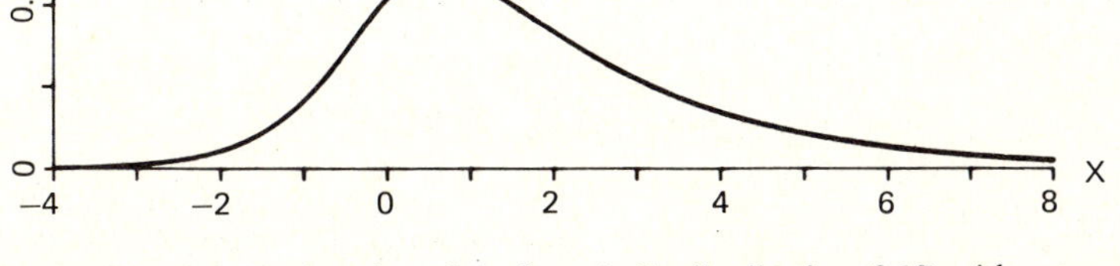

Fig. 1. The probability (density) function of the hyperbolic distribution (2.10) with parameters $\omega = 0$, $\delta_0 = 1$, $\alpha_0 = 1$, $\beta_0 = 0.5$.

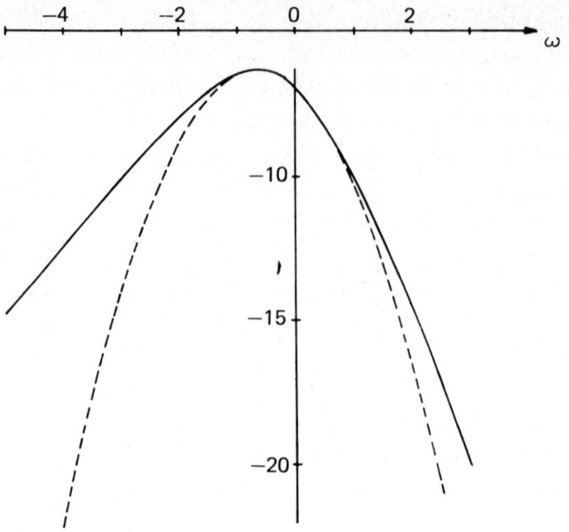

Fig. 2. The observed log-likelihood function, l, for a sample of 5 observations from the hyperbolic distribution of Fig. 1, and the approximating normal log-likelihood function.

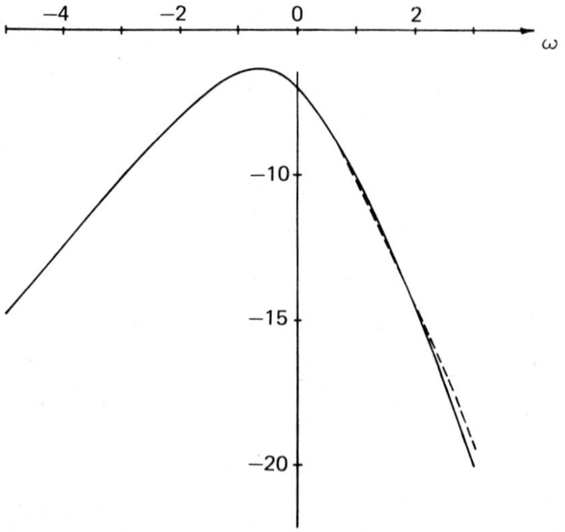

Fig. 3. The observed log-likelihood function, l, and the approximating hyperbolic log-likelihood function $\tilde{l}$ with parameters $\tilde{x}=-0.29, \delta=1.11, \alpha=3.69, \beta=1.14$, as determined from (2.6) and (2.7) (l: —— ; $\tilde{l}$:-----).

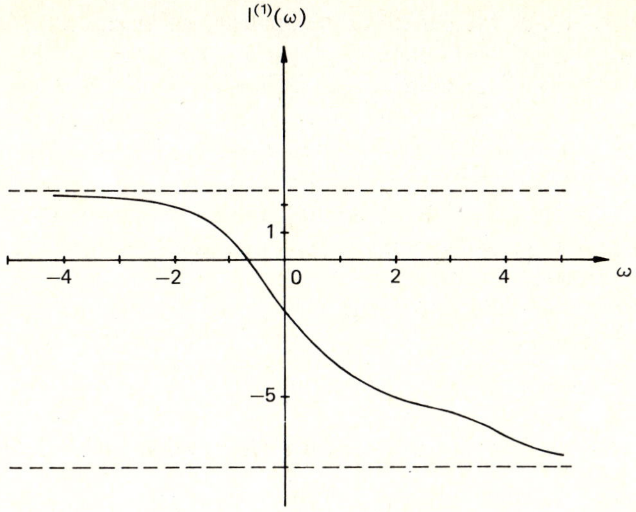

Fig. 4. The observed score function $l^{(1)}$.

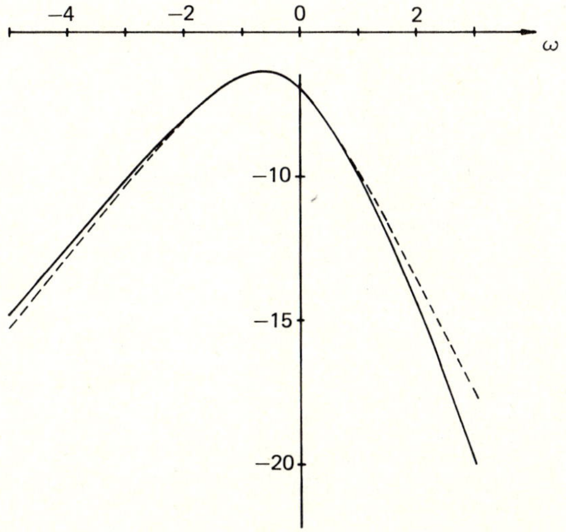

Fig. 5. The observed log-likelihood function l and the approximating hyperbolic log-likelihood function $\tilde{l}$ with parameters $\tilde{x}=-0.39, \delta=1.12, \alpha=3.42, \beta=0.776$, as determined from (2.8) and (2.9) (l: —— ; $\tilde{l}$:-----).

corresponding values of the distribution function of the approximating hyperbolic distribution are 0.000082 and 0.999955, respectively.

The score function $l^{(1)}$ is presented in Fig. 4 while Fig. 5 shows the hyperbolic approximation to the log-likelihood function l obtained from (2.9). This approximation is also rather satisfactory, though not as good as that of Fig. 3. It should be noted that the hyperbolic score function $\tilde{l}^{(1)}$, given in (2.3), is antisymmetric around the point $(\omega, \tilde{l}^{(1)}(\omega))=(\tilde{x}, -\beta)$. Comparing this to the observed score function in Fig. 4 gives a good impression of the possibilities and limitations for the fitting of a hyperbolic likelihood in this case.

Finally, the approximation determined by (2.4) and (2.5) is close to that of Fig. 5, the fitted parameters being $\tilde{x}=-0.41$, $\delta=1.07$, $\alpha=3.28$, $\beta=0.736$.

3. Hyperbolic approximations to multidimensional likelihood functions

If an observed log-likelihood function $l(\omega)$ for an r-dimensional vector parameter ω is approximately concave and if the level sets $\{\omega: l(\omega)=c\}$ of l are nearly of ellipsoidal shape then it may be possible to construct a useful hyperbolic approxima-

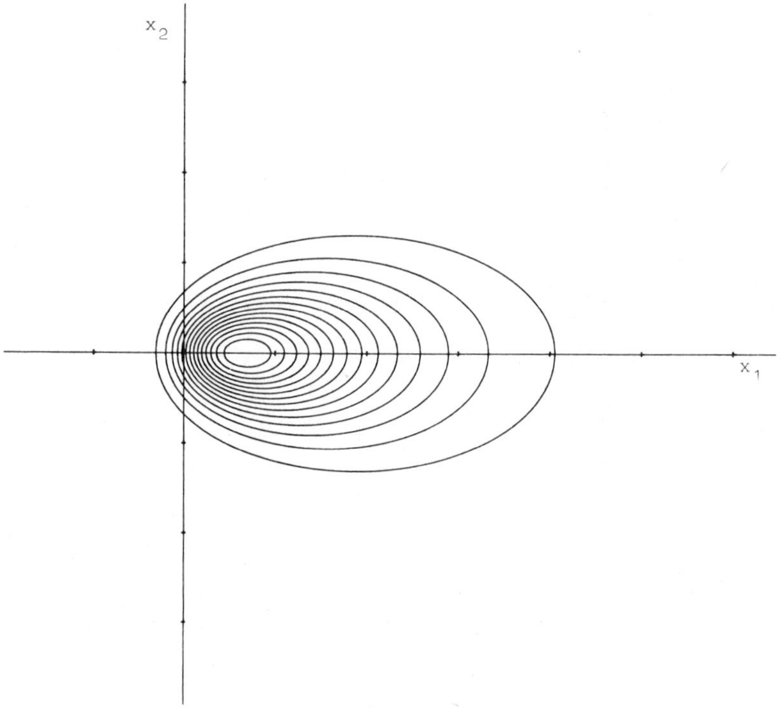

Fig. 6. Level curves of the two-dimensional hyperbolic distribution with parameters: $\omega=(0,0)$, $\delta=0.5$, $\epsilon_{11}=\epsilon_{22}=1$, $\epsilon_{12}=\epsilon_{21}=0$, $\alpha=5$ and $\beta=(4,0)$.

tion

$$\tilde{l}(\omega) = -\alpha\sqrt{\delta^2 + (\tilde{x}-\omega)\Delta^{-1}(\tilde{x}-\omega)'} + \beta(\tilde{x}-\omega)' \tag{3.1}$$

to l. The function $\tilde{l}$ is the log-likelihood function corresponding to a single observation $\tilde{x}$ from the r-dimensional hyperbolic distribution with translation parameter ω and fixed values for the other parameters, i.e. δ, Δ (a symmetric, positive definite $r \times r$ matrix, having determinant $|\Delta| = 1$), α, and β (an $1 \times r$ vector). The level sets $\{\omega : \tilde{l}(\omega) = c\}$ are ellipsoids and $\tilde{l}$ is strictly concave. The aim of the following remarks is to indicate the feasibility and practicality of approximation by a multivariate hyperbolic likelihood. A more complete discussion of the question will not be given here, and only the case $r=2$ will be considered.

A comprehensive account of the various possible shapes that a two-dimensional hyperbolic distribution can take has been given by Blæsild (1978) and that account may, of course, be equally viewed as a description of the possible shapes of the hyperbolic log-likelihood function (3.1). Figures 6 and 7 show two examples of the family of levels sets of $\tilde{l}$.

In the two-dimensional case, the total number of free parameters in $\tilde{l}$ available for the fitting purpose, i.e. the number of free variables in $\tilde{x}$, δ, Δ, α and β, is equal to 8.

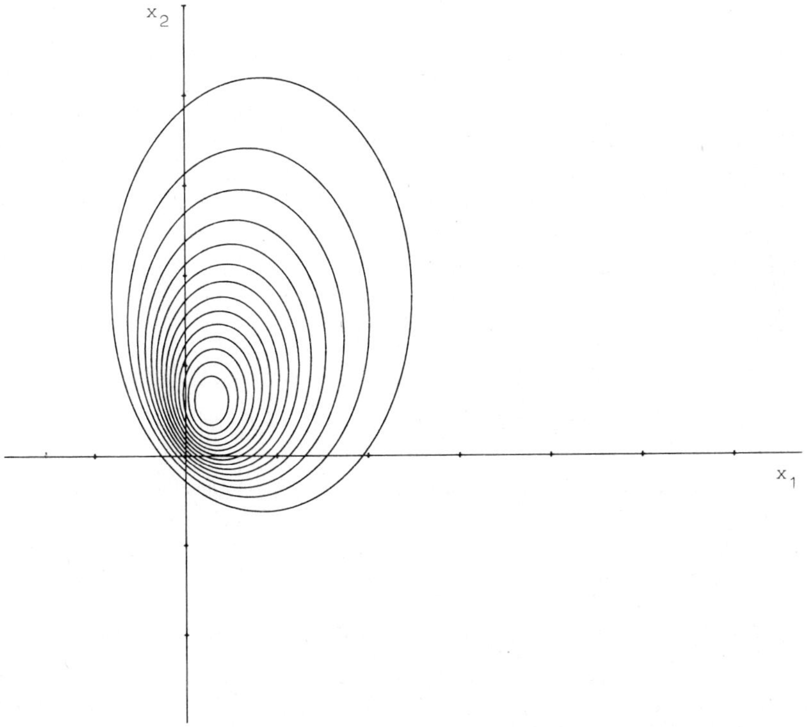

Fig. 7. Level curves of the two-dimensional hyperbolic distribution with parameters: $\omega=(0,0)$, $\delta=0.5$, $\epsilon_{11}=1$, $\epsilon_{22}=1.41$, $\epsilon_{12}=\epsilon_{21}=-0.64$, $\alpha=5$ and $\beta=(4,4)$.

As in one dimension, one may follow a variety of ways in selecting the values of these parameters. Only one type of method, which allows a direct utilization of the results of Section 2, will be discussed here.

For any fixed value of ω_2 the function $\tilde{l}(\cdot,\omega_2)$ is a one-dimensional hyperbolic log-likelihood function, and similarly for fixed ω_1. It is natural therefore to fit a hyperbolic likelihood to each of $l(\cdot,\hat{\omega}_2)$ and $l(\hat{\omega}_1,\cdot)$. This may be done by any of the procedures discussed in Section 2 and it leads to precisely 8 equations for the parameters of $\tilde{l}$, obtained by expressing the parameters of the one-dimensional hyperbolic log-likelihood functions $\tilde{l}(\cdot,\hat{\omega}_2)$ and $\tilde{l}(\hat{\omega}_1,\cdot)$ in terms of the parameters of $\tilde{l}$. Denoting the elements of Δ by ϵ_{ij} and letting $\kappa=\sqrt{\alpha^2-\beta\beta'}$, the equations are

$$\tilde{x}-\delta\kappa^{-1}\beta\Delta=\hat{\omega},$$

$$\delta^2(\epsilon_{11}^{-1},\epsilon_{22}^{-1})+(\epsilon_{11}^{-2},\epsilon_{22}^{-2})(\tilde{x}_2-\hat{\omega}_2)^2,(\tilde{x}_1-\hat{\omega}_1)^2)'=d^2,$$

$$\alpha\left(\sqrt{\epsilon_{11}},\sqrt{\epsilon_{22}}\right)=a,\qquad \beta=b, \tag{3.2}$$

where $d^2=(d_1^2,d_2^2)$, $a=(a_1,a_2)$ and $b=(b_1,b_2)$ are known constant vectors; in fact, (d_1,a_1,b_1) and (d_2,a_2,b_2) are the fitted (δ,α,β) vectors from $\tilde{l}(\cdot,\hat{\omega}_2)$ and $\tilde{l}(\hat{\omega}_1,\cdot)$, respectively. By means of the first vector equation in (3.2) the quantities $\tilde{x}_1-\hat{\omega}_1$ and $\tilde{x}_2-\hat{\omega}_2$ may be eliminated from the second vector equation. This yields

$$\delta^2\left\{(\epsilon_{11}^{-1},\epsilon_{22}^{-1})+\kappa^{-2}(\epsilon_{11}^{-2},\epsilon_{22}^{-2})\left((\beta_1\epsilon_{12}+\beta_2\epsilon_{22})^2,(\beta_1\epsilon_{11}+\beta_2\epsilon_{21})^2\right)'\right\}=d^2.$$

From this equation the $\epsilon-s$ and $\beta-s$ can be eliminated using the third and fourth equations of (3.2) together with the relation $|\Delta|=1$. The expression in the curly brackets of (3.3) then depends on α only, and eliminating δ^2 in the obvious way one arrives at a fairly simple equation for α which in general has to be solved by numerical iteration. More specifically, the equation is of the form $Q(\alpha^2)=0$ where Q is generally a polynomial of degree four; however, if either b_1 or b_2 equals 0 then Q reduces to a second degree polynomial. Once α is determined it is immediate to find the remaining parameters.

References

Bagnold, R.A. and Barndorff-Nielsen, O. (1979). The pattern of natural size distributions. Research Report No. 47, Dept. Theor. Statist., Aarhus University. (To appear in *Sedimentology*.)

Barndorff-Nielsen, O. (1978). Hyperbolic distributions and distributions on hyperbolae. *Scand. J. Statist.* **5**, 151–157.

Barndorff-Nielsen, O. (1979). Models for non-Gaussian variation with applications to turbulence. *Proc. R. Soc. London A* **368**, 501–520.

Blæsild, P. (1978). On the two-dimensional hyperbolic distribution and some related distributions; with an application to Johannsen's bean data. Research Report No. 40, Dept. Theor. Statist., Aarhus University. (To appear in *Biometrika*.)

G. Kallianpur, P.R. Krishnaiah, J.K. Ghosh, eds., *Statistics and Probability: Essays in Honor of C.R. Rao*
© North-Holland Publishing Company (1982) 77–88

THE DEVELOPMENT OF POPULATION MODELS

M.S. BARTLETT
Emeritus Professor of Biomathematics,
University of Oxford, United Kingdom

1. Historical beginnings

Dr. Rao's own contributions to population biology are perhaps predominantly biometrical. I recall, in particular, his important paper to the Royal Statistical Society in 1948 on classification and discriminatory problems. However, he would certainly be familiar with the wider context of population models, without which indeed much of our biometrical work would be pointless. It is pertinent to go right back to the beginnings of probability and statistics in the 17th century, with the statistical side dating in effect from John Graunt's study of mortality in 1662, which led to formulations of the life table both by Graunt himself and by Edmund Halley. Smith and Keyfitz (1977) give evidence of a very much earlier Italian work in their historical compilation, *Mathematical Demography*, but it was the scientific interest aroused among his contemporaries by Graunt's publication which seems to have launched what might be called modern actuarial mathematics. An intriguing account of the communication of these ideas is given by Kendall (1961),—for example, the correspondence on life expectancies between Ludovic Huyghens and his more famous brother Christian.

The existence of such historical surveys as Smith and Keyfitz means that an article like my own can concentrate more on facets of personal interest, and it should certainly avoid excessive attention to actuarial demography [cf. also Pollard (1973)]. On a theoretical point, actuarial demography has moreover been formulated on a largely "deterministic" basis for large populations, whereas much of my own work has concentrated on stochastic models. This distinction can, however, be overemphasized; for example, when we speak of the survival of 50% of males after age x, this may be thought of in probability terms, *or* in terms of a large population. The important thing is to know when sampling effects are to be reckoned with, and when they can be neglected.

On the question of content, the actuarial interest in mortality and its association with such matters as annuities, presumably held up for a while the inclusion of questions of fertility and net changes in population size. The systematic development of such a more comprehensive formulation is largely due to Lotka, though such early papers as that by Euler (1760) must not be overlooked. What is rather remarkable is the failure, until quite recently, to have any very realistic self-consistent models incorporating both sexes; I will comment on some aspects of this development presently.

When we turn to stochastic models as such, it is a salutary revelation to such comparatively late comers as myself to learn how much more had already been accomplished than we at first realized. I was well aware of Francis Galton's interest as long ago as 1889 in the "extinction of surnames" problem, though even this knowledge was lacking in post-war research by physicists into extinction probabilities for their own branching processes. It became quite customary to refer to Galton's appeal to the mathematician Watson for assistance, and to the loose ends which he left dangling and only being resolved by Steffenson in 1930. Not until the note by Heyde and Seneta (1972, see also other references therein) did we realize that Bienaymé had already resolved the problem in 1845!

A smaller time-span, but nonetheless a slightly embarrassing one, was brought to light by Kendall's discovery of McKendrick's (1926) work on stochastic epidemic models, after Kendall and I and others had been formulating these anew. I think McKendrick himself did not make it easy for us when he wrote a more publicized deterministic version of epidemic theory with Kermack, and made no reference himself to his earlier work. This temporary neglect of McKendrick by my contemporaries was at least not restricted to theoretical epidemiologists, for the discovery of the negative binomial distribution was at the time usually attributed to Yule and Greenwood in connection with their work in 1920 on accident-proneness among munition workers. Polya's association with the distribution as a "contagion model" came even later, so McKendrick's discussion of this contagion model in 1914 was especially startling. I suppose these later realizations of prior work must not surprise us unduly when we remember Pearson's note on de Moivre's discovery of the normal or Gaussian distribution as a limit of the binomial (with "Stirling's approximation" for a factorial thrown in for good measure).

To conclude these introductory remarks on early pioneers, I have to bear in mind the wide field in which population models are relevant, including not only demography and epidemiology, but population ecology and population genetics. If we attempted to classify the models introduced as deterministic or stochastic, those in ecology would be deterministic [e.g., Volterra (1926)], those in epidemiology largely deterministic at first [though Ross (1910) includes some probabilistic considerations] and those in population genetics more mixed [Haldane (1924), deterministic; Wright (1931) used diffusion models in discussing random drift; Fisher (1930) made use of both deterministic and diffusion models, the latter being approximations to full stochastic formulations and used extensively since by Kimura (1964) and others].

2. Stochastic formulations in the last fifty years

In spite of the sporadic early appearance of stochastic formulations of population models, especially in relation to extinction probabilities, it is the systematic development of stochastic process theory by Kolmogorov and others in the last fifty years which has strongly interacted with the advance of stochastic models on a much broader front. My own interest in stochastic population models (as distinct from time-series) largely dates from the post-war period (see, however, an early paper on fluctuations in the theory of inbreeding in 1937), and was stimulated in part by Feller's writings (e.g. 1939). In my North Carolina lectures, I derived the solution of the simple birth-and-death process introduced by Feller by means of the operational equation I had developed for the characteristic function of Markov processes in general, a solution which Feller had failed to obtain, but which I later discovered had already been deduced by Palm, by similar methods to my own. Not long after returning to England, I joined, in 1949, with Kendall and Moyal in contributing to the Symposium on Stochastic Processes under the auspices of the Royal Statistical Society, in which Kendall concentrated on population models and, in particular, showed how stochastic fluctuations in population numbers involving a *continuous* parameter such as age can be rigorously handled, a problem I had raised but not solved in my North Carolina lectures.

This last problem had also arisen in connection with physical populations of particles, for example, when classified by their energy, workers in this area including Arley, Bhabha, Jànossy, Moyal and Ramakrishnan [for references, see, for example, Bartlett (1978)].

In discrete-time and digitalized versions, this particular difficulty does not arise, and of course discrete-time formulations are often convenient as an approximation convenient for numerical work, and familiar in demography and ecology in the matrix (deterministic) formulation by Leslie (1945).

This interest in stochastic models inadvertently rather dichotomized the study of population models into the more classical deterministic approach and the more modern stochastic approach. I have said before that no one wants to use the more complicated stochastic approach when it is not necessary, so let me say a little more about when stochastic effects are relevant.

There are in fact at least five ways in which deterministic models are inadequte, so I will list these before considering when deterministic models are sufficient:

(i) The first may sound "academic", but I think it is important to understand the correct specification of a population model, however much we may curtail this later for tractability and convenience. To take the simplest case, a population of N individuals, unclassified for age or sex or genetic composition, can only change by integers, so, if we are considering continuous time change, $\mathrm{d}N_t/\mathrm{d}t$ is a nonsense expression, however common it may be in physical, chemical or biological equations. Often we may be unable or even not wish to specify a stochastic model, on the grounds that we know it would be over-idealized. On the other hand, if we specify a deterministic model that is inconsistent with any stochastic model, we are as much at fault as specifying a stochastic model that implies consequences that are not intended (a practice that has occurred in physical literature). I have discussed this in the context of age-distributions for dividing cell populations, and in the context of two-sex population models with age-structure.

In the former case simplifying assumptions of negligible death rate and constant time-period of the mitotic cycle before cell-division led to the conclusion of an exponential ultimate age-distribution, in complete disregard of the non-existence of *any* continuous distribution if all the initial cells were started off at the same "age". In the correct formulation in stochastic terms, the number of individuals $N(u;t)$ of all existing individuals up to age u at time t can be averaged, and *then* differentiated with respect to u or to t. For example, we define the so-called first order density with respect to age:

$$f_1(u;t)=\mathrm{d}_u\{EN(u;t)\}/\mathrm{d}u$$

and then under appropriate conditions

$$f_1(u;t)\to C\exp\left\{K(t-u)-\int_0^u[\lambda(v)+\mu(v)]\mathrm{d}v\right\}, \tag{1}$$

with K the dominant root of the equation

$$\int_0^\infty 2\lambda(t)\mathrm{e}^{-Kt}\exp\left\{-\int_0^t[\lambda(v)+\mu(v)]\mathrm{d}v\right\}\mathrm{d}t=1, \tag{2}$$

where $\mu(v)$ is the age-dependent death rate and $\lambda(v)$ the rate of completion of the mitotic cycle depending on the cycle age t [for further details see, for example, Bartlett (1970)].

A rather analogous but even more complicated inconsistency arose with the deterministic formulation of a two-sex human population model by Das Gupta (1972). We define $M_t(u)$, $F_t(v)$ as the numbers of males and females with ages less than or equal to u and v, respectively, at time t. We are thus able to define in any stochastic formulation $m_t(u)$, $f_t(v)$ as the first-order densities

$$m_t(u)=E\{\mathrm{d}M_t(u)\}/\mathrm{d}u,\qquad f_t(v)=E\{\mathrm{d}F_t(v)\}/\mathrm{d}v.$$

In any deterministic version, $m_t(u)$ and $f_t(v)$ are (non-rigorously) written for $\mathrm{d}M_t(u)/\mathrm{d}u$ and $\mathrm{d}F_t(v)/\mathrm{d}v$.

It is known [e.g. Bartlett (1970)] that $m_t(u)$ and $f_t(v)$ satisfy the continuity equations

$$\begin{aligned}\frac{\partial m_t(u)}{\partial t}+\frac{\partial m_t(u)}{\partial u}+\mu_{\mathrm{m}}(u)m_t(u)&=0,\quad (u>0),\\ \frac{\partial f_t(v)}{\partial t}+\frac{\partial f_t(v)}{\partial v}+\mu_{\mathrm{f}}(v)f_t(v)&=0,\quad (v>0),\end{aligned} \tag{3}$$

where $\mu_{\mathrm{m}}(u)$, $\mu_{\mathrm{f}}(v)$ are male and female age-dependent death rates. Continuity equations like the above are rather curiously by Rubinow (1978) referred to as Scherbaum–Rasch–von Foerster equations, though in an Addendum, Rubinow notes that they had already appeared in the work of McKendrick (1926). These equations provide the solutions

$$\begin{aligned}m_t(u)&=p_{\mathrm{m}}(u)m_{t-u}(0),\quad (0<u<t),\\ f_t(v)&=p_{\mathrm{f}}(v)f_{t-v}(0),\quad (0<v<t),\end{aligned} \tag{4}$$

where

$$p_{\mathrm{m}}(u)=\exp\left\{-\int_0^u\mu_{\mathrm{m}}(v)\mathrm{d}v\right\},\qquad p_{\mathrm{f}}(v)=\exp\left\{-\int_0^v\mu_{\mathrm{f}}(u)\mathrm{d}u\right\}.$$

Deterministic models require the specification of $m_t(0)$, $f_t(0)$, which in the Das

Gupta model are assumed given by

$$m_t(0)=\int_0^\infty\int_0^\infty \Lambda_m\{m_t(u), f_t(v)\}\,du\,dv,$$

$$f_t(0)=\int_0^\infty\int_0^\infty \Lambda_f\{m_t(u), f_t(v)\}\,du\,dv, \tag{5}$$

where Λ_m and Λ_f are in general functionals of $m_t(u)$, $f_t(v)$ which specify the entire state of the population at time t.

To illustrate the difficulties arising, let us consider the specific model of Das Gupta's where Λ_m and Λ_f depend only on the numbers of males and females at ages u, v, respectively, and particularly in one model are assumed proportional to $[m_t(u)f_t(v)]^{1/2}$, this assumption being consistent with Das Gupta's requirement of being of degree one [not unreasonable in the human population case, though not appropriate in some animal populations—see model in point (ii) below]. The stochastic version of this assumption becomes nonsensical, for if we tried to write $[dM_t(u)dF_t(v)]^{1/2}$ with, for simplicity, no multiple births, $dM_t(u)dF_t(v)$ is either 0 or 1, and $[dM_t(u)dF_t(v)]^{1/2}=dM_t(u)dF_t(v)$. A further complication of second-order densities arising from the product $dM_t(u)dF_t(v)$ need not concern us here, nor the difficulty of finding general, as distinct from asymptotic, solutions of the (deterministic) equations.

The formulation difficulty at least is avoided by introducing what I called a weighting or "distance" function $w(u, v)$, so that the expression replacing $[m_t(u)f_t(v)]^{1/2}$ would be of the form

$$\left[\int w_m(u, v)m_t(v)\,dv\int w_f(v, u)f_t(u)\,du\right]^{1/2}$$

and now quite consistently in the stochastic analogue

$$\left[\int w_m(u, v)\,dM_t(v)\int w_f(v, u)\,dF_t(u)\right]^{1/2}.$$

(ii) An obvious consequence of point (i) above is that *small-scale* vicissitudes in the population size N cannot be covered by deterministic theory. This means that precise statistical inference of the kind available in some specialized contexts is never available, and no possibility exists of discussing in any satisfactory way the fit of deterministic models to data. We must not overemphasize this point, for the use of models for large-scale populations is vital whether or not such questions of goodness of fit can be answered. Nevertheless, precise stochastic models for inferring parameter values are invaluable where applicable. Examples include the estimation of parameter values for measles from family household data by Bailey (1957), and the use of what I have termed *stochastic analysis*, e.g. in a laboratory experiment on the mating of sheep blow-flies (cf. Bartlett (1978) §8.31).

This method of analysis is applicable to any pure birth (or death) homogeneous process for which the detailed times of births are available. The chance of an increment is in general $\lambda\{N_t\}\,dt$, say, during $t, t+dt$, and the time to the next event thus inversely proportional to $\lambda\{N_t\}$ when the population size is N_t. By transforming each time-interval appropriately, we are thus able to standardize the stochastic exponential intervals.

In the experiment referred to, mating pairs of flies were removed at the times of mating, which were recorded. The original deterministic model for the number of males M_t and females F_t remaining at time t was

$$\mathrm{d}M_t/\mathrm{d}t = -\alpha M_t F_t \tag{6}$$

with $F_t = F_0 + (M_t - M_0)$. The equivalent stochastic model, if we write N_t for the number of matings, is

$$\lambda\{N_t\} = (M_0 - N_t)(F_0 - N_t)/\theta, \tag{7}$$

where $\theta = 1/\alpha$. When standardizing the n intervals between matings by multiplying by $M_t F_t$ the subsequent interval, the log-likelihood function L takes the simple form (in spite of the non-linearity of the model)

$$L = \sum_r (-T_r/\theta - \log\theta) \tag{8}$$

where T_r are the standardized intervals. This leads to an immediate estimator $\bar{T} = \Sigma_r T_r/n$ for θ which is unbiassed with minimum variance. Moreover, plotting the cumulative sum of the T_r, this sum is a random walk conditioned (given $\bar{T}$) at 0 and n, enabling goodness-of-fit questions to be answered.

(iii) The small-scale vicissitudes of the population number N_t are also vital theoretically in certain contexts, as I have previously emphasized. The most familiar is perhaps the fate of mutant genes in a genetically-structured population; but the onset or not of a major epidemic following the introduction of one or two infected individuals is another important problem which may be adequately answered by treating the number S_t of susceptible individuals at the beginning of an epidemic as approximately constant. Then if the chance of an infection during $t, t+\mathrm{d}t$ is $\lambda I_t S_t \mathrm{d}t$ where I_t is the number of infected individuals, and $\mu I_t \mathrm{d}t$ is the chance of removal of the infected individuals, the process becomes a simple birth-and-death process with birth rate for new infected individuals λS_0. The chance of a major epidemic is then approximately, from the theory of the simple birth-and-death process,

$$1 - [\mu/(\lambda S_0)]^{I_0},$$

if $S_0 > \mu/\lambda$, and 0 if $S_0 < \mu/\lambda$.

(iv) While individual extinction occurrences are the most obvious consequence of stochastic fluctuations, it is possible for the latter in some circumstances to be more global in their effects, and markedly modify the behaviour of quite large communities. I have illustrated this in the context of measles endemicity by examining its permanence in relation to total community size, leading to the concept of a critical community size (see, e.g. Bartlett (1960)). Theories of island biogeography (MacArthur and Wilson (1967)) also make essential use of stochastic models; and so do some investigations involving spatial spread (e.g., Oaten (1977)).

(v) Finally, before I redress the balance by noting circumstances where stochastic effects are unimportant, let me just note the relevance of what might be termed *doubly stochastic* models, in which variable environments and the like are represented by parameter values which are themselves changing, not, say, periodically with the season, but stochastically with, say, the weather. It will be clear that, while such doubly stochastic models are at times very relevant, they do not by any means exhaust the relevance of stochastic models, as might inadvertently be implied in the literature [cf., for example, the remarks by May on stochastic versus deterministic models, in Anderson et al. (1979) p. 393].

3. Approximations to stochastic models for large population size

3.1. Deterministic models

(i) *Limit cycles in ecology*. When previously discussing population equations (see, e.g. Bartlett (1960) p. 31), I have noted that one form of representation is the direct stochastic formulation (for a vector $\boldsymbol{N}_t$ of various types of individual changing in continuous time t)

$$\mathrm{d}\boldsymbol{N}_t=\boldsymbol{f}(\boldsymbol{N}_t)\,\mathrm{d}t+\mathrm{d}\boldsymbol{Z}_t. \tag{9}$$

This form of representation obscures the complexity of the stochastic component $\mathrm{d}\boldsymbol{Z}_t$, but has the advantage of isolating the deterministic analogue

$$\mathrm{d}\boldsymbol{N}_t=\boldsymbol{f}(\boldsymbol{N}_t)\,\mathrm{d}t, \tag{10}$$

which simply ignores $\mathrm{d}\boldsymbol{Z}_t$. The justification for doing this will usually depend on its relative effect being small as the population size $\boldsymbol{N}_t$ increases, but without going into specific cases it is possible to indicate situations where we may expect the stochastic effects to be unimportant. Suppose equation (10) has equilibrium solutions, given by

$$\boldsymbol{f}(\boldsymbol{N}_t)=0 \tag{11}$$

which are unstable. Then the population will explode or become extinct (or a combination of both in the *vector* $\boldsymbol{N}_t$) and, apart perhaps from any stochastic effects due to the *initial* numbers being small, stochastic fluctuations are unlikely to disturb this evolutionary behaviour.

Even if there is global stability, but local instability, a situation conducive to *limit cycles*, this conclusion is still likely to be true. The importance of limit cycles with some kinds of non-linear models has been emphasized, among others, by May (1973). As non-linear equations are the rule rather than the exception in biology, such behaviour can be more prevalent than I myself realized when first studying population models, although I had noticed this theoretical possibility, admittedly in a crude way, when referring to oscillations in sheep blow-fly populations (Bartlett (1960) p. 50); thus I have been very interested to see the subsequent fitting of limit-cycle models to such data in a more detailed manner.

As one simple model, consider the population size N_t for a single type of individual following the logistic model

$$\mathrm{d}N_t=\lambda N_t(1-N_t/M)\,\mathrm{d}t, \tag{12}$$

but with a delayed regularity effect approximating to the more detailed effect of age-structure, so that the equation is rewritten

$$\mathrm{d}N_t = \lambda N_t(1 - N_{t-T}/M)\,\mathrm{d}t. \tag{13}$$

May (1978) notes that this model, first introduced into ecology by Hutchinson (1948), leads to a limit cycle if $\lambda T > \frac{1}{2}\pi$, and shows that the value $\lambda T = 2.1$ best fits the blow-fly data. For this value, the maximum value of N_t is much larger than the minimum value by a factor 84.1, with a cycle period 4.54 T. With still larger values of λT, the cycle period does not change markedly, but the oscillations become much more violent, e.g. for $\lambda T = 2.4$, the period is $5.11T$ but the ratio N(max) to N(min) has risen to 1040. Of course, in this latter case, if N(max) were only 10000, N(min) would be only 10, and stochastic effects might no longer be negligible.

(ii) *Human population models.* I have already remarked that the early human population models were usually deterministic, apart from the special problem of extinction. Workers like Lotka had covered the question of age-structure, but sex was included, if at all, in a linear way by introducing males and females, but allowing, say, only the females to be assigned a reproductive role. A partial relaxation of this restriction has been made by Goodman and others [for references, see, for example Sivamurthy (1979)], but a more consistent approach to the two-sex problem attempted by Das Gupta (1972) has already been commented on. Of course, there is hardly any limit to the complications possible in the search for greater realism, i.e. immigration, emigration, marriages, divorces, etc. An attempt to introduce marriages into the population model was made by Fredrikson in 1971. Thus, if we write as before, the males of age $(u, u+\mathrm{d}u)$ as $m_t(u)\mathrm{d}u$, and females $f_t(v)$, but restrict this population to the unmarried, and include an additional density $n_t(u, v)\mathrm{d}u\mathrm{d}v$ for the married population with a male partner $u, u+\mathrm{d}u$ and with a female partner $v, v+\mathrm{d}v$, then the continuity equations become:

$$\begin{aligned}
\frac{\partial m_t}{\partial t} + \frac{\partial m_t}{\partial u} &= -\mu_{\mathrm{m}} m_t + \int \mu_{\mathrm{f}} n_t(u, v)\,\mathrm{d}v \\
&\quad - \int \Phi_{\mathrm{mf}}\{m_t, f_t\}\,\mathrm{d}v, \quad (u>0) \\
\frac{\partial f_t}{\partial t} + \frac{\partial f_t}{\partial v} &= -\mu_{\mathrm{f}} f_t + \int \mu_{\mathrm{m}} n_t(u, v)\,\mathrm{d}u \\
&\quad - \int \Phi_{\mathrm{mf}}\{m_t, f_t\}\,\mathrm{d}u, \quad (v>0) \\
\frac{\partial n_t}{\partial t} + \frac{\partial n_t}{\partial u} + \frac{\partial n_t}{\partial v} &= -(\mu_{\mathrm{m}} + \mu_{\mathrm{f}}) n_t + \Phi_{\mathrm{mf}}\{m_t, f_t\},
\end{aligned} \tag{14}$$

the last term in each equation representing the marriage rate. The new births are given by, say,

$$\left\{\begin{aligned}
m_t(0) &= \iint \lambda_1(u, v) n_t(u, v)\,\mathrm{d}u\mathrm{d}v \\
f_t(0) &= \iint \lambda_2(u, v) n_t(u, v)\,\mathrm{d}u\mathrm{d}v
\end{aligned}\right\} \tag{15}$$

and the total male and female populations of age u and v, respectively, are given by

$$m_t(u)+\int n_t(u,v)\,\mathrm{d}v, \qquad f_t(v)+\int n_t(u,v)\,\mathrm{d}u.$$

Unfortunately, the solutions of these equations are not possible to develop in any simple manner, and as they still neglect such aspects as divorces, unmarried mothers, family planning, and so on, it seems doubtful whether the model possesses any advantages over reasonably formulated two-sex models not explicitly introducing marriages.

3.2. *Diffusion models*

Returning to the complete stochastic representation (9), an approximation less drastic than the deterministic is its replacement by a diffusion equation, which may be thought of as representing an approximating normal random drift of the correct variance about the deterministic path. For example, if the population size $\boldsymbol{N}_t$ is large, and a stable equilibrium point $\boldsymbol{N}$ in the deterministic equation exists, then deviations x from the equilibrium point may be studied on some appropriate quasi-continuous scale N_t/N for each component. The second-order moment terms (depending in general on x) now affect the solution, which in one variable is known as the Sewall Wright distribution, first arising in a genetical context. Under appropriate conditions it is also possible to derive higher-order approximations (e.g. Bartlett (1960) §4.3) than the normal, which results if the model is "linearized".

However, considerable care is necessary in checking when these approximations are justified. In particular, it has only recently been realized that the diffusion approximation for a growing population literally diffusing in space such as the dispersal of an advantageous gene over space or the spread of an epidemic is not applicable at the boundary, and other approximations must be employed for valid conclusions on such questions as the velocity of propagation (see Daniels (1977), Mollison (1977)).

3.3. *Branching processes*

A different approximation is involved in the use of branching process models, for example, in the increase in numbers of a mutant gene or the initial growth of an epidemic. Here the population size is regarded as virtually infinite, but the individuals of interest—mutants or infected individuals—are at first very scarce, so that diffusion models in relation to these individuals are inapplicable.

3.4. *Mixed models*

The case for deterministic models becomes of course stronger as the model becomes more complex, especially as even these simple diffusion and branching process stochastic approximations become inapplicable or intractable; the deterministic formulation can be quite complicated enough. It may then be useful to introduce stochastic elements in a partial way into the model, wherever stochastic variability cannot be neglected. Such *mixed models* have been employed by many workers, for example, Kendall using the deterministic solution for an epidemic *after* the initial uncertainty phase, Gani in bacteriophage models, Anderson in ecological models, etc. A quite early example was the treatment by Lea and Coulson (1949) of the limiting distribution of mutated bacteria, treating the normal bacteria as growing deterministically.

When complex non-linear models involving more than one type of individual arise in ecology with the possibility of more than one equilibrium point, there seems even more justification for simplifying the stochastic element to assisting passage from one equilibrium to another, rather analogously to the situation envisaged by Sewall Wright in population genetics when multiple equilibrium points exist.

3.5. "Doubly stochastic" models

We now come to what I have called "doubly stochastic" models, in which the parameters of the model are themselves stochastic. (This terminology seems to me consistent with Cox's doubly stochastic Poisson processes, in which the average rate of occurrence of events is not constant, but itself a random process.)

I think these stochastic models should be distinguished from those arising in a practice sometimes employed of replacing the stochastic model (9), where $\mathrm{d}\boldsymbol{Z}_t$ is a complicated and interrelated component of $\mathrm{d}\boldsymbol{N}_t$, by the deterministic equation plus a simple and crude "noise" element $\mathrm{d}\boldsymbol{Y}_t$, independent of the deterministic component. This may be adequate for linear time-series models in econometrics, but is dangerous for the non-linear models of biology, even if of occasional value [as noted in Subsection 3.4 above] in indicating the effect of simple random drift about the deterministic solution.

A familiar example of doubly stochastic models in recent years is, in the case of the simple branching process approximation in genetics, allowing the environment to be random. In the case of the extinction probability for mutant genes for discrete generations, the relevant theory has been investigated by Wilkinson and others (see Bartlett (1978) §2.31). It has some relevance to the efficiency of group selection under variable environmental conditions.

The theoretical problem is a particular case of absorption probabilities for Markov chain sequences with variable transition probability matrices. With this wider formulation, a transition matrix $\boldsymbol{M}$ governing the distribution vector $\boldsymbol{p}_t$ by the *forward* equation $\boldsymbol{p}_{t+1}=\boldsymbol{M}\boldsymbol{p}_t$ is replaced by a random matrix $\{\boldsymbol{M}\}$, say. Any absorption probability limit $\boldsymbol{P}$ (a vector depending on the initial state), which previously satisfied the *backward* equation $\boldsymbol{P}=\boldsymbol{P}\boldsymbol{M}$, now satisfies the more general equation $\boldsymbol{P}=\boldsymbol{P}E\{\boldsymbol{M}\}$.

4. Further remarks

The plethora of population models now to be found in the literature shows how far we have come even from the early post-war days of simple birth-and-death processes and the like, journals now devoting much of their space to this topic (especially of course *Theoretical Population Biology*). Physicists have become reconciled to the growing complexities of fundamental physics, with their concepts of "strangeness" and "charm" and so on, but I doubt if they would claim to compete with the complexities of the living world.

At the community level models are still somewhat tentative, and empirical study of species abundances and diversity [see, for example, Engen (1978)] will no doubt help to stimulate much more theoretical model-building in the future at this complex but more realistic level.

References

Anderson, R.M., Turner, B.D. and Taylor, L.R. (Eds.) (1979). *Population Dynamics*. Blackwell Scientific Publications, Oxford.

Bailey, N.T.J. (1957). *The Mathematical Theory of Epidemics*. Griffin, London.

Bartlett, M.S. (1937). Deviations from expected frequencies in the theory of inbreeding. *J. Genet.* **35,** 83.

Bartlett, M.S. (1946). *Stochastic Processes* (mimeographed notes of a course given at the Univ. of North Carolina in the Fall Quarter, 1946).

Bartlett, M.S. (1960). *Stochastic Population Models in Ecology and Epidemiology*. Chapman and Hall, London.

Bartlett, M.S. (1970). Age distributions. *Biometrics* **26**, 377–385.

Bartlett, M.S. (1973). A note on Das Gupta's two-sex population model. *J. Theor. Pop. Biol.* **4**, 418–424.

Bartlett, M.S. (1978). *An Introduction to Stochastic Processes*. Camb. Univ. Press, 3rd ed.

Bienaymé, I.J. (1845). De la loi de multiplication et de la durée des familles. *Soc. Philomath. Paris Extraits Ser.* **5**, 37.

Daniels, H.E. (1977). The advancing wave in a spatial birth process. *J. Appl. Prob.* **14**, 689–701.

Das Gupta, P. (1972). On two-sex models leading to stable populations *J. Theor. Pop. Biol.* **3**, 358–375.

Engen, S. (1978). *Stochastic Abundance Models*. Chapman and Hall, London.

Euler, L. (1760). Recherches générales sur la mortalité et la multiplication du genre humaine. *Histoires l'Acad. R. Sci. Belles Lettres* **16**, 144–164.

Feller, W. (1939). Die Grundlagen der Volterraschen Theorie des Kampfes ums Dasein Wahrscheinlichkeitstheoretischer Behandlung *Acta Biotheoretica* **5**, 11.

Fisher, R.A. (1930) *Genetical Theory of Natural Selection*. Oxford.

Fredrickson, A.G. (1971). A mathematical theory of age structure in sexual populations: random mating and monogamous marriage models. *Math. Biosci.* **10**, 117–143.

Galton, F. (1889). *Natural Inheritance*. Macmillan, London.

Graunt, J. (1662). *Natural and Political Observations Upon the Bills of Mortality*. London.

Greenwood, M. and Yule, G.U. (1920). An inquiry into the nature of frequency distributions representative of multiple happenings with particular reference to the occurrence of multiple attacks of disease or of repeated accidents. *J.R. Statist. Soc.* **83**, 255–279.

Haldane, J.B.S. (1924). A mathematical theory of natural and artificial selection. *Trans. Camb. Phil. Soc.* **23,** 19–41.

Heyde, C.C. and Seneta, E. (1972). Studies in the history of probability and statistics XXXI The simple branching process, a turning point test and a fundamental inequality: A historical note on I.J. Bienaymé. *Biometrika* **59**, 680–683.

Hutchinson, G.E. (1948). Circular causal systems in ecology. *Ann. N.Y. Acad. Sci.* **50,** 211–246.

Kendall, D.G. (1949). Stochastic process and population growth *J.R. Statist. Soc. B.* **11**, 230–264.

Kendall, M.G. (1961). Studies in the history of probability and statistics *X*, Where shall the history of statistics begin? *Biometrika* **47,** 447–449.

Kimura, M. (1964). Diffusion models in population genetics *J. Appl. Prob.* **1**, 177–232.

Lea, D.E. and Coulson, C.A. (1949). The distribution of the number of mutants in bacterial populations. *J. Genet.* **49**, 264.

Leslie, P.H. (1945). On the use of matrices in certain population mathematics. *Biometrika* **34**, 183–212.

Levin, S.A. (Ed.) (1978). *Studies in Mathematical Biology Part II Populations and Communities.* Math. Assoc. of America.

Lotka, A.J. (1939). A contribution to the theory of self-renewing aggregates with special reference to industrial replacement *Ann. Math. Statist.* **10**, 1–25.

Ludwig, D. (1974). *Stochastic Population Theories.* Springer-Verlag, New York.

MacArthur, R.H. and Wilson, E.O. (1967). *The Theory of Island Biogeography.* Princeton.

McKendrick, A.G. (1914). Studies on the theory of continuous probabilities, with special reference to its bearing on natural phenomena of a progressive nature. *Proc. Lond. Math. Soc.* **13**, 401–416.

McKendrick, A.G. (1926). Applications of mathematics to medical problems. *Proc. Edin. Math. Soc.* **44** 98–130.

May, R.M. (1973). *Stability and Complexity in Model Ecosystems.* Princeton.

May, R.M. (1978). Mathematical aspects of the dynamics of animal populations. (See Levin, S.A., pp. 317–366.)

Mollison, D. (1977). Spatial contact models for ecological and epidemiological spread. *J. R. Statist. Soc. B* **39**, 283–326.

Oaten, A. (1977). Optimal foraging in patches: a case for stochasticity. *Theor. Pop. Biol.* **12**, 263–285.

Pearson, K. (1924). Historical note on the origin of the normal curve of errors. *Biometrika* **16**, 402–404.

Pollard, J.H. (1973). *Mathematical Models for the Growth of Human Populations.* Cambridge.

Rao, C.R. (1948). The utilisation of multiple measurements in problems of biological classification. *J. R. Statist. Soc. B* **10**, 159.

Ross, R. (1910). *The Prevention of Malaria*, 2nd ed. Murray, London.

Rubinow, S.I. (1978). Age-structured equations in the theory of cell populations. (See Levin, S.A., pp. 389–410.)

Sivamurthy, M. (1979). Convergence of human age-sex distributions to an equilibrium-state age-sex distribution. *Theor. Pop. Biol.* **16**, 233–252.

Smith, D. and Keyfitz, N. (1977). *Mathematical Demography-Selected Papers.* Springer-Verlag, New York.

Volterra, V. (1926). Variazioni e fluttuazioni del numero d'individui in specie animali convivente. *Mem. Acad. Lincei Roma* **2**, 31–112.

Wright, S. (1931). Evolution in Mendelian populations. *Genetics* **16**, 97–159.

G. Kallianpur, P.R. Krishnaiah, J.K. Ghosh, eds., *Statistics and Probability: Essays in Honor of C.R. Rao*

A NOTE ON THE DIRICHLET PROCESS

D. BASU and R. C. TIWARI
The Florida State University, Tallahassee, FL, U.S.A.

Written mainly for its pedagogical interest, this expository note is concerned primarily with the question of existence of a Dirichlet process. A random probability measure on a measurable space $(\mathcal{X}, \mathcal{A})$ is a stochastic process $\{P(A): A \in \mathcal{A}\}$ – a collection of random variables indexed by the measurable sets in $\mathcal{X}$ – such that almost every realization of the process is a probability measure on $(\mathcal{X}, \mathcal{A})$. Given a finite measure α on $(\mathcal{X}, \mathcal{A})$, a Dirichlet process D^{α} is a random probability measure on $(\mathcal{X}, \mathcal{A})$ such that, for every partition $(A_1, A_2, \ldots, A_k)$ of $\mathcal{X}$ into a finite number of measurable sets, the joint distribution of the random variables $(P(A_1), P(A_2), \ldots, P(A_k))$ is a singular Dirichlet distribution with parameters $(\alpha(A_1), \alpha(A_2), \ldots, \alpha(A_k))$.

Part one of this article deals with the familiar case where $\mathcal{X}$ is a finite set. Properties of the k-dimensional Dirichlet distribution are so exposited as to motivate Blackwell's (1973) constructive definition of the Dirichlet process. In part two, the case where $(\mathcal{X}, \mathcal{A})$ is a Borel space is discussed in some detail.

1. Introduction

This report is concerned primarily with the question of existence of a Dirichlet process on a Borel space $(\mathcal{X}, \mathcal{A})$. Part one of this report deals with the familiar case when $\mathcal{X}$ is a finite set. Properties of the k-dimensional Dirichlet distribution are so exposited as to motivate Blackwell's (1973) constructive definition of the Dirichlet process. In part two, the case where $(\mathcal{X}, \mathcal{A})$ is a Borel space is discussed in some detail.

Section 2 deals with Bayesian (parametric) inference and the family of natural conjugate priors. Section 3 is devoted to some characterizations of the Dirichlet distribution and elucidations of its useful properties. In Section 4, the results of Section 3 are extended and it is shown that there exists a Dirichlet process on $(\mathcal{X}, \mathcal{A})$, when $\mathcal{X}$ is a finite or a countably infinite set.

In Section 5, some preliminary material on the Dirichlet process is presented. Some useful results on a Borel space are given in Section 6. Blackwell's (1973) construction of a Dirichlet process on a Borel space $(\mathcal{X}, \mathcal{A})$ is discussed in Section 7. Section 8 is devoted to the study of some properties of a Dirichlet process.

PART I: THE DIRICHLET DISTRIBUTION

2. Bayesian inference and natural conjugate priors

Let X be an observable random variable (r.v.) with a statistical model that is characterized by a probability density function (p.d.f.) $f(\cdot|\theta)$, where $\theta \in \Theta$ is the unknown parameter of interest. Before any data are collected, a Bayesian represents his prior opinion about θ by a distribution on Θ, called a prior distribution. If he observes X n-times and denotes by $\boldsymbol{D}_1^n=(x_1, x_2, \ldots, x_n)$ the data so obtained, then his opinion about θ is represented by the distribution of θ given $\boldsymbol{D}_1^n$, called the posterior distribution.

Let q be the prior p.d.f. of θ. The posterior p.d.f. of θ given $\boldsymbol{D}_1^n$, namely $q(\cdot|\boldsymbol{D}_1^n)$, is expressed by the relation

$$q(\theta|\boldsymbol{D}_1^n) \propto q(\theta)L(\theta|\boldsymbol{D}_1^n). \tag{2.1}$$

Here, $L(\theta|\boldsymbol{D}_1^n)$ is the likelihood function of θ at point $\boldsymbol{D}_1^n$, and the proportionality symbol, $\propto$, is used to indicate that the posterior p.d.f. of θ given $\boldsymbol{D}_1^n$ is equal to the right side of (2.1) divided by the factor $\int_\Theta L(\theta|\boldsymbol{D}_1^n)q(\theta)\mathrm{d}\theta$ which does not involve θ.

This is how new knowledge, obtained through data, may be combined with prior knowledge. The Bayesian continually updates his knowledge as more observations are taken. Clearly,

$$\begin{aligned} q(\theta|\boldsymbol{D}_1^{n+m}) &\propto q(\theta)L(\theta|\boldsymbol{D}_1^{n+m}) \\ &\propto q(\theta)L(\theta|\boldsymbol{D}_1^n)L(\theta|\boldsymbol{D}_{n+1}^{n+m}) \\ &\propto q(\theta|\boldsymbol{D}_1^n)L(\theta|\boldsymbol{D}_{n+1}^{n+m}). \end{aligned}$$

Thus, the opinion $q(\theta|\boldsymbol{D}_1^{n+m})$ based on data $\boldsymbol{D}_1^{n+m}$ may be regarded as the posterior based on data $\boldsymbol{D}_{n+1}^{n+m}$ and prior $q(\theta|\boldsymbol{D}_1^n)$. This process of updating opinion may go through many stages.

It is clear from the relation (2.1) that the change of opinion about θ after the data are obtained is effected through the likelihood function. In the context of a chosen statistical model, a Bayesian will regard the likelihood function as the sole reservoir of all the relevant information about the parameter that is contained in the data. This is usually stated as: *The Likelihood Principle*. Two sets of data generating equivalent likelihood functions contain the same relevant information about the parameter. Two likelihood functions are said to be equivalent if one of them is a constant multiple of the other, where the constant may depend on the data. [See Basu (1975) for more on the likelihood principle.]

In many situations, it is convenient to access the prior within a family, C, of distributions. The class C should be large enough to accommodate various shades of opinion about the parameter. Further, if $q \in C$ is a prior p.d.f. of θ, then the posterior p.d.f. $q(\theta|\boldsymbol{D}_1^n)$ of θ given the data $\boldsymbol{D}_1^n$ ought to be in a simple computable form. If $q(\theta|\boldsymbol{D}_1^n) \in C$ for all $q \in C$ and data $\boldsymbol{D}_1^n$, then C is called a *conjugate family of priors*.

It frequently happens that a conjugate family of priors naturally coexists with a given statistical model for the observable r.v. X. Suppose the model is such that there exists an $n_0 > 0$ with the property that for all data $\boldsymbol{D}_1^n$ with $n \geqslant n_0$ the induced likelihood function $L(\theta|\boldsymbol{D}_1^n)$ is integrable (with respect to some integrating measure μ) over the parameter space Θ. Consider then the family C_0 of p.d.f.'s $q(\theta)$ of the form:

$$q(\theta)=\frac{L(\theta|\boldsymbol{D}_1^n)}{\int_\theta L(\theta|\boldsymbol{D}_1^n)\,\mathrm{d}\mu(\theta)}.$$

If the prior p.d.f. $q(\theta)$ corresponds to a so-called prior data, $\boldsymbol{D}_1^n=(y_1, y_2,\dots, y_n)$, then with the current data $\boldsymbol{D}_1^m=(x_1, x_2,\dots, x_m)$. The posterior p.d.f. $q(\theta|\boldsymbol{D}_1^m)$ will correspond to the likelihood function $L(\theta|\boldsymbol{D}_1^{n+m})$, where $\boldsymbol{D}_1^{n+m}$ is the extended data $(y_1, y_2,\dots, y_n, x_1, x_2,\dots, x_m)$. Thus for each prior $q\in C_0$, the posterior $q(\cdot|\boldsymbol{D}_1^m)$ belongs to C_0 for all possible current data $\boldsymbol{D}_1^m$.

The natural conjugate family C_0 of prior distributions takes on a simple form when, irrespective of the sample size n, there exists a sufficient statistic $T=(T_1, T_2,\dots, T_k)$ of fixed and small dimension $k(k\geqslant 1)$. Then,

$$L(\theta|\boldsymbol{D}_1^n)\propto H_n(\theta, T_1, T_2,\dots, T_k), \tag{2.2}$$

where $T_1, T_2,\dots, T_k$ are functions of $\boldsymbol{D}_1^n$. In this case, the natural conjugate family C_0 of prior distributions is characterized by $k+1$ superparameters, namely, particular values of $T_1, T_2,\dots, T_k$ and n.

For example, suppose each observation on an observable r.v. X belongs to one of the $k+1$ mutually exclusive and collectively exhaustive categories. Let $p_i(0<p_i<1)$ be the probability that an observation belongs to the ith category, $i=1,2,\dots, k+1$, where $\Sigma_{i=1}^{k+1}p_i=1$. We may regard $(p_1, p_2,\dots, p_k)$ as the model parameters. Suppose X is observed n times and let $\boldsymbol{D}_1^n$ be the data $(x_1, x_2,\dots, x_n)$ collected. Furthermore, let n_i denote the number of x's that belong to the ith category, $i=1,2,\dots, k+1$. Then each n_i is a non-negative integer and $\Sigma_{i=1}^{k+1}n_i=n$. Also, since $\Sigma_{i=1}^{k+1}n_i=n$, we may regard $T=(n_1, n_2,\dots, n_k)$ as the k-dimensional sufficient statistic. Before the data are collected, $(n_1, n_2,\dots, n_k)$ are r.v.'s having a multinomial distribution with parameters n and $(p_1, p_2,\dots, p_k)$. The likelihood function $L(p_1, p_2,\dots, p_k|\boldsymbol{D}_1^n)\propto \Pi_{i=1}^k p_i^{n_i}(1-\Sigma_{i=1}^k p_i)^{n-\Sigma_{i=1}^k n_i}$, which is of the form (2.2) with $\theta=(p_1, p_2,\dots, p_k)$ and $T=(n_1, n_2,\dots, n_k)$.

The natural conjugate family of prior distributions for the parameters $(p_1, p_2,\dots, p_k)$ is then the family $\mathcal{C}_0$ of distributions with p.d.f.'s of the form

$$q(p_1, p_2,\dots, p_k)\propto \prod_{i=1}^k p_i^{a_i-1}\left(1-\sum_{i=1}^k p_i\right)^{a_{k+1}-1};\quad p_i>0,\quad i=1,2,\dots, k,$$

$\Sigma_{i=1}^k p_i<1$, and each a_i is a positive integer.

Clearly, for any prior p.d.f. $q(p_1, p_2,\dots, p_k)\in\mathcal{C}_0$ and any data $\boldsymbol{D}_1^n$ the posterior p.d.f.,

$$q(p_1, p_2,\dots, p_k|\boldsymbol{D}_1^n)\propto \prod_{i=1}^k p_i^{a_i+n_i-1}\left(1-\sum_{i=1}^k p_i\right)^{a_{k+1}+(n-\Sigma_{i=1}^k n_i)-1}$$

and so $q(p_1, p_2,\dots, p_k|\boldsymbol{D}_1^n)\in\mathcal{C}_0$.

The natural conjugate family $\mathcal{C}_0$ of prior distributions for the parameters $(p_1, p_2, \ldots, p_k)$ is a subfamily of the family of the Dirichlet distributions, defined in the next section.

3. The Dirichlet distribution

This section is devoted to the study of the family of the Dirichlet distributions, as the natural conjugate family for the parameters of a multinomial distribution, and its characterizations. The Dirichlet distribution is defined as follows.

Definition 3.1. Let $\alpha_i > 0, i=1,2,\ldots,k+1$. The r.v.'s $(Y_1, Y_2, \ldots, Y_k)$ are said to have a Dirichlet distribution with parameters $(\alpha_1, \alpha_2, \ldots, \alpha_{k+1})$, denoted by $(Y_1, Y_2, \ldots, Y_k) \sim D(\alpha_1, \alpha_2, \ldots, \alpha_{k+1})$, if the joint distribution of $(Y_1, Y_2, \ldots, Y_k)$ has the p.d.f. $f(y_1, y_2, \ldots, y_k) = \text{const}\, y_1^{\alpha_1 - 1} y_2^{\alpha_2 - 1}, \ldots, y_k^{\alpha_k - 1}(1 - y_1 - \cdots - y_k)^{\alpha_{k+1}-1}$, over the k-dimensional simplex S_k defined by the inequalities $y_i > 0, i=1,2,\ldots,k, \Sigma_{i=1}^k y_i < 1$.

More generally, in the above definition one may assume $\alpha_i \geqslant 0$ for each i, and $\Sigma_{i=1}^{k+1} \alpha_i > 0$. However, if $\alpha_i = 0$ for some i, then the corresponding $Y_i = 0$ with probability one.

For $k=1$, the Dirichlet distribution $D(\alpha_1, \alpha_2)$ for Y_1 is the familiar Beta distribution with parameters α_1 and α_2, $\text{Beta}(\alpha_1, \alpha_2)$. The proof of the following basic proposition has already been outlined in the previous section.

Proposition 3.1. *Let $D(\alpha_1, \alpha_2, \ldots, \alpha_{k+1})$ be the prior probability model for the parameters $(p_1, p_2, \ldots, p_k)$ in the statistical model of a $k+1$ valued r.v. X. Then, with n independent observations on X, giving rise to the sample frequencies $n_1, n_2, \ldots, n_{k+1}$ for the $k+1$ values of the r.v. X, the posterior distribution of $(p_1, p_2, \ldots, p_k)$ will be $D(\alpha_1 + n_1, \alpha_2 + n_2, \ldots, \alpha_{k+1} + n_{k+1})$.*

The rest of this section is devoted to some characterizations of the Dirichlet distribution and elucidations of some of its more useful properties.

First of all, note that if we define Y_{k+1} as $1 - \Sigma_{i=1}^k Y_i$ then the joint distribution of $(Y_1, Y_2, \ldots, Y_{k+1})$ is singular with respect to the $k+1$ dimensional Lebesgue measure λ_{k+1} on R_{k+1}. The support of this singular distribution is the k-dimensional simplex E_{k+1} defined by the inequalities $y_i > 0, i=1,2,\ldots,k+1, \Sigma_{i=1}^{k+1} y_i = 1$. The joint p.d.f. (with respect to the k-dimensional Lebesgue measure on E_{k+1}) of the $k+1$ variables may be neatly represented as const $\Pi_{i=1}^{k+1} y_i^{\alpha_i - 1}$.

The following result follows immediately.

Proposition 3.2. *If $i_1, i_2, \ldots, i_k$ is any sequence of distinct integers from the set $\mathcal{X} = \{1, 2, \ldots, k+1\}$ then $(Y_{i_1}, Y_{i_2}, \ldots, Y_{i_k}) \sim D(\alpha_{i_1}, \alpha_{i_2}, \ldots, \alpha_{i_{k+1}})$.*

A characterization of the Dirichlet distribution in terms of mutually independent Beta r.v.'s is given by

Proposition 3.3. *Let* $(Y_1, Y_2, \ldots, Y_k) \sim D(\alpha_1, \alpha_2, \ldots, \alpha_{k+1})$. *Let* $U_1 = Y_1$, *and* $U_i = Y_i/(1-Y_1-\ldots-Y_{i-1})$, $i=2,3,\ldots,k$. *Then* $U_i \sim Beta(\alpha_i, \Sigma_{j=i+1}^{k+1}\alpha_j)$, $i=1,2,\ldots,k$, *and* $U_1, U_2, \ldots, U_k$ *are mutually independent.*

Proof. The joint p.d.f. of $(Y_1, Y_2, \ldots, Y_k)$ is $f(y_1, y_2, \ldots, y_k) = \text{const}\ \Pi_{i=1}^{k} y_i^{\alpha_i - 1}(1-\Sigma_{i=1}^{k} y_i)^{\alpha_{k+1}-1}; (y_1, y_2, \ldots, y_k) \in S_k$. Consider the one–one transformation of S_k onto the k-dimensional cube $(0,1)^k$ given by the relation $y_1 = u_1$, $y_i = u_i \Pi_{j=1}^{i-1}(1-u_j)$, $i = 2, 3, \ldots, k$, the Jacobian of transformation being $\Pi_{i=1}^{k-1}(1-u_i)^{k-i}$. It follows then that the joint p.d.f. of $(U_1, U_2, \ldots, U_k)$ is $g(u_1, u_2, \ldots, u_k) = \text{const}\ \Pi_{i=1}^{k} u_i^{\alpha_i - 1}(1-u_i)^{\alpha_{i+1}+\alpha_{i+2}+\cdots+\alpha_{k+1}-1}$.

That the converse of Proposition 3.3 is true, is established by reversing the above chain of arguments.

Remark 3.1. As a by-product of the above proposition we immediately have that the r.v. $Y_1 \sim \text{Beta}(\alpha_1, \alpha - \alpha_1)$ where $\alpha = \Sigma_{i=1}^{k+1}\alpha_i$. This fact together with Proposition 3.2 then gives:

Corollary 3.1. *If* $(Y_1, Y_2, \ldots, Y_k) \sim D(\alpha_1, \alpha_2, \ldots, \alpha_{k+1})$, *then* $Y_i \sim Beta(\alpha_i, \alpha - \alpha_i)$, $i = 1, 2, \ldots, k$.

This corollary may be generalized by using the converse of Proposition 3.3 to:

Corollary 3.2. *If* $(Y_1, Y_2, \ldots, Y_k) \sim D(\alpha_1, \alpha_2, \ldots, \alpha_{k+1})$, *then* $(Y_1, Y_2, \ldots, Y_r) \sim D(\alpha_1, \alpha_2, \ldots, \alpha_r, \alpha - \Sigma_{i=1}^{r}\alpha_i)$.

A more general extension is then given by Proposition 3.2 and the converse of Proposition 3.3 as

Corollary 3.3. *For any subset* $\{i_1, i_2, \ldots, i_r\}$ *of* $\mathscr{X}$, $(Y_{i_1}, Y_{i_2}, \ldots, Y_{i_r}) \sim D(\alpha_{i_1}, \alpha_{i_2}, \ldots, \alpha_{i_r}, \alpha - \Sigma_{j=1}^{r}\alpha_{i_j})$.

The following proposition follows immediately from Proposition 3.3 and its converse.

Proposition 3.4. *Let* $(Y_1, Y_2, \cdots, Y_k) \sim D(\alpha_1, \alpha_2, \cdots, \alpha_{k+1})$. *Then, for any integer* r *such that* $2 \leq r \leq k$,

$$\left(\frac{Y_r}{1-Y_1-\cdots-Y_{r-1}}, \frac{Y_{r+1}}{1-Y_1-\cdots-Y_{r-1}}, \ldots, \frac{Y_k}{1-Y_1-\cdots-Y_{r-1}}\right)$$

is independent of $(Y_1, Y_2, \ldots, Y_{r-1})$. *Also,*

$$\left(\frac{Y_r}{1-Y_1-\cdots-Y_{r-1}}, \frac{Y_{r+1}}{1-Y_1-\cdots-Y_{r-1}}, \ldots, \frac{Y_k}{1-Y_1-\cdots-Y_{r-1}}\right)$$
$$\sim D(\alpha_r, \alpha_{r+1}, \ldots, \alpha_{k+1}).$$

The Dirichlet distribution can also be characterized in terms of mutually independent Gamma r.v.'s. This is given by the following

Proposition 3.5. *Let $Z_1, Z_2, \ldots, Z_{k+1}$ be mutually independent Gamma r.v.'s with the common scale parameter $\beta>0$ and possibly different shape parameters $\alpha_i>0, i=1,2,\ldots,k+1$. Let $Z=\Sigma_i Z_i$ and $Y_i=Z_i/Z, i=1,2,\ldots,k$. Then $(Y_1, Y_2, \ldots, Y_k)\sim D(\alpha_1, \alpha_2, \ldots, \alpha_{k+1})$. Also, $(Y_1, Y_2, \ldots, Y_k)$ is independent of Z.*

Proof. The joint p.d.f. of the Z_i's is

$$f(z_1, z_2, \ldots, z_{k+1}) \propto \exp\left\{-\beta \sum_i z_i\right\} \Pi_i z_i^{\alpha_i - 1}, \quad z_i>0, i=1,2,\ldots,k+1.$$

Consider the transformation $z=\Sigma_i z_i$, $y_i=z_i/z$, $i=1,2,\ldots,k$, the reverse transformation being $z_i=zy_i$, $i=1,2,\ldots,k$, and $z_{k+1}=z(1-\Sigma_{i=1}^k y_i)$. The Jacobian of transformation is z^k. It then follows that the joint p.d.f. of $(Z, Y_1, Y_2, \ldots, Y_k)$ is $g(z, y_1, y_2, \ldots, y_k) \propto \exp\{-\beta z\} z^{\alpha-1} \Pi_{i=1}^k y_i^{\alpha_i - 1}(1-\Sigma_{i=1}^k y_i)^{\alpha_{k+1}-1}$.

That $(Y_1, Y_2, \ldots, Y_k)$ is independent of Z in the above proposition can also be seen as follows. Regard the parameters $(\alpha_1, \alpha_2, \ldots, \alpha_{k+1})$, where each $\alpha_i>0$, as known constants and $\beta>0$ as the unknown parameter. With $(Z_1, Z_2, \ldots, Z_{k+1})$ as the sample, $Z=\Sigma_i Z_i$ is a complete sufficient statistic. The vector valued statistic $((Z_1/Z),(Z_2/Z),\ldots,(Z_k/Z))$ is scale invariant. Since $\beta>0$ is a scale parameter, it follows that $((Z_1/Z),(Z_2/Z),\ldots,(Z_k/Z))$ is ancillary statistic. Hence it follows (Basu (1955)) that $((Z_1/Z),(Z_2/Z),\ldots,(Z_k/Z))=(Y_1, Y_2, \ldots, Y_k)$ is independent of Z.

The converse of Proposition 3.5 may be stated as follows. The proof is omitted.

Proposition 3.6. *If Z is a r.v. having a Gamma distribution with shape parameter $\alpha_i>0$ and scale parameter $\beta>0$, denoted by $Z\sim G(\alpha, \beta)$, and if Z is independent of $(Y_1, Y_2, \ldots, Y_k)$, where $(Y_1, Y_2, \ldots, Y_k)\sim D(\alpha_1, \alpha_2, \ldots, \alpha_{k+1})$; then the r.v.'s $Z_i=ZY_i$, $i=1,2,\ldots,k$, and $Z_{k+1}=Z(1-\Sigma_{i=1}^k Y_i)$ are mutually independent with $Z_i\sim G(\alpha_i, \beta)$, $i=1,2,\ldots,k+1$.*

Remark 3.2. Proposition 3.4 can also be proved using the Basu theorem (Basu (1955)) and the Gamma characterization of the Dirichlet distribution.

For more on the Dirichlet distribution, see Wilks (1962).

4. Further properties of the Dirichlet distribution

Suppose A is any subset of the set $\mathcal{X}=\{1,2,\ldots,k+1\}$ and $(y_1, y_2, \ldots, y_{k+1})$ is any given point in the simplex E_{k+1}. Define $\mathrm{P}(A)=\Sigma_{i\in A} y_i$. Then, P is a probability measure on $\mathcal{X}$, and P is identified by the point $(y_1, y_2, \ldots, y_{k+1})$ in E_{k+1}. Thus, E_{k+1} represents a class of probability measures on $\mathcal{X}$. If $(Y_1, Y_2, \ldots, Y_{k+1})$ is a random point in E_{k+1} then $P(A)=\Sigma_{i\in A}Y_i$ is random probability measure of the set A, and P is a random probability measure on $\mathscr{X}$. A random probability measure on $\mathscr{X}$ can, therefore, be viewed as a probability measure on E_{k+1}.

From now on, we consider the particular case where the r.v.'s $(Y_1, Y_2, \ldots, Y_{k+1})$ have a singular Dirichlet distribution with parameters $(\alpha_1, \alpha_2, \ldots, \alpha_{k+1})$. To simplify matters, we introduce $k+1$ mutually independent Gamma r.v.'s $Z_1, Z_2, \ldots, Z_{k+1}$ with $Z_i \sim G(\alpha_i, \beta)$, $i=1,2,\ldots,k+1$, and regard $Y_i=Z_i/Z$, where $Z=Z(\mathscr{X})=\Sigma_i Z_i$. For any subset A of $\mathscr{X}$ we write $Z(A)=\Sigma_{i\in A}Z_i$, and $\alpha(A)=\Sigma_{i\in A}\alpha_i$. Then $P(A)=\Sigma_{i\in A}Y_i=(Z(A))/(Z(\mathscr{X}))$.

For any subsets A and B of $\mathscr{X}$, let $P(A|B)$ be the (random) conditional probability of A given B defined as

$$P(A|B)=\begin{cases} P(AB)/P(B), & \text{if } P(B)>0 \\ 0, & \text{if } P(B)=0. \end{cases}$$

Note that for any collection $(A_1, A_2, \ldots, A_n)$ of disjoint subsets of $\mathscr{X}$, the r.v.'s $Z(A_1), Z(A_2), \ldots, Z(A_n)$ are mutually independent and $Z(A_i)\sim G(\alpha(A_i), \beta)$, $i=1,2,\ldots,n$.

The following is a general property of the Dirichlet distribution.

Proposition 4.1. *Let $\mathscr{X}$ be partitioned into non-empty subsets $A_1, A_2, \ldots, A_{m+1}$, $1\leqslant m \leqslant k$. Then, $(P(A_1), P(A_2), \ldots, P(A_m))\sim D(\alpha(A_1), \alpha(A_2), \ldots, \alpha(A_{m+1}))$. Also, $(P(A_1), P(A_2), \ldots, P(A_m))$ is independent of $Z(\mathscr{X})$.*

This follows immediately from Proposition 3.5.

For $m=1$, the above proposition can be stated as: For any two disjoint subsets A_1 and A_2 of $\mathscr{X}$, $(Z(A_1))/(Z(A_1\cup A_2))$ is independent of $Z(A_1\cup A_2)$. Also, $(Z(A_1))/(Z(A_1\cup A_2))\sim \text{Beta}(\alpha(A_1), \alpha(A_2))$, and $Z(A_1\cup A_2)\sim G(\alpha(A_1\cup A_2), \beta)$. As a direct consequence of Proposition 4.1, we have the following:

Corollary 4.1. *The marginal distribution of the sum of any r, $1\leqslant r\leqslant k$, r.v.'s $Y_{i_1}, Y_{i_2}, \ldots, Y_{i_r}$ is Beta$(\alpha_{i_1}+\alpha_{i_2}+\cdots+\alpha_{i_r}, \alpha-\Sigma_{j=1}^r \alpha_{i_j})$, where $i_1, i_2, \ldots, i_r$ is any sequence of r distinct integers from $\mathscr{X}=\{1,2,\ldots,k+1\}$.*

For any subset B of $\mathscr{X}$ we denote by $\bar{B}$ the complement of B. The next result is a preliminary to Proposition 4.2.

Lemma 4.1. *Let B_1 and B_2 be any two subsets of $\mathscr{X}$. Then, the r.v.'s $P(B_1), P(B_2|B_1), P(B_2|\bar{B}_1)$ are mutually independent.*

Proof. It suffices to show that the r.v.'s $(Z(B_1))/(Z(\mathscr{X}))$, $(Z(B_1B_2))/(Z(B_1))$, $(Z(\bar{B}_1B_2))/(Z(\bar{B}_1))$ are mutually independent. Since the r.v.'s $Z(B_1B_2)$, $Z(B_1\bar{B}_2)$, $Z(\bar{B}_1B_2)$ and $Z(\bar{B}_1\bar{B}_2)$ are mutually independent, the pairs $(Z(B_1B_2), Z(B_1\bar{B}_2))$ and $(Z(\bar{B}_1B_2), Z(\bar{B}_1\bar{B}_2))$ of r.v.'s are independent both "within and between". Applying Proposition 4.1 to each of the two pairs we have then that the pairs $((Z(B_1B_2))/(Z(B_1)), Z(B_1))$ and $((Z(\bar{B}_1B_2))/(Z(\bar{B}_1)), Z(\bar{B}_1))$ of r.v.'s are independent both "within and between". Thus, the r.v.'s $(Z(B_1B_2))/(Z(B_1))$, $(Z(\bar{B}_1B_2))/(Z(\bar{B}_1))$, $Z(B_1)$ and $Z(\bar{B}_1)$ are mutually independent. Applying Proposition 4.1 to the last two r.v.'s we finally conclude that the r.v.'s $(Z(B_1B_2))/(Z(B_1))$, $(Z(\bar{B}_1B_2))/(Z(\bar{B}_1))$, $(Z(B_1))/(Z(\mathscr{X}))$, and $Z(\mathscr{X})$ are mutually independent.

For any subset B of $\mathscr{X}$, define B^t as B when $t=1$ and as $\bar{B}$ when $t=0$, $i=1,2,\ldots,n$. We now state Proposition 4.2.

Proposition 4.2. *For any collection $B_1, B_2, \ldots, B_n$ of subsets of $\mathscr{X}$, the (2^n-1) r.v.'s $P(B_1), \{P(B_2|B_1^{t_1})\}, \ldots, \{P(B_n|B_1^{t_1}B_2^{t_2}\ldots B_{n-1}^{t_{n-1}})\}$ are mutually independent with $P(B_1) \sim Beta(\alpha(B_1), \alpha(\bar{B}_1))$ and $P(B_{r+1}|B_1^{t_1}B_2^{t_2}\ldots B_r^{t_r}) \sim Beta(\alpha(B_1^{t_1}B_2^{t_2}\ldots B_r^{t_r}B_{r+1}), \alpha(B_1^{t_1}B_2^{t_2}\ldots B_r^{t_r}\bar{B}_{r+1}))$, $r=1,2,\ldots,n-1$.*

Proof. For $n=2$ the proof is established in Lemma 4.1. The rest follows by induction.

Proposition 4.3. *If for any collection $B_1, B_2, \ldots, B_n$ of subsets of $\mathscr{X}$ the (2^n-1) r.v.'s $P(B_1), \{P(B_2|B_1^{t_1})\}, \ldots, \{P(B_n|B_1^{t_1}B_2^{t_2}\ldots B_{n-1}^{t_{n-1}})\}$ are mutually independent with $P(B_1) \sim Beta(\alpha(B_1), \alpha(\bar{B}_1))$ and $P(B_{r+1}|B_1^{t_1}B_2^{t_2}\ldots B_r^{t_r}) \sim Beta(\alpha(B_1^{t_1}B_2^{t_2}\ldots B_r^{t_r}B_{r+1}), \alpha(B_1^{t_1}B_2^{t_2}\ldots B_r^{t_r}\bar{B}_{r+1}))$, $r=1,2,\ldots,n-1$; then the joint distribution of 2^n r.v.'s $\{P(B_1^{t_1}B_2^{t_2}\ldots B_n^{t_n})\}$ is singular Dirichlet with parameters $\{\alpha(B_1^{t_1}B_2^{t_2}\ldots B_n^{t_n})\}$.*

Proof. Let $\{Y_{t_1t_2\ldots t_n}\}$ be a collection of 2^n r.v.'s having a singular Dirichlet distribution with parameters $\{\alpha(B_1^{t_1}B_2^{t_2}\ldots B_n^{t_n})\}$. Let $\mathfrak{y}=\{(t_1t_2\ldots t_n)\}$ be the set consisting of 2^n points. Define $C_i=\{(t_1t_2\ldots t_n): t_i=1\}, i=1,2,\ldots,n$, and for any subset C of $\mathfrak{y}$ write

$$Q(C)=\sum_{(t_1t_2,\ldots,t_n)\in C} Y_{t_1t_2\ldots t_n}.$$

Then Q is a random probability measure on $\mathfrak{y}$, and the joint distribution of 2^n r.v.'s $\{Q(C_1^{t_1}, C_2^{t_2}, \ldots, C_n^{t_n})\}$ is singular Dirichlet with parameters $\{\alpha(B_1^{t_1}, B_2^{t_2}, \ldots, B_n^{t_n})\}$. Furthermore, it follows from Proposition 4.2 that the (2^n-1) r.v.'s $Q(C_1), \{Q(C_2,|C_1^{t_1})\}, \ldots, \{Q(C_n|C_1^{t_1}C_2^{t_2}\ldots C_{n-1}^{t_{n-1}}\}$ are mutually independent with $Q(C_1) \sim$ Beta$(\alpha(B_1), \alpha(\bar{B}_1))$ and $Q(C_{r+1}|C_1^{t_1}C_2^{t_2}\ldots C_r^{t_r}) \sim$ Beta$(\alpha(B_1^{t_1}B_2^{t_2}\ldots B_r^{t_r}B_{r+1}), \alpha(B_1^{t_1}B_2^{t_2}\ldots B_r^{t_r}\bar{B}_{r+1}))$, $r=1,2,\ldots,n-1$. Thus, the joint distribution of (2^n-1) r.v.'s $\{P(B_1), P(B_2|B_1), P(B_2|\bar{B}_1), \ldots\}$ is the same as the joint distribution of (2^n-1) r.v.'s $\{Q(C_1), Q(C_2|C_1), Q(C_2|\bar{C}_1), \ldots\}$. It then follows that the joint distribution of 2^n r.v.'s $\{P(B_1^{t_1}B_2^{t_2}\ldots B_n^{t_n})\}$ is the same as the joint distribution of 2^n r.v.'s $\{Q(C_1^{t_1}C_2^{t_2}\ldots C_n^{t_n})\}$.

Remark 4.1. If the collection $\{(B_1^{t_1}B_2^{t_2}\ldots B_n^{t_n})\}$ is such that every single point subset of $\mathscr{X}=\{1,2,\ldots,k+1\}$ appears in the collection (in other words, $B_1, B_2, \ldots, B_n$ is a separating sequence), then the random probability measure P on $\mathscr{X}$ is a Dirichlet process (see Definition 5.1).

Up to this stage, only finite dimensional Dirichlet distributions were considered. A more general Dirichlet distribution may be defined as follows.

Let $\{\alpha_n\}$ be a sequence of numbers satisfying $\alpha_i > 0$ for each i and $\Sigma\alpha_i < \infty$. A sequence $\{Y_n\}$ of r.v's such that $0 < Y_i < 1$ for each i and $\Sigma Y_i = 1$ is said to have a Dirichlet distribution with parameters $\{\alpha_n\}$ if for each $k, (Y_1, Y_2, \ldots, Y_k) \sim D(\alpha_1, \alpha_2, \ldots, \alpha_k, \Sigma_{i=k+1}^{\infty}\alpha_i)$.

In the above definition one may assume $\alpha_i \geqslant 0$ for each i and $0 < \Sigma\alpha_i < \infty$. However, if $\alpha_i = 0$ for some i, then the corresponding $Y_i = 0$ with probability one.

The Dirichlet process on a countable infinite set $\mathcal{X} = \{1, 2, \ldots\}$ may now be defined as follows. Let $\{\alpha_n\}$ be a convergent sequence of non negative numbers. Consider the separating sequence $\{B_n\}$ of sets in $\mathcal{X}$ where $B_n = \{n\}$. Consider a sequence of mutually independent Beta r.v.'s $\{P(B_1), P(B_2|B_1), P(B_2|\bar{B}_1), \ldots\}$, where $P(B_1) \sim \text{Beta}(\alpha(B_1), \alpha(\bar{B}_1))$, $P(B_2|B_1) \sim \text{Beta}(\alpha(B_1B_2), \alpha(B_1\bar{B}_2))$, $P(B_2|\bar{B}_1) \sim \text{Beta}(\alpha(\bar{B}_1B_2), \alpha(\bar{B}_1\bar{B}_2))$, and so on. Then from Remark 4.1 it defines a Dirichlet distribution on $\{1, 2, \ldots, n\}$ for every n in a consistent manner.

In Part II we demonstrate how this constructive approach to the Dirichlet process also works in the case where $\mathcal{X}$ is a Borel space.

PART II: THE DIRICHLET PROCESS

5. Dirichlet process preliminaries

In the Bayesian analysis of non-parametric problems there is a sequence $\{X_n\}$ of independent identically distributed (i.i.d.) random variables with a common unknown distribution P, that is, given $P = \mathrm{P}$ the X_n's are i.i.d. P. Here P is regarded as the parameter and belongs to $\mathcal{P}$, the class of all probability measures on a given space $(\mathcal{X}, \mathcal{Q})$. A prior for P is a probability measure on $(\mathcal{P}, \sigma(\mathcal{P}))$, where $\sigma(\mathcal{P})$ is the smallest σ-field of subsets of $\mathcal{P}$ such that the map $\mathrm{P} \to \mathrm{P}(A)$ from $\mathcal{P}$ into $[0,1]$ is $\sigma(\mathcal{P})$-measurable $\forall A \in \mathcal{Q}$. This prior may be viewed as a stochastic process $\{P(A) : A \in \mathcal{Q}\}$ whose sample functions are probability measures on $(\mathcal{X}, \mathcal{Q})$. As in the parametric case, a class of processes satisfying the following properties is desired:

(I) It is wide enough to accommodate various shades of opinion about P.

(II) If a prior is selected from this class, then the posterior distribution given a sample of observations from P is manageable analytically, and it belongs to the class, i.e. the class is closed under "the Bayesian operation".

The class of Dirichlet process introduced by Ferguson (1973) is especially convenient since it satisfies the properties (I) and (II).

Let us look back at the definition of a random probability measure as given in the abstract of the paper. A random probability measure P on an arbitrary measurable space $(\mathcal{X}, \mathcal{A})$ may be viewed as a measurable map from a probability space $(\Omega, \mathcal{F}, \mu)$ to the space $(\mathcal{P}, \sigma(\mathcal{P}))$. It may also be regarded as a transition function from $(\Omega, \mathcal{F})$ into $(\mathcal{X}, \mathcal{A})$. In otherwords, $P(\cdot, \cdot)$ is a measurable map from $\Omega \times \mathcal{A}$ into $[0,1]$ such that (i) for every ω in Ω, $P(\omega, \cdot)$ is a probability measure on $(\mathcal{X}, \mathcal{A})$, and (ii) for every set A in $\mathcal{A}$, $P(\cdot, A)$ is a measurable function on $(\Omega, \mathcal{F})$, i.e. $P(\cdot, A)$ is a random variable with values in $[0,1]$. The distribution of P, namely μP^{-1}, is the prior probability measure on $(\mathcal{P}, \sigma(\mathcal{P}))$. Therefore, this paper can be thought of as dealing with a class of random probability measures, with a class of stochastic processes, or with a class of prior probabilities. The Dirichlet process is defined as follows.

Definition 5.1. Let α be a finite measure on $(\mathcal{X}, \mathcal{A})$. A Dirichlet process D^α is a random probability measure on $(\mathcal{X}, \mathcal{A})$ such that, for every partition $(A_1, A_2, \ldots, A_k)$ of $\mathcal{X}$ into a finite number of measurable sets, the joint distribution of random variables $(P(A_1), P(A_2), \ldots, P(A_k))$ is singular Dirichlet with parameters $(\alpha(A_1), \alpha(A_2), \ldots, \alpha(A_k))$.

Ferguson (1973) shows through the Kolmogorov extension theorem that there exists a probability measure on $([0,1]^{\mathcal{A}}, \sigma([0,1]^{\mathcal{A}}))$ yielding the above finite dimensional Dirichlet distributions. Here $[0,1]^{\mathcal{A}}$ is the product space having for each of its factors the closed unit interval $[0,1]$, there being as many factors as elements of $\mathcal{A}$. Also, $\sigma([0,1]^{\mathcal{A}})$ is the product σ-field for $[0,1]^{\mathcal{A}}$, the σ-field generated by the measurable cylinders having a finite base. Viewing $[0,1]^{\mathcal{A}}$ as a class of set functions defined on $\mathcal{A}$ with values in $[0,1]$, each set in $\sigma([0,1]^{\mathcal{A}})$ may be defined by restrictions on a countable collection $\{p(A_n); n=1,2,\ldots\}$, where $\{A_n\}$ is a given countable subset of $\mathcal{A}$ and p denotes an element of $[0,1]^{\mathcal{A}}$. Observe that with $\mathcal{A}$ uncountable the single point sets in $[0,1]^{\mathcal{A}}$ are not in $\sigma([0,1]^{\mathcal{A}})$. Also, the class $\mathcal{P}$ of all probability measures on $(\mathcal{X}, \mathcal{A})$ does not belong to $\sigma([0,1]^{\mathcal{A}})$; it is not determined by a countable number of restrictions when $\mathcal{A}$ is uncountable. Thus a statement like "a Dirichlet process gives probability one to the class $\mathcal{P}$" is not meaningful.

Berk and Savage (1979) discuss other technical problems relating to measurability in addition to some fundamental difficulties with Ferguson's definition of a Dirichlet process. However, as proved by Blackwell (1973), none of these difficulties arise when $(X, \mathcal{A})$ is a Borel space. A full discussion on Blackwell's construction is given in Section 7. Some basic results on a Borel space is given in the next section.

6. Some useful results on a Borel space

Let $(\mathcal{X}, \mathcal{A})$ be a Borel space — a complete separable metric space, $\mathcal{X}$, with $\mathcal{A}$ being the σ-field generated by the open subsets of $\mathcal{X}$. Since $\mathcal{X}$ is a separable metric space, $\mathcal{A}$ is countably generated. We may, therefore, assume that there exists a countable field $\mathcal{B}=\{B_1, B_2, \ldots\}$ such that its Borel extension is $\mathcal{A}$. The family $\mathcal{B}$ forms a separating sequence, i.e. for any distinct points x_1 and x_2 in $\mathcal{X}$ there exists a set $B_n \in \mathcal{B}$ which contains either x_1 or x_2 but not both.

Consider all sequences $t=(t_1, t_2, \ldots)$ such that each $t_i=0$ or 1. Let $T=\{t\}$ be the class of all such sequences and $\mathcal{T}$ be the σ-field for T — the σ-field generated by the cylinders having a finite base, the so called Kolmogorov's sets. Then $(T, \mathfrak{T})$ is a Borel space.

Consider the map $\xi: \mathfrak{X} \to T$ defined by $\xi(x) = (\xi_1(x), \xi_2(x), \ldots)$, where ξ_i is the indicator of B_i, $i = 1, 2, \ldots$. Notice that ξ is a measurable map since each coordinate is measurable. Also, ξ is one–one since any two x's that agree on all ξ_i's are the same. Then we have the following

Lemma 6.1. *$\xi(\mathfrak{X})$ is a Borel subset of T.*

The proof of the above lemma is a direct consequence of *The Kuratowski Theorem* (Parthasarathy (1967), Theorem 3.9). If ρ is a one–one measurable map from a Borel subset E_1 of a complete separable metric space into another complete separable metric space with $\rho(E_1) = E_2$, then E_2 is a Borel set. Also, the map ρ from E_1 onto E_2 is one–one and bimeasurable.

Let $[0,1]^\infty$ be the set of all sequences $(w_1, w_2, \ldots)$ with $0 \leqslant w_n \leqslant 1$ for each n. Note that $([0,1]^\infty, \sigma([0,1]^\infty))$ is a Borel space. Consider the map $\eta: \mathfrak{P} \to [0,1]^\infty$ defined as $\eta(\mathrm{P}) = \{\mathrm{P}(B_1), \mathrm{P}(B_2), \ldots\}$. The map is one–one since $\mathfrak{B}$ is a field. And it is measurable since each of its coordinates is a measurable map of $\mathfrak{P}$ into $[0,1]$. Let $\mathfrak{S}$ be the range of the map η. From Kuratowski's theorem it then follows:

Lemma 6.2. *$\mathfrak{S}$ is a Borel subset of $[0,1]^\infty$, and the map η from $\mathfrak{P}$ onto $\mathfrak{S}$ is one–one and bimeasurable.*

For the remainder of this section we need only to assume $\mathfrak{A}$ is countably generated and contains the single point sets. Let $\mathfrak{B}$ be a countable field that generates $\mathfrak{A}$.

The following result is a preliminary to Proposition 6.1.

Lemma 6.3. *Let* P *be a probability measure on $(\mathfrak{X}, \mathfrak{A})$. Then, for every $x \in \mathfrak{X}$ we have*

$$\inf_{n: x \in B_n} \mathrm{P}(B_n) = \mathrm{P}(\{x\}).$$

Proof. Let $C_1, C_2, \ldots$ be an enumeration of sets in $\mathfrak{B}$ that contain x. Then, $\bigcap_{n=1}^\infty C_n = \{x\}$. Defining $D_n = C_1 C_2 \ldots C_n$, we have $\mathrm{P}(D_n) \downarrow \mathrm{P}(\{x\})$.

Consider the set of all pairs (P, x) such that $\mathrm{P} \in \mathfrak{P}$ and $x \in \mathfrak{X}$, that is, the product space $\mathfrak{P} \times \mathfrak{X}$. Equip this with product σ-field $\sigma(\mathfrak{P}) \times \mathfrak{A}$. Let $E = \{(\mathrm{P}, x): \mathrm{P}(\{x\}) > 0\}$. The following result is useful.

Proposition 6.1. $\mathrm{E} \in \sigma(\mathfrak{P}) \times \mathfrak{A}$.

Proof. It suffices to show that the map $(\mathrm{P}, x) \to \mathrm{P}(\{x\})$ from $\mathfrak{P} \times \mathfrak{X}$ into $[0,1]$ is $\sigma(\mathfrak{P}) \times \mathfrak{A}$-measurable. Consider the map $H: R_2 \to R_1$ defined as

$$H(a, b) = \begin{cases} a, & \text{if } b \neq 0, \\ 1, & \text{if } b = 0. \end{cases}$$

Now, the map $P \to P(B_n)$ is $\sigma(\mathcal{P})$-measurable $\forall n$, and the map $x \to \xi_n(x)$ is $\mathcal{Q}$-measurable $\forall n$. Also, observe that H is a measurable map from R_2 into R_1. Therefore, the map $(P, x) \to H(P(B_n), \xi_n(x))$ is $\sigma(\mathcal{P}) \times \mathcal{Q}$-measurable $\forall n$, and so is $\inf_n H(P(B_n), \xi_n(x))$. Note that $\inf_n H(P(B_n), \xi_n(x)) = \inf_{n: x \in B_n} P(B_n) = P(\{x\})$, where the last equality follows from Lemma 6.3. Thus the map $(P, x) \to P(\{x\})$ is $\sigma(\mathcal{P}) \times \mathcal{Q}$-measurable.

For $P \in \mathcal{P}$, let $E_P = \{x : P(\{x\}) > 0\}$ be the P-section of E. E_P is the discrete mass points of P. Also, if P is a discrete probability measure, then E_P is the support of P. We have the following:

Proposition 6.2. *The map* $\psi : \mathcal{P} \to [0, 1]$ *defined as* $\psi(P) = P(E_P)$, *the discrete mass of* P, *is* $\sigma(\mathcal{P})$*-measurable.*

Note that the maps $P \to P_d$, the discrete part of P, and $P_d \to P_d(\mathcal{X})$ are measurable. Thus the map $P \to P_d(\mathcal{X}) = P(E_P)$ is $\sigma(\mathcal{P})$-measurable.

Corollary 6.1. *The class* $\mathcal{P}_0$ *of all discrete probability measures on* $(\mathcal{X}, \mathcal{Q})$ *is* $\sigma(\mathcal{P})$*-measurable.*

Observe that $\mathcal{P}_0 = \{P \in \mathcal{P} : \psi(\mathcal{P}) = 1\} \in \sigma(\mathcal{P})$. For further details refer to Dubins and Freedman (1964).

7. Existence of a dirichlet process

We proceed to prove the existence of a Dirichlet process D^α on a Borel space $(\mathcal{X}, \mathcal{Q})$ corresponding to any finite measure α on $\mathcal{Q}$. Choose and fix a countable field $\mathcal{B} = \{B_1, B_2, \ldots\}$ of sets in $\mathcal{X}$ such that $\mathcal{B}$ is a generator of the σ-field $\mathcal{Q}$. The map $x \to \boldsymbol{\xi}(x) = \{\xi_1(x), \xi_2(x), \ldots\}$, where ξ_n is the indicator of B_n, is then a one–one bimeasurable map of $(\mathcal{X}, \mathcal{Q})$ into $(T, \mathcal{T})$. A probability measure Q on $(T, \mathcal{T})$ defines a probability measure $P = Q\boldsymbol{\xi}$ on $(\mathcal{X}, \mathcal{Q})$ provided $Q[\boldsymbol{\xi}(\mathcal{X})] = 1$.

To simplify our notations we denote a typical point $(t_1, t_2, \ldots, t_n)$ of the product space $T_n = \{0, 1\}^n$ by s_n. By s_n0 we denote the point in T_{n+1} that is obtained by augmenting s_n by 0, that is, $s_n0 = (t_1, t_2, \ldots, t_n, 0)$, and similarly for s_n1. Finally, we denote by $[s_n]$ the cylinder set of all points in T whose first n coordinates form the vector s_n. For example, [0] is the set of all $t \in T$ such that $t_1 = 0$.

It is easily seen that a probability measure Q on $(T, \mathcal{T})$ is uniquely defined by a sequence of blocks ω of numbers in the closed unit interval [0, 1]:

$$\omega = \{u, (u_0, u_1), (u_{00}, u_{01}, u_{10}, u_{11}), \ldots\}, \tag{7.1}$$

where $u = Q([1])$, $u_0 = Q([0, 1] \| [0])$, $u_1 = Q([1, 1] \| [1])$, $u_{00} = Q([0, 0, 1] \| [0, 0])$ and so on, a typical term of the $(n+1)$th block of the sequence of blocks being

$$u_{s_n} = Q([s_n 1] \| [s_n]), \quad s_n \in T_n.$$

Let Ω denote the space of all sequence of blocks ω with its coordinates lying in [0, 1].

The probability measure on $(T, \mathfrak{T})$ that coexists with each $\omega \in \Omega$ is denoted by Q_ω. If Ω is equipped with the product σ-field $\sigma(\Omega)$, then the map $\omega \to Q_\omega$ defines a transition function from $(\Omega, \sigma(\Omega))$ to $(T, \mathfrak{T})$. If $(\Omega, \sigma(\Omega))$ is equipped with a probability measure μ, then we have a random probability measure on $(T, \mathfrak{T})$ which we denote by Q_μ. How do we choose μ so that $Q_\mu \boldsymbol{\xi}$ is a Dirichlet process on $(\mathfrak{X}, \mathcal{Q})$ with parameter α?

For an arbitrary but fixed n, consider the partition $\{B^{s_n}: s_n \in T_n\}$ of $\mathfrak{X}$, where by B^{s_n} we denote the set $B_1^{t_1} B_2^{t_2} \dots B_n^{t_n}$. If P_μ is D^α on $(\mathfrak{X}, \mathcal{Q})$, then the joint distribution of the 2^n r.v.'s $\{P_\mu(B^{s_n}): s_n \in T_n\}$ is singular Dirichlet with parameters $\{\alpha(B^{s_n}): s_n \in T_n\}$. Invoking Proposition 4.2 we then have

$$P_\mu(B_1),\ P_\mu(B_2|\bar{B}_1),\ P_\mu(B_2|B_1),\ P_\mu(B_3|\bar{B}_1\bar{B}_2), \dots \tag{7.2}$$

are mutually independent random variables with

$$P_\mu(B_1) \sim \mathrm{Beta}\big(\alpha(B_1), \alpha(\bar{B}_1)\big), \quad \text{and}$$

$$P_\mu(B_{m+1}|B^{s_m}) \sim \mathrm{Beta}\big(\alpha(B^{s_m 1}), \alpha(B^{s_m 0})\big), \quad s_m \in T_m, \quad m = 1, 2, \dots n. \tag{7.3}$$

Observe that the map $\boldsymbol{\xi}: \mathfrak{X} \to T$ transforms B_1 to [1], B^{s_m} to $[s_m]$, $P_\mu(B_1)$ to $Q_\mu([1])$, and so on. It is, therefore, clear that if under μ the coordinates of ω are mutually independent and are distributed as

$$u \sim \mathrm{Beta}\big(\alpha(B_1), \alpha(\bar{B}_1)\big), \quad \text{and}$$

$$u_{s_n} \sim \mathrm{Beta}\big(\alpha(B^{s_n 1}), \alpha(B^{s_n 0})\big), \quad n = 1, 2, \dots, \tag{7.4}$$

then (7.2) and (7.3) hold true for all n.

Theorem 7.1 (Blackwell (1973)). *If under μ coordinates of ω are mutually independent and* (7.4) *holds, then P_μ is D^α on* $(\mathfrak{X}, \mathcal{Q})$.

Proof. Since (7.2) and (7.3) hold, it follows (Remark 4.1) that $P_\mu(B_n) \sim \mathrm{Beta}(\alpha(B_n), \alpha(\bar{B}_n))$ for all n. Since $\mathfrak{B}$ is a field, $\mathfrak{X} = B_n$ some n. Therefore, $P_\mu(\mathfrak{X}) = 1$ a.s. $[\mu]$. This proves that P_μ is a random probability measure on $(\mathfrak{X}, \mathcal{Q})$.

To prove that $P_\mu(A) \sim \mathrm{Beta}(\alpha(A), \alpha(\bar{A}))$ for each $A \in \mathcal{Q}$ we proceed as follows: The map $\mathrm{P} \to (\mathrm{P}(B_1), \mathrm{P}(B_2), \dots)$ from $\mathfrak{P}$ to $\mathfrak{S}$ is one–one and bimeasurable (Lemma 6.2). For any $A \in \mathcal{Q}$, the map $\mathrm{P} \to \mathrm{P}(A)$ from $\mathfrak{P}$ to $[0,1]$ is measurable. Hence there exists a measurable map $h_A: \mathfrak{S} \to [0,1]$ such that $h_A(\mathrm{P}(B_1), \mathrm{P}(B_2), \dots) = \mathrm{P}(A)$ for all $\mathrm{P} \in \mathfrak{P}$. For each n, the joint distribution of $P_\mu(B_1), P_\mu(B_2), \dots, P_\mu(B_n)$ is well defined in terms μ. And for different n these joint distributions are mutually consistent. The Kolmogorov extension theorem, therefore, guarantees that the joint distribution of the whole sequence $P_\mu(B_1), P_\mu(B_2), \dots,$ is well defined. If we denote this joint distribution by Π_μ, then $P_\mu(A) \sim \Pi_\mu h_A^{-1}$.

Consider now the hypothetical situation where we might have started with $\mathfrak{B}^* = \{A, B_1, B_2, \dots\}$ as generator of $\mathcal{Q}$. Proceeding as before, we would then have defined a random probability measure P_μ^* on $(\mathfrak{X}, \mathcal{Q})$. Under μ, the random probability measures P_μ and P_μ^* on $(\mathfrak{X}, \mathcal{Q})$ are the same, and therefore the joint distribution of $(P_\mu(A), P_\mu(B_1), \ldots)$ is the same as the joint distribution of $(P_\mu^*(A), P_\mu(B_1), \ldots)$. For the random probability measure P_μ^* it is clear that $P_\mu^*(A) \sim \text{Beta}(\alpha(A), \alpha(\bar{A}))$ and $(P_\mu^*(B_1), P_\mu^*(B_2), \ldots) \sim \Pi_\mu$. Therefore, $\Pi_\mu h_A^{-1}$ is $\text{Beta}(\alpha(A), \alpha(\bar{A}))$. This proves that $P_\mu(A) \sim \text{Beta}(\alpha(A), \alpha(\bar{A}))$ for all $A \in \mathcal{A}$.

The above argument goes through, word for word, for an arbitrary measurable partiation $(A_1, A_2, \ldots, A_k)$ of $\mathcal{X}$ leading to the conclusion that the r.v.'s $(P_\mu(A_1), P_\mu(A_2), \ldots, P_\mu(A_k))$ have a singular Dirichlet distribution with parameters $(\alpha(A_1), \alpha(A_2), \ldots, \alpha(A_k))$. This proves that P_μ is D^α on $(\mathcal{X}, \mathcal{A})$.

8. Support of the Dirichlet process

The existence theorem of the previous section may be restated as:

Theorem 8.1. *If $(\mathcal{X}, \mathcal{A})$ is a Borel space, then, for each finite measure α on $(\mathcal{X}, \mathcal{A})$, there exists a probability measure D^α on $(\mathcal{P}, \sigma(\mathcal{P}))$ such that with $P \sim D^\alpha$, the r.v.'s $(P(A_1), P(A_2), \ldots, P(A_k))$ have singular Dirichlet distribution with parameters $(\alpha(A_1), \alpha(A_2), \ldots, \alpha(A_k))$ for any measurable partition $(A_1, A_2, \ldots, A_k)$ of $\mathcal{X}$.*

Let $\mathcal{P}_0$ be the family of discrete probability measure on $(\mathcal{X}, \mathcal{A})$. That $\mathcal{P}_0$ belongs to $\sigma(\mathcal{P})$ has been noted in Corollary 6.1.

Theorem 8.2. *If $\mathrm{P} \sim D^\alpha$, then almost every realization of* P *is a discrete probability measure on $(\mathcal{X}, \mathcal{A})$, that is,*

$$D^\alpha(\mathcal{P}_0) = 1.$$

Historical Note: Ferguson (1973) gave a rather involved argument to prove this result. Blackwell (1973) and Blackwell and MacQueen (1973) gave alternative arguments for the same result. The proof given here is a streamlined version of an ingenious argument given by Berk and Savage (1979).

Consider the pair (P, X) of random entities such that (i) $P \sim D^\alpha$ and (ii) $X|P \sim P$, that is, conditional on $P = \mathrm{P}$, the probability distribution of X on $(\mathcal{X}, \mathcal{A})$ is P. Let Δ^α denote the joint distribution of (P, X) on the product space $(\mathcal{P} \times \mathcal{X}, \sigma(\mathcal{P}) \times \mathcal{A})$. The marginal distribution of X is then easily verified to be the normalized measure $\bar{\alpha} = \alpha/\alpha(\mathcal{X})$. It is well known (see Ferguson (1973)) that $P|X = x \sim D^{\alpha + \delta_x}$, where δ_x denotes the degenerate probability measure with its whole mass concentrated at x.

We have noted earlier (Proposition 6.1) that $E = \{(\mathrm{P}, x) : \mathrm{P}(\{x\}) > 0\}$ belongs to $\sigma(\mathcal{P}) \times \mathcal{A}$. The following proposition is a preliminary to the proof of Theorem 8.2.

Proposition 8.1. $\Delta^\alpha(E) = 1$.

Proof. Writing E^x for the x-section of E, we have

$$\Delta^\alpha(E) = \int D^{\alpha + \delta_x}(E^x) \, d\bar{\alpha}(x).$$

Now, E^x is the set of all $\mathrm{P} \in \mathcal{P}$ such that $\mathrm{P}(\{x\}) > 0$. Under the distribution $D^{\alpha + \delta_x}$, the random variable $P(\{x\})$ is positive with probability one— this is because the $\alpha + \delta_x$ measure of the set $\{x\}$ is positive. Therefore, $D^{\alpha + \delta_x}\{\mathrm{P} : \mathrm{P}\{x\} > 0\} = 1$ for all x. In other words, $\Delta^\alpha(E) = 1$.

Proof of Theorem 8.2. Consider now the P-section $E_{\mathrm{P}}=\{x:\mathrm{P}(\{x\})>0\}$ of the set E. Since X, given $P=\mathrm{P}$, is distributed as P we have

$$\Delta^{\alpha}(E)=\int\psi(\mathrm{P})\,\mathrm{d}D^{\alpha}(\mathrm{P}),$$

where $\psi(\mathrm{P})=\mathrm{P}(E_{\mathrm{P}})$ is the discrete mass of P. Since $\Delta^{\alpha}(E)=1$, we at once have $\psi(P)=1$ a.s. $[\mathrm{D}^{\alpha}]$. But $\{\mathrm{P}:\psi(\mathrm{P})=1\}=\mathcal{P}_0$.

Let $\mathcal{V}=\{V\}$ be the collection of all open sets in $\mathcal{X}$. Since $\mathcal{X}$ is a separable metric space, there exists a countable subcollection $\{V_1,V_2,\dots\}$ of open sets such that every V contains some V_n. Let $\mathcal{P}'$ be the collection of all $\mathrm{P}\in\mathcal{P}$ such that $\mathrm{P}(V)>0$ for all $V\in\mathcal{V}$. Similarly, let $\mathcal{P}_n=\{\mathrm{P}:\mathrm{P}(V_n)>0\}$. It is then clear that $\mathcal{P}'=\bigcap_{n=1}^{\infty}\mathcal{P}_n$.

Theorem 8.3. *If $\alpha(V)>0$ for all $V\in\mathcal{V}$, then $D^{\alpha}(\mathcal{P}')=1$.*

Proof. Since $P(V_n)\sim\mathrm{Beta}(\alpha(V_n),\alpha(\bar{V}_n))$ and $\alpha(V_n)>0$ it follows that $P(V_n)>0$ a.s. $[D^{\alpha}]$, that is, $D^{\alpha}(\mathcal{P}_n)=1$. Therefore, $D^{\alpha}(\mathcal{P}')=1$.

The set $\mathcal{P}_0\cap\mathcal{P}'$ is the collection of all discrete probability measure P on $(\mathcal{X},\mathcal{A})$ such that the mass points of P are everywhere dense in $\mathcal{X}$. Putting Theorems (8.2) and (8.3) together we finally have:

Theorem 8.4. *If $P\sim D^{\alpha}$ and the α-measure of every open subset of $\mathcal{X}$ is positive, then for almost every realization* P *of P it is true that* P *is discrete with its mass points everywhere dense in $\mathcal{X}$.*

Further properties of the Dirichlet process will be discussed in a forthcoming note.

Acknowledgement

The authors are greatly indebted to Professor David Blackwell for the benefit of some prolonged discussions and consultations during the Fall and Winter Quarters of 1978–1979 when he was visiting the Florida State University. The authors are also indebted to Professor J. Sethuraman for his useful discussions.

References

Basu, D. (1955). On statistics independent of a complete sufficient statistic. *Sankhyā A*, **15**, 337–380.

Basu, D. (1975). Statistical information and likelihood. *Sankhyā A* **37**, 1–71.

Berk, R.H. and Savage, I.R. (1979). Dirichlet process produce discrete measures: An elementary proof. *Contributions to Statistics. Jaroslav Hájek Memorial Volume*, pp. 25–31. Academia, North-Holland, Prague.

Blackwell, D. (1973). Discreteness of Ferguson Selections. *Ann. Statist.* **1**, 356–358.

Blackwell, D. and MacQueen, J.B. (1973). Ferguson distributions via Pólya urn schemes. *Ann. Statist.* **1**, 353–355.

Dubins, L. and Freedman, D. (1964). Measurable sets of measures. *Pacific J. Math.* **14**, 1211–1222.

Ferguson, T.S. (1973). A Bayesian analysis of some nonparametric problems. *Ann. Statist.* **1**, 209–230.

Parthasarathy, K.R. (1967). *Probability Measures on Metric Spaces*. Academic Press, New York.

Wilks, S.S. (1962). *Mathematical Statistics*. Wiley, New York.

G. Kallianpur, P.R. Krishnaiah, J.K. Ghosh, eds., *Statistics and Probability: Essays in Honor of C.R. Rao*
© North-Holland Publishing Company (1982) 105–108

STRONG CONSISTENCY OF LEAST SQUARES ESTIMATES AND CONDITIONAL CONSISTENCY

B.R. BHAT
Karnatak University, Dharwar, India

Criteria for strong consistency of least squares estimates under general Gauss–Markov setup are derived. Concept of conditional consistency is introduced and illustrated by means of two examples.

1. Introduction

Consistency is perhaps the most fundamental property of any good estimator. If T_n is an estimator of θ based on n observations, whose true value is θ_0, then consistency of T_n demands that T_n should converge to θ_0 as $n \to \infty$ when P_{θ_0} is the underlying probability measure. If A is the set of convergence of T_n to θ_0, and $P_{\theta_0}(A)=1$, then we have strong or almost sure (a.s.) consistency of T_n. If $P_{\theta_0}(A)>0$, we have consistency conditional on A. If there exists a sequence $\{b_n\}$ such that $b_n \uparrow \infty$ and $b_n(T_n - \theta_0)$ is bounded in probability, then we have b_n-consistency [see Bhat (1979)].

Consider the sequence $\{y_n\}$ of random variables following the probability law P_θ, $\theta \in \Theta$. The sample space, Ω, consists of all infinite sequences of y_n's, with $y_n \in R$ $(n=1,2,\ldots)$. When Θ is finite, Chao (1971) has proved the equivalence of the following:

(i) there exists a sequence of strongly consistent estimator of θ;

(ii) maximum likelihood estimator is strongly consistent;

(iii) P_{θ_1} and P_{θ_2} are mutually singular $(\theta_1 \neq \theta_2)$.

From (i) and (iii) it is clear that the existence of an a.s. consistent estimator is related to the mutual singularity of the underlying probability measures on the infinite dimensional sample space Ω. In the case of independent identically-distributed observations, (iii) holds true if Kullback–Leibler information on θ relative to θ_0 is positive, although P_θ and P_{θ_0} are mutually absolutely continuous on any finite dimensional subspace of Ω (cf. Neveu (1975)). But the result is not obvious for the non-identically distributed or correlated cases. In this context, a.s. consistency of least squares estimators of estimable linear parametric functions under the general Gauss–Markoff set-up are discussed in Section 2.

Even though Chao has established the equivalence of (i) and (iii) for finite Θ, the result is true for Θ infinite also, in the sense that a.s. consistent estimator exists iff for each θ_0, there exists a measurable set A_{θ_0} such that $P_{\theta_0}(A_{\theta_0})=1$ and $P_\theta(A_{\theta_0})=0$ $(\theta \neq \theta_0)$. (Personal communication from J. Sethuraman.) Evidently, Ω is equal to the disjoint union of A_θ, $\theta \in \Theta$ a.s. If A_{θ_0} is such that $P_{\theta_0}(A_{\theta_0})>0$ and $P_\theta(A_{\theta_0})=0$ $(\theta \neq \theta_0)$, and $UA_{\theta_0}=A$, a measurable subset of Ω, then one can define A to be the consistency set of the parameter θ. Then, conditional on A, a.s. consistent estimators exist. Unconditionally, there exist estimators which are partially consistent (on A), but there exists no a.s. consistent estimator. This problem is discussed using two illustrative examples in Section 3.

2. Consistency of least squares estimates

Theory of linear estimation has been developed by Rao [see (1973)] and others. Even under the general Gauss–Markoff set-up involving correlated observations it is a well-discussed topic. The latter model is found appropriate particularly in the areas of econometrics and time-series analysis.

Consider the model $(\boldsymbol{Y}_n, \boldsymbol{X}_n\boldsymbol{\beta}, \boldsymbol{\Sigma}_n)$ where $\boldsymbol{\beta}'=(\beta_1,\ldots,\beta_m)$ is a vector of unknown parameters and $\boldsymbol{X}_n$ $(n\times m)$ and $\boldsymbol{\Sigma}_n$ $(n\times n)$ are matrices of known terms. For the time being we assume that $\boldsymbol{\Sigma}_n$ and $\boldsymbol{X}_n'\boldsymbol{\Sigma}_n^{-1}\boldsymbol{X}_n=\boldsymbol{A}_n$ to be non-singular for all n. The normal equations to estimate $\boldsymbol{\beta}$ are:

$$\boldsymbol{A}_n\boldsymbol{\beta}=\boldsymbol{X}_n'\boldsymbol{\Sigma}_n^{-1}\boldsymbol{y}_n=\boldsymbol{c}_n, \quad \text{say.}$$

The estimator $\hat{\boldsymbol{\beta}}_n$ of $\boldsymbol{\beta}$ based on n observations $\boldsymbol{y}_n$ $(y_1,\ldots,y_n)$ is given by:

$$\hat{\boldsymbol{\beta}}_n=\boldsymbol{A}_n^{-1}\boldsymbol{c}_n.$$

This generalized least squares estimator is unbiased and its covariance matrix is $\boldsymbol{A}_n^{-1}$. If $\boldsymbol{y}_n$ has n-variate normal distribution $\hat{\boldsymbol{\beta}}_n$ is the maximum likelihood estimate of $\boldsymbol{\beta}$ and is m-variate normal with mean $\boldsymbol{\beta}$ and covariance matrix $\boldsymbol{A}_n^{-1}$. We are interested in studying the consistency of this estimator as $n\to\infty$.

A sufficient condition for weak consistency of $\hat{\boldsymbol{\beta}}_n$ is that the variance of each of its components $\to 0$ or, equivalently $\boldsymbol{A}_n^{-1}\to\boldsymbol{0}$, as $n\to\infty$. This follows from Chebyshev's inequality. Then every estimable linear parametric function, can be estimated consistently.

If we know that the distribution of $\boldsymbol{y}_n$, and hence that of $\hat{\boldsymbol{\beta}}_n$ is normal, then the distance between the two Gaussian measures corresponding to two $\boldsymbol{\beta}$-values which differ only in the ith coordinate is proportional to ith diagonal term of $\boldsymbol{A}_n$. [see Rao (1973) last chapter]. If the ith diagonal term of $\boldsymbol{A}_n\to\infty$ as $n\to\infty$, then the two Gaussian measures on the space of infinite sequences are mutually singular and conversely. Then and only then, the ith coordinate of $\boldsymbol{\beta}$ is estimable consistently. If all the coordinates of $\boldsymbol{\beta}$ are to be estimable consistently and hence all their linear combinations, all the diagonal terms of $\boldsymbol{A}_n$ should tend to infinity as $n\to\infty$. Equivalently, $\boldsymbol{A}_n^{-1}\to\boldsymbol{0}$ as $n\to\infty$. Hence we have the following theorem.

Theorem. *Every estimable linear parametric function of $\boldsymbol{\beta}$ will be consistently estimable, if $\boldsymbol{x}_{in}'\boldsymbol{\Sigma}_n^{-1}\boldsymbol{x}_{in}\to\infty$ as $n\to\infty$ $(i=1,2,\ldots,m)$, where $(\boldsymbol{x}_{1n},\ldots,\boldsymbol{x}_{mn})=\boldsymbol{X}_n$. The condition is also necessary if $\boldsymbol{y}_n$ is normal.*

Corollary 1. *If y_i's are independent with means βx_i $(i=1,2,\ldots,n)$ and variance unity, the least squares (l.s.) estimator $\hat{\beta}_n$ of β will be consistent if $\Sigma_1^N x_i^2\to\infty$ as $n\to\infty$.*

If $x_i=\mathrm{O}(1/i)$, then $\hat{\beta}_n$ will not be consistent.

Corollary 2. *If A_n^{-1} converges to a matrix $\boldsymbol{B}$ of rank k $(<m)$, there exist $(m-k)$ linearly independent estimable functions of β_i's, which are estimable consistently and the remaining k functions cannot be consistently estimated.*

Proof. We note that $\boldsymbol{B}$ may be reduced to a diagonal form. This diagonal matrix may be viewed as the limiting covariance matrix of a linear transformation of $\hat{\boldsymbol{\beta}}_n$. Since $(m-k)$ diagonal terms are zero, the corresponding linear functions can be consistently estimated, while the remaining k linear functions cannot be consistently estimated.

Corollary 3. *Even if $\boldsymbol{A}_n$ is singular and so also is $\boldsymbol{\Sigma}_n$, every estimable linear parametric function can be consistently estimated if its variance $\to 0$ as $n \to \infty$.*

Corollary can be proved using generalised inverses or inverse partitioned matrix [see Rao (1973), p. 298].

In the above corollaries, the sufficient conditions stated are also necessary, if the underlying distributions are Gaussian. In the latter case we also have strong consistency of estimates.

3. Conditional consistency

Example 1. Consider sequences of length, n, from a Markov chain (M.C.) with state space 0, 1, 2, 3, with the 0th observation $y_0 = 3$. One step transition probability matrix is given by:

$$\boldsymbol{P} = \begin{pmatrix} 1 & 0 & 0 & 0 \\ 0 & \theta & 1-\theta & 0 \\ 0 & 1-\theta & \theta & 0 \\ 1-2\theta & \frac{1}{2}\theta & \frac{1}{2}\theta & \theta \end{pmatrix}, \quad 0 \leqslant \theta \leqslant \tfrac{1}{2}.$$

The M.C. has two ergodic classes $\{0\}$ and $\{1,2\}$. State 3 is transient. Hence for large n, either $y_n \equiv 0$ or $y_n \in \{1,2\}$.

Let $p_k(\theta)$ be the probability of absorption into $\{0\}$ at the kth step. Then,

$$p_k(\theta) = \theta^{k-1}(1-2\theta) \quad (k=1,2,\ldots),$$

$$\sum_1^{\infty} p_k(\theta) = (1-2\theta)/(1-\theta),$$

the latter being the probability of ultimate absorption in $\{0\}$. If the absorption takes place at the kth step, for $n > k$, there will be no change in the likelihood function and hence the probability measures for any two different values of θ, will be mutually absolutely continuous. There exists no consistent estimator of θ, as the set of sequences with ultimate absorption in $\{0\}$ has positive probability.

For sequences which enter the ergodic class $\{1,2\}$ at the kth step, likelihood function is given by

$$(1-2\theta)^{k-1}(\tfrac{1}{2}\theta)\theta^{n_{11}+n_{22}}(1-\theta)^{n_{12}+n_{21}},$$

where n_{ij} is the number of transitions from i to j. In this case, there exist strongly-consistent estimators, say, the (unconditional) maximum likelihood estimator (MLE). If A is the set of all such sequences, $P_\theta(A)=\theta/(1-\theta)$, which is positive except in the extreme case, when $\theta=0$. A is the consistency set. Conditional on A, there exist strongly consistent estimators $\hat{\theta}_n$, converging to θ_0 under $P_{\theta_0}(\cdot/A)$ probability measure. It is more useful and informative to conclude that a given estimator is at least partially consistent, instead of simply asserting that it is not strongly consistent.

In the above example, A and A^c are measurable sets with $P_\theta(A)$ depending on θ. Hence the unconditional likelihood function and the likelihood function conditional on A, are not the same. Consequently, the unconditional MLE and conditional MLE are not the same. The unconditional MLE will be partially consistent, conditional one will be a.s. consistent. In the following example (due to Adke), $P_\theta(A)$ does not depend on θ and the two MLE's will be the same. This simple example brings out clearly the need and usefulness of the concepts of consistency set and conditional consistency.

Example 2. A box contains coins, two-thirds of which are possibly biased, with a common probability θ of success in a single toss and the remaining ones are unbiased. A coin is picked up at random and tossed n times. The problem is to obtain an estimator of θ.

The likelihood function is $(\frac{1}{2})^n$ if the unbiased coin is picked up and is $\theta^r(1-\theta)^{n-r}$ otherwise, where r is the number of successes observed. The unconditional likelihood function is:

$$\tfrac{1}{3}\cdot(\tfrac{1}{2})^n+\tfrac{2}{3}\theta^r(1-\theta)^{n-r}.$$

The MLE of θ is r/n. This converges to θ, with probability $\frac{2}{3}$, i.e. on the set A, where we pick up the biased coin at the first instance. It converges to $\frac{1}{2}$ on A^c, whatever be θ. Thus A is the consistency set.

References

Bhat, B.R. (1979). Strong consistency of maximum likelihood estimator for dependent observations. *J. Ind. Statist. Assoc.* **17**, pp. 27–39.

Chao, Min Te (1971). A note on statistical consistency. *Tankang J. Math. Taipeh* **2**.

Neveu, J. (1975). *Discrete Parameter Martingales*. North-Holland, New York.

Rao, C.R. (1973). *Linear Statistical Inference and Its Applications*. Wiley, New York.

G. Kallianpur, P.R. Krishnaiah, J.K. Ghosh, eds., *Statistics and Probability: Essays in Honor of C.R. Rao*
© North-Holland Publishing Company (1982) 109–124

ASYMPTOTIC THEORY OF SELECTION PROCEDURES AND OPTIMALITY OF GUPTA'S RULES*

Peter J. BICKEL
University of California at Berkeley, Berkeley, CA, U.S.A.

Joseph A. YAHAV**
The Hebrew University, Jerusalem, Israel

By using an asymptotic theory introduced in Bickel and Yahav (1977), we show that subject to a lower bound on the probability of correct selection, the appropriate Gupta (1965) rules are asymptotically optimal for a wide class of loss functions and Bayes prior distributions.

1. Introduction

In our paper Bickel–Yahav (1977) (which we abbreviate B–Y) we studied the problem of selecting a set of populations with "large" means from a set of normal populations with common known variance, say 1, and differing means. We considered several sets of assumptions concerning our knowledge of the set of population means including:

(i) The means are known up to a permutation.

(ii) There is a possibly unknown Bayes prior distribution on the vector of means with independent, identically distributed components which are bounded above.

In this context we restricted ourselves to invariant rules and chose as our selection criterion, maximization of the expected average mean among the populations selected, subject to a positive lower bound on the probability of correct selection, i.e. of including in the selected set the population with the highest mean.

We showed that the optimal rules selected the populations corresponding to the r largest observations where r depended only on the values of the ordered observations.

We studied the behavior of these optimal rules as the number k of populations tended to ∞ in such a way that the empirical distribution of the population means stabilized [as always happens in case (ii)] to a limiting d.f. and showed (under regularity conditions) that the optimal rules asymptotically select a fixed proportion

*This research was partially supported by National Science Foundation grant MCS79-03716 and Office of Naval Research contract N00014-75-C-0444. Work performed at Brookhaven National Laboratory under contract EY-76-C-02-0016 with the U.S. Department of Energy.

**Visiting Statistician at Brookhaven National Laboratory, Summer 1979.

of the populations where the proportion depends on the bound for the probability of correct selection and on the limiting d.f.

In this paper we show that:

(a) The same rules also minimize the asymptotic risk for a wide class of smooth "monotone" loss functions within the class of procedures with probability of correct selection bounded below as specified. In the process of doing so we obtain a simple expression for the asymptotic risk.

(b) Gupta's rule with minimum probability of correct selection equal to the specified lower bound, is asymptotically optimal within the same class of procedures and for the same class of loss functions for essentially any prior for which the empirical d.f. of the means tends to a *fixed* d.f. with prior probability, 1, and whose essential supremum is finite. This can be viewed as a universal asymptotic optimality property since Gupta's rule is uniquely determined by the minimum probability of correct selection desired.

2. The model and notation

As in B–Y, let $\pi_1,\ldots,\pi_k$ be normal populations with common variance 1 and means $\mu_1,\ldots,\mu_k$. To begin with we suppose the parameter vector $\boldsymbol{\mu}$ is known up to a permutation of its coordinates, i.e. the parameter space is a single orbit of the permutation group. Our decision space A is the set of 2^k-1 non-empty subsets of $\{\pi_1,\ldots,\pi_k\}$. We consider measures of loss l which are invariant and monotone in the sense of Eaton (1967). That is if g is a permutation of $(\mu_1,\ldots,\mu_k)$ when acting on the parameter space and of $\{\pi_1,\ldots,\pi_k)$ when acting on the decision space we must have,

$$l(g\boldsymbol{\mu},g^{-1}S)=l(\boldsymbol{\mu},S),\qquad l(\boldsymbol{\mu},S)\geqslant l(\boldsymbol{\mu},S'),$$

whenever $\mu_i\geqslant\mu_j$, S and S' have the same number of elements, $\pi_i\in S'$, $\pi_j\in S$, $S-\{\pi_j\}=S'-\{\pi_i\}$.

We observe one observation X_i from each population π_i, $i=1,\ldots,k$, and let $Z_1\geqslant\cdots\geqslant Z_k$ be the corresponding order statistics. Let $D_1,\ldots,D_k$ be the antiranks defined by

$$X_{D_i}=Z_i,\quad i=1,\ldots,k.$$

Also let $\boldsymbol{\mu}^*=(\mu_{(1)},\ldots,\mu_{(k)})$ where $\mu_{(1)}\geqslant\cdots\geqslant\mu_{(k)}$ be the vector of ordered μ_1.

We restrict ourselves to the class of invariant decision procedures. Such a procedure d has constant probability of selecting the population with the largest mean as well as constant risk, $R(\boldsymbol{\mu}^*,d)$.

It follows from Theorem 4.1 of Eaton (1967) that for invariant, monotone loss functions the admissible invariant procedures are monotone, i.e. are of the form

$$d(\boldsymbol{x})=\{\pi_{D_1},\ldots,\pi_{D_{r(Z,U)}}\},$$

where $\boldsymbol{Z}=(Z_1,\ldots,Z_k)$, U is uniform on $(0,1)$ and independent of $\boldsymbol{X}$ and r takes on values $1,\ldots,k$. We can write the risk of a monotone rule suggestively as follows. Let, for all y,

$$\hat{G}_k(y,\lambda)=\frac{1}{[\lambda k]}\sum_{i=1}^{[\lambda k]} I_{[\mu_{D_i}\leqslant y]}, \quad \frac{1}{k}\leqslant\lambda\leqslant 1$$

$$=\hat{G}_k\left(y,\frac{1}{k}\right), \quad 0\leqslant\lambda<\frac{1}{k}. \tag{2.1}$$

$$F_k(y)=\frac{1}{k}\sum_{i=1}^{k} I_{[\mu_i\leqslant y]}. \tag{2.2}$$

We can make an obvious correspondence between

$$l\left(\boldsymbol{\mu}^*,\{\pi_{D_1},\dots,\pi_{D_r}\}\right)$$

and a functional

$$L\left(F_k,\hat{G}_k\left(\cdot,\frac{r}{k}\right),\frac{r}{k}\right),$$

defined on d.f.'s concentrating on at most k and r points, respectively. Then, for a monotone invariant rule d and corresponding $r=r(\boldsymbol{Z},U)$,

$$R(\boldsymbol{\mu},d)=E_{\boldsymbol{\mu}^*}L\left(F_k,\hat{G}\left(\cdot,\frac{r}{k}\right),\frac{r}{k}\right). \tag{2.3}$$

We now consider the asymptotic behavior of R of type (2.3) and the probability of correct selection as $k\to\infty$ in such a way that $F_k\xrightarrow{\mathcal{L}}F$.

3. Asymptotic behavior of $\hat{G}_k$

The motivation of this type of asymptotics, as we indicated in B–Y, is that for k large one would expect F_k to be fairly stable. For example, if $\mu_1,\dots,\mu_k$ are the values of a sample of size k from F then by the Glivenko–Cantelli theorem, $F_k\xrightarrow{\mathcal{L}}F$ with probability 1.

We begin by showing that under this assumption and a further regularity condition if $r/k\to\lambda$, $0\leqslant\lambda\leqslant 1$ then,

$$G_k\left(\cdot,\frac{r}{k}\right)\xrightarrow{\mathcal{L}}G(\cdot,\lambda) \tag{3.1}$$

where

$$G(y,\lambda)=\frac{1}{\lambda}\int_{H^{-1}(1-\lambda)}^{\infty}F(y|x)\,\mathrm{d}H(x), \quad \lambda>0, \tag{3.2}$$

$$G(\cdot,0)=\delta_{\{\mathrm{ess\,sup}\,F\}},$$

$$H(x)=\int_{-\infty}^{\infty}\Phi(x-\mu)\,\mathrm{d}F(\mu), \tag{3.3}$$

and $F(\cdot|x)$ is the conditional distribution of $\mu|X=x$ when $\mu\sim H$ and $X|\mu$ has a

$N(\mu,1)$ distribution. The heuristics behind this convergence is easy if one assumes $\mu_1,\dots,\mu_k$ are the values of a sample of size k from a distribution F. In that case, the X_i are i.i.d. with marginal distribution H, and given $\mathbf{Z}$, $\mu_{D_1},\dots,\mu_{D_k}$ are independent with μ_{D_i} having distribution $F(\cdot|Z_i)$. But for $i\leqslant r$, $Z_i\approx H^{-1}(1-i/k)$ and hence the empirical distribution of $\mu_{D_1},\dots,\mu_{D_r}$ should be approximately that of r independent random variables with distributions $F(\cdot|H^{-1}(1-i/k))$. Approximation (3.2) follows readily. Here is our rigorous development.

Condition 1 (A1 of B–Y).

$$F_k \xrightarrow{\mathfrak{L}} F \tag{3.4}$$

where F is the d.f. of a proper probability measure which is:

(i) *Bounded above; i.e.* $\exists y_0<\infty$ *such that* $F(y_0)=1$.

(ii) *Continuous.*

(iii) *Strictly increasing on* $\{y\colon 0<F(y)<1\}$.

Without loss of generality we take $\operatorname{ess\,sup} F=0$.

The set $\{\mu_1,\dots,\mu_k\}$ which determines F_k typically depends on k and we strictly should write $\{\mu_{1k},\dots,\mu_{kk}\}$. We do not do this here and in the sequel to avoid unnecessarily complicating an already cumbersome notation.

The second assumption requires more notation. For $1\leqslant h\leqslant k$, $0\leqslant i\leqslant[k/h]-1$, let

$$\mu(i,h)=h^{-1}\sum_{j=ih+1}^{(i+1)h}\mu_j,$$

$$S^2(i,h)=\sum_{j=ih+1}^{(i+1)h}(\mu_j-\mu(i,h))^2,$$

$$M(\varepsilon,h)=\max\left\{S^2(i,h)\colon 0\leqslant i\leqslant\left[(1-\varepsilon)\frac{k}{h}\right]-1\right\}.$$

Condition 2 (A3 of B–Y). *Suppose that for every* $\varepsilon>0$ *we can find a sequence* $\{h(k)\}$ *such that for every* $\delta>0$,

$$k\bar{e}^{\delta h}\to 0 \tag{3.5}$$

and

$$M(\varepsilon,h)\to 0. \tag{3.6}$$

We note here that (3.5) and (3.6) are readily satisfied if for some $M(\varepsilon)$ independent of k, $1\leqslant j\leqslant(1-\varepsilon)k$, $(\mu_j-\mu_{j+1})\leqslant(M(\varepsilon))/k$. Then $M(\varepsilon,h)=O(h^3k^{-2})$ and $h(k)\sim k^{\alpha}$, where $0<\alpha<\frac{2}{3}$, is a sequence which satisfies (3.5) and (3.6). In particular, if the μ_j are the order statistics of a sample of size k from F, where F has a density bounded away from 0 on compacts contained in the set $0<F<1$, the conditions 1 and 2 are satisfied almost surely.

Since we are only interested in invariant procedures we will in the sequel without loss of generality take $\boldsymbol{\mu}=\boldsymbol{\mu}^*$. Also for simplicity we write P for $P_{\boldsymbol{\mu}}$ (which depends on k) throughout. Thus, for example, $W_k \xrightarrow{P} 0$ for a sequence W_k of random variables such that W_k is a function of $(X_1,\dots,X_k)$ means $P_{\boldsymbol{\mu}}[|W_k|\geqslant\varepsilon]\to 0$, $\forall\varepsilon>0$.

Theorem 3.1. *Under Conditions* 1 *and* 2

$$\sup_y \sup_{\varepsilon\leqslant\lambda\leqslant 1-\varepsilon} |\hat{G}_k(y,\lambda)-G(y,\lambda)| \xrightarrow{P} 0. \tag{3.7}$$

The proof proceeds by a series of lemmas.

Let, for $k\lambda\geqslant 1$,

$$G_k(y,\lambda)=E(\hat{G}_k(y,\lambda)|\mathbf{Z})=\frac{1}{[k\lambda]}\sum_{i=1}^{[k\lambda]} P[\mu_{D_i}\leqslant y|\mathbf{Z}]. \tag{3.8}$$

Lemma 3.1. *Under Conditions* 1 *and* 2

$$\sup_{\varepsilon\leqslant\lambda\leqslant 1-\varepsilon} |G_k(y,\lambda)-G(y,\lambda)| \xrightarrow{P} 0. \tag{3.9}$$

Proof. Write, for $k\lambda\geqslant 1$,

$$\begin{aligned}\hat{G}_k(y,\lambda)&=\frac{1}{[k\lambda]}\sum_{j=1}^{[k\lambda]}\sum_{i=1}^{k} I_{[\mu_{D_j}\leqslant y]} I_{[D_j=i]}\\ &=\frac{1}{[k\lambda]}\sum_{i=1}^{k} I_{[\mu_i\leqslant y]}\sum_{j=1}^{[k\lambda]} I_{[D_j=i]}\\ &=\frac{1}{[k\lambda]}\sum_{i=1}^{k} I_{[\mu_i\leqslant y]} I_{[X_i\geqslant Z_{[k\lambda]}]}.\end{aligned} \tag{3.10}$$

Thus,

$$G_k(y,\lambda)=\frac{1}{[k\lambda]}\sum_{i=1}^{k} I_{[\mu_i\leqslant y]} A\left(\frac{i}{k},\lambda\right) \tag{3.11}$$

where A is defined as in B–Y by

$$A(\lambda_1,\lambda_2)=\sum_{j=1}^{[\lambda_2 k]} P[D_j=\langle\lambda_1 k\rangle|\mathbf{Z}] \tag{3.12}$$

where $\langle t\rangle$ = smallest integer $\geqslant t$. Now, for every $\delta>0$, by Lemma 4.3 of B–Y,

$$\sup\left\{\left|A\left(\frac{i}{k},\lambda\right)-\left(1-\Phi\left(H^{-1}(1-\lambda)-F^{-1}\left(1-\frac{i}{k}\right)\right)\right)\right|:\right.$$

$$\left.\varepsilon\leqslant\lambda\leqslant 1-\varepsilon, 1\leqslant i\leqslant(1-\delta)k\right\}\xrightarrow{P}0. \tag{3.13}$$

Therefore, since $\Sigma_{i=(1-\delta)k}^{k} I_{[\mu_i\leqslant y]}A((i/k),\lambda)\leqslant\delta k$ for every $\delta>0$,

$$\sup\left\{\left|\frac{1}{[k\lambda]}\sum_{i=1}^{k} I_{[\mu_i\leqslant y]}\left[A\left(\frac{i}{k},\lambda\right)-\left(1-\Phi\left(H^{-1}(1-\lambda)-F^{-1}\left(1-\frac{i}{k}\right)\right)\right)\right]\right|:\right.$$

$$\left.\varepsilon\leqslant\lambda\leqslant 1-\varepsilon\right\}\xrightarrow{P}0. \tag{3.14}$$

Now,

$$\sum_{i=1}^{k} I_{[\mu_i\leqslant y]}\left(1-\Phi\left(H^{-1}(1-\lambda)-F^{-1}\left(1-\frac{i}{k}\right)\right)\right)=$$

$$=\Sigma\left\{\left(1-\Phi\left(H^{-1}(1-\lambda)-F^{-1}\left(1-\frac{i}{k}\right)\right)\right):\frac{i}{k}\geqslant 1-F(y)\right\}$$

$$+\mathrm{O}(k|F_k(y)-F(y)|), \tag{3.15}$$

where O is independent of λ. From (3.13)–(3.15) we get,

$$\sup\left\{\left|G_k(y,\lambda)-\frac{1}{[k\lambda]}\Sigma\left\{\left[1-\Phi\left(H^{-1}(1-\lambda)-F^{-1}\left(1-\frac{i}{k}\right)\right)\right]:\right.\right.\right.$$

$$\left.\left.\left.F^{-1}\left(1-\frac{i}{k}\right)\leqslant y\right\}\right|:\varepsilon\leqslant\lambda\leqslant 1-\varepsilon\right\}\xrightarrow{P}0. \tag{3.16}$$

Finally, by standard arguments in view of the continuity of F^{-1} and H^{-1} the second term in (3.16) tends uniformly for $\varepsilon\leqslant\lambda\leqslant 1-\varepsilon$ to,

$$\frac{1}{\lambda}\int_{-\infty}^{y}\left(1-\Phi\left(H^{-1}(1-\lambda)-x\right)\right)\mathrm{d}F(x).$$

A standard Fubini argument shows that,

$$G(y,\lambda)=\frac{1}{\lambda}\int_{-\infty}^{y}\left(1-\Phi\left(H^{-1}(1-\lambda)-x\right)\right)\mathrm{d}F(x) \tag{3.17}$$

and we thus have for each y.

$$\sup\{|G_k(y,\lambda)-G(y,\lambda)|:\varepsilon\leqslant\lambda\leqslant 1-\varepsilon\}\xrightarrow{P}0. \tag{3.18}$$

Lemma 3.2. *Under Conditions* 1 *and* 2 *for every* y *and* $0<\lambda<1$,

$$G_k(y,\lambda)-\hat{G}_k(y,\lambda)\xrightarrow{P}0. \tag{3.19}$$

Proof. Since, for $\lambda k \geqslant 1$,

$$\hat{G}_k(y,\lambda)=\frac{1}{[k\lambda]}\sum_{i=1}^{k} I_{[\mu_i\leqslant y]}I_{[X_i\geqslant Z_{[k\lambda]}]}, \tag{3.20}$$

$$E\Big[\big(\hat{G}_k(y,\lambda)-G_k(y,\lambda)\big)^2/\boldsymbol{Z}\Big]=$$

$$=\frac{1}{[\lambda k]^2}\sum_{i=1}^{k}\sum_{l=1}^{k} I_{[\mu_i\leqslant\mu]}I_{[\mu_l\leqslant\mu]}\{P[X_i\geqslant Z_{[\lambda k]},X_l\geqslant Z_{[\lambda k]}|\boldsymbol{Z}]$$

$$-P[X_i\geqslant Z_{[\lambda k]}|\boldsymbol{Z}]\cdot P[X_l\geqslant Z_{[\lambda k]}|\boldsymbol{Z}]\}$$

$$\leqslant\frac{k}{[\lambda k]^2}+2\frac{1}{[\lambda k]^2}\sum_{i\leqslant l} I_{[\mu_i\leqslant y]}\{P[X_i\geqslant Z_{[\lambda k]},X_l\geqslant Z_{[\lambda k]}|\boldsymbol{Z}]$$

$$-P[X_i\geqslant Z_{[\lambda k]}|\boldsymbol{Z}]\cdot P[X_l\geqslant Z_{[\lambda k]}|\boldsymbol{Z}]\}. \tag{3.21}$$

As in B–Y we define, for $i=1,2,\ldots,[k/h]-1$,

$$V_k(i,h)=(Z_{ih+1},Z_{ih+2},\ldots,Z_{i(h+1)}) \tag{3.22}$$

and $\nu_i^{(k)}(\cdot)$ to be the empirical distribution of $V_k(i,h)$ (i.e., assigning mass $1/h$ to each member of $V_k(i,h)$). We then have

$$[\lambda k]^{-2}\sum_{i\leqslant l} I_{[\mu_i\leqslant y]}\{P[X_i\geqslant Z_{[\lambda k]},X_l\geqslant Z_{[\lambda k]}|\boldsymbol{Z}]-P[X_i\geqslant Z_{[\lambda k]}|\boldsymbol{Z}]\cdot P[X_l\geqslant Z_{[\lambda k]}|\boldsymbol{Z}]\}$$

$$\leqslant[\lambda k]^{-2}\sum_{a}\sum_{\substack{(i,l)\in V_k(a,h)\\ i\leqslant l}}\{P[X_i\geqslant Z_{[\lambda k]},X_l\geqslant Z_{[\lambda k]}|\boldsymbol{Z}]$$

$$-P[X_i\geqslant Z_{[\lambda k]}|\boldsymbol{Z}]\cdot P[X_l\geqslant Z_{[\lambda k]}|\boldsymbol{Z}]\}$$

$$+[\lambda k]^{-2}\sum_{a\neq b}\sum_{\substack{i\in V_k(a,h)\\ l\in V_k(b,h)}}\{P[X_i\geqslant Z_{[\lambda k]},X_l\geqslant Z_{[\lambda k]}|\boldsymbol{Z}]$$

$$-P[X_i\geqslant Z_{[\lambda k]}|\boldsymbol{Z}]\cdot P[X_l\geqslant Z_{[\lambda k]}|\boldsymbol{Z}]\}. \tag{3.23}$$

The first term of the r.h.s. of this inequality is less than $(kh)/[\lambda k]^2$ and hence for h, k such that $h/k\to 0$ as $k\to\infty$ this term converges to zero. Now,

$$[\lambda k]^{-2}\sum\{P[X_i\geqslant Z_{[\lambda k]},X_l\geqslant Z_{[\lambda k]}|\boldsymbol{Z}]-P[X_i\geqslant Z_{[\lambda k]}|\boldsymbol{Z}]P[X_l\geqslant Z_{[\lambda k]}|\boldsymbol{Z}]:$$

$$i\in V_k(a,h),l\in V_k(b,h),a\neq b\}$$

$$=[\lambda k]^{-2}\sum\{P[X_i\geqslant H^{-1}(1-\lambda),X_l\geqslant H^{-1}(1-\lambda)|\boldsymbol{Z}]$$

$$-P[X_i\geqslant H^{-1}(1-\lambda)|\boldsymbol{Z}]P[X_l\geqslant H^{-1}(1-\lambda)|\boldsymbol{Z}]:$$

$$i\in V_k(a,h),l\in V_k(b,h),a\neq b\}+R_k(\lambda) \tag{3.24}$$

where

$$\sup\{|R_k(\lambda)|: \varepsilon\leqslant\lambda\leqslant 1-\varepsilon\} \xrightarrow{P} 0.$$

To see this, note first that

$$|P[X_i\geqslant Z_{[\lambda k]}|\boldsymbol{Z}]-P[X_i\geqslant H^{-1}(1-\lambda)|\boldsymbol{Z}]|\leqslant$$
$$\leqslant P[H^{-1}(1-\lambda)\leqslant X_i\leqslant Z_{[\lambda k]}|\boldsymbol{Z}]+P[Z_{[\lambda k]}\leqslant X_i\leqslant H^{-1}(1-\lambda)|\boldsymbol{Z}]. \quad (3.25)$$

Then, by Shorack (1973), we can bound the right-hand side of (3.25) by $2P[|X_i-H^{-1}(1-\lambda)|\leqslant\delta|\boldsymbol{Z}]+o_p(1)$, $\forall\delta>0$, uniformly for $\varepsilon\leqslant\lambda\leqslant 1-\varepsilon$, $1\leqslant i\leqslant k$. We can apply Theorem A.1 of B–Y and the boundedness of the normal density to conclude that

$$\sup\{P[|X_i-H^{-1}(1-\lambda)|\leqslant\delta|\boldsymbol{Z}]: 1\leqslant i\leqslant(1-\delta)k\}\leqslant M\delta+R_k(\lambda) \quad (3.26)$$

where

$$\sup\{R_k(\lambda): \varepsilon\leqslant\lambda\leqslant 1-\varepsilon\}=o_p(1).$$

We conclude from (3.25) and (3.26) that

$$\sup\{|P[X_i\geqslant Z_{[\lambda k]}|\boldsymbol{Z}]-P[X_i\geqslant H^{-1}(1-\lambda)|\boldsymbol{Z}]|:$$
$$1\leqslant i\leqslant(1-\delta)k,\ \varepsilon\leqslant\lambda\leqslant 1-\varepsilon\}=o_p(1). \quad (3.27)$$

A similar argument applies to $P[X_i\geqslant Z_{[\lambda k]}, X_l\geqslant Z_{[\lambda k]}|\boldsymbol{Z}]$. We can then proceed to get rid of $\delta>0$ as in Lemma 3.1, and (3.24) follows.

By a generalization of Lemma A.1 in B–Y, if $i\in V_k(a,h)$, $j\in V_k(b,h)$, $a\neq b$,

$$P[X_i\geqslant H^{-1}(1-\lambda), X_j\geqslant H^{-1}(1-\lambda)|\boldsymbol{Z}]$$
$$=E\{P[X_i\geqslant H^{-1}(1-\lambda)|V_k(a,h)]\cdot P[X_j\geqslant H^{-1}(1-\lambda)|V_k(b,h)]|\boldsymbol{Z}\}, \quad (3.28)$$

and hence the terms in the summation on the right-hand side of (3.24) are just,

$$\mathrm{Cov}(P[X_i\geqslant H^{-1}(1-\lambda)|V_k(a,h)], P[X_l\geqslant H^{-1}(1-\lambda)|V_k(b,h)]|\boldsymbol{Z}).$$

By (A.20) of B–Y, if $W(i,l,\lambda,a,b)$ is the foregoing expression, for every $\delta>0$,

$$\max\{|W(i,l,\lambda,a,b)|: 1\leqslant i\leqslant(1-\delta)k, 1\leqslant l\leqslant(1-\delta)k, \varepsilon\leqslant\lambda\leqslant 1-\varepsilon, a\neq b\} \xrightarrow{P} 0, \quad (3.29)$$

since the $\nu_a^{(k)}(H^{-1}(1-\lambda))$ are constant given $\boldsymbol{Z}$.

The lemma now follows by dispensing with the $\delta>0$ restriction in the standard way.

Lemma 3.3. *Under Conditions* 1 *and* 2

$$\sup_{\varepsilon\leqslant\lambda\leqslant 1-\varepsilon} |\hat{G}_k(y,\lambda)-G_k(y,\lambda)| \xrightarrow{P} 0.$$

Proof. Given $\varepsilon=\lambda_1\leqslant\lambda_2\leqslant\cdots\leqslant\lambda_p=1-\varepsilon$

$$\sup_{\varepsilon\leqslant\lambda\leqslant 1-\varepsilon}\{|\hat{G}_k(y,\lambda)-G_k(y,\lambda)|\}\leqslant\max_{1\leqslant j\leqslant p}\{|\hat{G}_k(y,\lambda_j)-G_k(y,\lambda_j)|\}$$

$$+\max_{1\leqslant j\leqslant p-1}\left\{\sup_{\lambda_j\leqslant\lambda<\lambda_{j+1}}\{|\hat{G}_k(y,\lambda)-\hat{G}_k(y,\lambda_j)|\}\right\}$$

$$+\max_{1\leqslant j\leqslant p-1}\left\{\sup_{\lambda_j\leqslant\lambda<\lambda_{j+1}}\{|G_k(y,\lambda)-G_k(y,\lambda_j)|\}\right\}. \tag{3.30}$$

The last two terms are bounded, using the monotonicity of G_k and $\hat{G}_k$ in λ for fixed y, by $2\max_{1\leqslant j\leqslant p-1}\{G_k(y,\lambda_{j+1})-G_k(y,\lambda_j)\}$ and this term is bounded by

$$2\max_{1\leqslant j\leqslant p-1}\{G(y,\lambda_{j+1})>(y,\lambda_j)\}+4\sup_{\varepsilon\leqslant\lambda\leqslant 1-\varepsilon}\{|G_k(y,\lambda)-G(y,\lambda)|\}. \tag{3.31}$$

We then have,

$$\sup_{\varepsilon\leqslant\lambda\leqslant 1-\varepsilon}\{|\hat{G}_k(y,\lambda)-G_k(y,\lambda)|\}\leqslant\max_{1\leqslant j\leqslant p}\{|\hat{G}_k(y,\lambda_j)-G_k(y,\lambda_j)|\}$$
$$+2\max_{1\leqslant j\leqslant p-1}\{G(y,\lambda_{j+1})-G(y,\lambda_j)\}$$
$$+4\sup_{\varepsilon\leqslant\lambda\leqslant 1-\varepsilon}\{|G_k(y,\lambda)-G(y,\lambda)|\}. \tag{3.32}$$

By the uniform continuity of G we can choose $\lambda_1,\ldots,\lambda_p$ so that the middle term of the r.h.s. of (3.32) is arbitrarily small. We can then apply Lemma 3.2 to the first term and Lemma 3.1 to the third term and conclude the proof of the lemma.

Proof of Theorem. By Lemmas 3.1–3.3,

$$\sup\{|\hat{G}_k(y,\lambda)-G(y,\lambda)|:\varepsilon\leqslant\lambda\leqslant 1-\varepsilon\}\xrightarrow{P}0. \tag{3.33}$$

Since $G(y,\lambda)$ is by I uniformly continuous for $\varepsilon\leqslant\lambda\leqslant 1-\varepsilon$, $y\in R$ we can find for each $\delta>0$, a partition $y_1<\cdots<y_p$ such that, if $y_0=-\infty$, $y_{p+1}=\infty$,

$$\max_{1\leqslant j\leqslant p+1}\sup\{|G(y,\lambda)-G(y_{j-1},\lambda)|:y_{j-1}\leqslant y\leqslant y_j,\ \varepsilon\leqslant\lambda\leqslant 1-\varepsilon\}\leqslant\delta.$$

We can then use these partitions in a standard way as in the proof of Polya's theorem to conclude that we can take the sup over y in (3.33) as well. The theorem is proved.

4. Asymptotic behavior of the risk and optimality in a Bayesian Framework

In various cases of interest the loss function l can be written as the sum of three components,

$$l(\mu^*,\{\pi_{D_1},\ldots,\pi_{D_r}\})=l_1(\mu^*,\{\pi_{D_1},\ldots,\pi_{D_r}\})+b\frac{r}{k}+cI_{[\mu_{(1)}\notin\{\pi_D1,\ldots,\pi_{D_r}\}]} \tag{4.1}$$

where b, $c \geqslant 0$,

$$l_1\left(\mu^*, \{\pi_{D_1}, \ldots, \pi_{D_r}\}\right) = L_1\left(F_k, \hat{G}_k\left(\cdot \frac{r}{k}\right)\right) \tag{4.2}$$

and L_1 is a functional defined on $\mathfrak{F} \times \mathfrak{F}$ where $\mathfrak{F}$ is the set of all d.f.'s carried by some fixed set (possibly R). The third component, of course, gives a risk component corresponding to the probability of incorrect selection. The first two, typically lead to large losses if too many populations are selected. The loss function given by (2.1) in B–Y corresponds to having an L_1 equal to

$$L_{11}(F, G) = \operatorname{ess\,sup} F - \int x \, \mathrm{d}G.$$

Gupta (1965) considers the "average rank of the set of populations selected" as a measure of gain. When translated into a measure of loss this is equivalent (up to an additive constant) to

$$l_1\left(\mu^*, \{\pi_{D_1}, \ldots, \pi_{D_r}\}\right) = \frac{1}{r} \sum_{i=1}^{r} D_i$$

which in turn is equivalent to having an L_1 equal to

$$L_{12}(F, G) = \int_{-\infty}^{\infty} (1 - F(x)) \, \mathrm{d}G(x).$$

In both of these cases, $b = 0$. On the other hand, Gupta and Hsu (1978) take $L_1 = 0$, $b = 1$. We shall assume:

Condition 3. *$L_1(\cdot, \cdot)$ is continuous (in the usual sense of weak convergence) and bounded on $\mathfrak{F} \times \mathfrak{F}$ at all F, $G \in \mathfrak{F}$ which are continuous.*

Condition4. *If F, G_1, $G_2 \in \mathfrak{F}$ and G_1 is stochastically smaller than G_2, then*

$$L_1(F, G_1) \geqslant L_1(F, G_2).$$

Condition 4 corresponds to monotonicity in l_1 and also drives us to selecting smaller sets of populations. It holds for both L_{11} and L_{12}. Condition 3 holds for L_{12} in general and for L_{11} if all distributions in F are carried by a compact. In the model we've been considering so far the optimal invariant procedure for l as in (4.1), (4.2) is obtained by selecting the populations corresponding to the top $r_k^*(\boldsymbol{Z})$ order statistics where $r_k^*(\boldsymbol{Z})$ minimizes as a function of r,

$$V_k(r|\boldsymbol{Z}) = E\left(L_1\left(F_k, G_k\left(\cdot, \frac{r}{k}\right)\right) \middle| \boldsymbol{Z}\right) + b\frac{r}{k} + cP[X_1 < Z_r | \boldsymbol{Z}]. \tag{4.3}$$

Let

$$V(\lambda, F) = L_1(F, G(\cdot, \lambda)) + b\lambda + c\Phi\left(H^{-1}(1-\lambda)\right). \tag{4.4}$$

Theorem 4.1. *If Conditions 1–4 hold and $V(\lambda, F)$ is uniquely minimized for $0 < \lambda < 1$ by $\lambda_0(F)$, then*

$$\frac{r_k^*(\boldsymbol{Z})}{k} \xrightarrow{P} \lambda_0(F) \tag{4.5}$$

and

$$V_k(r_k^*(Z)|Z) \xrightarrow{P} V(\lambda_0(F), F). \tag{4.6}$$

Moreover, if $\{d_k\}$ *is any sequence of invariant monotone procedures with corresponding numbers of population selected* $r_k(Z)$ *such that*

$$\frac{r_k(Z)}{k} \xrightarrow{P} \lambda_0(F), \tag{4.7}$$

then

$$R(\mu^*, d_k) = EV_k(r_k(Z)|Z) \to V(\lambda_0(F), F) \tag{4.8}$$

and hence $\{d_k\}$ *is asymptotically optimal.*

The following lemma is basic. It consists of a special case of Lemma 4.3 of B–Y and an immediate consequence of Theorem 3.1.

Lemma 4.1. *Under Conditions* 1 *and* 2

$$\sup\{|P[X_1 < Z_{[\lambda k]}|Z] - \Phi(H^{-1}(1-\lambda))| : \lambda \leqslant 1-\varepsilon\} \xrightarrow{P} 0. \tag{4.9}$$

If, in addition, Condition 2 *holds,*

$$E(L_1(F_k, \hat{G}_k(\cdot, \lambda))|Z) = L_1(F, G(\cdot, \lambda)) + R_k(\lambda, Z)$$

where

$$\sup\{|R_k(\lambda, Z)| : \varepsilon \leqslant \lambda \leqslant 1-\varepsilon\} \xrightarrow{P} 0.$$

Proof of Theorem 4.1. By Lemma 4.1 and Conditions 1–3

$$V_k(\lambda k|Z) = V(\lambda, F) + R(\lambda, Z) \tag{4.11}$$

where

$$\sup\{|R(\lambda, Z)| : \varepsilon \leqslant \lambda \leqslant 1-\varepsilon\} \xrightarrow{P} 0.$$

Fix ε such that, $\varepsilon < \lambda_0 < 1-\varepsilon$. Then by Condition 4

$$\inf\{V_k(r|Z) : 1 \leqslant r \leqslant k\varepsilon\} \geqslant E[L_1(F_k, \hat{G}_k(\cdot, 0))|Z] + cP[X_1 < Z_{\langle k\varepsilon \rangle}|Z]. \tag{4.12}$$

But $\mathfrak{L}(\hat{G}_k(\cdot,0), \delta_{\{0\}}) \xrightarrow{P} 0$ for $\mathfrak{L}$, the Levy metric. This follows since by Lemma 5.1 of the next section,

$$\mu_{D_1} \underset{P}{\to} 0.$$

By the continuity of L_1, and (4.9), the right-hand side of (4.12) thus tends to the limit,

$$L_1(F, \delta_{\{0\}}) + c\Phi(H^{-1}(1-\varepsilon)). \tag{4.13}$$

It may be readily shown that as $\lambda \to 0$, $G(\cdot, \lambda)$ tends in law to $G(\cdot, 0) = \delta_{\{0\}}$. Therefore, as $\lambda \to 0$,

$$V(\lambda, F) \to L_1(F, \delta_{\{0\}}) + c$$

which is $> V(\lambda_0, F)$ by assumption. Thus $\exists \varepsilon = \varepsilon_0 > 0$ such that

$$L(F, \delta_{\{0\}}) + c\Phi(H^{-1}(1-\varepsilon_0)) > V(\lambda_0, F) \tag{4.14}$$

and by using (4.11)–(4.14) we conclude that

$$P[r^*(\boldsymbol{Z}) \geqslant k\varepsilon_0] \to 1. \tag{4.15}$$

By a similar argument,

$$P[r^*(\boldsymbol{Z}) \leqslant (1-\varepsilon_1)k] \to 1 \quad \text{for } \varepsilon_1 > 0.$$

Applying (4.11) again the theorem follows.

We now give a more satisfactory Bayesian formulation. Suppose the set of means $\mu_1, \ldots, \mu_k$ of our populations can vary freely in a broader model and have a Bayes prior distribution such that $\mu_1, \ldots, \mu_k$ are i.i.d. with distribution F which is bounded above and below and in addition has a density bounded away from 0 on its convex support. Since the Bayes prior is exchangeable, if l is monotone, the Bayes rule is monotone, invariant. Suppose now that l is of the form (4.1), for $\mathcal{F} = \{$all d.f. with the same carrier as $F\}$. We can then write the Bayes risk of any monotone invariant rule d_k as,

$$R(d_k) = E\{E_{\boldsymbol{\mu}^*}(V_k(r_k(\boldsymbol{Z})|\boldsymbol{Z}))\}$$

where the second expectation is taken according to the prior distribution on $\boldsymbol{\mu}^*$. If F is as above, it follows that Conditions 1 and 2 hold (with probability 1) for F_k corresponding to $\boldsymbol{\mu}^*$. If, in addition, Conditions 3 and 4 are valid we conclude from Theorem 4.1 that

$$\liminf_k R(d_k) \geqslant V(\lambda_0(F), F),$$

with equality holding for any sequence of procedures such that (4.17) holds (where P can now be taken as the probability induced on R^∞ by the Bayes prior and the normal kernel). Since equality can be achieved for instance by $r_k(\boldsymbol{Z}) = \langle k\lambda_0(F)\rangle$ for all $\boldsymbol{Z}$, $V(\lambda_0(F), F)$ is the asymptotic Bayes risk.

This argument clearly does not use the independence of the μ_i in an essential way. The conclusion continues to hold if for k observations we have an exchangeable prior distribution P_k on $(\mu_1, \ldots, \mu_k)$ such that the resulting empirical distributions F_k obey Conditions 3 and 4 in probability for a suitable unique F.

5. Asymptotic optimality of Gupta's and other rules in a frequentist framework

We now consider the problem of minimizing

$$E[L_1(F_k, G_k(\cdot, r_k(\boldsymbol{Z})/k))] + b\frac{r}{k} \tag{5.1}$$

subject to the probability of correct selection

$$P\left[X_1 \geqslant Z_{\langle r_k(\boldsymbol{Z})\rangle}\right] \geqslant \gamma > 0 \tag{5.2}$$

among monotone invariant rules.

It is clear, under Conditions 1–4, that by varying c from 0 to ∞ in the loss function l given by (4.1) we can make the optimal $\lambda_0(F)$ (which also depends on c) vary from 0 to 1, and the asymptotic probability of correct selection of the optimal rule $1-\Phi(H^{-1}(1-\lambda_0(F)))$ vary continuously from 0 to 1. It follows by a standard argument from Theorem 4.1 that under Conditions 1–4 any sequence of monotone invariant rules d_k with

$$\frac{r_k(\boldsymbol{Z})}{k} \xrightarrow{P} \lambda(\gamma, F) \tag{5.3}$$

where

$$1-\Phi\left(H^{-1}(1-\lambda(\gamma))\right) = \gamma$$

i.e. with

$$\lambda(\gamma, F) = 1 - H\left(\Phi^{-1}(1-\gamma)\right)$$

asymptotically minimizes (5.1) subject to (5.2) *whatever be* L_1 and b. This is not yet satisfactory since it appears that construction of an optimal rule requires knowledge of H. In B–Y we suggested a rule satisfying (5.3) which depended only on knowledge of $\operatorname{ess\,sup} F$. This rule is: d^0: Select π_i with $X_i - \operatorname{ess\,sup} F \geqslant \Phi^{-1}(1-\gamma)$. This rule, in fact, if $\mu_{(1)} = \operatorname{ess\,sup} F$ has probability of correct selection,

$$P_{\mu^*}\left[X_1 - \mu_{(1)} \geqslant \Phi^{-1}(1-\gamma)\right] = \gamma \tag{5.4}$$

for all k.

We show in B–Y that for this sequence of rules for any F satisfying Conditions 1 and 2 the number of populations selected r_k^0 has,

$$\frac{r_k^0(\boldsymbol{Z})}{k} \xrightarrow{P} \lambda(\gamma, F) \tag{5.5}$$

and thus d^0 is asymptotically optimal.

Gupta (1965) proposed rules of the form d^{G}: Select π_i with $X_i - Z_1 \geqslant c_k(1-\gamma)$ where $c_k(1-\gamma)$ is the $(1-\gamma)$ quantile of the distribution of $X_1 - Z_1$ for $X_1, \ldots, X_k$ i.i.d. $N(0,1)$. He showed that

$$\min_{\boldsymbol{\mu}} P_{\boldsymbol{\mu}^*}[X_1 - Z_1 \geqslant c_k(1-\gamma)] = \gamma. \tag{5.6}$$

We shall show that if $r_k^G(\boldsymbol{Z})$ is the number of populations selected by Gupta's rule and F is *any* distribution satisfying Conditions 1 and 2,

$$\frac{r_k^G(\boldsymbol{Z})}{k} \xrightarrow{P} \lambda(\gamma, F). \tag{5.7}$$

The following theorem is a consequence of (5.7).

Theorem 5.1. *Gupta's rules with probability of correct selection γ asymptotically minimize the risk* (5.1) *for all L_1 satisfying Conditions* 3 *and* 4 *among all monotone invariant rules with asymptotic probability of correct selection $\geqslant\gamma$ uniformly for F satisfying Conditions* 1 *and* 2.

The proof of (5.7) rests on the following lemma.

Lemma 5.1. *If F satisfies Condition* 1, *then*

$$Z_1 - \sqrt{2\log k} \xrightarrow{P} \operatorname{ess\,sup} F, \tag{5.8}$$

$$\mu_{D_1} \xrightarrow{P} \operatorname{ess\,sup} F. \tag{5.9}$$

Proof. Without loss of generality, as usual, we take $\operatorname{ess\,sup} F = 0$, $\mu_1 \geqslant \cdots \geqslant \mu_k$. Write,

$$X_i = \mu_i + \varepsilon_i, \quad i = 1, \ldots, k \tag{5.10}$$

where the ε_i are i.i.d. $N(0,1)$ and let

$$M_\lambda = \max\{\varepsilon_i \colon 1 \leqslant i \leqslant \lambda k\}, \quad 0 < \lambda \leqslant 1.$$

It is well known that, see, e.g. David (1970),

$$M_\lambda - \sqrt{2\log \lambda k} \xrightarrow{P} 0. \tag{5.11}$$

Since

$$\sqrt{2\log\lambda k} = \sqrt{2\log k}\left(1 + \left(1 + \frac{\log\lambda}{\log k}\right)^{1/2}\right) = \sqrt{2\log k} + \mathrm{o}(1),$$

$$M_\lambda - \sqrt{2\log k} \xrightarrow{P} 0. \tag{5.12}$$

However, since $\mu_i \leqslant 0$, $i = 1, \ldots, k$,

$$M_\lambda + \mu_{\langle\lambda k\rangle} \leqslant Z_1 \leqslant M_\lambda. \tag{5.13}$$

Now,

$$\mu_{\langle\lambda k\rangle} = F_k^{-1}\left((1-\lambda) - \frac{1}{k}\right) \to F^{-1}(1-\lambda),$$

by Condition 1 and since F is continuous and strictly increasing $F^{-1}(1-\lambda) \to 0$ as $\lambda \to 0$, (5.8) follows from (5.12) and (5.13).

Let, for $\lambda_1 < \lambda_2$,

$$Z(\lambda_1, \lambda_2) = \max\{X_i\colon \lambda_1 k \leqslant i < \lambda_2 k\}.$$

To prove (5.9), in view of Condition 1, it clearly suffices to show that,

$$P[Z(0,\lambda) > Z(\lambda,1)] \to 1 \tag{5.14}$$

for every $0 < \lambda \leqslant 1$. But by (5.8)

$$Z(\lambda,1) - \sqrt{2\log k} \to F^{-1}(1-\lambda) < 0$$

and

$$Z(0,\lambda) - \sqrt{2\log k} \xrightarrow{P} 0.$$

(5.14) follows. The lemma is proved.

Applying (5.11) with $\mu_1 = \cdots = \mu_k = 0$ we conclude that

$$c_k(1-\gamma) = -\sqrt{2\log k} + \Phi^{-1}(1-\gamma) + o(1). \tag{5.15}$$

Now,

$$\frac{r_k^{\mathrm{G}}(\boldsymbol{Z})}{k} = 1 - \hat{H}_k(c_k(1-\gamma) + Z_1) \tag{5.16}$$

where $\hat{H}_k$ is the empirical d.f. of $X_1, \ldots, X_k$.
By Lemma 5.1 and (5.13),

$$c_k(1-\gamma) + Z_1 \xrightarrow{P} \Phi^{-1}(1-\gamma) \tag{5.17}$$

and by Shorack (1973)

$$1 - \hat{H}_k(c_k(1-\gamma) + Z_1) \xrightarrow{P} 1 - H(\Phi^{-1}(1-\gamma))$$

and (5.7) and the theorem follows.

6. Discussion

(i) We can deduce from Theorems 4.1 and 5.1 that if we take a Bayesian point of view and a prior distribution making $\mu_1, \ldots, \mu_k$ independent with common distribution F with compact support and positive density on its convex support and take l given by (4.1) satisfying 2 and 4 then there is a rule of Gupta's type which is asymptotically optimal Bayes. (The boundedness of F from below is inessential and requires a condition like A.2 of B–Y.) Numerical results of Chernoff and Yahav (1977), and Gupta and Hsu (1978) using normal priors are in agreement with these results. Unfortunately our theory is not really applicable. In the Gupta–Hsu case the loss function is satisfactory ($L_{11} = 0$) but the unboundedness of the normal prior from above forces the optimal Bayes rule to take $r_k^*(\boldsymbol{Z})$ of smaller order than k and in fact yields asymptotic Bayes risk 0. This is true with the B–Y loss structure as well. In the Chernoff–Yahav case

$$I_{[\mu_{(1)} \notin \{\pi_{D_1} \cdots \pi_{D_r}\}]}$$

is not a component of the loss function but is replaced by $\mu_{(1)} - \max\{\mu_{D_i}\colon 1 \leqslant i \leqslant r\}$, $b=0$, and L_{11} is used. It is easy to see by using Lemma 5.1, that *even if* F is bounded above, this loss structure leads to optimal rules which asymptotically have $r_k^*(\boldsymbol{Z})$ of smaller order than k and hence also have asymptotic Bayes risk 0. These questions need further investigation.

(ii) There are many asymptotically equivalent types of rules both in the Bayesian and frequentist frameworks. We have mentioned the rules d^{G}, d^0 and rules taking a fixed proportion of the populations (in the Bayesian framework). Preliminary numerical calculations suggest, as one might expect, that d^0 is substantially better than d^{G} for moderate k if $\operatorname{ess\,sup} F$ is indeed known. Further numerical comparisons of these classes of rules are desirable.

(iii) We note that in the Bayesian framework where the prior distribution makes the μ_i i.i.d. according to F our results could be obtained more easily with methods such as those of Yang (1977) on concomitants of order statistics. His computations and those of David and Galambos (1974) are greatly simplified by the fact that given $\boldsymbol{Z}$, $\mu_{D_1}, \ldots, \mu_{D_k}$ are independent. Our Condition 2 arises precisely because in our framework we are studying the (discrete) conditional distribution of the μ_{D_i} given not only $\boldsymbol{Z}$ but also $\boldsymbol{\mu}^*$.

(iv) Conversely, our methods should yield results stronger than those of David and Galambos (1974) who dealt with concomitants of order statistics when the (μ_i, X_i) have bivariate normal distribution. They cannot simply yield those of Yang since he deals with (μ_i, X_i) where the conditional distribution of X given μ is not necessarily normal. Clearly, our results can be extended to such models and hence to general selection problems of the type dealt with by Gupta and others. The extension is not trivial since adequate substitutes need to be found for Condition 2 and Lemmas A.2–A.4 of B–Y.

References

[1] Bickel, P.J. and Yahav, J.A. (1977). On selecting a set of good populations. In: Gupta, S.S. and Moore, Eds., D.S., *Stat. Dec. Theory and Related Topics II*. Academic Press, New York, pp. 37–55.

[2] Chernoff, H. and Yahav, J.A. (1977). On subset selection problem employing a new criterion. In: Gupta, S.S. and Moore, D.S., Eds., *Stat. Dec. Theory and Related Topics II*. Academic Press, New York.

[3] David, H.A. (1970). *Order Statistics*. Wiley, New York.

[4] David, H.A. and Galambos, J. (1974). The asymptotic theory of concomitants of order statistics. *J. Appl. Probability* **11**, 762–770.

[5] Eaton, M.L. (1967). Some optimum properties of ranking procedures. *Ann. Math. Statist.* **38**, 124–137.

[6] Gupta, S.S. (1965). On some multiple decision (selection and ranking) rules. *Technometrics* **7**, 225–245.

[7] Gupta, S.S. and Hsu, J. (1978). On the performance of subset selection procedures. *Commun. Statist.* **B7**, 561–591.

[8] Shorack, G. (1973). Convergence of reduced empirical and quantile processes. *Ann. Statist.* **1**, 146–152.

[9] Yang, S.S. (1977). General distribution of the concomitants of order statistics. *Ann. Statist.* **5**, 996–1002.

G. Kallianpur, P.R. Krishnaiah, J.K. Ghosh, eds., *Statistics and Probability: Essays in Honor of C.R. Rao*
© North-Holland Publishing Company (1982) 125–140

$\mathcal{G}$-INVARIANT DESIGNS AND $\mathcal{G}$-BALANCED ARRAYS WITH APPLICATIONS

R.C. BOSE* and H.K. IYER
Statistics Department, Colorado State University, Fort Collins, CO, U.S.A.

1. Introduction

The concept of Orthogonal Arrays of strength d was introduced by C.R. Rao (1946). They have been found useful in the construction of fractional factorial designs, confounded symmetrical and asymmetrical factorials and many other useful combinatorial arrangements. Orthogonal arrays played an important role in the disproof of Euler's conjecture by Bose, et al. (1960) and also in the solution of Kirkman's schoolgirl problem by Ray-Chaudhuri and Wilson (1971). Chakravarthi (1956) defined partially balanced arrays of strength d in s symbols, and these were later renamed as balanced arrays by Srivastava. Balanced arrays in two symbols were extensively investigated by Srivastava (1972), Srivastava and Anderson (1970), Srivastava and Chopra (1971) and Srivastava and Ariyarathna (1979), as well as Yamamoto et al. (1975). These authors have used balanced arrays to obtain balanced designs of resolutions V and VII.

Rao (1973) has defined orthogonal arrays of the mixed type corresponding to asymmetrical factorial experiments. He called them "orthogonal arrays in a variable number of symbols". These were found to be useful in the construction of asymmetrical factorial fractions.

In the present paper, we introduce the wider class of $\mathcal{G}$-invariant designs corresponding to a suitable group $\mathcal{G}$. This class of designs has a subclass which we call the class of $\mathcal{G}$-balanced designs. Balanced designs and orthogonal designs are special cases of $\mathcal{G}$-balanced designs. We also introduce the concept of $\mathcal{G}$-balanced arrays of which balanced arrays in a variable number of symbols and orthogonal arrays in a variable number of symbols are special cases. A correspondence is established between $\mathcal{G}$-balanced designs of resolution $2t+1$ and $\mathcal{G}$-balanced arrays of strength $2t$.

The usefulness of our approach stems from the fact that the representation theory for the group $\mathcal{G}$ can be used to derive the characteristic polynomial of the information matrix corresponding to a general $\mathcal{G}$-balanced design. This would be very useful in obtaining optimal designs in various asymmetrical factorial experiments.

A sequel to this paper will deal with $2^m \times 3^n$ asymmetrical fractional factorials. We obtain E-optimal designs of resolution III and indicate the procedure for obtaining E-optimal designs of resolution V for various practical values of N where N is the number of runs.

*Research supported by the Air Force research grant AFOSR-77-3127.

We assume that the reader is familiar with the notion of general factorial effects and the use of the λ-operator in calculating the elements of the information matrix corresponding to a design T. We use the usual notation for factorial effects. For example, $F_1F_2F_3^2$ is a single degree of freedom component of the three factor interaction involving F_1, F_2 and F_3 and may be called the "linear by linear by quadratic" component of it. The reader is referred to Srivastava (1961) for details.

The vector $F^i=(F_0^i, F_1^i,\ldots, F_{s_i-1}^i)^{\mathrm{T}}$ denotes the vector of all factorial effects for factor i, f_j^i denotes the expected response for the jth level of factor i, $f^i=(f_0^i,\ldots, f_{s_i-1}^i)^{\mathrm{T}}$ and $F^i=\Delta_i f^i$. Also, we use D_i for Δ_i^{-1}, $\phi^i=(\phi_0^i,\ldots,\phi_{s_i-1}^i)^{\mathrm{T}}$ where $\phi_\alpha^i=F_\alpha^i/\pi_\alpha^i$ and $\Delta_i\Delta_i^{\mathrm{T}}=\mathrm{diag}(\pi_0^i,\pi_1^i,\ldots,\pi_{s_i-1}^i)$. Furthermore $f_{j_1}^1 f_{j_2}^2\ldots f_{j_m}^m$ denotes the expected response for the treatment combination using j_1th level of factor $1,\ldots, j_m$th level of factor m and a denotes the linear span of $\{f_{j_1}^1 f_{j_2}^2\ldots f_{j_m}^m \mid 0\leqslant j_r\leqslant s_r-1, 1\leqslant r\leqslant m\}$.

2. $\mathcal{G}$-invariant designs

2.1. *Definition of $\mathcal{G}$-invariant designs*

Consider m factors $F_1, F_2,\ldots, F_m$. Let $F_1, F_2,\ldots, F_{m_1}$ be at s_1-levels each; $F_{m_1+1},\ldots, F_{m_2}$ at s_2-levels each, $\ldots$; $F_{m_1+m_2+\cdots+m_{r-1}+1},\ldots, F_m$ at s_r-levels each, where $2\leqslant s_1<s_2<\cdots<s_r$ and $m=m_1+m_2+\cdots+m_r$.

Let us suppose that the vector, $\boldsymbol{\beta}$, of unknown parameters in the underlying linear model consists of all factorial effects involving t or less factors. Let $\mathcal{F}$ denote the set $\{F_1, F_2,\ldots, F_m\}$ of all factors. Let $\mathcal{F}_i$ denote the set $\{F_j \mid m_1+m_2+\cdots+m_{i-1}<j\leqslant m_1+m_2+\cdots+m_i\}$ of all factors which are at s_i-levels each. Let $S(\mathcal{F}_i)$ denote the full symmetric group on $\mathcal{F}_i$. Let $S=S(\mathcal{F}_1)\times S(\mathcal{F}_2)\times\cdots\times S(\mathcal{F}_r)$ be the direct product of $S(\mathcal{F}_1),\ldots, S(\mathcal{F}_r)$. Then S is a group of permutations acting on $\mathcal{F}$ such that $\mathcal{F}_i$ is mapped onto itself. The action of S on $\mathcal{F}$ naturally extends to an action on $\mathcal{A}$ as follows: For $\sigma\in S$, $\sigma(f_{j_1}^1 f_{j_2}^2\cdots f_{j_m}^m)=f_{j_1}^{\sigma(1)} f_{j_2}^{\sigma(2)}\cdots f_{j_m}^{\sigma(m)}$, where $\sigma(u)=v$ iff $\sigma(F_u)=F_v$, $1\leqslant u, v\leqslant m$.

In particular, σ acts on the set Ω of all factorial effects appearing in the model. Let $\Omega_0=\{\mu\}=\{F_0^1F_0^2,\ldots, F_0^m\}$, $\Omega_1,\Omega_2,\ldots,\Omega_q$ be the orbits of Ω under the action of S. The algebra, $\mathcal{A}$, regarded as a vector space over $\mathbb{R}$ has $V_{\Omega_0}, V_{\Omega_1},\ldots, V_{\Omega_r}$ as subspaces, where V_{Ω_i} is the $\mathbb{R}$-linear span of elements of Ω_i. In fact, $V_\Omega=V_{\Omega_0}\oplus\cdots\oplus V_{\Omega_q}$ (direct sum).

Let $G(\Omega_0), G(\Omega_1),\ldots, G(\Omega_q)$ be subgroups of $O(V_{\Omega_0}), O(V_{\Omega_1}),\ldots, O(V_{\Omega_q})$, respectively, where $O(V_{\Omega_i})$ is the real orthogonal group on V_{Ω_i}, $1\leqslant i\leqslant q$. Let $\mathcal{G}=G(\Omega_0)\times G(\Omega_1)\times\cdots\times G(\Omega_q)$. $\mathcal{G}$ can be regarded as a group of $p\times p$ matrices with $p=|\Omega|$. Each element of $\mathcal{G}$ is then a matrix with a block diagonal form with blocks corresponding to the subspaces $V_{\Omega_0}, V_{\Omega_1},\ldots, V_{\Omega_q}$, respectively.

Let $\mathcal{M}$ denote the set of all $p\times p$ symmetric matrices. Then the information matrix M_T corresponding to a design T belongs to $\mathcal{M}$. The group $\mathcal{G}$ acts on $\mathcal{M}$ as follows: For

$$P\in\mathcal{G},\qquad M\in\mathcal{M},\qquad P(M)=P^{\mathrm{T}}MP.$$

Definition. A design $T(N\times m)$ is $\mathcal{G}$-invariant if and only if $P^{\mathrm{T}}M_TP=M_T\ \forall P\in\mathcal{G}$.

We observe that since $P^{\mathrm{T}}=P^{-1}$, the above condition on M_T is equivalent to requiring that M_T belong to the centralizer algebra of $\mathcal{G}$, in the algebra of all $p\times p$ matrices.

2.2 *Orthogonal designs*

Suppose we take $G(\Omega_i)=O(V_{\Omega_i})$, $i=0,1,2,\ldots,q$. Then $\mathcal{G}=O(V_{\Omega_0})\times O(V_{\Omega_1})\times\cdots\times O(V_{\Omega_q})$. We now investigate the conditions for a design T to be $\mathcal{G}$-invariant, in this special case.

Lemma 2.1. *Let* $\mathcal{G}=O(V_{\Omega_0})\times O(V_{\Omega_1})\times\cdots\times O(V_{\Omega_q})$ *be a group of* $p\times p$ *matrices where* $p=\dim(V_{\Omega_0}\oplus V_{\Omega_1}\oplus\cdots\oplus V_{\Omega_q})$. *Let* M *be any* $p\times p$ *matrix such that* $MP=PM$ *for every* P *in* $\mathcal{G}$. *Then* M *must be a diagonal matrix such that* M *restricted to* V_{Ω_i} *is a scalar matrix for each* $i=0,1,2,\ldots,q$.

Proof. The proof is straightforward and is left to the reader.

Remarks. A design T is said to be *orthogonal* if the corresponding information matrix M_T is diagonal. We state here without proof that the requirement for M_T to be diagonal automatically implies that M_T restricted to V_{Ω_i} must be a scalar matrix for each $i=0,1,2,\ldots,q$. The proof of this statement will be contained in the proof of the following theorem. Thus a design T is $\mathcal{G}$-*invariant* with $\mathcal{G}=O(V_{\Omega_0})\times\cdots\times O(V_{\Omega_q})$ if and only if T is orthogonal.

2.3. *Necessary and sufficient conditions for a design to be orthogonal of resolution 2t+1*

Let the group S(defined in Subsection 2.1) act on the v-tuples of $\{F_1, F_2,\ldots, F_m\}$ where $v=t+u$, $u=\min(t, m-t)$ and t is the maximum number of factors involved in a factorial effect occurring in the underlying linear model. We recall that a design T is of resolution $2t+1$ if it allows the estimation of all factorial effects involving t or less factors assuming that the remaining effects are all zero. Let O denote the set of all v-tuples whose components are distinct elements of $\mathcal{F}$. Let $O_1, O_2,\ldots, O_d$ be the orbits of O under S. It is clear that each orbit, O_i, is characterized by a vector $(w_1^i, w_2^i,\ldots, w_r^i)$ where w_j^i is the number of factors in a v-tuple in O_i which are at s_j-levels each.

Theorem 2.1. *Let* T *be a design of resolution* $2t+1$ $(t\geqslant 1)$. *Suppose that* M_T *is a diagonal matrix. Then the following holds*:

Let T_1 *be an* $N\times v$ *matrix whose columns are the same as the* v *columns of* $T(N\times m)$ *corresponding to factors* $F_{i_1}, F_{i_2},\ldots, F_{i_v}$, *respectively. Each of the* $s_{i_1}\times s_{i_2}\times\cdots\times s_{i_v}$ *possible vectors which can occur as rows of* T_1, *do occur, each with the same frequency* λ_{O_i} *which depends only on the orbit* O_i *to which* $(F_{i_1}, F_{i_2},\ldots, F_{i_v})$ *belongs.*

Proof. Let $\{i_1, i_2, \ldots, i_t\} = I$ be a subset of $I_0 = \{1, 2, \ldots, m\}$. Let $\{i_{t+1}, i_{t+2}, \ldots, i_{t+u}\}$ be a subset of $I_0 - I$, where $u = \min(t, m-t)$. Consider the vector $\boldsymbol{F}^{i_1, i_2, \ldots, i_v} = \boldsymbol{F}^{i_1} \times \boldsymbol{F}^{i_2} \times \cdots \times \boldsymbol{F}^{i_v}$ (Kronecker product), $v = t+u$. Let $\boldsymbol{\epsilon}$ denote a vector of size $(s_{i_1} s_{i_2} \cdots s_{i_v}) \times 1$, whose rows correspond to the rows of $\boldsymbol{F}^{i_1, i_2, \ldots, i_v}$ and the element of $\boldsymbol{\epsilon}$ corresponding to $F^{i_1}_{\alpha_1} F^{i_2}_{\alpha_2}, \ldots, F^{i_v}_{\alpha_v}$ be $\epsilon(F^{i_1}_{\alpha_1} F^{i_2}_{\alpha_2}, \ldots, F^{i_t}_{\alpha_t}, F^{i_{t+1}}_{\alpha_{t+1}}, \ldots, F^{i_v}_{\alpha_v})$. Thus $\boldsymbol{\epsilon} = \lambda(\boldsymbol{\phi}^{i_1} \times \boldsymbol{\phi}^{i_2} \times \cdots \times \boldsymbol{\phi}^{i_v})$ where

$$\lambda(\boldsymbol{x}) = \begin{pmatrix} \lambda(x_1) \\ \lambda(x_2) \\ \vdots \\ \lambda(x_n) \end{pmatrix}.$$

for any vector $\boldsymbol{x} = (x_1, x_2, \ldots, x_n)'$ with $x_1, x_2, \ldots, x_n \in \mathcal{Q}$. Thus $\boldsymbol{\epsilon} = \lambda(\boldsymbol{\phi}^{i_1} \times \cdots \times \boldsymbol{\phi}^{i_v}) = (D^{\mathrm{T}}_{i_1} \times \cdots \times D^{\mathrm{T}}_{i_v}) \lambda(\boldsymbol{f}^{i_1} \times \cdots \times \boldsymbol{f}^{i_v})$. Under the hypothesis of the theorem, we must have:

$$\boldsymbol{\epsilon} = \begin{pmatrix} \lambda(\phi^{i_1}_0 \phi^{i_2}_0 \cdots \phi^{i_v}_0) \\ 0 \\ 0 \\ \vdots \\ 0 \end{pmatrix} = \begin{pmatrix} N \\ 0 \\ 0 \\ \vdots \\ 0 \end{pmatrix}.$$

So

$$\lambda(\boldsymbol{f}^{i_1} \times \boldsymbol{f}^{i_2} \times \cdots \times \boldsymbol{f}^{i_v}) = \left(D^{\mathrm{T}}_{i_1} \times \cdots \times D^{\mathrm{T}}_{i_v}\right)^{-1} \boldsymbol{\epsilon} = \left(\Delta^{\mathrm{T}}_{i_1} \times \cdots \times \Delta^{\mathrm{T}}_{i_v}\right) \boldsymbol{\epsilon}.$$

The first row of Δ_i is $(1/s_i, 1/s_i, \ldots, 1/s_i)$ for $1 \leq i \leq m$. Hence the first column of $\Delta^{\mathrm{T}}_{i_1} \times \cdots \times \Delta^{\mathrm{T}}_{i_v}$ is

$$(s_{i_1} s_{i_2} \cdots s_{i_v})^{-1} \cdot \begin{pmatrix} 1 \\ 1 \\ 1 \\ \vdots \\ 1 \end{pmatrix}.$$

Hence

$$\lambda(\boldsymbol{f}^{i_1} \times \cdots \times \boldsymbol{f}^{i_v}) = (s_{i_1} s_{i_2}, \ldots, s_{i_v})^{-1} \cdot \begin{pmatrix} N \\ N \\ \vdots \\ N \end{pmatrix}.$$

Thus, if T_1 is an $N \times v$ matrix whose v columns are the columns of T corresponding to factors $F_{i_1}, F_{i_2}, \ldots, F_{i_v}$, then every one of the $s_{i_1} s_{i_2} \cdots s_{i_v}$ possible vectors that can occur as rows of T_1, does occur as a row of T_1 with the same frequency $(N/s_{i_1} s_{i_2} \cdots s_{i_v})$. We observe that this number depends only on the orbit to which $(F_{i_1}, F_{i_2}, \ldots, F_{i_v})$ belongs. This establishes the theorem.

Theorem 2.2 (Converse of Theorem 2.1). *Let T be an $N\times m$ array which satisfies the conclusion of Theorem 2.1. Then M_T is a diagonal matrix and M_T restricted to V_{Ω_i} is a scalar matrix for each $i=0,1,2,\ldots,q$.*

Proof. The proof is left to the reader.

2.4. $\mathcal{G}$-balanced designs

Let the group S act naturally on the vector space V_Ω. For $\sigma\in S$, we then have $\sigma(F_{\alpha_1}^{i_1}F_{\alpha_2}^{i_2},\ldots,F_{\alpha_t}^{i_t})=F_{\alpha_1}^{\sigma i_1}F_{\alpha_2}^{\sigma i_2},\ldots,F_{\alpha_t}^{\sigma i_t}$. Hence, S leaves invariant the subspaces $V_{\Omega_0},V_{\Omega_1},\ldots,V_{\Omega_q}$ of V_Ω. Thus S can be regarded as a group of $p\times p$ matrices with a block diagonal form corresponding to the decomposition $V_\Omega=V_{\Omega_0}\oplus V_{\Omega_1}\oplus\cdots\oplus V_{\Omega_q}$. Let $\mathcal{G}$ be any subgroup of S, also regarded as $p\times p$ matrices, when convenient.

Definition 2.4.1. A design $T(N\times m)$ is called *$\mathcal{G}$-balanced* iff M_T is invariant under the action of $\mathcal{G}$, i.e.

$$P^{\mathrm{T}}M_TP=M_T,\quad \text{for all } P\in\mathcal{G}.$$

In the special case when $\mathcal{G}=S$, a $\mathcal{G}$-balanced design, T, is said to be *balanced*.

We remark that the above definition of a balanced design coincides with the usual notion of a balanced design. [see Srivastava (1970)]

Lemma 2.2. *A necessary and sufficient condition for T to be $\mathcal{G}$-balanced is that*

$$M_T(\theta_1,\theta_2)=M_T(\sigma\theta_1,\sigma\theta_2)$$

for every $\sigma\in\mathcal{G}$ and all $\theta_1,\theta_2\in\Omega$.

Proof. Let σ be an element of $\mathcal{G}$. In the matrix representation of $\mathcal{G}$ σ is represented by the matrix P given by,

$$\begin{aligned} P(\omega_1,\omega_2)&=1,\quad \text{if } \sigma(\omega_1)=\omega_2,\quad \omega_1,\omega_2\in\Omega,\\ &=0,\quad \text{otherwise.}\end{aligned}$$

Hence

$$\begin{aligned}(P^{\mathrm{T}}M_TP)(\theta_1,\theta_2)&=\sum_{\omega_1,\omega_2\in\Omega}P^{\mathrm{T}}(\theta_1,\omega_1)M_T(\omega_1,\omega_2)P(\omega_2,\theta_2)\\ &=M_T(\sigma^{-1}\theta_1,\sigma^{-1}\theta_2).\end{aligned}\tag{3.4.1}$$

If T is $\mathcal{G}$-balanced, we must have $P^{\mathrm{T}}M_TP=M_T$ for all $P\in\mathcal{G}$. This implies that (using Definition 2.4.1) $M_T(\theta_1,\theta_2)=M_T(\sigma^{-1}\theta_1,\sigma^{-1}\theta_2)$ or equivalently that,

$$M_T(\sigma\theta_1,\sigma\theta_2)=M_T(\theta_1,\theta_2)\quad \forall\,\sigma\in\mathcal{G},\ \theta_1,\theta_2\in\Omega.$$

The converse is obvious.

Theorem 2.3. *Suppose $T(N\times m)$ is $\mathcal{G}$-balanced. Then the following holds: Let $T_{i_1,i_2,\ldots,i_v}$ be an $N\times v$ matrix whose kth column is the i_kth column of T, $i_1,i_2,\ldots,i_v$ being distinct indices from $\{1,2,\ldots,m\}$. Let $I_l=\{0,1,2,\ldots,s_l-1\}$, $1\leqslant l\leqslant m$. Then for*

$\mathbf{x}\in I_{i_1}\times I_{i_2}\times\cdots\times I_{i_v}$, the frequency with which $\mathbf{x}$ occurs as a row of $T_{i_1,i_2,\ldots,i_v}$ is the same as that for $T_{\sigma i_1,\sigma i_2,\ldots,\sigma i_v}$ for every $\sigma\in\mathcal{G}$. (This frequency will be denoted by $\lambda_{O_i,\mathbf{x}}$ if $(i_1,i_2,\ldots,i_v)$ belongs to the orbit O_i where $O_1,O_2,\ldots,O_d$ are all the $\mathcal{G}$-orbits of $O=\{(j_1,j_2,\ldots,j_v)|1\leqslant j_1,\ldots,j_v\leqslant m;\ j_i$'s are all distinct$\}$.)

Proof. In our earlier notation, $u=\min\{t,m-t\}$ and $v=t+u$. Choose $i_1,i_2,\ldots,i_v$ to be distinct indices from $\{1,2,\ldots,m\}$. As before, let $\boldsymbol{\epsilon}$ be the vector given by

$$\boldsymbol{\epsilon}=\boldsymbol{\epsilon}_{i_1,i_2,\ldots,i_v}=\begin{bmatrix}\epsilon\left(F_0^{i_1}F_0^{i_2}\cdots F_0^{i_t},\ F_0^{i_{t+1}}F_0^{i_{t+2}}\cdots F_0^{i_v}\right)\\ \vdots\\ \epsilon\left(F_{\alpha_1}^{i_1}F_{\alpha_2}^{i_2}\cdots F_{\alpha_t}^{i_t},\ F_{\alpha_{t+1}}^{i_{t+1}}\cdots F_{\alpha_v}^{i_v}\right)\\ \vdots\\ \epsilon\left(F_{s_{i_1}-1}^{i_1}F_{s_{i_2}-1}^{i_2}\cdots F_{s_{i_t}-1}^{i_t},\ F_{s_{i_{t+1}}-1}^{i_{t+1}}\cdots F_{s_{i_v}-1}^{i_v}\right)\end{bmatrix}.$$

Then

$$\begin{aligned}\boldsymbol{\epsilon}&=\lambda(\boldsymbol{\phi}^{i_1}\times\boldsymbol{\phi}^{i_2}\times\cdots\times\boldsymbol{\phi}^{i_v})\\&=\left(D_{i_1}^{\mathrm{T}}\times D_{i_2}^{\mathrm{T}}\times\cdots\times D_{i_v}^{\mathrm{T}}\right)\lambda(\boldsymbol{f}^{i_1}\times\boldsymbol{f}^{i_2}\times\cdots\times\boldsymbol{f}^{i_v}).\end{aligned}$$

Let $\sigma\boldsymbol{\epsilon}=\sigma\boldsymbol{\epsilon}_{i_1,i_2,\ldots,i_v}=\boldsymbol{\epsilon}_{\sigma i_1,\sigma i_2,\ldots,\sigma i_v}$, for $\sigma\in\mathcal{G}$. Since T is $\mathcal{G}$-balanced, M_T is invariant under $\mathcal{G}$ and so $\sigma(\boldsymbol{\epsilon})=\boldsymbol{\epsilon}$. However, $\sigma(\boldsymbol{\epsilon})=(D_{i_1}^{\mathrm{T}}\times\cdots\times D_{i_v}^{\mathrm{T}})\lambda(\boldsymbol{f}^{\sigma i_1}\times\cdots\times\boldsymbol{f}^{\sigma i_v})$. Hence, we conclude that

$$\lambda(\boldsymbol{f}^{\sigma i_1}\times\cdots\times\boldsymbol{f}^{\sigma i_v})=\lambda(\boldsymbol{f}^{i_1}\times\cdots\times\boldsymbol{f}^{i_v}),\quad\forall\sigma\in\mathcal{G}.$$

This establishes the theorem.

Theorem 2.4. *Suppose that T $(N\times m)$ is a design satisfying the conclusion of Theorem 2.3, then T is $\mathcal{G}$-balanced.*

Proof. The proof is left to the reader.

Various authors [e.g., Srivastava (1970), Yamamoto et al. (1975)] have proved these two theorems in particular instances, for the case when $\mathcal{G}=S$, i.e. for balanced designs.

2.5. *Orthogonal and balanced arrays*

We continue using the notation introduced in Sub-sections 2.1 and 2.3. O denotes the set of all v-tuples whose elements are distinct and belong to $\{F_1,F_2,\ldots,F_m\}$. $O_1,O_2,\ldots,O_d$ are the orbits of O under S. We will also denote by O the set of all v-tuples whose elements are distinct and belong to $\{1,2,\ldots,m\}$ when convenient.

Definition 2.5.1. An array T $(N\times m)$ is said to be an orthogonal array with m constraints, N runs, strength v and $(s_1,s_2,\ldots,s_r)$ symbols if the following conditions hold:

(i) Columns $1,2,\ldots,m_1$ use the symbols in I_{s_1}, columns $m_1+1,\ldots,m_1+m_2$ use the symbols in I_{s_r}, columns $m_1+m_2+\cdots+m_{r-1}+1,\ldots,m_1+m_2+\cdots+m_r=m$

use the symbols in I_{s_r}, where $2 \leq s_1 < s_2 < \cdots < s_r$ and $m_1, m_2, \ldots, m_r \geq 1$, and $I_{s_l} = \{0, 1, \ldots, s_l - 1\}$ for $l = 1, 2, \ldots, r$.

(ii) For any $N \times v$ matrix T_1 whose columns occur as the i_1th, i_2th,..., i_vth columns of T respectively, all possible $1 \times v$ vectors which may occur as rows of T_1, do occur as rows of T_1 with the same frequency λ_{O_i} which depends only on the orbit O_i to which $(i_1, i_2, \ldots, i_v)$ belongs. The constants $\lambda_{O_1}, \lambda_{O_2}, \ldots, \lambda_{O_d}$ are called the *parameters* of the orthogonal array T.

An array T satisfying (i) of Definition 2.5.1 is said to be of type $s_1^{m_1} \times s_2^{m_2} \times \cdots \times s_r^{m_r}$.

Definition 2.5.2. An array T $(N \times m)$ is said to be a $\mathcal{G}$-*balanced* array with m constraints, N runs, strength v and $(s_1, s_2, s_3, \ldots, s_r)$ symbols if the following conditions hold:

(i) This condition is the same as condition (i) of Definition 2.5.1.

(ii) Let $T_{i_1, i_2, \ldots, i_v}$ denote a $N \times v$ matrix whose columns are respectively the i_1th, i_2th,..., i_vth columns of T, where $i_1, i_2, \ldots, i_v$ are distinct set of indices from $\{1, 2, \ldots, m\}$. Let $\boldsymbol{x}$ be any element of $I_{s_{i_1}} \times \cdots \times I_{s_{i_v}}$ (Cartesian product of sets). Then the frequency of $\boldsymbol{x}$ as a row of $T_{i_1, i_2, \ldots, i_v}$ is the same as that for $T_{\sigma i_1, \sigma i_2, \ldots, \sigma i_v}$ for every $\sigma \in \mathcal{G}$.

As mentioned in the statement of Theorem 2.3, this frequency will be denoted by $\lambda_{O_i, \boldsymbol{x}}$ if O_i is the orbit to which $(i_1, i_2, \ldots, i_v)$ belongs, where $O_1, O_2, \ldots, O_d$ are all the $\mathcal{G}$-orbits of O.

2.6. Example

Let T be an array of type $2^3 \times 3$ given by

$$T = \begin{bmatrix} 1 & 0 & 1 & 0 & 1 & 1 & 0 & 0 \\ 0 & 1 & 1 & 0 & 1 & 0 & 1 & 0 \\ 1 & 1 & 1 & 1 & 0 & 0 & 0 & 1 \\ 2 & 2 & 1 & 0 & 2 & 1 & 1 & 0 \end{bmatrix}^{\mathrm{T}}.$$

Let $\mathcal{G}$ = the symmetric group on $\{1,2\}$. We then have, $O_1 = \{(1,2), (2,1)\}$, $O_2 = \{(1,3), (2,3)\}$, $O_3 = \{(1,4), (2,4)\}$, $O_4 = \{(3,1), (3,2)\}$, $O_5 = \{(3,4)\}$, $O_6 = \{(4,1), (4,2)\}$, $O_7 = \{(4,3)\}$. Let $\boldsymbol{x}_1 = (0,0)$, $\boldsymbol{x}_2 = (0,1)$, $\boldsymbol{x}_3 = (1,0)$, $\boldsymbol{x}_4 = (1,1)$, $\boldsymbol{y}_1 = (0,0)$, $\boldsymbol{y}_2 = (0,1)$, $\boldsymbol{y}_3 = (0,2)$, $\boldsymbol{y}_4 = (1,0)$, $\boldsymbol{y}_5 = (1,1)$, $\boldsymbol{y}_6 = (1,2)$, $\boldsymbol{z}_1 = (0,0)$, $\boldsymbol{z}_2 = (0,1)$, $\boldsymbol{z}_3 = (1,0)$, $\boldsymbol{z}_4 = (1,1)$, $\boldsymbol{z}_5 = (2,0)$, $\boldsymbol{z}_6 = (2,1)$. We then have

$\lambda_{O_1, \boldsymbol{x}_i} = 2$ for $i = 1,2,3,4$;

$\lambda_{O_2, \boldsymbol{x}_1} = 1$; $\lambda_{O_2, \boldsymbol{x}_2} = 3$; $\lambda_{O_2, \boldsymbol{x}_3} = 2$; $\lambda_{O_2, \boldsymbol{x}_4} = 2$;

$\lambda_{O_3, \boldsymbol{y}_1} = 2$; $\lambda_{O_3, \boldsymbol{y}_2} = 1$, for $i = 2,3$; $\lambda_{O_3, \boldsymbol{y}_4} = 0$, $\lambda_{O_3, \boldsymbol{y}_5} = \lambda_{O_3, \boldsymbol{y}_6} = 2$;

$\lambda_{O_4, \boldsymbol{x}_i} = 1$, for $i = 1$; 2 for $i = 2$; 3 for $i = 3$; 2 for $i = 4$;

$\lambda_{O_6, \boldsymbol{z}_i} = 2$, for $i = 1,4,6$; 0 for $i = 1$; 1 for $i = 3,5$,

$\lambda_{O_5, \boldsymbol{y}_i} = 0$, for $i = 1$; 2 for $i = 2,4,6$; 1 for $i = 3,5$;

$\lambda_{O_7, \boldsymbol{z}_i} = 0$, for $i = 1$; 2 for $i = 2,3,6$; 1 for $i = 4,5$.

Thus, T is $\mathcal{G}$-balanced but clearly not balanced.

With T as above, it is easily seen that M_T has the form

$$\begin{bmatrix} a_0 & a_1 & a_1 & a_2 & a_3 & a_4 \\ a_1 & a_5 & a_6 & a_7 & a_8 & a_9 \\ a_1 & a_6 & a_5 & a_7 & a_8 & a_9 \\ a_2 & a_7 & a_7 & a_{10} & a_{11} & a_{12} \\ a_3 & a_8 & a_8 & a_{11} & a_{13} & a_{14} \\ a_4 & a_9 & a_9 & a_{12} & a_{14} & a_{15} \end{bmatrix}.$$

2.7. *Summary*

We have introduced the concept of $\mathcal{G}$-invariant designs and shown that orthogonal designs and balanced designs occur as special cases. We have also introduced the related idea of $\mathcal{G}$-balanced arrays and these include orthogonal arrays and balanced arrays of general type. The relationship between $\mathcal{G}$-balanced designs and $\mathcal{G}$-balanced arrays were also established. Particular cases of these results can be found in Srivastava (1970) and Yamamoto et al. (1975).

3. The characteristic polynomial of the information matrix for a $\mathcal{G}$-balanced design

We consider a design of the type $s_1^{m_1} \times s_2^{m_2} \times \cdots \times s_r^{m_r}$. We use the notation introduced earlier and we have the decomposition

$$V_\Omega = V_{\Omega_0} + V_{\Omega_1} + \cdots + V_{\Omega_q}.$$

$\mathcal{G}$ is a subgroup of the group S introduced earlier, represented as $p \times p$ matrices where $p = |\Omega|$. Suppose T $(N \times m)$ is a $\mathcal{G}$-balanced design. Then the elements of M_T are expressible in terms of the parameters $\lambda_{O_i, x}$ of the design T. We wish to find the characteristic polynomial of M_T as a function of the design parameters of T. To this effect we note that $PM_T = M_T P$ for all $P \in \mathcal{G}$. Thus M_T belongs to the algebra of all $p \times p$ matrices that commute with all the elements of $\mathcal{G}$. We will use some of the basic theorems of group representation theory and indicate a general procedure for transforming M_T to a simpler form, from which the characteristic polynomial of M_T may be obtained more easily.

3.1. *Results from group representation theory*

We state several results below the proofs of which can be found in any standard book on group representation theory [e.g., Ledermann (1977)].

Definition 3.1.1. Let $\mathcal{G}$ be a group of $n \times n$ matrices over a field K. $\mathcal{G}$ is said to be reducible if there exists a matrix P over K such that $P^{-1}AP$ has the form

$$\left[\begin{array}{c|c} C & \mathbf{0} \\ \hline E & D \end{array}\right],$$

with C of size $n_1 \times n_1$ and D of size $n_2 \times n_2$, $n_1 + n_2 = n$ and $\mathbf{0}$ being the zero matrix of size $n_1 \times n_2$, for every $A \in \mathcal{G}$ $(n_1, n_2 > 0)$.

Definition 3.1.2. Let $\mathcal{G}$ be a group of $n \times n$ matrices over a field K. $\mathcal{G}$ is said to be irreducible if $\mathcal{G}$ is not reducible.

Definition 3.1.3. Let $\mathcal{G} = \{A_1, A_2, \ldots, A_g\}$ be a group of $n \times n$ matrices over K. $\mathcal{G}$ is said to be completely reducible if there exists a nonsingular matrix P over K such that

$$P^{-1}A_iP = \begin{bmatrix} B_{i1} & & & \mathbf{0} \\ & B_{i2} & & \\ & & \ddots & \\ \mathbf{0} & & & B_{il} \end{bmatrix} = \operatorname{diag}(B_{i1}, \ldots, B_{il}),$$

with each B_{ij} of size $n_j \times n_j$, $j = 1, 2, \ldots, l$, $n_1 + n_2 + \cdots + n_l = n$, for every $i = 1, 2, \ldots, g$, and furthermore $\mathcal{G}_j = \{B_{ij} | i = 1, 2, \ldots, g\}$ is an irreducible group of matrices for every $j = 1, 2, \ldots, l$.

Theorem 3.1. *Let $\mathcal{G}$ be a finite group of $n \times n$ matrices over a field K of characteristic zero. Then $\mathcal{G}$ is completely reducible.*

Theorem 3.2. *Let $\mathcal{G} = \{A_1, A_2, \ldots, A_g\}$ be a group of $n \times n$ matrices over an algebraically closed field K. Let*

$$A_i = \operatorname{diag}(\underbrace{B_{i1}, B_{i1}, \ldots, B_{i1}}_{e_1}, \underbrace{B_{i2}, \ldots, B_{i2}}_{e_2}, \ldots, \underbrace{B_{il}, B_{il}, \ldots, B_{il}}_{e_l})$$

for $i = 1, 2, \ldots, g$, where B_{ij} is an $f_j \times f_j$ matrix for $j = 1, 2, \ldots, l$ and $n = e_1 f_1 + e_2 f_2 + \cdots + e_l f_l$. Also let $\mathcal{G}_j = \{B_{ij} | i = 1, 2, \ldots, g\}$ be an irreducible group of matrices for each $j = 1, 2, \ldots, l$. Suppose T is an $n \times n$ matrix over K such that $TA_i = A_iT$ for $i = 1, 2, \ldots, g$. Then T has the form,

$$T = \begin{bmatrix} T_{11} & & & \mathbf{0} \\ & T_{22} & & \\ & & \ddots & \\ \mathbf{0} & & & T_{ll} \end{bmatrix} = \operatorname{diag}(T_{11}, T_{22}, \ldots, T_{ll}),$$

with T_{rr} of size $e_r f_r \times e_r f_r$, $r = 1, 2, \ldots, l$, and furthermore,

$$T_{rr} = M_r \times I_{f_r}, \quad \text{(Kronecker product)},$$

where M_r is an arbitrary matrix of size $e_r \times e_r$ and I_{f_r} is an $f_r \times f_r$ identity matrix, $r = 1, 2, \ldots, l$.

Corollary 3.1. *With the same set-up as in Theorem 3.2, we have*

$$\begin{aligned} P_T(x) &= \text{the characteristic polynomial of } T \\ &= \{p_{M_1}(x)\}^{f_1}\{p_{M_2}(x)\}^{f_2},\ldots,\{p_{M_l}(x)\}^{f_l} \\ &= \prod_{i=1}^{l} \{p_{M_i}(x)\}^{f_i} \end{aligned}$$

where $p_A(x)$ denotes the characteristic polynomial of the square matrix A.

3.2. Applications

It is clear now how to apply the above results in calculating $p_{M_T}(x)$ for any $\mathcal{G}$-balanced design T. By Definition 3.1.3, the group $\mathcal{G}$ of $p \times p$ matrices is completely reducible and a matrix P can be found which effects this complete reduction. This matrix P must also transform M_T to its reduced form by virtue of Theorem 3.2 and by Corollary 3.1, the characteristic polynomial of M_T can be calculated.

Thus, the problem has been reduced to that of finding the matrix P which effects the complete reduction of $\mathcal{G}$. This is a standard problem in group representation theory and has been solved at least in theory. The reader is referred to Robinson (1961) and Littlewood (1958). Especially in the case of balanced designs, one can use the theory of symmetric groups and their representations, as these have been fully worked out. The reader is referred to Miller (1972) for information about symmetric groups.

It should be pointed out the Yamamoto et al. (1965), have approached this problem, with respect to the 2^n-factorial and the 3^n-factorial cases, via association schemes and their algebras. Our approach here is in some sense dual to theirs and makes use of the group theory that has been well established.

We shall now give examples which will illustrate the general technique discussed in the previous section.

3.3. Example

Consider a 2^4-factorial experiment in which the underlying model involves the general mean, main effects and two factor interactions only. The factors may be denoted by F_1, F_2, F_3, F_4. We then have

$$\begin{aligned} \Omega = \{ & F_0^1F_0^2F_0^3F_0^4,\ F_1^1F_0^2F_0^3F_0^4,\ F_0^1F_1^2F_0^3F_0^4,\ F_0^1F_0^2F_1^3F_0^4,\ F_0^1F_0^2F_0^3F_1^4, \\ & F_1^1F_1^2F_0^3F_0^4,\ F_1^1F_0^2F_1^3F_0^4,\ F_1^1F_0^2F_0^3F_1^4,\ F_0^1F_1^2F_1^3F_0^4,\ F_0^1F_1^2F_0^3F_1^4, \\ & F_0^1F_0^2F_1^3F_1^4 \}. \end{aligned}$$

Thus $|\Omega| = 11$. Let $\mathcal{G}$ denote the symmetric group on $\{F_1, F_2, F_3, F_4\}$. A $\mathcal{G}$-invariant design is therefore just a balanced design. We have,

$$\Omega_0 = \{F_0^1F_0^2F_0^3F_0^4\},$$
$$\Omega_1 = \{F_1^1F_0^2F_0^3F_0^4,\ F_0^1F_1^2F_0^3F_0^4,\ F_0^1F_0^2F_1^3F_0^4,\ F_0^1F_0^2F_0^3F_1^4\},$$
$$\Omega_2 = \Omega - (\Omega_0 \cup \Omega_1).$$

Then $\Omega_0, \Omega_1, \Omega_2$ are the orbits of Ω under $\mathcal{G}$. Using Lemma 2.2, we see that if T is $\mathcal{G}$-invariant, then M_T must have the form

$$M_T = \left[\begin{array}{c|cccc|cccccc}
a_0 & a_1 & a_1 & a_1 & a_1 & a_2 & a_2 & a_2 & a_2 & a_2 & a_2 \\
\hline
a_1 & a_3 & a_4 & a_4 & a_4 & a_5 & a_5 & a_5 & a_6 & a_6 & a_6 \\
a_1 & a_4 & a_3 & a_4 & a_4 & a_5 & a_6 & a_6 & a_5 & a_5 & a_6 \\
a_1 & a_4 & a_4 & a_3 & a_4 & a_6 & a_5 & a_6 & a_5 & a_6 & a_5 \\
a_1 & a_4 & a_4 & a_4 & a_3 & a_6 & a_6 & a_5 & a_6 & a_5 & a_5 \\
\hline
a_2 & a_5 & a_5 & a_6 & a_6 & a_7 & a_8 & a_8 & a_8 & a_8 & a_9 \\
a_2 & a_5 & a_6 & a_5 & a_6 & a_8 & a_7 & a_8 & a_8 & a_9 & a_8 \\
a_2 & a_5 & a_6 & a_6 & a_5 & a_8 & a_8 & a_7 & a_9 & a_8 & a_8 \\
a_2 & a_6 & a_5 & a_5 & a_6 & a_8 & a_8 & a_9 & a_7 & a_8 & a_8 \\
a_2 & a_6 & a_5 & a_6 & a_5 & a_8 & a_9 & a_8 & a_8 & a_7 & a_8 \\
a_2 & a_6 & a_6 & a_5 & a_5 & a_9 & a_8 & a_8 & a_8 & a_8 & a_7
\end{array}\right],$$

where the a_i's can all be expressed as functions of the parameters of the design T.

From Theorem 2.3, it follows that the parameters of the design T are λ_0, λ_1, λ_2, λ_3 and λ_4 where λ_i is the number of rows of T of strength i, $i=0,1,2,3,4$. We recall that a vector is of strength i if exactly i of its components are non-zero. Thus a_i is a function of $\lambda_0, \lambda_1, \lambda_2, \lambda_3, \lambda_4$ for $i=0,1,2,\ldots,9$.

Consider the representation of $\mathcal{G}$ on V_{Ω_1}. It can be shown using elementary character theory for S_4 that this representation is reducible and that a matrix P_1 which effects this reduction is given by

$$P_1 = \begin{bmatrix}
\frac{1}{2} & \frac{1}{2} & \frac{1}{2} & \frac{1}{2} \\
\frac{1}{\sqrt{2}} & -\frac{1}{\sqrt{2}} & 0 & 0 \\
\frac{1}{\sqrt{6}} & \frac{1}{\sqrt{6}} & \frac{-2}{\sqrt{6}} & 0 \\
\frac{1}{\sqrt{12}} & \frac{1}{\sqrt{12}} & \frac{1}{\sqrt{12}} & \frac{-3}{\sqrt{12}}
\end{bmatrix}.$$

Similarly, elementary representation theory of S_4 enables one to find a matrix P_2 (6×6) which completely reduces the representation of $\mathcal{G}$ on V_{Ω_2}. In fact, one such P_2

is given by

$$P_2=\operatorname{diag}\left(\frac{1}{\sqrt{6}},\frac{1}{2},\frac{1}{\sqrt{12}},\frac{1}{\sqrt{6}},\frac{1}{2},\frac{1}{\sqrt{12}}\right)\begin{bmatrix} 1 & 1 & 1 & 1 & 1 & 1\\ 0 & 1 & 1 & -1 & -1 & 0\\ 2 & -1 & 1 & -1 & 1 & -2\\ 1 & 1 & -1 & 1 & -1 & -1\\ 0 & -1 & 1 & 1 & -1 & 0\\ -2 & 1 & 1 & 1 & 1 & -2 \end{bmatrix}.$$

Let

$$P=\left[\begin{array}{c|c|c} 1 & 0 & 0\\ \hline 0 & P_1 & 0\\ \hline 0 & 0 & P_2 \end{array}\right].$$

Then we have, PM_TP^{T} is equal to

$$\left[\begin{array}{c|cccc|cccccc}
a_0 & 2a_1 & 0 & 0 & 0 & \sqrt{6}\,a_2 & 0 & 0 & 0 & 0 & 0\\ \hline
2a_1 & a_3+3a_4 & 0 & 0 & 0 & \sqrt{6}\,\theta_4 & 0 & 0 & 0 & 0 & 0\\
0 & 0 & a_3-a_4 & 0 & 0 & 0 & \sqrt{2}\,\theta_5 & 0 & 0 & 0 & 0\\
0 & 0 & 0 & a_3-a_4 & 0 & 0 & 0 & \sqrt{2}\,\theta_5 & 0 & 0 & 0\\
0 & 0 & 0 & 0 & a_3-a_4 & 0 & 0 & 0 & \sqrt{2}\,\theta_5 & 0 & 0\\ \hline
\sqrt{6}\,a_2 & \sqrt{6}\,\theta_4 & 0 & 0 & 0 & \theta_1 & 0 & 0 & 0 & 0 & 0\\
0 & 0 & \sqrt{2}\,\theta_5 & 0 & 0 & 0 & \theta_2 & 0 & 0 & 0 & 0\\
0 & 0 & 0 & \sqrt{2}\,\theta_5 & 0 & 0 & 0 & \theta_2 & 0 & 0 & 0\\
0 & 0 & 0 & 0 & \sqrt{2}\,\theta_5 & 0 & 0 & 0 & \theta_2 & 0 & 0\\
0 & 0 & 0 & 0 & 0 & 0 & 0 & 0 & 0 & \theta_3 & 0\\
0 & 0 & 0 & 0 & 0 & 0 & 0 & 0 & 0 & 0 & \theta_3
\end{array}\right]$$

where

$$\theta_1=a_7+4a_8+a_9,\qquad \theta_2=a_7-a_9,\qquad \theta_3=a_7+a_0-2a_8$$
$$\theta_4=a_5+a_6,\qquad \theta_5=a_5-a_6.$$

Rearranging rows and columns we have that M_T is equivalent to the matrix,

$$M_T^*=\left[\begin{array}{c|c|c|c|c}
\begin{matrix} a_0 & 2a_1 & \sqrt{6}\,a_2\\ 2a_1 & a_3+3a_4 & \sqrt{6}\,\theta_4\\ \sqrt{6}\,a_2 & \sqrt{6}\,\theta_4 & \theta_1 \end{matrix} & \mathbf{0} & \mathbf{0} & \mathbf{0} & \mathbf{0}\\ \hline
\mathbf{0} & \begin{matrix} a_3-a_4 & \sqrt{2}\,\theta_5\\ \sqrt{2}\,\theta_5 & \theta_2 \end{matrix} & \mathbf{0} & \mathbf{0} & \mathbf{0}\\ \hline
\mathbf{0} & \mathbf{0} & \begin{matrix} a_3-a_4 & \sqrt{2}\,\theta_5\\ \sqrt{2}\,\theta_5 & \theta_2 \end{matrix} & \mathbf{0} & \mathbf{0}\\ \hline
\mathbf{0} & \mathbf{0} & \mathbf{0} & \begin{matrix} a_3-a_4 & \sqrt{2}\,\theta_5\\ \sqrt{2}\,\theta_5 & \theta_2 \end{matrix} & \mathbf{0}\\ \hline
\mathbf{0} & \mathbf{0} & \mathbf{0} & \mathbf{0} & \begin{matrix} \theta_3 & \mathbf{0}\\ \mathbf{0} & \theta_3 \end{matrix}
\end{array}\right].$$

Let

$$R_1=\begin{bmatrix} a_0 & 2a_1 & \sqrt{6}\,a_2 \\ 2a_1 & a_3+3a_4 & \sqrt{6}\,\theta_4 \\ 6a_2 & \sqrt{6}\,\theta_4 & \theta_1 \end{bmatrix},$$

$$R_2=\begin{bmatrix} a_3-a_4 & \sqrt{2}\,\theta_5 \\ \sqrt{2}\,\theta_5 & \theta_2 \end{bmatrix}.$$

Then $p_{M_T}(x)=p_{R_1}(x)\{p_{R_2}(x)\}^3(x-\theta_3)^2$. Since R_1 is only of size 3×3, the characteristic polynomial of R_1 is easily calculated and hence $p_{M_T}(x)$ can be explicitly written down.

We remark that Srivastava and Chopra (1971a) have obtained $p_{M_T}(x)$ when T is a balanced design of type 2^m of resolution 5 and Yamamoto et al. (1976) have obtained $p_{M_T}(x)$ when T is a balanced design of type 2^m of resolution $2l+1$ $(l\leqslant\frac{1}{2}m)$ and Kuwada (1977a) when T is a balanced design of type 3^m and of resolution 5, using association schemes and association algebras. In later work, yet to be published, Kuwada (1980), has also obtained $p_{M_T}(x)$ when T is a balanced design of type $2^m\times3^n$ and of resolution 5. The advantage of our approach is that it gives us a general procedure for calculating $p_{M_T}(x)$ for the more general *$\mathcal{G}$-balanced designs* using only the representation theory for the group $\mathcal{G}$ which has been worked out in fair detail in group theoretical literature.

To illustrate this point, we present below an example of $\mathcal{G}$-balanced designs which are not balanced and calculate the corresponding characteristic polynomial.

3.4. Example

Consider a factorial experiment involving 4 factors F_1, F_2, F_3, F_4 each at 2 levels and in which all 2-factor and higher order interactions are zero. Let S be the symmetric group on $\{F_1, F_2, F_3, F_4\}$ and let $\mathcal{G}$ be the subgroup of S of order 4 generated by the permutation $(F_1\ F_2\ F_3\ F_4)$. Suppose we wish to consider $\mathcal{G}$-balanced designs T. Then M_T must have the form:

$$M_T=\left[\begin{array}{c|cccc} a_0 & a_1 & a_1 & a_1 & a_1 \\ \hline a_1 & a_2 & a_3 & a_4 & a_3 \\ a_1 & a_3 & a_2 & a_3 & a_4 \\ a_1 & a_4 & a_3 & a_2 & a_3 \\ a_1 & a_3 & a_4 & a_3 & a_2 \end{array}\right],$$

where a_0, a_1, a_2, a_3, a_4 can be expressed as functions of the parameters of the design T. Observe that a balanced design is a special case of $\mathcal{G}$-balanced design here. If we were seeking an optimal $\mathcal{G}$-balanced design with $\mathcal{G}$ as above, we would be working with a larger class of designs than balanced designs.

We have $\Omega=\{F_0^1F_0^2F_0^3F_0^4,\ F_1^1F_0^2F_0^3F_0^4,\ F_0^1F_1^2F_0^3F_0^4,\ F_0^1F_0^2F_1^3F_0^4,\ F_0^1F_0^2F_0^3F_1^4\}$ and $\Omega_0=\{F_0^1F_0^2F_0^3F_0^4\}$, $\Omega_1=\Omega-\Omega_0$ are the two orbits of Ω under $\mathcal{G}$. $\mathcal{G}$ has a 5-dimensional representation on V_Ω and can be completely reduced by the matrix

$$P=\text{diag}(1,\tfrac{1}{2},\tfrac{1}{2},\tfrac{1}{2},\tfrac{1}{2})\begin{bmatrix} 1 & 0 & 0 & 0 & 0 \\ 0 & 1 & 1 & 1 & 1 \\ 0 & 1 & \mathrm{i} & -1 & -\mathrm{i} \\ 0 & 1 & -1 & 1 & -1 \\ 0 & 1 & -\mathrm{i} & 1 & -\mathrm{i} \end{bmatrix}$$

which is also Hermitian.

$$PM_TP^{-1}=PM_T\bar{P}^t \quad (\bar{P}=\text{complex conjugate of } P)$$

$$=\begin{pmatrix} a_0 & 2a_1 & 0 & 0 & 0 \\ 2a_1 & \theta_1 & 0 & 0 & 0 \\ 0 & 0 & \theta_2 & 0 & 0 \\ 0 & 0 & 0 & \theta_3 & 0 \\ 0 & 0 & 0 & 0 & \theta_2 \end{pmatrix}.$$

So $P_{M_T}(x)=\{(a_0-x)(\theta_1-x)-4a_1^2\}(\theta_2-x)^2(\theta_3-x)$. Thus $P_{M_T}(x)$ has been obtained explicitly and can be used in a search for an optimal $\mathcal{G}$-balanced design of resolution 3 and type 2^4.

4. Final Remarks

In a sequel to this paper, we consider balanced designs of the $2^m \times 3^n$ series of resolution III and obtain E-optimal designs for various values of m and n for values of N, the number of runs, within a practical range. We also consider resolution V designs of the $2^m \times 3^n$ series and indicate the method of obtaining E-optimal designs by considering specific examples.

References

Addelman, S. (1962). Orthogonal main-effect plans for asymmetrical factorial experiments. *Technometrics* **4**, 21–46.

Anderson, D.A. and Srivastava, J.N. (1972). Resolution IV designs of $2^m \times 3$ series. *J. R. Statist. Soc. (B)* **34**, 377–384.

Ariyarathna, M.W. (1979). Optimal balanced resolution V designs of the 3^m series. *Ph.D. dissertation*, Colorado State Univ., CO.

Bellman, R. (1970). *Introduction to Matrix Analysis*. McGraw-Hill Book Company, New York.

Bose, R.C. (1947). Mathematical theory of the symmetrical factorial design. *Sankhya* **8**, 107–166.

Bose, R.C., Shrikhande, S.S. and Parker, E.T. (1960). Further results on the construction of mutually orthogonal Latin Squares and the falsity of Euler's Conjecture. *Canad. J. Math.* **12**, 189–203.

Bose, R.C. and Srivastava, J.N. (1964a). Analysis of irregular factorial design. *Sankhya A*, **26**, 117–144.

Bose, R.C. and Srivastava, J.N. (1964b). Multidimensional partially balanced designs and their analysis, with applications to partially balanced factorial fractions. *Sankhya A*, **26**, 145–168.

Chakravarti, I.M. (1956). Fractional replication in asymmetrical factorial designs and partially balanced arrays. *Sankhya* 17, 143–164.

Chakravarti, I.M. (1961). On some methods of constructions of partially balanced arrays. *Ann. Math. Statist.* **32**, 1181–1185.

Hoke, A.T. (1975). The characteristic polynomial of the information matrix for second-order models. *Ann. Statist.* **3**, 780–786.
Kishen, K. and Srivastava, J.N. (1960). Mathematical theory of confounding in symmetrical and asymmetrical factorial designs. *J. Ind. Soc. Agr. Statist.* **11**, 73–110.
Kuwada, M. (1977b). Balanced arrays of strength 4 and balanced fractional 3^m factorial designs. (To appear.)
Ledermann, W. (1977). *Introduction to Group Characters.* Cambridge Univ. Press, Cambridge.
Littlewood, D. (1958). *Theory of Group Characters.* 2nd ed., Oxford Univ. Press (Clarendon), London and New York.
Margolin, B.H. (1967). Systematic methods for analyzing 2^n3^m factorial experiments with applications. *Technometrics* **9**, No. 2, 245–259.
Margolin, B.H. (1968). Orthogonal main-effect 2^n3^m designs and two factor interaction aliasing. *Technometrics* **10**, No. 3, 559–573.
Margolin, B.H. (1969a). Results on factorial designs of resolution IV for the 2^n and 2^n3^m series. *Technometrics* **11**, No. 3, 431–444.
Margolin, B.H. (1969b). Orthogonal main-effect plans permitting estimation of all two-factor interactions for the 2^n3^m factorial series of designs. *Technometrics* **11**, No. 4, 747–762.
Miller, W., Jr. (1972). *Symmetry Groups and Their Applications.* Academic Press, New York.
Morrison, M. (1956). Fractional replication for mixed series. *Biometrics* **12**, 1–19.
Rao, C.R. (1946). Hypercubes of strength d leading to confounded designs in factorial experiments, *Bull. Calcutta Math. Soc.* **38**, 67–68.
Rao, C.R. (1973). Some combinatorial problems of arrays and applications to design of experiments. In *A Survey of Combinatorial Theory*, Srivastava, J.N. Ed., pp. 349–359. North-Holland Publishing Co., Amsterdam.
Rafter, J.A. and Seiden, E. (1974). Contributions to the theory and construction of balanced arrays. *Ann. Statist.* **2**, 1256–1273.
Ray-Chaudhuri, D.K. and Wilson, R.M. (1971). A Solution of Kirkman's school girl problem, Proc. Symp. in pure mathematics, *Combinatorics Am. Math. Soc.* **19**, 187–204.
Robinson, G. de B. (1961). *Representation Theory of the Symmetric Group.* Univ. of Toronto Press, Toronto.
Shirakura, T. (1975). On balanced arrays of 2 symbols, strength $2l$, m constraints and index set $\{\mu_0, \mu_1, \ldots, \mu_{2l}\}$ with $\mu_l = 0$. *J. Jpn. Statist. Soc.* **5**, No. 2, 53–56.
Shirakura, T. (1976). Optimal balanced fractional 2^m factorial designs of resolution VII, $6 \leqslant m \leqslant 8$. *Ann. Statist.* **4**, No. 3, 515–531.
Shirakura, T. (1976). Balanced fractional 2^m factorial designs of even resolution obtained from balanced arrays of strength $2l$ with index $\mu_l = 0$. *Ann. Statist.* **4**, No. 4, 723–735.
Shirakura, T. (1976). A note of the norm of alias matrices in fractional replication. *Austral. J. Statist.* **18**, No. 3, 158–160.
Shirakura, T. and Kuwada, M. (1975). Note on balanced fractional 2^m factorial designs of resolution $2l+1$. *Ann. Inst. Statist. Math.* **27**, 377–386.
Srivastava, J.N. (1961). *Contributions to the Construction and Analysis of Designs.* Univ. of North Carolina, Institute of Statistics, NC. mimeo. Series No. 301.
Srivastava, J.N. (1970). Optimal balanced 2^m fractional factorial designs. In *S.N. Roy Memorial Volume.* pp. 689–706. Univ. of North Carolina and Indian Statistical Institute, NC.
Srivastava, J.N. (1972). Some general existence conditions for balanced arrays of strength t and 2 symbols. *J. Comb. Theory* **12**, 198–206.
Srivastava, J.N. and Anderson, D.A. (1969). Fractional factorial designs for estimating main effects orthogonal to two factor interactions, 3^n and $2^m \times 3^n$ series. ARL Tech. Rep., Aerospace Res. Labs., Wright Patterson Air Force Base, No. 690123iii.
Srivastava, J.N. and Anderson, D.A. (1970). Some basic properties of multidimensioal partially balanced designs. *Ann. Math. Statist.* **40**, 1438–1445.
Srivastava, J.N. and Anderson, D.A. (1971). Factorial association schemes with applications to the construction of multidimensional partially balanced designs. *Ann. Math. Statist.* **42**, 1167–1181.
Srivastava, J.N. and Anderson, D.A. (1972). Resolution IV designs of $2^m \times 3$ series. *J.R. Statist. Soc.* (*B*) **34**, 377–384.

Srivastava, J.N. and Chopra, D.V. (1971a). On the characteristic roots of the information matrix of 2^m balanced factorial designs of resolution V with applications. *Ann. Math. Statist.* **42**, 722–734.

Srivastava, J.N. and Chopra, D.V. (1973b). Balanced arrays and orthogonal arrays. In *A Survey of Combinatorial Theory*, J.N. Srivastava, Ed. pp. 411–428. North-Holland Publishing Co., Amsterdam.

Yamamoto, S. (1964). Some aspects for the composition of relationship algebras of experimental designs. *J. Sci. Hiroshima Univ. Ser. A-I* **28**, No. 2, 167–197.

Yamamoto, S., Fujii, Y. and Hamada, N. (1965). Composition of some series of association algebras. *J. Sci. Hiroshima Univ., Ser. A-I* **29**, 181–215.

Yamamoto, S., Shirakura, T. and Kuwada, M. (1975). Balanced arrays of strength $2l$ and balanced fractional 2^m factorial designs. *Ann. Inst. Statist. Math.* **27**, 143–157.

Yamamoto, S. and Shirakura, T. (1976). Characteristic polynomials of the information matrices of balanced fractional 2^m factorial designs of higher $(2l+1)$ resolution. In *Essays in Prob. and Statistics*, S. Ikeda et al. Ed., pp. 73–94.

Yamamoto, S. and Tamari, F. (1976). An existence condition for multidimensional partially balanced association schemes. In *Essays in Prob. and Statistics, Birthday volume in honor of Prof. J. Ogawa*, S. Ikeda et al., Ed. pp. 65–71.

G. Kallianpur, P.R. Krishnaiah, J.K. Ghosh, eds., *Statistics and Probability: Essays in Honor of C.R. Rao*

A PROOF OF THE CENTRAL LIMIT THEOREM MOTIVATED BY THE CRAMÉR–RAO INEQUALITY

Lawrence D. BROWN*
Cornell University, Ithaca, NY, U.S.A.

This paper contains a proof of the classical central limit theorem. This proof is based on the Fisher information of the sequence of normalized sums. It is motivated by the Cramér–Rao inequality and involves a simple eigenfunction property of Hermite polynomials.

1. The central limit theorem

Let $X_1, X_2, \ldots$ be independent identically distributed random variables possessing a mean, μ, and a variance, σ^2. Define the normalized sums

$$S_n = \left(\sum_{i=1}^{n} X_i - n\mu \right) \Big/ \sigma\sqrt{n}. \tag{1.1}$$

The central limit theorem (for identically distributed variables) states that S_n tends to a standard normal variable in distribution; in formal terms, that $F_{S_n}(s) \to \Phi(s)$ as $n \to \infty$ for each $s \in R$, where $F_{S_n}(s) = \Pr\{S_n \leqslant s\}$ and

$$\Phi(s) = \int_{-\infty}^{s} \varphi(t)\,\mathrm{d}t = \int_{-\infty}^{s} (2\pi)^{-1/2} \mathrm{e}^{-t^2/2}\,\mathrm{d}t.$$

In this paper we present what we believe is a new proof of this theorem. Our proof is based on the concept of Fisher information. It is motivated by a form of the Cramér–Rao inequality; although that inequality plays no role in the formal proof.

2. Information

Let Z be a random variable with probability density f. Suppose f is continuously differentiable with a bounded derivative. The information of Z is defined as

$$I(Z) = E\left(\left(f'(Z)/f(Z) \right)^2 \right). \tag{2.1}$$

This is, of course, the Fisher information at $\theta = 0$ of the location family of densities $f_\theta(z) = f(z - \theta)$. Note that

$$0 < I(Z) \leqslant \infty, \tag{2.2}$$

and

$$I(cZ) = I(Z)/c^2. \tag{2.3}$$

*Work supported in part by N.S.F. research grant MCS78-24175.

One form of the Cramér–Rao inequality states that $I(Z)\geqslant(\operatorname{Var} Z)^{-1}$ with equality if and only if Z is a normal variable. [This result is true with no regularity conditions other than the differentiability of f assumed above. See, e.g. Pitman (1979, p. 37).] In the setting of Section 1 suppose $I(X_i)<\infty$ but X_i is not a normal variable. Then it will be seen that $I(S_n)$ is a strictly decreasing sequence. If it could be shown that $\lim_{n\to\infty} I(S_n)=1\equiv\operatorname{Var} S_n$ it would follow that S_n tends in distribution (and in density) to a standard normal variable. This is the motivation for the proof which follows; although the proof proceeds somewhat differently.

Let X, Y be independent variables. Let $Z=X+Y$. It can be shown via the Cauchy–Schwarz inequality that

$$I(Z)\leqslant\min(I(X), I(Y)). \tag{2.4}$$

This result will be used later. (A detailed proof is given in Lemma 5.5.)

Now, suppose Z_1, Z_2 are independent random variables each with finite information. Let $S=(Z_1+Z_2)/\sqrt{2}$ and let $\rho_{Z_i}=f'_{Z_i}/f_{Z_i}$ and $\rho_S=f'_S/f_S$. Then $f_S(s)=2\int(f_{Z_1}((s+t)/\sqrt{2})f_{Z_2}((s-t)/\sqrt{2}))\,\mathrm{d}t$. Further calculation yields

$$\begin{aligned}\rho_S(s)&=(1/\sqrt{2})\int(\rho_{Z_1}((s+t)/\sqrt{2})+\rho_{Z_2}((s-t)/\sqrt{2}))f_{T|S}(t|s)\,\mathrm{d}t\\&=(1/\sqrt{2})E\big(\rho_{Z_1}(Z_1)+\rho_{Z_2}(Z_2)|Z_1+Z_2=\sqrt{2}\,s\big),\end{aligned} \tag{2.5}$$

where $T=(Z_1-Z_2)/\sqrt{2}$ and $f_{T|S}$ is the indicated conditional density. It follows that

$$\begin{aligned}(I(Z_1)+I(Z_2))/2-I(S)&=\tfrac{1}{2}(E(\rho^2_{Z_1})+E(\rho^2_{Z_2})-E(E^2(\rho_{Z_1}+\rho_{Z_2}|Z_1+Z_2=\sqrt{2}\,s))\\&=\tfrac{1}{2}\iint(\rho_{Z_1}(z_1)+\rho_{Z_2}(z_2)-\sqrt{2}\,\rho_S((Z_1+Z_2)/\sqrt{2}))^2\\&\qquad\times f_{Z_1}(z_1)f_{Z_2}(z_2)\,\mathrm{d}z_1\,\mathrm{d}z_2.\end{aligned} \tag{2.6}$$

In particular, $\tfrac{1}{2}(I(Z_1)+I(Z_2))-I(S)\geqslant 0$. (It can also be shown that equality holds only if Z_1 and Z_2 are normal variables.)

3. A basic lemma involving Hermite polynomials

The expression (2.6) motivates the study of a lower bound for integrals of the form $\int(v(z_1)+v(z_2)-w((z_1+z_2)/\sqrt{2}))^2 f_1(z_1)f_2(z_2)\,\mathrm{d}z_1\,\mathrm{d}z_2$. The following lemma provides such a bound when $f_1=f_2$ is a normal density and v and w are arbitrary functions.

Lemma 3.1. *Let $\varphi_{\beta^2}(z)=\beta^{-1}\varphi(z/\beta)$. Suppose $\int v^2(z)\varphi_{\beta^2}(z)\,\mathrm{d}z<\infty$. Then*

$$\begin{aligned}\Delta(v,w)&=\iint(v(z_1)+v(z_2)-w((z_1+z_2)/\sqrt{2}))^2\\&\qquad\times\varphi_{\beta^2}(z_1)\varphi_{\beta^2}(z_2)\,\mathrm{d}z_1\,\mathrm{d}z_2\\&\geqslant\inf\left\{\int(v(z)-(az+b))^2\varphi_{\beta^2}(z)\,\mathrm{d}z\;:(a,b)\in R^2\right\}\end{aligned} \tag{3.1}$$

Proof. Set $\beta^2=1$ for convenience – there is no loss of generality. Define the Hermite polynomials, H_k, by

$$(-1)^k H_k(z)\varphi(z)=\frac{d^k}{dz^k}\varphi(z), \quad k=0,1\ldots. \tag{3.2}$$

Recall that these polynomials are characterized by being the unique polynomials of degree k with positive leading coefficient satisfying

$$\int H_k(z)H_l(z)\varphi(z)\,dz=\begin{matrix}0, & k\neq l\\ k!, & k=l\end{matrix}. \tag{3.3}$$

(See, e.g. Feller (1966, p. 524). The relation (3.3) is provable from (3.2) by repeated integration by parts.)

For given v the expression $\Delta(v,w)$ is minimized by the choice

$$\begin{aligned} w_v(s)&=\int (v((s+t)/\sqrt{2})+v((s-t)/\sqrt{2}))\varphi(t)\,dt\\ &=2\int v((s+t)/\sqrt{2})\varphi(t)\,dt \end{aligned} \tag{3.4}$$

since $S=(Z_1+Z_2)/\sqrt{2}$ and $T=(Z_1-Z_2)/\sqrt{2}$ are independent standard normal variables. Furthermore

$$\begin{aligned} \Delta_{\min}(v)&=\Delta(v,w_v)\\ &=2\int v^2(z)\varphi(z)\,dz+2\left(\int v(z)\varphi(z)\,dz\right)^2\\ &\quad-\int w_v^2(s)\varphi(s)\,ds. \end{aligned} \tag{3.5}$$

Now, suppose $v=H_k$. Then w_v is a polynomial of degree k with positive leading coefficient and

$$\begin{aligned} &\int H_m(z)w_v(z)\varphi(z)\,dz=\\ &=2\iint H_m(z)H_k((z+t)/\sqrt{2})\varphi(t)\varphi(z)\,dt\,dz\\ &=2^{1-m/2}\iint H_k^{(m)}((z+t)/\sqrt{2})\varphi(z)\varphi(t)\,dz\,dt\\ &=2^{1-m/2}\int H_k^{(m)}(w)\varphi(w)\,dw\\ &=\begin{matrix}0 & m\neq k\\ 2^{1-k/2}k!, & m=k,\end{matrix} \end{aligned} \tag{3.6}$$

where we have used integration by parts (m times) and then the change of variables $w=(z+t)/\sqrt{2}$. Consequently, if $v=H_k$ then $w_v=2^{1-k/2}H_k$.

In general $v(z)=\sum_{k=0}^{\infty}\alpha_k H_k(z)$ with $\int v^2(z)\varphi(z)\mathrm{d}z=\sum_{k=0}^{\infty}\alpha_k^2 k!$ by basic Hilbert space theory. From (3.6) $w_v(z)=\sum_{k=0}^{\infty}2^{1-k/2}\alpha_k H_k(z)$. Consequently, from (3.5),

$$\begin{aligned}\Delta_{\min}(v) &= \sum_{k=2}^{\infty}(2-2^{1-k/2})\alpha_k^2 k!\\ &\geqslant \sum_{k=2}^{\infty}\alpha_k^2 k!\\ &= \int\left(v(t)-\alpha_0-\alpha_1 t\right)^2\varphi(t)\,\mathrm{d}t.\end{aligned}\tag{3.7}$$

This is equivalent to (3.1), the desired result.

The preceding proof can be summarized by saying that the map $v\to w_v$ of $L_2\to L_2$ has eigenfunctions $\{H_k\}$ with corresponding eigenvalues $2^{1-k/2}$. The lemma then follows from the simple fact (3.5).

4. The proof

Let $X_1, X_2,\ldots$ be as in Section 1. Without loss of generality assume $\mu=0$ and $\sigma^2=1$. Let $Y_1^{(\tau)}, Y_2^{(\tau)},\ldots$ be independent identically distributed normal variables with mean 0 and variance τ^2, and let

$$Z_i^{(\tau)}=X_i+Y_i^{(\tau)},\qquad S_n^{(\tau)}=\sum_{i=1}^{n}Z_i^{(\tau)}/\sqrt{n},$$

and

$$Q_k^{(\tau)}=S_{2^k}^{(\tau)}.$$

Let

$$Q_k=S_{2^k}=\sum_{i=1}^{2^k}X_i/2^{k/2}.$$

Then, $I(Q_k^{(\tau)})<\infty$ by (2.2). Also,

$$f_{Q_k^{(\tau)}}(q)\geqslant\zeta_\tau\varphi_{\tau^2/2}(q).\tag{4.1}$$

(See Lemma 5.1 for a formal proof involving only Chebyshev's inequality and the fact that $\operatorname{Var}S_n\equiv 1$.)

From (2.6), (3.1) and (4.1)

$$\begin{aligned}&I(Q_k^{(\tau)})-I(Q_{k+1}^{(\tau)})=\\ &=\iint\left(\rho_{Q_k^{(\tau)}}(q_1)+\rho_{Q_k^{(\tau)}}(q_2)-\sqrt{2}\,\rho_{Q_{k+1}^{(\tau)}}((q_1+q_2)/\sqrt{2})\right)^2\\ &\qquad\times f_{Q_k^{(\tau)}}(q_1)f_{Q_k^{(\tau)}}(q_2)\,\mathrm{d}q_1\,\mathrm{d}q_2\\ &>\zeta_\tau^2\inf\left\{\int\left(\rho_{Q_k^{(\tau)}}(q)-aq-b\right)^2\varphi_{\tau^2/2}(q)\,\mathrm{d}q;\ (a,b)\in R^2\right\}.\end{aligned}\tag{4.2}$$

Hence $I(Q_k^{(\tau)})$ is a decreasing sequence; and

$$\lim_{k\to\infty}\int\left(\rho_{Q_k^{(\tau)}}(q)-(q-b_k)/a_k^2\right)\varphi_{\tau^2/2}(q)\,\mathrm{d}q=0 \tag{4.3}$$

for suitably chosen (a_k, b_k) (depending also on τ) since $I(Q_k^{(\tau)})>0$. (Actually, $I(Q_k^{(\tau)})\geqslant 1+\tau^2$ as noted in Section 2.)

The expression (4.3) yields

$$\lim_{k\to\infty}\left(f_{Q_k^{(\tau)}}(q)-c_k \mathrm{e}^{-(q-b_k)^2/2a_k^2}\right)=0 \tag{4.4}$$

uniformly on compact subsets for suitable $\{c_k\}$, since $\rho_{Q_k^{(\tau)}}(q)=(\mathrm{d}/\mathrm{d}q)\ln f_{Q_k^{(\tau)}}(q)$. Now,

$$\lim_{B\to\infty}\int_{-B}^{B} q^m f_{Q_k^{(\tau)}}(q)\,\mathrm{d}q=$$

$$=\lim_{B\to\infty}\int_{-B}^{B} q^m \varphi_{1+\tau^2}(q)\,\mathrm{d}q \quad \text{for } m=0,1,2, \tag{4.5}$$

uniformly in k. (See Lemma 5.2, whose proof involves only Chebyshev's inequality and a truncation argument for the case $m=2$). Hence, $c_k\to(2\pi(1+\tau^2))^{-1/2}$, $b_k\to 0$, $a_k^2\to 1+\tau^2$ and

$$\lim_{k\to\infty} f_{Q_k^{(\tau)}}(q)=\varphi_{1+\tau^2}(q) \tag{4.6}$$

uniformly on compact subsets of R.

The expression (4.6) shows that $Q_k^{(\tau)}$ tends to a normal $(0,1+\tau^2)$ variable in distribution (and in density). Letting $\tau\to 0$ yields the result that $Q_k=S_{2^k}$ tends in distribution to a standard normal variable since for all $\epsilon>0$

$$\Phi(\epsilon/\tau)F_{Q_k}(q-\epsilon)\leqslant F_{Q_k^{(\tau)}}(q)$$

$$\leqslant\Phi(\epsilon/\tau))F_{Q_k}(q+\epsilon)+\Phi(-\epsilon/\tau). \tag{4.7}$$

Finally, the fact that the binary sequence $S_1, S_2, S_4, S_8,\ldots$ tends to the standard normal distribution implies that the full sequence $S_1, S_2, S_3, S_4, S_5,\ldots$ also does. This fact is formally proved in Lemma 5.4. This completes the proof of the central limit theorem.

5. Technical lemmas

This section contains statements and proofs of the three elementary technical lemmas quoted in Section 4.

Lemma 5.1. *Let* $Z=Z_i^{(\tau)}$ *be defined as in Section* 4. *Then there is a constant* $\zeta_\tau=3(32)^{-1/2}\mathrm{e}^{-4/\tau^2}$ *such that*

$$f_Z(z)\geqslant\zeta_\tau\varphi_{\tau^2/2}(z). \tag{5.1}$$

The inequality (5.1) *also applies to* $Q_k^{(\tau)}$ *since* $Q_k^{(\tau)}=Q_k+Y^{(\tau)}$.

Proof. $\Pr(|X|<2)\geqslant\frac{3}{4}$ by Chebyshev's inequality. Hence

$$f_Z(z)\geqslant\int_{-2}^{2}\varphi_{\tau^2}(z-x)\,\mathrm{d}F_X(x)$$
$$\geqslant\tfrac{3}{4}\min\{\varphi_\tau(z-x):|x|<2\}$$
$$=\tfrac{3}{4}\varphi_\tau(|z|+2)\geqslant\zeta_\tau\varphi_{\tau^2/2}(z).$$

Lemma 5.2. *Suppose* $X_1, X_2,\ldots$ *are independent identically distributed random variables with mean* 0 *and variance* σ^2. *Let* $S_n=\Sigma_{i=1}^n X_i/\sqrt{n}$. *Then*

$$\lim_{B\to\infty}\int_{-B}^{B}s^m f_{S_n}(s)\,\mathrm{d}s=$$
$$=\int s^m\varphi_{\sigma^2}(s)\,\mathrm{d}s \quad \textit{for } m=0,1,2 \tag{5.2}$$

uniformly in n.

Proof. Assume $\sigma^2=1$ without loss of generality. For each $c>0$ define $C(c)$ by

$$\int_{C(c)}^{\infty}xf_X(x)\,\mathrm{d}x=\int_{-\infty}^{-c}|x|f_X(x)\,\mathrm{d}x.$$

Let

$$V_c(X)=\begin{cases}X, & -c\leqslant X\leqslant C(c),\\ 0, & \text{otherwise},\end{cases}$$

and $W_c(X)=X-V_c(X)$. Then $E(V_c)=0$, $\lim_{c\to\infty}\operatorname{Var}(V_c)=\sigma^2$ and $\lim_{c\to\infty}\operatorname{Var}(W_c)=0$. Also $M_{c4}=E(V_c^4)\leqslant\max(c^4, C^4(c))<\infty$. Now define $V_{ci}=V_c(X_i)$, $\tilde{V}_{cn}=\Sigma_{i=1}^n V_{ci}/\sqrt{n}$ and similarly for $W_{ci}, \tilde{W}_{cn}$.

Then

$$E(\tilde{V}_{cn}^4)=\frac{M_c^4}{n}+3n(n-1)\sigma^4/n^2$$
$$\leqslant M_{c4}+3\sigma^4,$$

and $\operatorname{Var}\tilde{W}_{cn}\equiv\operatorname{Var}W_c\to 0$ as $c\to\infty$. Hence

$$E(S_n^2,|S_n|>B)=$$
$$=E((\tilde{V}_{cn}+\tilde{W}_{cn})^2,|S_n|>B)$$
$$\leqslant 2\,E(\tilde{V}_{cn}^2+\tilde{W}_{cn}^2,|S_n|>B)$$
$$\leqslant 2\{E(\tilde{V}_{cn}^4)^{1/2}(\Pr(|S_n|>B))^{1/2}+E(\tilde{W}_{cn}^2)\}$$
$$\leqslant 2\{(M_{c4}+3\sigma^4)^{1/2}\sigma/B+\operatorname{Var}(W_c)\}. \tag{5.3}$$

The right side of (5.3) can be made arbitrarily small by first choosing c large so that $\operatorname{Var}(W_c^2)$ is small and then choosing B large so that $2(M_{c4}+3\sigma^4)^{1/2}\sigma/\mathrm{B}$ is small.

The desired result, (5.2), for $m=2$ then follows since

$$E(S_n^2, |S|<B)=\sigma^2-E(S_n^2, |S|>B).$$

Equation (5.2) for $m=0$, 1 is easier to establish since no truncation argument is needed; we omit the details.

The next lemma is needed in the proof of Lemma 5.4.

Lemma 5.3. *Let $R_1, R_2, \ldots$ be independent random variables each with variance* 1 *and with $F_{R_i}(r)\to\Phi(r)$ for all $r\in R$. Then there exist variables Δ_i such that*

$$R_i=N_i+\Delta_i$$

where N_i are independent standard normal random variables and Δ_i satisfy $\operatorname{Var}\Delta_i\to 0$. [*Note that the pair (N_i, Δ_i) may be dependent.*]

Proof. Let U_i be independent uniform variables on $(0,1)$ and write $R_i=F_{R_i}^{-1}(U)$, $N_i=\Phi^{-1}(U_i)$, $\Delta_i=\Delta(U_i)=R_i-N_i$. Then

$$\operatorname{Var}\Delta_i\leqslant 2(\operatorname{Var}R_i+\operatorname{Var}N_i)=4, \tag{5.4}$$

and $\Delta_i\to 0$ in probability. Consequently, $\operatorname{cov}(\Delta_i, N_i)\to 0$, as $i\to\infty$, by the Cauchy–Schwartz inequality. Thus

$$\begin{aligned}\operatorname{Var}\Delta_i&=\operatorname{Var}R_i-\operatorname{Var}N_i-2\operatorname{cov}(\Delta_i, N_i)\\&=-2\operatorname{cov}(\Delta_i, N_i)\to 0, \quad \text{as } i\to\infty.\end{aligned}$$

Lemma 5.4. *Let S_n and $Q_k=S_{2^k}$ be as in Section* 1. *Suppose $F_{Q_k}(q)\to\phi(q)$. Then $F_{S_n}(s)\to\phi(s)$.*

Proof. Let n have binary expansion $n=\sum_{k=0}^{[\log_2 n]} i_k 2^k$. Then

$$S_n=\sum_{k=0}^{[\log_2 n]} i_k(2^k/n)^{1/2}\hat{Q}_k,$$

where the $\hat{Q}_k$ are independent and have the distribution of Q_k, respectively. Then

$$S_n=N+\sum_{k=0}^{K} i_k(2^k/n)^{1/2}\Delta_k+\sum_{K+1}^{[\log_2 n]} i_k(2^k/n)^{1/2}\Delta_k \tag{5.5}$$

with Δ_k as in Lemma 5.3, and $N=\sum_{k=0}^{[\log_2 n]} i_k(2^k/n)^{1/2}N_k$, a standard normal variable. Now, $\operatorname{Var}(\sum_{k=0}^{K} i_k(2^k/n)^{1/2}\Delta_k)\leqslant 4\sum_{k=0}^{K} i_k(2^k/n)\to 0$ as $n\to\infty$ for fixed K by (5.4). And, $\operatorname{Var}(\sum_{K+1}^{[\log_2 n]} i_k(2^k/n)^{1/2}\Delta_k)\leqslant\sup\{\operatorname{Var}\Delta_k\colon K\geqslant K+1\}\to 0$ as $K\to\infty$ by Lemma 5.3 (uniformly in n). The lemma therefore follows upon letting $n\to\infty$ then $K\to\infty$ in the preceding.

We conclude by verifying eq. (2.4).

Lemma 5.5. *If X, Y are independent random variables with continuously differentiable densities and $Z = X + Y$ then*

$$I(Z) \leqslant \min(I(X), I(Y)).$$

Proof. Assume $I(X) \leqslant I(Y)$. Then $f_Z(z) = \int f_X(z-t) f_Y(t)\,\mathrm{d}t$ and, using the Cauchy – Schwartz inequality

$$I(Z) = \int \frac{(f_Z'(z))^2}{f_Z(z)}\,\mathrm{d}z = \int \left\{ \frac{\left(\int (f_X'(z-t)/f_X(z-t)) f_X(z-t) f_Y(t)\,\mathrm{d}t \right)^2}{f_Z(z)} \right\} \mathrm{d}z$$

$$\leqslant \int \left\{ \int \left(\frac{f_X'(z-t)}{f_X(z-t)} \right)^2 f_X(z-t) f_y(t)\,\mathrm{d}t \right\} \mathrm{d}z$$

$$= \iint \left(\frac{f_X'(x)}{f_X(x)} \right)^2 f_X(x) f_Y(t)\,\mathrm{d}t\,\mathrm{d}x = I(X).$$

This completes the proof.

Acknowledgement

I would like to thank A. Zaman, who suggested a fresh look at the central limit theorem.

References

Feller, W. (1966). *An Introduction to Probability Theory and Its Applications*, Vol. II, Wiley, New York.
Pitman, E.J.G. (1979). *Some Basic Theory for Statistical Inference.* Chapman and Hall Ltd., London.

G. Kallianpur, P.R. Krishnaiah, J.K. Ghosh, eds., *Statistics and Probability: Essays in Honor of C.R. Rao*

MINIMAX ESTIMATORS IN LINEAR MODELS UNDER NORMALIZED QUADRATIC LOSS FUNCTIONS

O. BUNKE and M. MÖHNER
Humboldt University Berlin,
Berlin, German Democratic Republic

1. Introduction

This paper deals with the estimation of l-dimensional linear parameters $\gamma=\gamma(\vartheta)=CX\beta$ in a linear model for the observable vector y:

$$y=X\beta+\mathcal{E}. \tag{1}$$

X is a known $n\times k$ matrix of rank k. The vector $\vartheta=(\beta,\sigma)$ is an unknown parameter. $\mathcal{E}$ is a random vector with unknown p.d., which has expectation $E\mathcal{E}=0$ and covariance matrix $D\mathcal{E}=\sigma^2\Lambda$ with a known positive definite matrix Λ. The use of quadratic loss functions

$$L(\vartheta,\bar{\gamma})=\sigma^{-2}\|\bar{\gamma}-\gamma(\vartheta)\|^2, \qquad \|\gamma\|^2:=\gamma'H\gamma, \tag{2}$$

with some positive definite "weighting" matrix H is standard in the linear theory, as comprehensively presented in Rao (1973, 1976) and Humak (1977). Here the minimax property of the BLUE

$$\hat{\gamma}=CX\hat{\beta}, \qquad \hat{\beta}=(X'\Lambda^{-1}X)^{-1}X'\Lambda^{-1}y$$

has been proved (Bunke, 1975). Some minimax results for a more general loss of the type $G(\sigma^{-1}\|\bar{\gamma}-\gamma(\vartheta)\|)$ are reviewed in Humak (1977), but some restrictions on the class of admitted p.d.'s for $\mathcal{E}$ are needed.

Unbounded loss functions like (2) have been criticized because of the unjustified too heavy weighting of large estimation errors $\bar{\gamma}-\gamma$ in the corresponding risk $E\|\bar{\gamma}-\gamma\|^2$ and some other more realistic loss functions have been proposed.

Moreover, estimation errors which are small relative to the magnitude of γ may be taken to be negligible. Therefore a normalized loss like

$$L_1(\vartheta,\bar{\gamma})=\frac{\|\bar{\gamma}-\gamma(\vartheta)\|^2}{h\sigma^2+\|\gamma(\vartheta)\|^2}, \quad \text{with } h>0 \tag{3}$$

or (with componentwise normalization)

$$L_2(\vartheta,\bar{\gamma})=\sum_{j=1}^{l}\frac{|\bar{\gamma}_j-\gamma_j(\vartheta)|^2}{h_j\sigma^2+|\gamma_i(\vartheta)|^2}, \quad \text{with } h_j>0, \quad j=1,\ldots,l \tag{4}$$

deserve special interest. Both losses are bounded, for large γ only the magnitude of

$\bar{\gamma}-\gamma$ relative to γ has influence. The constant h is introduced to prevent unjustified high losses for small γ. While some other more realistic but more complicated loss functions are conceivable, the relatively simple loss functions (3) and (4) may be preferred to (2).

The main result of this paper is the interesting structure of minimax estimators with respect to the normalized risks

$$r_i(\beta,\sigma,\bar{\gamma}):=E_\vartheta L_i(\gamma(\beta),\bar{\gamma}[y]), \quad i=1,2, \tag{5}$$

namely essentially as shrunken BLUE's.

2. Minimax estimators for loss L_1

As the following theorem shows, certain estimators of the form

$$\tilde{\gamma}=RCX\hat{\beta} \tag{6}$$

are minimax with respect to the risk r_1, that is:

$$\sup_\beta r_1(\beta,\sigma,\tilde{\gamma})\leqslant \sup_\beta r_1(\beta,\sigma,\bar{\gamma}) \tag{7}$$

for all estimators $\bar{\gamma}$ and all $\sigma>0$.

Theorem 1. *The estimator* (6) *with*

$$R:=\sqrt{h}/\left(\sqrt{h}+\sqrt{t}\right),$$

$$t:=\mathrm{tr}\left[HCX(X'\Lambda^{-1})^{-1}X'C'\right]$$

is minimax in the sense (7).

Proof. Analogous to the proof in Bunke (1975) it suffices to show for a model with fixed $\sigma^2=\sigma_0^2>0$ and the loss

$$L[\beta,\bar{\gamma}]:=\frac{\|\bar{\gamma}-\gamma(\beta)\|^2}{h\sigma_0^2+\|\gamma(\beta)\|^2},$$

(i) that the risk $r[\beta,\tilde{\gamma}]:=E_{\beta,\sigma_0^2}L(\beta,\tilde{\gamma}[y])$ is constant and

(ii) that under a normal distribution for y, $\tilde{\gamma}$ is Bayes with respect to the risk r and some prior density ξ.

We use

$$\xi(\beta)\propto\varphi\left(\beta|0,\rho\sigma_0^2[X'\Lambda^{-1}X]^{-1}\right)\left(h\sigma_0^2+\|\gamma(\beta)\|^2\right) \tag{8}$$

where $\varphi(\beta|\mu,B)$ denotes the density of the p.d. $\mathfrak{N}(\mu,B)$ and

$$\rho:=(R^{-1}-1)^{-1}. \tag{9}$$

We have

$$r[\beta,\tilde{\gamma}]=\left(h\sigma_0^2+\|\gamma(\beta)\|^2\right)^{-1}(1-R)^2\left[\|\gamma(\beta)\|^2+\left\{R(1-R)^{-1}\right\}^2\sigma_0^2 t\right]$$
$$=t\left(\sqrt{t}+\sqrt{h}\right)^{-2} \tag{10}$$

which is independent of β.

The posterior density for β under a normal distribution for y may be easily calculated [as in Humak (1977)].

$$\xi(\beta|y)\propto\xi(\beta)\varphi(y|X\beta,\sigma_0^2\Lambda)$$
$$\propto\varphi\left(\beta|R\hat{\beta},R\sigma_0^2[X'\Lambda^{-1}X]^{-1}\right)\left(h\sigma_0^2+\|\gamma(\beta)\|^2\right). \tag{11}$$

Thus, the posterior risk is

$$E(L[\beta,\gamma]|y)\propto\int\|\gamma-\gamma(\beta)\|^2\varphi\left(\beta|R\hat{\beta},R\sigma_0^2[X'\Lambda^{-1}X]^{-1}\right)\mathrm{d}\beta \tag{12}$$

and it is minimized for $\gamma=CX(R\hat{\beta})$, the expectation of $\gamma(\beta)$ under the density φ in the integrand. Therefore $\tilde{\gamma}$ is Bayes.

3. Minimax estimators for loss L_2

For the loss (4), we are only able to derive minimax estimators for β under the additional assumption of a diagonal matrix

$$D=(X'\Lambda^{-1}X)^{-1}=\mathrm{Diag}[d_1,\ldots,d_k]. \tag{13}$$

They are of the form

$$\tilde{\beta}=K\hat{\beta} \tag{14}$$

with

$$K=\mathrm{Diag}[R_1,\ldots,R_k] \tag{15}$$

Theorem 2. *The estimator* (14) *with*

$$R_j:=\sqrt{h_j}/\left(\sqrt{h_j}+\sqrt{d_j}\right),\quad j=1,\ldots,k \tag{16}$$

fulfills

$$\sup_\beta r_2(\beta,\sigma,\tilde{\beta})\leqslant\sup_\beta r_2(\beta,\sigma,\bar{\beta}), \tag{17}$$

for all estimators $\bar{\beta}$ and all $\sigma>0$.

Proof. As in Theorem 1, we use a loss

$$L[\beta,\bar{\beta}]:=L_2(\beta,\sigma_0^2,\bar{\beta})$$

for fixed $\sigma^2=\sigma_0^2>0$.

We have a constant risk for the estimator (14):

$$r[\beta,\tilde{\beta}]=\sum_{j=1}^{k} E_{\beta,\sigma_0^2}|\tilde{\beta}_j[y]-\beta_j|^2\left(h_j\sigma_0^2+\beta_j^2\right)^{-1}$$

$$=\sum_{j=1}^{k} d_j\left(\sqrt{d_j}+\sqrt{h_j}\right)^{-2}. \tag{18}$$

We use the prior density

$$\xi(\beta)\propto\varphi(\beta|0,\sigma_0^2AD)\prod_{j=1}^{k}\left(h_j\sigma_0^2+\beta_j^2\right), \tag{19}$$

with

$$A:=\mathrm{Diag}[\rho_1,\ldots,\rho_k], \tag{20}$$

$$\rho_i:=\left(R_j^{-1}-1\right)^{-1},\quad j=1,\ldots,k. \tag{21}$$

The posterior risk is then

$$E\left(L[\beta,\bar{\beta}]|y\right)\propto\sum_{j=1}^{k}\int|\beta_j-\bar{\beta}_j|^2\varphi(\beta|K\hat{\beta},\sigma_0^2KD)\prod_{i\neq j}\left(h_i\sigma_0^2+\beta_i^2\right)\mathrm{d}\beta$$

$$=\sum_{j=1}^{k}\int|\beta_j-\bar{\beta}_j|^2\varphi(\beta_j|R_j\hat{\beta}_j,\sigma_0^2R_jd_j)\,\mathrm{d}\beta_j$$

$$\times\left\{\prod_{i\neq j}\int\varphi(\beta_i|R_i\hat{\beta}_i,\sigma_0^2R_id_i)(h_i\sigma_0^2+\beta_i^2)\,\mathrm{d}\beta_i\right\} \tag{22}$$

and each term is minimized by $\bar{\beta}_j=R_j\hat{\beta}_j$. Therefore the estimator $\tilde{\beta}=K\hat{\beta}$ is Bayes.

4. Numerical risk comparisons

To give impressions on the performance of the minimax estimator $\tilde{\gamma}$ in comparison with other estimators we calculated risk functions in a very simple model. We assume a three-dimensional observation, y, with

$$y\sim\mathcal{N}(\mu,I_3) \tag{23}$$

and as loss for estimating the vector μ:

$$L(\mu,\bar{\mu})=\frac{(\bar{\mu}-\mu)'(\bar{\mu}-\mu)}{1+\mu'\mu}. \tag{24}$$

We compare the following estimators:

BLUE;	$\hat{\mu}=y$,
Minimax:	$\tilde{\mu}=(1+\sqrt{3})^{-1}y$,
Other shrunken BLUE's:	$\mu_1^*=0.2\times y$,
	$\mu_2^*=0.6\times y$,
Stein estimator:	$\bar{\mu}=[1-(y'y)^{-1}]y$

and obtain the following graphical picture (Fig. 1) for their risk functions for μ:

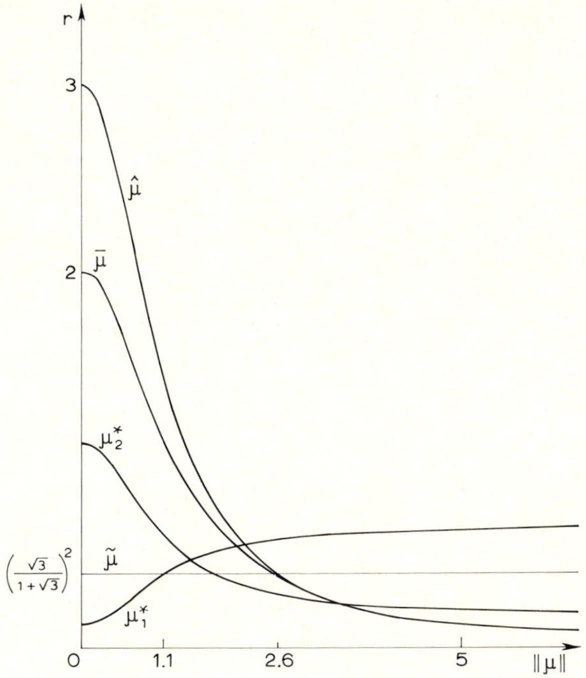

Fig. 1.

References

Bunke, O. (1975). Least squares as robust and minimax estimates. *Math. Operationsforsch. Statist.* **6**, 687–688.

Huber, P.J. (1977). Robust methods of estimation of regression coefficients. *Math. Operationsforsch. Statist., Ser. Statistics*, **8**, 41–54.

Humak, K.M.S. (1977). *Statistische Methoden der Modellbildung. Band* 1. *Statistische Inferenz für lineare Parameter*. Akademie-Verlag, Berlin.

Rao, C.R. (1973). *Linear Statistical Inference and its Applications*. 2nd. ed., Wiley, New York.

Rao, C.R. (1976). Estimation of parameters in a linear model. *Ann. Statist.* **4**, 1023–1037.

G. Kallianpur, P.R. Krishnaiah, J.K. Ghosh, eds., *Statistics and Probability: Essays in Honor of C.R. Rao*

BIVARIATE NEGATIVE BINOMIAL–POISSON AND NEGATIVE BINOMIAL–BERNOULLI MODELS WITH AN APPLICATION TO ACCIDENT DATA*

T. CACOULLOS and H. PAPAGEORGIOU
University of Athens, Athens, Greece

1. Introduction and summary

Discrete compound (generalized) distributions arise in several areas of application, e.g. ecology, genetics, physics, etc. (Feller [4, ch. 12]). The following example serves as an illustration and at the same time it has motivated the present investigation. Let X denote the number of injury accidents in a given locality during a certain period of time and Z the corresponding total number of fatalities or fatal accidents. In such a case, Z has the compound distribution from generalizing X by the distribution of Z_i, the number of fatalities or fatal accidents per accident $i=1,\ldots, X$. Thus,

$$Z=Z_1+Z_2+\cdots+Z_X.$$

The joint distribution of X and Z was considered by Leiter and Hamdan [9] to analyze the accident data recorded during 639 days (in 1969 and 1970) along a 50-mile stretch of highway in eastern Virginia. They suggested a Poisson–Poisson (P–P) model to generate the joint distribution of the number X of daily injury accidents and the number Z of daily fatalities and a Poisson–Bernoulli model when Z denotes the daily number of fatal accidents. Instead of the P–P model, the authors [1] considered a Poisson–Binomial (P–B) model, where the Z_i are i.i.d. binomial variables with parameters n and p. This implies the rather unrealistic assumption that the n passengers involved in a car accident play the role of n independent Bernoulli trials.

Both models, the P–P and the P–B, were fitted by the method of maximum likelihood to three sets of accident data collected over the 22-month period in Virginia [9]. However, though the fit was generally good (according to the χ^2 criterion), it was felt that the bivariate (X, Z) accident data were not adequately described by either of the preceding two models. This raised the question of finding a more appropriate bivariate model, that would provide a better fit. Apparently, as seen in the sequel, this has been achieved by the Negative Binomial–Poisson model, which is the object of this paper. A similar remark applies to the Negative Binomial–Bernoulli model as an alternative to the Poisson–Bernoulli model.

In the present bivariate Negative Binomial–Poisson, NB–P, and NB–Bernoulli models, we have assumed that the number of accidents X follows a negative

*This research was partially sponsored by the National Research Foundation of Greece, under Grant No. 067/79.

binomial distribution, because it is generally accepted (see [8]) that accident data are more adequately described by a negative binomial distribution (i.e. a Poisson with the parameter λ varying according to a gamma distribution) than by a Poisson distribution.

Various properties of the bivariate (X, Z) structure are examined, including the derivation of the conditional probability generating function (p.g.f.) of X given Z, using the basic relation between Bell polynomials and the derivatives of a composite function (see, e.g. Riordan [12, p. 34]).

Parameter estimators were derived using the methods of moments and "even points," the "double-zero proportion" method, and an estimation procedure using the marginal means and the ratio of the first two frequencies. For comparison purposes, NB–P and NB–Bernoulli models were fitted to the same sets of accident data used in [9]. This, as expected, resulted in a significant reduction of the corresponding χ^2 values.

2. Bell polynomials

For our purposes we require the following preliminaries concerning Bell polynomials.

The so called Bell polynomials $A_n(f; g_1,\ldots, g_n)$ may be associated with the nth derivative of the composite function $A(t)=f(g(t))$ as follows. Let $D_t=\mathrm{d}/\mathrm{d}t$, $D_u=\mathrm{d}/\mathrm{d}u$ and

$$A_n=D_t^nA(t),\qquad f_n=D_u^tf(u)|_{u=g(t)},\qquad g_n=D_t^ng(t).$$

Then (see, e.g. Riordan [12, p. 35])

$$A_n\equiv A_n(f; g_1,\ldots, g_n)=\sum_{\pi(n)}\frac{n!f_k}{k_1!\cdots k_n!}\left(\frac{g_1}{1!}\right)^{k_1}\cdots\left(\frac{g_n}{n!}\right)^{k_n},\tag{2.1}$$

where the summation extends over all partitions $\pi(n)$ of n, i.e. over all non-negative integers $k_1,\ldots, k_n$ such that $k_1+2k_2+\cdots+nk_n=n$; $k=k_1+k_2+\cdots+k_n$ denotes the number of parts in a given partition. (2.1) is known as di Bruno's formula.

In the special case $f_k=(\nu)_k$, where $(\nu)_k=\nu(\nu-1)\cdots(\nu-k+1)$, $(\nu)_0\equiv1$, we have the polynomials

$$C_{n,\nu}(g_1,\ldots, g_n)=A_n((\nu)_k; g_1,\ldots, g_n).\tag{2.2}$$

When $g_k=\theta^kc$, then

$$C_{n,\nu}(\theta c,\theta^2c,\ldots,\theta^nc)=\theta^nS_{n,\nu}(c)=\theta^n\sum_{k=0}^{n}(\nu)_kS(n,k)c^k\tag{2.3}$$

where $S(n,k)$ denotes a Stirling number of the second kind. Similarly

$$C_{n,-\nu}(-c,\ldots,-c)=S_{n-\nu}(-c)=\sum_{k=0}^{n}(-1)^k(-\nu)_kS(n,k)c^k,\tag{2.4}$$

or

$$S_{n,-\nu}(-c)=\sum_{k=0}^{n}\nu^{(k)}S(n,k)c^{k}, \tag{2.5}$$

where $\nu^{(k)}=\nu(\nu+1)\cdots(\nu+k-1)$, $\nu^{(0)}\equiv 1$.

The Bell polynomials have already been used in expressing the probabilities and moments of a univariate generalized discrete random variable [2].

3. The Negative Binomial–Poisson model

Suppose that the number of accidents X recorded at a specific location in a given time interval has a negative binomial distribution with probability function

$$h_1(x)=\binom{N+x-1}{N-1}P^xQ^{-(N+x)},\quad x=0,1,2,\ldots,\quad Q-P=1. \tag{3.1}$$

Let Z_i be the number of fatalities resulting from the ith accident and suppose that Z_i has a Poisson distribution with parameter λ; then

$$P(Z_i=n)=\frac{e^{-\lambda}\lambda^n}{n!},\quad n=0,1,2,\ldots. \tag{3.2}$$

Now if the Z_i are assumed to be mutually independent, then the conditional distribution of

$$Z=Z_1+Z_2+\cdots+Z_x, \tag{3.3}$$

the total number of fatalities, given that $X=x$ accidents have occurred, is Poisson with parameter λx. That is,

$$h(z/x)=\frac{e^{-\lambda x}(\lambda x)^z}{z!},\quad z=0,1,2,\ldots. \tag{3.4}$$

Hence, the joint distribution of the number of accidents X and the corresponding number of fatalities Z is given by

$$\begin{aligned} h(x,z)&=h(z/x)h_1(x)\\ &=\binom{N+x-1}{N-1}P^xQ^{-(N+x)}\frac{e^{-(\lambda x)}(\lambda x)^z}{z!}, \end{aligned} \tag{3.5}$$

and the Negative Binomial–Poisson probability function of Z (see Johnson and Kotz [6, p. 187] and in their notation Poisson $(\theta)\underset{\theta/\lambda}{\wedge}$ Negative Binomial (N,P)) is

$$\begin{aligned} h_2(z)&=\sum_{x=0}^{\infty}h(x,z)\\ &=\frac{\lambda^z}{z!}\sum_{x=0}^{\infty}e^{-(\lambda x)}x^z\binom{N+x-1}{N-1}P^xQ^{-(N+x)}. \end{aligned} \tag{3.6}$$

Also, it can be easily seen (e.g. by conditioning on X) that the p.g.f. of (X, Z) is given by

$$H_{X,Z}(u, v) = \left[Q - Pu\exp\{\lambda(v-1)\}\right]^{-N}, \tag{3.7}$$

since the p.g.f. of X is

$$H_X(u) = (Q - Pu)^{-N}. \tag{3.8}$$

The marginal p.g.f. for Z is

$$H_Z(v) = \left[Q - P\exp\{\lambda(v-1)\}\right]^{-N}. \tag{3.9}$$

3.1. *Moments and cumulants*

The cumulant generating function of (X, Z), readily obtained from (3.7), is given by

$$K(t_1, t_2) = -N\log\left[Q - Pe^{t_1}e^{\lambda(e^{t_2}-1)}\right]. \tag{3.10}$$

Upon differentiating (3.10) i times with respect to t_1, and j times with respect to t_2 and changing the order of differentiation on the right-hand side and putting $t_1 = t_2 = 0$, we obtain the recurrence for the cumulants κ_{ij} of (X, Z):

$$\kappa_{i+1,j} = PQ\frac{\partial}{\partial P}\kappa_{i,j}. \tag{3.11}$$

Similarly

$$\kappa_{i,j+1} = \lambda PQ\sum_{\nu=1}^{j}\binom{j}{\nu}\frac{\partial}{\partial P}\kappa_{i,\nu} + \lambda\kappa_{i+1,0} \quad (j \geqslant 1). \tag{3.12}$$

Moreover, from the factorial cumulant generating function of (X, Z),

$$L(t_1, t_2) = -N\log\left[Q - P(1+t_1)e^{\lambda t_2}\right], \tag{3.13}$$

the following recurrences for the factorial cumulants $\kappa_{(i,j)}$ are easily derived

$$\kappa_{(i+1,j)} = PQ\frac{\partial}{\partial P}\kappa_{(i,j)} - i\kappa_{(i,j)}, \tag{3.14}$$

$$\kappa_{(i,j+1)} = \lambda PQ\frac{\partial}{\partial P}\kappa_{(i,j)}, \tag{3.15}$$

$$\kappa_{(i,0)} = (i-1)!NP^i, \tag{3.16}$$

$$\kappa_{(i,1)} = i!NP^i\lambda Q. \tag{3.17}$$

From the known relationships between factorial cumulants, ordinary cumulants, factorial moments and central moments (see, e.g. [3, Ch. 9]), we find the low-order moments

$$\begin{aligned} E(X) &= \kappa_{(1,0)} = \kappa_{1,0} = NP, \qquad \mathrm{Var}(X) = \kappa_{(2,0)} + \kappa_{(1,0)} = \kappa_{2,0} = NPQ,\\ E(Z) &= \kappa_{(0,1)} = \kappa_{0,1} = NP\lambda, \qquad \mathrm{Var}(Z) = \kappa_{(0,2)} + \kappa_{(0,1)} = \kappa_{0,2} = NP\lambda(\lambda Q + 1),\\ \mathrm{Cov}(X, Z) &= \kappa_{(1,1)} = \kappa_{1,1} = NPQ\lambda. \end{aligned} \tag{3.18}$$

It is noticed that

$$\mathrm{Corr}(X, Z) = \left(\frac{\lambda Q}{1+\lambda Q}\right)^{1/2}$$

is independent of N, but dependent on the basic parameter Q in the NB. Clearly, we have $\mathrm{Cov}(X, Z) > E(Z)$, whereas, if X is a Poisson with parameter λ, then $\mathrm{Cov}(X, Z) = E(Z) = \lambda E(Z_i)$ (see [1]) and the $\mathrm{Corr}(X, Z)$ does not depend on λ (for any Z_i). In fact, the following general result is true.

Proposition 3.1. *Let*

$$Z = Z_1 + \cdots + Z_X$$

where the Z_i are i.i.d. r.v.'s independent of the r.v. X.

(i) *If $E(Z_i) > 0$, then* $\mathrm{Cov}(X, Z) \gtreqless E(Z)$ *according to whether* $\mathrm{Var}(X) \gtreqless E(X)$, *respectively*;

(ii) *if $E(Z_i) < 0$, then the inequality signs are reversed*;

(iii) *if $E(Z_i) = 0$, then* $\mathrm{Cov}(X, Z) = E(Z) = 0$.

Proof. This is immediate from the relations

$$E(Z) = E(Z_i)E(X), \qquad E(XZ) = E(Z_i)E(X^2),$$

from which

$$\mathrm{Cov}(X, Z) = E(Z_i)\,\mathrm{Var}(X). \tag{3.19}$$

Relation (3.19) and the formula

$$\begin{aligned}\mathrm{Var}(Z) &= E\{\mathrm{Var}(Z|X)\} + \mathrm{Var}\{E(Z|X)\}\\ &= E(X)\,\mathrm{Var}(Z_i) + \{E(Z_i)\}^2\,\mathrm{Var}(X)\end{aligned}$$

can be used for the computation of $\mathrm{Corr}(X, Z)$.

3.2. Conditionality, regression

As regards the conditional distribution of Z given $X = x$, this is clearly the x-fold convolution of the distribution of Z_1, with p.g.f.

$$H_{Z/X=x}(v) = \exp\{\lambda x(v-1)\}. \tag{3.20}$$

To derive the conditional distribution of X given $Z = z$, it is convenient to use the following formula (see, e.g. [13]),

$$H_{X/Z=z}(u) = H_v^{(z)}(u, 0) \Big/ H_v^{(z)}(1, 0), \tag{3.21}$$

where we set

$$H_v^{(z)}(\alpha, \beta) = \left.\frac{\partial^z H(u, v)}{\partial v^z}\right|_{\substack{u=\alpha\\ v=\beta}}. \tag{3.22}$$

Consequently, by applying (3.21) to $H(u,v)$ of (3.7) in the role of $A(t)$ of (2.1) with $f(u)=u^{-N}$, $g(t)=Q-Pu\exp\{\lambda(t-1)\}$ and by simplifying by (2.1)–(2.5), we obtain

$$H_{X/Z=z}(u)=\left[\frac{1-\theta u}{1-\theta}\right]^{-N}\frac{S_{z,-N}(-ct)}{S_{z,-N}(-c)} \tag{3.23}$$

where $\theta=(P/Q)e^{-\lambda}$, $c=\theta/(1-\theta)$, $t=u[(1-\theta u)/(1-\theta)]^{-1}$. Note that the first factor on the right-hand side of (3.23) is the p.g.f. of the conditional distribution of X given $Z=0$, i.e.

$$H_{X/Z=0}(u)=\frac{H(u,0)}{H(1,0)}=\left[\frac{1-\theta u}{1-\theta}\right]^{-N}, \tag{3.24}$$

and the second factor is a mixture of z negative binomial distributions the kth negative binomial having parameters θ and k (and shifted to the right by k), $k=1,\ldots,z$.

The regression of X on Z is easily obtained from (3.23) as

$$E(X/Z=z)=Nc+\frac{1}{1-\theta}\frac{\sum_{k=0}^{z} kN^{(k)}S(z,k)c^k}{\sum_{k=0}^{z} N^{(k)}S(z,k)c^k}. \tag{3.25}$$

In particular,

$$E(X/Z=0)=Nc,$$

$$E(X/Z=1)=Nc+\frac{1}{1-\theta},$$

$$E(X/Z=2)=Nc+\frac{1}{1-\theta}\left[1+\frac{Nc+c}{Nc+c+1}\right].$$

These regression functions may be used, e.g. for estimating the number X of accidents given the number Z of fatalities, since the latter is usually more available than X.

4. The Negative Binomial–Bernoulli model

Here the analogous results of Section 3 are given for the NB–Bernoulli distribution. Details are omitted.

The p.g.f. of (X,Z) is

$$G(u,v)=(Q-Pqu-Ppuv)^{-N}, \tag{4.1}$$

where $p=1-q=P[Z_i=1]$ is the Bernoulli parameter. This is seen to be a special case of the bivariate negative binomial distribution with p.g.f.

$$G_0(u,v)=(\alpha_0-\alpha_1 u-\alpha_2 v-\alpha_3 uv)^{-N}. \tag{4.2}$$

It should also be noted that Z, like X, is negative binomial with parameter pP, whereas the conditional distribution of X given $Z=z$ is a negative binomial shifted to the right by z, with p.g.f.

$$G_{X|Z=z}(u)=u^{z}\left[\frac{Q-Pqu}{Q-Pq}\right]^{-(N+z)}. \tag{4.3}$$

The regression of X on Z is easily verified to be

$$E(X|Z=z)=z+(N+z)\frac{Pq}{Q-Pq}. \tag{4.4}$$

5. Estimation

5.1. *Negative Binomial–Poisson*

(a) *Method of moments.* Let $\bar{x}, \bar{z}$ be the marginal means and s_{xz} the unbiased estimator of the covariance. Then moment estimators are simply derived as:

$$\hat{\lambda}=\frac{\bar{z}}{\bar{x}}, \tag{5.1}$$

$$\hat{P}=\frac{s_{xz}}{\bar{z}}-1, \tag{5.2}$$

$$\hat{N}=\frac{\bar{x}}{\hat{P}}. \tag{5.3}$$

(b) *Method of "even points".* Patel [11] used this procedure in order to estimate the parameters of a univariate Hermite distribution. Papageorgiou and Kemp [10] introduced a bivariate version of the "even-points" method and the technique was illustrated for a variety of bivariate generalized Poisson distributions.

In particular, since

$$H(1,1)+H(-1,-1)=2\left[\sum_{i=0}^{\infty}\sum_{j=0}^{\infty}P_{2i,2j}+\sum_{i=0}^{\infty}\sum_{j=0}^{\infty}P_{2i+1,2j+1}\right], \tag{5.4}$$

denoting by See and Soo the sums of observed frequencies at points $(2i,2j)$ and $(2i+1,2j+1)$ in a bivariate sample of size n, for the bivariate NB–P model we have the estimation equation

$$H(1,1)+H(-1,-1)=1+\{Q+Pe^{-2\lambda}\}^{-N}=\frac{2(\text{See}+\text{Soo})}{n}. \tag{5.5}$$

Consequently, even-point estimators can be obtained from the following system of equations

$$-N\log\{Q+Pe^{-2\lambda}\}=\log\left\{\frac{2(\text{See}+\text{Soo})}{n}-1\right\}, \tag{5.6}$$

$$\bar{x}=NP, \tag{5.7}$$

$$\bar{z}=NP\lambda. \tag{5.8}$$

Combining equations (5.6) and (5.7) gives

$$\frac{P}{\log\{P(1+e^{-2\lambda})+1\}}=\frac{\bar{x}}{-\log\left\{\frac{2(\text{See}+\text{Soo})}{n}-1\right\}}, \tag{5.9}$$

which can be solved iteratively with respect to P (we used Newton–Raphson).

(c) *The "double-zero proportion" method*. This method was introduced by Holgate [5], who obtained estimators for the parameters of a bivariate Poisson distribution using the marginal means and the proportion of observations in the (0,0) cell ($f_{0,0}$). For the bivariate NB–P model

$$H(0,0)=f_{0,0}=Q^{-N}. \tag{5.10}$$

Hence, parameter estimators can be derived from the following system of equations

$$\log(f_{0,0})=-N\log(1+P), \tag{5.11}$$

$$\bar{x}=NP, \tag{5.12}$$

$$\bar{z}=NP\lambda. \tag{5.13}$$

N may be eliminated between equations (5.12) and (5.11) giving the equation

$$\frac{P}{\log(1+P)}=\frac{\bar{x}}{-\log(f_{0,0})}, \tag{5.14}$$

which can be solved iteratively with respect to P. Provided $\bar{x}>-\log(f_{0,0})$, there is always a unique solution [6, p. 131].

(d) *An estimation procedure using a ratio of frequencies*. In the univariate case when the first two frequencies were large in comparison with the remaining, use was made of an estimation method using the first moment(s) and the ratio of the first two frequencies (see, e.g. [7]).

This procedure can also be applied in bivariate situations. Thus for the bivariate NB–P model, parameter estimators are derived using the marginal means and the ratio.

$$\frac{f_{1,0}}{f_{0,0}}=\frac{NPe^{-\lambda}}{Q}. \tag{5.15}$$

In particular

$$\lambda^*=\frac{\bar{z}}{\bar{x}}, \tag{5.16}$$

$$P^*=\bar{x}e^{-\lambda^*}\frac{f_{0,0}}{f_{1,0}}-1, \tag{5.17}$$

$$N^*=\frac{\bar{x}}{P^*}. \tag{5.18}$$

5.2. *Negative Binomial–Bernoulli*

For this model, the moment estimators of the parameters N, P and p are

$$\tilde{p}=\frac{\bar{z}}{\bar{x}}, \qquad \tilde{P}=\frac{s_{xz}}{\bar{z}}-1, \qquad \tilde{N}=\frac{\bar{x}}{\tilde{P}}. \tag{5.19}$$

Even-point estimators are solutions of the system

$$\bar{x}=NP, \qquad \bar{z}=NPp, \tag{5.20}$$

$$-\log\left\{\frac{2(\text{See}+\text{Soo})}{n}-1\right\}=N\log\{1+2P(1-p)\}. \tag{5.21}$$

Double-zero proportion estimators can be obtained by combining (5.20) with (5.11).

Ratio-of-frequencies estimators are derived by using (5.20) and

$$\frac{f_{1,0}}{f_{0,0}}=\frac{P}{Q}qN. \tag{5.22}$$

6. An application to accident data

For comparison purposes, the NB–P and NB–Bernoulli distributions were fitted to the same sets of accident data used by [9].

Recurrences for the probabilities were calculated by differentiating the p.g.f. once with respect to a generating variable and then equating coefficients.

Thus for the NB–P probabilities P_{ij} the following recurrences were obtained:

$$(i+1)P_{i+1,j}=\frac{Pe^{-\lambda}}{Q}(N+i)\sum_{k=0}^{j}\frac{\lambda^k}{k!}P_{i,j-k}, \tag{6.1}$$

$$(j+1)P_{i,j+1}=\frac{Pe^{-\lambda}}{Q}\left[(j+1)\sum_{k=0}^{j+1}\frac{\lambda^k}{k!}P_{i-1,j+1-k}+N\lambda\sum_{k=0}^{j}\frac{\lambda^k}{k!}P_{i-1,j-k}\right], \tag{6.2}$$

with

$$P_{0,0}=H(0,0)=Q^{-N}.$$

Similarly, the recurrences for the NB–Bernoulli probabilities p_{ij} are:

$$(i+1)p_{i+1,j}=\frac{P}{Q}(N+i)(qp_{i,j}+pp_{i,j-1}), \tag{6.3}$$

$$(j+1)p_{i,j+1}=\frac{P}{Q}[q(j+1)p_{i-1,j+1}+p(N+j)p_{i-1,j}], \tag{6.4}$$

with

$$p_{0,0}=G(0,0)=Q^{-N}.$$

Table 1
Estimated values of the NB–P parameters

NB–P distribution		N	P	λ[a]
	EP	3.716015	0.232046	0.067151
639 days	DZP	5.800666	0.148653	0.067151
	RF	5.237675	0.164631	0.067151
	EP	3.893513	0.225929	0.074919
349 days	DZP	5.707780	0.154115	0.074919
	RF	5.036314	0.174663	0.074919
	EP	3.521162	0.238949	0.057377
290 days	DZP	5.963775	0.141082	0.057377
	RF	5.520613	0.152407	0.057377

[a]Notice that all estimation procedures give the same estimate for λ, i.e. $\bar{z}/\bar{x}$ (as in the P–P model).

Table 2
Estimated values of the NB–Bernoulli parameters

NB–Bernoulli distribution		N	P	p[a]
	EP	3.352314	0.257221	0.063521
639 days	DZP	5.800666	0.148653	0.063521
	RF	5.181818	0.166406	0.063521
	EP	3.559489	0.247130	0.071661
349 days	DZP	5.707780	0.154115	0.071661
	RF	5.017413	0.175321	0.071661
	EP	3.130927	0.268732	0.053279
290 days	DZP	5.963775	0.141082	0.053279
	RF	5.412969	0.155438	0.053279

[a]Notice that all estimation procedures give the same estimate for p, i.e. $\bar{z}/\bar{x}$ (as in the P–Bernoulli model).

Moment estimators were not considered since, for most sets of data $0<s_{xz}<\bar{z}$, which would result in a negative value for P. This is explained by the fact that $\mathrm{Cov}(X, Z)>E(Z)$ (see Proposition 3.1) for both models.

Table 1 gives the even-point (EP), double-zero proportion (DZP) and ratio-of-frequencies (RF) estimates of the NB–P parameters for the entire study (639 days) and the years 1969 (349 days) and 1970 (290 days). Table 2 gives the corresponding estimates for the NB–Bernoulli model.

Tables 3 and 4 give the observed and expected frequencies, for the NB–P and NB–Bernoulli models, when even-point, double-zero proportion or ratio-of-frequencies parameter estimates are used based on the entire set of data (639 days). For comparison purposes, the χ^2 values and the corresponding significance levels for

Table 3
Observed (first entry) and expected frequencies when the NB–P distribution is fitted by the method of EP (second entry), DZP (third entry) or RF (fourth entry) to the number of injury accidents X and the number of fatalities Z for the entire study (639 days)

	X/Z	0	(Z) 1	2	Total/EP/DZP/RF
	0	286 294.26 286.00 287.63	— — — —	— — — —	286 294.26 286.00 287.63
	1	198 192.57 200.75 199.13	17 12.93 13.48 13.37	1 0.43 0.45 0.45	216 205.93 214.68 212.95
	2	82 79.97 82.61 82.09	10 10.74 11.09 11.02	0 0.72 0.75 0.74	92 91.43 94.45 93.85
(X)	3	24 26.83 25.99 26.18	5 5.41 5.24 5.27	1 0.54 0.53 0.53	30 32.78 31.76 31.98
	4	13 7.93 6.92 7.13	1 2.13 1.86 1.91	0 0.29 0.25 0.26	14 10.35 9.03 9.30
	5	1 2.16 1.64 1.74	0 0.72 0.55 0.58	0 0.12 0.09 0.10	1 3.00 2.28 2.42
	Total	604	33	2	639
	EP	603.72	31.93	2.10	637.75
	DZP	603.91	32.22	2.07	638.20
	RF	603.90	32.15	2.08	638.13

all models and all three sets of accident data are shown in Table 5. Maximum likelihood (ML) estimates were used for the P–P, P–B ($n=4$) and P–Bernoulli models. It should be noticed that the comparative performance of the preceding models (as judged by the χ^2 criterion) is preserved when the models are fitted to the individual data for the years 1969 (349 days) and 1970 (290 days).

Table 4
Observed (first entry) and expected frequencies when the NB–Bernoulli distribution is fitted by the method of EP (second entry), DZP (third entry) or RF (fourth entry) to the number of injury accidents X and the number of fatal accidents Z for the entire study (639 days)

		(Z)		
	X/Z	0	1	Total/EP/DZP/RF
	0	286 296.65 286.00 287.80	– – – –	286 296.65 286.00 287.80
	1	198 190.54 201.06 199.25	18 12.93 13.64 13.52	216 203.47 214.70 212.77
	2	82 79.44 82.86 82.28	10 10.78 11.24 11.16	92 90.22 94.10 93.44
(X)	3	24 27.16 26.11 26.32	6 5.53 5.31 5.36	30 32.69 31.42 31.68
	4	13 8.26 6.96 7.19	1 2.24 1.89 1.95	14 10.50 8.85 9.14
	5	1 2.33 1.65 1.76	0 0.79 0.56 0.60	1 3.12 2.21 2.36
	Total	604	35	639
	EP	604.38	32.27	636.65
	DZP	604.64	32.64	637.28
	RF	604.60	32.59	637.19

Comparing the χ^2 values* and Tables 3 and 4 with the corresponding tables given by Leiter and Hamdan [9] or Cacoullos and Papageorgiou [1], we conclude that the negative binomial as the distribution of X is more appropriate than the Poisson for the analysis of such types of accident data in a bivariate set-up.

*The degrees of freedom (D.F.) given by Leiter and Hamdan [9] should be decreased by two, the number of estimated parameters.

Table 5
χ^2 values and corresponding significance levels (S.L.)

Accident – fatalities data

Days	Model	P–P (ML)	P–B (ML)	NB–P (EP)	NB–P (DZP)	NB–P (RF)
	χ^2	20.52	21.10	9.52	9.94	9.65
639	D.F.	13	13	12	12	12
	S.L.	0.08	0.07	0.66	0.62	0.65
	χ^2	14.51	14.82	7.90	8.27	8.05
349	D.F.	13	13	12	12	12
	S.L.	0.34	0.32	0.79	0.76	0.78
	χ^2	11.55	15.97	8.58	8.99	8.85
290	D.F.	10	10	9	9	9
	S.L.	0.32	0.10	0.48	0.44	0.45

Accident – fatal accidents data

Days	Model	P–Bernoulli (ML)	NB–Bernoulli (EP)	NB–Bernoulli (DZP)	NB–Bernoulli (RF)
	χ^2	19.12	8.16	8.32	7.99
639	D.F.	8	7	7	7
	S.L.	0.02	0.32	0.31	0.34
	χ^2	12.54	5.69	6.02	5.79
349	D.F.	8	7	7	7
	S.L.	0.13	0.58	0.54	0.57
	χ^2	6.67	3.25	3.78	3.58
290	D.F.	6	5	5	5
	S.L.	0.35	0.66	0.58	0.61

References

[1] Cacoullos, T. and Papageorgiou, H. (1980). On some bivariate probability models applicable to traffic accidents and fatalities. *Int. Statist. Rev.* **48**, 345–356.

[2] Charalambides, Ch.A. (1977). On the generalized discrete distributions and the Bell polynomials. *Sankhyā Ser. B* **39**, 36–44.

[3] David, F.N. and Barton, D.E. (1962). *Combinatorial Chance*. Griffin, London.

[4] Feller, W. (1957). *An Introduction to Probability Theory and its Applications*. Vol. 1. Wiley, New York.

[5] Holgate, P. (1964). Estimation for the bivariate Poisson distribution. *Biometrika* **51**, 241–245.

[6] Johnson, N.L. and Kotz, S. (1969). *Distributions in Statistics: Discrete Distributions*. Houghton Mifflin Co., Boston, IL.

[7] Katti, S.K. and Gurland, J. (1961). The Poisson Pascal distribution. *Biometrics* **17**, 527–538.

[8] Kemp, C. D. (1970). "Accident proneness" and discrete distribution theory. In Patil, G.P., (Ed.), *Random Counts in Biomedical and Social Sciences*, The Pennsylvania State University Press, PA, pp. 41–65.

[9] Leiter, R.E. and Hamdan, M.A. (1973). Some bivariate probability models applicable to traffic accidents and fatalities. *Int. Statist. Rev.* **41**, 87–100.

[10] Papageorgiou, H. and Kemp, C. D. (1977). *Even point estimation for bivariate generalized Poisson distributions*. Paper presented at the 10th European Meeting of Statisticians, Leuven, Belgium. Statistics Reports and Preprints, No. 29, School of Mathematics, University of Bradford, Bradford, England.

[11] Patel, Y.C. (1976). Even-point estimation and moment estimation in Hermite distribution. *Biometrics* **32**, 865–873.

[12] Riordan, J. (1958). *An Introduction to Combinatorial Analysis*. Wiley, New York.

[13] Subrahmaniam, K. (1966). A test for "intrinsic" correlation in the theory of accident proneness. *J. Roy. Statist. Soc. Ser. B* **28**, 180–189.

G. Kallianpur, P.R. Krishnaiah, J.K. Ghosh, eds., *Statistics and Probability*: *Essays in Honor of C.R. Rao*

SYMMETRIES (GROUP OF AUTOMORPHISMS) OF DESARGUESIAN FINITE PROJECTIVE AND AFFINE PLANES AND THEIR ROLE IN STATISTICAL MODEL CONSTRUCTION*

I.M. CHAKRAVARTI and Catherine T. BURTON
Department of Statistics, University of North Carolina at Chapel Hill, Chapel Hill, NC, U.S.A.

1. Introduction

Weyl (1952, 1964) defined the symmetries of a configuration in space by its group of automorphisms and used this concept to analyze symmetries in nature and art. Following Weyl, we define the symmetries of a block design (more generally, an incidence structure) by its group of automorphisms.

An *incidence structure* [see, for instance, (Dembowski, 1968)] is defined as a triple $S=(\mathcal{P}, \mathcal{B}, \mathcal{I})$ where $\mathcal{P}, \mathcal{B}, \mathcal{I}$ are sets with

$$\mathcal{P}\cap\mathcal{B}=\emptyset \quad \text{and} \quad \mathcal{I}\subseteq\mathcal{P}\times\mathcal{B}.$$

The elements of $\mathcal{P}$ are called points, those of $\mathcal{B}$ blocks, and those of $\mathcal{I}$ flags. Only finite incidence structures are considered here. This means that $\mathcal{P}$, $\mathcal{B}$ and $\mathcal{I}$ are finite sets. Incidence structures with equally many points on every block and equally many blocks through every point are called *tactical configurations*. A tactical configuration is called a *balanced incomplete block design* (BIB design) if every pair of points is incident with an equal number of blocks, say λ (>0) blocks. The parameters of the BIB design are v, b, r, k, λ, where $v=|\mathcal{P}|$, $b=|\mathcal{B}|$, r is the number of blocks through a point and k is the number of points in a block.

Let $S=(\mathcal{P}, \mathcal{B}, \mathcal{I})$ and $T=(\mathcal{Q}, \mathcal{C}, \mathcal{J})$ be two incident structures. An *incidence preserving* map of S into T is a mapping ϕ of $\mathcal{P}\cup\mathcal{B}$ into $\mathcal{Q}\cup\mathcal{C}$ such that pIB implies $p\phi JB\phi$ for all $p\in\mathcal{P}$ and $B\in\mathcal{B}$. Such a mapping is called a *homomorphism* if $\mathcal{P}\phi\subseteq\mathcal{Q}$ and $\mathcal{B}\phi\subseteq\mathcal{C}$ and an anti-homomorphism if $\mathcal{P}\phi\subseteq C$ and $\mathcal{B}\phi\subseteq\mathcal{Q}$. An *epimorphism* of S onto T is a homomorphism ϕ with $\mathcal{P}\phi=\mathcal{Q}$ and $\mathcal{B}\phi=C$ and an *isomorphism* is a one–one epimorphism whose inverse is likewise incidence-preserving (and hence also an isomorphism). For special classes of incidence structures (for designs, for instance) the condition that ϕ^{-1} be incidence-preserving is superfluous. Anti-epimorphisms and anti-isomorphisms are defined in the similar manner. S and T are dual to each other if there exists an anti-isomorphism of S onto T. *Automorphisms* (anti-automorphisms) are isomorphisms (anti-isomorphisms) of an incidence structure onto itself and a *polarity* is an anti-automorphism of order 2. The set of all automorphisms and anti-automorphisms of S is a group denoted by $A(S)$, and the

*This research was supported by the National Science Foundation under Grant MCS 78-01434.

automorphisms form a subgroup Aut(S) which is of index 2 if S is self-dual and coincides with $A(S)$ if S is not self-dual.

2. Finite planes and their collineations

Finite projective planes and finite affine planes [see, for instance, (Hall, 1959; Dembowski, 1968; Carmichael, 1937)] are balanced incomplete block designs with parameters $v=b=s^2+s+1$, $r=k=s+1$, $\lambda=1$ and $v=s^2$, $b=s^2+s$, $r=s+1$, $k=s$, $\lambda=1$, respectively.

An automorphism of a plane is called a *collineation* and it is customary to call the group of automorphisms of a plane the *group of collineations*. A projective Desarguesian plane PG(2, s) of order $s=p^r$, where p is a prime and r is an integer, (that is with $s+1$ points on a line) has a collineation group PC(2, s) of order $r(s^2+s+1)(s^2+s)s^2(s-1)^2$ and an affine Desarguesian plane EG(2, s) has a collineation group EC(2, s) of order $rs^2(s^2-1)(s^2-s)$, [see, for instance (Carmichael, 1937; Hall, 1959)].

Finite projective geometries PG(d, s) and finite affine geometries EG(d, s), where $s=p^r$, p a prime and r an integer and $d>2$, can be only Desarguesian. A collineation of projective or affine geometry is a permutation of its points, which maps lines into lines and hence every subspace is mapped into a subspace. A balanced incomplete block design constructed from a finite geometry by taking points of the geometry as the points of the design and subspaces of the geometry as blocks of the design will then have as its full group of automorphisms the full collineation groups PC(d, s) and EC(d, s) of the geometry. The full collineation groups PC(d, s) and EC(d, s) of the geometries PG(d, s) and EG(d, s) respectively, are known [see, for instance, (Carmichael, 1937)].

The collineation groups of finite non-Desarguesian planes have been studied to a certain extent [see, for instance, (Dembowski, 1968)] by different authors.

3. Deficiency of a linear model for a BIBD. First step towards modification

Groups of automorphisms of a known design and other classical groups have been used by combinatorial mathematicians to distinguish one design from the other or to construct new designs. However, to the best of our knowledge, the symmetries (group of automorphisms) of a block design have hardly been taken into consideration by a statistician in constructing statistical models and analysis of observations from such a design. Thus the linear model $E(\boldsymbol{Y})=\mu \boldsymbol{j}+A'\boldsymbol{\theta}+L'\boldsymbol{\beta}$, $\operatorname{Var}\boldsymbol{Y}=\sigma^2 I$, (where $\boldsymbol{Y}=(Y_1, Y_2, \ldots, Y_n)^T$ denotes the vector of observations arising from the $n=bk$ experimental units of a BIB design with parameters $v, b, r, k, \lambda, \boldsymbol{\theta}$ the vector of treatment effects, $\boldsymbol{\beta}$ the vector of block effects and μ a constant) commonly used by a statistician does not discriminate between two BIB designs having the same set of numerical values for the parameters v, b, r, k, λ but different groups of symmetries (automorphisms).

This is the first of a series of papers in which we will report the results of our research efforts towards letting the group of automorphisms of a design play a role in modifying the commonly used linear model, in successive steps. As a first step, we will assume that the distribution of $Y=(Y_1, Y_2,\ldots, Y_n)^T$ from a BIB design is such that its covariance matrix $\operatorname{Var} Y=\Sigma$ is invariant under the action of the group of automorphisms of the design. This means that Σ belongs to the commutant algebra (centralizer ring) corresponding to the group of automorphisms G represented as the group of matrices π_G permuting the flags (the incident point-block pairs) of the design. If t is the dimension of this *commutant algebra* and V_i, $i=1,2,\ldots,t$ is a linear basis of this algebra, then our assumption, that $P_g \Sigma P_g^{-1}=\Sigma$ for every P_g belonging to the π_G, implies that $\Sigma=c_1V_1+\cdots+c_tV_t$.

4. Relationship algebra of James. Commutant algebra

James (1957) defined the relationship matrices B, T, I, G of a block design as $B=(b_{ij})$, $b_{ij}=1$ if the experimental units i and j are in the same block, $b_{ij}=0$, otherwise, $T=(t_{ij})$, $t_{ij}=1$ if the experimental units i and j receive the same treatment, $t_{ij}=0$, otherwise. I is the $n\times n$ identity matrix and $G=(g_{ij})$, $g_{ij}=1$ for all pairs (i, j), $i, j=1,2,\ldots,n$. The $n=bk$ experimental units are the n flags (the incident point-block pairs) of the BIB design (v, b, r, k, λ). If the BIB design is asymmetric, $v<b$, the matrices I, G, B, T, BT, TB, BTB form a linear basis of a seven-dimensional non-commutative algebra R_7 (James, 1957). If the design is symmetric, $v=b$, $BTB=\lambda G+(k-\lambda)B$ and (I, G, B, T, BT, TB) is the linear basis of a six-dimensional algebra R_6 (Mann, 1960).

James (1957, pp. 1001–1002) mentioned "For certain designs, the relationship algebra is the commutator algebra (it is customary to call it a *commutant* algebra or a centralizer ring) of the representation of a group expressing the symmetry of the experimental design. Such will be the subject of a further paper". (James told the author on October 1, 1979, that he had earlier obtained some results in this area but never published them.)

Sysoev and Shaikin (1978, II, p. 1029) considered the commutant algebra corresponding to the group $\mathcal{G}$ of all those permutations g of the experimental unit numbers (flags) of a BIB design under which $b_{ij}=b_{g(i)g(j)}$ and $t_{ij}=T_{g(i)g(j)}$ for all pairs (i, j) and every g in $\mathcal{G}$. They mentioned: "In concluding this section we note *without proof* the following fact. For symmetric schemes $\mathrm{PG}(m, q)$, the R-equivalence classes coincide with equivalence classes in the sense (23), i.e. the extended relationship algebra $\mathcal{P}_7$ is a commutator (commutant) algebra. In the special case of block-schemes $\mathrm{PG}(2, q)$ for which $\lambda=1$, the commutator (commutant) algebra coincides with the relationship algebra R_6. The relationship algebra R_7 is a commutator (commutant) algebra in the case of asymmetric block schemes with $\lambda=1$, generated by the finite Euclidean geometries $\mathrm{EG}(2, q)$." The above statements were made *without proof* and the authors have found no reference, if a proof existed, in the literature.

5. Symmetries of PG(2,2) and the dimension and basis of the corresponding commutant algebra

In this paper, we illustrate the use of Schur's theorem in deriving the dimension and a linear basis of the commutant algebra corresponding to the collineation group PC(2,2) (represented as a group of matrices permuting the 21 flags (incident point-line pairs)) of the finite projective plane PG(2,2) with 3 points on a line. We quote from Wielandt (1964, p. 80) the following theorem due to Schur (1933). "Theorem 28.4. If a transitive permutation group $\mathcal{G}$ is regarded as a matrix group $\mathcal{G}^*$, then the matrices which commute with all the matrices of $\mathcal{G}^*$ form a ring $V=V(\mathcal{G})$. We call V the centralizer ring corresponding to $\mathcal{G}$. V is a vector space over the complex number field which has the matrices $B(\Delta)$ corresponding to the orbits Δ of $\mathcal{G}_1$ (stabilizer-subgroup of $\mathcal{G}$ fixing the element "1") as a linear basis. In particular, the dimension of V coincides with the number k of orbits of $\mathcal{G}_1$." The matrix $\mathcal{B}(\Delta)=(v^{\Delta}_{\alpha,\beta})$, $\alpha,\beta=1,2,\ldots,n$ corresponding to the orbit Δ, is defined as

$$v^{\Delta}_{\alpha,\beta}=\begin{cases}1, & \text{if there exists } g\in\mathcal{G},\ \delta\in\Delta \text{ with } 1^g=\beta,\ \delta^g=\alpha,\\ 0, & \text{otherwise.}\end{cases}$$

$\Omega=\{1,2,\ldots,n\}$ is the set of n elements on which every permutation g of $\mathcal{G}$ acts.

Hall (1959, Example 4, pp. 21–23) gave the group of permutations $\mathcal{G}$ (and its subgroups) on the seven letters A, B, C, D, E, F, G which permute among themselves the columns of the following diagram:

	(1)	(2)	(3)	(4)	(5)	(6)	(7)
	A	B	C	D	E	F	G
$\mathcal{D}_1$:	B	C	D	E	F	G	A
	D	E	F	G	A	B	C

If the seven letters are regarded as points and the columns as lines, the diagram represents the finite projective plane PG(2,2) with seven points and the group $\mathcal{G}$ is its collineation group PC(2,2). $|\mathcal{G}|=168$ and the subgroup K of $\mathcal{G}$, which fixes the flag $(1,A)\equiv\{1\}$ is our stabilizer $\mathcal{G}_{\{1\}}$. From Hall (1959),

$$K=\{I,(CE)(FG),(CF)(EG),(CG)(EF),(BD)(EF),(BD)(CFGE),$$

$$(BD)(CEGF),(BD)(CG)\}.$$

The action of each one of the permutations of K on the flags (incident letter-column pairs) of $\mathcal{D}_1$ gives us eight diagrams including $\mathcal{D}_1$ (action of I on $\mathcal{D}_1$). For each diagram we note also the permutation of columns induced by the given permutation of letters.

$\mathfrak{D}_2$:

	(1)	(2)	(4)	(3)	(7)	(6)	(5)	No. of flags fixed
	Ⓐ	Ⓑ	E	D	C	G	F	= 5
(CE)(FG)	Ⓑ	E	D	C	G	F	A	
(34)(57)	Ⓓ	C	G	F	A	Ⓑ	E	

$\mathfrak{D}_3$:

	(1)	(6)	(3)	(4)	(7)	(2)	(5)	No. of flags fixed
	Ⓐ	B	F	Ⓓ	G	C	E	= 5
(CF)(EG)	Ⓑ	F	Ⓓ	G	C	E	A	
(26)(57)	Ⓓ	G	C	E	A	B	F	

$\mathfrak{D}_4$:

	(1)	(6)	(4)	(3)	(5)	(2)	(7)	No. of flags fixed
	Ⓐ	B	G	D	F	E	C	= 5
(CG)(EF)	Ⓑ	G	D	F	E	C	Ⓐ	
(26)(34)	Ⓓ	F	E	C	Ⓐ	B	G	

$\mathfrak{D}_5$:

	(1)	(3)	(2)	(6)	(5)	(4)	(7)	No. of flags fixed
	Ⓐ	D	C	B	F	E	Ⓖ	= 5
(BD)(EF)	D	C	B	F	E	G	Ⓐ	
(23)(46)	B	F	E	G	Ⓐ	D	Ⓒ	

$\mathfrak{D}_6$:

	(1)	(3)	(6)	(2)	(7)	(4)	(5)	No. of flags fixed
	Ⓐ	D	F	B	C	G	E	= 1
(BD)(CFGE)	D	F	B	C	G	E	A	
(2463)(57)	B	C	G	E	A	D	F	

$\mathfrak{D}_7$:

	(1)	(4)	(2)	(6)	(7)	(3)	(5)	No. of flags fixed
	Ⓐ	D	E	B	G	C	F	= 1
(BD)(CEGF)	D	E	B	G	C	F	A	
(2364)(57)	B	G	C	F	A	D	E	

$\mathfrak{D}_8$:

	(1)	(4)	(6)	(2)	(5)	(3)	(7)	No. of flags fixed
	Ⓐ	D	G	B	Ⓔ	F	C	= 5
(BD)(CG)	D	G	B	E	Ⓕ	C	Ⓐ	
(24)(36)	B	E	F	C	Ⓐ	D	G	

In this case it is simple to find the orbits of $K=\mathcal{G}_{\{1\}}$. There are six of them, viz.,

$$\Delta_1=\{(1,A)\}\equiv\{1\},$$
$$\Delta_2=\{(1,B),(1,D)\},$$
$$\Delta_3=\{(2,B),(6,B),(3,D),(4,D)\},$$
$$\Delta_4=\{(2,C),(2,E),(6,F),(6,G),(3,C),(3,F),(4,E),(4,G)\},$$
$$\Delta_5=\{(5,E),(7,C),(7,G),(5,F)\},$$
$$\Delta_6=\{(5,A),(7,A)\}.$$

Alternatively, one can use the following theorem [Theorem VII on p. 191 of Burnside (1911)] to get the number of orbits of K ($\equiv\mathcal{G}_{\{1\}}$), the stabilizer. "The sum of the numbers of symbols left unchanged by each of the permutations of a permutation group of order N is tN, where t is the number of transitive sets (orbits) in which the group permutes the symbols. The sum of the squares of the numbers of symbols left unchanged by each of the permutations of a *transitive group* of order N is sN where s is the number of transitive sets in which a subgroup leaving one symbol unchanged (stabilizer) permutes the symbols."

In our case, the number of flags left fixed by the elements of K are, respectively, 21, 5, 5, 5, 5, 1, 1, 5 in the diagrams $\mathcal{D}_1,\mathcal{D}_2,\ldots,\mathcal{D}_8$. Hence the number of orbits of the stabilizer $K=(21+5\times5+2\times1)/|K|=48/8=6$.

The six incidence matrices $B(\Delta_i)\equiv B_i$, $i=1,2,\ldots,6$ corresponding to the six orbits can be easily derived. These provide a linear basis of the commutant algebra corresponding to the group $\mathcal{G}\equiv\mathrm{PC}(2,2)$ represented as a group $\mathcal{G}^*$ of permutation matrices acting on the incident letter-column pairs. Hence, if we assume that the covariance matrix $\Sigma_{21\times21}$ of the observations $Y_1,Y_2,\ldots,Y_{21}$ from this BIB design ($v=b=7$, $r=k=3$, $\lambda=1$) is invariant under the action of the permutation matrices

$$\Sigma=\sum_{i=1}^{6}c_iB_i$$

where the B_i's are known.

However, an alternative linear basis of the six-dimensional commutant algebra is the six relationship matrices of James, I, G, B, T, BT and TB. This is because these matrices are linearly independent and invariant under the action of the permutation matrices of $\mathcal{G}^*$.

6. Decompositions into direct sums of irreducible representations of the permutation representations of PC(2,2) and EC(2,3)

Further insight into the structure of the permutation representation of a collineation group of a design and that of the corresponding commutant algebra can be obtained by deriving first the decomposition of the permutation-representation into a direct-sum of its irreducible representations over an algebraically closed field (in our case the field of complex numbers). We have derived these decompositions for the two groups PC(2,2) and EC(2,3) and obtained the structures of the corresponding commutant algebras. [Details of proofs which involve results from the theory of group representations are reported in Burton's dissertation (1980) and will be published in a mathematical journal.]

Here we give a brief outline of an approach and comment on the statistical nature of the decompositions. Both PC(2,2) and EC(2,3) are flag-transitive collineation groups and $|PC(2,2)|=168$ and $|EC(2,3)|=432$. First, using a computer, the numbers of flags left fixed by each element of the full collineation groups PC(2,2) and EC(2,3) were obtained. Next, using these counts and the tables of group characters given by Littlewood (1950), the decompositions of the permutation representations (permutations of the set of flags) of the two groups, into direct sums of their irreducible representations were obtained. Then using Schur's lemma, the structures of the corresponding commutant algebras were derived.

We recall PC(2,2) is the collineation group of the design $v=b=7$, $r=k=3$, $\lambda=1$ and this design has 21 flags (point-block incident pairs). A 21×21 matrix D belonging to the algebra of matrices generated by the permutation matrices representing PC(2,2) has then the representation

$$D=D_1\oplus I_2\otimes D_6\oplus D_8,$$

where D_i is a real (orthogonal) square matrix with i rows and I_j is the identity matrix of order j. This means D can be seen pictorially as

```
        1  [ *                                  ]
           [     ******                         ]
           [     ******                         ]
        6  [     ******                         ]
           [     ******                         ]
           [     ******                         ]
           [     ******                         ]
           [           ******                   ]
           [           ******                   ]
D=         [           ******                   ]
        6  [           ******                   ]
           [           ******                   ]
           [           ******                   ]
           [                 ********           ]
           [                 ********           ]
           [                 ********           ]
        8  [                 ********           ]
           [                 ********           ]
           [                 ********           ]
           [                 ********           ]
           [                 ********           ]
```

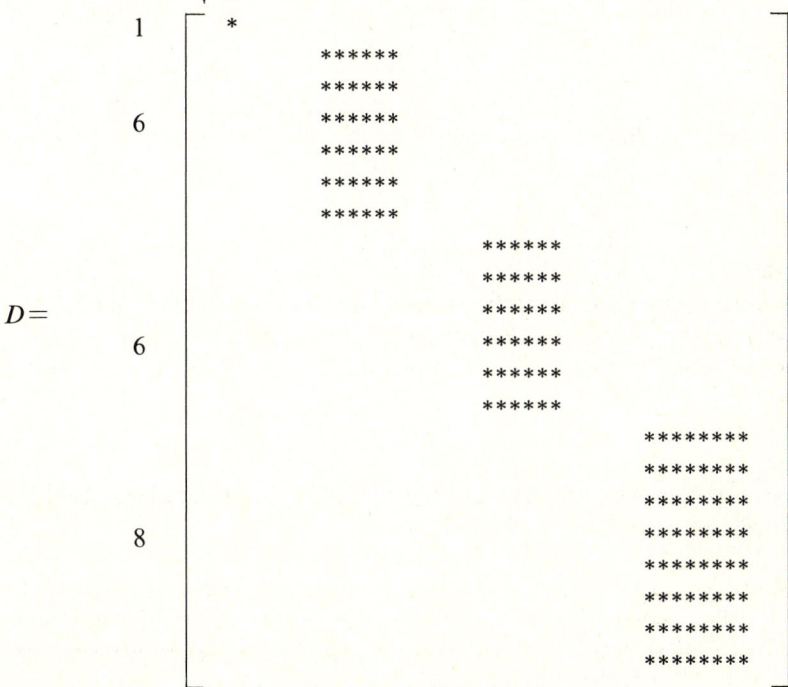

The partition $21=1+6+6+8$ corresponds to the partition of the degrees of freedom in the usual analysis of variance of such a design, viz., 1 for the constant, 6 for treatments, 6 for blocks and 8 for the error. Let C denote a matrix belonging to the corresponding commutant algebra. Then

$$C=a\oplus\begin{pmatrix} b_{11} & b_{12} \\ b_{21} & b_{22} \end{pmatrix}\otimes I_6\otimes cI_8$$

$$=\begin{pmatrix} a & & & \\ & b_{11}I_6 & b_{12}I_6 & \\ & b_{21}I_6 & b_{22}I_6 & \\ & & & cI_8 \end{pmatrix}$$

where the parameters $(a, b_{11}, b_{12}, b_{21}, b_{22}, c)$ are real.

EC(2, 3) is the collineation group of the design $v=9, b=12, r=4, k=3, \lambda=1$. This design has 36 flags. The 36-dimensional space of the permutation representation of this group, is a direct sum in which the one-dimensional real space occurs *once*, a three-dimensional real space occurs *once*, an eight-dimensional real space occurs *twice* and a sixteen-dimensional real space occurs *once*. In symbols

$$D=D_1\oplus D_3\oplus I_2\otimes D_8\oplus D_{16}$$

where D is a matrix belonging to the algebra of permutation-matrices (36×36) representing this group. The partition $36=1+3+8+8+16$ corresponds to the partition of the degrees of freedom in the conventional analysis of variance for such a resolvable BIB design, namely, 1 for the constant, 3 for the replications, 8 for the treatments, 8 for the blocks within replications and 16 for the error. If C is a matrix belonging to the corresponding commutant algebra, then C has the structure

$$C=aI_1\oplus bI_3\oplus\begin{pmatrix} c_{11}c_{12} \\ c_{21}c_{22} \end{pmatrix}\otimes I_8\oplus dI_{16}.$$

The parameters $(a, b, c_{11}, c_{12}, c_{21}, c_{22}, d)$ are real. $(I, G, B, T, BT, TB$ and $BTB)$ form a linear basis in this case.

We thus conclude that under the assumption that the covariance matrix Σ of the observations $(Y_1, Y_2, \ldots, Y_n)$ belong to the commutant algebra corresponding to the group of matrices permuting the incident point-line pairs of PG(2, 2) (EG(2, 3)), Σ has the form $c_1I+c_2G+c_3B+C_4T+c_5BT+c_6TB$ $(d_1I+d_2G+d_3B+d_4T+d_5BT+d_6TB+d_7BTB)$ instead of $\Sigma=\sigma^2I$ as is assumed in the conventional analysis of variance.

7. Conclusion

Efficient estimation of parameters of this modified linear model, occurring both in the expectation and the covariance matrix of the vector of observations, has been worked out and will be published in a separate communication.

In conclusion, we point out that we have proved that:

(i) The *dimension* of the commutant algebra corresponding to the full collineation group PC(2, p^r) (represented as a permutation group of matrices permuting the flags (the incident point-line pairs) of PG(2, p^r) is *six* for every prime p and every integer r. The relationship matrices I, G, B, T, BT and TB defined by James (1957) form a linear basis for this commutant algebra.

(ii) The dimension of the commutant algebra corresponding to the full collineation group EC(2, p^r) (represented as a permutation group of matrices acting on the flags) of EG(2, p^r) is *seven* for every prime p and every integer r. The matrices I, G, B, T, BT, TB, BTB form a linear basis of this commutant algebra.

(iii) The dimension of the commutant algebra corresponding to the full collineation group PC(3, p^r) (represented as a permutation group of matrices acting on the flags – the incident point-plane pairs) of PG(3, p^r) is *seven* for every prime p and every integer r. The relationship matrices I, G, B, T, BT, TB and S form a linear basis of this commutant algebra. The matrix $S=(s_{ij})$ is defined by $s_{ij}=1$ if $(BT)_{ij}=(TB)_{ij}=1$, $s_{ij}=0$ otherwise (Sysoev and Shaikin, 1976).

(iv) The dimension of the commutant algebra corresponding to the full collineation group EC(3, p^r) (represented as a permutation group of the matrices acting on the flags – incident point-plane pairs) of EG(3, p^r) is *eight* for every prime p and every integer r. The relationship matrices I, G, B, T, BT, TB, BTB and S form a linear basis of this algebra.

The proofs which make use of Theorem 28.4 (due to Schur) in Wielandt (1964, p. 80), are presented in the doctoral dissertation of Burton and will be published elsewhere.

Note added in proof

The authors have now shown that the results in (iii) and (iv) hold respectivily for PC(d, p^r) and EC(d, p^r) for every $d \geqslant 3$.

References

Burnside, W. (1911). *Theory of Groups of Finite Order.* Cambridge University Press (Dover Publications, New York, 1955, 2nd ed.).

Burton, C.T. (1980). Automorphism groups of balanced incomplete block designs and their use in statistical model construction and analysis. Dissertation, Department of Statistics, University of North Carolina at Chapel Hill, NC.

Carmichael, R.D. (1937). *Introduction to the Theory of Groups of Finite Order.* Dover Publications, New York.

Dembowski, P. (1968). *Finite Geometries.* Springer-Verlag, New York.

Hall Jr., M. (1959). *The Theory of Groups.* Macmillan, New York.

James, A.T. (1957). The relationship algebra of an experimental design. *Ann. Math. Statist.* **28**, 993–1002.

Mann, H.B. (1960). The algebra of a linear hypothesis. *Ann. Math. Statist.* **31**, 1–15.

Sysoev, L.P. and Shaikin, M.E. (1976). Algebraic methods of investigating the correlation connections in incompletely balanced block schemes for experiment design. I. Characterization of covariance matrices in the case of asymmetric block schemes. *Avtomatika i Telemekhanika* (in Russian) No. 7, 57–67. (Translated as *Automation and Remote Control* **37** (1976) No. 5, 696–704.) II. Relationship algebras and characterization of covariance matrices in the case of symmetric block schemes. *Avtomatika i Telemechanika* (in Russian) No. 7, 57–67. (Translated as *Automation and Remote Control* **37** (1976) No. 7, 1022–1031.)

Weyl, H. (1964). *Symétrie et mathématique moderne* (originally published in English under the title *Symmetry*, Princeton University Press, 1952). Flammarion, Paris.

Wielandt, H. (1964). *Finite Permutation Groups*. Academic Press, New York.

G. Kallianpur, P.R. Krishnaiah, J.K. Ghosh, eds., *Statistics and Probability: Essays in Honor of C.R. Rao*
© North-Holland Publishing Company (1982) 179–192

DEFICIENCY FOR MULTIPARAMETER TESTING PROBLEMS

Tapas K. CHANDRA and J.K. GHOSH
Indian Statistical Institute, India

1. Introduction

Consider a simple null hypothesis H_0: $\theta=\theta_0$, a composite alternative H_1: $\theta\in\Theta_1$, two test statistics $T_{1,n}$ and $T_{2,n}$ and critical regions of the form $\{T_{i,n}>t_{i,n}\}$. Let $\alpha_{i,n}$ and $\beta_{i,n}(\theta)$, $i=1,2$, denote the error of first kind and the power function, respectively. One way of comparing the two tests under a fixed $\theta\in\Theta_1$ is to choose $t_{i,n}$ such that $\beta_{1,n}(\theta)=\beta_{2,n}(\theta)=\beta$ (β being fixed and strictly between zero and one) and then compare the sample sizes n_1, n_2 necessary to get (approximately) the same error of first kind; specifically one considers the limit of n_1/n_2 as the common (approximate) value of $\alpha_{i,n}$ goes to zero. This is a rough description of the approach due to Bahadur (1967, 1971) and Cochran (1952). In the case when the limit of n_1/n_2 turns out to be one, Chandra and Ghosh (1978) have studied the limiting behaviour of (n_1-n_2). In the next paragraph, we will summarise most of what we need from this paper.

Assume that for $i=1,2$,

$$\log\alpha_{i,n}(\theta,\beta)=-na(\theta;\beta)+n^{1/2}b_i(\theta,\beta)$$
$$+(\log n)c_i(\theta;\beta)+d_i(\theta,\beta)+o(1). \tag{1.1}$$

Then [vide Lemma 2.3.1 and its proof, Chandra and Ghosh (1978, pp. 261, 273)] there is an extension $\alpha_{i,x}$ of $\alpha_{i,n}$ defined for all non-negative real x such that $\alpha_{i,x}$ is continuous and decreasing in x for sufficiently large x and satisfies (1.1) with x in place of n. The first two properties of $\alpha_{i,x}$ ensure that for sufficiently small δ, $n_1(\delta)$ and $n_2(\delta)$ exist such that $\alpha_{i,x}=\delta$ if and only if $x=n_i(\delta)$. Theorem 2.3.1 of Chandra and Ghosh (1978, p. 261) then ensures that the limit of (n_1-n_2) as $\delta\to 0$ exists and equals $\lambda=(d_1-d_2)/a$ if $b_1=b_2$ and $c_1=c_2$. (If $b_1\neq b_2$ or $c_1\neq c_2$, the limit exists but is not finite.) This limit, finite or not, is the approximate Bahadur–Cochran deficiency (of $T_{1,n}$ with respect to $T_{2,n}$) to be abbreviated henceforth as deficiency. Another simple interpretation of λ is provided by the relation $\alpha_{2,n}=(\exp(-a\lambda)+o(1))\alpha_{1,n}$ showing how large $\alpha_{1,n}$ is compared with $\alpha_{2,n}$.

In the subsequent pages we calculate this quantity in some common multiparameter multivariate problems. For these problems Proposition 2.5.1, a one-dimensional large deviation result (and hence Theorem 2.5.1) of Chandra and Ghosh (1978) requires substantial modifications; one has to replace the techniques of Bahadur and Ranga Rao (1960) by those of Borovkov and Rogozin (1965).

In Section 2 we have a sample of size n from a k-variate normal population with the mean vector, θ, and identity as the dispersion matrix; the null hypothesis H_0: $\theta=0$ is tested against H_1: $\theta_i>0$, $i=1,2,\ldots,k$. Here $T_{2,n}$ is the likelihood ratio test and $T_{1,n}$ is the likelihood ratio test against the unrestricted alternative H_1': $\theta\neq 0$. If $\theta_i>0$, $i=1,\ldots,k$, the deficiency equals $(2\log 2/\|\theta\|^2)k$ and what is more illuminating in this case, $\alpha_{2,n}=(2^{-k}+o(1))\alpha_{1,n}$. A similar result holds for $k=2$ when the alternative is restricted to a non-convex set $\{(\theta_1,\theta_2)$: $\theta_1>0$ and $\theta_2>0$ or $\theta_1<0$ and $\theta_2<0\}$.

In Section 3 we will consider a sample of size n from a bivariate normal population with the mean vector, θ, and dispersion matrix identity; the null hypothesis is H_0: $\theta=0$ to be tested against H_1: $\theta\neq 0$. Here $T_{1,n}$ is the likelihood ratio statistic and

$$T_{2,n}=\int f_\theta \pi(\theta)\,\mathrm{d}\theta/f_0 \tag{1.2}$$

where f_θ is the joint density under θ and π is a prior density (we will refer to it as the Bayes test). Assuming that $\pi(\theta)$ is continuous and positive everywhere, we first approximate $T_{2,n}$ and then approximate $t_{2,n}$. Under additional conditions, $\alpha_{2,n}$ is evaluated; the value of $\alpha_{1,n}$ remains the same as in Section 2 and so the deficiency can be calculated. It turns out that the deficiency of the likelihood ratio test is less than

$$\|\theta\|^{-2}\log\{(\pi\partial^2 g/\partial\phi^2)(\|\theta\|^2,\phi_0)\} \tag{1.3}$$

where $-\frac{1}{2}g$ is the logarithm of the prior density written in terms of the polar coordinates $\|\theta\|^2$ and ϕ, and ϕ_0 is the value of ϕ at which, for fixed $\|\theta\|$, g attains its minimum. Thus the deficiency at θ depends on the curvature of g on the circle that passes through θ and has origin at the centre. If the prior density is normal with a zero mean vector, the variances equal to σ^2 and correlation coefficient $\rho\neq 0$, the deficiency is

$$\frac{1}{\|\theta\|^2}\left\{-|\rho|\frac{(\theta_1\mp\theta_2)^2}{(1-p^2)\sigma^2}+\log\left(\frac{\pi|\rho|\,\|\theta\|^2}{\sigma^2(1-\rho^2)}\right)\right\}, \tag{1.4}$$

accordingly as ρ is positive or negative.

In Section 4 we extend the results of Section 3 to linear exponential densities

$$f_\theta(x)=\exp\left\{\sum_{i=1}^{k}\theta_i x_i-C(\theta)\right\} \tag{1.5}$$

with respect to an absolutely continuous measure, with θ varying over the natural parameter space which is assumed to be an open set. We test H_0: $\theta=0$ against H_1: $\theta\neq 0$ and define $T_{1,n}, T_{2,n}$ as before. Under the assumptions that π is positive everywhere and twice continuously differentiable we show that an approximate Bayes test is given by

$$\langle\hat\theta,\bar X\rangle-C(\hat\theta)-n^{-1}(\sigma(\hat\theta)-\log\pi(\hat\theta))>k_{2,n}' \tag{1.6}$$

where $\hat{\theta} \equiv \hat{\theta}(\bar{X})$ is the maximum likelihood estimator and σ is defined by (4.4). [This follows from (4.5) which may be regarded as a refinement of the main result of Schwarz (1978), under conditions stronger than his.] Under rather stringent conditions, $\alpha_{2,n}$ has been found. By using Theorem 1 of Woodroofe (1978), the deficiency can then be readily computed and bounded by a certain integral involving the curvature of $\sigma(\hat{\theta}(x)) - \log \pi(\hat{\theta}(x))$ for fixed $\phi(x) = \langle \hat{\theta}(x), x \rangle - C(\hat{\theta}(x))$. Section 3 contains the special case where M_n [see (4.13)] is zero dimensional. Results of the same type for composite hypotheses have been obtained similarly, but they require even more stringent conditions and so are not reported here.

The remarks of Section 4 are intended to clarify the technical assumptions. In particular, the final remark indicates how our results are related to Theorem 1 of Woodroofe (1978) and can be used to get large deviation probabilities for a class of statistics which includes the likelihood ratio criterion. (In the subsequent sections we will adhere to the notations introduced above unless otherwise stated.)

2. Normal with restricted mean vector

We will first find the size of the unrestricted likelihood ratio test. The critical region of this test is $\{\|\bar{X}\|^2 > k_{1,n}\}$ where $k_{1,n}$ is such that the power at θ is β. Let z_β be defined by,

$$\int_{z_\beta}^{\infty} (2\pi)^{-1/2} \exp(-x^2/2)\, dx = \beta.$$

Since $\|\bar{X}\|^2$ when suitably normalised has an Edgeworth expansion (in powers of $n^{-1/2}$), there exists a constant $d(k)$ (free from n) such that,

$$k_{1,n} = \|\theta\|^2 + 2\|\theta\| z_\beta n^{-1/2} + 2\|\theta\| d(k) n^{-1} + o(n^{-1}); \tag{2.1}$$

$(d(k) = (z_\beta^2 + k - 1)/2\|\theta\|)$. Now,

$$\alpha_{1,n} = \frac{(nk_{1,n})^{(k/2)-1} \exp(-\tfrac{1}{2} nk_{1,n})}{2^{k/2}\Gamma(\tfrac{1}{2}k)} \int_0^{\infty} \frac{2}{n} \exp(-\tfrac{1}{2}u)\left(1 + \frac{u}{k_{1,n}}\right)^{(k/2)-1} du$$

$$= \frac{(\tfrac{1}{2} nk_{1,n})^{(k/2)-1} \exp(-\tfrac{1}{2} nk_{1,n})}{\Gamma(\tfrac{1}{2}k)} (1 + o(1)). \tag{2.2}$$

We now come to the case of the restricted likelihood ratio test. Its critical region is of the following form:

$$\{\|\bar{X}\|^2 > k_{2,n}, \bar{X}_i > 0, i = 1, \ldots, k\} \cup \left\{ \sum_{i \notin J} \bar{X}_i^2 > k_{2,n}, \bar{X}_i < 0 \text{ iff } i \in J \right\},$$

where the union is taken over all non-empty proper subsets J of $\{1, \ldots, k\}$. As before, we determine $k_{2,n}$ such that the power of this test at θ is β. One sees immediately from the well-known asymptotic estimate of the tail of the standard normal distribution that,

$$P_\theta(\|\bar{X}\|^2 > k_{2,n}) - \beta \to 0, \quad \text{as } n\to\infty,$$

at an exponential rate. Consequently,

$$k_{2,n} - k_{1,n} = o(n^{-i}), \quad i = 1,2,\dots .$$

It is then evident that

$$P_{\theta=0}(\|\bar{X}\|^2 > k_{2,n}, \bar{X}_i > 0, i=1,\dots,k) = (2^{-k} + o(1))\alpha_{1,n}$$

and that for any J

$$\begin{aligned} P_{\theta=0}\Big(\sum_{i\notin J} \bar{X}_i^2 > k_{2,n}, \bar{X}_i < 0 \text{ iff } i\in J\Big) &= \\ &= 2^{-j} \;\; (\text{right side of (1.2) with } k \text{ replaced by } (k-j)) \\ &= o(\alpha_{1,n}), \end{aligned}$$

j being the number of elements of J. Thus

$$\alpha_{2,n} = (2^{-k} + o(1))\alpha_{1,n}$$

implying that deficiency of the unrestricted likelihood ratio test with respect to the restricted likelihood ratio test is $(2\log 2\|\theta\|^{-2})k$.

We next consider the above problem except that the new alternative hypothesis specifies a non-convex set of θ. Specifically let $k=2$ and H_1 be $\{\theta\colon \theta_1\theta_2 > 0\}$. Then one has $\alpha_{2,n} = (2^{-1} + o(1))\alpha_{1,n}$ and so the deficiency is $\|\theta\|^{-2}\log 2$.

3. Comparison of Bayes and likelihood ratio tests for the mean vector of a bivariate normal population

To get an asymptotic expansion for $\alpha_{2,n}$, it is convenient to write the Bayes critical region in the form

$$2n^{-1}\log\big((2\pi)^{-1} nT_{2,n}\big) > k_{2,n} \tag{3.1}$$

where $k_{2,n}$ is such that the power at (θ_1, θ_2) is β. [The reason for writing the critical region in this form will become clear from (3.2) below.] Now

$$\begin{aligned} T_{2,n} &= \exp(\tfrac{1}{2}n\|\bar{X}\|^2)\int \exp(\tfrac{1}{2}n\|\theta - \bar{X}\|^2 + \log\pi(\theta))\,d\theta \\ &= \exp(\tfrac{1}{2}n\|\bar{X}\|^2)\Big\{\int_0^\delta\int_0^{2\pi} 2^{-1}\exp\big(-\tfrac{1}{2}nr + \log\pi(\bar{X}_1 + r^{1/2}\cos\phi, \\ &\qquad \bar{X}_2 + r^{1/2}\sin\phi)\big)\,dr\,d\phi + O\big(\exp(-\tfrac{1}{2}n\delta)\big)\Big\} \\ &= 2\pi n^{-1}\exp\big(\tfrac{1}{2}n\|\bar{X}\|^2 + \log\pi(\bar{X}_1, \bar{X}_2)\big)(1 + o(1)), \end{aligned} \tag{3.2}$$

where the o(1) term is uniformly small on compact sets of $\bar{X}$. [Here we assume that $\pi(\theta_1, \theta_2)$ is positive and continuous everywhere.] We therefore consider an approximate Bayes test whose critical region is

$$T'_{2,n} \equiv \{\|\bar{X}\|^2 + 2n^{-1}\log\pi(\bar{X}_1, \bar{X}_2)\} > k'_{2,n} \tag{3.3}$$

where $k'_{2,n}$ is determined so that its power at (θ_1, θ_2) is $\beta + o(n^{-1/2})$. To get an expansion for $k'_{2,n}$, observe that

$$T'_{2,n} = \|\theta\|^2 + 2n^{-1}\log\pi(\theta_1, \theta_2)$$
$$+ n^{-1/2}(2\|\theta\|U + n^{-1/2}(U^2 + V^2)) + o_p(n^{-1})$$

where U, V are (appropriate linear) functions of $\bar{X}_1$ and $\bar{X}_2$ and are i.i.d. $N(0;1)$. Clearly then the statistic $T'_{2,n}$ has an Edgeworth expansion (in powers of $n^{-1/2}$). There exists, therefore, a constant, d, such that

$$n^{1/2}(k'_{2,n} - \|\theta\|^2 - 2n^{-1}\log\pi(\theta_1, \theta_2))/2\|\theta\| = z_\beta + n^{-1/2}d + o(n^{-1/2}),$$

that is,

$$k'_{2,n} = \|\theta\|^2 + 2\|\theta\| z_\beta n^{-1/2} + 2(\|\theta\| d + \log\pi(\theta_1, \theta_2))n^{-1} + o(n^{-1}) \tag{3.4}$$

[vide Lemma 2.5.1 of Chandra and Ghosh (1978)]. It is to be noted that d does not depend on the prior π; in fact d can alternatively be determined from the condition that,

$$\text{Prob}(U + (U^2 + V^2)/(2\|\theta\| n^{+1/2}) > z_\beta + n^{-1/2}d) = \beta + o(n^{-1/2}),$$

where U, V are i.i.d. $N(0;1)$ $(d = (z_\beta^2 + 1)/2\|\theta\|)$.

To evaluate the size $\alpha'_{2,n}$ of the approximate test, we apply the standard polar transformation and get

$$\alpha'_{2,n} = n(4\pi)^{-1}\exp(-nr_{0,n}/2)\left\{\int_0^\delta \int_{A_n} \exp(-nr/2)\,d\phi\,dr\right.$$
$$\left. + O(\exp(-\tfrac{1}{2}n\delta))\right\}$$

where

$$r_{0,n} = \inf\{r: 0 < r < \infty, 0 < \phi < 2\pi, r - n^{-1}g(r,\phi) > k'_{2,n}\},$$
$$g(r,\phi) = -2\log\pi(r^{1/2}\cos\phi, r^{1/2}\sin\phi), \tag{3.5}$$
$$A_n = \{\phi: 0 < \phi < 2\pi, r_{0,n} + r - n^{-1}g(r_{0,n} + r, \phi) > k'_{2,n}\}.$$

We now want to replace A_n by a suitable (approximating) interval of ϕ. Note that $\{r_{0,n}\}$ is bounded and the infimum is attained on the boundary and hence:

(i) $r_{0,n} > 0$ and $r_{0,n} - k'_{2,n} \to 0$ $(n \to \infty)$.

We now assume that:

(ii) there exists a unique ϕ, say $\phi_{0,n}$, such that,

$$r_{0,n} - n^{-1}g(r_{0,n}, \phi) = k'_{2,n},$$

and that $\phi_{0,n}\to\phi_0$ (say);

(iii) g is twice continuously differentiable with respect to r and ϕ; and finally

(iv) $(\partial^2 g/\partial\phi^2)(\|\theta\|^2,\phi_0)>0$.

Then by (i), (ii) and (iii) and equation (3.4), one can conclude that

$$\begin{aligned} r_{0,n}=\|\theta\|^2+2\|\theta\|z_\beta n^{-1/2}+2\|\theta\|\,\mathrm{d}n^{-1}\\ +2n^{-1}(\log\pi(\theta_1,\theta_2)-\log\pi(\|\theta\|\cos\phi_0,\|\theta\|\sin\phi_0))\\ +\mathrm{o}(n^{-1})\end{aligned}$$

and that

$$\begin{aligned} r_{0,n}+r-n^{-1}g(r_{0,n}+r,\phi)=k'_{2,n}+r\big(1-n^{-1}\partial g/\partial\phi(1+\mathrm{o}_r(1))\big)\\ -(2n)^{-1}(\phi-\phi_{0,n})^2\partial^2 g/\partial\phi^2(1+\mathrm{o}_r(1)),\end{aligned}$$

where $\mathrm{o}_r(1)$ denotes a term which goes to zero (uniformly in n) as $r\to0$; here we have used the fact that $\partial g/\partial\phi$ vanishes at $(r_{0,n},\phi_{0,n})$. Above and henceforth all derivatives of g are evaluated at $(r_{0,n},\phi_{0,n})$. Next we approximate A_n by the interval $|\phi-\phi_{0,n}|\le t_n(nr)^{1/2}$ where

$$t_n=\big(2(1-n^{-1}\partial g/\partial\phi)/(\partial^2 g/\partial\phi^2)\big)^{1/2}.$$

In fact A_n contains and is contained in intervals of the type $\phi_{0,n}\pm t_n(nr)^{1/2}(1+\mathrm{o}_r(1))$. Thus

$$\begin{aligned}\alpha'_{2,n}&=(2\pi)^{-1/2}t_n\exp(-\tfrac12 nr_{0,n})(1+\mathrm{o}(1))\\ &=\exp\big\{-\tfrac12 n\|\theta\|^2-\|\theta\|z_\beta n^{-1/2}-\|\theta\|d\\ &\quad-\tfrac12\log t-\log\pi(\theta_1,\theta_2)\\ &\quad+\log\pi(\|\theta\|\cos\phi_0,\|\theta\|\sin\phi_0)+\mathrm{o}(1)\big\}\end{aligned}\tag{3.6}$$

where

$$t=\pi(\partial^2 g/\partial\phi^2)(\|\theta\|^2,\phi_0).$$

We now return to the (exact) Bayes test with power β. Fix $\delta>0$ and choose a compact set C such that

$$\begin{aligned}&\text{(a)}\quad P_\theta\{(\bar X_1,\bar X_2)\notin C\}=\mathrm{o}(n^{-1/2})\\ &\text{(b)}\quad P_{\theta=0}\{(\bar X_1,\bar X_2)\notin C\}=\mathrm{O}\big(\exp(-\tfrac12 n(\|\theta\|^2+\delta))\big).\end{aligned}\tag{3.7}$$

Then for any $\varepsilon>0$ the set

$$\{2n^{-1}\log((2\pi)^{-1}nT_{2,n})>k_{2,n}\}\cap\{(\bar X_1,\bar X_2)\in C\}$$

lies between the sets

$$\{T'_{2,n} > k_{2,n} \pm 2\|\theta\|\varepsilon n^{-1}\} \cap \{(\bar{X}_1, \bar{X}_2) \in C\}.$$

Also (3.2) and (3.7a) imply that $T_{2,n}$ and $T'_{2,n}$ (in appropriate normalised form) have the same Edgeworth expansion up to $o(n^{-1/2})$. From this one derives as before that,

$$k_{2,n} = k'_{2,n} + o(n^{-1}). \tag{3.8}$$

It follows from the above facts and the expansion for the size of the approximate test that

$$\begin{aligned}&(\text{the right side of (3.6) with } d \text{ replaced by } (d+\varepsilon)) + O\big(\exp\big(-\tfrac{1}{2}n(\|\theta\|^2+\delta)\big)\big)\\ &\quad \leqslant \alpha_{2,n}\\ &\quad \leqslant (\text{the right side of (3.6) with } d \text{ replaced by } (d-\varepsilon))\\ &\qquad + O\big(\exp\big(-\tfrac{1}{2}n(\|\theta\|^2+\delta)\big)\big),\end{aligned}$$

which shows that (3.6) can be taken as the expansion for $\alpha_{2,n}$ as well.

By Theorem 2.3.1, p. 261 of Chandra and Ghosh (1978), it follows that the deficiency of the likelihood ratio test with respect to the Bayes test is

$$\|\theta\|^{-2}(2(-\log \pi(\|\theta\|\cos\phi_0, \|\theta\|\sin\phi_0) + \log \pi(\theta_1, \theta_2)) + \log t). \tag{3.9}$$

From our assumptions, it follows that for fixed $\|\theta\|$, $-\log \pi(\|\theta\|\cos\phi_0, \|\phi\|\sin\phi_0)$ is minimised at ϕ_0. Hence the deficiency (3.9) is less than $\|\theta\|^{-2}\log t$.

If instead of a unique $\phi_{0,n}$ we have $\phi^1_{0,n}, \ldots, \phi^J_{0,n}$ satisfying similar assumptions and converging to $\phi_0, \ldots, \phi^J_0$, respectively, then the deficiency will be a sum of J terms, obtained by replacing ϕ_0 by $\{\phi^j_0\}$ in (3.9).

In case the prior is bivariate normal with the mean vector 0, variances equal to σ^2 and correlation coefficient ρ, one can directly work with the Bayes test (instead of the approximate test) and easily verify that:

(a) the critical region of the Bayes test is

$$\|\bar{X}\|^2 + 2\rho\big(n\sigma^2(1-\rho^2)+1\big)^{-1}\bar{X}_1\bar{X}_2 > k_{2,n};$$

(b) $J=2$, $\phi^1_{0,n} = \frac{1}{4}\pi$, $\phi^2_{0,n} = \frac{5}{4}\pi$ and

$$r_{0,n} = \frac{k_{2,n}}{1+|\rho|\big(n\sigma^2(1-\rho^2)+1\big)^{-1}};$$

(c) the deficiency (3.9) is as given in (1.4).

For priors whose densities are not positive everywhere, the deficiency may be finite or infinity and in case it is finite, the value may considerably differ from (3.9). For example, if the prior is Lebesgue measure on the positive quadrant and is zero elsewhere, then it can be shown that the deficiency at θ (lying in the positive quadrant) is $-4\log 2\|\theta\|^{-2}$, which is same as the negative of the deficiency of

Section 2 (with $k=2$). On the other hand, if the prior is degenerate at θ, then

$$\alpha_{2,n} = \frac{\exp\left(-\frac{1}{2}\left(n^{1/2}\|\theta\|+z_\beta\right)^2\right)}{(2\pi n)^{1/2}\|\theta\|}(1+o(1))$$

and so the deficiency (at θ) is ∞.

4. Bayes test for the exponential family

For the exponential family (1.5), assume that the natural parameter space $\boldsymbol{\Theta}$ is open, $C(\theta)$ is strictly convex and that for some n_0, $\bar{X}_n$ lies in the set $(\nabla C)(\boldsymbol{\Theta})$ of possible expectations of the family almost surely (under θ_0) for all $n \geqslant n_0$. (The symbols ∇ and ∇^2 will denote gradient and Hessian, respectively.) The last condition ensures that the maximum likelihood estimator $\hat{\theta} \equiv \hat{\theta}(\bar{X}_n)$ given by the (unique) solution of the equation

$$(\nabla C)(\hat{\theta}) = \bar{X}_n \tag{4.1}$$

is well defined. Assume without loss of generality that:

$$\theta_0 = 0, \qquad C(0) = 0, \qquad \nabla C(0) = 0. \tag{4.2}$$

Let $I(\theta;\theta')$ be the Kullback–Leibler information number of θ, with respect to θ' and $\Sigma(\theta)$, the dispersion matrix of the family under θ:

$$\begin{aligned} I(\theta;\theta') &= \langle \theta-\theta', \nabla C(\theta)\rangle - C(\theta) + C(\theta'), \\ \Sigma(\theta) &= \nabla^2 C(\theta). \end{aligned} \tag{4.3}$$

Now consider the Bayes testing procedure and observe that

$$\begin{aligned} T_{2,n} &= \int \exp\left(n\langle\theta,\bar{X}\rangle - nC(\theta) + \log\pi(\theta)\right)\mathrm{d}\theta \\ &= \exp\left(n\phi(\bar{X}) - \sigma(\hat{\theta})\right)\Bigg(\int_{0<t(\theta)\leqslant\delta} \exp\Big(nt(\theta) \\ &\qquad + \log\pi(\hat{\theta}+B^{-1}\theta)\Big)\mathrm{d}\theta + \mathrm{O}(\exp(-n\delta))\Bigg). \end{aligned}$$

Here B is a matrix such that $B^{\mathrm{T}}B \equiv \Sigma(\hat{\theta})$ and

$$\begin{aligned} \phi(x) &= \sup_\theta \{\langle\theta, x\rangle - C(\theta)\} \\ &= \langle\hat{\theta}(x), x\rangle - C\left(\hat{\theta}(x)\right), \\ \sigma(\theta) &= \tfrac{1}{2}\log(\det(\Sigma(\theta))), \\ t(\theta) &= \langle B^{-1}\theta, \bar{X}\rangle - C(\hat{\theta}+B^{-1}\theta) + C(\hat{\theta}). \end{aligned} \tag{4.4}$$

One now applies the standard polar transformation and observes that the set $\{0<t(\theta)\leqslant\delta\}$ lies between two sets of the form $\{\|\theta\|^2\leqslant\delta'\}$ $(\delta'=0(\delta))$ (use the convexity of $t(\theta)$). Thus $T_{2,n}$ becomes:

$$(2\pi)^{k/2}n^{-k/2}\exp\big(n\phi(\bar{X})-\sigma(\hat{\theta})+\log\pi(\hat{\theta})(1+o(1))\big), \tag{4.5}$$

where the o(1) term goes to zero as $n\to\infty$ uniformly over compact sets of $\bar{X}$.

We therefore consider the approximate Bayes test which rejects H_0 if and only if $\bar{X}\in S_n$:

$$S_n=\{x:\phi(x)-n^{-1}G(\theta(x))>k'_{2,n}\}, \tag{4.6}$$

$$G(\theta)=\sigma(\theta)-\log\pi(\theta). \tag{4.7}$$

The constant $k'_{2,n}$ is determined so that the power at θ is $\beta+o(n^{-1/2})$. It follows from Theorem 2 of Bhattacharya and Ghosh (1978) that under θ, $\phi(\bar{X})$ has an Edgeworth expansion in powers of $n^{-1/2}$. One gets, as in Sections 2 and 3, by expanding ϕ a constant d (free from n) such that

$$\begin{aligned}P_\theta\{n^{1/2}(\langle\bar{X}-E_\theta(\bar{X}),\theta\rangle+\langle\hat{\theta}-\theta,\bar{X}-E_\theta(\bar{X})\rangle\\ -\tfrac{1}{2}\langle\hat{\theta}-\theta,\Sigma(\theta)(\hat{\theta}-\theta)\rangle)>z_\beta+n^{-1/2}d\}=\beta+o(n^{-1/2}).\end{aligned} \tag{4.8}$$

It also follows that

$$\begin{aligned}k'_{2,n}=I(\theta;\theta_0)+|\theta|z_\beta n^{-1/2}+\|\theta\|\,dn^{-1}\\ -(\log\pi(\theta)-\sigma(\theta))n^{-1}+o(n^{-1})\end{aligned} \tag{4.9}$$

where

$$|\theta|=(\langle\theta,\Sigma(\theta)\theta\rangle)^{1/2}. \tag{4.10}$$

To evaluate the size $\alpha'_{2,n}$ of the approximate test, we assume as in Woodroofe (1978) that there exists an integer n_1 such that the vector of the sample totals has a bounded continuous density for all $n\geqslant n_1$. Let

$$\phi_{\min}=\inf\{\phi(x):\phi(x)-n^{-1}G_1(x)\geqslant k'_{2,n}\} \tag{4.11}$$

where the subscript 1 in G_1 refers to the composition of G with $\hat{\theta}$ as a function of $\bar{X}$ (i.e., $G_1(x)=G(\hat{\theta}(x))$); similar conventions will be used below for other functions also. Now, by using the techniques of Borovkov and Rogozin (1965) [see also Proposition 1, Section 2 of Woodroofe (1978)], we get:

$$\begin{aligned}\alpha'_{2,n}=(n/2\pi)^{k/2}\Big\{\int_{A_{n,\delta}}\exp(-n\phi(x)-\sigma_1(x))\,dx\\ +O(\exp(-n(\phi_{\min}+\delta)))\Big\}(1+o(1))\end{aligned} \tag{4.12}$$

where

$$A_{n,\delta}=\{x|\phi(x)-n^{-1}G_1(x)\geqslant k'_{2,n},\ 0\leqslant\phi(x)-\phi_{\min}\leqslant\delta\}.$$

Note that $A_{n,\delta}$ is compact and so on it $\sigma_1(x)$ is bounded above.

To evaluate the integral on the right side of (4.12), we need to make a few assumptions. The following remark explains why they are plausible. Let

$$M_n = \{x \in S_n | \phi(x) = \phi_{\min}\}. \tag{4.13}$$

Remark 4.1. Since ϕ is strictly convex and its global minimum is attained outside S_n, M_n is a subset of ∂S_n, the boundary of S_n. Using Lagrangian multipliers, one expects that M_n may alternatively be obtained from the dual problem of minimising G_1 subject to the condition that $\phi = \phi_{\min}$. Noticing that $\phi_{\min}$ converges to $I(\theta; \theta_0)$ as $n \to \infty$, one can also consider the problem of minimising G_1 subject to the condition that $\phi = I(\theta; \theta_0)$. Let M^* be the set of points where this restricted minimum of G_1 is attained. One would then expect that M_n will "converge" to a (unique) subset M of M^*; since ϕ is strictly convex, M_n and M^* are (non-empty and) compact. We assume below that M is also compact. Let M be a $(k-r-1)$-dimensional submanifold of the $(k-1)$-dimensional manifold $N = \{x | \phi(x) = I(\theta; \theta_0)\}$. (In Section 3, M_n and M are zero-dimensional.) Get a finite open cover $U_1, \ldots, U_m$ of M satisfying the following conditions: for a fixed U_i, there exists a coordinate system $\eta_i(x) = (\eta_i^{(1)}(x), \ldots, \eta_i^{(k)}(x))$ such that:

(a) $x \to \eta_i(x)$ is a diffeomorphism on U_i;

(b) $\eta_i^{(k)} = \phi$; and

(c) $\eta_i^{(1)}, \ldots, \eta_i^{(k-1)}$ are local coordinates for the manifold N and

$$M \cap U_i = \{x \in N \cap U_i | \eta^{(1)}(x) = \cdots = \eta^{(k-r-1)}(x) = 0\}.$$

In view of Remark 4.3 below, we may assume without loss of generality that there is only one such coordinate system η which is a diffeomorphism on a neighbourhood V of M. Put

$$\eta^1 = (\eta^{(1)}, \ldots, \eta^{(k-r-1)}) \quad \text{and} \quad \eta^2 = (\eta^{(k-r)}, \ldots, \eta^{(k-1)}). \tag{4.14}$$

Since M_n converges to M, it seems reasonable to assume that for all sufficiently large n, M_n is also $(k-r-1)$-dimensional and that it is in fact of the form

$$M_n = \{x \in V | \eta^{(k)}(x) = \phi_{\min}, \eta^2(x) = C_n(\eta^1(x), \eta^{(k)}(x))\} \tag{4.15}$$

where

$$C_n(\eta^1(x), \eta^{(k)}(x)) \in R^r \text{ tends to zero uniformly in } V. \tag{4.16}$$

Then on M the gradient $\nabla_{(\eta^1, \eta^2)} G_1$ is zero and the Hessian $\nabla^2_{(\eta^1, \eta^2)} G_1$ is positive semidefinite. We assume below that the Hessian $\nabla^2_{\eta^2} G_1$ of G_1 with respect to η^2 is positive definite on V and hence on M_n for all sufficiently large n.

We now state these assumptions more formally. Assume that on $A_{n,\delta}$ there exists a one-to-one thrice continuously differentiable transformation $x \to \eta(x)$ such that

(a) $\eta^{(k)} = \phi$;

(b) equations (4.15) and (4.16) hold for all sufficiently large n;

(c) the Hessian $\nabla^2_{\eta^2} G_1(\eta^1, C_n(\eta^1, \eta^{(k)}), \phi_{\min})$ of G_1 with respect to η^2 is positive definite for all η;

(d) the elements of the Jacobian matrix are bounded. Here and in the following we use the same notations σ_1, G_1, etc., even after a change of variables.

[The assumption that a set M exists to which M_n converges will be made later, vide (4.20).]

If I_n denotes the integral on the right side of (4.12), then I_n can be written as:

$$\int_{\phi_{\min}}^{\phi_{\min}+\delta} \exp(-n\eta^{(k)}) \int_{D_{n,\delta}(\eta^{(k)})} \left(\int_{B_{n,\delta}(\eta^1,\eta^{(k)})} \exp(-\sigma_1(\eta)) J(\eta)\, d\eta^2 \right) d\eta^1\, d\eta^{(k)},$$

where $J(\eta)$ is the Jacobian of the transformation and

$$\begin{aligned} D_{n,\delta}(\eta^{(k)}) &= \text{the projection of } A_{n,\delta}(\eta^{(k)}) \text{ to } \eta^1 \text{ space}, \\ B_{n,\delta}(\eta^1,\eta^{(k)}) &= \text{the section at } \eta^1 \text{ of } A_{n,\delta}(\eta^{(k)}), \\ A_{n,\delta}(\eta^{(k)}) &= \text{the section at } \eta^{(k)} \text{ of the image } \eta(A_{n,\delta}) \text{ of } A_{n,\delta} \text{ under } \eta. \end{aligned} \tag{4.17}$$

For any $\eta \in \eta(A_{n,\delta})$ one gets, by expanding G_1 around $(\eta^1, C_n(\eta^1,\eta^{(k)}), \phi_{\min}) \in \eta(M_n)$,

$$\begin{aligned} \phi(x) - n^{-1}G_1(x) = k'_{2,n} + (\eta^{(k)} - \phi_{\min}) \left(1 - n^{-1} \frac{\partial G_1}{\partial \eta^{(k)}} (1+o(1)) \right) \\ - \frac{1}{2n} \left\langle \eta^2 - C_n(\eta^1,\eta^{(k)}), (\nabla^2_{\eta^2} G_1 + o(1))(\eta^2 - C_n(\eta^1,\eta^{(k)})) \right\rangle, \end{aligned}$$

where the o(1) terms go to zero uniformly in n and η^1 as $\delta \to 0$ and all the derivatives of G_1 are evaluated at $(\eta^1, C_n(\eta^1,\eta^{(k)}), \phi_{\min})$. (Here we have used the fact that $\partial G_1/\partial \eta^{(i)} = 0$ for $i = (k-r), \ldots, (k-1)$, that $\phi(x)$ attains its infimum on the boundary of S_n and that the following inclusion holds because of assumption (d):

$$\eta(A_{n,\delta}) \subset (\eta(M_n))^{\delta'}, \quad \delta' = 0(\delta), \quad \delta' \text{ free from } n, \tag{4.18}$$

the set on the right side being the δ'-neighbourhood of $\eta(M_n)$). By assumption (c) one therefore gets:

$$\begin{aligned} &\int_{B_{n,\delta}(\eta^1,\eta^{(k)})} \exp(-\sigma_1(\eta)) J(\eta)\, d\eta^2 = \\ &= (1+o(1)) \int_{\langle \eta^2 - C_n(\eta^1,\eta^{(k)}), \nabla^2_{\eta^2} G_1 (\eta^2 - C_n(\eta^1,\eta^{(k)})) \rangle \leq a_n} \exp(-\sigma_1) J\, d\eta^2 \\ &= \frac{\exp(-\sigma_1) J}{\left(\det(\nabla^2_{\eta^2} G_1)\right)^{1/2}} \cdot \frac{(\pi a_n)^{r/2}}{\Gamma(\frac{1}{2}r+1)} \cdot (1+o(1)), \end{aligned} \tag{4.19}$$

where $a_n = 2n(\eta^{(k)} - \phi_{\min})(1 - n^{-1}\partial G_1/\partial \eta^{(k)})$ and all the functions are evaluated at $(\eta^1, C_n(\eta^1,\eta^{(k)}), \phi_{\min})$.

Now we note that,

$$D_{n,\delta}(\phi_{\min}) = \{\eta^1 | (\eta^1, C_n(\eta^1,\eta^{(k)}), \phi_{\min}) \in \eta(M_n)\}.$$

We shall denote the set on the right side by $\eta^1(M_n)$. Let M be a non-empty compact subset of R^k such that as $n \to \infty$

$$\text{the Hausdorff distance of } M_n \text{ and } M \to 0 \tag{4.20}$$

and

$$\lambda\left(\eta^1(M_n)\Delta\eta^1(M)\right) \to 0, \tag{4.21}$$

where λ is $(k-r-1)$-dimensional Lebesgue measure, Δ denotes symmetric difference and $\eta^1(M)$ is the projection of M to η^1 space. We note that the strict convexity of ϕ, the fact that the origin lies outside S_n and assumption (d) imply that:

$$\begin{aligned}&\text{the Hausdorff distance of } D_{n,\delta}\left(\eta^{(k)}\right) \text{ and } \eta^1(M_n)\\ &\to 0 \text{ as } \delta \to 0 \text{ uniformly in } n \text{ and } \eta^{(k)}.\end{aligned} \tag{4.22}$$

We also assume that as $\delta \to 0$,

$$\lambda\left(D_{n,\delta}(\eta^{(k)})\Delta\eta^1(M_n)\right) \to 0 \text{ uniformly in } n \text{ and } \eta^{(k)}. \tag{4.23}$$

Note that under (4.20) and (4.22), assumption (c) can be deduced [using compactness of $\eta(M)$] from assumption (c′) below:

(c′) the Hessian $\nabla^2_{\eta^2}G_1(\eta^1, 0, I(\theta;\theta_0))$ is positive definite for all $\eta^1 \in \eta^1(M)$.

We now replace assumption (c) by (stronger) assumption (c′). We finally assume that:

(e) the Jacobian determinant of the transformation $\eta(x)$ is positive and bounded away from zero on $\eta(M)$.

Using (4.22) and (4.23) and eq. (4.19), the integral I_n can be reduced to:

$$n^{-1}(2\pi)^{r/2}\exp(-n\phi_{\min})(1+o(1))\int_{\eta^1(M_n)} \frac{\exp(-\sigma_1)J}{\left(\det\left(\nabla^2_{\eta^2}G_1\right)\right)^{1/2}} \cdot d\eta^1.$$

Finally by using (4.20) and (4.21) we get

$$\begin{aligned}\alpha'_{2,n} &= n^{(k/2)-1}(2\pi)^{-(k-r)/2}\exp(-n\phi_{\min})\\ &\quad\times(1+o(1))\int_{\eta^1(M)} \frac{\exp(-\sigma_1)J}{\left(\det\left(\nabla^2_{\eta^2}G_1\right)\right)^{1/2}} \cdot d\eta^1\end{aligned} \tag{4.24}$$

where in the integrand, all functions are evaluated at $(\eta^1, 0, I(\theta;\theta_0))$. It can now be seen as in Section 3 that the size of the Bayes test has the same asymptotic expansion as that of $\alpha'_{2,n}$.

Remark 4.2. Let $C_{n,\eta^{(k)}}$ and C be non-empty compact subsets of R^{k-r-1} such that as $n \to \infty$:

(i) the Hausdorff distance of $C_{n,\eta^{(k)}}$ and C tends to zero uniformly in $0 \leqslant \eta^{(k)} \leqslant \delta$;

(ii) the Hausdorff distance of $\partial C_{n,\eta^{(k)}}$ and ∂C tends to zero uniformly in $0 \leqslant \eta^{(k)} \leqslant \delta$ and

(iii) $\lambda(\partial C) = 0$; λ being $(k-r-1)$-dimensional Lebesgue measure. Then

$$\lambda\left(C_{n,\eta^{(k)}}\Delta C\right) \to 0 \quad (n \to \infty)$$

uniformly in $0 \leqslant \eta^{(k)} \leqslant \delta$. This result may be used to check conditions like (4.21) and (4.23).

Remark 4.3. Suppose our assumptions do not hold for S_n but there exists a finite open cover $U_1, \ldots, U_m$ of $\{\phi(x) \leqslant I(\theta; \theta_0) + \delta\}$ and the assumptions are true for $S_{n,i} = S_n \cap U_i$. Then we can write S_n as a finite disjoint union of sets $S'_{n,j}$ for each of which our assumptions hold. In this case final result (4.24) remains true.

Remark 4.4. If instead of assumption (e) we assume

$$J(\eta^1, \eta^2, \eta^{(k)}) \sim J_1(\eta^1, \eta^{(k)}) \psi(\eta^2) \quad \text{as } |\eta^2| \to 0,$$

where $\psi(C \cdot \eta^2) = C^\gamma.\ \psi(\eta^2)$ then (4.24) holds with a new integrand.

Example. Consider a trivariate normal population with mean $\theta = (\theta_1, \theta_2, \theta_3)$ and dispersion matrix identity. Let the prior density be

$$\pi(\mathrm{d}\theta) = \exp(-\tfrac{1}{2} a \theta_3^2)\, \mathrm{d}\theta, \quad a > 0.$$

The critical region of the Bayes test is

$$\| \overline{X} \|^2 - b_3 \overline{X}_3^2 \geqslant k_{2,n}$$

where

$$b_n = \frac{n+a-1}{n(n+a)} \cdot \frac{a}{2},$$

$$k_{2,n} = \| \theta \|^2 + 2\| \theta \| z_\beta n^{-1/2} + 2 \| \theta \| \mathrm{d} n^{-1} + \mathrm{o}(n^{-1}),$$

d being a suitable constant. The size of the test is

$$\alpha_{2,n} = \frac{n^{3/2}}{(2\pi)^{3/2}} \cdot \iiint \mathrm{e}^{-(n/2) r} \tfrac{1}{2} r^{1/2} \sin\phi_1 \, \mathrm{d}\phi_1 \, \mathrm{d}\phi_2 \, \mathrm{d}r$$

the integral being taken over the region

$$r(1 - b_n \sin^2 \phi_1) \geqslant k_{2,n}, \quad 0 < r < \infty, 0 < \phi_1 < \pi, 0 < \phi_2 < 2\pi.$$

Here $M_n = \{(\phi_1, \phi_2) \colon \phi_1 = 0 \text{ or } \phi_1 = \pi\}$ (free from n) and is of dimension 1; $r_{0,n}$, the smallest r in the region, is $k_{2,n}$ and $\gamma = 1$ (see Remark 4.5). One can verify that

$$\alpha_{2,n} = \frac{2 \exp(-\tfrac{1}{2} n k_{2,n}) n^{1/2}}{(2\pi k_{2,n})^{1/2}} (1 + \mathrm{o}(1)).$$

Remark 4.5. To see how our results are related to Theorem 1 of Woodroofe (1978), consider instead of S_n a set $S = \{x | \phi(x) - G(x) \geqslant C\}$ where G is thrice continuously differentiable. Let

$$M = \{x \in \partial S | \phi(x) = \phi_{\min}\},$$

$$\phi_{\min} = \inf\{\phi(x) | x \in S\}$$

and suppose that η can be defined with properties analogous to (a), (b), (c), (d) and (e). (In particular, $\eta^2=0$ on M.) Assume furthermore that:

$$1-\frac{\partial G}{\partial \eta^{(k)}}>0 \quad \text{on } M.$$

Then it can be shown that,

$$P_{\theta_0}(S)=\frac{n^{((k-r)/2)-1}\exp(-n\phi_{\min})}{(2\pi)^{(k-r)/2}}(1+o(1))$$

$$\times\int_{\eta(M)}\frac{\exp(-\sigma)J(1-\partial G/\partial\eta^{(k)})^{r/2}}{\left(\det\left(\nabla^2_{\eta^2}G\right)\right)^{1/2}}\cdot d\eta^1 \tag{4.25}$$

where in the integrand all functions are evaluated at $(\eta^1, 0, \phi_{\min})$ (definitions of σ, J etc., are obvious) and the dimension of M is $(k-r-1)$.

In Theorem 1 of Woodroofe, G is his ϕ_0 and one can take η^2 as his $\hat{\omega}_0$; here $\partial G/\partial\eta^{(k)}=0$ on M.

The assumptions made here can be relaxed as in Remark 4.3. In fact both (4.24) and (4.25) can be suitably modified to cover cases where M is a finite disjoint union of manifolds of different dimensions.

References

Bahadur, R.R. (1967). Rates of convergence of estimates and test statistics. *Ann. Math. Statist.* **38**, 303–324.

Bahadur, R.R. (1971). *Some Limit Theorems in Statistics*. SIAM, Philadelphia, PA.

Bahadur, R.R. and Ranga Rao, R. (1960). On deviations of the sample mean. *Ann. Math. Statist.* **31**, 1015–1027.

Bhattacharya, R.N. and Ghosh, J.K. (1978). On the validity of the formal Edgeworth expansion. *Ann. Statist.* **6**, 434–451.

Borovkov, A.A. and Rogozin, B.A. (1965). On the multi-dimensional central limit theorem. *Theor. Probability Appl.* **10**, 55–62.

Chandra, T.K. and Ghosh, J.K. (1978). Comparison of tests with same Bahadur-efficiency. *Sankhyā Ser. A* **40**, 253–277.

Chochran, W.G. (1952). The χ^2-goodness of fit test. *Ann. Math. Statist.* **23**, 493–507.

Schwarz, G. (1978). Estimating the dimension of a model. *Ann. Statist.* **6**, 461–464.

Woodroofe, M. (1978). Large deviations of likelihood ratio statistics with applications to sequential analysis. *Ann. Statist.* **6**, 72–84.

G. Kallianpur, P.R. Krishnaiah, J.K. Ghosh, eds., *Statistics and Probability: Essays in Honor of C.R. Rao*
© North-Holland Publishing Company (1982) 193–196

A REMARK ON THE CRAMÉR–RAO INEQUALITY

S.D. CHATTERJI
Département de Mathématique, Ecole Polytechnique Fédérale de Lausanne, Lausanne, Switzerland

1. Introduction

The Cramér–Rao inequality states that, under suitable regularity conditions, the variance of any estimator cannot be smaller than a certain quantity (given in terms of an expression containing the reciprocal of Fisher information: cf. (5) infra). Various authors [1,3] have pointed out that quite acceptable (and even better) inequalities of the above type hold under very mild conditions. The purpose of the present note is to indicate a general reasoning which contains the above-mentioned inequalities as well as the generalized Bhattacharya lower bounds as in [2].

2. Cramér–Rao bounds

Let (X,Σ) be a measurable sample space and $\{P_\alpha\}$, $\alpha\in I$, be a family of probability measures on it, indexed by some parameter set I and dominated by some σ-finite positive measure μ. Let $P_\alpha(\mathrm{d}x)=p(x,\alpha)\mu(\mathrm{d}x)$ and let $f\colon X\to\mathbb{R}$ be a measurable function such that

$$\int f(x)P_\alpha(\mathrm{d}x)=g(\alpha),\quad \alpha\in I,$$

i.e. f is an unbiased estimator of the real-valued quantity $g(\alpha)$. We shall assume that f is square-integrable with respect to all the P_α and obtain a lower bound for $V_{\alpha_0}(f)$, the variance of f under the probability measure P_{α_0}.

Let ν be some complex measure on a suitable σ-algebra on I such that $\nu(I)=0$ and such that p is $X\times I$ measurable and

$$\int\int |f(x)|\,p(x,\alpha)|\nu|(\mathrm{d}\alpha)\mu(\mathrm{d}x)<\infty.$$

This can always be arranged by taking ν to be some discrete measure concentrated at a finite number of points of I. Then by Fubini's theorem,

$$\begin{aligned}\int_I g(\alpha)\nu(\mathrm{d}\alpha)&=\int_X f(x)\cdot\left\{\int_I p(x,\alpha)\nu(\mathrm{d}\alpha)\right\}\mu(\mathrm{d}x)\\ &=\int_X f(x)K_\nu(x)\mu(\mathrm{d}x)\end{aligned}\tag{1}$$

where $K_\nu(x)=\int_I p(x,\alpha)\nu(\mathrm{d}\alpha)$. By the same argument (using now that $\nu(I)=0$), we have

$$0=\int_I \nu(\mathrm{d}x)=\int_X K_\nu(x)\mu(\mathrm{d}x). \tag{2}$$

Here Fubini's theorem is automatically applicable since

$$\int_{X\times I} p(x,\alpha)\mu(\mathrm{d}x)|\nu|(\mathrm{d}x)=|\nu|(I)<\infty.$$

If

$$\{x|K_\nu(x)\neq 0\}\subset\{x|p(x,\alpha_0)>0\}=X_0, \quad \mu \text{ a.e.,} \tag{3}$$

then from (1) and (2) we get,

$$\begin{aligned}\int_I g(\alpha)\nu(\mathrm{d}\alpha)&=\int_X \{f(x)-g(\alpha_0)\}K_\nu(x)\mu(\mathrm{d}x)\\ &=\int_{X_0}\{f(x)-g(\alpha_0)\}\sqrt{p(x,\alpha_0)}\cdot\{K_\nu(x)/p(x,\alpha_0)\}\\ &\qquad\cdot\sqrt{p(x,\alpha_0)}\,\mu(\mathrm{d}x)\end{aligned}$$

whence an application of the Cauchy–Schwarz inequality yields immediately that

$$V_{\alpha_0}(f)\geqslant\left|\int_I g(x)\nu(\mathrm{d}\alpha)\right|^2\Big/\int_X |K_\nu(x)/p(x,\alpha_0)|^2 P_{\alpha_0}(\mathrm{d}x). \tag{4}$$

If we can find ν such that (3) holds and the denominator is a finite non-zero quantity, then (4) gives a lower bound; taking the sup over all such ν's on the right-hand side of (4) gives even better lower bounds. The methods of [1] and [3] are clearly special cases of the reasoning indicated.

If I is a non-degenerate interval of $\mathbb{R}$ and $\nu_{\alpha'}$ is the obvious signed measure (concentrated at α_0 and α') such that

$$\int_I g(\alpha)\nu_{\alpha'}(\mathrm{d}\alpha)=\{g(\alpha')-g(\alpha_0)\}/(\alpha'-\alpha_0),$$

then the use of $\nu_{\alpha'}$ in (4) followed by a passage to the limit of α' to α_0 gives the usual Cramér–Rao inequality

$$V_{\alpha_0}(f)\geqslant|g'(\alpha_0)|^2/I(\alpha_0) \tag{5}$$

where $I(\alpha_0)=\int|(\partial \ln p)/(\partial\alpha)\,(x,\alpha_0)|^2 P_{\alpha_0}(\mathrm{d}x)$ (Fisher information) provided that certain conditions are satisfied. For instance, the following conditions would suffice:

(i) $g'(\alpha_0)$ exists;

(ii) $p(x,\alpha)>0$ μ a.e. for α near α_0;

(iii) For an x-set of full μ-measure and for α near α_0, $(\partial p/\partial\alpha)(x,\alpha)$ exists and

$$\left|\frac{\partial p}{\partial\alpha}(x,\alpha)\right|\leqslant M(x)p(x,\alpha_0)\quad\text{where } \int M(x)\mu(\mathrm{d}x)<\infty;$$

(iv) $(\partial p/\partial\alpha)(x,\alpha_0)\neq 0$ on an x-set of positive μ-measure.

Another possible way of deriving (5) from (4) under certain conditions is via distribution theory. Take $\nu_n(d\alpha)=\varphi_n'(\alpha)d\alpha$ where φ_n is a suitable sequence of test functions (infinitely differentiable with the compact support contained in the interior of I) such that as $n\to\infty$,

$$\int_I g(\alpha)\nu_n(d\alpha)=-\int_I g'(\alpha)\varphi_n(\alpha)d\alpha\to -g'(\alpha_0),$$

$$\int_I p(x,\alpha)\nu_n(d\alpha)=-\int_I \frac{\partial p}{\partial\alpha}(x,\alpha)\varphi_n(\alpha)d\alpha\to -\frac{\partial p}{\partial\alpha}(x,\alpha_0).$$

This method may be more useful if I is an open set in $\mathbb{R}^d$.

3. Bhattacharya bounds

The procedure here is exactly as in Section 2; we simply take ν to be vector-valued (say with values in $\mathbb{R}^k$) and use a standard inequality concerning Grammians in place of the Cauchy–Schwarz inequality.

Let f, g, etc. be as in Section 2 but take ν to be a $\mathbb{R}^k$-valued (column vectors) measure on a suitable σ-algebra on I. Assume (3) as before and consider the Grammian G of the $(k+1)$ elements of the Hilbert space $L^2(P_{\alpha_0})$ given by

$$f-g(\alpha_0) \quad\text{and}\quad K_\nu/p(\cdot,\alpha_0),$$

i.e. the covariance matrix of the above $(k+1)$-dimensional random vector in the probability space (X,Σ,P_{α_0}).

Then $G=\det A$ where [from (1) and (2)]

$$A=\left(\begin{array}{c|c} V_{\alpha_0}(f) & \gamma' \\ \hline \gamma & B\end{array}\right),$$

$$\gamma=\int_I g(\alpha)\nu(d\alpha),\ (\text{a } k\times 1 \text{ column vector}),$$

$$\gamma'=\text{transpose of }\gamma,$$

$$B=\int_X \{K_\nu(x)K_\nu'(x)/p^2(x,\alpha_0)\}P_{\alpha_0}(dx).$$

From a standard identity concerning determinants [cf. 4, p. 46 (II)]

$$G=(\det B)\cdot\{V_{\alpha_0}(f)-\gamma' B^{-1}\gamma\} \tag{6}$$

provided that B is nonsingular. Actually, in the present situation, (6) simply gives the well-known fact concerning Grammians that $G/\det B$ equals the square of the distance of $f-g(\alpha_0)$ from the subspace generated by $K_\nu/p(\cdot,\alpha_0)$. Since $G\geqslant 0$ we obtain from (6) the generalized Bhattacharya bounds:

$$V_{\alpha_0}(f)\geqslant\gamma' B^{-1}\gamma. \tag{7}$$

Let I be a non-degenerate interval of $\mathbb{R}$ and ν an $\mathbb{R}^k$ valued discrete measure of finite support such that $\int_I g(\alpha)\nu(d\alpha)$ is a vector of k divided differences around α_0, i.e. the coordinates are respectively:

$$\frac{g(\alpha_1)-g(\alpha_0)}{\alpha_1-\alpha_0},\ldots,\sum_{j=0}^{k}\left\{\frac{g(\alpha_j)}{\prod_{i\neq j}(\alpha_j-\alpha_i)}\right\},$$

where $\alpha_1\cdots\alpha_k$ are k points in I. This choice of ν in (7) gives the result of [2]. By letting α_1, α_2, etc. tend to α_0, (7) yields the usual Bhattacharya bounds (under some conditions quite analogous to those in Section 2) viz. (7) with

$$\gamma'=(g'(\alpha_0),\ldots,g^{(k)}(\alpha_0)),$$
$$B=(b_{ij})$$

where
$$b_{ij}=\int_X\left\{\frac{\partial^i p}{\partial\alpha^i}(x,\alpha_0)\frac{\partial^j p}{\partial\alpha^j}(x,\alpha_0)/p(x,\alpha_0)\right\}\mu(dx).$$

4. Further remarks

If f is a vector-valued estimator with values in $\mathbb{R}^d$ of a vector quantity $g(\alpha)$, then we can proceed (as in Section 2) with a scalar measure ν with $\nu(I)=0$ and apply the matrix considerations of Section 3 to obtain

$$\det\{\tilde{V}_{\alpha_0}(f)-b^{-1}\gamma'\gamma\}\geqslant 0 \tag{8}$$

where γ is now the $(1\times d)$ row vector $\int_I g(\alpha)\nu(d\alpha)$, $b=\int_X\{K_\nu^2(x)/p^2(x,\alpha_0)\}P_{\alpha_0}(dx)$ and $\tilde{V}_{\alpha_0}(f)$ is the $(d\times d)$ covariance matrix of f under the probability measure P_{α_0}. If we take ν to be a vector valued measure also, then proceeding as before but taking now tensor integrals $\int_I g(\alpha)\otimes\nu(d\alpha)$ obtain other determinantal inequalities like (8). Passage to limit as in Sections 2–3 in case I is a domain in some $\mathbb{R}^q$ will give inequalities like those in [5, p. 194]. Finally, we remark that the considerations of Section 2 extend readily to other convex loss functions in place of the quadratic function. Similar generalizations of the considerations of Section 3 or the preceding seem difficult to obtain.

References

[1] Chapman, D.C. and Robbins, H. (1951). Minimum variance estimation without regularity assumptions. *Ann. Math. Statist.* **22**, 581–586.
[2] Fraser, D.A.S. and Guttman, I. (1952). Bhattacharya bound without regularity assumptions. *Ann. Math. Statist.* **23**, 629–632.
[3] Keifer, J. (1952). On minimum variance estimators. *Ann. Math. Statist.* **23**, 627–629.
[4] Gantmacher, F.R. (1959). *The Theory of Matrices*, Vol. 1. Chelsea, New York. (Translated from Russian by K.A. Hirsch.)
[5] Zacks, S. (1971). *The Theory of Statistical Inference.* Wiley, New York.

G. Kallianpur, P.R. Krishnaiah, J.K. Ghosh, eds., *Statistics and Probability: Essays in Honor of C.R. Rao*

RANDOMIZATION AND CONCOMITANT VARIABLES IN THE DESIGN OF EXPERIMENTS

D.R. COX
Department of Mathematics, Imperial College, London, U.K.

General comments are made on experimental design in the presence of concomitant observations. The circumstances under which appreciable non-orthogonality may arise are investigated. It is shown that under some circumstances repeated rerandomization, until a nearly balanced design emerges, is justifiable from the point of view of pure randomization theory.

1. Introduction

A central theme in statistical studies of experimental design is the control of error via information about the experimental units available before the assignment of treatments. Control is usually achieved by a judicious combination of balancing (blocking or stratification), adjustment (for example by analysis of covariance or by fitting some suitable special model) and randomization. Sometimes the initial information worth including is quite limited and then, at least for fairly simple treatment structures, a satisfactory design can usually be found by conventional methods.

It is useful to draw a non-rigid distinction between blocking and stratification as devices for the control of error. In stratification the levels correspond to factors classifying the experimental units (e.g., by age and sex), whereas in blocking the levels correspond to groups of supposedly similar individuals not necessarily meaningfully labelled. Further, there may well be a large number of replications of each treatment within a stratum, whereas normally the number of replications per block will be small.

We consider experiments comparing t treatments $T_1, \ldots, T_t$, equally replicated, interest attaching to

$$\operatorname*{ave}_{u,v} \operatorname{var}(\hat{\tau}_u - \hat{\tau}_v), \tag{1}$$

where $\hat{\tau}_u$ is the estimated mean response for T_u. For simple experiments with one concomitant variable initially available per experimental unit, there will be a choice between using this variable to form blocks or as a basis for adjustment. Cox (1957) showed that with the criterion (1) it is only in very small experiments or when the relation between response and the concomitant variable is very strong that the choice is critical.

Consideration of a second use of concomitant variables, namely to detect interactions, tends to strengthen the case for a direct use of the concomitant variable, as against its indirect use to form blocks or strata.

Now if we use randomization followed by adjustment by analysis of covariance, there will be some inflation of variance arising from non-orthogonality, the amount depending on the lack of balance in the design realized and on the number of concomitant variables.

If there are initially available a considerable number of concomitant observations per experimental unit, it will often be a good idea to reduce the concomitant variables at the start to a small number of possibly derived variables. If, however, this cannot be done, stratification will typically be possible only for a few of the variables; the conventional approach is to randomize and to adjust for the remaining variables by analysis of covariance. We investigate this in more detail below.

The issues are of particular importance in connection with clinical trials where there has been some controversy over the best procedure when there are many prognostic (concomitant) variables. In clinical trials treatment assignment is nearly always sequential, a complication we largely ignore.

2. Theoretical development

Consider first an experiment with $n=2r$ experimental units to compare two treatments T_1 and T_2, each treatment being assigned to r units. Let a $p\times 1$ vector z of concomitant variables be available on each unit. Assume that linear methods of analysis based on the method of least squares are used.

Then if $\hat{\tau}_2-\hat{\tau}_1$ is the adjusted difference between treatments after allowing for linear regression on z, we have under the usual second-order assumptions about error that

$$\operatorname{var}(\hat{\tau}_2-\hat{\tau}_1)=\sigma^2\left\{\frac{2}{r}+(\bar{z}_2-\bar{z}_1)^{\mathrm{T}}S_z^{-1}(\bar{z}_2-\bar{z}_1)\right\}. \tag{2}$$

Here σ^2 is the residual variance per observation, $\bar{z}_1$ and $\bar{z}_2$ are the vector means of the concomitant variables for T_1 and T_2 in the design chosen and S_z is the matrix of sums of squares and products of z within treatments, eliminating block differences where appropriate.

More generally, if there are t treatments, each assigned to r experimental units,

$$\operatorname{ave}\operatorname{var}(\hat{\tau}_u-\hat{\tau}_v)=\sigma^2\left\{\frac{2}{r}+\frac{2}{r(t-1)}\operatorname{tr}\left(B_zS_z^{-1}\right)\right\}, \tag{3}$$

where B_z is the matrix of sums of squares and products between treatments, i.e. of r times the sum of squares and products of deviations of means.

If treatment assignment is randomized

$$E_{\mathrm{R}}(S_z/d_w)=\Omega_z,$$

where E_{R} denotes expectation over the randomization, d_w is the degrees of freedom within treatments and Ω_z is the "finite population" covariance matrix, within blocks or strata where appropriate. Further, the randomization distribution of $\bar{z}_2-\bar{z}_1$ is such that

$$E_{\mathrm{R}}(\bar{z}_2-\bar{z}_1)=0, \qquad E_{\mathrm{R}}\{(\bar{z}_2-\bar{z}_1)(\bar{z}_2-\bar{z}_1)^{\mathrm{T}}\}=2\Omega_z/r.$$

Approximate multivariate normality will hold as $n \to \infty$ for fixed p, under weak conditions.

It follows that

$$\tfrac{1}{2}r(\bar{z}_2-\bar{z}_1)^{\mathrm{T}}\Omega_z^{-1}(\bar{z}_2-\bar{z}_1)=\tfrac{1}{2}r\,\|\bar{z}_2-\bar{z}_1\|^2_{\Omega_z}, \tag{4}$$

say, has expectation p and asymptotically a chi-squared distribution with p degrees of freedom. Hence also

$$\tfrac{1}{2}r\,\|\bar{z}_2-\bar{z}_1\|^2_{S_z/d_w} \tag{5}$$

has approximately the same properties.

Thus, if we write W_p for a chi-squared random variable with p degrees of freedom, from (2) we have approximately

$$\operatorname{var}(\hat{\tau}_2-\hat{\tau}_1)=\frac{2\sigma^2}{r}\left(1+\frac{W_p}{d_w}\right); \tag{6}$$

the same formula follows from the more general form (3). The value of W_p depends on the particular design produced by the randomization.

If the numbers of blocks and treatments are small compared with n, d_w in (6) can be replaced by n. If we take expectations over the randomization, the inflation factor in (6) becomes

$$(1+p/d_w)\simeq(1+p/n), \tag{7}$$

in the special case just mentioned.

Despite the simplicity of these formulae, interpretation needs a little care.

3. Preliminary discussion

An initial interpretation of (7) when $d_w \simeq n$ is that, because $1+p/n \simeq n/(n-p)$, non-orthogonality is equivalent to a loss of p observations. This would suggest that so long, say, as $p<\frac{1}{10}n$ the effect of non-orthogonality is unimportant.

There are two reasons why this is oversimplified. First, in the final analysis it may well be that adjustment is carried out not for the full p concomitant variables but rather for a subset of p_0 variables which appear to show fairly clear connection with the response; some form of ridge regression is an alternative. The effect is likely to be to reduce the inflation factor $1+W_p/d_w$, although presumably the factor will typically exceed $1+W_{p_0}/d_w$.

The second effect works in the opposite direction. It is no defence of a particular experiment which is seriously unbalanced to show that the realized design is atypical. As a general guide in design, it is reasonable to require that most of the designs produced by the randomization are satisfactory. If $c^{*2}_{p\varepsilon}$ is the upper ε point of the chi-squared distribution with p degrees of freedom, we need not $p<\frac{1}{10}n$ but $c^{*2}_{p\varepsilon}<\frac{1}{10}n$ for perhaps $\varepsilon=0.01$, an appreciably more stringent requirement. For $\varepsilon=0.01$, $p=1,5,10,50,100$ the minimum values of n are $67,151,232,762,1358$. Thus, quite large values of n are needed if the chance of appreciable imbalance is to be kept at 0.01.

The above discussion is for linear methods of analysis. It is likely that similar conclusions apply more broadly, for example to the maximum likelihood analysis of models for binary or survival data. There is, however, the technical difficulty that if the adjustments are relatively small their study requires greater refinement than the "first order" asymptotic theory of maximum likelihood estimation.

4. Randomization

So far we have assumed that a standard scheme of randomization is used, taking account of blocking or stratification as appropriate. There are two difficulties.

One is that adjustment by analysis of covariance does not have an exact second-order randomization theory. This can be overcome by weighted randomization (Cox 1956) or by reliance on asymptotic arguments.

A more serious point concerns the circumstances under which arguments based on randomization are convincing for the interpretation of the particular experimental arrangement actually used. For this it is necessary that the arrangement will not be distinguishable in a relevant respect from the reference set of arrangements used in the randomization calculations. A precise formulation will not be attempted.

This raises special difficulties in the present instance because the quantity W_p does classify the possible designs into sets of differing precisions, at least under the associated standard linear model. This reinforces the desirability of rejecting any arrangement for which W_p/d_w is appreciable; if the proportion of arrangements so rejected is not small, the whole relevance of the randomization scheme is suspect.

Of course, quite generally when randomization is used, the individual arrangements have distinguishing features. It is always possible that rational arguments will be produced for regarding the arrangement used as atypical, in which case an ad hoc analysis has to be considered. There is, however, an onus on those suggesting additional complications to show that substantial benefit results. The wide success of randomization depends on the empirical fact that often, especially when the randomization set contains many arrangements, there are indeed appreciable numbers of designs that can reasonably be regarded as equivalent. Even then, however, designs having some feature in an extreme form are best rejected, the effect on the randomization distribution being small provided that the proportion of rejected arrangements is small. In the present instance one would normally want to discard arrangements with $W_p/d_w > \frac{1}{10}$; see Section 3 for the conditions under which this will involve only a small proportion of arrangements.

There is, however, another possible procedure for randomization. Suppose that we randomize conditionally on $\|\bar{z}_2 - \bar{z}_1\|_{\Omega_z}$ being small. That is, we randomize, compute $\|\bar{z}_2 - \bar{z}_1\|_{\Omega_z}$ or, more generally $\operatorname{tr}(B_z\Omega_z^{-1})$, accept the design if $\|\bar{z}_2 - \bar{z}_1\|_{\Omega_z} < a$, for suitable a, and otherwise reject the design and rerandomize. In some special cases, notably when the z's identify a system of blocking, there may exist a large number of arrangements with $\|\bar{z}_2 - \bar{z}_1\|_{\Omega_z} = 0$ and an exact second moment randomization theory may hold. In general, however, we must argue approximately as follows.

Under the usual randomization model the response observed on unit i is $\xi_i + \tau_{[i]}$, where $T_{[i]}$ is the treatment applied to that unit; in addition, a vector of concomitant variables is available on each unit. Then under complete randomization the joint distribution of $(\bar{\xi}_2 - \bar{\xi}_1, \bar{z}_2 - \bar{z}_1)$, the differences between the means for T_2 and T_1, has zero expectation and covariance matrix

$$\frac{2}{r}\begin{pmatrix} \sigma_\xi^2 & \Omega_{\xi z} \\ \Omega_{\xi z}^{\mathrm{T}} & \Omega_z \end{pmatrix},$$

say, with elements defined by the finite population of ξ's and z's. To the extent that a multivariate normal approximation is adequate, the distribution of $\bar{\xi}_2 - \bar{\xi}_1$, when randomization is constrained so that $\| \bar{z}_2 - \bar{z}_1 \|_{\Omega_z}$ is small, is normal with zero mean and variance

$$2\sigma_{\xi \cdot z}^2 / r,$$

in the usual notation. Finally, $\sigma_{\xi \cdot z}^2$ is consistently estimated from the empirical variance eliminating treatment effects and linear regression on z.

Now there are several requirements for the precise scheme of rerandomization:

(i) We want

$$\tfrac{1}{2} r \| \bar{z}_2 - \bar{z}_1 \|_{\Omega_z}^2 / d_w \leqslant b \tag{8}$$

for some suitable b, e.g. $b \leqslant \frac{1}{20}$, so that negligible inflation of variance occurs.

(ii) We want the selected design to satisfy

$$\tfrac{1}{2} r \| \bar{z}_2 - \bar{z}_1 \|_{\Omega_z}^2 \leqslant c^{*2}_{p,1-\alpha} \tag{9}$$

for some fairly small, but not very small, value of α, so that $\| \bar{z}_2 - \bar{z}_1 \|$ lies near the centre of its probability distribution under full randomization.

(iii) Except where special arguments can be employed, we want α defined by (9) not to be too small, so that the number of rerandomizations, geometrically distributed with mean $1/\alpha$, is not excessive.

(iv) If n_{R} is the number of distinct designs in the full randomization set, the mean in the constrained randomization set is approximately αn_{R} and this should be at least, say, 50–100 if an effective analysis based on the randomization distribution is to be in principle possible.

In very small experiments, having satisfied (8) and (9), condition (iv) will not be achieved. Then it seems necessary to abandon reliance on randomization theory, to assume a physical model for the random variation and to choose a systematic design; see Cox (1951) for the case where the z's determine a polynomial trend with "time".

Of course, if $p \ll n$ the much simpler procedure of Section 3, randomization with rejection of occasional extreme arrangements, will be satisfactory.

In clinical trials there is the further complication that entry is nearly always sequential. A scheme of "minimization" (Pocock 1979) could be used in which the assignment probabilities for a new individual with given z are, subject to concealed assignment, biased to the treatment leading to greater balance. The effect of such schemes on randomization analysis has not been fully discussed, although it is clear in a general way that any time trends in response should be eliminated from error.

References

Cox, D.R. (1951). Some systematic experimental designs. *Biometrika* **38**, 312–323.

Cox, D.R. (1956). A note on weighted randomization. *Ann. Math. Statist.* **27**, 1144–1151.

Cox, D.R. (1957). The use of a concomitant variable in selecting an experimental design. *Biometrika* **44**, 150–158.

Pocock, S.J. (1979). Allocation of patients to treatment in clinical trials. *Biometrics* **35**, 183–197.

G. Kallianpur, P.R. Krishnaiah, J.K. Ghosh, eds., *Statistics and Probability: Essays in Honor of C.R. Rao*
© North-Holland Publishing Company (1982) 203–207

A NOTE ON MULTIPLICITY THEORY

Harald CRAMÉR
University of Stockholm, Stockholm, Sweden

The multiplicity theory for purely non-deterministic stochastic processes was introduced in 1960 almost simultaneously by Hida for a one-dimensional normal (Gaussian) process, and by the present author for a general class of finite-dimensional vector processes. We both based our work on the theory of Hilbert space, Hida using self-reproducing kernels, and I, the class of selfadjoint linear transformations. (References to the literature of the subject will be found in my paper in Sankhyā A 40, pp. 91–115.)

In some unpublished university lectures during the years 1970–1972, I pointed out that it is by no means necessary to use these fairly complicated parts of Hilbert space theory for the deduction of the basic propositions of multiplicity theory. In fact, only the simplest properties of projection operators are required for the proof of a theorem in Hilbert space geometry which directly leads to the relevant properties of stochastic processes. In the present note I shall give the proof of this theorem, and then briefly indicate its application to stochastic processes.

We consider a separable Hilbert space, H, with points or elements $x, y, z, \ldots$, inner product (x, y) and norm

$$\|x\| = (x, x)^{1/2}.$$

Suppose a family of subspaces of H is given, i.e. $H(t)$, defined for all real t and such that $H(u) \subset H(t)$ for $u < t$. Suppose further that

$$H(-\infty) = 0, \qquad H(+\infty) = H, \qquad \text{and } H(t-0) = H(t),$$

for all finite t, where $H(-\infty)$ is the intersection, and $H(+\infty)$ the closure of the $H(t)$ for all t, while $H(t-0)$ is the closure of the $H(u)$ for all $u < t$.

By P_t we denote the projection operator in H with range $H(t)$, so that $H(t) = P_t H$. Then for $a < b$ the difference $P_b - P_a$ is the projection on the orthogonal complement $H(b) \ominus H(a)$, and may be conceived as the value of a projection measure for the interval $(a, b]$. In the usual way this may be extended to a projection measure $P(S)$ defined for all Borel sets S on the real line, and uniquely determined by the $H(t)$ family.

For any element y in H we denote by $C(y)$ the cyclic subspace of H spanned by the projections $P_u y$ for all real u, and by $C(y, t)$ the subspace of $C(y)$ spanned by the $P_u y$ for all $u \leqslant t$. Then obviously

$$C(y, t) = P_t C(y).$$

For any fixed y the squared norm

$$\|P_t y\|^2 = G(t)$$

is a real-valued never decreasing function of t, which is everywhere continuous to the left and is such that $G(-\infty)=0$ and $G(+\infty)=\|y\|^2$. We will call $G(t)$ the spectral function of y, and say that it determines the spectral type of y. Every element z in $C(y)$ can be expressed as an integral with respect to t

$$z = \int_{-\infty}^{\infty} h(t)\, dP_t y,$$

where $h(t) \in L_2 G(t)$ is a complex-valued function of t, and the correspondence between z and $h(t)$ determines an isomorphy between $C(y)$ and $L_2 G(t)$. It follows that every element in the subspace $C(y, s)$ can be expressed by an integral of the same form, with the upper limit replaced by s.

In the set of all elements y in H, and in the set of the corresponding spectral functions $G(t)$, we define a partial ordering by saying that y_1 dominates y_2, and G_1 dominates G_2, and writing

$$y_1 > y_2 \quad \text{and} \quad G_1 > G_2$$

whenever the function $G_2(t)$ is absolutely continuous with respect to $G_1(t)$. If $G_1 > G_2$, and at the same time $G_2 > G_1$, we say that G_1 and G_2 are equivalent, and similarly for y_1 and y_2. The set of all G equivalent to a given G_1 is called the equivalence class of G_1.

Let y_1 be an arbitrary non-zero element in H, with spectral function $G_1(t)$. Consider the orthogonal complement

$$H_1 = H \ominus C(y_1).$$

If $H_1 \neq 0$, which means that H_1 does not reduce to the zero element, we take a non-zero element y_2 in H_1, with spectral function $G_2(t)$. Then $y_2 \in H$ and $y_2 \perp C(y_1)$. From elementary properties of the projection family P_t it then follows that $P_t y_2 \perp C(y_1)$ for all t, and thus

$$C(y_2) \perp C(y_1).$$

Then consider the orthogonal complement

$$H_2 = H \ominus [C(y_1) \oplus C(y_2)],$$

where the expression between the brackets is the vector sum of the mutually orthogonal cyclic subspaces $C(y_1)$ and $C(y_2)$. If $H_2 \neq 0$, take a non-zero element y_3 in H_2, with spectral function $G_3(t)$, and proceed further in the same way.

As H is separable this procedure must be terminated after a finite or enumerable number of steps, and we finally obtain

$$H = C(y_1) \oplus C(y_2) \oplus \cdots \oplus C(y_M),$$

where the $C(y_n)$ are all mutually orthogonal, and M is a finite integer or equal to $\aleph_0$. Applying the projection P_t to this relation, we obtain for all t

$$H(t) = C(y_1, t) \oplus C(y_2, t) \oplus \cdots \oplus C(y_M, t). \tag{1}$$

We have thus expressed H as the vector sum of the mutually orthogonal cyclic subspaces $C(y_n)$, and the subspaces $H(t)=P_tH$ as the vector sum of the corresponding $C(y_n, t)=P_tC(y_n)$.

Neither the individual components $C(y_n, t)$ in (1) nor their number M is uniquely determined. We now proceed to show that by an appropriate transformation it is possible to obtain a similar representation possessing at least a certain degree of uniqueness.

By means of the y_n occurring in (1) we define a new element z_1 in H by writing

$$z_1=a_1y_1+a_2y_2+\cdots+a_My_M,$$

where the a_n are positive constants. If M is finite, we take $a_n=1$ for all n; if M is infinite we let the a_n decrease to zero sufficiently rapidly to make the series convergent in norm. As the $C(y_n)$ are mutually orthogonal the spectral function of z_1 is

$$F_1(t)=a_1^2G_1(t)+\cdots+a_M^2G_M(t). \tag{2}$$

We can now easily show that, for an arbitrary z in H with spectral function $F(t)$ we have

$$F_1>F \quad \text{and thus} \quad z_1>z,$$

so that z_1 is an element of maximal spectral type in H.

In fact, if S is a Borel set such that $F_1(t)$ does not increase over S, then according to (2), no $G_n(t)$ can increase over S, and the relation (1) then shows that the projection measure $P(S)$ defined by the P_t family is equal to zero, which implies that the spectral function F of any given z does not increase over S, and so F is absolutely continuous with respect to F_1.

If the orthogonal complement

$$H_1=H\ominus C(z_1)$$

does not reduce to the zero element, we can repeat the same procedure, finding an element z_2 of maximum spectral type in H_1, with spectral function $F_2(t)$, and so on. As before, this procedure must terminate after a finite or enumerable number of steps. We thus obtain a finite or enumerable sequence of elements $z_1, z_2, \ldots, z_N$ in H, such that the corresponding cyclic subspaces are mutually orthogonal, and we have for all t

$$H(t)=C(z_1, t)\oplus C(z_2, t)\oplus\cdots\oplus C(z_N, t) \tag{3}$$

with

$$z_1>z_2>\cdots>z_N, \qquad F_1>F_2>\cdots>F_N. \tag{4}$$

Neither the z_n nor the F_n are uniquely determined here, but it follows from the procedure used that the number N and the equivalence classes of the spectral functions F_n are uniquely determined by the given family of subspaces $H(t)$ and the corresponding projections P_t. We can thus state the following proposition:

In any two representations of $H(t)$ in the form of (3), *with the z_n and their spectral functions F_n satisfying* (4), *the number N of terms will have the same value, while the z_n and the F_n will be pairwise equivalent.*

In the sense implied by this statement, (3) may be regarded as a canonical representation. The number N will be called the multiplicity of the given family of subspaces $H(t)$.

In order to show the application of the above statement to stochastic processes, we consider the set of all complex-valued random variables x defined on a given probability space, and such that $Ex=0$, $E|x|^2<\infty$. It is well known that this set becomes a Hilbert space if we define the norm of x and the inner product of two variables x and y by the relations

$$\|x\|^2=E|x|^2, \qquad (x,y)=Ex\bar{y},$$

and regard x and y as identical if $E|x-y|^2=0$.

We now consider a finite-dimensional stochastic vector process

$$\boldsymbol{x}(t)=(x_1(t),\ldots,x_q(t))$$

with the real-valued parameter t. Let $H(\boldsymbol{x})$ be the Hilbert space spanned by the $x_n(u)$ for $n=1,\ldots,q$ and all real u, while $H(\boldsymbol{x},t)$ is the subspace obtained when the u are restricted to values $\leqslant t$. Obviously the $H(\boldsymbol{x},t)$ will be a never decreasing family of subspaces in $H(\boldsymbol{x})$, and as before, we denote by P_t the projection operator with range $H(\boldsymbol{x},t)$. We will impose the following two conditions on the vector process $\boldsymbol{x}(t)$:

(i) The quadratic mean limits $x_n(t\pm 0)$ exist for all n and t, and $x_n(t-0)=x_n(t)$.

(ii) $H(\boldsymbol{x},-\infty)=0$, which means that the intersection of all $H(\boldsymbol{x},t)$ only contains the random variable almost everywhere equal to zero.

It is known that the condition (i) implies that the Hilbert space $H(\boldsymbol{x})$ is separable, and that $H(\boldsymbol{x},t-0)=H(\boldsymbol{x},t)$, while (ii) implies that the $\boldsymbol{x}(t)$ process is purely non-deterministic.

For any element z in $H(\boldsymbol{x})$ the projections

$$z(t)=P_t z$$

determine a stochastic process with orthogonal increments, and the corresponding spectral function is

$$E|z(t)|^2=F(t).$$

The general proposition on Hilbert space given above can now be applied, and shows that the given $\boldsymbol{x}(t)$ process has a uniquely determined multiplicity N, which may be finite or infinite, and that we can find a sequence $z_1,\ldots,z_N$ of random variables in $H(\boldsymbol{x})$ such that the corresponding cyclic subspaces are mutually orthogonal, and we have for all t

$$H(\boldsymbol{x},t)=C(z_1,t)\oplus\cdots\oplus C(z_N,t), \tag{5}$$

while the spectral functions F_n of the z_n satisfy the relation

$$F_1>F_2>\cdots>F_N, \tag{6}$$

and the representation is canonical in the sense given above.

Any component $x_i(t)$ is an element of $H(\boldsymbol{x}, t)$, and is thus, by (5), a sum of elements in the $C(z_n, t)$, so that we have the representation

$$x_i(t)=\sum_{n=1}^{N}\int_{-\infty}^{t} g_{in}(t,u)\,\mathrm{d}z_n(u), \tag{7}$$

or in matrix form

$$\boldsymbol{x}(t)=\int_{-\infty}^{t} G(t,u)\,\mathrm{d}\boldsymbol{z}(u), \tag{8}$$

where the $z_n(u)$ are mutually orthogonal processes with orthogonal increments, with spectral functions F_n satisfying (6), while for any fixed t we have $g_{in}(t,u)\in L_2F_n(u)$. Further, $\boldsymbol{x}(t)$ and $\boldsymbol{z}(u)$ are column vectors, $\boldsymbol{G}(t,u)$ is the matrix $\{g_{in}(t,u)\}$ with $i=1,\ldots,q$ and $n=1,\ldots,N$.

It follows from the general Hilbert space theorem that, in any two representations of $x(t)$ in the form (7) or (8), with spectral functions F_n satisfying (6), the multiplicity N will have the same value, and the F_n will be pairwise equivalent.

The relations (7) and (8) are the basic representation formulas of stochastic multiplicity theory. They show how the $\boldsymbol{x}(t)$ vector process may be regarded as linearly built up by N-dimensional mutually orthogonal innovation elements $\mathrm{d}\boldsymbol{z}(u)=(\mathrm{d}z_1(u),\ldots,\mathrm{d}z_N(u))$ acting during the entire past up to the instant t. The best linear prediction (in the least square sense) of $\boldsymbol{x}(t)$, based on the development of the process up to the instant $s<t$ is obtained by replacing the upper limit of integration by s in (7) or (8).

The covariance matrix of the (t) process is

$$\boldsymbol{R}(s,t)=E\boldsymbol{x}(s)\boldsymbol{x}(t)^*=\{r_{ij}(s,t)\}$$

with

$$r_{ij}(s,t)=Ex_i(s)\overline{x_j(t)}\,.$$

It is easily seen that the multiplicity N of $\boldsymbol{x}(t)$ and the equivalence classes of its spectral functions are uniquely determined by the $\boldsymbol{R}(s,t)$ matrix. Suppose, in fact, that the processes

$$\boldsymbol{x}(t)=(x_1(t),\ldots,x_q(t)),$$
$$\boldsymbol{y}(t)=(y_1(t),\ldots,y_q(t)),$$

have the same $\boldsymbol{R}(s,t)$ matrix. Any linear relation between the random variables $x_i(t_k)$ which holds with probability one, will then imply the same relation between the $y_i(t_k)$, and conversely. It follows that we can define an isometric transformation from $h(\boldsymbol{x})$ to $h(\boldsymbol{y})$ by taking

$$Vx_i(t)=y_i(t)$$

and extending the transformation V linearly and by convergence in norm. For the projections P_t^x and P_t^y associated respectively with the subspaces $H(\boldsymbol{x},t)$ and $H(\boldsymbol{y},t)$ we shall then have

$$P_t^y=VP_t^xV^{-1}.$$

Thus there is a unique mutual correspondence between the $H(\boldsymbol{x},t)$ and $H(\boldsymbol{y},t)$ families, and it follows that the two processes have the same multiplicity properties.

G. Kallianpur, P.R. Krishnaiah, J.K. Ghosh, eds., *Statistics and Probability: Essays in Honor of C.R. Rao*

CLASSIFICATION INTO TWO MULTIVARIATE NORMAL POPULATIONS WITH KNOWN MEANS

Somesh DAS GUPTA
University of Minnesota, Minneapolis, MN, U.S.A.

1. Introduction

We consider the problem of classifying a (randomly observed) experimental unit, ω, into one of two populations Π_1 and Π_2. It is assumed that a random vector $X(\omega)$: $p\times 1$ is distributed as $N_p(\mu_i, \Sigma)$ in Π_i; furthermore, μ_1 and μ_2 are assumed to be known, but Σ is an unknown positive-definite matrix. Based on random training samples from Π_1 and Π_2 we may get a statistic S: $p\times p$, where nS is distributed as the Wishart distribution $\mathcal{W}_p(n; \Sigma)$, $n\geqslant p$.

If Σ were known, the admissible Bayes and minimax rule δ classifies ω into Π_1 iff

$$(x-\mu_1)'\Sigma^{-1}(x-\mu_1)<(x-\mu_2)'\Sigma^{-1}(x-\mu_2). \tag{1}$$

The plug-in rule $\hat{\delta}$ (obtained by replacing Σ by S) classifies ω into Π_1 iff

$$T\equiv(\mu_2-\mu_1)'S^{-1}\left(x-\tfrac{1}{2}[\mu_1+\mu_2]\right)<0. \tag{2}$$

The likelihood-ratio rule suggested by Anderson [1] also turns out to be the same as $\hat{\delta}$. In this note we shall study the rule $\hat{\delta}$.

2. Probabilities of correct classification

Let

$$\Delta=\left[(\mu_1-\mu_2)'\Sigma^{-1}(\mu_1-\mu_2)\right]^{1/2}. \tag{3}$$

Without loss of generality, we may assume that

$$\mu_1=0, \qquad \mu_2=(\Delta,0,\ldots,0)', \qquad \Sigma=I_p. \tag{4}$$

Let $A=nS=[a_{ij}]$, $A^{-1}=[a^{ij}]$. Then it can be seen that

$$T/n=\Delta\left[x_{1.(2)}-\tfrac{1}{2}\Delta\right]/A_{11.(2)} \tag{5}$$

where

$$A=\left[\begin{array}{c|c} A_{11} & A_{12} \\ \hline A_{21} & A_{12} \end{array}\right]\begin{matrix}1\\p-1\end{matrix}, \qquad A_{11.2}=A_{11}-A_{12}A_{22}^{-1}A_{21},$$
$$\qquad\; 1 \quad p-1 \tag{6}$$

$$X_{1\cdot(2)}=X_1-A_{12}A_{22}^{-1}X_{(2)}, \qquad X=(X_1, X_2,\ldots, X_p)'=(X_1, X'_{(2)})'. \tag{7}$$

The conditional distribution of $[X_{1.(2)}-\frac{1}{2}\Delta]$, given $X'_{(2)}A_{22}^{-1}X_{(2)}=v$, is given by

$$N((-1)^i\Delta/2, 1+v) \tag{8}$$

independent of $A_{11.(2)}$, if ω comes from Π_i. Note that $X'_{(2)}A_{22}^{-1}X_{(2)}$ and $A_{11.(2)}$ are distributed independently as $f_{p-1,n-p+2}$ and χ^2_{n-p+1}, respectively, where $f_{a,b}$ is the distribution of the ratio of independent χ^2_a and χ^2_b. Thus the probabilities of correct classification are given by

$$\begin{aligned}P(T\leqslant 0|\Pi_1)&=P(T\geqslant 0|\Pi_2)\\ &=\mathcal{E}\Phi(\tfrac{1}{2}\Delta\beta^{1/2})\end{aligned} \tag{9}$$

where Φ is the c.d.f. of $N(0,1)$ and $\beta^{1/2}$ is distributed as the square root of a $\beta(\frac{1}{2}(n-p+2), \frac{1}{2}(p-1))$ variate.

3. Optimum properties of $\hat{\delta}$

Consider an orthogonal matrix L: $p\times p$ such that the first row is proportional to $(\mu_2-\mu_1)'$. Define

$$Y=L(x-\mu_1), \qquad B=LAL'=[b_{ij}], \qquad \Gamma=L\Sigma L'. \tag{10}$$

Then $Y\sim N_p(0,\Gamma)$ in Π_1 and $Y\sim N_p(\nu,\Gamma)$ in Π_2, where

$$\nu=(c,0,\ldots,0)', \quad c=[(\mu_1-\mu_2)'(\mu_1-\mu_2)]^{1/2}. \tag{11}$$

Consider now the classification problem in terms of Y and B. The problem is invariant under transformations:

$$Y\to TY, \qquad B\to TBT' \tag{12}$$

where

$$T=\left[\begin{array}{c|c} 1 & t'_{(1)} \\ \hline 0 & T_{22} \end{array}\right]\begin{matrix}1\\p-1\end{matrix},$$
$$1 \quad p-1 \tag{13}$$

is a non-singular matrix. A set of maximal invariant statistics is given by

$$U=Y_{1.(2)}, \qquad V=Y'_{(2)}B_{22}^{-1}Y_{(2)}, b_{11.(2)} \tag{14}$$

where

$$B=\left[\begin{array}{c|c} b_{11} & B_{12} \\ \hline B_{21} & B_{22} \end{array}\right]\begin{array}{l} 1 \\ p-1 \end{array}, \qquad b_{11.(2)}=b_{11}-B_{12}B_{22}^{-1}B_{21},$$
$$\begin{array}{cc} 1 & p-1 \end{array} \tag{15}$$

$$Y_{1.(2)}=Y_1-B_{12}B_{22}^{-1}Y_{(2)}, \qquad Y=(Y_1,Y_2,\ldots,Y_p)'=(Y_1Y_{(2)}')'. \tag{16}$$

Note that $b_{11.(2)}$ is independent of U and V, and $b_{11.(2)}\sim\gamma_{11.(2)}\chi^2_{n-p+1}$, where

$$\gamma_{11.(2)}=\Gamma_{11}-\Gamma_{12}\Gamma_{22}^{-1}\Gamma_{21}, \qquad \Gamma=\left[\begin{array}{c|c} \gamma_{11} & \Gamma_{12} \\ \hline \Gamma_{21} & \Gamma_{22} \end{array}\right]\begin{array}{l} 1 \\ p-1 \end{array}.$$
$$\begin{array}{cc} 1 & p-1 \end{array} \tag{17}$$

Note that

$$\Delta^2=(\mu_1-\mu_2)'\Sigma^{-1}(\mu_1-\mu_2)=\nu'\Gamma^{-1}\nu=c^2/\gamma_{11.(2)}. \tag{18}$$

Hence

$$\gamma_{11.(2)}=c^2/\Delta^2. \tag{19}$$

Also

$$\begin{aligned}(\mu_2-\mu_1)A^{-1}(X-\tfrac{1}{2}[\mu_1+\mu_2])&=\\ =\nu'B^{-1}[Y-\tfrac{1}{2}\nu]&=c(Y_{1.(2)}-\tfrac{1}{2}c)/b_{11.(2)}.\end{aligned} \tag{20}$$

The distribution of U, given $V=v$, is given by

$$N(\theta,\gamma_{11.(2)}(1+v)), \tag{21}$$

where $\theta=0$ in Π_1 and $\theta=c$ in Π_2.

Consider a prior distribution which sets $\gamma_{11.(2)}=\gamma_0$ and attaches equal probabilities to Π_1 and Π_2. Then the unique (a.e.) Bayes rule (with zero–one loss function) in the class of all invariant rules classifies ω into Π_1 iff

$$U<\tfrac{1}{2}c. \tag{22}$$

Thus this invariant admissible Bayes rule coincides with $\hat{\delta}$. Since for a fixed Δ, the probabilities of correct classification for $\hat{\delta}$ are equal, it can be easily seen that $\hat{\delta}$ is invariant minimax.

Kiefer and Schwarz [3] derived a class of Bayes rules (not invariant) for this problem. Such a Bayes rule decides ω in Π_1 iff

$$\begin{aligned}&\operatorname{etr}[(\mu_2-\mu_1)'X][1+(X-\mu_1)'A^{-1}(x-\mu_1)]^{-r/2}\\ &\qquad\times[1+(X-\mu_2)'A^{-1}(X-\mu_2)]^{r/2}>k.\end{aligned} \tag{23}$$

The rule $\hat{\delta}$ does not provide a similar test if this decision problem is reviewed as a hypothesis testing problem. Such a similar region (for ω in Π_1) is given by

$$\sqrt{n-p+1}\, Y_{1.(2)} \Big/ \left[\left(1+Y'_{(2)}B_{22}^{-1}Y_{(2)}\right)b_{11.(2)}\right]^{1/2} < k, \tag{24}$$

since the statistics in the left-hand side of (24) is distributed as Student's t_{n-p+1}.

Note: The above problem was also considered by Geisser [2] from the viewpoint of predictive discrimination; he used an improper prior distribution for Σ.

References

[1] Anderson, T.W. (1958). *An Introduction to Multivariate Statistical Analysis*. Wiley, New York.

[2] Geisser, S. (1966). Predictive discrimination. In: Ed., Krishnaiah, P.R., *Multivariate Analysis*, I. Academic Press, New York.

[3] Kiefer, J. and Schwartz, R. (1965). Admissible Bayes character of T^2-, R^2-, and other fully invariant tests for classical multivariate normal problems. *Ann. Math. Statist.* **36**. 747–770.

G. Kallianpur, P.R. Krishnaiah, J.K. Ghosh, eds., *Statistics and Probability: Essays in Honor of C.R. Rao*

SOME FORMULAS USEFUL FOR COVARIANCE ESTIMATION WITH GAUSSIAN LINEAR COMPONENT MODELS

A.P. DEMPSTER*
Department of Statistics, Harvard University, Cambridge, MA, U.S.A.

The posterior distribution of components in Gaussian linear models depends on the prior covariance structure (or scale) of the components. Formulas are given which describe how such posterior distributions change as the covariances change. The formulas apply when prior variances are infinite under stated conditions. The derivations are shown to be trivial in terms of SWP operators.

1. Introduction

Manipulations with complicated analytic representations of likelihood can sometimes be drastically reduced by using an elegant expression for the array of kth derivatives of log likelihood due to Sundberg (1974). Although the Sundberg theory nominally applies to incomplete data from exponential families, it also applies to Gaussian linear models where unobserved components in the linear structure play the role of missing data. This extension to variance components and factor models was pointed out by Dempster et al. (1977), and further illustrated by Dempster et al. (1979a) and by Dempster et al. (1979b).

The Sundberg formula asserts that the array of kth derivatives of log likelihood with respect to the natural parameters of the exponential family is a difference of two arrays of kth cumulants of the complete data sufficient statistics, where the first array is calculated using the marginal sampling distribution of the sufficient statistics, while the second array uses the conditional distribution given the incomplete data. In the case of Gaussian component models, the main computational task in applying incomplete data technology is to find the required conditional distributions. The computations are formally identical to the well-understood computations of least squares, but deserve reconsideration here for two reasons. First, they are equivalent to very large least squares problems when the data set is large. Second, certain iterative techniques, such as those required to find a maximum likelihood estimate or to integrate over a posterior distribution, require repeated evaluations of the conditional moments for different sets of parameter values. The formulas derived in this paper are intended to aid in the process of repeated computation by showing how to modify the conditional distribution in place, thus avoiding recomputation from scratch.

*Research was supported by NSF Grant MCS77-27119.

2. Preliminaries

Observable processes are often modelled as sums of independent Gaussian components. Such models can be represented in the general form

$$\boldsymbol{Y}=\boldsymbol{X}\boldsymbol{\beta}, \tag{2.1}$$

where $\boldsymbol{Y}$ is an $n\times 1$ vector of observables, $\boldsymbol{X}$ is an $n\times p$ given matrix, and $\boldsymbol{\beta}$ is a $p\times 1$ vector of unknown linear effects.

Component structure arises through partitioning the columns of $\boldsymbol{X}$ and the corresponding rows of $\boldsymbol{\beta}$ into subsets, so that (2.1) can be written in the form

$$\boldsymbol{Y}=\sum_{i=1}^{k} \boldsymbol{X}_i\boldsymbol{\beta}_i. \tag{2.2}$$

The Gaussian assumption asserts that $\boldsymbol{\beta}$ is *a priori* multivariate normal $N(\boldsymbol{\mu},\boldsymbol{\Sigma})$ distributed. When the components $\boldsymbol{\beta}_i$ are independent $N(\boldsymbol{\mu}_i,\boldsymbol{\Sigma}_i)$, then $\boldsymbol{\Sigma}$ has a block-diagonal form with the $\boldsymbol{\Sigma}_i$ forming the diagonal blocks.

Depending on circumstances, the parameters $\boldsymbol{\Sigma}_i$ may be of direct interest, or it may be that estimation of the $\boldsymbol{\Sigma}_i$ is carried out primarily so that inferences may be drawn about $\boldsymbol{\beta}$ and $\boldsymbol{\mu}$ given estimated values for the $\boldsymbol{\Sigma}_i$.

Examples of models of the form (2.2) are many and varied. Specializing to $k=2$, and setting $\boldsymbol{\mu}_2=\boldsymbol{0}$, $\boldsymbol{\Sigma}_2=\sigma^2\boldsymbol{I}$, $\boldsymbol{X}_2=\boldsymbol{I}$ and $\boldsymbol{\beta}_2=\boldsymbol{e}$ yields the familiar representation

$$\boldsymbol{Y}=\boldsymbol{X}_1\boldsymbol{\beta}_1+\boldsymbol{e} \tag{2.3}$$

for linear models with a homoscedastic Gaussian error term $\boldsymbol{e}$. Another example of current interest to the author is a time series model of the form

$$\boldsymbol{Y}=\boldsymbol{\beta}_1+\boldsymbol{\beta}_2+\boldsymbol{\beta}_3+\boldsymbol{\beta}_4+\boldsymbol{\beta}_5, \tag{2.4}$$

where the component time series $\boldsymbol{\beta}_i$ denote respectively, a constant term, a fixed seasonal pattern which repeats from year to year, a component depending on exogenous events, a stochastic non-seasonal component, and finally a stochastic seasonal component representing slow variation of a seasonal pattern. Another large class of models of the form (2.2) consists of hierarchical variance component models, arising for example, from multistage sample surveys.

Gaussian component models typically contain two types of components, traditionally called fixed effects and random effects. While fixed effects can be estimated by maximum likelihood, random effects are generally estimated from their conditional means given the data and the parameters of the model, as discussed, for example, by Harville (1977). As shown by Lindley and Smith (1972) the two principles of estimation can be unified by treating the fixed effects as limiting cases of random effects when the associated scale factors $\boldsymbol{\Sigma}_i$ tend to infinity. The unified approach is adopted here because of its simplicity.

Within the established conventions, any model of the form (2.1) with $\boldsymbol{\beta}$ having a $N(\boldsymbol{\mu},\boldsymbol{\Sigma})$ distribution can be formally rewritten as a different model of the same general form but with $\boldsymbol{\mu}=\boldsymbol{0}$. Specifically, if $\boldsymbol{Y}=\boldsymbol{X}\boldsymbol{\beta}$ where $\boldsymbol{\beta}$ is $N(\boldsymbol{\mu},\boldsymbol{\Sigma})$, then $\boldsymbol{Y}=\boldsymbol{X}\boldsymbol{\mu}+\boldsymbol{X}(\boldsymbol{\beta}-\boldsymbol{\mu})$, where the fixed effect $\boldsymbol{\mu}$ is $N(\boldsymbol{0},\boldsymbol{\Omega})$ for large $\boldsymbol{\Omega}$ and $(\boldsymbol{\beta}-\boldsymbol{\mu})$ is $N(\boldsymbol{0},\boldsymbol{\Sigma})$. Thus, $\boldsymbol{\beta}$ is replaced by $(\boldsymbol{\mu},\boldsymbol{\beta}-\boldsymbol{\mu})$ with zero means.

The remainder of Section 2 develops elementary facts about the model (2.1) where $\boldsymbol{\beta}$ has an initial $N(\mathbf{0}, \boldsymbol{\Sigma})$ distribution. Since $\boldsymbol{X}$ is known, the relation $\boldsymbol{Y}=\boldsymbol{X\beta}$ implies that the initial joint distribution of $\boldsymbol{Y}$ and $\boldsymbol{\beta}$ is

$$\left[\frac{\boldsymbol{Y}}{\boldsymbol{\beta}}\right] \sim N\left(\left[\frac{\mathbf{0}}{\mathbf{0}}\right], \left[\begin{array}{c|c} \boldsymbol{Q} & \boldsymbol{X\Sigma} \\ \hline \boldsymbol{\Sigma X}^{\mathrm{T}} & \boldsymbol{\Sigma} \end{array}\right]\right), \tag{2.5}$$

where

$$\boldsymbol{Q}=\boldsymbol{X\Sigma X}^{\mathrm{T}}. \tag{2.6}$$

In Sections 2 and 3, it is assumed that $\boldsymbol{\Sigma}$ is positive definite and finite, that $p \geqslant n$, and that $\boldsymbol{X}$ has rank n, whence it follows that $\boldsymbol{Q}$ has rank n. In Section 4, the assumption that $\boldsymbol{\Sigma}$ is finite is relaxed.

From (2.5), the conditional distribution of $\boldsymbol{\beta}$ given $\boldsymbol{Y}$ is

$$(\boldsymbol{\beta}|\boldsymbol{Y}) \sim N(\hat{\boldsymbol{\beta}}, \boldsymbol{C}), \tag{2.7}$$

where

$$\hat{\boldsymbol{\beta}}=\boldsymbol{U}^{\mathrm{T}}\boldsymbol{Y}, \tag{2.8}$$

$$\boldsymbol{U}=\boldsymbol{Q}^{-1}\boldsymbol{X\Sigma}, \tag{2.9}$$

and

$$\begin{aligned} \boldsymbol{C} &= \boldsymbol{\Sigma}-\boldsymbol{\Sigma X}^{\mathrm{T}}\boldsymbol{Q}^{-1}\boldsymbol{X\Sigma} \\ &= (\boldsymbol{I}-\boldsymbol{U}^{\mathrm{T}}\boldsymbol{X})\boldsymbol{\Sigma}. \end{aligned} \tag{2.10}$$

Note that $\boldsymbol{C}$ has rank $p-n$ because the relation $\boldsymbol{Y}=\boldsymbol{X\beta}$ restricts $\boldsymbol{\beta}$ to a $(p-n)$-dimensional subspace of p-space.

To recapitulate, the model $\boldsymbol{Y}=\boldsymbol{X\beta}$ where $\boldsymbol{\beta}$ is $N(\mathbf{0}, \boldsymbol{\Sigma})$ distributed has a likelihood function depending on $\boldsymbol{\Sigma}$. According to Sundberg (1974), the derivatives of log likelihood with respect to the natural parameters (in this case the elements of $\boldsymbol{\Sigma}^{-1}$) are simply expressible in terms of the moments of the prior $N(\mathbf{0}, \boldsymbol{\Sigma})$ distribution of $\boldsymbol{\beta}$ and corresponding moments of the posterior $N(\hat{\boldsymbol{\beta}}, \boldsymbol{C})$ distribution of $\boldsymbol{\beta}$ defined by (2.7)–(2.10). The formulas in Section 3 are aids in the recomputation of (2.9) and (2.10) as aspects of $\boldsymbol{\Sigma}$ change.

3. Formulas

The columns of $\boldsymbol{U}$ may be partitioned into $\boldsymbol{U}_i$ corresponding to the partition of $\boldsymbol{X}$ into $\boldsymbol{X}_i$ for $i=1,2,\ldots,k$. While $\boldsymbol{\Sigma}$ correspondingly partitions into block diagonal form with $\boldsymbol{\Sigma}_i$ on the diagonals, the posterior covariance $\boldsymbol{C}$ is no longer block diagonal, so the notation $\boldsymbol{C}_{ii}$ will be used for the diagonal blocks and $\boldsymbol{C}_{ij}$ will be used for the off-diagonal blocks of posterior covariances between $\boldsymbol{\beta}_i$ and $\boldsymbol{\beta}_j$.

The formulas listed below relate the components $\boldsymbol{U}_i$ and $\boldsymbol{C}_{ij}$ of $\boldsymbol{U}$ and $\boldsymbol{C}$ to the components of $\boldsymbol{U}_i^*$ and $\boldsymbol{C}_{ij}^*$ of $\boldsymbol{U}^*$ and $\boldsymbol{C}^*$ when $\boldsymbol{\Sigma}_1$ is replaced by $\boldsymbol{\Sigma}_1^*$ while $\boldsymbol{\Sigma}_2,\ldots,\boldsymbol{\Sigma}_k$ are unchanged. It is sufficient to treat the case $k=2$ since the last $k-1$ components can be aggregated. The results are:

$$(C_{11}^*)^{-1}-(\Sigma_{11}^*)^{-1}=C_{11}^{-1}-\Sigma_{11}^{-1}, \tag{3.1}$$

$$U_1^*(C_{11}^*)^{-1}=U_1C_{11}^{-1}, \tag{3.2}$$

$$(C_{11}^*)^{-1}C_{12}^*=C_{11}^{-1}C_{12}, \tag{3.3}$$

$$-(Q^*)^{-1}-U_1^*(C_{11}^*)^{-1}U_1^{*\mathrm{T}}=-Q^{-1}-U_1C_{11}^{-1}U_1^{\mathrm{T}}, \tag{3.4}$$

$$U_2^*-U_1^*(C_{11}^*)^{-1}C_{12}^*=U_2-U_1C_{11}^{-1}C_{12}, \tag{3.5}$$

$$C_{22}^*-C^{*T}_{12}(C_{11}^*)^{-1}C_{12}^*=C_{22}-C_{12}^{\mathrm{T}}C_{11}^{-1}C_{12}. \tag{3.6}$$

These formulas are derived in the Appendix. In addition to the assumptions that $p\geqslant n$, that X has rank n, and that Σ and Σ^* have rank p, it is further required that C_{11} and C_{11}^* have full rank.

To facilitate use of formulas (3.1)–(3.6) in sequence for computing starred from unstarred quantities, they can easily be recast and extended as follows:

$$C_{11}^*=\left(C_{11}^{-1}+(\Sigma_1^*)^{-1}-\Sigma_1^{-1}\right)^{-1}, \tag{3.7}$$

$$U_1^{*\mathrm{T}}=A_{11}U_1^{\mathrm{T}}, \tag{3.8}$$

$$C_{12}^*=A_{11}C_{12}, \tag{3.9}$$

$$(Q^*)^{-1}=Q^{-1}+U_1B_{11}U_1^{\mathrm{T}}, \tag{3.10}$$

$$U_2^*=U_2-U_1B_{11}C_{12}, \tag{3.11}$$

$$C_{22}^*=C_{22}-C_{12}^{\mathrm{T}}B_{11}C_{12}, \tag{3.12}$$

$$\hat{\beta}_1^*=A_{11}\hat{\beta}_1, \tag{3.13}$$

and

$$\hat{\beta}_2^*=\hat{\beta}_2-\hat{\beta}_1^{\mathrm{T}}B_{11}C_{12}, \tag{3.14}$$

where

$$A_{11}=C_{11}^*C_{11}^{-1}, \tag{3.15}$$

and

$$B_{11}=C_{11}^{-1}-(C_{11}^*)^{-1}C_{11}^{-1}(C_{11}^*)^{-1}. \tag{3.16}$$

The formulas (3.7)–(3.16) are most useful when the blocks Σ_i all have small dimensions and hence the formulas are applied repeatedly using each of $\Sigma_1, \Sigma_2, \ldots, \Sigma_k$ in place of Σ_1. One common situation is that $\Sigma_i=\sigma_i\Omega_i$ where σ_i is a scale parameter and Ω_1 is known. In this case, β_i can be replaced by a set of linear combinations γ_i whose prior covariance is $\sigma_i I$, and the formulas (3.7)–(3.16) can be applied in the new coordinates with each Σ_i having dimension 1×1, so that even a small matrix inversion is avoided in (3.7).

Different subsets of the formulas may be used in different circumstances, depending on which arrays are continuously updated in a process of repeatedly altering components $\boldsymbol{\Sigma}_1, \boldsymbol{\Sigma}_2, \ldots$. For example, only $\boldsymbol{Q}^{-1}$ may be retained throughout, and $\boldsymbol{U}_1$ recomputed at each step from (2.9), with other desired quantities computed from $\boldsymbol{Q}^{-1}$, $\boldsymbol{X}$, $\boldsymbol{Y}$, and $\boldsymbol{\Sigma}$ after all components of $\boldsymbol{\Sigma}$ have been modified. Alternatively, we may wish to retain only $\boldsymbol{U}$ throughout, and compute other quantities such as $\boldsymbol{Q}^{-1} = \boldsymbol{U}\boldsymbol{\Sigma}^{-1}\boldsymbol{U}^{\mathrm{T}}$ from $\boldsymbol{U}$, $\boldsymbol{X}$, $\boldsymbol{Y}$, and $\boldsymbol{\Sigma}$. A quite different approach is to use only formulas (3.7), (3.9), (3.12), (3.13), (3.14), which means retaining the current versions of $\boldsymbol{C}$ and $\hat{\boldsymbol{\beta}}$ throughout the calculations.

Finally, it is shown in the Appendix that

$$\det \boldsymbol{Q}^* = \frac{\det \boldsymbol{C}_{11}}{\det \boldsymbol{\Sigma}} \times \frac{\det \boldsymbol{\Sigma}^*}{\det \boldsymbol{C}_{11}^*} \times \det \boldsymbol{Q} \tag{3.17}$$

which is useful for computing values of the likelihood as $\boldsymbol{\Sigma}$ changes.

4. Extension to infinite variances

In the Appendix it is shown that the arrays $\boldsymbol{Q}^{-1}$, $\boldsymbol{U}$, $\boldsymbol{C}$ and $\hat{\boldsymbol{\beta}}$ have unique finite limits as $\boldsymbol{\Sigma}_1 \to \infty$ if and only if $\boldsymbol{X}_1$ has rank equal to its number of columns, say p_1, while the remaining set of columns $\left[\boldsymbol{X}_2 \,\vdots\, \boldsymbol{X}_3 \,\vdots\, \cdots \,\vdots\, \boldsymbol{X}_k\right]$ of $\boldsymbol{X}$ have rank n. The notation $\boldsymbol{\Sigma}_1 \to \infty$ is taken to mean that $\boldsymbol{a}\boldsymbol{\Sigma}_1\boldsymbol{a}^{\mathrm{T}} \to \infty$ for all $\boldsymbol{a} \neq \boldsymbol{0}$. It is assumed that the components $\boldsymbol{\Sigma}_i$ of $\boldsymbol{\Sigma}$ have full rank.

Since (2.6) can be written,

$$\boldsymbol{Q} = \sum_{i=1}^{k} \boldsymbol{X}_i \boldsymbol{\Sigma}_i \boldsymbol{X}_i^{\mathrm{T}}; \tag{4.1}$$

it is evident that $\boldsymbol{Q}$ becomes infinite in the precise sense that $\boldsymbol{b}\boldsymbol{Q}\boldsymbol{b}^{\mathrm{T}} \to \infty$ except for $\boldsymbol{b}$ in an $(n-p_1)$-dimensional subspace of n-space, whence the corresponding $\boldsymbol{Q}^{-1}$ has a finite limit with rank $n-p_1$.

Explicit expressions for the limiting $\boldsymbol{Q}^{-1}$, $\boldsymbol{U}$, and $\boldsymbol{C}$ follow immediately from the relation (A.4). These expressions depend on

$$\boldsymbol{Q}(-) = \sum_{i=2}^{k} \boldsymbol{X}_i \boldsymbol{\Sigma}_i \boldsymbol{X}_i^{\mathrm{T}} \tag{4.2}$$

which does not depend on $\boldsymbol{\Sigma}_1$. For example

$$\lim_{\boldsymbol{\Sigma}_1 \to \infty} \boldsymbol{Q}^{-1} = \boldsymbol{Q}(-)^{-1} - \boldsymbol{Q}(-)^{-1}\boldsymbol{X}_1\left(\boldsymbol{X}_1^{\mathrm{T}}\boldsymbol{Q}(-)^{-1}\boldsymbol{X}_1\right)^{-1}\boldsymbol{X}_1^{\mathrm{T}}\boldsymbol{Q}(-)^{-1}, \tag{4.3}$$

but the remaining formulas for $\boldsymbol{U}$ and $\boldsymbol{C}$ are omitted to save space. From (4.3) it is easily checked that

$$\lim_{\boldsymbol{\Sigma}_1 \to \infty} \boldsymbol{Q}^{-1}\boldsymbol{X}_1 = \boldsymbol{0} \tag{4.4}$$

in accordance with the fact that $\lim \boldsymbol{Q}^{-1}$ has rank $n-p_1$. Note that (2.9) cannot be used in the limit to find $\boldsymbol{U}_1$ since (2.9) becomes the product of a zero factor (4.4) and an infinite factor $\boldsymbol{\Sigma}_1$.

The discussion of the preceding paragraph shows that it is natural to expand the class of models with all components of $\boldsymbol{\Sigma}$ finite, to include limiting models where some components of $\boldsymbol{\Sigma}$ are infinite, the rule being that the matrix of columns of $\boldsymbol{X}$ corresponding to infinite variances must have rank equal to its number of columns, and the matrix of remaining columns must have rank n. These rank conditions are already familiar to users of linear models. The first condition reflects the well-known fact that the design matrix corresponding to unconstrained fixed effects must have full rank if the fixed effects are to be estimable. The second condition is automatically satisfied in the case of model (2.3) since the residuals $\boldsymbol{e}$ are never taken to be fixed effects and $\boldsymbol{X}_2=\boldsymbol{I}$ always has rank n.

Continuity considerations imply that the relations, formulas and algorithms derived in Sections 2 and 3 hold automatically in the expanded class of models, subject only to the condition that the limits of all the quantities appearing in any relation remain finite. In particular, the relations (3.7)–(3.14) can be used to go back and forth among any pair of models in the expanded class.

Appendix

The computing operations SWP (for sweep) and RSW (for reverse sweep) provide compact derivations of the formulas of this paper. For completeness SWP and RSW are defined and their properties are sketched below, but see also Dempster (1969).

Suppose that $\boldsymbol{M}$ is an $r\times r$ symmetric matrix whose (i, j) element is denoted by m_{ij}. Then for any i on $1\leqslant i\leqslant r$, SWP[i] $\boldsymbol{M}$ is another $r\times r$ symmetric matrix $\boldsymbol{M}^*$ with elements m_{ij}^*, where

$$\left.\begin{aligned} m_{ii}^* &= -\frac{1}{m_{ii}}, & & \\ m_{ij}^* &= m_{ji}^* = m_{ij}m_{ii}^{-1}, & & \text{for } j\neq i, \\ m_{jk}^* &= m_{jk} - m_{ji}m_{ii}^{-1}m_{ik} & & \text{for } j\neq i,\ k\neq i. \end{aligned}\right\} \tag{A.1}$$

The elementary sweep operations SWP[1], SWP[2], ..., SWP[r] are commutative, and hence we can define SWP[$\boldsymbol{u}$] for any subset $\boldsymbol{u}$ of the integers $1,2,\ldots,r$ to be the result of applying SWP[i] for each $i\in\boldsymbol{u}$, using any order of application. The end results of SWP[$\boldsymbol{u}$] is best displayed in terms of a partition of $(1,2,\ldots,r)$ into $\mathbf{1}=(1,2,\ldots,s)$ and $\mathbf{2}=(s+1,\ldots,r)$ and a corresponding partition of $\boldsymbol{M}$. It is easily seen that

$$\mathrm{SWP}[\mathbf{1}]\left[\begin{array}{c|c} \boldsymbol{M}_{11} & \boldsymbol{M}_{12} \\ \hline \boldsymbol{M}_{12}^{\mathrm{T}} & \boldsymbol{M}_{22} \end{array}\right] = \left[\begin{array}{c|c} -\boldsymbol{M}_{11}^{-1} & \boldsymbol{M}_{11}^{-1}\boldsymbol{M}_{12} \\ \hline \boldsymbol{M}_{12}^{\mathrm{T}}\boldsymbol{M}_{11}^{-1} & \boldsymbol{M}_{22}-\boldsymbol{M}_{12}^{\mathrm{T}}\boldsymbol{M}_{11}^{-1}\boldsymbol{M}_{12} \end{array}\right], \tag{A.2}$$

for any s on $1\leqslant s\leqslant r$.

RSW in the inverse operation of SWP. From (A.1) it is easily seen that RSW[i] is identical to SWP[i] except for a change of sign in the second line of (A.1). RSW operations commute with each other, and with SWP operations. RSW[$\boldsymbol{u}$] is defined analogously to SWP[$\boldsymbol{u}$] and the analog of (A.2) is identical to (A.2) except for a change of sign in the 12 position.

It follows from the above that the successive application of SWP[1], SWP[2], ..., SWP[r] to $\boldsymbol{M}$ yields $-\boldsymbol{M}^{-1}$, whatever the order of application. An important byproduct of these equations is det $\boldsymbol{M}$, which is found simply by multiplying the pivot elements in each of the r steps, where the pivot element associated with SWP[i] is defined to be the (i, i) element of the matrix operated upon at that step.

The operation of finding the conditional distribution (2.7) from the marginal distribution (2.5) can be expressed in SWP terms as

$$\left[\begin{array}{c|c} \boldsymbol{Q} & \boldsymbol{X\Sigma} \\ \hline \boldsymbol{\Sigma X}^{\mathrm{T}} & \boldsymbol{\Sigma} \end{array}\right] \overset{\mathrm{SWP}[1,2,\ldots,n]}{\longrightarrow} \left[\begin{array}{c|c} -\boldsymbol{Q}^{-1} & \boldsymbol{U} \\ \hline \boldsymbol{U}^{\mathrm{T}} & \boldsymbol{C} \end{array}\right]. \tag{A.3}$$

In fact, (A.3) is an operator equivalent o f (2.9) and (2.10).

The formulas of Sections 3 and 4 follow from the basic relation

$$\left[\begin{array}{c|c|c} -\boldsymbol{Q}^{-1} & \boldsymbol{U}_1 & \boldsymbol{U}_2 \\ \hline \boldsymbol{U}_1^{\mathrm{T}} & \boldsymbol{C}_{11} & \boldsymbol{C}_{12} \\ \hline \boldsymbol{U}_2^{\mathrm{T}} & \boldsymbol{C}_{12}^{\mathrm{T}} & \boldsymbol{C}_{22} \end{array}\right] \overset{\mathrm{SWP}[n+1,n+2,\ldots,n+p_1]}{\longrightarrow}$$

$$\left[\begin{array}{c|c|c} -\boldsymbol{Q}(-)^{-1} & \boldsymbol{Q}(-)^{-1}\boldsymbol{X}_1 & \boldsymbol{Q}(-)^{-1}\boldsymbol{X}_2\boldsymbol{\Sigma}_2 \\ \hline \boldsymbol{X}_1^{\mathrm{T}}\boldsymbol{Q}(-)^{-1} & -\boldsymbol{\Sigma}_1-\boldsymbol{X}_1^{\mathrm{T}}\boldsymbol{Q}(-)^{-1}\boldsymbol{X}_1 & -\boldsymbol{X}_1^{\mathrm{T}}\boldsymbol{Q}(-)^{-1}\boldsymbol{X}_2\boldsymbol{\Sigma}_2 \\ \hline \boldsymbol{\Sigma}_2\boldsymbol{X}_2^{\mathrm{T}}\boldsymbol{Q}(-)^{-1} & -\boldsymbol{\Sigma}_2\boldsymbol{X}_2^{\mathrm{T}}\boldsymbol{Q}(-)^{-1}\boldsymbol{X}_1 & \boldsymbol{\Sigma}_2-\boldsymbol{\Sigma}_2\boldsymbol{X}_2^{\mathrm{T}}\boldsymbol{Q}(-)^{-1}\boldsymbol{X}_2\boldsymbol{\Sigma}_2 \end{array}\right]. \tag{A.4}$$

Relation (A.4) is a consequence of the relation

$$\left[\begin{array}{c|c|c} \boldsymbol{Q} & \boldsymbol{X}_1\boldsymbol{\Sigma}_1 & \boldsymbol{X}_2\boldsymbol{\Sigma}_2 \\ \hline \boldsymbol{\Sigma}_1\boldsymbol{X}_1^{\mathrm{T}} & \boldsymbol{\Sigma}_1 & \boldsymbol{0} \\ \hline \boldsymbol{\Sigma}_2\boldsymbol{X}_2^{\mathrm{T}} & \boldsymbol{0} & \boldsymbol{\Sigma}_2 \end{array}\right] \overset{\mathrm{SWP}[n+1,n+2,\ldots,n+p_1]}{\longrightarrow}$$

$$\left[\begin{array}{c|c|c} \boldsymbol{Q}(-) & \boldsymbol{X}_1 & \boldsymbol{X}_2\boldsymbol{\Sigma}_2 \\ \hline \boldsymbol{X}_1^{\mathrm{T}} & -\boldsymbol{\Sigma}_1^{-1} & \boldsymbol{0} \\ \hline \boldsymbol{\Sigma}_2\boldsymbol{X}_2^{\mathrm{T}} & \boldsymbol{0} & \boldsymbol{\Sigma}_2 \end{array}\right], \tag{A.5}$$

which is an immediate consequence of the definitions of $\boldsymbol{Q}$, $\boldsymbol{Q}(-)$, and SWP. The left side of (A.4) results from applying SWP[$1,2,\ldots,n$] to the left side of (A.5), and similarly for the right sides of (A.4) and (A.5), whence (A.4) follows from (A.5) because SWP[$1,2,\ldots,n$] and SWP[$n+1,n+2,\ldots,n+p_1$] commute.

Formulas (3.1)–(3.6) are immediate consequences of applying (A.4) first to the starred quantities and then to the unstarred quantities. Note that $\boldsymbol{\Sigma}_1$ or $\boldsymbol{\Sigma}_1^*$ appear on the right side of (A.4) only in the middle diagonal part, while the remaining parts of the right side of (A.4) is identical for starred and unstarred quantities. Formulas

(3.2)–(3.6) merely express these identical right side parts in terms of the left sides according to the definition of SWP[$n+1, n+2, \ldots, n+p_1$]. Formula (3.1) follows because $-C_{11}^{-1} = -\Sigma_1^{-1} - X_1^{\mathrm{T}} Q(-)^{-1} X_1$, whence $C_{11}^{-1} - \Sigma_1^{-1} = X_1^{\mathrm{T}} Q(-) X_1$, where the right side does not depend on Σ_1 and so must equal the same expression in starred terms.

In the limit as $\Sigma_1 \to \infty$, the right side of (A.4) approaches an obvious limit, where the only change is that $\Sigma_1^{-1} \to \mathbf{0}$ in the middle diagonal part. Thus, as remarked in Section 4, the limiting expressions for Q^{-1}, U and C can be immediately written down, as illustrated in (4.3).

Finally, note that (3.17) follows from the fact that

$$\det\left[\begin{array}{c|c} Q & X_1\Sigma_1 \\ \hline \Sigma_1 X_1^{\mathrm{T}} & \Sigma_1 \end{array}\right],$$

can be represented in either of the forms $\det Q \times \det C_{11}$ or $\det \Sigma_1 \times \det Q(-)$.

References

Dempster, A.P. (1969). *Elements of Continuous Multivariate Analysis*. Addison-Wesley, Reading, MA.

Dempster, A.P., Laird, N.M. and Rubin, D.B. (1977). Maximum likelihood from incomplete data via the EM algorithms. *J. Roy. Statist. Soc. Ser. B* **39**, 1–38.

Dempster, A.P., Selwyn, Murray R. and Patel, Chandu, M. (1979a). Algorithms for mixed model analysis with two variance components. Submitted.

Dempster, A.P., Rubin, D.B. and Tsutakawa, R.K. (1979b). Estimation in covariance components models. Submitted.

Harville, David A. (1977). Maximum likelihood approaches to variance component estimation and to related problems. *J. Amer. Statist. Assoc.* **72**, 320–340.

Lindley, D.V. and Smith, A.F.M. (1972). Bayes estimates for the linear model. *J. Roy. Statist. Soc. Ser. B.* **34**, 1–41.

Sundberg, R. (1974). Maximum likelihood theory for incomplete data from an exponential family. *Scand. J. Statist.* **1**, 49–58.

G. Kallianpur, P.R. Krishnaiah, J.K. Ghosh, eds., *Statistics and Probability: Essays in Honor of C.R. Rao*

CONNECTEDNESS AND ZERO VARIANCE IN SAMPLING DESIGNS

S.W. DHARMADHIKARI
Southern Illinois University, Carbondale, IL, U.S.A.

If samples are considered as blocks and units are considered as treatments, the definition of a connected sampling design becomes quite straightforward. This has been done by Wynn (1976) and by Patel and Dharmadhikari (1977). Wynn considered designs of fixed sample size and gave a characterization of connected sampling designs in terms of the variance of the Horvitz–Thompson estimator. In this paper we give a different characterization of connectedness without requiring the design to be of fixed sample size. We also make some remarks which show an interplay between sampling designs and incomplete block designs.

1. Preliminaries

Let $\mathcal{U}=\{1,2,\ldots,N\}$ be a population of N identifiable units. Let S be a class of subsets of $\mathcal{U}$ whose union is $\mathcal{U}$ and $P=\{p(s), s\in S\}$ be such that $p(s)>0$ for all s and $\Sigma_{s\in S}p(s)=1$. The pair (S,P) is thus a sampling design for $\mathcal{U}$. Let $\boldsymbol{Y}=(Y_1,\ldots,Y_N)$ be a vector of real-valued measurements on the units in $\mathcal{U}$. We consider the problem of estimating the total $\tau(\boldsymbol{Y})=\Sigma_{i=1}^N Y_i$ in an unbiased and linear manner. A linear estimator T has the form

$$T(s)=\Sigma_{i\in s}b(s,i)Y_i, \quad s\in S. \tag{1}$$

For T to be unbiased for $\tau(\boldsymbol{Y})$ we must have

$$\sum_{s:i\in s} b(s,i)p(s)=1, \quad i=1,\ldots,N, \tag{2}$$

We say that units i and j are *connected* if we can find units $i_1,\ldots,i_{n-1}$ and samples $s_1,\ldots,s_n$ such that $i\in s_1$, $i_1\in s_1\cap s_2,\ldots,i_{n-1}\in s_n\cap s_{n-1}$ and $j\in s_n$. It is clear that "being connected" is an equivalence relation. Therefore $\mathcal{U}$ splits into equivalence classes $\mathcal{U}_1,\ldots,\mathcal{U}_t$. The design is said to be *connected* if $t=1$ or, equivalently, if i and j are connected for every choice of i, j. The subsets $\mathcal{U}_h$ will be called *components*.

When we have t components we can express S as the disjoint union of t subclasses $S_1,\ldots,S_t$ such that the samples in S_h consist of units from $\mathcal{U}_h$ only. We will write

$$\alpha_h=\sum_{s\in S_h} p(s).$$

We will also use the standard symbol π_i for the inclusion probability of the unit i.

2. Necessary and sufficient conditions for zero variance at a given point

In this section we first assume that $\boldsymbol{k}=(k_1,\ldots,k_N)$ is a given point in R^N with $k_i \neq 0$ for all i and $\tau(\boldsymbol{k})=\Sigma_{i=1}^N k_i \neq 0$. Suppose that we can find a linear estimator T of the form (1) which is unbiased for $\tau(\boldsymbol{Y})$ and which has zero variance at $\boldsymbol{k}$. Then in addition to (2), we must have

$$\sum_{i\in s} b(s,i)k_i = \tau(\boldsymbol{k}). \tag{3}$$

Multiplying (3) by $p(s)$ and summing over $s \in S_h$, we get

$$\sum_{s\in S_h} \sum_{i\in s} b(s,i)k_i p(s) = \tau(\boldsymbol{k})\alpha_h. \tag{4}$$

Write $w_h(\boldsymbol{k})=\Sigma_{i\in U_h} k_i$. Then, multiplying (2) by k_i and summing over $i \in \mathfrak{U}_h$, we get

$$\sum_{i\in \mathfrak{U}_h} \sum_{s:i\in s} b(s,i)p(s)k_i = w_h(\boldsymbol{k}). \tag{5}$$

However, the left sides of relations (4) and (5) are the same. Therefore, we must have

$$w_h(\boldsymbol{k}) = \alpha_h \tau(\boldsymbol{k}), \quad h=1,\ldots,t. \tag{6}$$

We have thus obtained a necessary condition (6) for the existence of an unbiased linear estimator with zero variance at $\boldsymbol{k}$. We now proceed to prove that this condition is also sufficient.

Suppose that (6) holds. We want to obtain $b(s,i)$ so that (2) and (3) hold. Since zero variance at $\boldsymbol{k}$ is the same as zero variance at $c\boldsymbol{k}$ with $c \neq 0$, we may assume that $\tau(\boldsymbol{k})=1$. Suppose we take

$$b(s,i) = \xi_i + a_s k_i.$$

Condition (3) will be satisfied if we take

$$a_s = \Big(1 - \sum_{j\in s} \xi_j k_j\Big)\Big/\delta(s),$$

where $\delta(s) - \Sigma_{i\in s} k_i^2$. Therefore, we take

$$b(s,i) = \frac{k_i}{\delta(s)} + \xi_i - \frac{k_i}{\delta(s)} \sum_{j\in s} k_j \xi_j. \tag{7}$$

The quantities ξ_i in (7) remain to be determined. They must be so chosen that the $b(s,i)$ satisfy the condition (2) for unbiasedness. The resulting linear equations can be written as

$$A\boldsymbol{\xi} = \boldsymbol{d}, \tag{8}$$

where the entries in the $N \times N$ matrix A and in the N-vector $\boldsymbol{d}$ are given by

$$a_{ii} = \pi_i - k_i^2 \sum_{s:i\in s} [p(s)/\delta(s)],$$

$$a_{ij} = -k_i k_j \sum_{s:i,j\in s} [p(s)/\delta(s)], \quad i \neq j,$$

and

$$d_i = 1 - k_i \sum_{s:i\in s} [p(s)/\delta(s)].$$

We have to show that the system (8) has a solution if (6) holds.

Suppose $\mathfrak{U}$ has t components $\mathfrak{U}_1,\ldots,\mathfrak{U}_t$. Then $a_{ij}=0$ whenever i and j belong to different components. Therefore, system (8) splits into subsystems

$$A_h\xi_h = \boldsymbol{d}_h, \quad h=1,\ldots,t. \tag{9}$$

The matrix A is symmetric. Further

$$\begin{aligned}\sum_{j=1}^{N} a_{ij}k_j &= \pi_i k_i - k_i \sum_{j=1}^{N} k_j^2 \sum_{s:i,j\in s} [p(s)/\delta(s)] \\ &= \pi_i k_i - k_i \sum_{s:i\in s} \frac{p(s)}{\delta(s)} \sum_{j\in s} k_j^2 \\ &= \pi_i k_i - k_i \sum_{s:i\in s} p(s) \\ &= \pi_i k_i - k_i \pi_i = 0.\end{aligned}$$

Thus, $A\boldsymbol{k}=\boldsymbol{0}$. Therefore if $K=\text{diag}(k_1,\ldots,k_N)$, then the matrix $C=KAK$ is such that C is symmetric and $\Sigma_{j=1}^{N} c_{ij}=0$ for all i. Furthermore, $c_{ij}\neq 0$ whenever $a_{ij}\neq 0$, which happens when there is a sample that contains both i and j. The proof of Theorem 1 of Patel and Dharmadhikari (1977) now shows that the rank of A_h is N_h-1 where N_h is the number of units in $\mathfrak{U}_h$. The rows of $\boldsymbol{A}_h$ therefore satisfy only one linear constraint. Since $\boldsymbol{k}'A=0$, the system (9) is consistent whenever $\Sigma_{i\in\mathfrak{U}_h} k_i d_i = 0$. However,

$$\begin{aligned}\sum_{i\in\mathfrak{U}_h} k_i d_i &= \sum_{i\in\mathfrak{U}_h} k_i\Big\{1 - k_i \sum_{s:i\in s} [p(s)/\delta(s)]\Big\} \\ &= \sum_{i\in\mathfrak{U}_h} k_i - \sum_{s\in S_h} \frac{p(s)}{\delta(s)} \sum_{i\in s} k_i^2 \\ &= w_h(\boldsymbol{k}) - \sum_{s\in S_h} p(s) \\ &= w_h(\boldsymbol{k}) - \alpha_h,\end{aligned}$$

which is zero because (6) is assumed to hold and $\tau(\boldsymbol{k})=1$. Given a solution $\boldsymbol{\xi}$ of system (8), we can use (7) to get the required estimator. We have proved the following theorem.

Theorem 1. *Suppose $k_i\neq 0$ for all i and $\Sigma_{i=1}^{N} k_i \neq 0$. A necessary and sufficient condition for the existence of an unbiased linear estimator having zero variance at $\boldsymbol{k}$ is that*

$$\sum_{i\in\mathfrak{U}_h} k_i = \Big(\sum_{s\in S_h} p(s)\Big)\Big(\sum_{i=1}^{N} k_i\Big), \quad h=1,\ldots,t.$$

So far we have assumed that $k_i \neq 0$ for all i and $\tau(\boldsymbol{k}) \neq 0$. We now briefly mention the modifications required when these conditions fail.

(a) Suppose that $k_i \neq 0$ for all i, but $\tau(\boldsymbol{k})=0$. Then condition (6) needs no modification. In other words, we require that $w_h(\boldsymbol{k})=0$ for $h=1,\ldots,t$.

(b) Suppose that $k_i=0$ for some i. Observe that condition (3) puts no restriction on the $b(s,i)$ if $k_i=0$. For such units i, we can therefore take $b(s,i)=\pi_i^{-1}$. Next we consider the remaining set of units, namely,

$$\mathfrak{U}'=\{i\in\mathfrak{U}: k_i\neq 0\}.$$

We look at the components $\mathfrak{U}'_h$, $h=1,\ldots,t'$ with reference to the reduced population $\mathfrak{U}'$. Now S'_h, $w'_h(\boldsymbol{k})$ and α'_h can be defined in the same way as before. Condition (6) then has to be replaced by

$$w'_h(\boldsymbol{k})=\alpha'_h\tau(\boldsymbol{k}), \quad h=1,\ldots,t'. \tag{10}$$

If $\tau(\boldsymbol{k})=0$, then condition (10) enables us to prove a modification of Theorem 1. However, if $\tau(\boldsymbol{k})\neq 0$, then an additional condition is required, namely,

$$\text{Every } s\in S \text{ contains some unit } i \text{ for which } k_i\neq 0. \tag{11}$$

To see the necessity of condition (11) observe that (3) cannot hold if $k_i=0$ for all $i\in s$ and $\tau(\boldsymbol{k})\neq 0$.

3. Characterization of connectedness

Theorem 1 gives us the following characterization of connectedness.

Theorem 2. *A sampling design is connected if, and only if, for every given* $\boldsymbol{k}$ *satisfying* $k_i\neq 0$ *for all i and* $\sum_{i=1}^N k_i\neq 0$, *the design admits an unbiased linear estimator which has zero variance at* $\boldsymbol{k}$.

Proof. Suppose the design is connected. Then $t=1$ and the condition of Theorem 1 is trivially satisfied for all $\boldsymbol{k}$ and hence the required linear estimator exists.

Conversely, suppose the design is not connected. Then we can easily choose $\boldsymbol{k}$ such that $k_i\neq 0$ for all i, $\tau(\boldsymbol{k})\neq 0$ and $w_h(\boldsymbol{k})$ are not proportional to α_h. Condition (6) would then fail. Therefore an unbiased linear estimator with zero variance at $\boldsymbol{k}$ cannot exist. The proof is complete.

4. Concluding remarks

Wynn (1976) gives a characterization of connectedness applicable to fixed sample size designs. Our results are free from this restriction. The statement of Theorem 2 is similar to the result in the theory of incomplete block designs which says that a design is connected if, and only if, every treatment contrast is estimable.

If $\boldsymbol{k}=(1,0,\ldots,0)$, then we can have zero variance at $\boldsymbol{k}$ if, and only if, $\pi_1=1$. So if $\pi_i<1$ for all i, then an unbiased linear estimator cannot have zero variance along any coordinate axis. There may also be other points at which zero variance is unattainable. For example, let $N=3$ and take one of the two samples $\{1,2\}$, $\{1,3\}$ with equal probability. Then an unbiased linear estimator with zero variance at $(0, k_2, k_3)$ is obtainable if, and only if, $k_2=k_3$.

We also remark that all the results of this paper apply to *homogeneous* linear estimators only. If non-homogeneous estimators are accepted, then zero variance is attainable at every $\boldsymbol{k}$; see Basu (1971).

Suppose $\nu_i(s)=1$ or 0 according as $i \in s$ or $i \in \mathfrak{U}-s$. Write $\nu=\sum_{i=1}^{N}\nu_i$ for the effective sample size. Define the matrix $C=(c_{ij})$ as follows

$$c_{ii}=E(\nu_i)-E(\nu_i/\nu) \quad \text{and} \quad c_{ij}=-E(\nu_i\nu_j/\nu), \qquad i \neq j.$$

The c_{ij}'s are precisely the values of the a_{ij}'s when $k_i=1/N$ for all i. The matrix C may be called the C-matrix of the sampling design because of its close resemblance with the C-matrix of an incomplete block design. In particular, C is nonnegative definite and consequently acts as a covariance matrix. It would be interesting to find some variables related to the sampling design whose covariance matrix is a multiple of C. Similarly, it would be nice if the characteristic roots of C have some simple interpretation. These problems are being investigated.

References

Basu, D. (1971). An essay on the logical foundations of survey sampling, Part I. In *Foundations of Statistical Inference*. pp. 203–242. Holt, Rinehart and Winston, Toronto.

Patel, H.C. and Dharmadhikari, S.W. (1977). On linear invariant unbiased estimators in survey sampling. *Sankhyā C* **39**, 21–27.

Wynn, H.P. (1976). Connected finite population sampling plans. *Biometrika* **63**, 208–210.

G. Kallianpur, P.R. Krishnaiah, J.K. Ghosh, eds., *Statistics and Probability: Essays in Honor of C.R. Rao*

A POTENTIAL THEORETIC APPROACH TO MARTINGALE THEORY

J.L. DOOB
University of Illinois at Urbana-Champaign, Urbana, IL, U.S.A.

The purpose of this paper is to outline certain parallel aspects of potential theory and martingale theory in order to show that a potential theoretic approach to martingale theory is likely to be fruitful.

1. Potential theory

Some aspects of potential theory will be described in this section; corresponding aspects of martingale theory will be described in Section 2.

1.1. Superharmonic functions

In each context of potential theory there is a basic space, S, and a class of "superharmonic" functions defined on subsets of S. A superharmonic function is defined as an extended real valued function satisfying certain finiteness and smoothness conditions and also satisfying a system of average inequalities. If these inequalities are equalities, or under certain related conditions, the superharmonic function is called "harmonic".

Example (A). Potential theory of Laplace's equation. In this context S is an open subset of $\mathbb{R}^N$. A superharmonic function is a function u defined on an open subset B of S and satisfying the following conditions.

(A.1) $-\infty < u \leqslant +\infty$; $u < +\infty$ on a dense subset of B.

(A.2) u is lower semicontinuous.

(A.3) If ξ is a point of B and if $L(u, \xi, r)$ is the average of u over the sphere of center ξ and radius r, with r so small that the sphere and its interior are in S, then $u(\xi) \geqslant L(u, \xi, r)$.

A function u and its negative $-u$ are both superharmonic if and only if u satisfies Laplace's equation $\Delta u = 0$, and u is then called "harmonic".

Example (B). Potential theory of the heat equation. In this context S is an open subset of $\mathbb{R}^{N+1}$ and a point of S is written $(t, \xi_1, \ldots, \xi_N)$ to distinguish the first coordinate from the others. A superharmonic function (sometimes called "superparabolic" in this context) is a function defined on an open subset of B of S, satisfying (A.1) and (A.2) and a certain different average inequality from (A.3) which need not be stated explicitly here. A function u and its negative are both superharmonic if and only if u satisfies the heat equation $\partial u/\partial t = \Delta_\xi u$, and u is then called "harmonic" or sometimes "parabolic".

1.2. The fine topology of a potential theory

The fine topology of the space S is the coarsest topology making every superharmonic function continuous. In Examples (A) and (B) countable subsets of S have no limit points in the fine topology. In Example (B) a hyperplane $t=$const. has no limit point in the fine topology.

1.3. Small sets in potential theory

In each potential theoretic context a subset A of S is called "polar" if there is a superharmonic function which is identically $+\infty$ on A. The polar sets are negligible in most potential theoretic operations and a relation on a subset B of S true up to a polar subset of B is said to be true quasi everywhere on B. A polar set is known to have no fine topology limit points. A subset of S is called "semipolar" if it is a countable union of sets each of which has no fine topology limit points. A polar set is therefore semipolar. In Example (A) the semipolar sets are the same as the polar sets but in Example (B) a hyperplane $t=$const. is a semipolar set which is not polar.

1.4. Convergence of an upward directed family of superharmonic functions

Owing to the type of smoothness condition imposed on superharmonic functions, the limit of an increasing sequence of superharmonic functions satisfies this smoothness condition. The Lebesgue dominated convergence theorem can then be applied to show that the limit function satisfies the superharmonic function average inequality in the specified context, and the limit function is therefore superharmonic if the appropriate finiteness condition is satisfied. More generally, let $\{u_\alpha, \alpha\in I\}$ be an upward directed family of superharmonic functions (that is if α_1 and α_2 are in I, there is an index α_3 in I such that $u_{\alpha_1}\leqslant u_{\alpha_3}$ and $u_{\alpha_2}\leqslant u_{\alpha_3}$). Then it is known that the limit u of this upward directed family of functions, that is $u=\sup_\alpha u_\alpha$, is superharmonic if u satisfies the finiteness conditions imposed on superharmonic functions in the given context.

1.5. Convergence of a downward directed family of superharmonic functions

If $\{u_\alpha, \alpha\in I\}$ is a family of positive superharmonic functions the function $u=\inf_\alpha u_\alpha$ is not necessarily superharmonic but is very nearly so in the following sense. The function u satisfies the superharmonic function average inequality but not the smoothness conditions imposed on a superharmonic function. However a certain smoothing operation changes u to a superharmonic function $\hat{u}$ for which $\hat{u}\leqslant u$ with equality up to a semipolar set. This theorem is actually a convergence theorem because the minimum of finitely many superharmonic functions is superharmonic so if $\{u_\alpha, \alpha\in I\}$ is replaced by the family of finite minima $u_{\alpha_1}\wedge\cdots\wedge u_{\alpha_k}$ the new family of superharmonic functions is directed downwards and has u as limit. In Examples (A) and (B) $\hat{u}(\xi)=\liminf_{\eta\to\xi} u(\eta)$.

2. Martingale theory

Let $(\Omega, \mathcal{F}, \mathcal{F}(\cdot), P)$ be a probability space provided with a filtration $\mathcal{F}(\cdot)$. That is Ω is a space, $\mathcal{F}$ is a σ algebra of subsets of Ω, P is a probability measure on $\mathcal{F}$, and to each t in $\mathbb{R}^+$ corresponds a sub-σ algebra $\mathcal{F}(t)$ of $\mathcal{F}$ such that $s<t$ implies that $\mathcal{F}(s)\subset\mathcal{F}(t)$. It is supposed that the probability measure P on $\mathcal{F}$ is complete, that $\mathcal{F}(0)$ contains the P null sets and that $\mathcal{F}(t)=\bigcap_{\varepsilon>0}\mathcal{F}(t+\varepsilon)$ for all t. The product space $\mathbb{R}^+\times\Omega$ is the counterpart of S in the potential theory context. A stochastic process $\{x(t),\ t\in\mathbb{R}^+\}$ is a function $(t,\omega)\circ\!\!\to x(t,\omega)$ on $\mathbb{R}^+\times\Omega$ such that each function $x(t,\cdot)$, also written $x(t)$, is $\mathcal{F}$ measurable. All processes will be supposed adapted to the filtration $\mathcal{F}(\cdot)$ in the sense that $x(t)$ is $\mathcal{F}(t)$ measurable for all t.

2.1. *Supermartingales*

A supermartingale is an adapted stochastic process $\{x(t),\ t\in\mathbb{R}^+\}$ on $(\Omega, \mathcal{F}, \mathcal{F}(\cdot), P)$ for which $E\{|x(t)|\}<\infty$ for all t and for which $s<t$ implies that

$$E\{x(t)|\mathcal{F}(s)\}\leqslant x(s) \quad \text{a.s.} \tag{2.1}$$

A martingale is a process satisfying the same conditions except that there is equality in (2.1). We shall call a "(super)martingale standard" if almost all its sample functions are right continuous with left limits at all points. A standard (super)martingale is the probabilistic counterpart of a (super)harmonic function.

2.2. *Small sets in martingale theory*

A subset A of $\mathbb{R}^+\times\Omega$ is called "evanescent" if there is a P null subset Λ of Ω such that $A\subset\mathbb{R}^+\times\Lambda$. The evanescent sets are the probabilistic counterparts of the polar sets of potential theory and a relation on a subset B of $\mathbb{R}^+\times\Omega$ true up to an evanescent set will be said to be true quasi everywhere on B. Optional times (=stopping times) are defined as usual in terms of $\mathcal{F}(\cdot)$ and a subset of $\mathbb{R}^+\times\Omega$ will be called "semipolar" if it is contained in a countable union of graphs $\{(t,\omega): t=T(\omega)<+\infty\}$ of optional times T. Here we will not discuss the counterpart on $\mathbb{R}^+\times\Omega$ of the fine topology of potential theory.

Two stochastic processes $x(\cdot)$ and $y(\cdot)$ will be identified if they are equal quasi everywhere on $\mathbb{R}^+\times\Omega$, that is if $x(t,\omega)=y(t,\omega)$ up to an evanescent set. Thus the counterpart of a (super)harmonic function is an equivalence class of (super) martingales but we shall adopt the usual loose language which ignores this refinement.

2.3. *Convergence of an upward directed family of supermartingales*

According to a theorem of Meyer the limit $x(\cdot)$ of an increasing sequence of standard supermartingales is equal quasi everywhere to a standard supermartingale if $E\{|x(0)|\}<\infty$, and this theorem is easily extended to an upward directed set of supermartingales. Thus the exact counterpart of the potential theory theorem on the limit of an upward directed set of superharmonic functions is true. The proof of Meyer's theorem is however much deeper than that of its potential theory counterpart.

2.4. *Convergence of a downward directed family of supermartingales*

The following theorem will be proved elsewhere. Let $\{x_\alpha(\cdot), \alpha \in I\}$ be a family of standard supermartingales. There is then a supermartingale $x(\cdot)$ with the following properties.

(C.1) Almost every $x(\cdot)$ sample function has right and left limits at all points.

(C.2) For each α in I, $x(\cdot) \leqslant x_\alpha(\cdot)$ quasi everywhere on $\mathbb{R}^+ \times \Omega$. (The exceptional evanescent set depends on α.)

(C.3) There is a sequence $\alpha_0, \alpha_1, \ldots$ such that $x(\cdot) = \inf_j x_{\alpha_j}(\cdot)$ quasi everywhere on $\mathbb{R}^+ \times \Omega$.

(C.4) If $\hat{x}(t) = x(t+)$ the process $\hat{x}(\cdot)$ is a standard supermartingale. Moreover $\hat{x}(\cdot) \leqslant x(\cdot)$ quasi everywhere on $\mathbb{R}^+ \times \Omega$, with equality up to a semipolar subset of $\mathbb{R}^+ \times \Omega$. This result is a close counterpart of the potential theory theorem on the limit of a downward directed set of superharmonic functions. The process $x(\cdot)$, uniquely determined up to an evanescent set, will be written $\operatorname{ess}^*_\alpha \inf x_\alpha(\cdot)$.

3. Sweeping in potential theory and martingale theory

We make a few remarks on sweeping (balayage) in the two theories to illustrate further the tight correspondence between potential theory and martingale theory. If v is a strictly positive superharmonic function on S and if A is a subset of S the reduction R_v^A of v on A is defined as the infimum of the class of positive superharmonic functions u on S such that $u \geqslant v$ on A. This reduction need not be superharmonic but according to Section 1 its smoothing $\hat{R}_v^A$ is. The martingale theory counterpart of this reduction is the reduction $R_{y(\cdot)}^A$ of a strictly positive standard supermartingale on a subset A of $\mathbb{R}^+ \times \Omega$, defined as the essential* infimum of the class of positive standard supermartingales $x(\cdot)$ such that $x(\cdot) \geqslant y(\cdot)$ on A. This reduction need not be a standard supermartingale but according to Section 2 its smoothing $\hat{R}_{y(\cdot)}^A$ is. In both theories the reduction has certain desirable properties not possessed by its smoothing, as illustrated below. To make the potential theory more concrete we consider Example (B). In this context let Γ be the class of compact subsets of S and consider the set function $A \circ\!\!\to \phi(A) = R_v^A$. This function from sets A into functions on S is not additive but is strongly subadditive, that is

(R.1) $\phi(A \cup B) + \phi(A \cap B) \leqslant \phi(A) + \phi(B)$

and ϕ also has the following two properties:

(R.2) If A is an increasing sequence of subsets of S with union A then $\lim_{n\to\infty} \phi(A_n) = \phi(A)$.

(R.3) If v is finite valued and if A is a decreasing sequence of sets in Γ with intersection A then $\lim_{n\to\infty}\phi(A_n)=\phi(A)$.
These properties of ϕ imply that, for v finite valued, for each point ξ the set function $A\circ\!\!\rightarrow\phi(A)(\xi)=R_v^A(\xi)$ is a Choquet capacity relative to Γ and that for every Borel or even analytic subset A of S the value $\phi(A)(\xi)$ is the supremum [infimum] of the values of $\phi(\cdot)(\xi)$ on compact subsets [open supersets] of A. The smoothed reduction $\hat{R}_v^A$ satisfies (R.1) and (R.2) but not (R.3).

In the martingale theory context we write S for $\mathbb{R}^+\times\Omega$ from now on. Suppose that $y(\cdot)$ is a strictly positive standard supermartingale of class $\boldsymbol{D}$, that is the class of random variables

$$\{y(T): T \text{ finite and optional}\}$$

is uniformly integrable. Suppose also, as is true in many contexts, that the given filtered probability space has the property that almost every sample function of a standard supermartingale is lower semicontinuous. Define a subset of $\mathbb{R}^+\times\Omega$ as compact if the set of values of t with (t,ω) in the set is compact for each ω. Let Γ be the class of compact progressively measurable subsets of S. Then if $\phi(A)=R_{y(\cdot)}^A$, so that ϕ is a function from subsets A of S into (equivalence classes of) supermartingales, the same technique used in the potential theory context to prove (R.1), (R.2), (R.3) can be used to prove that these relations hold quasi everywhere on S in the martingale theory context. The exceptional set in (R.1) depends on A and B and there is a corresponding qualification for (R.2) and (R.3). It is possible to invoke Choquet capacity theory to obtain the counterpart of the potential theory results just described, but the technical details and results are omitted here.

G. Kallianpur, P.R. Krishnaiah, J.K. Ghosh, eds., *Statistics and Probability: Essays in Honor of C.R. Rao*

ASYMPTOTIC CONFIDENCE-INTERVALS FOR THE VARIANCE IN THE LINEAR REGRESSION MODEL

Hilmar DRYGAS
Gesamthochschule Kassel, D-3500, Kassel, Federal Republic of Germany

Consider a sequence $\varepsilon_1, \varepsilon_2, \ldots$ of i.i.d. random variables on the probability space $(\Omega, \mathrm{F}, \mathrm{P})$ such that:

$$E\varepsilon_i = 0, \qquad E\varepsilon_i^2 = 1, \quad i = 1, 2, \ldots. \tag{1}$$

Consider, moreover, a sequence $x_1, x_2, \ldots$ of non-stochastic $1 \times k$-vectors and let,

$$\eta_i = x_i\theta + \sigma\varepsilon_i; \quad i = 1, 2, \ldots, \tag{2}$$

where θ is an unknown $k \times 1$-vector and $\sigma > 0$ is an unknown parameter, too.

It is the purpose of this note to find asymptotic confidence-intervals for the unknown variance σ^2 which are valid even in the non-normal case.

In the sequel we use $y_n = (\eta_1, \ldots, \eta_n)'$, $X_n = (x_1', x_2', \ldots, x_n')'$ and $u_n = (\varepsilon_1, \ldots, \varepsilon_n)'$. Then

$$y_n = X_n\theta + \sigma u_n \tag{3}$$

describes the usual linear regression model with n observations and

$$Ey_n = X_n\theta, \qquad \operatorname{Cov} y_n = \sigma^2 I_n \tag{4}$$

where I_n denotes the $n \times n$ unit-matrix. Let:

$$E\varepsilon_i^4 = \beta, \tag{5}$$

then $\beta \geqslant 1$ (with equality if and only if $\mathrm{P}(\varepsilon_i = 1) = \mathrm{P}(\varepsilon_i = -1) = \frac{1}{2}$) and $\beta = 3$ in the normal case.

The basis for constructing asymptotic confidence-intervals for σ and σ^2 will be the following theorem:

Theorem 1. *If $P_n y$ denotes the projection of $y \in \mathbb{R}^n$ onto* $\operatorname{im}(X_n)$, *the column-space of X_n, then if $\beta > 1$,*

$$(\sigma^{-1}\| y_n - P_n y_n \| - \eta^{1/2})2(\beta - 1)^{-1/2}, \tag{6}$$

converges in distribution towards the normal distribution $N(0,1)$.

Proof. Evidently $y_n - P_n y_n = \sigma(u_n - P_n u_n)$ and

$$\| u_n - P_n u_n \|^2 = \| u_n \|^2 - \| P_n u_n \|^2.$$

Now, therefore,

$$\sigma^{-1}\| y_n - P_n y_n \| - n^{1/2} =$$
$$=\left(\left[\| u_n \|^2 - \| P_n u_n \|^2\right] - n\right)$$
$$\times\left(\left(\| u_n \|^2 - \| P_n u_n \|^2\right)^{1/2} + n^{1/2}\right)^{-1}. \tag{7}$$

But $\{\varepsilon_i^2;\ i=1,2,\dots\}$ is a sequence of i.i.d. random variables. Thus, since $\mathrm{Var}(\varepsilon_i^2)=(\beta-1)$,

$$(\beta-1)^{-1/2} n^{-1/2} \sum_{i=1}^{n} (\varepsilon_i^2 - 1) =$$
$$=(\beta-1)^{-1/2} n^{-1/2}(\| u_n \|^2 - n)$$
$$=(\beta-1)^{-1/2} n^{-1/2} \| u_n \|^2 - (\beta-1)^{-1/2} n^{1/2}, \tag{8}$$

converges in distribution towards the standard normal distribution $N(0,1)$. Since $E\| P_n u_n \|^2 = \mathrm{tr}(P_n) \leqslant k$, it follows from Chebyshev's inequality*, that as n tends to infinity $n^{-1/2}\| P_n u_n \|^2 \to 0$ in probability. Thus,

$$(\beta-1)^{-1/2}\left[n^{-1/2}\| u_n - P_n u_n \|^2 - n^{1/2}\right] =$$
$$=(\beta-1)^{-1/2}\left[\sigma^{-2} n^{-1/2} \| y_n - P_n y_n \|^2 - n^{1/2}\right] \tag{9}$$

converges in distribution towards the standard normal distribution, too. Finally, by the (weak) law of large numbers $n^{-1}\| u_n \|^2 \to 1$ in probability. From this and (7), and a well-known theorem (see, e.g. Rao [2, pp. 122, 124]), the assertion of the theorem follows.

This theorem already allows the construction of confidence-intervals for σ. Let $\Phi(x)$ denote the distribution function of the standard normal distribution, then if $\alpha = \Phi^{-1}(\frac{1}{2}(1+\delta))$ $(0<\delta<1)$ by Theorem 1:

$$\lim_{n\to\infty} P\left(-\alpha \leqslant \left[\sigma^{-1}\|(I_n - P_n) y_n\| - n^{1/2}\right] 2(\beta-1)^{-1/2} \leqslant \alpha\right) =$$
$$= \lim_{n\to\infty} P\left(-\tfrac{1}{2}(\beta-1)^{1/2}\alpha + n^{1/2} \leqslant \sigma^{-1}\|(I-P_n)y_n\| \leqslant \tfrac{1}{2}(\beta-1)^{1/2}\alpha + n^{1/2}\right)$$
$$=\delta. \tag{10}$$

This shows that

$$J_n = \left[2\|(I_n - P_n)y_n\| / \left(2n^{1/2} + \alpha(\beta-1)^{1/2}\right),\right.$$
$$\left. 2\|(I_n - P_n)y_n\| / \left(2n^{1/2} - \alpha(\beta-1)^{1/2}\right)\right] \tag{11}$$

*This inequality is sometimes also called Markov's inequality.

is an asymptotic confidence-interval for σ^2 at the confidence-level δ whereas

$$\tilde{J}_n = \Big[4\|(I_n - P_n)y_n\|^2/\big(2n^{1/2} + \alpha(\beta-1)^{1/2}\big)^2,\quad 4\|(I_n - P_n)y_n\|^2/\big(2n^{1/2} - \alpha(\beta-1)^{1/2}\big)^2\Big] \tag{12}$$

is an asymptotic confidence-interval for σ^2 at the confidence-level δ. These two formulae would also apply if $\beta=1$. However, in this case

$$\|(I_n - P_n)y_n\|^2 = \sigma^2\|(I_n - P_n)u_n\|^2 = n\sigma^2 - \sigma^2\|P_n u_n\|^2$$

with probability one.

Usually β is unknown and has to be replaced by an estimator β_n in such a way that $\{\beta_n\}$ is a consistent sequence of estimators of β. In the sequel we will construct such a sequence. It is remarkable, however, that we have to assume the existence of the sixth moments $E(\varepsilon_i^6)$ to establish this consistency. Let:

$$\hat{\beta}_n = \begin{cases} n\|(I_n - P_n)y_n\|^{-4} \sum_{i=1}^{n} \big(e_i'(I_n - P_n)y_n\big)^4 \\ \quad \text{if } y_n \notin \operatorname{im}(X_n), \\ 3 \quad \text{otherwise,} \end{cases} \tag{13}$$

where e_i denotes the ith unit-vector in $\mathbb{R}^n$ and let, moreover,

$$\tilde{\beta}_n = \begin{cases} \hat{\beta}_n & \text{if } \hat{\beta}_n > 1, \\ 3 & \text{if } \hat{\beta}_n = 1. \end{cases} \tag{14}$$

Remark 1. If y_n has an absolutely continuous distribution with respect to the Lebesgue measure, then

$$\tilde{\beta}_n = n\|(I_n - P_n)y_n\|^{-4} \sum_{i=1}^{n} \big(e_i'(I_n - P_n)y_n\big)^4$$

with probability one, if $n>k+1$. Indeed $\operatorname{im}(X_n)$ is a hyperplane in $\mathbb{R}^n$ and has probability zero if $n>k$. $\hat{\beta}_n=1$ can happen if and only if $(e_i'(y_n - P_n y_n))^2$ is constant for all i. This means that if $P_n = (p_{ij}^{(n)})$, then

$$\eta_1 - \sum_{k=1}^{n} p_{ik}^{(n)}\eta_k = \pm\left(\eta_i - \sum_{k=1}^{n} p_{ik}^{(n)}\eta_k\right) \quad \text{for all } i.$$

If this equation is not trivially met, then it defines a hyperplane which has probability zero. The equation is trivial if and only if,

$$1 - p_{ii} \pm p_{1i} = 1 - p_{11} \pm p_{i1} = p_{ik} \mp p_{1k} = 0,$$

for $k \neq i, 1$. $p_{1i} = p_{i1}$ implies $p_{ii} = p_{11}$ and $p_{1i} = \mp(1-p_{11})$. From the fact that P_n is idempotent, we get $p_{11} = p_{11}^2 + (n-1)(1-p_{11})^2$. This yields $p_{11}=1$ or $p_{11}=(n-1)n^{-1}$. In the first case we get $P_n = I_n$ (impossible if $n>k$) while in the second case $\operatorname{Rank}(X_n) = \operatorname{tr}(P_n) = n-1$ (impossible if $n>k+1$). The triviality of the equations in question results in very special designs, namely the projection on the orthogonal complement of $(-1, 1'_m, -1'_{n-m-1})'$, where 1_k is all one vector of dimension k, i.e.

$$P_n = I_n - n^{-1}\left(\begin{array}{c|c|c} 1 & 0 & 0 \\ \hline 0 & I_m & 0 \\ \hline 0 & 0 & -I_{n-m-1} \end{array}\right) 1_n 1'_n$$

$$\times \left(\begin{array}{c|c|c} -1 & 0 & 0 \\ \hline 0 & I_m & 0 \\ \hline 0 & 0 & -I_{n-m-1} \end{array}\right), \tag{15}$$

for a suitable integer m.

Theorem 2. *If $E(\varepsilon_i^6)$ exists, then $\{\tilde{\beta}_n\}$ is a consistent sequence of estimators of $\beta > 1$.*

Proof. First of all we show that:

$$\hat{\mu}_4^{(n)} = n^{-1} \sum_{i=1}^{n} (e_i'(I_n - P_n) y_n)^4, \tag{16}$$

is a consistent sequence of estimator of $\beta\sigma^4$. Indeed:

$$\begin{aligned} \hat{\mu}_4^{(n)} &= \sigma^4 n^{-1} \sum_{i=1}^{n} (e_i'(I_n - P_n) u_n)^4 = \sigma^4 n^{-1} \sum_{i=1}^{n} (e_i' u_n)^4 \\ &\quad - 4\sigma^4 n^{-1} \sum_{i=1}^{n} (e_i' u_n)^3 (e_i' P_n u_n) + 6\sigma^4 n^{-1} \sum_{i=1}^{n} (e_i' u_n)^2 (e_i' P_n u_n)^2 \\ &\quad - 4\sigma^4 n^{-1} \sum_{i=1}^{n} (e_i' u_n)(e_i' P_n u_n)^3 + \sigma^4 n^{-1} \sum_{i=1}^{n} (e_i' P_n u_n)^4. \end{aligned} \tag{17}$$

Now $e_i' u_n = \varepsilon_i$ and therefore $n^{-1}\Sigma_{i=1}^n \varepsilon_i^4 \xrightarrow{p} E(\varepsilon_i^4) = \beta$ by the weak law of large numbers. Moreover,

$$\begin{aligned} E(e_i' P_n u_n)^4 &= \left(E(e_i' P_n u_n)^2\right)^2 + \operatorname{Var}\left((e_i' P_n u_n)^2\right) \\ &= \|P_n e_i\|^4 + 2\|P_n e_i\|^4 + (\beta - 3) \sum_{j=1}^{n} \left(p_{ij}^{(n)}\right)^4 \\ &= 3p_{ii}^2 + (\beta - 3) \sum_{j=1}^{n} p_{ij}^4, \end{aligned} \tag{18}$$

if $P_n = (p_{ij}^{(n)})$. (The upper script n is omitted for brevity.) Now $p_{ii}, p_{ij}^2 \leqslant 1$ and therefore $p_{ij}^2 \leqslant p_{ii}$, $\Sigma_{j=1}^n p_{ij}^4 \leqslant \Sigma_{j=1}^n p_{ij}^2 = p_{ii}$. Therefore,

$$E\left(n^{-1} \sum_{i=1}^{n} (e_i' P_n u_n)^4\right) \leqslant n^{-1} \beta \operatorname{tr}(P_n) \to 0, \tag{19}$$

as n tends to infinity. This implies that the term in (19) under the expectation sign converges towards zero in probability by Chebyshev's inequality*. The third term in (17) can be estimated by Cauchy–Schwarz inequality: it is less or equal to a constant times the root of the first and the last term in (17). Therefore, this term converges to zero in probability, too. A similar argument holds for the second term: by Cauchy–Schwarz inequality its absolute value is less than or equal to $4\sigma^4(n^{-1}\sum_{i=1}^n \varepsilon_i^6)^{1/2}\|P_n u\| n^{-1/2}$. Since $E(\varepsilon_i^6)<\infty$ by assumption the second term will as well tend to zero in probability if n tends to infinity. The term before the last term can be estimated by

$$4\sigma^4\left(n^{-1}\sum_{i=1}^n \varepsilon_i^2\right)^{1/2}\left(n^{-1}\sum_{i=1}^n\left(\sum_{j=1}^n p_{ij}\varepsilon_j\right)^6\right)^{1/2}. \tag{20}$$

Now

$$\begin{aligned}E\left(\left(\sum_{j=1}^n p_{ij}\varepsilon_j\right)^6\right) &= E(\varepsilon_j^6)\sum_{j=1}^n p_{ij}^6 \\ &\quad + 15\beta\sum_{j=1}^n\sum_{k=1}^n p_{ij}^4 p_{ik}^2 + 20\left(E(\varepsilon_i^3)\right)^2 \sum_{\substack{j,k=1\\ k\neq j}}^n p_{ij}^3 p_{ik}^3 \\ &\quad + 120 \sum_{\substack{j,k,l\\ j\neq k\neq l}} p_{ij}^2 p_{ik}^2 p_{il}^2.\end{aligned} \tag{21}$$

An upper bound for $E(n^{-1}\sum_{i=1}^n(\sum_{j''1}^n p_{ij}\varepsilon_j)^6)$ is therefore,

$$n^{-1}\operatorname{tr}(P_n)\left[E(\varepsilon_i^6)+15\beta+20\left(E(\varepsilon_i^3)\right)^2+120\right]\to 0, \tag{22}$$

as n tends to infinity. Thus $\{\hat\mu_4^{(n)}\}$ is a consistent sequence of estimators of μ_4. It is well known (see Drygas [2]) that $\{\hat\sigma_n^2\}=\{n^{-1}\|(I_n-P_n)y_n\|^2\}$ is a consistent sequence of estimators of σ^2.

Now let $\delta<\sigma^2$. Then $P(\hat\sigma_n^2\leqslant\delta)\leqslant P(|\hat\sigma_n^2-\sigma^2|\geqslant\sigma^2-\delta)\to 0$ as $n\to\infty$. Finally,

$$\begin{aligned}P(|\hat\beta_n-\beta|\geqslant\eta) &\leqslant P(|\hat\beta_n-\beta|\geqslant\eta,\ \hat\sigma_n^2>\delta)+P(\hat\sigma_n^2\leqslant\delta) \\ &= P(|\hat\mu_4^{(n)}-\beta\hat\sigma_n^4|\geqslant\eta\hat\sigma_n^4,\ \hat\sigma_n^2>\delta)+P(\hat\sigma_n^2\leqslant\delta) \\ &\leqslant P(|\hat\mu_4^{(n)}-\beta\hat\sigma_n^4|\geqslant\eta\delta^2)+P(\hat\sigma_n^2\leqslant\delta) \\ &\to 0,\end{aligned} \tag{23}$$

as $n\to\infty$. Now if $\beta\neq 1$, $1+\delta<\beta$, then:

$$\begin{aligned}P(|\tilde\beta_n-\beta|\geqslant\eta) &= P(|\tilde\beta_n-\beta|\geqslant\eta,\ \hat\beta_n\geqslant 1+\delta)+P(|\tilde\beta_n-\beta|\geqslant\eta,\ \hat\beta_n<1+\delta) \\ &\leqslant P(|\hat\beta_n-\beta|\geqslant\eta)+P(|\hat\beta_n-\beta|\geqslant\beta-1-\delta) \\ &\to 0\end{aligned} \tag{24}$$

as $n\to\infty$. This finishes the proof of the theorem.

*This inequality is sometimes also called Markov's inequality.

We now get immediately:

Theorem 3. *If $\beta>1$ and $E(\varepsilon_i^6)<\infty$, then*

$$A_n=\Big[2\|(I_n-P_n)y_n\|/\big(2n^{1/2}+\alpha(\tilde\beta_n-1)\big)^{1/2},\quad 2\|(I_n-P_n)y_n\|/\big(2n^{1/2}-\alpha(\tilde\beta_n-1)\big)^{1/2}\big)\Big] \tag{25}$$

is an asymptotic confidence for σ at the level δ, where $\alpha=\Phi^{-1}(\frac{1}{2}(1+\delta))$. Similarly,

$$\tilde A_n=\Big[4\|(I_n-P_n)y_n\|^2/\big(2_n^{1/2}+\alpha(\tilde\beta_n-1)^{1/2}\big)^2,\quad 4\|(I_n-P_n)y_n\|^2/\big(2_n^{1/2}-\alpha(\tilde\beta_n-1)^{1/2}\big)^2\Big], \tag{26}$$

is an asymptotic confidence-interval for σ^2 at the confidence-level δ.

Now

$$E\|(I_n-P_n)y_n\|^2=\sigma^2(n-\operatorname{tr}P_n)$$

and

$$\operatorname{Var}\|(I_n-P_n)y_n\|^2=\sigma^4\Big\{2(n-\operatorname{tr}P_n)+(\beta-3)\sum_{i=1}^n m_{ii}^2\Big\},$$

where $M_n=I_n-P_n=(m_{ij}^{(n)})$ (the upper script n is omitted for brevity). Therefore, one is tempted to use the standardized variable,

$$\big[\sigma^{-2}\|(I_n-P_n)y_n\|^2-(n-\operatorname{tr}P_n)\big]\Big[2(n-\operatorname{tr}P_n)+(\beta-3)\sum_{i=1}^n m_{ii}^2\Big]^{-1/2}, \tag{27}$$

as an approximation of a standard normal variable. Indeed from Cauchy–Schwarz inequality and $0\leqslant m_{ii}\leqslant 1$, it follows that,

$$(n-\operatorname{tr}P_n)^2/n^2\leqslant\Big(\sum_{i=1}^n m_{ii}^2\Big)\Big/n\leqslant(n-\operatorname{tr}P_n)/n, \tag{28}$$

showing that $\lim_{n\to\infty}(\Sigma_{i=1}^n m_{ii}^2)/n=1$. From this it follows that

$$\big[\sigma^{-1}\|(I_n-P_n)y_n\|-(n-\operatorname{tr}P_n)^{1/2}\big]\Big[2-2n^{-1}\operatorname{tr}P_n+(\beta-3)n^{-1}\sum_{i=1}^n m_{ii}^2\Big]^{-1/2} \tag{29}$$

converges in distribution towards the standard normal distribution. This result will again give rise to the construction of asymptotic confidence-intervals for σ and σ^2. Only a thorough asymptotic analysis can show which one of the asymptotic confidence-intervals is the better one. Again, of course, β may be replaced by $\tilde\beta_n$, where $\{\tilde\beta_n\}$ is a consistent sequence of estimators of β.

References

[1] Drygas, H. (1972). The estimation of residual variance in regression analysis. *Math.-Operat. Statist.* **3**, 373–388.

[2] Drygas, H. (1975). A note on a paper by T. Kloek concerning the consistency of the residual variance estimator in the linear model. *Econometrica* **43**, 175.

[3] Rao, C.R. (1973). *Linear Statistical Inference and Its Applications*. 2nd ed. Wiley, New York.

G. Kallianpur, P.R. Krishnaiah, J.K. Ghosh, eds., *Statistics and Probability: Essays in Honor of C.R. Rao*
© North-Holland Publishing Company (1982) 241–247

ON THE WAY TOWARDS A NON-PARAMETRIC MULTIDIMENSIONAL TEST OF NORMALITY

Daniel DUGUÉ
Institut de Statistique, Université de Paris, Paris, France

It is a pleasure to dedicate this paper to my good friend Professor C.R. Rao who happened to be present during discussions on this topic twice before. The first time was in June 1968 at the Dayton meeting on Multivariate Analysis organised by Dr. P.R. Krishnaiah, when he was in the chair as I presented my paper quoted in Section 2. In December 1977, when I attended the ISI session in Delhi during which Dr. Rao was elected President of ISI, a participant asked me if I knew a multivariate non-parametric test of normality. At that time my answer was "no".

Our paths crossed each other many times; I may reminisce that the first time was in Cambridge in Gonville and Caius College where both of us were invited by dear Sir Ronald Fisher. This was in the early fifties. I hope we will have many occasions in the future to meet again.

Notations. As I do in papers dealing with analogous problems, I shall write A_s for an $s\times s$ matrix with $a_{ij}=0$ $(i<j)$ and 1 $(i\geqslant j)$; B_s will be ${}^tA_sA_s$ with $b_{ij}=\min(s-i+1, s-j+1)$ where ${}^tA_s$ denotes the transpose of A_s; M_s is an $s\times s$ matrix with $m_{ij}=1$. I_s is the $s\times s$ unit matrix.

1. A probability law analogous to the multinomial law

Let us consider $n=sr$ observations of a uniform random variable on [0, 1]. Divide the unit interval into s segments, each of which except the last one has $(r-1)$ observations and the end point of each of which is an observation and the last segment containing the remaining r observations. Let us take the lengths $x_1,\dots,x_{s-1},x_s$ $(x_1+\cdots+x_s=1)$ of the segments which are random variables whose elementary probability law is

$$\frac{(sr)!}{(r-1)!^{s-1}r!}x_1^{r-1}\cdots x_{s-1}^{r-1}(1-(x_1+\cdots+x_{s-1}))^r\,\mathrm{d}x_1\cdots\mathrm{d}x_{s-1}=$$

$$=f(x_1,\dots,x_{s-1})\,\mathrm{d}x_1\cdots\mathrm{d}x_{s-1}.$$

Put $\sqrt{sr}\,(x_i-(1/s))=y_i$ and so $x_i=(1/s)+(y_1/\sqrt{sr})$. We get $g(y_1,\dots,y_{s-1})$ $\mathrm{d}y_1\cdots\mathrm{d}y_{s-1}$. Through Stirling approximation it is easy to see that:

$$\lim_{r\to\infty} g(y_1,\dots,y_{s-1})=\frac{1}{(2\pi)^{(s-1)/2}}s^{s/2}$$

$$\times\exp-\tfrac{1}{2}\left[s(y_1^2+\cdots+y_{s-1}^2)+s(y_1+\cdots+y_{s-1})^2\right].$$

The $(s-1)\times(s-1)$ matrix of the quadratic form appearing in the exponent is $s[I_{s-1}+M_{s-1}]$. The covariance matrix of $Y_1,\dots,Y_{s-1}$ which is a normal variable is $\{s[I_{s-1}+M_{s-1}]\}^{-1}=(1/s)[I_{s-1}-(1/s)M_{s-1}]$. The covariance matrix of $Y_1,\dots,Y_s$

with $Y_s = 1-(Y_1 + \cdots + Y_{s-1})$ will be, as it is easy to see, the preceeding one with sth row equal to: $E[Y_sY_j]$ and the sth column equal to $E[Y_iY_s]$. It is found easily that this matrix is $(1/s)[I_s - (1/s)M_s]$.

1.1. Extension to $\mathbb{R}^p$

Suppose now that we have a p-dimensional random variable in $\mathbb{R}^p$ whose probability law is continuous. Consider the cumulated p-dimensional histogram of frequency over $n = rs^p$ results. We can divide $\mathbb{R}^p$ (whose coordinates are denoted by $x^{(1)}, \ldots, x^{(p)}$) into s^p cells in such way that we have the same number r of results in each cell. First divide the coordinate $x^{(1)}$ by $x_1^{(1)}, \ldots, x_{s-1}^{(1)}$, to have s 1-regions and we can choose the $x_i^{(1)}$ such that in each 1-region we have the same number of results n/s; now each 1-region will be divided into s 2-regions (s^2 2-regions in the whole) according to $x^{(2)}$, where we have n/s^2 results in each 2-region and so on to get the p-regions. The p-regions are the s^p cells $C_{i_1 \cdots i_p}$ each of the indexes $i_1, \ldots, i_p$ varying from 1 to s. This way of dividing $\mathbb{R}^p$ will be called the equipartition of $\mathbb{R}^p$ according to the histogram of cumulated frequencies $H_n(x^{(1)}, \ldots, x^{(p)})$. The probability of the cell $C_{i_1 \cdots i_p}$ is a random number $P_{i_1 \cdots i_p}$ which tends in probability towards $1/s^p$ when n increases to infinity. $\sqrt{rs^p}\,(P_{i_1 \cdots i_p} - (1/s^p))$ plays the part of y. When r increases indefinitely (whatever s may be), if the initial probability law is continuous, this s^p dimensional variable tends in law towards a normal law with zero mean and covariance matrix $(1/s^p)[I_{s^p} - (1/s^p)M_{s^p}]$ which is equal to $((1/s)I_s)^{\otimes p} - ((1/s^2)M_s)^{\otimes p}$.

2. An extension of a result on the *p*-dimensional Von Mises–Smirnoff law

Now take the difference:

$$\sqrt{n}\left[H_n(x^{(1)}, \ldots, x^{(p)}) - F(x^{(1)}, \ldots, x^{(p)})\right],$$

n being rs^p. In the cell $C_{i_1 \cdots i_p}$ the increment of this difference is $\sqrt{n}\,[(1/s^p) - P_{i_1 \cdots i_p}] = Y_{i_1 \cdots i_p}$, coordinates of a vector Y.

The quadratic form:

$$Q_n^{(p)} = \sum_{k_1=1}^{s} \sum_{k_2=1}^{s} \cdots \sum_{k_p=1}^{s} \left(\sum_{i_1=1}^{k_1} \sum_{i_2=1}^{k_2} \cdots \sum_{i_p=1}^{k_p} Y_{i_1 \cdots i_p} \right)^2 s^{1/p}$$

will be obtained by saturation of the matrix $(1/s^p)^t A_s^{\otimes p} \cdot A_s^{\otimes p} = (1/s^p) B_s^{\otimes p}$.

The characteristic function of $Q_s^{(p)}$ is:

$$E\left[\exp iuQ_s^{(p)}\right] = E\left[\exp iu^tY \frac{1}{s^p} B_s^{\otimes p} Y\right].$$

With the same calculations as in Dugué [1, p. 298] we find:

$$E\left(\exp iuQ_s^{(p)}\right) = \det\left[I_s^{\otimes p} - \frac{2iu}{s^{2p}} B_s^{\otimes p}\left(I_s^{\otimes p} - \frac{1}{s^p} M_s^{\otimes p}\right)\right]^{-1/2}.$$

The same result as in this paper will hold. The only difference with it is that here the random element is the increment of the probability law $P_{i_1 \cdots i_p}$ and the certain element is the increment of the cumulated histogram and there is the other way around. And we have the result:

Theorem 1.

$$\lim_{s\to\infty} \lim_{n\to\infty} E\left[\exp \mathrm{i}uQ_s^{(p)}\right] =$$

$$= \left[\mathrm{i}2^{p-1}\frac{\mathrm{d}}{\mathrm{d}u}\prod_{k_1=1}^{\infty} \cdots \prod_{k_p=1}^{\infty}\left(1-\frac{2\mathrm{i}u}{\left(k_1-\frac{1}{2}\right)^2\cdots\left(k_p-\frac{1}{2}\right)^2\pi^{2p}}\right)\right]^{-1/2}.$$

This is a way to compare the cumulated histogram to a continuous theoretical law through the equipartition of $\mathbb{R}^p$ according to the histogram.

Let us write:

$$\mathrm{i}2^{p-1}\frac{\mathrm{d}}{\mathrm{d}u}\prod_{k_1=1}^{\infty} \cdots \prod_{k_p=1}^{\infty}\left(1-\frac{2\mathrm{i}u}{\left(k_1-\frac{1}{2}\right)^2\cdots\left(k_p-\frac{1}{2}\right)^2\pi^{2p}}\right) = \prod_{n=1}^{\infty}\left(1-\frac{2\mathrm{i}u}{\lambda_n^{(p)}}\right).$$

It is easily shown that the $\lambda_n^{(p)}$ are positive.

3. A non-parametric test of independence of multivariate variables

Let us now have a set of n independent pairs of results on two p-dimensional variables X and Y: $X_1, Y_1; X_2, Y_2; \ldots; X_n, Y_n$. How does one test the independence of X and Y?

In this case the probability law of X and Y is $F(x^{(1)}, \ldots, x^{(p)}; y^{(1)}, \ldots, y^{(p)})$. We will write it as $F(\{x^{(i)}\}; \{y^{(j)}\})$. The marginal laws are: $F(\{x^{(i)}\}; \{+\infty\})$ and $F(\{+\infty\}; \{y^{(j)}\})$ and in the case of independence we have

$$F(\{x^{(i)}\}; \{y^{(j)}\}) = F(\{x^{(i)}\}; \{+\infty\})F(\{+\infty\}; \{y^{(j)}\}).$$

We will use the same notation as in Hoeffding [4] and Dugué [2].

In $\mathbb{R}^{2p}$ we have an histogram of frequencies $H_n(\{x^{(i)}\}; \{y^{(j)}\})$ on n pairs of results. Exactly as in Section 1.1, $\mathbb{R}^{2p}$ will be divided in s^{2p} cells $C_{i_1 \cdots i_p; j_1 \cdots j_p}$. If we consider the vector Z with $2p$ indexes $Z_{i_1 \cdots i_p; j_1 \cdots j_p}$ whose components are the increments of

$$\sqrt{n}\left[H_n(\{x^{(i)}\}; \{y^{(j)}\}) - F(\{x^{(i)}\}; \{y^{(j)}\})\right]$$

in each cell; we will get n increasing indefinitely, s^{2p} components of a normal law with zero mean and covariance matrix $((1/s)I_s)^{\otimes 2p} - ((1/s^2)M_s)^{\otimes 2p}$. Let

$$D_n(\{x^{(i)}\}; \{y^{(j)}\}) = \sqrt{n}\left[H_n(\{x^{(i)}\}; \{y^{(j)}\}) - H_n(\{x^{(i)}\}; \{+\infty\})H_n(\{+\infty\}; \{y^{(j)}\})\right]$$

and

$$\Delta_n(\{x^{(i)}\};\{y^{(j)}\}) = \sqrt{n}\left[H_n(\{x^{(i)}\};\{+\infty\}) - F(\{x^{(i)}\};\{+\infty\})\right]$$

$$\times\left[H_n(\{+\infty\};\{y^{(j)}\}) - F(\{+\infty\};\{y^{(j)}\})\right].$$

It is easily seen that:

$$\sqrt{n}\left[H_n(\{x^{(i)}\};\{+\infty\}) - F(\{x^{(i)}\};\{+\infty\})\right]$$

converges in law towards a normal variable and

$$H_n(\{\infty\};\{y^{(j)}\}) - F(\{+\infty\};\{y^{(j)}\})$$

converges in law to zero and so the product Δ_n converges in law to zero.

$$\begin{aligned} D_n(\{x^{(i)}\};\{y^{(j)}\}) - \Delta_n(\{x^{(i)}\};\{y^{(j)}\}) = \\ = \sqrt{n}\left[H_n(\{x^{(i)}\};\{y^{(j)}\}) - F(\{x^{(i)}\};\{+\infty\})F(\{+\infty\};\{y^{(j)}\})\right] \\ - \sqrt{n}\left[H_n(\{x^{(i)}\};\{+\infty\}) - F(\{x^{(i)}\};\{+\infty\})\right]H_n(\{+\infty\};\{y^{(j)}\}) \\ - \sqrt{n}\,H_n(\{x^{(i)}\};\{+\infty\})\left[H_n(\{+\infty\};\{y^{(j)}\}) - F(\{+\infty\};\{y^{(j)}\})\right]. \end{aligned}$$

$D_n - \Delta_n$ and D_n converge in law to the same distribution.

Let us denote the increments in $C_{i_1\cdots i_p;\, j_1\cdots j_p}$ of:

$$D_n \quad \text{by } d^{(n)}_{i_1\cdots i_p;\, j_1\cdots j_p},$$

$$\Delta_n \quad \text{by } \delta^{(n)}_{i_1\cdots i_p;\, j_1\cdots j_p},$$

$$\sqrt{n}\left[H_n(\{x^{(i)}\};\{y^{(j)}\}) - F(\{x^{(i)}\};\{+\infty\})F(\{+\infty\};\{y^{(j)}\})\right] \quad \text{by } \alpha^{(n)}_{i_1\cdots i_p;\, j_1\cdots j_p},$$

$$\sqrt{n}\left[H_n(\{x^{(i)}\};\{+\infty\}) - F(\{x^{(i)}\};\{+\infty\})\right]H_n(\{+\infty\};\{y^{(j)}\}) \quad \text{by } \beta^{(n)}_{i_1\cdots i_p;\, j_1\cdots j_p}$$

and

$$\sqrt{n}\,H_n(\{x^{(i)}\};\{+\infty\})\left[H_n(\{+\infty\};\{y^{(j)}\}) - F(\{+\infty\};\{y^{(j)}\})\right] \quad \text{by } \gamma^{(n)}_{i_1\cdots i_p;\, j_1\cdots j_p}.$$

Of course $d^{(n)} - \delta^{(n)} = \alpha^{(n)} - \beta^{(n)} - \gamma^{(n)}$. Put

$$R_s^{(p)} = \sum_{k_1=1}^{s}\cdots\sum_{k_p=1}^{s}\sum_{l_1=1}^{s}\cdots\sum_{l_p=1}^{s}\left[\sum_{i_1=1}^{k_1}\cdots\sum_{i_p=1}^{k_p}\sum_{j_1=1}^{l_1}\cdots\sum_{j_p=1}^{l_p} d^{(n)}_{i_1\cdots i_p;\, j_1\cdots j_p}\right]^2\frac{1}{s^{2p}}$$

$R_s^{(p)}$ is a law free functional of H_n.

As Δ_n tends in law to 0, it can be proved without difficulty that in the bracket $d^{(n)}$ can be replaced by $d^{(n)}-\delta^{(n)}$ to calculate the limit of $R_s^{(p)}$ $(n\to\infty)$. So we consider

$$\sum_{k_1=1}^{s}\cdots\sum_{k_p=1}^{s}\sum_{l_1=1}^{s}\cdots\sum_{l_p=1}^{s}$$

$$\left[\sum_{i_1=1}^{k_1}\cdots\sum_{i_p=1}^{k_p}\sum_{j_1=1}^{l_1}\cdots\sum_{j_p=1}^{l_p}\left(\alpha^{(n)}_{i_1\cdots i_p;\,j_1\dots j_p}-\beta^{(n)}_{i_1\cdots i_p;\,j_1\cdots j_p}-\gamma^{(n)}_{i_1\cdots i_p;\,j_1\dots j_p}\right)\right]^2\frac{1}{s^{2p}}$$

$$=R_s'^{(p)}.$$

It is easy to see that the following equalities give the s^{2p} dimensional vectors with $2p$ indexes:

$$\sum_{i_1=1}^{k_1}\cdots\sum_{i_p=1}^{k_p}\sum_{j_1=1}^{l_1}\cdots\sum_{j_p=1}^{l_1}\alpha^{(n)}_{i_1\cdots i_p;\,j_1\cdots j_p}=A_s{}^{\otimes p}\otimes A_s{}^{\otimes p}Z,$$

$$\sum_{i_1=1}^{k_1}\cdots\sum_{i_p=1}^{k_p}\sum_{j_1=1}^{l_1}\cdots\sum_{j_p=1}^{l_1}\beta^{(n)}_{i_1\cdots i_p;\,j_1\cdots j_p}=A^{\otimes p}\otimes A^{\otimes p}\frac{1}{s^p}M_s{}^{\otimes p}Z,\qquad \begin{aligned}&k_1\dots k_p;\\&l_1\cdots l_p\\&=1,\dots,s.\end{aligned}$$

$$\sum_{i_1=1}^{k_1}\cdots\sum_{i_p=1}^{k_p}\sum_{j_1=1}^{l_1}\cdots\sum_{j_p=1}^{l_p}\gamma^{(n)}_{i_1\cdots i_p;\,j_1\cdots j_p}=A_s{}^{\otimes p}\frac{1}{s^p}M_s{}^{\otimes p}\otimes A_s{}^{\otimes p}Z,$$

Now we can write:

$$R_s'^{(p)}=\frac{1}{s^{2p}}{}^{\mathrm t}Z\,{}^{\mathrm t}\!\left[\left(A_s{}^{\otimes p}\left(I_s{}^{\otimes p}-\frac{1}{s^p}M_s{}^{\otimes p}\right)\right)^{\otimes 2}-\left(A_s{}^{\otimes p}\frac{1}{s^p}M_s{}^{\otimes p}\right)^{\otimes 2}\right]$$

$$\times\left[\left(A_s{}^{\otimes p}\left(I_s{}^{\otimes p}-\frac{1}{s^p}M_s{}^{\otimes p}\right)\right)^{\otimes 2}-\left(A_s{}^{\otimes p}\frac{1}{s^p}M_s{}^{\otimes p}\right)^{\otimes 2}\right]Z.$$

With the classical calculus of characteristic function of quadratic forms:

$$E\left[\exp iuR_s'^{(p)}\right]$$

$$=\det\left\{I_s{}^{\otimes 2p}-\frac{2iu}{s^{2p}}{}^{\mathrm t}\!\left[\left(A_s{}^{\otimes p}\left(I_s{}^{\otimes p}-\frac{1}{s^p}M_s{}^{\otimes p}\right)\right)^{\otimes 2}-\left(A_s{}^{\otimes p}\frac{1}{s^p}M_s{}^{\otimes p}\right)^{\otimes 2}\right]\right.$$

$$\times\left[\left(A_s{}^{\otimes p}\left(I_s{}^{\otimes p}-\frac{1}{s^p}M_s{}^{\otimes p}\right)\right)^{\otimes 2}-\left(A_s{}^{\otimes p}\frac{1}{s^p}M_s{}^{\otimes p}\right)^{\otimes 2}\right]$$

$$\left.\times\left[\left(\frac{1}{s}I_s\right)^{\otimes 2p}-\left(\frac{1}{s^2}M_s\right)^{\otimes 2p}\right]\right\}^{-1/2}.$$

By using the fact that $\det[1-AB]=\det[1-BA]$ and that $M_sM_s=sM_s$ we can write the determinant in the following way:

$$\det\left\{I_s^{\otimes 2p} - \frac{2iu}{s^{2p}}\left[A_s^{\otimes 2p}\left(I_s^{\otimes p} - \frac{1}{s^p}M_s^{\otimes p}\right)^{\otimes 2}\right]\left(\frac{1}{s}I_s\right)^{\otimes 2p} \times {}^{t}\left[A_s^{\otimes 2p}\left(I_s^{\otimes p} - \frac{1}{s^p}M_s^{\otimes p}\right)^{\otimes 2}\right]\right\}^{-1/2}$$

and then as

$$\det\left\{I_s^{\otimes 2p} - \frac{2iu}{s^{4p}}\left[B_s^{\otimes p}\left(I_s^{\otimes p} - \frac{1}{s^p}M_s^{\otimes p}\right)\right]^{\otimes 2}\right\}^{-1/2}.$$

Then with the notations of Section 2, we have the following.

Theorem 2.

$$\lim_{s\to\infty}\lim_{n\to\infty} E\left[\exp iuR_s^{(p)}\right] = \left[\prod_{l=1}^{\infty}\prod_{m=1}^{\infty}\left(1 - \frac{2iu}{\lambda_l^{(p)}\lambda_m^{(p)}}\right)\right]^{-1/2}$$

($R_s^{(p)}$ is a law free functional of H_n) if X and Y are independent, and have continuous laws.

4. A non-parametric multidimensional test of normality

Let us have a sequence of $2n$ independent results of p-dimensional variables, identically distributed with continuous laws: $X_1,\ldots, X_n, X_{n+1},\ldots, X_{2n}$, and consider now the n pairs

$$X_1 + X_{n+1}, X_1 - X_{n+1}; X_2 + X_{n+2}, X_2 - X_{n+2}; \ldots; X_n + X_{2n}, X_n - X_{2n}.$$

These pairs are independent and through Section 3 we can test the independence of the sum and of the difference. The following multivariate extension of a well-known univariate theorem of Darmois and Bernstein is proved in Fuchs [3]. If X and Y are independent and identically distributed p-dimensional random vectors such that $X+Y$ and $X-Y$ are independent then the common law of X and Y is normal.

Section 3 would give the non-parametric multidimensional test of normality we are looking for if we had the distribution function.

5. Two problems

(i) The first problem is to get the Fourier inverse of

$$\left[\prod_{l=1}^{\infty}\prod_{m=1}^{\infty}\left(1 - \frac{2iu}{\lambda_l^{(p)}\lambda_m^{(p)}}\right)\right]^{-1/2},$$

to obtain the distribution function, and to write down the tables.

(ii) The limiting laws only are considered in Theorems 1 and 2. Since an infinite set of results are not available in practical applications it would be important to know the rates of convergence in s and in n in these theorems.

Acknowledgement

I would like to propose these problems to the sagacity of the students of Professor Rao and my own (others also, of course).

References

[1] Dugué, D. (1969). Characteristic functions of random variables connected with Brownian motion and of the Von Mises multidimensional ω_n^2. In: P.R. Krishnaiah, Ed., *Multivariate Analysis* II (Academic Press, New York) pp. 289–301.

[2] Dugué, D. (1975). Sur des tests d'independence "independants de la loi". *C. R. Acad. Sci. Paris* **281**, 1103–1104.

[3] Fuchs, A. Lois de probabilité liées a la loi de Laplace–Gauss a k dimensions. *Publications de l'Institut de Statistique de l'Universite de Paris*, XII-2, pp. 117–129.

[4] Hoeffding, W. (1948). A nonparametric test of independence. *Ann. Math. Statist.* **19**, 546–557.

G. Kallianpur, P.R. Krishnaiah, J.K. Ghosh, eds., *Statistics and Probability: Essays in Honor of C.R. Rao*
© North-Holland Publishing Company (1982) 249–262

ASYMPTOTIC JOINT DISTRIBUTIONS OF FUNCTIONS OF THE ELEMENTS OF SAMPLE COVARIANCE MATRIX*

C. FANG
University of Pittsburgh and Carnegie-Mellon University, Pittsburgh, PA, U.S.A.

P.R. KRISHNAIAH
University of Pittsburgh, Pittsburgh, PA, U.S.A

1. Introduction

Several test statistics in multivariate analysis are based upon certain functions of the elements of the sample covariance matrix or sample correlation matrix. For example, the sample correlation coefficient, partial and multiple correlation coefficients, and various transformations of the sample correlation coefficients depend upon the elements of the sample covariance matrix and correlation matrix. The exact distributions of many of these statistics are quite complicated and so there is a need to derive asymptotic distributions of the above statistics. In this paper, we consider asymptotic joint distributions of functions of the elements of the non-central Wishart matrix and the associated non-central correlation matrix.

In Section 2 of this paper, we derive the asymptotic joint distribution of certain functions of the elements of the non-central Wishart matrix. The first term in the asymptotic expression is the multivariate normal density, whereas the second term involves partial derivatives of the multivariate normal density. Similar expressions are given for the case of the non-central correlation matrix. The method used involves expanding the functions in terms of Taylor series and computing the characteristic functions. Olkin and Siotani (1976) obtained the first term in the asymptotic joint distribution of functions of the elements of the central correlation matrix. Siotani and Hayakawa (1964) obtained the first terms in the asymptotic joint distributions of the partial and multiple correlation coefficients by expressing them as functions of the elements of the central Wishart matrix. Konishi (1979) obtained the first two terms in the asymptotic joint distribution of various functions of the central correlation matrix; these results are special cases of the results given in this paper. In Section 3 of this paper, we study the accuracy of the asymptotic

*This work is sponsored by the Air Force Office of Scientific Research under Contract F49620-79-0161.

expressions given in Section 2 for some special cases. Asymptotic expressions for the joint distributions of functions of the elements of the sample covariance matrix are given in Section 4 when the underlying distribution is not multivariate normal. Finally, applications of the results of this paper are discussed in Section 5.

2. Joint distribution of the functions of the elements of the non-central Wishart matrix

Let $\boldsymbol{X}_1,\ldots,\boldsymbol{X}_n$ be distributed independently as multivariate normal with mean vectors $\boldsymbol{\mu}_1,\ldots,\boldsymbol{\mu}_n$ and covariance matrix $\Sigma=(\sigma_{ij})$. Then, the distribution of $S=\Sigma_{j=1}^n \boldsymbol{X}_j\boldsymbol{X}_j'=(S_{ij})$ is known to be the non-central Wishart matrix with n degrees of freedom and $E(S/n)=\Sigma+(M/n)=\Omega$, with non-centrality parameter $M=\Sigma_{j=1}^n \boldsymbol{\mu}_j\boldsymbol{\mu}_j'=n(\nu_{ij})$. Now, let,

$$T_i(S/n)=T_i(s_{11},\ldots,s_{pp},s_{12},\ldots,s_{1p},s_{23},\ldots,s_{p-1,p}),$$

for $i=1,\ldots,k$, where $s_{ij}=S_{ij}/n$, are analytic functions in the neighborhood of $\Omega=(\omega_{ij})$. Also, let

$$a_{j_1j_2}^{(i)}=\left(\frac{1+\delta_{j_1j_2}}{2}\right)\frac{\partial}{\partial s_{j_1j_2}}T_i(S/n)\bigg|_{(S/n)=\Omega}, \tag{2.1}$$

$$a_{j_1j_2\cdot j_3j_4}^{(i)}=\left(\frac{1+\delta_{j_3j_4}}{2}\right)\left(\frac{1+\delta_{j_1j_2}}{2}\right)\frac{\partial^2}{\partial s_{j_3j_4}\partial s_{j_1j_2}}T_i(S/n)\bigg|_{(S/n)=\Omega}.$$

where δ_{hk} is given by:

$$\delta_{hk}=\begin{cases}1, & h=k,\\ 0, & h\neq k.\end{cases}$$

The Taylor expansion of $T_i(S/n)$ about Ω is

$$\begin{aligned}T_i(S/n)=T_i(\Omega)&+\sum_{j_1=1}^{p}\sum_{j_2=1}^{p}a_{j_1j_2}^{(i)}\big((S_{j_1j_2}/n)-\omega_{j_1j_2}\big)\\&+\tfrac{1}{2}\sum_{j_1,j_2,j_3,j_4}^{p}a_{j_1j_2\cdot j_3j_4}^{(i)}\big((S_{j_1j_2}/n)-\omega_{j_1j_2}\big)\\&\qquad\times\big((S_{j_3j_4}/n)-\omega_{j_3j_4}\big)\\&+\text{higher order terms,}\end{aligned} \tag{2.2}$$

where $\Sigma_{j_1,j_2,\ldots,j_u}^{p}$ denotes the summation over all values of $j_1, j_2,\ldots, j_u$ varying from 1 to p. Now, let:

$$L_i=\sqrt{n}\,\{T_i(S/n)-T_i(\Omega)\}. \tag{2.3}$$

Using (2.2), we obtain the following expression for the joint characteristic function of $\boldsymbol{L}'=(L_1,\ldots,L_k)$:

$$\psi(\boldsymbol{t})=E\left\{\exp\left(\mathrm{i}\sum_{j=1}^{k} t_j L_j\right)\right\}$$
$$=E_1(\boldsymbol{t})+E_2(\boldsymbol{t})+\mathrm{O}(n^{-1}), \tag{2.4}$$

where $\boldsymbol{t}'=(t_1,\ldots,t_k)$ and,

$$E_1(\boldsymbol{t})=E\left\{\exp\left[\mathrm{i}\sqrt{n}\sum_{g=1}^{k}\sum_{j_1,j_2}^{p} t_g a_{j_1j_2}^{(g)}\left((S_{j_1j_2}/n)-\omega_{j_1j_2}\right)\right]\right\}$$
$$=\exp\left[-\mathrm{i}\sqrt{n}\operatorname{tr}B\Omega\right]\left|I-\frac{2\mathrm{i}\,B\Sigma}{\sqrt{n}}\right|^{-n/2}\exp\left[\mathrm{i}\operatorname{tr}M\left(I-\frac{2\mathrm{i}\,B\Sigma}{\sqrt{n}}\right)^{-1}\frac{B}{\sqrt{n}}\right] \tag{2.5}$$

$$B=\sum_{i=1}^{k} t_i\left(a_{j_1j_2}^{(i)}\right)=\sum_{i=1}^{k} t_i A^{(i)},$$

$$E_2(\boldsymbol{t})=E\left\{\frac{\mathrm{i}\sqrt{n}}{2}\sum_{g=1}^{k}\sum_{j_1,j_2,j_3,j_4}^{p} t_g a_{j_1j_2\cdot j_3j_4}^{(g)}\left((S_{j_1j_2}/n)-\omega_{j_1j_2}\right)\left((S_{j_3j_4}/n)-\omega_{j_3j_4}\right)\right.$$
$$\left.\times\exp\left[\mathrm{i}\sqrt{n}\sum_{i_1=1}^{k}\sum_{j_1,j_2}^{p} t_{i_1}a_{j_1j_2}^{(i_1)}\left((S_{j_1j_2}/n)-\omega_{j_1j_2}\right)\right]\right\}. \tag{2.6}$$

Starting from (2.4), we obtain the following asymptotic expression for the joint characteristic function of $L_1,\ldots,L_k$:

$$\Psi(\boldsymbol{t})=\exp(-\tfrac{1}{2}t'Qt)\left\{1+\frac{\mathrm{i}}{2\sqrt{n}}\sum_{g=1}^{k} t_g h_1\right.$$
$$\left.+\frac{1}{\sqrt{n}}\sum_{i_1,i_2,i_3}^{k}\mathrm{i}^3 t_{i_1}t_{i_2}t_{i_3}(h_2+h_3)+\mathrm{O}(n^{-1})\right\} \tag{2.7}$$

where

$$Q=(Q_{i_1i_2}),\quad Q_{i_1i_2}=2\operatorname{tr}R^{(i_1)}R^{(i_2)}+4\operatorname{tr}R^{(i_1)}\psi^{(i_2)},$$
$$h_1=\sum_{j_1,j_2,j_3,j_4}^{p} a_{j_1j_2\cdot j_3j_4}^{(g)}\left(\sigma_{j_1j_3}\omega_{j_2j_4}+\sigma_{j_1j_4}\omega_{j_2j_3}+\sigma_{j_2j_3}\nu_{j_1j_4}+\sigma_{j_2j_4}\nu_{j_1j_3}\right),$$
$$h_2=\tfrac{4}{3}\operatorname{tr}R^{(i_1)}R^{(i_2)}R^{(i_3)}+4\operatorname{tr}R^{(i_1)}R^{(i_2)}\psi^{(i_3)}, \tag{2.8}$$
$$h_3=2\sum_{j_1,j_2,j_3,j_4}^{p} a_{j_1j_2\cdot j_3j_4}^{(i_1)}\left(\Xi_{j_1j_2}^{(i_2)}+\Upsilon_{j_1j_2}^{(i_2)}\right)\left(\Xi_{j_3j_4}^{(i_3)}+\Upsilon_{j_3j_4}^{(i_3)}\right)$$

and

$$R^{(i)}=A^{(i)}\Sigma,\qquad \psi^{(i)}=A^{(i)}M/n,$$
$$\Xi^{(i)}=\Sigma A^{(i)}\Omega,\qquad \Upsilon^{(i)}=\frac{M}{n}A^{(i)}\Sigma$$

and U_{ij} denotes the (i, j)th element of a matrix $U=(U_{ij})$. By inverting the above characteristic function, we obtain the following expression for the joint density of $L_1,\ldots, L_k$:

$$f(L_1,\ldots, L_k)=N(\boldsymbol{L},Q)\left\{1+\frac{1}{2\sqrt{n}}\sum_{g=1}^{k} H_g(\boldsymbol{L})h_1 + \frac{1}{\sqrt{n}}\sum_{i_1,i_2,i_3}^{k} H_{i_1i_2i_3}(\boldsymbol{L})(h_2+h_3)+\mathrm{O}(n^{-1})\right\}, \tag{2.9}$$

where

$$N(\boldsymbol{L},Q)=\left(\left(\sqrt{2\pi}\right)^k|Q|^{1/2}\right)^{-1}\exp(-\tfrac{1}{2}\boldsymbol{L}'Q^{-1}\boldsymbol{L}),$$

$$H_{j_1,\ldots,j_u}(\boldsymbol{L})N(\boldsymbol{L},Q)=(-1)^u\frac{\partial^u}{\partial L_{j_1},\ldots,\partial L_{j_u}}N(\boldsymbol{L},Q). \tag{2.10}$$

The correlation matrix is given by $\mathcal{R}=S_0^{-1/2}SS_0^{-1/2}=(r_{ij})$ where $S_0=\mathrm{diag}(S_{11},\ldots, S_{pp})$. Now, let

$$G_j(\mathcal{R})=G_j(r_{12}, r_{13},\ldots, r_{1p}, r_{23},\ldots, r_{2p},\ldots, r_{p-1,p}) \tag{2.11}$$

be an analytic function in the neighborhood of $\Omega_0^{-1/2}\Omega\Omega_0^{-1/2}=P^*=(\rho_{ij}^*)$ where $\Omega_0=\mathrm{diag}(\omega_{11},\ldots,\omega_{pp})$. If we denote $r_{k_1k_2}$ by $T_{k_1k_2}(S/n)$, eq. (2.11) can be written as:

$$G_j(\mathcal{R})=G_j(T_{12}(S/n),\ldots, T_{p-1,p}(S/n))=(G\circ T)_j(S/n). \tag{2.12}$$

Now, let

$$c_{k_1k_2}^{(j)}=\left.\frac{\partial}{\partial r_{k_1k_2}}G_j(\mathcal{R})\right|_{\mathcal{R}=P^*},$$

$$c_{k_1k_2\cdot k_3k_4}^{(j)}=\left.\frac{\partial^2}{\partial r_{k_3k_4}\partial r_{k_1k_2}}G_j(\mathcal{R})\right|_{\mathcal{R}=P^*}$$

in eq. (2.1). Then we obtain:

$$\begin{aligned} a_{j_1j_2}^{(j)} &= \left.\left(\frac{1+\delta_{j_1j_2}}{2}\right)\left(\frac{\partial}{\partial s_{j_1j_2}}\right)(G\circ T)_j(S/n)\right|_{(S/n)=\Omega} \\ &= \tfrac{1}{2}\sum_{k_1\neq k_2} c_{k_1k_2}^{(j)}\zeta_{j_1j_2}^{(k_1k_2)}, \end{aligned} \tag{2.13}$$

$$\begin{aligned} a_{j_1j_2\cdot j_3j_4}^{(j)} &= \left.\tfrac{1}{4}(1+\delta_{j_3j_4})(1+\delta_{j_1j_2})\frac{\partial^2}{\partial s_{j_3j_4}\partial s_{j_1j_2}}(G\circ T)_j(S/n)\right|_{(S/n)=\Omega} \\ &= \sum_{k_1<k_2}\sum_{k_3<k_4} c_{k_1k_2\cdot k_3k_4}^{(j)}\zeta_{j_1j_2}^{(k_1k_2)}\zeta_{j_3j_4}^{(k_3k_4)} \\ &\quad + \sum_{k_1<k_2} c_{k_1k_2}^{(j)}\zeta_{j_1j_2\cdot j_3j_4}^{(k_1k_2)} \end{aligned} \tag{2.14}$$

where

$$\zeta_{j_1 j_2}^{(k_1 k_2)} = \begin{cases} \left(2\sqrt{\omega_{k_1k_1}\omega_{k_2k_2}}\right)^{-1}, & j_1=k_1, j_2=k_2 \text{ or } j_1=k_2, j_2=k_1, \\ \dfrac{-\rho^*_{k_1k_2}}{2\omega_{k_1k_1}}, & j_1=j_2=k_1, \\ \dfrac{-\rho^*_{k_1k_2}}{2\omega_{k_2k_2}}, & j_1=j_2=k_2, \\ 0, & \text{otherwise}, \end{cases} \tag{2.15}$$

$$\zeta_{j_1j_2\cdot j_3j_4}^{(k_1k_2)} = \begin{cases} -\left(4\omega_{k_1k_1}^{1/2}\omega_{k_2k_2}^{3/2}\right)^{-1} & \text{for any 3 values of } j_1, j_2, j_3, j_4 \text{ equal to } k_2, \\ -\left(4\omega_{k_1k_1}^{3/2}\omega_{k_2k_2}^{1/2}\right)^{-1} & \text{for any 3 values of } j_1, j_2, j_3, j_4 \text{ equal to } k_1, \\ \dfrac{\rho^*_{k_1k_2}}{4\omega_{k_1k_1}\omega_{k_2k_2}} & \text{for } j_1=j_2=k_1, j_3=j_4=k_2 \text{ or } j_1=j_2=k_2, j_3=j_4=k_1, \\ \dfrac{3\rho^*_{k_1k_2}}{4\omega^2_{k_ik_i}} & \text{for all } j_1=j_2=j_3=j_4=k_i, \quad i=1,2, \\ 0 & \text{otherwise}. \end{cases}$$

By substituting (2.13) and (2.14) in eq. (2.9), we obtain the asymptotic density for functions of the elements of correlation matrix.

3. An empirical study on the accuracy of the asymptotic expressions

In this section, we study the accuracy of the asymptotic expressions derived in Section 2 for some special cases.

We will first study the accuracy of the approximation for the distribution of the non-central chi-square. Let $y_i = S_{ii}/\sigma_{ii}$ for $i=1,2,\ldots, p$. Then y_1 is distributed as the non-central chi-square distribution with n degrees of freedom and with the non-centrality parameter $\gamma = n\nu_{11}/\sigma_{11}$. Now let

$$\exp(-\gamma/2)\sum_{j=0}^{\infty}\frac{\gamma^j}{j!2^{(n/2)+2j}\Gamma(\tfrac{1}{2}n+j)}\int_0^u x^{(n/2)+j-1}\exp(-\tfrac{1}{2}x)\,\mathrm{d}x=\beta \tag{3.1}$$

and u is given by

$$\frac{1}{2^{n/2}\Gamma(\tfrac{1}{2}n)}\int_0^u \exp(-\tfrac{1}{2}x)x^{(n/2)-1}\,\mathrm{d}x=0.95. \tag{3.2}$$

The left side of (3.1) is the probability integral of y_1. Also the left side of (3.1) is equivalent to the left side of (3.2) when $\gamma=0$. Table 1 given below compares the exact values of β with the corresponding values obtained by using the asymptotic expression (2.9).

Table 1
Comparison of the asymptotic expression with exact expression for the non-central chi-square distribution

n	u	γ	O(1)	$O(n^{-1/2})$	$O(1)+O(n^{-1/2})$	Exact
5	11.071	0	0.9726	−0.0358	0.9368	0.95
		16.47	0.1163	−0.0091	0.1072	0.10
		2.67	0.7727	0.0245	0.7972	0.80
10	18.307	0	0.9684	−0.0260	0.9424	0.95
		20.53	0.1132	−0.0082	0.1050	0.10
		3.71	0.7820	0.0159	0.7979	0.80
20	31.41	0	0.9644	−0.0186	0.9458	0.95
		26.13	0.1104	−0.0071	0.1033	0.10
		5.18	0.7880	0.0105	0.7985	0.80
30	43.773	0	0.9623	−0.0153	0.9470	0.95
		30.38	0.1089	−0.0064	0.1025	0.10
		6.31	0.7906	0.0083	0.7988	0.80

The values in the column "O(1)" give the values of β when the first term in the asymptotic expression (2.9) is taken whereas the column "$O(1)+O(n^{-1/2})$" gives the values of β when the first two terms in (2.9) are taken. The column "$O(n^{-1/2})$" gives the contribution of the second term in (2.9) to the value of β. The exact values given in the last column are taken from Owen (1962).

We will compare the asymptotic expansion given by (2.9) for the distribution of y_1+y_2 with the corresponding exact expression. When $\rho=\rho_{12}=0$ the distribution of y_1+y_2 is the non-central chi-square with $2n$ degrees of freedom and with $\gamma_1+\gamma_2$ as the non-centrality parameter where $\gamma_1=n\nu_{11}/\sigma_{11}$ and $\gamma_2=n\nu_{22}/\sigma_{22}$. When $\rho_{12}\neq 0$, the distribution of y_1+y_2 is the same as the distribution of a quadratic form in the non-central case. Now, let:

$$\Pr[y_1+y_2\leqslant u|\rho,\gamma_1,\gamma_2]=\beta \tag{3.3}$$

where

$$\Pr[y_1+y_2\leqslant u|\rho=0,\gamma_1=\gamma_2=0]=0.95. \tag{3.4}$$

In Table 2, the entries in the column "$O(1)+O(n^{-1/2})$" represent the approximate value of β when the first two terms in the asymptotic expression (2.9) are taken. The entries in the columns "O(1)" and "$O(n^{-1/2})$" represent respectively the contribution of the first term and the second term to the above approximate value of β. The entries in the column "exact values" are computed by Monte Carlo methods using the IMSL subroutine GGNSM. In computing the simulated values, 5000 trials are performed.

We now discuss the distribution of the ratio $F_0=y_1/y_2$. The distribution of F_0 is known (Bose, 1935) for $\gamma_1=\gamma_2=0$ to be:

$$f(F_0)=\frac{2^n(1-\rho^2)^{n/2}\Gamma\left[\frac{1}{2}(n+1)\right]F_0^{n-1}(1+F_0^2)}{\sqrt{\pi}\,\Gamma\left(\frac{1}{2}n\right)\left[1+(1+F_0^2)^2-4\rho^2F_0^2\right]^{(n+1)/2}}. \tag{3.5}$$

Table 2
Comparison of the asymptotic expression with exact expression for the distribution of sum of correlated chi-square variables

n	u	ρ	γ_1	γ_2	O(1)	$O(n^{-1/2})$	$O(1)+O(n^{-1/2})$	Exact
10	31.41	0	0	0	0.9644	−0.0186	0.9458	0.95
		0.5	0	0	0.9467	−0.0230	0.9237	0.9364
		0.5	4	2	0.7353	0.0238	0.7591	0.7574
		0.9	0	0	0.9100	−0.0192	0.8908	0.9032
		0.9	4	2	0.7092	0.0340	0.7432	0.7362
25	67.505	0	0	0	0.9600	−0.0119	0.9481	0.95
		0.2	0	0	0.9570	−0.0125	0.9444	0.9454
		0.8	0	0	0.9142	−0.0126	0.9016	0.9018
		0.5	4	2	0.5730	0.0289	0.6019	0.5960

Finney (1938) showed that:

$$\Pr\{F_0 \leqslant t\} = 1 - I_x(\tfrac{1}{2}n, \tfrac{1}{2}n) \tag{3.6}$$

where

$$x = \tfrac{1}{2}\left[1 - \frac{(t - t^{-1})}{\left\{(t + t^{-1})^2 - 4\rho^2\right\}^{1/2}}\right],$$

$$I_x(a, b) = \frac{1}{\beta(a, b)} \int_0^x y^{a-1}(1-y)^{b-1}\, \mathrm{d}y.$$

Krishnaiah et al. (1965) obtained an alternative expression for $\Pr[F_0 \leqslant t]$ when $\gamma_1 = \gamma_2 = 0$ and also gave tables for this distribution. In Table 3, we give the values of $\Pr[F_0 \leqslant u | \rho, \gamma_1, \gamma_2] = \beta$ by using the asymptotic expressions (2.9), where u is the 75% critical value of the central F distribution with (n, n) degrees of freedom. In

Table 3
Comparison of the asymptotic expression with exact expression for the distribution of the ratio of correlated chi-square variables

n	u	ρ	γ_1	γ_2	O(1)	$O(n^{-1/2})$	$O(1)+O(n^{-1/2})$	Exact[a]
24	1.3214	0	0	0	0.7844	−0.0370	0.7474	0.75*
		0.2	0	0	0.7892	0.0373	0.7519	0.7543*
		0.5	0	0	0.8183	−0.0385	0.7798	0.7816*
		0.8	0	0	0.9053	−0.0356	0.8697	0.8675*
		0	12	12	0.7981	−0.0378	0.7604	0.7566
		0.8	12	12	0.8444	−0.0389	0.8055	0.7942
40	1.2397	0	0	0	0.7758	−0.0272	0.7486	0.75*
		0.2	0	0	0.7804	−0.0274	0.7530	0.7544*
		0.5	0	0	0.8093	−0.0285	0.7808	0.7818*
		0.8	0	0	0.8968	−0.0272	0.8696	0.8683*
		0	20	20	0.7893	−0.0278	0.7615	0.7564
		0.8	20	20	0.8352	−0.0290	0.8062	0.8038

[a] See formula (3.6) for entries with "*".

the last column, the entries with * are obtained by using the formula (3.6) and the remaining entries in this column are obtained by simulation for 5000 trials.

Next, we study the accuracy of the asymptotic expression for the case of the distribution of the sample correlation coefficient $r_{12}=r$. When $\gamma_1=\gamma_2=0$, the distribution of r was first found by Fisher (1915), and Hotelling (1953) has expressed the distribution in terms of the hypergeometric function. When $\rho\neq 0$, the distribution of r is complicated and the cumulative distribution of r has been tabulated by David (1938). Pillai (1946) suggested the transformations $g(r)=(r-\rho)/(1-r\rho)$ that renders the distribution

$$\frac{g(r)\sqrt{n-2}}{\sqrt{1-(g(r))^2}}$$

close to the Student's t distribution. In Table 4 we compare the asymptotic expressions for

$$\Pr\{r\leqslant u|\rho,\gamma_1,\gamma_2\} \quad \text{and} \quad \Pr\{g(r)\leqslant u|\rho,\gamma_1,\gamma_2\}$$

Table 4
Comparison of the asymptotic expression with exact expression for the distributions of functions of the sample correlation coefficient

n	ρ	u	γ_1	γ_2	Stat.	O(1)	$O(n^{-1/2})$	$O(1)+O(n^{-1/2})$	Exact[a]
49	0.5	0.5	0	0	r	0.5000	−0.0142	0.4858	0.4856*
		0	0	0	$g(r)$	0.5000	−0.0142	0.4858	0.4856*
		0.5	24.5	12.25	r	0.8748	0.0106	0.8854	0.8818
		0	24.5	12.25	$g(r)$	0.8948	−0.0088	0.8860	0.8818
		0.5	24.5	24.5	r	0.9235	0.0123	0.9358	0.9370
		0	24.5	24.5	$g(r)$	0.9438	−0.0077	0.9361	0.9370
		0.35	0	0	r	0.0808	0.0156	0.0964	0.0966*
		−0.1818	0	0	$g(r)$	0.1016	−0.0063	0.0952	0.0966*
		0.35	24.5	12.25	r	0.4486	−0.0064	0.4422	0.4486
		−0.1818	24.5	12.25	$g(r)$	0.4491	−0.0068	0.4422	0.4486
		0.35	24.5	24.5	r	0.5568	−0.0044	0.5524	0.5586
		−0.1818	24.5	24.5	$g(r)$	0.5573	−0.0049	0.5524	0.5586
24	0.8	0.7	0	0	r	0.0868	0.0349	0.1217	0.1183*
		−0.2273	0	0	$g(r)$	0.1328	−0.0175	0.1152	0.1183*
		0.7	12	6	r	0.8341	0.0208	0.8549	0.8434
		−0.2273	12	6	$g(r)$	0.8800	−0.022	0.8576	0.8434
		0.7	12	12	r	0.9105	0.0295	0.9400	0.9370
		−0.2273	12	12	$g(r)$	0.9600	−0.0180	0.9420	0.9370
		0.55	0	0	r	0.0003	0.0022	0.0025	0.0098*
		−0.4464	0	0	$g(r)$	0.0144	−0.0030	0.0114	0.0098*
		0.55	12	6	r	0.3870	−0.0069	0.3801	0.3680
		−0.4464	12	6	$g(r)$	0.3924	−0.0124	0.3800	0.3680
		0.55	12	12	r	0.5534	−0.0057	0.5478	0.5388
		−0.4464	12	12	$g(r)$	0.5547	−0.0069	0.5477	0.5388

[a]See the tables in David (1938) for exact values with "*".

with the corresponding exact values. The exact values with * are taken from the tables of David (1938) and the remaining exact values are obtained by simulation with 5000 trials.

4. Asymptotic distributions of functions of the elements of the sample covariance matrix for non-normal populations

Let $X=[\boldsymbol{X}_1,\ldots,\boldsymbol{X}_n]$ where the $p\times 1$ random vectors $\boldsymbol{X}_1,\ldots,\boldsymbol{X}_n$ are distributed independently. Also, let $S=XX'=(S_{ij})$ be the sample sums of squares and cross-products matrix such that $E(S/n)=\Omega=(\omega_{ij})$, $s_{ij}=S_{ij}/n$ and

$$Y=\sqrt{n}\left(\frac{S}{n}-\Omega\right)=(y_{ij}).$$

In addition, let the functions

$$T_i(S/n)=T_i(s_{11},\ldots,s_{pp},s_{12},\ldots,s_{1p},s_{23},\ldots,s_{p-1,p})$$

be analytic in the neighborhood of Ω for $i=1,2,\ldots,k$. The Taylor expansion of $L_i=\sqrt{n}(T_i(S/n)-T_i(\Omega))$ about Ω is:

$$L_i=\sum_{j_1,j_2} a^{(i)}_{j_1j_2}y_{j_1j_2}+\frac{1}{2\sqrt{n}}\sum_{j_1,j_2}\sum_{j_3,j_4} a^{(i)}_{j_1j_2\cdot j_3j_4}y_{j_1j_2}y_{j_3j_4}$$
$$+\mathrm{O}(n^{-1}).$$

Hence,

$$\mathrm{u}_i=E(L_i)=\frac{1}{2\sqrt{n}}\sum_{j_1,j_2}\sum_{j_3,j_4} a^{(i)}_{j_1j_2\cdot j_3j_4}\kappa(j_1j_2,j_3j_4)+\mathrm{O}(n^{-1}),$$

$$\mathrm{u}_{ij}=E(L_iL_j)=\sum_{j_1,j_2}\sum_{j_3,j_4} a^{(i)}_{j_1j_2}a^{(j)}_{j_3j_4}\kappa(j_1j_2,j_3j_4)+\mathrm{O}(n^{-1}),$$

$$\mathrm{u}_{ijl}=E(L_iL_jL_l)=\sum_{j_1,j_2}\sum_{j_3,j_4}\sum_{j_5,j_6} a^{(i)}_{j_1j_2}a^{(j)}_{j_3j_4}a^{(l)}_{j_5j_6}\kappa(j_1j_2,j_3j_4,j_5j_6)$$
$$+\mathrm{O}(n^{-1}), \tag{4.1}$$

where

$$\kappa(j_1 j_2, j_3 j_4) = E(y_{j_1 j_2} y_{j_3 j_4}),$$

$$\kappa(j_1 j_2, j_3 j_4, j_5 j_6) = E(y_{j_1 j_2} y_{j_3 j_4} y_{j_5 j_6}).$$

Now, let $\kappa^{(i)}_{j_1 j_2, \ldots, j_u}$ denote the cumulant of order u for $\boldsymbol{X}_i$, where $j_1, j_2, \ldots, j_u = 1, \ldots, p$. Then:

$$\kappa(j_1 j_2, j_3 j_4) = \overline{\kappa_{j_1 j_2 j_3 j_4}} + \sum^{4} \overline{\kappa_{j_1} \kappa_{j_2 j_3 j_4}} + \left(\overline{\kappa_{j_1 j_3} \kappa_{j_2 j_4}} + \overline{\kappa_{j_1 j_4} \kappa_{j_2 j_3}}\right) + \sum^{4} \overline{\kappa_{j_1 j_3} \kappa_{j_2} \kappa_{j_4}}, \tag{4.2}$$

$$\begin{aligned} \kappa(j_1 j_2, j_3 j_4, j_5 j_6) = \frac{1}{\sqrt{n}} \Bigg(& \overline{\kappa_{j_1 j_2 j_3 j_4 j_5 j_6}} + \sum^{6} \overline{\kappa_{j_1 j_2 j_3 j_4 j_5} \kappa_{j_6}} + \sum^{10} \overline{\kappa_{j_1 j_2 j_3} \kappa_{j_4 j_5 j_6}} \\ & + \sum^{12} \overline{\kappa_{j_1 j_2 j_3 j_5} \kappa_{j_4 j_6}} + \sum^{12} \overline{\kappa_{j_1 j_2 j_3 j_5} \kappa_{j_4} \kappa_{j_6}} + \sum^{48} \overline{\kappa_{j_1 j_2 j_3} \kappa_{j_4 j_5} \kappa_{j_6}} \\ & + \sum^{8} \overline{\kappa_{j_1 j_3 j_5} \kappa_{j_2} \kappa_{j_4} \kappa_{j_6}} + \sum^{8} \overline{\kappa_{j_1 j_4} \kappa_{j_2 j_5} \kappa_{j_3 j_6}} + \sum^{24} \overline{\kappa_{j_1 j_4} \kappa_{j_2 j_5} \kappa_{j_3} \kappa_{j_6}} \Bigg). \end{aligned} \tag{4.3}$$

The expressions under "——— " represent the average values over n samples. For instance:

$$n\, \overline{\kappa_{j_1 j_3} \kappa_{j_2 j_4}} = \sum_{i=1}^{n} \kappa^{(i)}_{j_1 j_3} \kappa^{(i)}_{j_2 j_4}.$$

The summations in eqs. (4.2) and (4.3) are over the possible ways of grouping the subscripts and the number of terms resulting is written over Σ. Eqs. (4.2) and (4.3) coincide with the expressions of Kaplan (1952) when $\boldsymbol{X}_1, \ldots, \boldsymbol{X}_n$ are identically distributed.

So the cumulants of $\boldsymbol{L} = (L_1, \ldots, L_k)$ are:

$$\begin{aligned} \mathcal{K}_i &= \mathfrak{u}_i, \\ \mathcal{K}_{ij} &= \mathfrak{u}_{ij} + O(n^{-1}), \\ \mathcal{K}_{ijl} &= \mathfrak{u}_{ijl} - (\mathfrak{u}_i \mathfrak{u}_{jl} + \mathfrak{u}_j \mathfrak{u}_{il} + \mathfrak{u}_l \mathfrak{u}_{ij}) + O(n^{-1}) \end{aligned} \tag{4.4}$$

for $i, j, l = 1, \ldots, k$. An equivalent equation [see Kendall and Stuart (1961)]:

$$\begin{aligned} & \kappa(j_1 j_2, j_5 j_6)\kappa(j_3 j_4, j_7 j_8) + \kappa(j_1 j_2, j_7 j_8)\kappa(j_3 j_4, j_5 j_6) = \\ & = -\kappa(j_1 j_2, j_3 j_4)\kappa(j_5 j_6, j_7 j_8) + O(n^{-1}) \end{aligned} \tag{4.5}$$

is used in calculating $\mathfrak{u}_i \mathfrak{u}_{jl}$, $\mathfrak{u}_j \mathfrak{u}_{il}$, $\mathfrak{u}_l \mathfrak{u}_{ij}$ in eq. (4.4).

The approximated characteristic function of $\boldsymbol{L}$ is,

$$E[\exp(\mathrm{i}\,\boldsymbol{t}'\boldsymbol{L})]=\exp\left(-\tfrac{1}{2}\boldsymbol{t}'Q\boldsymbol{t}+\mathrm{i}\sum_{g=1}^{k} t_g\mathcal{K}_g +\mathrm{i}^3\sum_{g,j,l}^{k}\frac{1}{\delta(g,j,l)}t_gt_jt_l\mathcal{K}_{gjl}\right)+\mathrm{O}(n^{-1})$$

$$=\exp(-\tfrac{1}{2}\boldsymbol{t}'Q\boldsymbol{t})\left\{1+\mathrm{i}\sum_{g=1}^{k} t_g\mathcal{K}_g +\mathrm{i}^3\sum_{g,j,l}^{k}\frac{1}{\delta(g,j,l)}t_gt_jt_l\mathcal{K}_{gjl}\right\}+\mathrm{O}(n^{-1}), \tag{4.6}$$

where $Q=(\mathcal{K}_{ij})$ and,

$$\delta(g,j,l)=\begin{cases}3! & g=j=l,\\ 2! & \text{any two values of } g,\ j,\ l \text{ are equal,}\\ 1 & g\neq j\neq l.\end{cases}$$

By inverting eq. (4.6), we obtain the following asymptotic expression for the joint density of $L_1,\ldots,L_k$:

$$f(L_1,\ldots,L_k)=N(\boldsymbol{L},Q)\left\{1+\sum_{g=1}^{k}H_g(\boldsymbol{L})\mathcal{K}_g +\sum_{g,j,l}^{k}\frac{1}{\delta(g,j,l)}H_{gjl}(\boldsymbol{L})\mathcal{K}_{gjl}\right\}+\mathrm{O}(n^{-1}), \tag{4.7}$$

where $N(\boldsymbol{L},Q)$, $H_g(\boldsymbol{L})$ and $H_{gjl}(\boldsymbol{L})$ are as defined in eq. (2.10).

5. Applications of the distributions of functions of the elements of the covariance matrix and correlation matrix

In this section, we discuss some applications of the results of Section 2 in simultaneous tests of hypotheses on the elements of the covariance matrix and correlation matrix.

Let $M=0$, H_{ij}: $\sigma_{ij}=\sigma_{0ij}$, A_{ij}: $\sigma_{ij}\neq\sigma_{0ij}$, H_{ij}^*: $\rho_{ij}=\rho_{0ij}$ and A_{ij}^*: $\rho_{ij}\neq\rho_{0ij}$. We will first discuss the problem of testing the hypotheses H_{ij} simultaneously against the alternative hypotheses A_{ij}. In this case, the hypothesis H_1 is accepted if

$$b_1\leqslant s_{ij}-\sigma_{0ij}\leqslant a_1 \quad \text{for } i<j,$$

$$b_2\leqslant\frac{s_{ii}}{\sigma_{0ii}}\leqslant a_2 \tag{5.1}$$

for $i, j=1,2,\dots, p$, where

$$\Pr[b_1 \leqslant s_{ij}-\sigma_{0ij} \leqslant a_1, i<j=1,2,\dots, p;\ b_2 \leqslant s_{ii}/\sigma_{0ii} \leqslant a_2, i=1,\dots, p|H_1]=(1-\alpha), \tag{5.2}$$

and $H_1=\bigcap_{i\leqslant j} H_{ij}$. For practical purposes, we may choose the constants a_1 and b_1, a_2 and b_2 such that $a_1=-b_1$ and $a_2=1/b_2$. We can propose similar procedures for testing the hypotheses H_{ij} against one-sided alternatives.

Next, consider the problem of testing the hypotheses $H_{11},\dots, H_{pp}$ simultaneously against $A_{11},\dots, A_{pp}$. In this case, we accept H_{ii} if

$$b_2 \leqslant \frac{s_{ii}}{\sigma_{011}} \leqslant a_2$$

and reject it otherwise where

$$\Pr\left[b_2 \leqslant \frac{s_{ii}}{\sigma_{0ii}} \leqslant a_2;\ i=1,\dots, p \middle| \bigcap_{i=1}^{p} H_{ii}\right]=(1-\alpha). \tag{5.3}$$

Also, consider the problem of testing the hypotheses H_{ij}^* $(i, j=1,2,\dots, p)$ simultaneously against the alternatives A_{ij}^*. In this case, we accept or reject H_{ij}^* according as:

$$(r_{ij}-\rho_{0ij})^2 \lesseqgtr c_\alpha$$

where

$$\Pr\left[(r_{ij}-\rho_{0ij})^2 \leqslant c_\alpha;\ i<j=1,2,\dots, p|H^*\right]=(1-\alpha)$$

and $H^*=\bigcap_{i<j} H_{ij}^*$.

The results of Section 2 are useful in computing approximate values of the critical values associated with the above tests. The results of Section 2 are also useful in finding approximate critical values of the tests of Krishnaiah (1975) for testing the hypothesis $\sigma_{11}=\cdots=\sigma_{pp}$ against different alternatives when the correlation matrix is known as well as the procedure of Roy and Bargmann (1958) for testing the hypothesis that the covariance matrix of multivariate normal is diagonal.

In the applications discussed above, we assumed that the matrix $S=(S_{ij})$ is the central Wishart matrix. But situations arise where the model itself is not correct. For example, if we assume that $\boldsymbol{X}_1,\dots, \boldsymbol{X}_n$ are distributed independently and identically as multivariate normal with a common mean vector, $\boldsymbol{\mu}$, and covariance matrix, Σ, then S is the central Wishart matrix when $S=\Sigma_{j=1}^n(\boldsymbol{X}_j-\bar{\boldsymbol{X}}_.)(\boldsymbol{X}_j-\bar{\boldsymbol{X}}_.)'$ and $n\bar{\boldsymbol{X}}_.=\Sigma_{j=1}^n \boldsymbol{X}_j$. But, if the mean vectors are given by $E(\boldsymbol{X}_j)=\boldsymbol{\mu}_j$, then S is the non-central Wishart matrix. So, the results given in Section 2 for the case of the non-central Wishart matrix and non-central correlation matrix are useful in studying the robustness of several test procedures on the elements of Σ if the assumption of common mean vector for $\boldsymbol{X}_1,\dots, \boldsymbol{X}_n$ is violated.

Next, consider the model,

$$X_{1j}=u_j+\delta_j, \qquad X_{2j}=v_j+\varepsilon_j,$$

when $u_j = \alpha + \beta v_j$ $(j=1,2,\ldots,n)$ and α, β are unknown constants. Also, we assume that v_j's are distributed independently and identically as normal with a common mean μ and variance σ^2. In addition, ε_j's and δ_j's are distributed as normal and:

$$E(\varepsilon_j)=E(\delta_j)=0;$$

$$\operatorname{cov}(\varepsilon_j,\delta_j)=0, \qquad \operatorname{cov}(v_j,\varepsilon_j)=\operatorname{cov}(v_j,\delta_j)=0;$$

$$\operatorname{Var}(\varepsilon_j)=\sigma_\varepsilon^2, \qquad \operatorname{var}(\delta_j)=\sigma_\delta^2.$$

We also assume that the random vectors $(X_{1j}, X_{2j}, u_j, v_j, \varepsilon_j, \delta_j)$ are distributed independent of each other for different values of j. When $\lambda=\sigma_\delta^2/\sigma_\varepsilon^2$ is known, the maximum likelihood estimate of β is known [see Kendall and Stuart (1973, ch. 29) to be

$$\hat{\beta}=\frac{(S_{11}-\lambda S_{22})+\{(S_{11}-\lambda S_{22})^2+4\lambda S_{12}^2\}^{1/2}}{2S_{12}}$$

where $S=(S_{ij})=\Sigma_{j=1}^n(\boldsymbol{X}_j-\overline{\boldsymbol{X}}_.)(\boldsymbol{X}_j-\overline{\boldsymbol{X}}_.)'$, $\boldsymbol{X}_j'=(X_{1j}, X_{2j})$, and $n\overline{\boldsymbol{X}}_.=\Sigma_{j=1}^n\boldsymbol{X}_j$. The results in Section 2 of this paper are useful in obtaining an asymptotic expression for the distribution of $\hat{\beta}$.

Now, let $\boldsymbol{Y}_j'=(Y_{1j}, Y_{2j})$ $(j=1,2,\ldots,n)$ be distributed independently as a bivariate normal with mean vector (μ_{1j}, μ_{2j}) and covariance matrix $\Sigma=\sigma^2 I$ where I is an identity matrix. Also, let $\mu_{1j}=\alpha+\beta\mu_{2j}$. Then, the maximum likelihood estimate $\hat{\hat{\beta}}$ of β is known [e.g., see Anderson (1976)] to be

$$\hat{\hat{\beta}}=\frac{S_{11}-S_{22}+\{(S_{11}-S_{22})^2+4S_{12}^2\}^{1/2}}{2S_{12}},$$

where $S=(S_{ij})=\Sigma_{j=1}^n(\boldsymbol{Y}_j-\overline{\boldsymbol{Y}}_.)(\boldsymbol{Y}_j-\overline{\boldsymbol{Y}}_.)'$ and $n\overline{\boldsymbol{Y}}_.=\Sigma_{j=1}^n\boldsymbol{Y}_j$. Since S is distributed as the non-central Wishart matrix, an asymptotic expression for the distribution of $\hat{\hat{\beta}}$ can be obtained as a special case of the results of Section 2. Here, we note that Kunimoto (1980) has recently obtained an asymptotic expression for the distribution of $\hat{\hat{\beta}}$.

References

[1] Anderson, T.W. (1976). Estimation of linear functional relationships: approximate distributions and connections with simultaneous equations in econometrics (with discussion). *J. Roy. Statist. Soc. Ser. B* **38**, 1–36.

[2] Bose, S.S. (1935). On the distribution of the ratio of two samples drawn from a given bivariate correlated population. *Sankhya* **2**, 65–72.

[3] David, F.N. (1938). *Tables of the Ordinates and Probability Integral of the Distribution of the Correlation Coefficient in Small Samples*. Biometrika Office, London.

[4] Finney, D.J. (1938). The distribution of the ratio of estimates of the two variances in a sample from a normal bivariate population. *Biometrika* **30**, 190–192.

[5] Fisher, R.A. (1915). Frequency distribution of the values of the correlation coefficient in samples from an indefinitely large population. *Biometrika* **10**, 507–521.

[6] Hotelling, H. (1953). New light on the correlation coefficient and its transforms. *J. Roy. Statist. Soc. Ser. B.* **15**, 193–225.

[7] Kaplan, E.L. (1952). Tensor notation and the sampling cumulants of k-statistics. *Biometrika* **39**, 319–323.

[8] Kendall, M.G. and Stuart, A. (1961). *The Advanced Theory of Statistics*, Vol. 1. Griffin, London.

[9] Kendall, M.G. and Stuart, A. (1973). *The Advanced Theory of Statistics*, Vol. 2. Griffin, London.

[10] Konishi, S. (1979). Asymptotic expressions for the distributions of statistics based on the sample correlation matrix in principle components analysis. *Hiroshima Math.* **9**, 647–700.

[11] Krishnaiah, P.R., Hagis Jr, P. and Steinberg, L. (1965). Test for the equality of standard deviations in a bivariate normal population. *Trabajos Estadist.* **17**, 3–15.

[12] Krishnaiah, P.R. (1975). Tests for the equality of the covariance matrices of correlated multivariate normal populations. In: Srivastava, J.N., Ed., *A Survey of Statistical Design and Linear Models*. North-Holland, Amsterdam, pp. 355–366.

[13] Kunitomo, N. (1980). Asymptotic expansion of the distributions of estimators in a linear functional relationship and simultaneous equations. *J. Am. Statist. Assoc.* **75**, 693–700.

[14] Olkin, I. and Siotani, M. (1976). Asymptotic distribution of functions of a correlation matrix. In: Ikeda, S. et al., Eds., *Essays in Probability and Statistics in Honor of J. Ogawa*. Shiko Tsusho Co. Ltd., Tokyo, pp. 231–251.

[15] Owen, D.B. (1962). *Handbook of Statistical Tables*. Addison Wesley, Reading, MA.

[16] Pillai, K.C.S. (1946). Confidence interval for the correlation coefficient. *Sankhya* **7**, 415–522.

[17] Roy, S.N. and Bargman, R.E. (1958). Tests of multiple independence and the associated confidence bounds. *Ann. Math. Statist.* **29**, 491–503.

[18] Siotani, M. and Hayakawa, T. (1964). Asymptotic distributions of functions of Wishart matrix. *Proceedings of Inst. Statist. Math.* **12**, 191–198.

G. Kallianpur, P.R. Krishnaiah, J.K. Ghosh, eds., *Statistics and Probability*: *Essays in Honor of C.R. Rao*

THE EARLY USE OF STOCHASTIC METHODS: AN HISTORICAL NOTE ON McKENDRICK'S PIONEERING PAPERS

J. GANI
CSIRO Division of Mathematics and Statistics, Canberra, Australia

A summary of McKendrick's early papers on mathematical methods applied to biological and medical problems is presented. It is pointed out that he solved the time-dependent Poisson, as well as simple birth and birth–death models, and pioneered work on two-dimensional spatial problems. McKendrick achieved a very practical balance between theory and applications; his work still bears reading.

Introduction

Workers in epidemic theory often have occasion to quote McKendrick (also spelt M'Kendrick) as the earliest user of stochastic methods in their field. However, despite the fact that McKendrick's very original paper on medical problems (1926) is referred to in Bartlett (1960) and Bailey (1975), details of his contributions, particularly those of his earlier paper (1914) are not often given in the literature. Irwin, in his presidential address to the Royal Statistical Society (1963), has indeed described at some length McKendrick's most important results of 1926. However, the priority of this work and details of his 1914 paper are still not fully appreciated, and his stochastic approach to spatial problems has not as yet received adequate recognition. What we will attempt in this note is a review of his papers, with a brief commentary on his most important ideas.

1. Early results and one-dimensional models

In his earlier paper McKendrick (1914) considers the problem of leucocytes which absorb $x=0,1,2,\ldots$ microorganisms in solution after colliding with them. If v_x denotes the (mean) number of leucocytes after x collisions in continuous time $t\geqslant 0$, this is shown to satisfy the equations

$$\mathrm{d}v_0/\mathrm{d}t=-\phi v_0, \qquad \mathrm{d}v_x/\mathrm{d}t=(v_{x-1}-v_x)\phi, \quad (x\geqslant 1)$$

where, in effect, the probability of a collision in the time interval $(t, t+\delta t)$ is $\phi\,\delta t+\mathrm{o}(\delta t)$. This describes the Poisson process, and McKendrick considers it in detail, particularly for ϕ time-dependent. It is possibly the earliest appearance of the process in the literature in this form.

McKendrick then proceeds to derive equations for the pure birth process, and a particular birth–death process. Since these are also considered in his later paper, we shall not describe their details here. He formulates very clearly for the first time the differential equations for the deterministic general epidemic which he later elaborated with Kermack in their famous paper of 1927. Finally, he concludes his work with the case where v_x tends to a continuous function $v(x)$ of x, and derives a wave equation, as well as several diffusion equations for v. These may well be their earliest derivation in a biological context.

In his second paper (1926), McKendrick considers two types of one dimensional problems, which he represents by discrete passages in continuous time $t \geqslant 0$ from compartment x to $x+1$, or x to $x-1$ and $x+1$ along the line. The first, which he refers to as irreversible, may describe the number of individuals v_x of grade x at time t who have experienced x attacks of a disease (the common cold, for example) and for which the probability of a further attack in $(t, t+\delta t)$ is given by

$$\Pr\{\text{individual of grade } x \text{ passes to grade } x+1 \text{ in } (t, t+\delta t)\} = f_{t,x}\,\delta t + o(\delta t).$$

Ordinarily, one would proceed to consider the probability $\Pr\{v_x = n \text{ at time } t \mid \text{initial conditions}\}$; but McKendrick effectively studies the semi-stochastic model in which v_x can be regarded as the mean number of individuals of grade x at time t. For these he derives the differential equation,

$$dv_x/dt = f_{t,x-1}v_{x-1} - f_{t,x}v_x, \tag{1.1}$$

which he solves for $f_{t,x} = b + cx$, when it reduces to

$$dv_x/dt = (b + c(x-1))v_{x-1} - (b+cx)v_x. \tag{1.2}$$

These are the equations which McKendrick had already derived in a different context in his earlier paper (1914). The reader will have recognised, in (1.2), the stochastic linear birth model, with birth parameter $\lambda_x = b + cx$, for which the p.g.f. $\phi(z,t) = \sum_{x=0}^{\infty} v_x z^x$ satisfies the partial differential equation

$$\partial\phi/\partial t = cz(z-1)\,\partial\phi/\partial z + b(z-1)\phi. \tag{1.3}$$

For $v_0 = 1$ at time $t = 0$, all other v_x being zero, its solution is readily found to be

$$\phi(z,t) = \left\{ e^{ct}\left(1 - z(1 - e^{-ct})\right)\right\}^{-b/c}. \tag{1.4}$$

The solution to (1.2) is derived by McKendrick in a slightly different form, and this predates Furry's (1937) treatment of the linear birth process, which he outlined in a physical context.

If c/b tends to zero, this p.g.f. will tend to $e^{-bt(1-z)}$, and McKendrick notes that v_x then takes the Poisson form. Hence his conclusion that a Poisson distribution for observed cases of illness means that there is no infectious process involved. This was the case, for example, in the number of houses in Luckau, Hanover, in which x cases of cancer had occurred in the period 1875–1898, and for which the estimate of $c/b = -0.009$ was considered negligible.

Observed no. of cases x	Observed no. of houses	Calculated no. of houses
0	64	65
1	43	40
2	10	12
3	2	2.5
4	1	0.4

McKendrick concludes: "Thus the figures afford no evidence that the occurrence of a late case was influenced by previous cases".

The reversible case is that in which, in addition to the $f_{t,x}$ of (1.1), there are also probabilities

$$\Pr\{\text{individual of grade } x \text{ passes to grade } x-1 \text{ in } (t, t+\delta t)\} = f'_{t,x}\,\delta t + o(\delta t).$$

These lead, when $f_{t,x} = f_x$, $f'_{t,x} = f'_x$ to the equation

$$dv_x/dt = f_{x-1}v_{x-1} - f_x v_x + f'_{x+1}v_{x+1} - f'_x v_x \tag{1.5}$$

or when $f_x = f'_x = f$ for simplicity, to

$$dv_x/dt = f(v_{x-1} - 2v_x + v_{x+1}). \tag{1.6}$$

Once again, this equation had been considered in McKendrick (1914). The reader will have recognized the stochastic birth–death model of the simplest kind, with $\lambda_x = \mu_x = f$, for which the p.g.f. $\phi(z,t) = \Sigma_{x=0}^{\infty} v_x z^x$ satisfies the partial differential equation

$$\partial\phi/\partial t = f(z-1)^2\,\partial\phi/\partial z, \tag{1.7}$$

whose solution for $v_1 = 1$ at $t=0$, all other v_x being zero, is

$$\phi(z,t) = (1+ft)^{-1}(ft + z(1-ft))(1 - ft(1+ft)^{-1}z)^{-1}. \tag{1.8}$$

McKendrick obtains the solution to (1.5) both in this simple case, and also when $f_x = f'_x = b + cx$.

2. Two-dimensional models

Let (x, y) represent a position on the square lattice $x, y = 0, 1, 2, \ldots$; then once again if $v_{x,y}$ is regarded as the mean number of individuals at x, y at time $t \geqslant 0$, and

$$\Pr\{\text{individual at } x, y \text{ passes to } x+1, y \text{ in time } \delta t\} = f_{x,y}\,\delta t + o(\delta t),$$

$$\Pr\{\text{individual at } x, y \text{ passes to } x, y+1 \text{ in time } \delta t\} = g_{x,y}\,\delta t + o(\delta t)$$

the equation

$$dv_{x,y}/dt = f_{x-1,y}v_{x-1,y} - f_{x,y}v_{x,y} + g_{x,y-1}v_{x,y-1} - g_{x,y}v_{x,y} \tag{2.1}$$

can be readily derived.

If $f_{x,y}=b+cx$ and $g_{x,y}=\beta+\gamma y$, the joint p.g.f., $\phi(z,w,t)=\sum_{x,y=0}^{\infty} v_{x,y} z^x w^y$ satisfies the partial differential equation

$$\partial\phi/\partial t=\{b(z-1)+\beta(w-1)\}\phi+\{c(z^2-z)\,\partial\phi/\partial z+\gamma(w^2-w)\,\partial\phi/\partial w\}. \quad (2.2)$$

The solution of this for $v_{00}=1$ at $t=0$, all other $v_{x,y}$ being zero, is

$$\phi(z,w,t)=\{e^{ct}(1-z(1-e^{-ct}))\}^{-b/c}\{e^{\gamma t}(1-w(1-e^{-\gamma t}))\}^{-\beta/\gamma} \quad (2.3)$$

as might be expected from the independence of the horizontal and vertical movements. Once again as c/b and γ/β tend to zero, $v_{x,y}$ tends to the bivariate Poisson form.

McKendrick also considered a third direction of movement, namely from x, y to $x+1$, $y+1$, diagonally across compartments. For this, by analogy with the earlier cases, he derives the equation

$$\begin{aligned} dv_{x,y}/dt=\ & f_{x-1,y}v_{x-1,y}-f_{x,y}v_{x,y} \\ & +g_{x,y-1}v_{x,y-1}-g_{x,y}v_{x,y} \\ & +h_{x-1,y-1}v_{x-1,y-1}-h_{x,y}v_{x,y}, \end{aligned} \quad (2.4)$$

where the last two terms refer to such diagonal motion. Among the other topics of interest discussed in his paper are:

(i) the summation of $v_{x,y}$ along the diagonals $z=x+y$ and $z=x-y$;

(ii) some restricted cases where $v_{x,y}=0$ for $x=0$, $y=1,2,\ldots$; here x denotes infections arising from outside a household, and y infections arising within the household;

(iii) some generalizations where $f_{x,y}$, $g_{x,y}$, $h_{x,y}$ indicate movement horizontally, vertically and diagonally from x, y to $x+1$, y, from x, y to x, $y+1$ and x, y to $x+1$, $y+1$, respectively, and $f'_{x,y}$, $g'_{x,y}$, $h'_{x,y}$ are the analogous probabilities for movement from x, y to $x-1$, y, from x, y to x, $y-1$, and x, y to $x-1$, $y-1$;

(iv) the consideration of continuous variables $v(x,y)$, analogous to the discrete $v_{x,y}$, leading to partial differential equations for them; and lastly

(v) an example written with Dr. Kermack, his lifelong collaborator, on the course of an epidemic where $v_{t,\theta}$ now denotes the number of cases of age θ at time t. Professor Bartlett has pointed out to me that this is probably the first instance of the derivation of the well-known partial differential equation referred to in Rubinow (1978)

$$\partial v_{t,\theta}/\partial t+\partial v_{t,\theta}/\partial\theta=-f_{t,\theta}v_{t,\theta}.$$

Here $f_{t,\theta}\,\delta t+o(\delta t)$ is the probability of dying at age θ in time $(t, t+\delta t)$.

3. The scope of McKendrick's work

One is struck throughout the papers by the breadth of McKendrick's knowledge, and his grasp of the relevant factors in biological and epidemic processes. While his treatment of the problems may be only semi-stochastic, he appears to have been familiar far earlier than his contemporaries with the equations for Poisson, birth and birth–death type processes, of which he obtained the solutions. He also anticipated some of the early results in diffusion equations and in spatial stochastic models.

But perhaps his greatest contribution of all is the familiarity he shows with medical literature, and his facility in passing from mathematical theory to practical applications. It is perhaps fitting that we should end this note with his treatment of some statistics collected by the Australian Government (1919) concerning influenza epidemics which had occurred in ships on their outward voyage to Australia. The data recorded was

Observed no. of cases x	Number of epidemics with x cases
$x=1$	34
2	15
3	5
4	3
5	8
6	1
7	3
8	4
9	2
10 or more	17

While he was unable to fit these figures by means of the equations he had derived for "restricted cases", he noted that the J character of the distribution was evident, as indeed his theory indicated. He concludes on a characteristically cautious note, illustrative of his understanding of the medical as well as the statistical background to epidemic theory:

"The conclusion is suggested that as our limited experience of epidemic influenza is based upon statistics which may relate only to a small selected minority of the total number of epidemics, it may be in no sense representative, and may even be misleading."

McKendrick's work still bears rereading: it is rich in ideas, contains many unexpected results, and is full of practical good sense. He remains an inspiring model on which modern mathematical epidemiologists could well pattern themselves.

References

Bailey, N.T.J. (1975). *The Mathematical Theory of Infectious Diseases and its Applications.* Griffin, London.

Bartlett, M.S. (1960). *Stochastic Population Models.* Methuen, London.

Commonwealth of Australia Quarantine Service (1919). Publication 18.

Furry, W.J. (1937). On fluctuation phenomena in the passage of high energy electrons through lead. *Phys. Rev.* **52**, 569–581.

Irwin, J.O. (1963). The place of mathematics in medical and biological statistics. *J. R. Statist. Soc. A* **126**, 1–44.

Kermack, W.O. and McKendrick, A.G. (1927). Contributions to the mathematical theory of epidemics. *Proc. R. Soc. A.* **115**, 700–721.

McKendrick, A.G. (1914). Studies on the theory of continuous probabilities, with special reference to its bearing on natural phenomena of a progressive nature. *Proc. London Math. Soc.* **13**, 401–416.

McKendrick, A.G. (1926). Applications of mathematics to medical problems. *Proc. Edinburgh Math. Soc.* **44**, 98–130.

Rubinow, S.I. (1978). Age-structured equations in the theory of cell populations. In *Studies in Mathematical Biology, Part II: Populations and Communities*, Levin, S.A., Ed. MAA Studies in Mathematics, Vol. 16, Math. Assoc. Amer., Washington, DC.

G. Kallianpur, P.R. Krishnaiah, J.K. Ghosh, eds., *Statistics and Probability: Essays in Honor of C.R. Rao*

PROJECTION PLOTS FOR DISPLAYING CLUSTERS

R. GNANADESIKAN, J.R. KETTENRING and J.M. LANDWEHR
Bell Laboratories, Murray Hill, NJ, U.S.A.

1. Introduction

The veritable explosion in the number of clustering algorithms, as well as the widespread use of cluster analysis in diverse fields, are well-recognized phenomena (Blashfield and Aldenderfer, 1978). The use of any clustering method necessarily confronts the user with a set of "clusters" whether or not such clusters are meaningful. Thus, from a data-analytic viewpoint, there is a crucial need for procedures that facilitate the interpretation of the results and enable a sifting of useful findings from less important ones and even methodological artifacts (see, e.g. Gnanadesikan et al., 1978; Rao, 1977). Formal tests for the statistical significance of clusters have been proposed (e.g., Baker and Hubert, 1975; Hubert, 1974) but informal more data-oriented methods that make fewer assumptions are also needed.

The present paper is concerned with graphical displays that can help elucidate the cluster structure in a body of data. Specifically the methods, to be used *after* a cluster analysis of "objects" into mutually exclusive groups has been performed, yield various two-dimensional orthogonal projections of the objects. The two dimensions for the display are chosen to satisfy specific criteria that are meaningful for studying separations amongst the objects or the clusters. Also, although the displays themselves are only two dimensional, information regarding additional dimensions is coded in the pictures so as to help assimilate what is not being seen directly. Visually simple representations, such as the ones to be described here, can indeed be helpful in understanding and interpreting the results of cluster analyses and, as such, they are in the spirit of the graphical methods discussed by Rao (1952, Section 9c.1) for recognizing groups on the basis of measurements on two or three correlated variables.

Section 2 describes the essential details of the different projection techniques. Section 3 illustrates the use of the methods in two real data examples and Section 4 consists of concluding discussion.

2. Methodology

Conventional methods for finding projections of multivariate data based upon principal components and discriminant analyses can be usefully modified for purposes of studying the structure of clusters. The two classes of procedures described in this section are intended either for examining the local neighborhood around a particular cluster or for displaying all of the clusters simultaneously so as to reveal their separations.

For the local neighborhood procedure, let: $x_1,\ldots,x_l$ be the $(p\times 1)$ observation vectors in the cluster of interest, with centroid at m; $x_1^*,\ldots,x_{n-l}^*$ the observations outside the cluster; and $y_j = x_j^* - m$. A standard principal components analysis based on $\Sigma y_j y_j'$ would pick out directions in which the x^*'s are maximally dispersed about m; but this is not quite what is needed. To focus on the local neighborhood aspect, the emphasis should be on x^*'s which are near m. This suggests using the $(p\times p)$ matrix

$$A = \sum_{j=1}^{n-l} w_j y_j y_j' \tag{1}$$

where w_j is a nonincreasing function of $d_j = \| y_j \|$.

From (1), it follows that

$$\Sigma\lambda_j(A) = \Sigma w_j d_j^2 \tag{2}$$

where $\lambda_j(A)$ is the jth largest eigenvalue of A. If $e_1(A),\ldots,e_p(A)$ are orthonormal eigenvectors corresponding to the λ's, then each e_i defines a direction along which a portion $\lambda_i = \Sigma_j w_j \| e_i' y_j \|^2$ of the total local neighborhood distances, given by (2), is accounted for. In particular, (e_1, e_2) defines a two-dimensional orthogonal projection which maximizes the local neighborhood distances. This projection is of interest because it shows those two dimensions in which the cluster is "best" separated from other nearby observations, and displays of this type are illustrated in Section 3. Study of the elements of e_1 and e_2 may indicate variables which are largely responsible for the separation which is attained.

A possible form for the weights is given by $w_j = d_j^{-q}$. Fig. 1 shows the effect of different q-values, with the displayed circle representing the unit sphere. The case of $q=0$ corresponds to the standard principal components analysis mentioned earlier. With $q=2$, only orientation information is being considered since the analysis boils down to doing principal components after moving the y's radially outward or inward so that they lie on the unit sphere centered at $\mathbf{0}$. Any choice of $q>2$ effectively pushes the x^*'s near m farther out and brings the others in closer. The choice used in this paper is $q=4$. This emphasizes the x^*'s nearest to m as can be seen from (2), which now reduces to $\Sigma\lambda_j = \Sigma d_j^{-2}$.

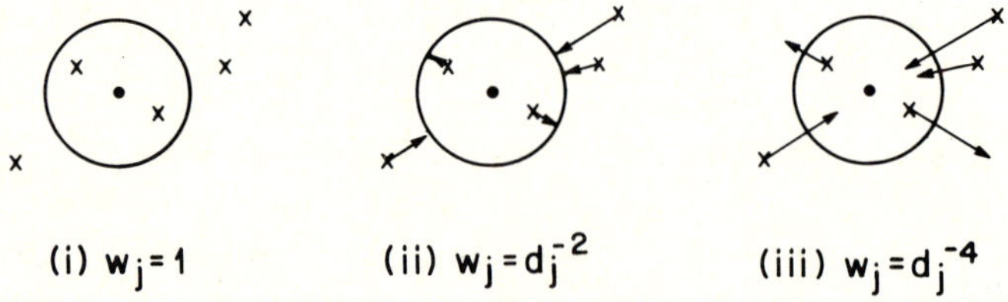

Fig. 1. Effect of weighting function.

The main features of the local neighborhood projection technique are that it treats one cluster at a time, is object oriented in the sense that it focuses on $\boldsymbol{x}^*$'s which are near the cluster of interest, and parallels principal components analysis in terms of computations. The second projection approach described below treats all clusters simultaneously, is cluster oriented, and resembles discriminant analysis in its computations.

For the all-cluster projection plots, let $\boldsymbol{x}_{ij}$ be the jth observation in the ith cluster, $j=1,\ldots,n_i$, $i=1,\ldots,g$; $\bar{\boldsymbol{x}}_i$ and $\bar{\boldsymbol{x}}$ the ith group and the overall centroids; and $n=\Sigma n_i$. The usual between-clusters covariance matrix can be expressed in two algebraically equivalent ways:

$$\boldsymbol{B}=\frac{1}{(g-1)}\sum_i n_i(\bar{\boldsymbol{x}}_i-\bar{\boldsymbol{x}})(\bar{\boldsymbol{x}}_i-\bar{\boldsymbol{x}})' \tag{3}$$

and

$$\boldsymbol{B}=\frac{1}{n(g-1)}\sum_{i<j} n_i n_j(\bar{\boldsymbol{x}}_i-\bar{\boldsymbol{x}}_j)(\bar{\boldsymbol{x}}_i-\bar{\boldsymbol{x}}_j)'. \tag{4}$$

Note in both (3) and (4) how the cluster sizes enter as weights.

As for $\boldsymbol{A}$, the eigenvectors of $\boldsymbol{B}$ can be used to define orthogonal projections. In particular $\boldsymbol{e}_1(\boldsymbol{B})$ and $\boldsymbol{e}_2(\boldsymbol{B})$ determine the plane for which the cluster centroids are maximally separated in the sense of between clusters mean squares (i.e., $\lambda_1(\boldsymbol{B})+\lambda_2(\boldsymbol{B})$). This projection, however, has the drawback that it will be largely determined by outlying clusters if there are any.

To give more consideration to the less extreme clusters, $\boldsymbol{B}$ can be modified by including additional weights w_i in (3) to obtain

$$\boldsymbol{B}^*=\frac{1}{(g-1)}\sum_i w_i n_i(\bar{\boldsymbol{x}}_i-\bar{\boldsymbol{x}})(\bar{\boldsymbol{x}}_i-\bar{\boldsymbol{x}})' \tag{5}$$

or w_{ij} in (4) to produce

$$\boldsymbol{B}^{**}=\frac{1}{n(g-1)}\sum_{i<j} w_{ij} n_i n_j(\bar{\boldsymbol{x}}_i-\bar{\boldsymbol{x}}_j)(\bar{\boldsymbol{x}}_i-\bar{\boldsymbol{x}}_j)' \tag{6}$$

where w_i and w_{ij} are nonincreasing functions of $d_{i0}=\|\bar{\boldsymbol{x}}_i-\bar{\boldsymbol{x}}\|$ and $d_{ij}=\|\bar{\boldsymbol{x}}_i-\bar{\boldsymbol{x}}_j\|$, respectively. Since $\bar{\boldsymbol{x}}$ is not of direct interest, $\boldsymbol{B}^{**}$ may have some advantages over $\boldsymbol{B}^*$.

From (3)–(6), it follows that

$$\sum\lambda_i(\boldsymbol{B})=\frac{1}{(g-1)}\sum n_i d_{i0}^2=\frac{1}{n(g-1)}\sum_{i<j} n_i n_j d_{ij}^2,$$

$$\sum\lambda_i(\boldsymbol{B}^*)=\frac{1}{(g-1)}\sum w_i n_i d_{i0}^2, \tag{7}$$

and

$$\sum\lambda_i(\boldsymbol{B}^{**})=\frac{1}{n(g-1)}\sum_{i<j} w_{ij} n_i n_j d_{ij}^2. \tag{8}$$

By analogy with the local neighborhood procedure, the choices $w_i = d_{i0}^{-4}$ and $w_{ij} = d_{ij}^{-4}$ simplify (7) and (8) to

$$(g-1)\sum \lambda_i(\boldsymbol{B}^*) = \sum n_i d_{i0}^{-2}$$

and

$$n(g-1)\sum \lambda_i(\boldsymbol{B}^{**}) = \sum_{i<j} n_i n_j d_{ij}^{-2}.$$

Clearly, these particular weights are very sensitive to configurations where either $\bar{\boldsymbol{x}}_i$ and $\bar{\boldsymbol{x}}$, or $\bar{\boldsymbol{x}}_i$ and $\bar{\boldsymbol{x}}_j$, are close.

The main reason for limiting attention to orthogonal transformations and projections in the $\boldsymbol{A}$- and $\boldsymbol{B}$-type procedures is that the cluster structure is often obtained in the context of a particular Euclidean coordinate system or metric. However, when the natural metric for the clusters is not simple Euclidean distance, more powerful displays may be attainable. For example, it may be appropriate to incorporate an estimate of the within-cluster variability,

$$\boldsymbol{W} = \frac{1}{(n-g)} \sum_{i=1}^{g} \sum_{j=1}^{n_i} (\boldsymbol{x}_{ij} - \bar{\boldsymbol{x}}_i)(\boldsymbol{x}_{ij} - \bar{\boldsymbol{x}}_i)',$$

directly into the analysis by rescaling the data by $\boldsymbol{W}^{-1/2}$ before calculating $\boldsymbol{A}$, $\boldsymbol{B}$, $\boldsymbol{B}^*$, or $\boldsymbol{B}^{**}$. The plot resulting from the rescaled $\boldsymbol{B}$ matrix is the usual discriminant analysis display.

A potential problem with the projection plots is that failure to account for more than two dimensions could result in incorrect and misleading impressions. For example, an observation could appear in the plot to be in the middle of a cluster and yet actually be far removed from it in the $(p-2)$-dimensional space not shown. To avoid giving such impressions, the size of the plotting character can be made to decrease with the distance of the observation from the plane of the projection. By studying the empirical distribution of these distances, one should be able to determine how to do this so that the smallest sizes are still legible and the intermediate sizes discriminate among the rest of the distances. For the applications in Section 3, the size decreases linearly with the squared distance to the plane but no characters are drawn smaller than 20% of the largest size.

3. Examples

The first data set involves 48 subsidiaries of a single parent company and is taken from Chen et al. (1975). Their goal was to cluster the subsidiaries based on the local environments in which they operate, and 13 clusters were formed using seven environmental-type variables. The clusters are denoted A through M; the objects within a cluster are numbered so that, for example, A1 refers to the first object in cluster A.

The local neighborhood plot for cluster J is shown in Fig. 2. Cluster J contains four subsidiaries which are located in Illinois, New Jersey, New York and Southern

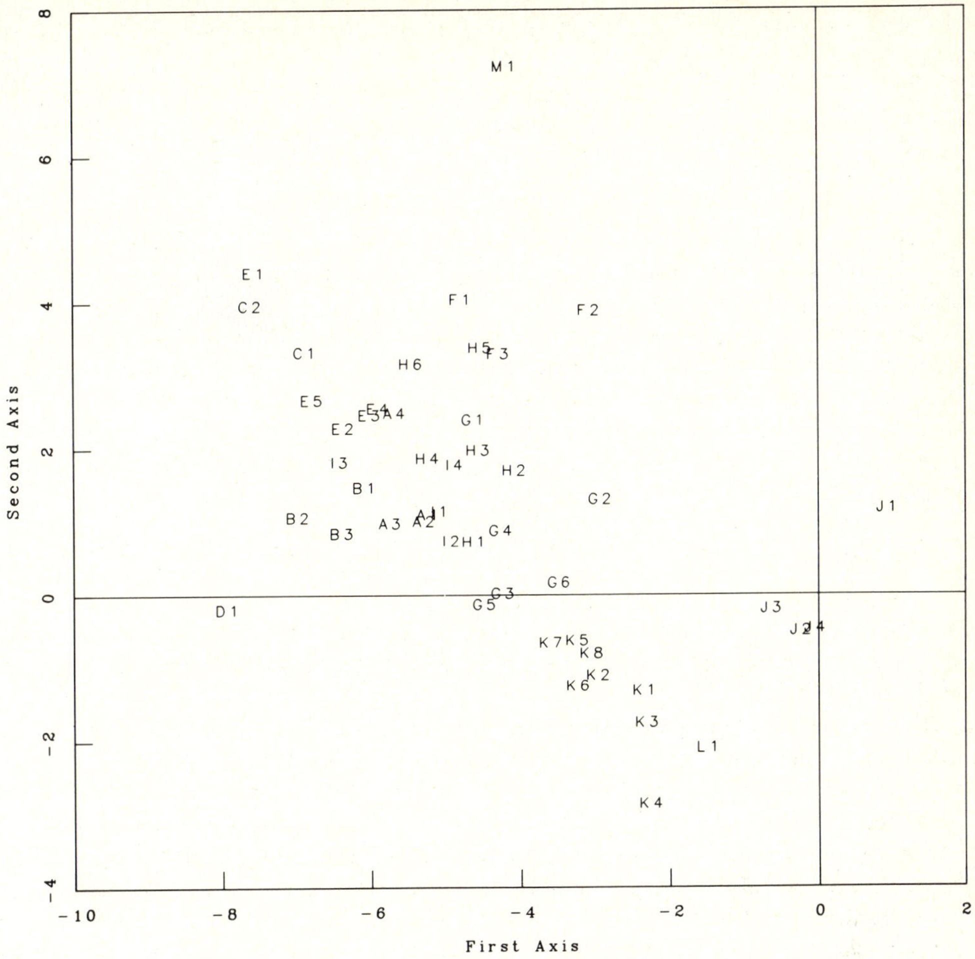

Fig. 2. Local neighborhood plot using equal character sizes (subsidiaries example).

California, all of which are highly industrialized, urbanized, and prosperous. Note that the plot itself is square with equal ranges on both axes. Thus distance as perceived by the eye corresponds to ordinary Euclidean distance in the space of these two components. Fig. 2 shows that the four members of cluster J are definitely separated in these dimensions from their nearest neighbors, which are the members of cluster K and the singleton cluster L1. Cluster K contains regions such as Pennsylvania, Maryland and Connecticut. It is interesting that this cluster is near to, yet distinct from, J, as these regions intuitively seem moderately industrialized and urbanized, but not quite as much as the members of J. Subsidiary L1 represents a small part of Indiana.

Fig. 3 shows the advantage of using variable character sizes, as discussed in Section 2. Evidently subsidiary L1 is not as near to cluster J as it appeared to be in

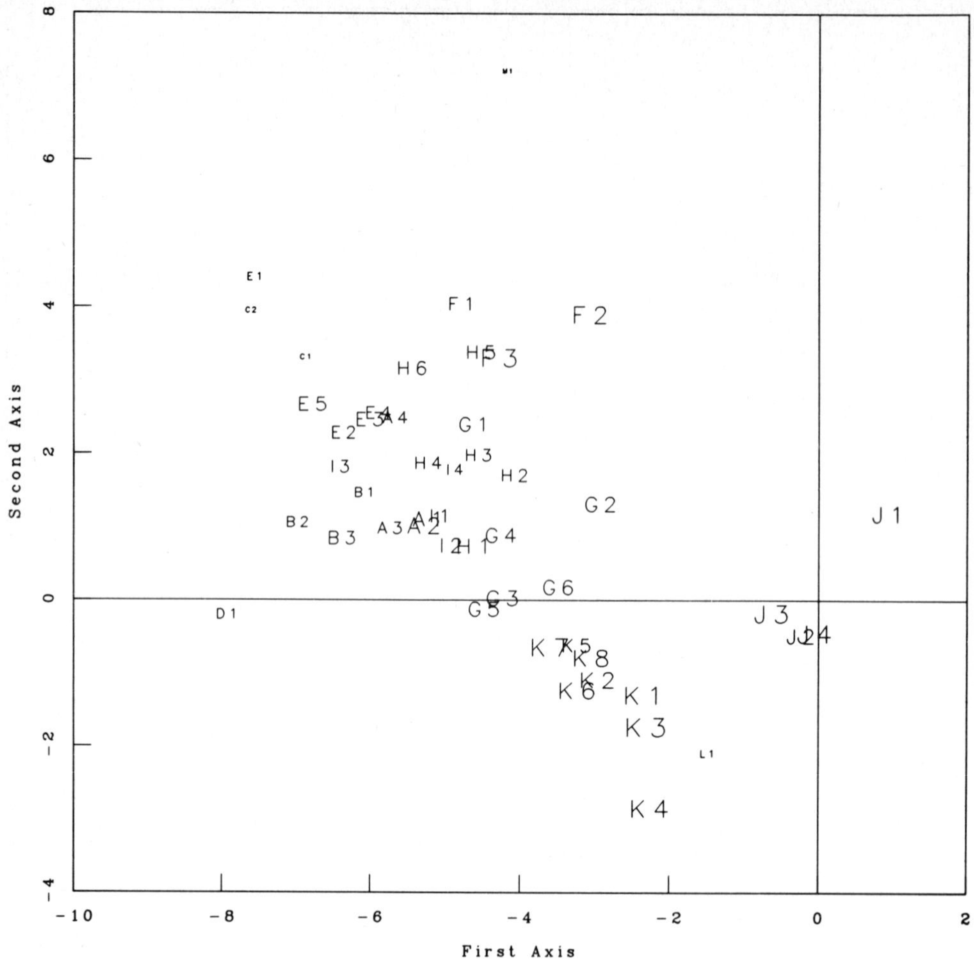

Fig. 3. Local neighborhood plot using variable character sizes (subsidiaries example).

Fig. 2. Moreover, the largeness of J1–J4 is reassuring since this means that the members of J are all close to the plane of display. For this plot $(\lambda_1+\lambda_2)/(\Sigma\lambda_i)$ is 87%, a high figure of merit.

Fig. 4 shows the ***B***-type plot for these data. Its main feature is the separation of the outlying singleton cluster M1 (Nevada) from the others (also note its position and size in the top central part of Figs. 2 and 3). This is exactly the situation in which a projection based on ***B**** may be helpful, as shown in Fig. 5. Now M1 is very small but not on the extreme edge. Clusters J and K are intermingled with each other, but the existence and structure of clusters G and C, in particular, are more apparent than before. Note that E1 and C1, representing Mississippi and North Dakota, are small and close together in Fig. 4, but not apparently so near in Fig. 5. This points out that adjacent objects with small-sized labels are not necessarily near

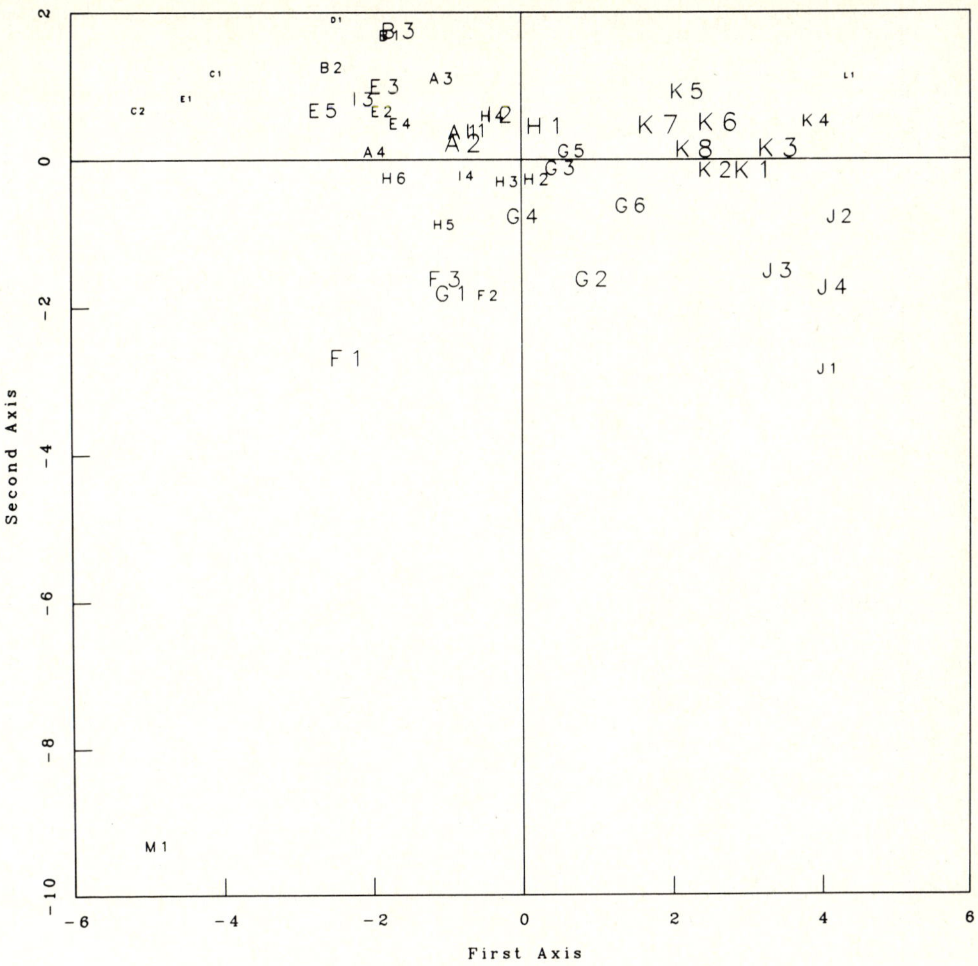

Fig. 4. ***B***-type plot (subsidiaries example).

each other in the entire space, as they can be far away from the plane in many different directions. Although the ***B****** plot does exhibit some interesting features of the data, it does not give a clear overall impression that there are 13 distinct clusters. This suggests that there is probably no two-dimensional projection of these data which would make all 13 clusters appear distinct.

The next series of projection plots uses a different set of data. One hundred ($=n$) tone ringer signals – electronically generated alternatives to the ordinary telephone bell – were rated by 43 ($=p$) subjects as to how much they liked each signal (Bricker, 1971). For present purposes, the tones are clustered into five groups, labeled A through E.

The projection derived from ***B*** is shown in Fig. 6. The impression is not one of strong clusters. However, Fig. 7, displaying the two-dimensional view of the

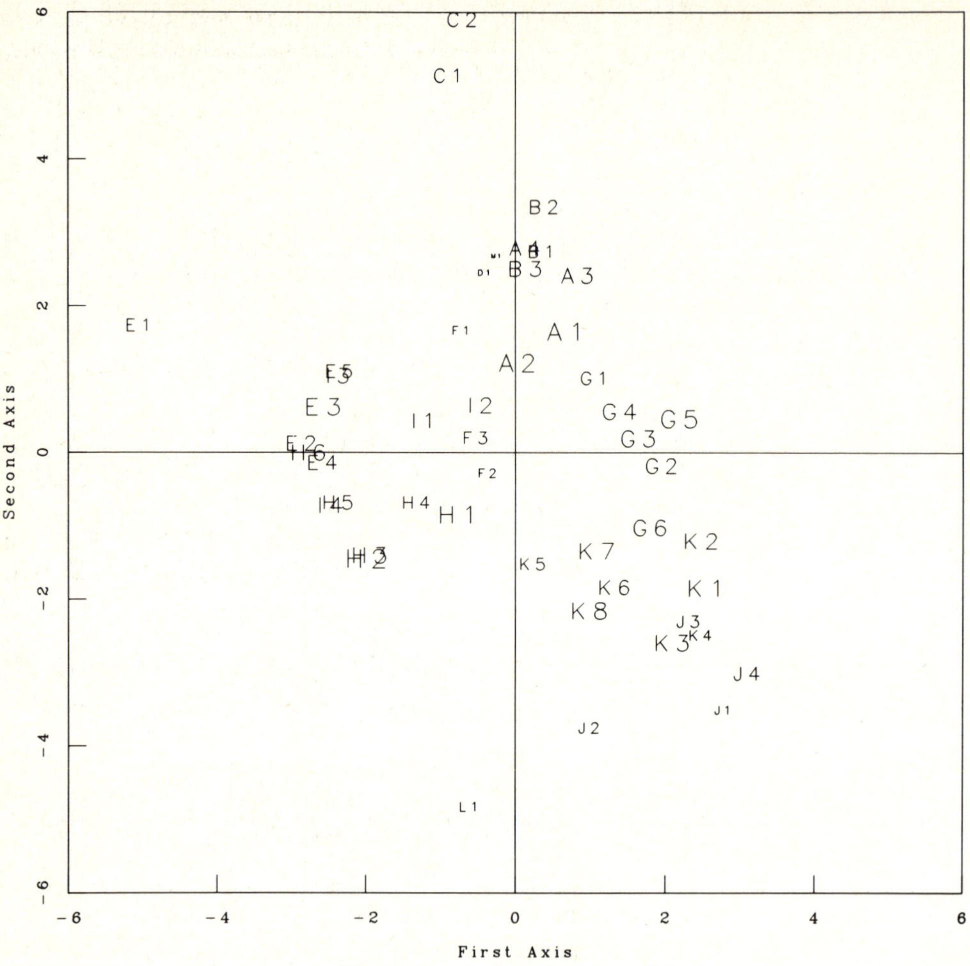

Fig. 5. $\boldsymbol{B}^*$-type plot (subsidiaries example).

43-dimensional space obtained by scaling by $\boldsymbol{W}^{-1/2}$ before calculating $\boldsymbol{B}$, does show much stronger evidence for these clusters. Cluster D, which contains the so-called "favorites" liked by many subjects, is very well separated. Furthermore, there is somewhat better separation between A and E, and also between B (a cluster of least-liked signals) and C. Most "variables" (people) here have medium-sized elements in the two eigenvectors going into this plot, and a noticeable fraction ($\approx \frac{1}{4}$) of the variables have relatively large elements.

The plot based on scaling by $\boldsymbol{W}^{-1/2}$ before calculating $\boldsymbol{B}^{**}$ is shown in Fig. 8. It differs very little from Fig. 7.

For this example, unlike the previous one, there are no clusters which require much down-weighting to achieve a balanced view of the separations among the clusters. The effect of rescaling by $\boldsymbol{W}^{-1/2}$ does, however, have a dramatic impact.

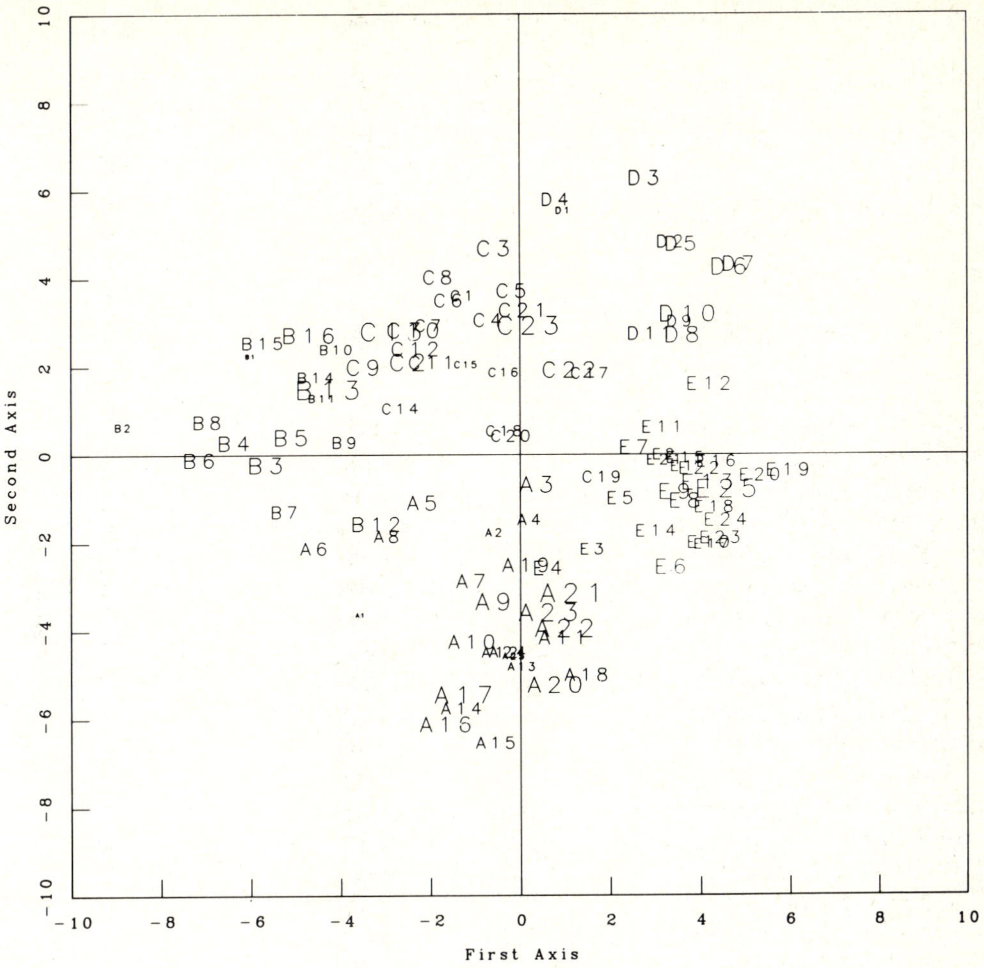

Fig. 6. ***B*-type plot (tone ringers example).**

4. Conclusions

This paper has described methods for obtaining two-dimensional orthogonal projections of objects that have been clustered into mutually exclusive sets. The choice of a projection is in part governed by whether the focus is on (i) a specific cluster and the desire is to study objects not in that cluster but in the immediate vicinity of it, or (ii) all clusters and the objective is to study them in terms of the relative spreads among the cluster centroids.

The essential features of the projection plots are:

(i) the choice of a matrix for performing an eigenanalysis;

(ii) the choice of a monotone nonincreasing function of distance for calculating weights for the data employed in forming this matrix; and

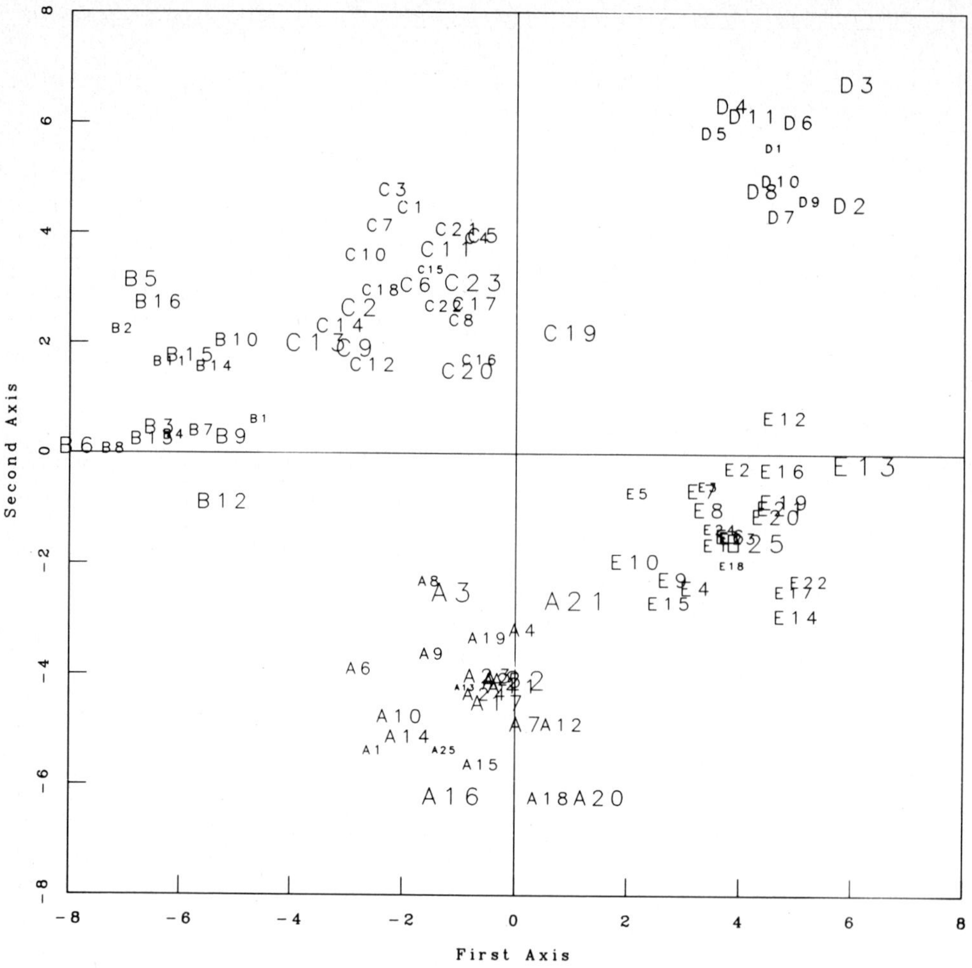

Fig. 7. $\boldsymbol{W}^{-1/2}$-scaled $\boldsymbol{B}$-type plot (tone ringers example).

(iii) the choice of label size for an object so as to indicate its distance away from the plane of projection.

The eigenanalyses are reminiscent of standard multivariate ones, such as principal components and multi-group discriminant analysis. The procedure of weighting observations so as to emphasize nearby ones is similar in spirit to ideas for developing robust estimates of multivariate dispersion (see, e.g., Gnanadesikan and Kettenring, 1972; Huber, 1977; Maronna, 1976).

In practice, trying a variety of projections and comparing the results would be a fruitful strategy. Since the computations involved are straightforward, one can indeed adopt such a strategy. Other authors (e.g., Friedman and Tukey, 1974; Friedman and Rafsky, 1979) have proposed different schemes for obtaining two-dimensional representations of the objects, and the area of multidimensional scaling

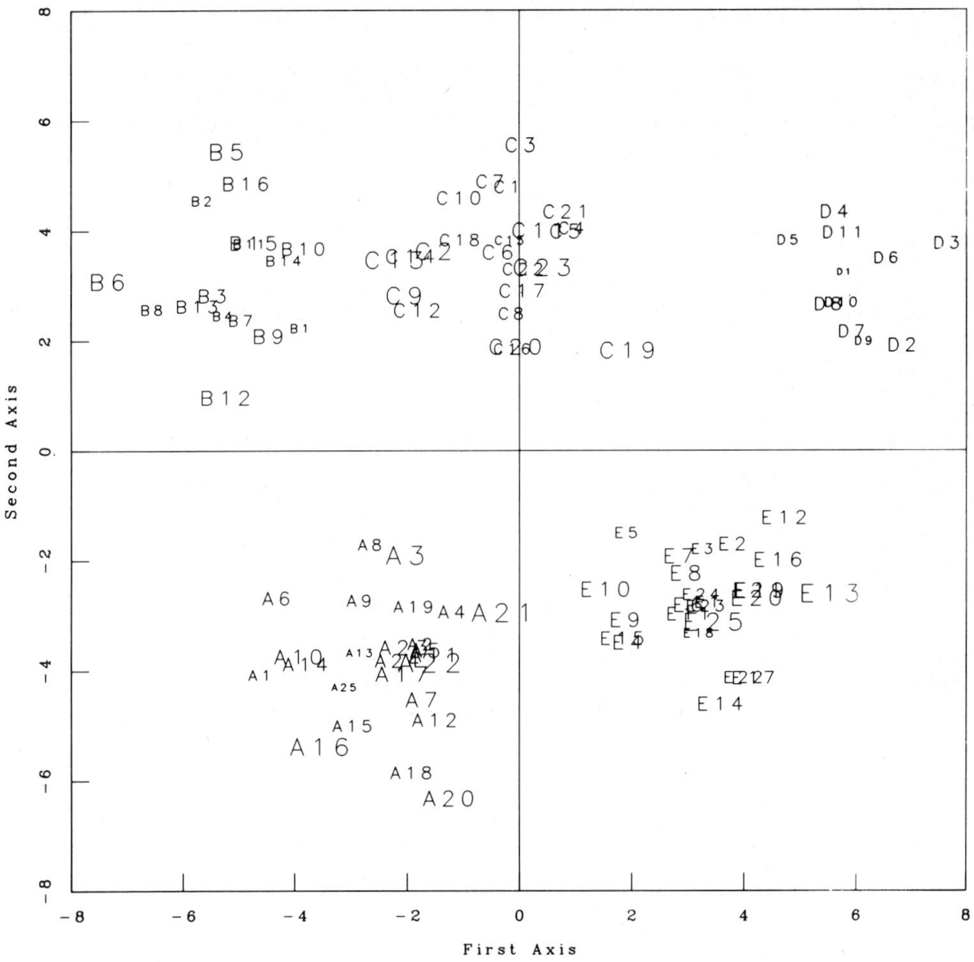

Fig. 8 $W^{-1/2}$-scaled B^{**}-type plot(tone ringers example).

shares similar concerns. (See also Andrews et al., 1973, for related ideas in a different context.) Some of these alternatives involve more complicated, iterative, algorithms than the ones used in this paper. Also some of them (e.g., Friedman and Tukey, 1974) are cluster-seeking techniques rather than post-clustering displays of the data. It would, however, be interesting to compare the relative performances of the different projection techniques systematically using both simulated and real data.

Within the context of the methods proposed in this paper, there are also several questions that can and should be investigated further. For instance, only one choice of the nonincreasing function of distance has been used for determining weights. Clearly other choices are feasible, and investigating the sensitivities of the different functions would be worthwhile. Although such investigations would shed more light on the methods and enhance their value, their very simplicity merits describing them for use by a wider audience at this stage.

References

Andrews, D.F., Gnanadesikan, R. and Warner, J.L. (1973). Methods for assessing multivariate normality. In Krishnaiah, P.R., Ed., *Multivariate Analysis-III*. Academic Press, New York, pp. 95–116.

Baker, F.B. and Hubert, L.J. (1975). Measuring the power of hierarchical cluster analysis. *J. Am. Statist. Assoc.* **70**, 31–38.

Blashfield, R.K. and Aldenderfer, M.S. (1978). The literature of cluster analysis. *Multiv. Behav. Res.* **13**, 271–295.

Bricker, P.D. (1971). Listener evaluation of simulated telephone calling signals. *Bell System Tech. J.* **50**, 1559–1578.

Chen, H., Dunn, D.M. and Landwehr, J.M. (1975). Grouping companies based on their operating environment. *Proc. Bus. Econ. Statist. Sect. Am. Statist. Assoc.*, pp. 278–283.

Friedman, J.H. and Rafsky, L.C. (1979). Multivariate generalizations of the Wald–Wolfowitz and Smirnov two-sample tests. *Ann. Statist.* **7**, 697–717.

Friedman, J.H. and Tukey, J.W. (1974). A projection pursuit algorithm for exploratory data analysis. *IEEE Trans. Computers* **C-23**, 881–890.

Gnanadesikan, R. and Kettenring, J.R. (1972). Robust estimates, residuals, and outlier detection with multiresponse data. *Biometrics* **28**, 81–124.

Gnanadesikan, R., Kettenring, J.R. and Landwehr, J.M. (1978). Interpreting and assessing the results of cluster analyses. *Bull. Int. Statist. Inst.* **47**(2), 451–463.

Huber, P.J. (1977). Robust covariances. In: Gupta, S.S. and Moore, D.S., Eds., *Statistical Decision Theory and Related Topics II*. Academic Press, New York, pp. 165–191.

Hubert, L. (1974). Approximate evaluation techniques for single-link and complete-link hierarchical clustering procedures. *J. Am. Statist. Assoc.* **69**, 698–704.

Maronna, R.A. (1976). Robust *M*-estimators of multivariate location and scatter. *Ann. Statist.* **4**, 51–67.

Rao, C.R. (1952). *Advanced Statistical Methods in Biometric Research*. Wiley, New York.

Rao, C.R. (1977). Cluster analysis applied to a study of race mixture in human populations. In: Van Ryzin, J., Ed., *Classification and Clustering*. Academic Press, New York, pp. 175–197.

G. Kallianpur, P.R. Krishnaiah, J.K. Ghosh, eds., *Statistics and Probability: Essays in Honor of C.R. Rao*

LIKELIHOOD PRINCIPLE AND RANDOMISATION

V.P. GODAMBE
University of Waterloo, Waterloo, Ontario, Canada

The likelihood principle and randomisation are studied in relation to the survey-sampling setup. A basic difficulty that arises in a thorough going implementation of the likelihood principle is illustrated. The meaning and the validity of the likelihood principle are discussed.

1. Introduction and notation

A survey-population, $\mathcal{P}$, is a collection of a finite number, say N, of *labelled* individuals i.

Definition 1.1. A survey-population $\mathcal{P}=\{i\}$, $i=1,\ldots,N$.

With each individual i of the population $\mathcal{P}$ is associated a variate value y_i, $i=1,\ldots,N$.

Definition 1.2. The population vector $\boldsymbol{y}=(y_1,\ldots,y_N)$.

The population vector $\boldsymbol{y}$ is supposed to be *unknown* and a complete survey of $\mathcal{P}$ to ascertain $\boldsymbol{y}$ is assumed to be impossible. Hence we are interested in inferring about the unknown population vector $\boldsymbol{y}$ or some function of $\boldsymbol{y}$ say the population mean.

Definition 1.3. The population mean $\bar{Y}(\boldsymbol{y})=\Sigma_1^N y_i/N$.

To infer about $\bar{Y}$ the values y_i are ascertained for individuals i belonging to a subset (sample) selected from $\mathcal{P}$.

Definition 1.4. Any subset s of $\mathcal{P}$, $s\subset\mathcal{P}$, is called a sample.

Now the sample s is selected by using a sampling design p.

Definition 1.5. If $S=\{s: s\subset\mathcal{P}\}$, a sampling design $p: S\to[0,1]$ such that for all $s\in S$, $\Sigma_S\, p(s)=1$.

In other words, s is selected by giving all samples S *preassigned* probabilities of selection $p(s)$, $s\in S$. Thus a sampling design is a *mode of randomisation*.

2. Sample space, parameter space and class of distributions

In the above (Def. 1.4), we have called s a subset of $\mathcal{P}$, a sample. This was done to keep up with the traditional notation. However, in the present case the total data consists of the selected subset s and the values $y_i : i \in s$, ascertained through the survey. Hence:

Definition 2.1. A "data-point" or a "sample-point" or for short simply "data" $\equiv (s, y_i : i \in s)$, where $s \in S$ (Def. 1.5) and $\boldsymbol{y}$ is given by Definition 1.2.

Therefore, what in the general statistical theory is called the *sample-space*, in the present context is given by the set of points

$$\{(s, y_i : i \in s) : A_i \leqslant y_i \leqslant A_i', i \in s, s \in S\}, \tag{2.1}$$

where A_i and A_i', $i = 1, \ldots, N$ are some known constants determined by the prior knowledge concerning the population $\mathcal{P}$ or more specifically, concerning the population vector (Def. 1.2) $\boldsymbol{y}$. (For instance if y_i denotes the income of the household i and the statistician knows a priori, i.e. before the conduct of the present survey, that no household in $\mathcal{P}$ has income less than \$2000 or greater than \$50000, he/she may put $A_i = 2000$ and $A_i' = 50000$, $i = 1, \ldots, N$.) To simplify the notation, however, we may replace in (2.1) A_i by $-\infty$ and A_i' by $+\infty$. Thus:

Definition 2.2.* The sample-space Υ is given by

$$\Upsilon = \{(s, y_i : i \in s), -\infty < y_i < \infty, i \in s, s \in S\}. \tag{2.2}$$

Of course, now the domain of the population vector $\boldsymbol{y}$ (Def. 1.2) is the entire R^N. To distinguish y_i of $\boldsymbol{y}$ from the y_i of the data (Def. 2.1) we would replace $\boldsymbol{y}$ by $\boldsymbol{y}' = (y_1', \ldots, y_i', \ldots, y_N')$. The purpose of this replacement will soon be obvious. Now the *probability distribution* on the sample-space Υ in (2.2) is completely determined by the sampling design p (Def. 1.5) and the population vector $\boldsymbol{y}'$. This "probability" is denoted by $\text{Prob}_{\boldsymbol{y}'}(\cdot)$:

$$\text{Prob}_{\boldsymbol{y}'}(s, y_i : i \in s) = \begin{cases} p(s), & \text{if } y_i = y_i', i \in s, \\ 0, & \text{otherwise}, \end{cases} \quad \boldsymbol{y}' \in R^N. \tag{2.3}$$

For *simplicity, on the left-hand side of* (2.3) *we have suppressed the notation for the sampling design p*. For a fixed sampling design, it is clear from (2.3) that $\boldsymbol{y}'$ completely determines the probability distribution "Prob" on the sample-space Υ in (2.2). Thus in the present context $\boldsymbol{y}'$ could be said to be the *parameter* and R^N the *parameter-space*. Hence, from (2.2) and (2.3), symbolically we can write the triple of sample-space, parameter space and class of distributions as follows:

$$(\Upsilon, R^N, \mathbb{P}), \mathbb{P} = \{\text{Prob}_{\boldsymbol{y}'} : \boldsymbol{y}' \in R^N\}. \tag{2.4}$$

*In this connection we refer to the discussion at the end of Section 8.

Following the usual notation of mathematical statistics, a general version of a triple (2.4) above, would be defined as an *experiment* M in the next section.

3. Likelihood principle

As it was mentioned in the last section a triple of sample-space, parameter space, and the class of distributions is called an experiment.

Definition 3.1. If $X=\{x\}$ is a sample-space $\Omega=\{\theta\}$ is a parameter space and P_θ is the probability distribution defined on X when θ obtains, then $M\equiv(X,\Omega,\mathbb{P})$ where $\mathbb{P}=\{P_\theta:\theta\in\Omega\}$, is called an experiment.

Remark 3.1. We consider experiments M with *discrete* distributions $\mathbb{P}$, only.

The outcome x of the experiment M is supposed to enable the statistician to make some inference about the unknown parameter θ. Now the likelihood principle relates to two experiments with a *common* parameter space.

Definition 3.2. Likelihood principle (LP). If for any two experiments M^r, $r=1,2$, $M^r=(X^r,\Omega,\mathbb{P}^r)$ where $X^r=\{x^r\}$, $\Omega=\{\theta\}$, $\mathbb{P}^r=\{P_\theta^r,\theta\in\Omega\}$, for some $x_0^1\in X^1$ and $x_0^2\in X^2$

$$P_\theta^1(x_0^1)=\text{constant}\,(x_0^1,x_0^2)P_\theta^2(x_0^2),\quad \theta\in\Omega, \tag{3.1}$$

then the two inferences about θ, one based on (M^1 and x_0^1) and the other based on (M^2 and x_0^2) must be identical. (Note as the notation in (3.1) indicates the "constant" of proportionality between $P_\theta^1(x_0^1)$ and $P_\theta^2(x_0^2)$ depends on x_0^1 and x_0^2).

The above formulation of the likelihood principle* (LP) closely follows Birnbaum (1962).

One feature of the above Definition 3.2 of LP is that it does *not* make any mention of the prior knowledge about the unknown parameter θ, on the part of the statistician performing the experiment M. The underlying *assumption* being that the pair (M and x) itself provides some inference about the unknown parameter θ and LP refers to this part of inference only. To avoid the above assumptions one may have

Definition 3.3. Modified likelihood principle (L′P′). If at any time a statistician can perform either of the two experiments M^1 and M^2 having a common parameter space Ω, $M^r=(X^r,\Omega,\mathbb{P}^r)$. $r=1,2$ such that $x_0^1\in X^1$ and $x_0^2\in X^2$ satisfy (3.1) of Definition 3.2 then the statistician's inference about θ if M^1 is performed and x_0^1 is observed should be the same as in case M^2 is performed and x_0^2 is observed.

*The original version due to Barnard (1947) and Barnard, Jenkins and Winsten (1962) is considerably restrictive.

Now for the convenience of discussion here, we would call a statistician a strict Bayesian (SB) if for him/her (i) all "prior knowledge" concerning an unknown parameter θ is *necessarily equivalent* to a prior distribution, say ξ on the parameter space Ω; (ii) at any given time, ξ is *common* for different experiments having the common parameter space $\Omega=\{\theta\}$. However, the prior distribution ξ may vary from one statistician to another; (iii) given any experiment M and an observation x, any inference about θ must depend on M and x only through the Bayes posterior distribution (of θ) conditional on x.

It is easy to see that the inferences of any SB-statistician satisfy the modified likelihood principle L′P′ (Def. 3.3).

4. Irrelevance of randomisation

The implications of the likelihood principle (LP or L′P′) (Defs. 3.2 and 3.3), in the context of survey-sampling are far reaching: LP or L′P′ imply that *after* the data $(s, y_i : i\in s)$ (Def. 2.1) is obtained any inference about the population vector $\boldsymbol{y}'$ must be *regardless* of what sampling design, (Def. 1.5), was used to select the sample (Def. 1.4), s! (Godambe 1966). At least apparently this conclusion contradicts the traditional sampling practice which involves use of sophisticated sampling designs (i.e., different modes of randomisation) to reduce the variance of some estimate. But before we discuss this practical contradiction we go through some details of logical implications of LP or L′P′ mentioned above.

Let p_1 and p_2 be two sampling designs (Def. 1.5) such that for a specified sample (Def. 1.4) s, $p_1(s)>0$ and $p_2(s)>0$. Let Prob^1 and Prob^2 denote the respective probabilities of the data $(s, y_i : i\in s)$ when sampling designs p_1 and p_2 are used to select s. Then from (2.3), introducing the prefix r for the sampling design p and the corresponding $\text{Prob}_y(\cdot)$, we have for $r=1,2$,

$$\text{Prob}^r_{y'}(s, y_i : i\in s)=\begin{cases} p_r(s), & \text{if } y_i=y'_i : i\in s, \\ 0, & \text{otherwise}, \end{cases} \qquad \boldsymbol{y}'\in R^N. \tag{4.1}$$

Hence,

$$\text{Prob}^r_{y'}(s, y_i : i\in s)=\text{constant}.\ \text{Prob}^2_{y'}(s, y_i : i\in s), \tag{4.2}$$

$\boldsymbol{y}'\in R^N$ where constant$=p_1(s)/p_2(s)$. Now consider the two experiments (Def. 3.1) M^r, $r=1,2$ obtained by putting in (2.4) for $\mathbb{P}, \mathbb{P}^r=(\text{Prob}^r_{y'} : \boldsymbol{y}'\in R)$ where $\text{Prob}^r_{y'}$ is as given by (4.1) above. Next substituting the experiments M^r, $r=1,2$ just mentioned in the Definition 3.2 and writing in it for x_0^1 and x_0^2, $(s, y_i : i\in s)$, from (3.1) and (4.2) we have the implication of LP or L′P′ (Defs. 3.2, 3.3): the inference about $\boldsymbol{y}'$ based on the data $(s, y_i : i\in s)$ should *not* depend whether s was selected by the sampling design p_1 or p_2.

Of course *before* drawing a sample the statistician could have preferences between different sampling designs based on the prior knowledge concerning the population $\mathcal{P}$: some samples (Def. 1.4) could be expected to give more accurate estimates than others. More formally, a strict Bayesian statistician – SB (Section 3) may choose a sampling design which minimises "a priori expectation" of some "posterior risk". But all these considerations can only lead to some set S_1 of samples such that each $s \in S_1$ is most preferable a priori, the samples $s \notin S_1$ being undesirable. [Ericson (1969), Godambe (1969), Zacks (1969).] All these considerations, therefore, cannot make "randomisation" in the usual sense of the word *compelling*; any (purposive) sampling design (Def. 1.5) choosing a sample say s_0, $s_0 \in S_1$ with probability 1, i.e. $p(s_0)=1$ would be *optimum*. It is true that some writers (Ericson 1969) have interpreted "randomisation" as means to justify the use of some exchangeable prior distribution within a *relaxed* Bayesian framework. But here we are concerned with the viewpoint of a *strict Bayesian* (SB) of Section 3. The only randomisation that is permissible (*yet superfluous*) from likelihood principle (LP, L′P′, Defs. 3.2, 3.3) considerations or from a strict Bayesian (SB) point of view is drawing with arbitrarily chosen probabilities a sample, s, from the set S_1 of all the *most preferable* samples a priori; that is drawing a sample s with *any* sampling design p such that $\Sigma_{S_1} p(s)=1$. For some additional related remarks we refer to Section 8.

To emphasize we again say, *after* the data $(s, y_i : i \in s)$ is at hand any inference about the population vector $\boldsymbol{y}'$ (Def. 1.2) which is consistent with the likelihood principle (LP, L′P′) must be *independent* of the sampling design used to select s: That is *the inference must depend exclusively on the data* $(s, y_i : i \in s)$, (Def. 2.1), *and the prior knowledge concerning the population* $\mathcal{P}$.

5. The prior knowledge

Here we would consider the *prior knowledge* about the unknown parameter, population vector (Def. 1.2), $\boldsymbol{y}'=(y'_1,\dots, y'_i,\dots, y'_N)$, which is equivalent to a specified class C of (prior) distributions ξ on the parameter space R^N. For some *known* positive numbers x_i, $i=1,\dots, N$, denoting by $\varepsilon_\xi(\cdot)$ the expectation under ξ,

$$C=\left\{\xi:\begin{array}{ll}\text{(i)} & y_1,\dots, y_N \text{ are independent when distributed as } \xi,\\ \text{(ii)} & \varepsilon_\xi(y_i)=\beta x_i, \quad i=1,\dots, N,\\ \text{(iii)} & \varepsilon_\xi(y_i-\beta x_i)^2=\sigma^2 f(x_i), \quad i=1,\dots, N, \text{ where}\\ & 0<\beta,\ \sigma^2<\infty, \text{ the function } f \text{ being known.}\end{array}\right\} \tag{5.1}$$

We would consider two cases of (5.1): *Case I* is obtained by putting in (5.1), $f(x_i)=x_i$ and for *Case II* we put $f(x_i)=x_i^2$. In each case we would try to obtain some plausible point estimate of the population mean, (Def. 1.3), $\overline{Y}=\Sigma_{i=1}^N y_i/N$ based on the data $(s, y_i : i \in s)$, (Def. 2.1). In this connection we refer to the last sentence of Section 4.

Case I. Here as said above we assume in (5.1), $f(x_i)=x_i$. Consider the class of all *linear unbiased* estimates, i.e. estimates, e, such that $e(s, y_i : i \in s)=\Sigma_{i\in s}\beta(s, i)y_i$, $\beta(s, i)$ being some real function of s and i for which conditional on the given s, $\varepsilon(e-\overline{Y})=0$ for all $\xi \in C$ (for simplicity we write ε for ε_ξ). In this class of unbiased linear estimate e, given s, $\varepsilon(e-\overline{Y})^2$ is minimised for all $\xi \in C$ for the estimate $e=\bar{e}$

given by,

$$\bar{e}(s, y_i: i \in s) = \bar{X}\hat{\beta}, \tag{5.2}$$

where $\bar{X} = \Sigma_1^N x_i / N$ and

$$\hat{\beta} = \sum_{i \in s} y_i \Big/ \sum_{i \in s} x_i, \tag{5.3}$$

(Royall 1971). Actually in (5.2) $\hat{\beta}$ is the usual *least squares* estimate of β. Similarly if we add to (i) of (5.1) assumption of *normality* of $y_1, \ldots, y_n$ and obtain from C a single *mixture distribution* say η, using the mixing distribution $\mathrm{d}\alpha\,\mathrm{d}\beta/\sigma$, then with respect to η, $\hat{\beta}$ given by (5.3) and $\bar{e}$ given by (5.2) are *Bayes estimates* for β and $\bar{Y}$, respectively, i.e. they are the corresponding means of the Bayes posterior distributions condition on the data $(s, y_i: i \in s)$ (Ericson 1969).

Case II. Here as said before we assume in (5.1), $f(x_i) = x_i^2$. Now as above the *least-squares* estimate of β and the *Bayes estimate* of β (for the latter assuming normality and the *mixture* distribution mentioned before) are given by

$$\hat{\hat{\beta}} = \frac{1}{n} \sum_{i \in s} \frac{y_i}{x_i}, \tag{5.4}$$

where n is the total number individuals i belonging to the subset s in the data $(s, y_i: i \in s)$ (Royall 1971, Ericson 1969). The estimate of $\bar{Y}$ that we would consider here is

$$\bar{\bar{e}}(s, y_i: i \in s) = \bar{X}\hat{\hat{\beta}} \tag{5.5}$$

where as before $\bar{X} = \Sigma_1^N x_i / N$.

Remark 5.1. Though $\bar{\bar{e}}$ in (5.5) has the same form as $\bar{e}$ in (5.2) *unlike* the latter the former is neither the least-square estimator nor the Bayes estimate of $\bar{Y}$. (See Remark 7.1 to follow in this connection.)

Here we emphasize that both the estimates $\bar{e}$ in (5.2) and $\bar{\bar{e}}$ in (5.5) are obtained exclusively on the basis of the corresponding data $(s, y_i: i \in s)$ and the prior knowledge; particularly, the estimates are obtained *regardless* of the consideration as to what sampling design p was used to select s. Hence, the uses of estimates $\bar{e}$ and $\bar{\bar{e}}$ in Cases I and II, respectively, are *consistent* with the likelihood principle LP or L′P′ (Defs. 3.2, 3.3), (see Section 4).

6. Randomisation provided by the Midzuno sampling design

We would call C in (5.1) a *model* of prior knowledge, i.e. a formalisation of prior knowledge. In this sense the model generally would *not* be exact. It is, therefore, realistic to consider the consequences of possible departures from the model C in (5.1) on our estimates given by (5.2) and (5.5). We would discuss the following departures from (ii) and (iii) of (5.1): Replace (ii) and (iii) by (ii)′ $\varepsilon_\xi(y_i) = \beta x_i + a_i$

and (iii)′ $\varepsilon_\xi(y_i-\beta x_i-a_i)^2=\sigma^2 f(x_i)$. Now following C in (5.1) we write for a given $\boldsymbol{a}=(a_1,\ldots,a_i,\ldots,a_N)$,

$$C(\boldsymbol{a})=\left\{\xi:\begin{array}{ll}\text{(i)} & \text{as (i) of (5.1)}\\ \text{(ii)} & \varepsilon(y_i)=\beta x_i+a_i,\quad i=1,\ldots,N\\ \text{(iii)} & \varepsilon(y_i-\beta x_i-a_i)^2=\sigma^2 f(x_i),\quad i=1,\ldots,N\\ & \text{where } 0<\beta,\sigma^2<\infty,\text{ the function,}\\ & f,\text{ being known.}\end{array}\right\} \tag{6.1}$$

Further we have,

$$C(D)=\bigcup_{\boldsymbol{a}\in D} C(\boldsymbol{a}). \tag{6.2}$$

Thus, $C(D)$ defines the (extended) model incorporating the possible departures given by $\boldsymbol{a}\in D$; D would be specified later but here we mention that it includes the 'origin' hence $C(D)\supseteq C$ in (5.1).

Remark 6.1. To avoid the possible confusion we emphasize that throughout the discussion we do *not* consider any probability distributions on the set D mentioned above in (6.2). It is very difficult if not impossible to think of a realistic prior distribution on $\boldsymbol{a}=(a_1,\ldots,a_N)$ partly because usually the survey population (Def. 1.1) size N, is very large (see Section 8).

Now as in Section 5, here Case I is given by $f(x_i)=x_i$ in (6.1). Thus for Case I, when model $C(\boldsymbol{a})$ in (6.1) obtains, following the discussion in Section 5 we obtain analogous least-squares and Bayes estimate $\hat{\beta}_{\boldsymbol{a}}$ of β as

$$\hat{\beta}_{\boldsymbol{a}}=\sum_{i\in s}(y_i-a_i)\Big/\sum_{i\in s}x_i=\hat{\beta}-\sum_{i\in s}a_i\Big/\sum_{i\in s}x_i, \tag{6.3}$$

where $\hat{\beta}$ is given by (5.3) of Section 5. Similarly corresponding to $\bar{e}$ of (5.2) we now, from (6.3), have for the estimate of the population mean, (Def. 1.3) $\bar{Y}$. the estimate

$$\bar{e}_{\boldsymbol{a}}=\bar{X}\hat{\beta}_{\boldsymbol{a}}+\frac{1}{N}\sum_1^N a_i=\bar{e}-\bar{X}\sum_{i\in s}a_i\Big/\sum_{i\in s}x_i+\frac{1}{N}\sum_1^N a_i. \tag{6.4}$$

Using (6.4) we write,

$$\bar{\Delta}(s,\boldsymbol{a})=\bar{e}-\bar{e}_{\boldsymbol{a}}=\bar{X}\sum_{i\in s}a_i\Big/\sum_{i\in s}x_i-\frac{1}{N}\sum_1^N a_i. \tag{6.5}$$

It is clear from the above discussion that if the departures from the model C in (5.1) correspond to the model $C(D)$ in (6.2), the least-squares and Bayes considerations in Section 5 would *not* provide a *unique* estimate for β or $\bar{Y}$; for, the estimates given by (6.3) and (6.4) depend on $\boldsymbol{a}$. Hence, when the model $C(D)$ obtains we ask the

Question:* Can we select a sample (Def. 1.4) s, with a sampling design (mode of randomisation) p (Def. 1.5), such that with overwhelmingly large probability (frequency) we arrive at an s for which in (6.5), $\bar{\Delta}\simeq 0$, i.e. $\bar{e}-\bar{e}_{\boldsymbol{a}}\simeq 0$? (6.6)

Let us suppose that the above question could (under some conditions) be answered *affirmatively* with a sampling design say p'. Then, if the model $C(D)$ obtains we could draw a sample s with p' and use the estimate $\bar{e}$ in (5.2) as a good approximation for $\bar{e}_{\boldsymbol{a}}$ in (6.4) which depends upon the unknown $\boldsymbol{a}$. Indeed this is possible if the set D in (6.2) is restrictively defined for the population $\mathcal{P}$ (Def. 1.1) under study as follows:

$$D=\left\{\boldsymbol{a}: \operatorname*{Max}_{i\in\mathcal{P}}|a_i|\leqslant K \text{ (known)}\right\}. \tag{6.7}$$

Evidently this would hold in many practical situations.

Now we consider a sampling design obtained by Midzuno sampling scheme which is as follows: From the population $\mathcal{P}=\{i\}$, $i=1,\ldots,N$, first draw is made giving each individual i probability of selection $x_i/N\bar{X}$, $i=1,\ldots,N$ [we assume in (5.1) and (6.1) all $x_i>0$]. If the individual i_0 is thus selected, from the remaining individuals, i.e. from $\mathcal{P}-\{i_0\}$, $n-1$ individuals are drawn by simple random sampling without replacement. Here, simple calculations would give,

$$p(s)=\begin{cases}\left({}^{N-1}C_{n-1}N\bar{X}\right)^{-1}\sum\limits_{i\in s}x_i, & \text{if } s \text{ contains } n \text{ individuals,}\\ 0, & \text{otherwise,}\end{cases}\qquad s\in S \tag{6.8}$$

(Def. 1.5). If the sampling design p, given by (6.8), is employed the expectation (E) and variance (V) of $\bar{\Delta}$ in (6.5) are given by

$$E_{\boldsymbol{a}}(\bar{\Delta})=0, \quad \text{for all } \boldsymbol{a}\in D \text{ in (6.7)}$$

and (6.9)

$$V_{\boldsymbol{a}}(\bar{\Delta})\leqslant\frac{\bar{X}}{x_m}\left(\frac{1}{n}-\frac{1}{N}\right)\frac{N}{N-1}K, \quad \text{for } \boldsymbol{a}\in D \text{ in (6.7)},$$

where $x_m=\mathrm{Min}(x_1,\ldots,x_N)$. *Thus* $V_{\boldsymbol{a}}(\bar{\Delta})$ *in* (6.9) *can be sufficiently small for suitably large sample size* n, assuming a sufficiently large population, i.e. large N.

Thus we are now in the position to answer the question (6.6). If the sampling design p (Def. 1.5) in (6.8) with n sufficiently large is used to select a sample (Def. 1.4), with a very large frequency (probability) one would get a sample, s, for which $\bar{\Delta}(s,\boldsymbol{a})$ in (6.5) is very small for all $\boldsymbol{a}\in D$ in (6.7). Hence the estimate $\bar{e}$ in (5.2) which was found *appropriate* under the model C in (5.1), *whatever the sampling design*,

*There are several previous ideas related to this explicitly formulated question. For instance, Yates (1960), Royall and Herson (1973) have emphasized the idea of balancing the sample. Godambe (1966, 1969), Scott et al. (1978) suggest choosing a sampling design to make, some Bayes or other estimate based exclusively on the prior knowledge and the data, (frequency) unbiased.

continues to be appropriate or nearly so under the extended model $C(D)$ defined by (6.2) and (6.7) *provided* the sample s is drawn using the sampling design given by (6.8) with suitably large n. A more pronounced effect of the sampling design is numerically illustrated in the next section.

7. Randomisation with specified inclusion probabilities

In Section 6 we discussed, in relation to the models (5.1) and (6.2) Case I, given by $f(x_i)=x_i$. Here, in a similar manner, we discuss Case II given by $f(x_i)=x_i^2$. Following the discussion in Sections 5 and 6 it will be clear that, when $f(x_i)=x_i^2$ under the model $C(\boldsymbol{a})$ in (6.1) the least-squares and the Bayes estimate (the latter under the assumption of normality and mixture as in Sections 5,6) for β are the same: $\hat{\hat{\beta}}_{\boldsymbol{a}}$ is obtained by replacing in $\hat{\hat{\beta}}$ of (5.4) y_i by y_i-a_i. That is,

$$\hat{\hat{\beta}}_{\boldsymbol{a}}=\frac{1}{n}\sum_{i\in s}\frac{y_i-a_i}{x_i}. \tag{7.1}$$

Similarly now under the model $C(\boldsymbol{a})$ the estimate for the population mean $\overline{Y}$, say $\bar{\bar{e}}_{\boldsymbol{a}}$ is obtained by replacing in $\bar{\bar{e}}$ of (5.5) $\hat{\hat{\beta}}$ by $\hat{\hat{\beta}}_{\boldsymbol{a}}$ in (7.1). Thus:

$$\bar{\bar{e}}_{\boldsymbol{a}}=\overline{X}\hat{\hat{\beta}}_{\boldsymbol{a}}+\frac{1}{N}\sum_{1}^{N}a_i. \tag{7.2}$$

Now analogous to (6.5) of Section 6, we write

$$\bar{\bar{\Delta}}(s,\boldsymbol{a})=\bar{\bar{e}}-\bar{\bar{e}}_{\boldsymbol{a}}=\frac{\overline{X}}{n}\sum_{i\in s}\frac{a_i}{x_i}-\frac{1}{N}\sum_{1}^{N}a_i. \tag{7.3}$$

From (7.2) it follows that under model $C(D)$ of (6.2) we do *not* have a *unique* estimate for $\overline{Y}$ (Def. 1.3), corresponding to $\bar{\bar{e}}$ of (5.5). In view of the similar non-uniqueness for Case I, $f(x_i)=x_i$, in Section 6 we raised the question (6.6). The question we now ask is obtained by replacing in (6.6) $\overline{\Delta}$ by $\bar{\bar{\Delta}}$ of (7.3):

> *Question:* Can we select a sample (Def. 1.4) s, with a sampling design (mode of randomisation) p (Def. 1.5.), such that with overwhelmingly large probability (frequency) we arrive at an s for which in (7.3), $\bar{\bar{\Delta}}\simeq 0$ and hence $\bar{\bar{e}}-\bar{\bar{e}}_{\boldsymbol{a}}\simeq 0$? (7.4)

To construct a sampling design providing a positive answer to the question (7.4) we need to define for any sampling design p the inclusion probabilities $\Pi_i(p)$

$$\Pi_i(p)=\sum_{s\ni i}p(s), \tag{7.5}$$

for individuals $i=1,\ldots,N$. Now consider a sampling design p for which in (7.5)

$$\Pi_i(p)\propto x_i,\quad i=1,\ldots,N. \tag{7.6}$$

For a design satisfying (7.6) it is easy to see that the expectation (E) of $\bar{\bar{\Delta}}$ in (7.3) is zero; specifically

$$E_a(\bar{\bar{\Delta}})=0 \quad \text{for all } \boldsymbol{a}\in D \text{ in (6.7)}. \tag{7.7}$$

Actually it can be easily shown that for a stratified unequal probability sampling design, p, satisfying (7.6), with the number of strata equal to the sample size n,

$$V_a(\bar{\bar{\Delta}})\leqslant 2\left(\frac{1}{n}-\frac{1}{N}\right)\left(\frac{K}{x_m}\right)^2, \quad \boldsymbol{a}\in D \text{ in (6.7)}, \tag{7.8}$$

where as before, $x_m=\min(x_1,\ldots,x_N)$ and for all $\boldsymbol{a}\in D$, and $a_i/x_i\leqslant K/x_m$, $i=1,\ldots,N$. That is for sufficiently large n and N

$$V_a(\bar{\bar{\Delta}})\leqslant \text{a small constant}, \quad \boldsymbol{a}\in D. \tag{7.9}$$

Now the sampling design (Def. 1.5) described in the above paragraph because of (7.7) and (7.9) would produce with large frequency (probability) samples s (Def. 1.4) for which $\bar{\bar{\Delta}}=\bar{\bar{e}}-\bar{\bar{e}}_a$ in (7.3) is nearly zero for all $\boldsymbol{a}\in D$. This answers the question (7.4).

As we have noted in Section 5, when the model C in (5.1) with $f(x_i)=x_i^2$, obtains *whatever the sampling design* an *appropriate* estimate for the population mean $\bar{Y}$ is given by the estimate $\bar{\bar{e}}$ in (5.5). From the preceding paragraph we note that even under the extended model $C(D)$ given by (6.2) and (6.7) the estimate $\bar{\bar{e}}$ continues to be *nearly* appropriate *provided* the sample (Def. 1.4) s is selected with the above described sampling design.

In a numerical example we had $N=1000$. The variations of x_i were between 10 and 50 and those of $|a_i|$ were between 1 and 100. That is $10\leqslant x_i\leqslant 50$, $1\leqslant|a_i|\leqslant 100$, $i=1,\ldots,1000$; $\beta=5$. For this example, the mean squared errors $E_a(\bar{\bar{\Delta}})^2$ were computed on the basis of two sampling designs: (I) This design p, is the same stratified sampling as the one described in this section, with $n=100$. Here the variance $E_a(\bar{\bar{\Delta}}^2)=V_a(\bar{\bar{\Delta}})=0.24$. (II) This is usual simple random sampling without replacement, with $n=100$. Here we had $E_a(\bar{\bar{\Delta}})^2=2.30$. The difference between the two mean squared errors namely 0.24 and 2.30 clearly illustrates how the sampling design p tends to produce more often samples with small $\bar{\bar{\Delta}}$ than the design given by the simple random sampling. Following are the actual estimates for 25 samples, each of size $n=100$.

Sampling design p			Simple random sampling		
$\bar{\bar{e}}$	$\bar{\bar{e}}_a$	$\bar{\bar{\Delta}}$	$\bar{\bar{e}}$	$\bar{\bar{e}}_a$	$\bar{\bar{\Delta}}$
5.07	5.44	−0.36	4.86	2.73	2.13
4.91	4.23	0.68	5.08	6.29	−1.20
5.07	4.38	0.69	5.11	2.31	2.80
5.05	5.27	0.21	5.05	4.38	0.66
4.92	5.69	0.77	4.99	5.69	−0.69
5.12	5.50	0.38	5.00	3.08	1.91
4.97	4.49	0.47	5.03	3.71	1.31
4.86	5.49	0.63	4.91	6.56	−1.65

Sampling design p			Simple random sampling		
$\bar{\bar{e}}$	$\bar{\bar{e}}_a$	$\bar{\bar{\Delta}}$	$\bar{\bar{e}}$	$\bar{\bar{e}}_a$	$\bar{\bar{\Delta}}$
4.98	4.77	0.20	5.05	4.70	0.34
5.05	4.81	0.24	5.06	3.98	1.08
4.90	5.55	−0.64	5.05	5.45	−0.40
5.09	5.28	−0.18	5.07	3.59	1.48
5.08	4.95	0.13	4.87	5.76	−0.89
5.12	4.45	0.67	5.11	6.73	−1.62
5.03	5.12	−0.09	5.17	4.23	0.93
4.87	4.79	0.08	5.02	7.83	−2.81
4.96	5.07	−0.10	5.06	4.71	0.34
5.02	6.04	−1.02	5.23	3.68	1.54
5.02	4.71	0.30	5.14	6.42	−1.28
4.93	4.76	0.17	4.87	5.39	−0.51
4.98	5.72	−0.73	4.93	3.14	1.79
4.99	4.71	0.28	4.92	5.40	−0.48
5.00	5.04	−0.04	5.06	7.64	−2.58
5.04	4.74	0.30	5.02	3.57	1.44
5.01	5.71	−0.70	5.03	4.15	0.88

Remark 7.1. Following Remark 5.1 we note that the least-squares and Bayes estimates for $\bar{Y}$ corresponding to (5.5) and (7.2) are respectively given by $\bar{\bar{e}}_1 = (n\bar{y} + \hat{\hat{\beta}}(N\bar{X} - \Sigma_{i\in s} x_i))/N$ and $\bar{\bar{e}}_{1\boldsymbol{a}} = (\Sigma_{i\in s}(y_i - a_i) + \hat{\hat{\beta}}_{\boldsymbol{a}}(N\bar{X} - \Sigma_{i\in s} x_i) + \Sigma_1^N a_i)/N$ where $\hat{\hat{\beta}}$ and $\hat{\hat{\beta}}_{\boldsymbol{a}}$ are the same as in (5.4) and (7.1). In contrast to the estimate $\bar{\bar{e}}$ discussed above it can be shown that for no fixed sample size design $E_{\boldsymbol{a}}(\bar{\bar{e}}_1 - \bar{\bar{e}}_{1\boldsymbol{a}}) = 0$, $\boldsymbol{a} \in D$ and hence the consequent problem concerning the robustness of $\bar{\bar{e}}_1$ when departures from (C) in (5.1) are characterized by $C(D)$ in (6.2).

8. A discussion

As we have seen in Section 4, a *strict* adherance to the likelihood principle LP or L′P′ of Section 3 precludes an interpretation for the randomisation as practised in survey-sampling. Particularly LP and L′P′ provides no interpretation for the sampling designs discussed in Sections 6 and 7. Even there is, at least, an apparent *conflict* with the likelihood principle LP involved here: For our conclusions at the end of Sections 6 and 7 are that if the model $C(D)$ given by (6.2) and (6.7), obtains, the estimates $\bar{e}$ in (5.2) and $\bar{\bar{e}}$ in (5.5) can be used (when $f(x_i) = x_i$ and $f(x_i) = x_i^2$, respectively) *only when* the sample s is drawn with correspondingly appropriate sampling design (mode of randomisation). Does this not mean that in the above two situations inferences *after* the data $(s, y_i : i \in s)$ (Def. 2.1), is obtained, *depend* on the sampling designs used? Indeed some supporters of the likelihood principles including some Bayesians, for whatever reasons they may have accepted simple random sampling (see Section 4) up to stratification, have *denounced* the use of additional or different *unequal probability sampling* involved by the sampling designs discussed in Sections 6 and 7. For instance we quote Basu (1971, p. 234)

> "However in survey practice, situations will occasionally arise where it will be necessary to insist upon a random sample. But this will only be to safeguard against some unknown biases. *In no situation is it possible to make any sense of unequal probability sampling.*"

(Italics due to the present author.)

One may view the above difficulty of relating the likelihood principle to randomisation as follows: Consider *all the experiments* (Def. 3.1) as in (2.4), obtained by varying the sampling designs p. They have the common sample-space Υ in (2.2). Now when the model $C(D)$ defined by (6.2) and (6.7) with $f(x_i)=x_i$ $[f(x_i)=x_i^2]$ obtains, *we assert* that only some *exceptional* points of Υ enable any inference at all about the unknown $\overline{Y}$. These exceptional points, referring to (6.5) [(7.3)] are characterized by $\bar{\Delta}=0$ $[\bar{\bar{\Delta}}=0]$ for all $\boldsymbol{a}\in D$. Hence one would have liked to choose from *all* the experiments mentioned above an *ideal experiment* (if it existed) for which,

$$\text{Probability}(\bar{\Delta}=0)=1. \qquad \left[\text{Probability}(\bar{\bar{\Delta}}=0)=1\right]. \tag{8.1}$$

The *choice* of such an experiment would be quite in line with the universal practice of the statisticians of choosing an experiment with fewer nuisance parameters. For not even a super ingenious Bayesian can find a reasonable joint prior distribution for hundreds of nuisance parameters which at least apparently are arbitrarily fixed constants (see Remark 6.1). Of course, the act of choosing an experiment from those that are available, does *not* contradict the likelihood principle. Now the above mentioned ideal experiment characterised by (8.1) is equivalent to an (*ideal*) *sampling design* (Def. 1.5) satisfying,

$$\sum_{S_0} p(s)=1, \quad \text{where } S_0=\left\{s:\bar{\Delta}(s,\boldsymbol{a})=0\left[\bar{\bar{\Delta}}(s,\boldsymbol{a})=0\right], \boldsymbol{a}\in D\right\}, \tag{8.2}$$

where $\bar{\Delta}$ and $\bar{\bar{\Delta}}$ are as before given by (6.5) and (7.3) and D by (6.7). We refer to Section 4, for some related comments. Now an ideal/sampling design satisfying (8.2) and hence an ideal experiment satisfying (8.1), obviously does *not* exist; for *no* sample (Def. 1.4) s, for all $\boldsymbol{a}\in D$ in (6.7),

$$\sum_{i\in s} a_i \Big/ \sum_{i\in s} x_i = \sum_1^N a_i \Big/ \sum_1^N x_i \left[\sum_{i\in s} a_i/x_i = n\sum_1^N a_i \Big/ \sum_1^N x_i\right].$$

But a good approximation to the ideal sampling design (8.2) is provided by the design described in Section 6 [Section 7]. For this latter sampling design, as we observed in Section 6 [Section 7], for suitably small ϵ and η if

$$S_0'(\boldsymbol{a})=\left\{s:|\bar{\Delta}(s,\boldsymbol{a})|\leqslant\varepsilon\left[|\bar{\bar{\Delta}}(s,\boldsymbol{a})|\leqslant\varepsilon\right]\right\},$$

$$\sum_{S_0'(\boldsymbol{a})} p(s)\geqslant 1-\eta, \quad \boldsymbol{a}\in D, \tag{8.3}$$

due to (6.9) [(7.7), (7.9)] assuming the sample size n to be sufficiently large.

In relation to (8.2) and (8.3) it is interesting to note that if D in (6.7) is replaced by some exceptional set D_1 one can find a sample s such that $|\bar{\Delta}(s, \boldsymbol{a})| < \varepsilon$ $[|\bar{\bar{\Delta}}(s, \boldsymbol{a})| < \varepsilon]$ for all $\boldsymbol{a} \in D_1$. For instance, assuming $x_i = \text{const.}$, $i = 1, \ldots, N$, if D_1 is such that for all $\boldsymbol{a} \in D_1$, $\Sigma_1^n a_i/n = \Sigma_1^N a_i/N$ then for $s = \{i\}$, $i = 1, 2, \ldots, n$, $\bar{\Delta}(s, \boldsymbol{a}) = \bar{\bar{\Delta}}(s, \boldsymbol{a}) = 0$. For any finite set D_1, *in principle* we can verify whether or not, for a given s, the *condition* $|\bar{\Delta}(s, \boldsymbol{a})| < \varepsilon$ $[|\bar{\bar{\Delta}}(s, \boldsymbol{a}| < \varepsilon]$ for all $\boldsymbol{a} \in D_1$ is satisfied. Yet to find a sample s, *if at all one exists*, satisfying the condition would generally, for large N and n, be a formidable job. Hence the recommendation to use a sampling design satisfying (8.3).*

Thus, from the discussion above and that in Section 4 we conclude that the employment of an ideal sampling design (8.2), if it existed would *not* have been in conflict with the likelihood principle. However, employment of a sampling design (8.3) which provides a "close approximation" to the "ideal" does *contradict* the likelihood principle. Yet the approximation (8.3) and ideal (8.2) in some situations (when n in (6.9) and (7.8) is sufficiently large) would be *practically indistinguishable*. This then could be considered as an essential *defect* of the likelihood principles LP, L′P′ (Defs. 3.2, 3.3).

Alternatively one may maintain that for the survey-sampling setup likelihood principle is *not* applicable for the following considerations. With the Definition 3.1, we called (2.4) an experiment; Υ in (2.2) being the sample-spaces, $R^N = \{\boldsymbol{y}'\}$ (Def. 1.2) being the parameter-space. Here one observes that when actually a specific parameter value say $\boldsymbol{y}'$ obtains, the only *logically possible* sample-points (Def. 2.1) are the ones given by $\{(s, y_i' : i \in s) : s \in S\}$. That is any point $(s, y_i : i \in s)$ with $y_i \neq y_i'$ for some $i \in s$, is a *logically impossible point*. For such a point, however, in (2.4) we have $\text{Prob}_{\boldsymbol{y}'}(s, y_i : i \in s) = 0$. Now the difficulties that arise due to equating such logically impossible points with zero probability points are discussed elsewhere (Godambe 1975). To avoid these difficulties one course of action would be to abandon the "fixed sample-space Υ in (2.2)" and say as suggested above that the sample-space depends on the parameter: Given $\boldsymbol{y}'$ the sample space is $\{(s, y_i' : i \in s) : s \in S\}$. Hence the likelihood principle (LP) (Def. 3.2) is *not* applicable to the survey-sampling set-up. The other possible course of action would be to limit applicability of likelihood principle to experiments (Def. 3.1) M for which $P_\theta > 0$, $x \in X$, $\theta \in \Omega$. With this limitation again application of the likelihood principle to the experiment (2.4) is ruled out.

Acknowledgements

The author acknowledges with pleasure some valuable discussions he had with Professors D. Blackwell and O. Kempthorne during his visits to Berkeley and Ames.

*Again for some exceptional set D_1, with a rare chance, the sample s, so drawn can be found to be "bad" in the sense, $|\bar{\Delta}(s, \boldsymbol{a})| > \varepsilon$ $[|\bar{\bar{\Delta}}(s, a)| > \varepsilon]$ for *all* $\boldsymbol{a} \in D_1$. Such a "bad" sample will then have to be replaced by another draw.

References

Barnard, G.A. (1947). Sequential Analysis (book review). *J. Amer. Statist. Assoc.* **42**, 658–664.

Barnard, G.A., Jenkins, G.M. and Winsten, C.B. (1962). Likelihood inference and time series (with discussion), *J. Roy. Statist. Soc. A* **125**, 321–372.

Basu, D. (1971). An Essay on the Logical Foundations of Survey-Sampling. In: *Foundations of Statistical Inference.* Godambe, V.P. and Sprott, D.A. Eds. Holt, Rinehart and Winston of Canada, Toronto.

Birnbaum, Allan (1962). On foundations of statistical inference (with discussion), *J. Amer. Statist. Assoc.* **57**, 269–326.

Ericson, W.A. (1969). Subjective Bayesian models in sampling finite populations (with discussion). *J. Roy. Statist. Soc. B* **31**, 195–233.

Godambe, V.P. (1966). A new approach to sampling from finite populations, *J. Roy. Statist. Soc. B* **17**, 269–278.

Godambe, V.P. (1969). Some aspects of theoretical developments in survey-sampling. In: *New Developments in Survey-Sampling.* Johnson, N.L. and Smith, H. Eds. Wiley Interscience, New York.

Godambe, V.P. (1975). A reply to my critics, *Sankhyā, Ser. C.* **37**, 53–76.

Royall, R.M. (1971). Linear regression models in finite population sampling theory. In: *Foundations of Statistical Inference.* Godambe, V.P. and Sprott, D.A. Eds. Holt, Rinehart and Winston of Canada, Toronto.

Royall, R.M. and Herson, J. (1973). Robust estimation in finite populations, I. *J. Amer. Statist. Assoc.* **68**, 880–889.

Scott, A., Brewer, K.R.W. and Ho, E.W.H. (1978). Finite population sampling and robust estimation, *J. Amer. Statist. Assoc.* **73**, 359–361.

Yates, F. (1960). *Sampling Methods for Censuses and Surveys*, 3rd Ed. Griffin, London.

Zacks, S. (1969). Bayes sequential designs for sampling finite populations, *J. Amer. Statist. Assoc.* **64**, 1342–49.

G. Kallianpur, P.R. Krishnaiah, J.K. Ghosh, eds., *Statistics and Probability: Essays in Honor of C.R. Rao*
© North-Holland Publishing Company (1982) 295–305

ON THE LEAST FAVORABLE CONFIGURATIONS IN CERTAIN TWO-STAGE SELECTION PROCEDURES*

Shanti S. GUPTA
Department of Statistics, Purdue University, U.S.A.

Klaus-J. MIESCKE
Department of Mathematics, University of Mainz, Federal Republic of Germany

The problem of finding the least favorable configuration for selecting the "best" of k populations, i.e. the one with the largest location parameter by use of six different two-stage selection procedures is considered. Each of the six procedures consists of a subset selection (screening) rule at the first stage followed by another rule based on (the first stage and) additional samples from the selected populations to decide finally which of the selected populations is the best. In the indifference-zone approach it is (or was) conjectured that the least favorable parameter configuration is of the slippage type. It is shown that this conjecture is true for four of these procedures. For a fifth procedure it is proved that at least a certain lower bound of the probability of a correct selection has this property which is analogous to the result of Tamhane and Bechhofer (1979) concerning the sixth procedure.

1. Introduction

Suppose we are given k normal populations $\pi_1, \ldots, \pi_k$ with different unknown means and a common (known or unknown) variance. If the experimenter's goal is to find that population having the largest mean by using suitably chosen samples, then a large variety of possible sampling plans and selection procedures can be found in the literature. In this paper we are dealing with the so-called two-stage procedures of the following type:

Stage 1: Take k independent samples $(X_{i1}, \ldots, X_{in})$ of size n, $i=1, \ldots, k$, from $\pi_1, \ldots, \pi_k$ and select a non-empty subset of these populations according to a pre-specified rule $S(\boldsymbol{X})$ where $\boldsymbol{X}=(X_1, \ldots, X_k)$ and $X_i = X_{i1} + \cdots + X_{in}$, $i=1, \ldots, k$. If the resulting subset consists of only one population, stop and decide that this is the population with the largest mean. Otherwise proceed to Stage 2.

Stage 2: Take additionally independent samples of size m $(Y_{i1}, \ldots, Y_{im})$ from those populations π_i selected in Stage 1. Among the selected populations decide finally in favor of that population yielding the largest Y_i (or $X_i + Y_i$), where $Y_i = Y_{i1} + \cdots + Y_{im}$, $i=1, \ldots, k$.

For convenience, let us represent the rules for Stage 1 in the form $S\colon \mathbb{R}^k \to \{s \mid \emptyset \neq s \subseteq \{1, \ldots, k\}\}$ where $i \in S$ means that the ith population is included in the subset of

*This research was supported by the Office of Naval Research contract N00014-75-C-0455 at Purdue University. Reproduction in whole or in part is permitted for any purpose of the United States Government.

selected populations. Moreover, let us represent the rules for Stage 2 in the form $d=\{d_s|\emptyset\neq s\subseteq\{1,\ldots,k\}\}$, where for every s, d_s: $\mathbb{R}^{2k}\to\{1,\ldots,k\}$ and $d_s(\boldsymbol{\xi},\boldsymbol{\eta})$ depends only on those ξ_i's and η_i's with $i\in s$.

Let us now study in more detail the four possible two-stage procedures (S_α, d_β), $\alpha,\beta=1,2$ which we get after combining any two of the different single-stage procedures given below [(S_3, d_1) and (S_3, d_2) will be discussed at the end of this section].

Stage 1: For $i\in\{1,\ldots,k\}$ let:
$i\in S_1(\boldsymbol{X})$ iff $X_i\geqslant\max_{j=1,\ldots,k}X_j-c$; $c>0$ fixed,
$i\in S_2(\boldsymbol{X})$ iff X_i is one of the t largest values of $X_1,\ldots,X_k$; $t\in\{2,\ldots,k-1\}$ fixed.
[$i\in S_3(\boldsymbol{X})$ iff $X_i\geqslant c_i$; $c_1,\ldots,c_k\in\mathbb{R}$ fixed.]

Stage 2: For $\emptyset\neq s\subseteq\{1,\ldots,k\}$ and $i\in s$ let:
$d_{1,s}(\boldsymbol{X},\boldsymbol{Y})=i$ iff $Y_i=\max_{j\in s}Y_j$,
$d_{2,s}(\boldsymbol{X},\boldsymbol{Y})=i$ iff $X_i+Y_i=\max_{j\in s}(X_j+Y_j)$.

A correct selection occurs whenever a procedure finally ends up with the population associated with the largest mean, which we may assume to be π_k without loss of generality, since we are obviously dealing with permutation-invariant two-stage procedures.

To implement such a procedure one usually wishes to guarantee that the probability of a correct selection is at least $P^*>k^{-1}$ over a certain set of parameter configurations. Now if the means ν_i, say, $i=1,\ldots,k$, are restricted to the condition $\nu_1,\ldots,\nu_{k-1}\leqslant\nu_k-\Delta$ with $\Delta\geqslant 0$ fixed, then it seems intuitively clear that the infimum of the probability of a correct selection should occur at parameter configurations $(\nu,\nu,\ldots,\nu,\nu+\Delta)$ with $\nu\in\mathbb{R}$, which are called the least favorable configurations. Since there is no proof for these conjectures till now in the literature except for the special case of $k=2$ populations for (S_1, d_2) (as we shall discuss below more explicitly) we have tried to fill this gap and solve the problems in a more general setup (without the assumption of normality). Briefly, we have been successful in proving the conjectures for procedures using S_2 and S_3 but not for those using S_1.

Discussion of different two-stage procedures:

(S_1, d_β): (S_1, d_2) has been studied by Alam (1970) and Tamhane and Bechhofer (1977, 1979). Alam has proved the conjecture of $k=2$ and his subsequent results are based on the assumption that the conjecture is true for all k. Tamhane and Bechhofer (1977, 1979), on the other hand, used lower bounds for the probability of a correct selection which assume their infima at the desired parameter configurations.

(S_1, d_1) has not been studied up to now. Surprisingly, it turns out that it is even difficult to prove the conjecture for this simpler procedure. Therefore, we propose a lower bound for the probability of a correct selection which appears to be quite good and which is minimal at the conjectured parameter configuration.

Remark. Recently Miescke and Sehr (1980) have proved that the conjecture holds true for (S_1, d_1) and (S_2, d_2) in case of $k=3$. The proof is nonstandard and uses geometrical arguments.

(S_2, d_β): (S_2, d_2) and (S_2, d_1) have been studied by Somerville (1971a) and (1974). In both papers he has claimed that the corresponding conjectures have been proved by Somerville (1954), Fairweather (1968) and Somerville (1971b).

Now, in the last paper Somerville (1971b) was shown to be in error by Carroll and Santner (1975), and, in fact, his method of proof does not even work for (S_2, d_1). Moreover, the "loss function approach" of Somerville (1954) and Fairweather (1968) is not applicable in our problem since the corresponding function, W, used there turns out to be an indicator function here which clearly does not have a continuous second derivative. Therefore, the conjectures for (S_2, d_1) and (S_2, d_2) remained totally unproved up to now.

(S_3, d_β): These types of procedures may be used when the k populations are compared with a predetermined standard value ν_0, say, for the means. They proceed in the same manner as the procedures discussed above, with the only difference that at Stage 1 $S_3(\boldsymbol{X})$ now may be empty, in which case we stop and decide that no population is better than the standard. The probability of this event is now desired to be at least β^*, say, if $\nu_1, \ldots, \nu_k \leqslant \nu_0$ and the probability of a correct selection is then studied over all parameter configurations with $\nu_1, \ldots, \nu_{k-1} \leqslant \nu_0 - \Delta$ and $\nu_0 < \nu_k$. We will show in this paper that the infimum occurs at the point $(\nu_0 - \Delta, \ldots, \nu_0 - \Delta, \nu_0)$.

Remark. Let us finally mention that S_1, S_2, S_3 and d are well-established, one-stage, multiple decision procedures, studied and used in a variety of papers which can not all be mentioned here. To give a few references, Gupta (1956, 1965) proposed and studied S_1, Bechhofer (1954) S_2, Dunnett (1955), Gupta and Sobel (1958) and Lehmann (1961) studied S_3 and Bahadur and Goodman (1952), Lehmann (1966) and Miescke (1979) investigated d.

2. General considerations concerning all procedures

Let X_i, Y_i, $i=1,\ldots,k$ be independent random variables where X_i and Y_i have distribution functions $F(\xi-\theta_i), \xi \in \mathbb{R}$ and $G(\eta-\mu_i), \eta \in \mathbb{R}$, $i=1,\ldots,k$. F and G are assumed to be known continuous functions and the θ_i's and μ_i's represent unknown location parameters. Since we restrict ourselves to two-stage procedures which are invariant under permutations of the k populations, we may assume, without loss of generality, that we have $\theta_1, \ldots, \theta_{k-1} \leqslant \theta_k$ and $\mu_1, \ldots, \mu_{k-1} \leqslant \mu_k$.

Now let S be any subset selection rule for Stage 1. Then using d_1 or d_2 in Stage 2 the probabilities of correct selections are as follows:

$$P_k(S, d_1) = \sum_{\tilde{s} \subseteq \{1,\ldots,k-1\}} P\{S(\boldsymbol{X}) = s\} \int_{\mathbb{R}} \prod_{i \in \tilde{s}} G(\eta + \mu_k - \mu_i)\, \mathrm{d}G(\eta), \tag{2.1}$$

$$P_k(S, d_2) = \sum_{\tilde{s} \subseteq \{1,\ldots,k-1\}} P\{S(\boldsymbol{X}) = s;\ X_i + Y_i < X_k + Y_k,\ i \in \tilde{s}\}, \tag{2.2}$$

with the understanding that here and in the sequel $s = \tilde{s} \cup \{k\}$, if both s and $\tilde{s}$ appear simultaneously. The product appearing in (2.1) is defined to be equal to one if $\tilde{s}$ is empty.

In the sequel let $|A|$ denote the size of any finite set A. Now we state our main result:

Theorem. *For every* $\delta, \Delta \geqslant 0$, $\beta \in \{1,2\}$ *and* $\theta_0, \mu_0 \in \mathbb{R}$ *the following holds:*

(i) *Subject to* $\theta_1, \ldots, \theta_{k-1} \leqslant \theta_k - \delta$ *and* $\mu_1, \ldots, \mu_{k-1} \leqslant \mu_k - \Delta$, $P_k(S_2, d_\beta)$ *assumes its minimal value at every parameter configuration* $(\theta, \ldots, \theta, \theta+\delta)$ *and* $(\mu, \ldots, \mu, \mu+\Delta)$ *with* $\delta, \mu \in \mathbb{R}$.

(ii) *Subject to the additional restrictions* $\theta_k \geqslant \theta_0$ *and* $\mu_k \geqslant \mu_0$, $P_k(S_3, d_\beta)$ *assumes its minimal value at every parameter configuration* $(\theta, \ldots, \theta, \theta+\delta)$ *and* $(\mu, \ldots, \mu, \mu+\Delta)$ *with* $\mu \geqslant \mu_0 - \Delta$ *and* $\theta \geqslant \theta_0 - \delta$.

Proof (first part). From expression (2.1) it is clear that $P_k(S, d_1)$ for every S is non-increasing in $\mu_1, \ldots, \mu_{k-1}$ and non-decreasing in μ_k. The same is seen to hold true for $P_k(S, d_2)$ since for every $\boldsymbol{\xi} \in \mathbb{R}^k$ and $\tilde{s} \subseteq \{1, \ldots, k-1\}$ $P\{S(\boldsymbol{X})=s;\ X_i + Y_i < X_k + Y_k,\ i \in \tilde{s} | \boldsymbol{X}=\boldsymbol{\xi}\}$ also has this property.

This accomplishes the first step towards a solution of our problem. We can assume from now on that $(\mu_1, \ldots, \mu_k) = (\mu, \ldots, \mu, \mu+\Delta)$ for some $\mu \in \mathbb{R}$, respectively, $\mu \geqslant \mu_0 - \Delta$ holds. Then (2.1) reduces to

$$P_k(S, d_1) = \sum_{\tilde{s} \subseteq \{1, \ldots, k-1\}} P\{S(\boldsymbol{X})=s\} \int_{\mathbb{R}} G(\eta+\Delta)^{|\tilde{s}|} \mathrm{d}G(\eta)$$

$$= \sum_{r=1}^{k} P\{k \in S(\boldsymbol{X}), |S(\boldsymbol{X})| = r\} \int_{\mathbb{R}} G(\eta+\Delta)^{r-1} \mathrm{d}G(\eta), \tag{2.3}$$

or, alternatively,

$$P_k(S, d_1) = P\{k \in S(\boldsymbol{X})\} \int_{\mathbb{R}} G(\eta+\Delta)^{k-1} \mathrm{d}G(\eta)$$

$$+ \sum_{r=1}^{k-1} P\{k \in S(\boldsymbol{X}), |S(\boldsymbol{X})| \leqslant r\} \int_{\mathbb{R}} G(\eta+\Delta)^{r-1} [1 - G(\eta+\Delta)] \mathrm{d}G(\eta). \tag{2.4}$$

(2.2) reduces to

$$P_k(S, d_2) = \sum_{\tilde{s} \subseteq \{1, \ldots, k-1\}} P\{S(\boldsymbol{X})=s;\ X_i + U_i < X_k + \Delta + U_k,\ i \in \tilde{s}\} \tag{2.5}$$

where $U_1, \ldots, U_k$ are independently and identically distributed random variables with distribution function G, which are also independent of $X_1, \ldots, X_k$. (End of proof's first part.)

Formulae (2.3)–(2.5) and the next lemma will be used repeatedly in Sections 3 and 4 when we give the second part of the proof, consisting of four versions corresponding to the four procedures under consideration.

Lemma. *For every* $A \subseteq B \subseteq \{1,\ldots,k-1\}$, $r \in \{0,1,\ldots,|A|\}$, $a_1,\ldots,a_{k-1} \in \mathbb{R}$ *and* $b_1,\ldots,b_{k-1} \in \mathbb{R}$,

$$P\{|\{i|i\in A,\ X_i \geqslant a_i\}| \leqslant r;\ X_j \leqslant b_j, j \in B\} \tag{2.6}$$

is non-increasing in θ_l, $l=1,\ldots,k-1$.

Proof. For $r=0$ the assertion is clearly true. For $r>0$ and $l\in A$, (2.6) is equal to

$$P\{|\{i|i\in A, i\neq l,\ X_i \geqslant a_i\}| \leqslant r-1;\ X_j \leqslant b_j, j\in B,\ j\neq l\}P\{X_l \leqslant b_l\}$$

$$+P\{|\{i|i\in A, i\neq l,\ X_i \geqslant a_i\}| = r;\ X_j \leqslant b_j, j\in B,\ j\neq l\}P\{X_l \leqslant \min(a_l, b_l)\}$$

which obviously is non-increasing in θ_l. Similarly one can prove the assertion in case of $l\in B\backslash A$, whereas in case of $l\notin B$ it is trivially true since in that case (2.6) does not even depend on θ_l.

3. The second part of the proof

3.1. *Case* (S_2, d_1)

For every fixed $t\in\{2,\ldots,k-1\}$, we have $|S_2(\boldsymbol{X})|=t$ with probability one, and thus (2.3) reduces to

$$P_k(S_2, d_1) = P\{k\in S_2(\boldsymbol{X})\}\int_{\mathbb{R}} G(\eta+\Delta)^{t-1}\mathrm{d}G(\eta). \tag{3.1}$$

Moreover, we have

$$\begin{aligned} P\{k\in S_2(\boldsymbol{X})\} &= P\{|\{i|X_k < X_i\}| \leqslant t-1\} \\ &= \int_{\mathbb{R}} P\{|\{i|X_i > \xi, i\neq k\}| \leqslant t-1 | X_k=\xi\}\,\mathrm{d}P\{X_k=\xi\} \\ &= \int_{\mathbb{R}} P\{|\{i|X_i > \xi+\theta_k, i\neq k\}| \leqslant t-1\}\,\mathrm{d}F(\xi). \end{aligned} \tag{3.2}$$

Since the integrand obviously is non-decreasing in θ_k and by the lemma is non-increasing in $\theta_1,\ldots,\theta_{k-1}$, the proof for (S_2, d_1) is completed.

3.2. *Case* (S_2, d_2)

Let $t \in \{2,\ldots,k-1\}$ be fixed. Then using the fact that $(U_1 - U_k, \ldots, U_{k-1} - U_k)$ is symmetrically distributed, from (2.5) we get

$$P_k(S_2, d_2) = \\ = \sum_{\substack{1 \leqslant i_1 < i_2 < \cdots < i_{t-1} \leqslant k-1 \\ \tilde{s} = \{i_1, \ldots, i_{t-1}\}}} P\{X_l < X_i, l \notin s, i \in s;\ U_j - U_k + X_{i_j} < X_k + \Delta, j = 1, \ldots, t-1\}$$

$$= \int \cdots \int_{\{a_1 < a_2 < \cdots < a_{t-1}\}} \int_{\mathbb{R}} \sum_{\substack{1 \leqslant i_1 < i_2 < \cdots < i_{t-1} \leqslant k-1 \\ \tilde{s} = \{i_1, \ldots, i_{t-1}\}}} \sum_{\pi} P\{X_l < X_i, \xi + \theta_k, l \notin s, i \in \tilde{s};\\ a_{\pi_j} + X_{i_j} < \xi + \theta_k + \Delta, j = 1, \ldots, t-1\} \, \mathrm{d}F(\xi) \, \mathrm{d}P\{U_j - U_k = a_j, j = 1, \ldots, t-1\}, \tag{3.3}$$

where in the second sum, π runs over all $(t-1)!$ permutations of $(1,2,\ldots,t-1)$. Since every probability term obviously is non-decreasing in θ_k, it follows that $P_k(S_2, d_2)$ also has this property.

Now we show that for every fixed $\xi \in \mathbb{R}$ and $a_1 < a_2 < \cdots < a_{t-1}$, the integrand is non-increasing in $\theta_1, \ldots, \theta_{k-1}$. For the sake of simplicity we prove it for θ_1. Let $b_j = \xi + \theta_k + \Delta - a_j$, $j = 1, \ldots, t-1$ and $\xi_k = \xi + \theta_k$.

$$\sum_{\substack{1 \leqslant i_1 < i_2 < \cdots < i_{t-1} \leqslant k-1 \\ \tilde{s} = \{i_1, \ldots, i_{t-1}\}}} \sum_{\pi} P\{X_l < X_i, \xi_k, l \notin s, i \in \tilde{s};\ X_{i_j} < b_{\pi_j}, j = 1, \ldots, t-1\}$$

$$= \int_{\mathbb{R}^{k-2}} \sum_{\substack{1 \leqslant i_1 < i_2 < \cdots < i_{t-1} \leqslant k-1 \\ \tilde{s} = \{i_1, \ldots, i_{t-1}\}}} \sum_{\pi} P\{X_l < X_i, \xi_k, l \notin s, i \in \tilde{s};\ X_{i_j} < b_{\pi_j},\\ j = 1, \ldots, t-1 | X_2 = \xi_2, \ldots, X_{k-1} = \xi_{k-1}\} \, \mathrm{d}P\{X_2 = \xi_2, \ldots, X_{k-1} = \xi_{k-1}\}. \tag{3.4}$$

Now let $\xi_2, \ldots, \xi_{k-1} \in \mathbb{R}$ be fixed and assume, without loss of generality, that $\xi_2 < \xi_3 < \cdots < \xi_{k-1}$ holds. Then the integrand reduces to

$$\sum_{\pi} P\{X_1, X_{k-t} < X_{k-t+1}, \xi_k;\ X_{k-t+1} \leqslant b_{\pi_1}, X_{k-t+2} \leqslant b_{\pi_2}, \ldots, X_{k-1} \leqslant b_{\pi_{t-1}} |\\ X_2 = \xi_2, \ldots, X_{k-1} = \xi_{k-1}\}\\ + \sum_{\pi} P\{X_{k-t+1} < X_1, \xi_k;\ X_1 \leqslant b_{\pi_1}, X_{k-t+2} \leqslant b_{\pi_2}, \ldots, X_{k-1} \leqslant b_{\pi_{t-1}} |\\ X_2 = \xi_2, \ldots, X_{k-1} = \xi_{k-1}\}. \tag{3.5}$$

Finally we have to distinguish between two cases according to whether $\xi_{k-t+1}<\xi_k$ or not. In case of $\xi_{k-t+1}<\xi_k$, (3.5) reduces to

$$\sum_{\pi} P\{X_1<X_{k-t+1}\leqslant b_{\pi_1};\quad X_{k-t+2}\leqslant b_{\pi_2},\ldots, X_{k-1}\leqslant b_{\pi_{t-1}}|X_2=\xi_2,\ldots, X_{k-1}=\xi_{k-1}\}$$

$$+\sum_{\pi} P\{X_{k-t+1}<X_1\leqslant b_{\pi_1};$$

$$X_{k-t+2}\leqslant b_{\pi_2},\ldots, X_{k-1}\leqslant b_{\pi_{t-1}}|X_2=\xi_2,\ldots, X_{k-1}=\xi_{k-1}\}$$

$$=\sum_{\pi} P\{X_1, X_{k-t+1}\leqslant b_{\pi_1}; X_{k-t+2}\leqslant b_{\pi_2},\ldots, X_{k-1}\leqslant b_{\pi_{t-1}}|X_2=\xi_2,\ldots, X_{k-1}=\xi_{k-1}\},\tag{3.6}$$

whereas if $\xi_{k-t+1}\geqslant\xi_k$, (3.5) reduces to

$$\sum_{\pi} P\{X_1, X_{k-t}<\xi_k; X_{k-t+1}\leqslant b_{\pi_1},$$

$$X_{k-t+2}\leqslant b_{\pi_2},\ldots, X_{k-1}\leqslant b_{\pi_{t-1}}|X_2=\xi_2,\ldots, X_{k-1}=\xi_{k-1}\}.\tag{3.7}$$

Since now these last terms are non-increasing in θ_1, the proof for (S_2, d_2) is completed.

3.3. *Case (S_3, d_1)*

For every fixed $c_1,\ldots,c_k\in\mathbb{R}$ and using (2.4), it suffices to show that for every $r\in\{1,\ldots,k\}$, $P\{k\in S_3(\boldsymbol{X}),|S_3(\boldsymbol{X})|\leqslant r\}$ is non-increasing in $\theta_1,\ldots,\theta_{k-1}$ and non-decreasing in θ_k. Now

$$P\{k\in S_3(\boldsymbol{X}),|S_3(\boldsymbol{X})|\leqslant r\}=P\{|\{i|X_i\geqslant c_i, i\neq k\}|\leqslant r-1\}P\{X_k\geqslant c_k\},\quad r=1,\ldots,k.\tag{3.8}$$

The first factor does not depend on θ_k and by the lemma is non-increasing in $\theta_1,\ldots,\theta_{k-1}$, whereas the second factor does not depend on $\theta_1,\ldots,\theta_{k-1}$ and is non-decreasing in θ_k. Thus the proof for (S_3, d_1) is completed.

3.4. *Case (S_3, d_2)*

For every fixed $c_1,\ldots,c_k$, from (2.5) we have

$$P_k(S_3, d_2)=\sum_{\tilde{s}\subseteq\{1,\ldots,k-1\}} P\Big\{\underset{i\in s}{X_i\geqslant c_i},\ \underset{j\notin s}{X_j<c_j};\ \underset{i\in\tilde{s}}{X_i+U_i}<X_k+\Delta+U_k\Big\}\tag{3.9}$$

which clearly is non-decreasing in θ_k since in every summand we have $s=\tilde{s}\cup\{k\}$ by our convention. Again, for sake of simplicity, it will be shown that $P_k(S_3, d_2)$ is non-increasing in θ_1. Now

$$P_k(S_3, d_2) = \sum_{\tilde{s}\subseteq\{2,\dots,k-1\}} P\Big\{ \underset{i\in s}{X_i \geq c_i},\ X_1 < c_1,\ \underset{\substack{j\notin s\\ j\neq 1}}{X_j < c_j};\ \underset{i\in\tilde{s}}{X_i + U_i} < X_k + \Delta + U_k \Big\}$$

$$+ \sum_{\tilde{s}\subseteq\{2,\dots,k-1\}} P\Big\{ \underset{i\in s}{X_i \geq c_i},\ X_1 \geq c_1,\ \underset{\substack{j\notin s\\ j\neq 1}}{X_j < c_j};$$

$$X_1 + U_1,\ \underset{i\in\tilde{s}}{X_i + U_i} < X_k + \Delta + U_k \Big\}$$

$$= \sum_{\tilde{s}\subseteq\{2,\dots,k-1\}} P\Big\{ \underset{i\in s}{X_i \geq c_i},\ X_1 < c_1,\ \underset{\substack{j\notin s\\ j\neq 1}}{X_j < c_j};$$

$$\underset{i\in\tilde{s}}{X_i + U_i} < X_k + \Delta + U_k \leq c_1 + U_1 \Big\}$$

$$+ \sum_{\tilde{s}\subseteq\{2,\dots,k-1\}} P\Big\{ \underset{i\in s}{X_i \geq c_i},\ \underset{\substack{j\notin s\\ j\neq 1}}{X_j < c_j};$$

$$c_1 + U_1,\ X_1 + U_1,\ \underset{i\in\tilde{s}}{X_i + U_i} < X_k + \Delta + U_k \Big\}. \qquad (3.10)$$

Clearly, every summand is non-increasing in θ_1 and, therefore, the proof for (S_3, d_2) is completed.

3.5. *Concluding remarks*

In the case of normal populations as described in Section 1 we have $X_i \sim N(n\nu_i, n\sigma^2)$ and $Y_i \sim N(m\nu_i, m\sigma^2)$, $i=1,\dots,k$, for some $\sigma^2 > 0$. Thus for every $\alpha \in \{1,2,3\}$ and $\beta \in \{1,2\}$, the probability that (S_α, d_β) finally leads to a decision in favor of population π_k can be represented by a certain function $H_{\alpha,\beta}(\nu_1, \nu_2, \dots, \nu_k)$. To prove the conjectures we could, alternatively, have tried to show that $H_{\alpha,\beta}$ is non-increasing in $\nu_1, \dots, \nu_{k-1}$ and non-decreasing in ν_k. However, this turns out to be a very difficult and cumbersome way.

4. A lower bound for (S_1, d_1)

To prove the conjecture for (S_1, d_1) in view of (2.4) it would suffice to show that for every $r \in \{1,\dots,k\}$, $P\{k \in S_1(X), |S_1(X)| \leq r\}$ is non-increasing in $\theta_1, \dots, \theta_{k-1}$ and non-decreasing in θ_k. For $r=1$ and $r=k$ this probability is equal to $P\{X_1, \dots, X_{k-1} < X_k - c\}$ and $P\{X_1, \dots, X_{k-1} \leq X_k + c\}$, respectively, where each of them clearly has the desired property. Thus the conjecture for $k=2$ is proved.

For $k>2$ it turns out to be rather difficult to prove the conjecture. Therefore, we derive a lower bound for the probability of a correct selection, which assumes its minimal value at the desired parameter configuration.

For $k>2$ and $r\in\{2,\ldots,k-1\}$ we have

$$P\{k\in S_1(\boldsymbol{X}),|S_1(\boldsymbol{X})|\leqslant r\}=$$

$$=\int_{\mathbb{R}} P\{k\in S_1(X_1,\ldots,X_{k-1},\xi),|S_1(X_1,\ldots,X_{k-1},\xi)|\leqslant r|X_k=\xi\}\,\mathrm{d}P\{X_k=\xi\}$$

$$\geqslant\int_{\mathbb{R}} P\{|\{i|X_i\geqslant\xi+\theta_k-c,\,i\neq k\}|\leqslant r-1;\,X_1,\ldots,X_{k-1}\leqslant\xi+\theta_k+c\}\,\mathrm{d}F(\xi). \tag{4.1}$$

The preceding inequality holds also for $r=1$ and follows from the fact that we have

$$\{k\in S(\boldsymbol{X}),|S(\boldsymbol{X})|\leqslant r\}=$$

$$=\left\{|\{i|X_i\geqslant X_k-c,\,i\neq k\}|\leqslant r-1,\,X_k=\max_{j=1,\ldots,k}X_j\right\}$$

$$\cup\bigcup_{l=1}^{k-1}\left\{|\{i|X_i\geqslant X_l-c,\,i\neq k,l\}|\leqslant r-2,\,X_k\geqslant X_l-c,\,X_l=\max_{j=1,\ldots,k}X_j\right\}. \tag{4.2}$$

Now the integrand in the last integral of (4.1) clearly is non-decreasing in θ_k and by our lemma it is non-increasing in $\theta_1,\ldots,\theta_{k-1}$. Thus, using (2.4), we get

$$P_k(S_1,d_1)\geqslant P\{Z_1,\ldots,Z_{k-1}\leqslant Z_k+\delta+c\}\int_{\mathbb{R}} G(\eta+\Delta)^{k-1}\mathrm{d}G(\eta)$$

$$+\sum_{r=1}^{k-1} P\{|\{i|Z_i\geqslant Z_k+\delta-c,\,i\neq k\}|\leqslant r-1;\,Z_1,\ldots,Z_{k-1}\leqslant Z_k+\delta+c\}$$

$$\times\int_{\mathbb{R}} G(\eta+\Delta)^{r-1}[1-G(\eta+\Delta)]\,\mathrm{d}G(\eta), \tag{4.3}$$

where $Z_1,\ldots,Z_k$ are independently and identically distributed random variables with distribution function F. Now the right-hand side of (4.3), being in a form similar to (2.4), can be brought into a form similar to that of (2.3). Then it is equal to

$$\sum_{r=1}^{k} P\{|\{i|Z_i\geqslant Z_k+\delta-c,\,i\neq k\}=r-1;\,Z_1,\ldots,Z_{k-1}\leqslant Z_k+\delta+c\}$$

$$\times\int_{\mathbb{R}} G(\eta+\Delta)^{r-1}\mathrm{d}G(\eta)$$

$$=\sum_{i=0}^{k-1}\int_{\mathbb{R}}\binom{k-1}{i}F(\xi+\delta-c)^{k-i-1}[F(\xi+\delta+c)-F(\xi+\delta-c)]^i\,\mathrm{d}F(\xi)$$

$$\times\int_{\mathbb{R}} G(\eta+\Delta)^i\,\mathrm{d}G(\eta). \tag{4.4}$$

Thus we finally arrive at the following result:

Corollary. *For* $k \geqslant 2$

$$P_k(S_1, d_1) \geqslant$$
$$\geqslant \int_{\mathbb{R}} \int_{\mathbb{R}} \{F(\xi+\delta-c)+[F(\xi+\delta+c)-F(\xi+\delta-c)]G(\eta+\Delta)\}^{k-1} \mathrm{d}F(\xi)\mathrm{d}G(\eta). \tag{4.5}$$

Note that for $k=2$ this lower bound for the probability of a correct selection is exact.

Acknowledgements

This research was supported by the Office of Naval Research Contract N00014-75-C-0455 at Purdue University.

References

Alam, K. (1970). A two sample procedure for selecting the population with the largest mean of k normal populations. *Ann. Inst. Statist. Math.* **22**, 127–136.

Bahadur, R.R. and Goodman, L. (1952). Impartial decision rules and sufficient statistics. *Ann. Math. Statist.* **23**, 553–562.

Bechhofer, R.E. (1954). A single-sample multiple decision procedure for ranking means of normal populations with known variances. *Ann. Math. Statist.* **25**, 16–39.

Carroll, R.J. and Santner, T.J. (1975). A note on a minimization result for sequential ranking procedures. Tech. Rep. No. 258, Dept. of Operations Research, College of Engineering, Cornell Univ., Ithaca, New York.

Dunnett, C.W. (1955). A multiple comparison procedure for comparing several treatments with a control. *J. Amer. Statist. Assoc.* **50**, 1096–1121.

Fairweather, W.R. (1968). Some extensions of Somerville's procedure for ranking means of normal populations. *Biometrika* **55**, 411–418.

Gupta, S.S. (1956). *On a decision rule for a problem in ranking means.* Mimeo. Ser. No. 150, Inst. of Statist., Univ. of North Carolina, Chapel Hill, NC.

Gupta, S.S. (1965). On some multiple decision (selection and ranking) rules. *Technometrics* **7**, 225–245.

Gupta, S.S. and Sobel, M. (1958). On selecting a subset which contains all populations better than a standard. *Ann. Math. Statist.* **29**, 274–281.

Lehmann, E.L. (1961). Some Model I problems of selection. *Ann. Math. Statist.* **32**, 990–1012.

Lehmann, E.L. (1966). On a theorem of Bahadur and Goodman. *Ann. Math. Statist.* **37**, 1–6.

Miescke, K.J. (1979). Identification and selection procedures based on tests. *Ann. Statist.* **7**, 207–219.

Miescke, K.J. and Sehr, J. (1980). On a conjecture concerning the least favorable configuration of certain two-stage selection procedures. *Commun. Statist.—Theor. Meth.* **A9**, 1609–1617.

Somerville, P.N. (1954). Some problems of optimum sampling. *Biometrika* **41**, 420–429.

Somerville, P.N. (1971a). A technique for obtaining probabilities of correct selection in a two-stage selection problem. *Biometrika* **58**, 615–623.

Somerville, P.N. (1971b). A generalization of a fundamental theorem of ranking and selection. *Biometrika* **58**, 227–228. Correction: (1976). *Biometrika* **63**, 412.

Somerville, P.N. (1974). On allocation of resources in a two-stage selection procedure. *Sankhyā Ser. B* **36**, 194–203.

Tamhane, A.C. and Bechhofer, R.E. (1977). A two-stage minimax procedure with screening for selecting the largest normal mean. *Commun. Statist.—Theor. Meth.* **A6**, 1003–1033.

Tamhane, A.C. and Bechhofer, R.E. (1979). A two-stage minimax procedure with screening for selecting the largest normal mean (II): An improved lower bound on the PCS and associated tables. *Commun. Statist.—Theor. Meth.* **A8**, 337–358.

G. Kallianpur, P.R. Krishnaiah, J.K. Ghosh, eds., *Statistics and Probability: Essays in Honor of C.R. Rao*

FITTING MULTIVARIATE ARMA MODELS

E.J. HANNAN
Australian National University, Canberra, Australia

1. Introduction

Consider observations, $y(t), t=1,2,\ldots,T$, which are r-vectors generated by a stationary, ergodic process with zero mean which is of autoregressive-moving average (ARMA) type, i.e.

$$\sum_0^p A(j)y(t-j)=\sum_0^q B(j)\varepsilon(t-j), \qquad A(0)=B(0), \qquad \mathcal{E}\{\varepsilon(s)\varepsilon(t)'\}=\delta_{st}\Sigma\ldots \tag{1}$$

If $\tilde{g}(z)=\Sigma A(j)z^j$, $\tilde{h}(z)=\Sigma B(j)z^j$ and,

$$\det(\tilde{g})\neq 0, \quad |z|\leqslant 1, \qquad \det(\tilde{h})\neq 0, \quad |z|<1, \tag{2}$$

then the $\varepsilon(t)$ are the linear innovations for $y(t)$. (See ref. [1], for example.) If $\mathcal{F}_t$ is the σ-algebra generated by $\varepsilon(s)$, $s\leqslant t$ and $\mathcal{E}\{\varepsilon(t)|\mathcal{F}_{t-1}\}=0$, then the best linear predictor is the best predictor and since eq. (1) is used for a model substantially for prediction, this is a natural condition.

For $r=1$, algorithms for fitting (1) are readily available, based on (possibly approximate) optimisation of a likelihood for the data, constructed as if that were Gaussian. When $r=1$ we will put $A(0)=B(0)=1$, $\Sigma=\sigma^2$. $A(j)=\alpha(j)$, $B(j)=\beta(j)$. We shall use $\hat{\Sigma}$, $\hat{A}(j)$ etc. for the maximum likelihood estimates. If $r=1$, p,q are known and either $\alpha(p)$ or $\beta(q)$ is non-zero then $\hat{\alpha}(j)$, $\hat{\beta}(j)$, $\hat{\sigma}^2\to\alpha(j)$, $\beta(j)$, σ^2. A.s. (See ref. [2] for details concerning the general case.) However p,q are not known in practice. If the true values are p_0, q_0 and $p>p_0, q>q_0$ is tried then the algorithm for optimising the likelihood will tend to break down for the following reason. It will tend to be true that the optimising values will be such that $\tilde{g}$, $\tilde{h}$ have a common zero (or more than one) that is very near to the unit circle. (We will explain this later; see also ref. [3]). In such circumstances, optimisation will be difficult. For example, one approximate procedure is to compute $\Sigma\,\varepsilon_\tau(t)^2$, where $\varepsilon_\tau(t)$ is computed on the basis of the vector of system parameters $\tau'=(\alpha(1),\ldots,\alpha(p),\beta(1),\ldots,\beta(q))$. Then τ is chosen to minimise $\Sigma\,\varepsilon_\tau(t)^2$. However if τ is such that $\tilde{h}$ has a zero near to the unit circle, the calculation of the $\varepsilon_\tau(t)$ will be very unstable in that it will be badly affected by the choice of initial values.

Problems of the kind just discussed are greatly magnified in the case $r>1$. This paper discusses these problems.

2. Parameterisation of vector ARMA models

By introducing $k(z)=\tilde{g}(z^{-1})^{-1}\tilde{h}(z^{-1})$, we then obtain

$$k(z)=\sum_{0}^{\infty} K(j)z^{-j}, \qquad K(0)=I, \qquad |z|\geqslant 1. \tag{3}$$

We may topologise the rational matrix functions $k(z)$ by means of the metric

$$\|k\|=\sum^{\infty}\|K(j)\|2^{-j}, \tag{4}$$

for example, where $\|K(j)\|$ is the Euclidean norm. Of course, also

$$\det(k)\neq 0, \quad |z|>1 \tag{5}$$

because of the second part of (2). It is well known (see ref. [1]) that k is uniquely determined, say from the spectrum of $y(t)$, and hence may be strongly consistently estimated. However, it is one thing to say that and another to construct an algorithm to estimate k. For this purpose we need coordinates, at least locally, i.e. in some neighbourhood of each k. To achieve this we may proceed as follows. If $k=g^{-1}h$ where g, h are matrices of polynomials, then k is said to have been given a matrix fraction description (m.f.d). If $g=ug_1$, $h=uh_1$, where u, g, h, are also polynomials, implies that $\det(u)$ is a non-zero constant (i.e., u is unimodular) then g, h are said to be left prime. This is so if and only if $[g(z):h(z)]$ is of rank r for all complex z. (See ref. [4] for details.) The determinant of g in any left prime m.f.d for k has a uniquely determined degree, n let us say, since any other left prime m.f.d, $g_1^{-1}h_1$, has $g_1=vg$, $h_1=vh$ for v unimodular. This integer n is called the order of the system or sometimes the McMillan degree. (See ref. [5] for further refs. and details.) Let $M(n)$ be the set of all k satisfying (3) for given n, but with (5) strengthened to $\det(k)\neq 0$, $|z|\geqslant 1$ (This is done purely to make $M(n)$ an open set.) Let $n=n_1+n_2+\cdots+n_r$, $n_j\geqslant 0$ be some partition of n. Consider $k\in M(n)$ having a left prime m.f.d with $g_{jj}(z)$ (the jth diagonal element of g) a monic polynomial of degree n_j and degree $(g_{ij})<n_j$, $i\neq j$. Call $U(\alpha)$ the set of all such k, where we have used α as a symbol for a typical partition $(n_1,\ldots,n_r)$. Then $M(n)$ is a union of the $U(\alpha)$ and each $U(\alpha)$ is dense in $M(n)$. If $\alpha_{ij}(u)$ is the coefficient of z^u in g_{ij} and $\kappa_{ij}(u)$ is the coefficient of z^u in $K(u)$, then the $\alpha_{ij}(u)$, $u=0,1,\ldots,n_j-1$, $j=1,\ldots,r$; $\kappa_{ij}(u)$, $u=1,\ldots,n_i$, $i, j=1,\ldots,r$ form a set of $2nr$ coordinates in $U(\alpha)$ and if ϕ_a is the mapping into $\mathfrak{R}^{2nr}$ instituted on $U(\alpha)$ by the coordinates then ϕ_α is a homeomorphism. Moreover $\phi_\beta\phi_\alpha^{-1}$, which maps $\phi_\alpha(U(\alpha)\cap U(\beta))$ onto $\phi_\beta(U(\alpha)\cap U(\beta))$, is real analytic so that $M(n)$ is an analytic manifold. (We refer the reader to ref. [5] for some further details and references to the literature.) Call $U_0(n)$ the $U(\alpha)$ for which $n_1=n_2=\cdots=n_s=n_{s+1}+1=\cdots n_r+1, n=rn_r+s, s<r$. For U_0 we may impose stronger degree conditions, namely

$$\begin{aligned} &\text{degree}\,(g_{ij})\leqslant n_i, \quad j\leqslant i; \qquad \text{degree}\,(g_{ij})<n_i, \quad j>i; \\ &\text{degree}\,(g_{ij})<n_j, \quad i\neq j; \qquad \text{degree}\,(h_{ij})\leqslant n_i, \quad j=1,\ldots,r. \end{aligned} \tag{5}$$

Thus, the set of all $k(z)$ may be parameterised by one integer, n, plus $2nr$

coordinates, given n, and it would seem, since $U_0(n)$ is dense in $M(n)$, that it would be sufficient to examine only $U_0(n)$ which is homeomorphic with an open set in $\mathcal{R}^{2nr}$. However things are not that simple, as we shall see.

Let the matrix Z be diagonal with z^{n_i} in the main diagonal and put $\tilde{g}(z)=Zg(z^{-1})$, $\tilde{h}(z)=Zh(z^{-1})$. Then $y(t)$ may be represented as the form of eq. (1) for these $\tilde{g}$, $\tilde{h}$. Let $k\in U_0(n)$ and g, h satisfy (5), then $[\tilde{g}:\tilde{h}]=Z[g(z^{-1}):h(z^{-1})]$ and $[\tilde{g}(0):\tilde{h}(0)]$ has $\tilde{g}(0)$ lower triangular with units in the main diagonal [see (5)]. Hence $\tilde{g}$, $\tilde{h}$ are left prime. Reversing this argument we see that the matrix, H, having as its ith row the coefficients of z^{n_i} in the ith row of $[\tilde{g}:\tilde{h}]$, is also of full rank. If $n=rn_r$ then $H=[A(n_r):B(n_r)]$. Thus if $M_0(p)$ is the set of all structures (1) for $p=q$, satisfying (2) with $|z|<1$ strengthened to $|z|\leqslant 1$, with $A(0)=B(0)=I_r$, $\tilde{g}$, $\tilde{h}$ left prime and $[A(p):B(p)]$ of full rank then $M_0(p)$ is just $U_0(n)$, $n=rp$ and is thus dense in $M(n)$, $n=rp$. In restricting attention to $M_0(p)$ we can be doing no real damage for the following reason. Adjoin to $M(n)$ those limit points obtained by allowing a zero of $\det(k)$ or a pole of k to be on the unit circle. Then ref. [5], calling this $M_c(n)$,

$$\overline{M_c(n)} = \bigcup_{m\leqslant n} M_c(m). \tag{6}$$

Thus, since $M_0(p)$ is dense in $M(rp)$, and $M(rp)$ is dense in the set of all $M(n)$, $n\leqslant rp$ then nothing would be lost in restricting attention to the $M_0(p)$, since it is almost certain, a priori, that the maximum of the likelihood over $M(m)$, $m\leqslant rp$ will be found in $M_0(p)$. Again, however, there are problems, which will be discussed further in the next section. Before doing that let us mention constraints. One way to reduce the problem of fitting is to use prior knowledge that imposes constraints so that the set of models to be considered becomes so circumscribed that fitting is no longer a problem. Here we have not discussed the state space representation of an ARMA system (see ref. [6], for example) but if the model is initially built in this form, often very many constraints are available. The use of constraints for the canonical forms associated with the partitions $n=\Sigma n_i$ is usually impossible because the constraints will refer to some original model having physical meaning and the functions mapping this into canonical form will usually be too complicated for the constraints to be recognisable in that form. This is not so true of $M_0(p)$ since the requirement that $[A(p):B(p)]$ be of full rank could usually be assumed to hold for the model as originally constructed. One form of constraint that can easily be applied is to require that some $A(j)$, $B(j)$ be null. Let $M(p,q)$ be the set of all $k(z)$ satisfying (1) and (2) with $A(0)=B(0)=I$, with $\tilde{g}$, $\tilde{h}$ left prime and $[A(p):B(q)]$ of full rank. This is a slight extension of a constrained subset of $M_0(p)$ if, say, $q<p$. An extreme case is $M(p,0)$, which is just the set of autoregressions of order p and for which there is no fitting problem of the kind met, in general.

3. Properties of the parameterisations relating to the unknown order

As shown in ref. [2], the maximum likelihood estimate of $k_0\in M(n_0)$ converges a.s. to k_0, uniformly on $|z|\geqslant 1$, and indeed $\det(k)=0$, $|z|=1$ may even be allowed. This assumes that n_0 is known. The same is true for $M_0(p_0)$ or $M(p_0,q_0)$ if p_0 or (p_0,q_0)

is known. However the proof does not use coordinates and acts as if the maximum of the likelihood had been located on $M(n_0)$, for example. It is the computation of the estimate, which needs coordinates, and the determination of n_0 that causes the trouble.

Let π_α be the mapping inverse to ϕ_α. Thus π_α sends any vector τ in $\mathcal{R}^{2nr}$ into $k=g^{-1}h$ with g, h in canonical form for $U(\alpha)$. Of course, k might not then satisfy the pole-zero requirements or the left prime condition but if $\tau \in \phi_\alpha U(\alpha)$ it will. Then [5],

$$\pi_\alpha \overline{\phi_\alpha U(\alpha)} = \bigcup_{\beta \leqslant \alpha} U_c(\beta) \tag{7}$$

whereby $\beta \leqslant \alpha$ we mean that β corresponds to n_i', $i=1,\dots,r$ with $n_i' \leqslant n_i$, $i=1,\dots,r$. By $U_c(\beta)$ we mean that we have adjoined limit points for which there may be poles or zeros on the unit circle. Of course *sets* of points in $\overline{\phi_\alpha U(\alpha)}$ may be mapped onto one $k \in \overline{M(n)}$. Such sets can be called "equivalence classes," if equivalence is taken to mean giving rise to the same k. The $\tau \in \phi_\alpha U(\alpha)$ are equivalence classes that are singletons. Of course $\pi_\alpha \overline{\phi_\alpha U(\alpha)}$ may be topologised via (4). Call T_p this topology. Let T_Q be the quotient topology on $\pi_\alpha \overline{\phi_\alpha U(\alpha)}$, i.e. the largest topology such that π_α is continuous. Then (see ref. [5]) T_p and T_q are equivalent on $U(\alpha)$ and on no larger set in $\pi\overline{\phi_\alpha(U(\alpha))}$. To see what this indicates consider a simple example where $n=r=1$ and $k_T(z)=(z+\alpha_T+\eta_T)/(z+\alpha_\tau)$, $|\alpha_\tau| \leqslant c < 1$, $|z| \geqslant 1$, $\eta_T \to 0$. Then $k_\tau(z)$ converges to $k_0(z) \equiv 1$, uniformly in $(z) \geqslant 1$. However, $(\alpha_T+\eta_T, \alpha_\tau)$ might stay outside of some open set containing $(0,0)$ and hence $k_T(z)$ does not converge in T_Q. The rather arbitrary nature of the sequence α_T indicates how troublesome the computation of $k_T(z)$ might be, if it is interpreted as the maximum likelihood solution. Indeed in this kind of situation, as we will see, α_T may converge towards ± 1 which causes great problems, as outlined in Section 1.

Returning to (7) assume that $U(\alpha)$ is being examined and $k_0 \notin U(\alpha)$. If $n_0=n$ then ϕ_α must map k_0 onto the point at infinity. The same will be true if $n_0<n$ and k_0 does not lie in any $U(\beta)$, $\beta \leqslant \alpha$. As we have already seen if $n_0<n$ and $k_0 \in U(\beta)$, $\beta \leqslant \alpha$ then $\hat{k}$ might not converge to k_0 in the quotient topology. Each of these possibilities indicates how an algorithm for the determination of $\hat{k}$ might break down. Of course there will be no true n_0 but in a practical situation the data may be well fitted by a k_0 for some particular n_0 and then analogous considerations will apply. All of this shows that it will be essential, in practice, to bound τ, a priori, when maximising any $U(\alpha)$ and also to bound the zeros of $k(z)$ away from the unit circle. (Since $k_T \to k_0$, a.s. in T_p this ensures that the poles are also bounded away from the unit circle.) Of couse, this amounts to the use of prior constraints and really depends on some knowledge that the data can be well represented by a k_0 bounded in this way. When this is done the only points which can be attained in the process of optimising the likelihood over $U(\alpha)$ are $k_0 \in U(\beta)$, $\beta \leqslant \alpha$. In the remainder of this Section we examine only the situation when the possibilities are restricted by the bounding we have described.

Since $M_0(p)$ is just a special $U(\alpha)$, all of the above considerations apply directly to it. For $M(p,q)$ the situation is examined in ref. [7] and again a full analogy with what has been said for $U(\alpha)$ holds. Now ϕ is defined so as to make $k \in M(p,q)$ into $(p+q)r^2$ dimensional Euclidean space via the elements of the $A(j)$, $B(j)$, $j>0$. The inverse mapping is defined as before as well as the topologies T_p, T_q and $T_p \subset T_Q$ and they are the same in $M(p,q)$ but in no larger set. Of course, the analogue of (7) is more complicated to describe.

4. The estimation of order

The remaining problem is the estimation of n [or of p or of (p,q)]. Akaike (see ref. [6] for further details and earlier references) has proposed, for special cases, the following method. Put

$$A(n) = T \log \det \hat{\Sigma}_n + nC_T, \quad n \leqslant N \tag{8}$$

where C_T is a chosen sequence of constraints. The estimate $\hat{n}$ is chosen so as to minimise $A(n)$. We have written $\hat{\Sigma}_n$ to indicate that $\hat{\Sigma}_n$ is computed assuming n to be the true order. In fact Akaike chose $C_T = 4r$, when $A(n)$ is called $AIC(n)$. This was done so as to minimise the error of prediction (in the mean square sense) when an account is taken of the errors in estimating k, but when it is not believed that $k_0(z)$ lies in any $M(n)$, though it will be well approximated by a $k \in M(n)$ for some sufficiently large n. If it is hypothesised that $k_0 \in M(n_0)$ then it seems that Akaike would use $C_T = 2r \log T$. [Now $A(n)$ is called $BIC(n)\cdot$] We further discuss the choice of C_T later. Here we consider $k_0 \in M(n_0)$, $n_0 \leqslant N$, in order to elucidate the situation. (For details see ref. [5].)

Let $k_0 \in U(\beta)$, $\beta \leqslant \alpha$. What is the image, $\phi_\alpha k_0$, in $\overline{\phi_\alpha U(\alpha)}$? It is an affine subspace of $\mathcal{R}^{2nr}$ of dimension $(n-n_0)r$, or at least the intersection of such a space with the set of $\tau \in \mathcal{R}^{2nr}$ for which the zero and poles of the corresponding k have modulus greater than unity. On the other hand

$$h - gk_0 = \sum_{-p+1}^{\infty} \Psi(i) z^{-i},$$

where $p = \max n_i$, $\beta = (n_1, n_2, \ldots, n_r)$. Then the elements of the $\Psi(i)$ are linear in $\tau \in \phi_\alpha U(\alpha)$, with coefficients determined by k_0. This set of linear functions of τ is of dimension $(n+n_0)r$. If we choose $(n-n_0)r$ linear functions of τ lying in the space $\phi_\alpha k_0$, which we call χ_j, $j=1,\ldots,(n-n_0)r$, and $(n+n_0)r$ linear functions of τ from among the components of $\Psi(i)$ which we call Ψ_j, $j=1,\ldots,(n+n_0)r$ then we get a set of $2nr$ linearly independent linear forms in τ.

Since $\hat{k} \to k_0$, a.s., then the vector $\hat{\Psi}$ composed of the $\hat{\Psi}_j$ converges to 0, a.s. However $\hat{\chi}$ does not behave in any such reasonable fashion as can be imagined from the fact that the likelihood is constant over the affine space in which χ lies, because all points in it give rise to the same transfer function, namely k_0.

The general situation can be seen by consideration of the simplest relevant case, where $r_1 = 1$, $n_0 = 0$, $n = 1$. Then the space over which optimisation is effected is that of all $k(z)$

$$k(z)=\frac{z+\beta}{z+\alpha}, \qquad |\beta|\leqslant 1-\delta, \quad \delta>0.$$

Then whether $A(n)$ is smaller at $n=1$ or $n=0$ is determined by the behaviour of

$$T\{\log\hat{\sigma}_0^2-\log\hat{\sigma}_1^2\}=\sup_{-1+\delta\leqslant\psi\leqslant 1-\delta}\frac{1-\psi^2}{\sigma^4}\left[T^{-1/2}\sum_1^T\varepsilon(t)\left\{\sum_0^\infty\psi^k\varepsilon(t-1-k)\right\}\right]^2+\mathrm{o}(1), \tag{9}$$

where the limit that is o(1) converges a.s. to zero. (See ref. [3].) Putting $\psi=\tanh\frac{1}{2}v$, $|v|\leqslant\log\{(2-\delta)/\delta\}=V$ the expression behind "sup" in the first term on the right converges weakly to $\phi(v)^2$ where $\phi(v)$ is a stationary Gaussian process with spectral density $(\cosh\pi\omega)^{-1}$. It is now evident that (9) will become indefinitely large as first $T\to\infty$ and then $\delta\to 0$ (i.e., $V\to\infty$) and that the point at which it will be optimised will tend to be for larger and larger $|v|$, i.e. $|\psi|$ near and nearer to 1, since ψ is here $\frac{1}{2}(\hat{\alpha}+\hat{\beta})$ and $(\hat{\alpha}-\hat{\beta})\to 0$ and $\hat{\alpha}$, $\hat{\beta}\to\pm 1$. This shows also that, for $T\to\infty$ and then $\delta\to 0$, $n=1$ will be preferred to $n=0$ unless $C_T\to\infty$. In general, the following can be shown [see ref. [8] for details and conditions]

(i) Consistency cannot hold in general unless $C_T/T\to 0$.

(ii) If,

$$\limsup_{T\to\infty} C_T/\{2\log\log T\}>1, \quad C_T/T\to 0,$$

then a.s. convergence of $\hat{n}$ to n_0 holds.

(iii) If,

$$\limsup C_T/\{2\log\log T\}<1,$$

then a.s. convergence of $\hat{n}$ to n_0 does not hold.

(iv) If,

$$C_T\uparrow\infty, \qquad C_T/T\to 0,$$

$\hat{n}$ converges to n_0 in probability.

(v) If,

$$\limsup_{T\to\infty} C_T<\infty,$$

then,

$$\lim_{T\to\infty} P\{n>n_0\},$$

converges to unity as the bounds on the vector τ are relaxed. This last result may only be relevant for *very* large values of T (see ref. [3]).

Thus n_0 can be consistently estimated. Of course, since there is no true n_0 these results, especially point (v), have to be interpreted with great care. We discuss this further in the next section.

The situation for $M_0(p)$ is, of course, fully analogous to that for $M(n)$. For $M(p,q)$ the situation is more complicated in that the dimension of the set ϕk_0,

$k_0 \in M(p_0, q_0)$, in $\overline{\phi M(p,q)}$ cannot be stated in simple terms. It is now easier to specify the ψ vector via the expression

$$\sum_1^\infty \Psi(j) z^j = \tilde{h}(z) - \tilde{g}(z) \sum_0^\infty K_0(j) z^j = \tilde{h} - \tilde{g}\tilde{k}_0,$$

let us say. Then $\psi_\infty = B_0 \tau + b_0$ where B_0 and b_0 depend only on K_0 and have infinitely many rows. We may determine the rank of B_0 by determining the dimension of its null space, i.e. of the set of solutions of $\tilde{h} - \tilde{g}\tilde{k}_0 = 0$, which implies $\tilde{g} = x\tilde{g}_0$, $\tilde{h} = x\tilde{h}_0$. If m is the degree of x then it is easily seen that $\min(p-p_0, q-q_0) \leqslant m \leqslant \max(p-p_0, q-q_0)$ so that the dimension of the null space is no less than $r^2 m$ in $(p-p_0, q-q_0)$. Thus the dimension of the ψ-space is no greater than $r^2\max(p+q_0, q+p_0)$. All the results of points (i) to (v) above continue to hold, with (v) holding as stated only if the maximum values, P, Q, examined satisfy $p_0 < P$, $q_0 < Q$. However, even if one, *but not both*, of the equalities $p_0 = P$, $q_0 = Q$ hold a modified form of (v) will hold showing that $\hat{n}$ is not consistent. (We emphasise again that $\hat{n}$ for the case covered by point (v) is not meant to be consistent).

Other criteria for estimating order have been suggested. We briefly consider one of these for the scalar case (see ref. [10] for related ideas). This is, essentially, to consider

$$B(n) = T \log \hat{\sigma}_n^2 - T \log\det(T^{-1}\hat{I}_n), \quad n \leqslant N, \ldots \tag{10}$$

where $\hat{I}_n$ is the information matrix estimated from the data. Then $\hat{n}$ is chosen to minimise $B(n)$. Since $\hat{I}_n$ will become singular for $n > n_0$ the second term will become large for $n > n_0$ and prevent over estimation. However the presence of the term $-nT\log T$ on the right side of (10) ensures that asymptotically this criterion will under estimate n. Indeed consider the case $n=1$ compared to $n=0$. When $n_0 = 1$ obtains then $\hat{I}_n$ will converge almost surely to a nonsingular matrix so that the term $-T\log T$ will dominate and eventually we will prefer $\hat{n}=0$. When $n_0 = 0$ is reached then it may be shown that

$$\log \hat{\sigma}_0^2 / \hat{\sigma}_1^2 = (\hat{\alpha} - \hat{\beta})^2 + o(T^{-1})$$

while

$$-\log\det(T^{-1}\hat{I}_1) = \log(1-\hat{\alpha}^2) + \log(1-\hat{\beta}^2) + \log(1-\hat{\alpha}\hat{\beta})^2 \quad -\log(\hat{\alpha}-\hat{\beta})^2. \tag{11}$$

If $|\hat{\beta}| \leqslant 1-\delta$ then since $(\hat{\alpha}-\hat{\beta}) \to 0$, a.s. the dominant terms in $T^{-1}\{B(0)-B(1)\}$ are

$$(\hat{\alpha}-\hat{\beta})^2 + \log(\hat{\alpha}-\hat{\beta})^2 + \log T$$

Since $(\hat{\alpha}-\hat{\beta})^2 = O(\log\log T/T)$ again the last term will eventually dominate and $\hat{n}=0$ will again be preferred. To avoid the underestimation (i.e., choice of $\hat{n}=0$ when $n_0=1$) one might eliminate the term $-n\log T$ from $B(n)$. Now, in our example, when $n_0=0$ certainly $\hat{n}=0$ will be chosen asymptotically. When $n_0=1$ holds then $\hat{\sigma}_1^2$ converges a.s. to $\sigma^2\{1+(\alpha-\beta)^2/(1-\alpha^2)\}$. Comparison of this with (11) shows that for α, β small and $\alpha \neq \beta$ but $(\alpha-\beta)$ small then $\hat{n}=0$ will asymptotically be preferred since $1+(\alpha-\beta)^2/(1-\alpha^2)$ will be near to 1 while $\hat{I}_1$ will be large so that

$\log\{\hat{\sigma}_1^2/\hat{\sigma}_0^2\hat{I}_1^{-1}\}$ will become large. Thus if the term $-n\log T$ is eliminated again one cannot in general get consistency. A third possibility is to consider $A(n)-T\log\det(\hat{I}_n)$. However again under some circumstances this will asymptotically, lead to under estimation, when $C_T/T\to 0$. The trouble is easy to see. Though the inclusion of the term $-T\log\det(\hat{I}_n)$ will act against over estimation, yet when the true k_0 is near to a k for which $n<n_0$ then this additional term will also become large at n_0 and lead to too small a value being chosen. Since under estimation of n_0 usually seems worse than over estimation it may be that formulae of the form (10) will not prove useful.

An alternative to order estimation via statistics such as $A(n)$ would be an hypothesis testing approach. In general it will be difficult, if not impossible, to obtain explicitly the asymptotic distribution of the likelihood ratio test statistic for testing $n=n_1$ against $n=n_0$ on the null hypothesis. The statistic is equivalent to

$$-T\{\log\det\hat{\Sigma}_{n_1}-\log\det\hat{\Sigma}_{n_0}\}.$$

For $r=1$, $n_1=1$, $n_0=0$ this statistic is described in (9) and its asymptotic distribution, on the null hypothesis, is described below that formula. For $r=1$, $p_0=1$, $q_0=0$; $p_1=2$, $q_1=1$ detailed calculation, that we do not present here, shows that the statistic is asymptotically distributed as

$$\sup_{-V\leqslant v\leqslant V}\{\phi(v)^2-\phi(v_0)^2\},\quad v_0=\log\{(1+\alpha_0)/(1-\alpha_0)\}$$

where V, $\phi(v)$ are as described below and and in (9). However other cases seem too complicated to be of any use. An alternative would be to construct a test statistic by choosing a suitable, prior, value for χ and optimising, for $n=n_1$, only over the ψ values for this value of χ. Of course k_0 will be set at the values estimated on the basis of $n=n_0$. Now the likelihood ratio test statistic will be found by standard methods and will evidently have a chi square distribution with $(n_1+n_0)r-2n_0r=(n_1-n_0)r$ degrees of freedom. For example, for $r=1$, $n_1=1$, $n_0=0$ then one may choose $\chi=\beta$, $\psi=\alpha-\beta$. Putting $\beta=0$ leads to the use of the first autocorrelation of the data as a test statistic. This is briefly discussed in ref. [3].

5. The consequences of overestimation of order

We have seen that if sufficient prior information is available reasonably to bound the estimate $\hat{k}$, a priori, (both by bounding $\hat{\tau}$ for each n and by bounding the zeros of $\det(\hat{k})$ away from the unit circle) then since the order of the process can be consistently estimated then multivariate ARMA models should be susceptible to fitting by reasonable algorithms. However, in practice the most commonly used criterion would be of the form of $AIC(n)$, which is not consistent. This is because one does not believe that there is any true n_0. Nevertheless it is apparent that there is some risk that $AIC(n)$ will over fit. This leads us to ask how important that would be. Consider the construction of a linear predictor using $\hat{k}$ for $n\geqslant n_0$. The best linear predictor is of the form $(I-k_0^{-1})y(T)$, where z^{-1} is now interpreted as the backward shift. If $\hat{k}$ is used instead of k_0 then the error is $(k_0^{-1}-\hat{k}^{-1})y(T)$. (We assume that the infinite past is used but this is inconsequential since terms for $t\leq 0$ contribute to $(k_0^{-1}-\hat{k}^{-1})$ a component that is, a.s., $O(\rho^T)$, $0<\rho<1$.) However,

$$(k_0^{-1}-\hat{k}^{-1})y(T)=\hat{h}^{-1}(\hat{h}-\hat{g}k_0)k_0^{-1}y(T)$$
$$=\hat{h}^{-1}(\hat{h}-\hat{g}k_0)\varepsilon(T)=\hat{h}^{-1}\left(\sum_{-p+1}^{\infty}\hat{\Psi}(i)z^{-1}\right)\varepsilon(T).$$

Now the coefficient matrices in the expression of $\hat{h}^{-1}$ and the $\hat{\Psi}(i)$ have elements also that decrease to zero as ρ^i, for some $0<\rho<1$. On the other hand, it is shown in refs. [8, 9] that the vector $\hat{\psi}$ essentially obeys the law of iterated logarithm, i.e.

$$\|\hat{\psi}\|=O\{(\log\log T/T)^{1/2}\}.$$

All elements of the $\Psi(i)$ are linear combinations of the elements of ψ with uniformly bounded coefficients. Hence,

$$\left\|\hat{h}^{-1}\left(\sum_{p+1}^{\infty}\hat{\Psi}(i)z^{-i}\right)\varepsilon(T)\right\|\leq C\left(\frac{\log\log T}{T}\right)^{1/2}\sum_{0}^{\infty}\rho_2^j\|\varepsilon(T-j)\|,\quad 0<\rho_2<1$$
$$=\left(\frac{\log\log T}{T}\right)^{1/2}w(T),$$

let us say, where $w(t)$ is a stationary scalar process. [The conditions of the theorems used here, see refs. [8, 9], require at least a finite fourth moment for the $\varepsilon(t)$.] These considerations show how fast the error in the linear predictor obtained from using $\hat{k}$ in place of k_0 converges to zero and how inconsequential, in a sense, is the use of $n>n_0$, provided that $n\leq N$, for some fixed N (of course the result would continue to hold if N increased sufficiently slowly with T). Of course, if $n<n_0$ the error in $(k_0^{-1}-\hat{k}^{-1})y(T)$ will not converge to zero, in general.

In the case where there is no true n_0 more accurate comparisons of the use of $AIC(n)$ for prediction have been made in ref. [11], to which we refer the reader.

Though prediction is a prime purpose of the estimation of ARMA systems it is not the only purpose. One might be concerned to estimate the structure in connection with problems of control. In that case over estimation of n_0 might be troublesome since undue effort might thereby be induced to stabilise some aspect of the system by relocating zeros or poles that are near to the unit circle, when those zeros and poles are spurious. In practice, of course, there will be no true n_0 but again trouble may be caused by estimating a system of too high order.

6. ARMAX models

If an additional term,

$$\sum_{1}^{s}\Delta(j)x(t-j),$$

is inserted on the right in (1), where the $x(t)$ are m-vectors generated by a process independent of the $\varepsilon(t)$ process then (1) is said to be ARMAX. Now k is replaced by $[k:l]$ where

$$l(z)=g^{-1}(z^{-1})\hat{d}(z^{-1}) \quad \text{and} \quad d(z)=\sum \Delta(j)z^{j}.$$

Then $[k:l]=g^{-1}[h:d]$ and the theory develops in almost the same way as before $M(p,q)$ being replaced by $M(p,q,s)$ wherein now $[A(p):B(q):\Delta(s)]$ is required to be of full rank. An alternative model is the transfer function model wherein the transfer functions, k, l, form $\varepsilon(t)$ and $x(t)$, respectively, to $y(t)$ are separately parameterised and two integer parameters, the McMillan degrees of k and l, respectively, are introduced. We will not go into the fairly obvious extensions of what has been said in this paper.

References

[1] Hannan, E.J. (1970). *Multiple Time Series*. Wiley, New York.

[2] Hannan, E.J., Dunsmuir, W. and Deistler, M. (1980). Estimation of vector Armax models. *J. Multivar. Anal.* **10**, 275–295.

[3] Hannan, E.J. (1982). Testing for autocorrelation and Akaike's criterion. In *Essays in Statistical Science* (*The Moran Festschrift*), J.M. Gani and E.J. Hannan, Eds. Applied Probability Trust, Sheffield.

[4] Rosenbrock, H.H. (1970). *State Space and Multivariable Theory*. Nelson, London.

[5] Deistler, M. and Hannan, E.J. (1981). Some properties of the parameterisation of ARMA systems of unknown order. *J. Multivar. Anal.* **11**.

[6] Akaike, H. (1976). Canonical correlation analysis of time series and the use of an information criterion. In *System Identification*: *Advances and Case Studies*. R.K. Mehra and D.G. Lainiotis, Eds. pp. 27–96. Academic Press, New York

[7] Deistler, M., Dunsmuir, W.T.M. and Hannan, E.J. (1978). Vector linear time series models: Corrections and extensions. *J. App. Prob.* **10**, 360–372.

[8] Hannan, E.J. (1981). Estimating the dimension of a linear system. *J. Multivar. Anal.* **11**.

[9] Hannan, E.J. (1980). The estimation of the order of an ARMA process. *Ann. Statist.* **8**, 1071–1081.

[10] Young, P., Jakeman, A. and McMurtrie, R. (1980). An instrumental variable method for model order identification. *Automatica* **16**, 281–294.

[11] Shibata, R. (1976) Selection of the order of an autoregressive model by Akaike's information criterion. *Biometrika* **63**, 117–126.

G. Kallianpur, P.R. Krishnaiah, J.K. Ghosh, eds., *Statistics and Probability: Essays in Honor of C.R. Rao*
© North-Holland Publishing Company (1982) 317–323

ESTIMATION IN THE PRESENCE OF A THRESHOLD THEOREM: PRINCIPLES AND THEIR ILLUSTRATION FOR THE TRAFFIC INTENSITY

C. C. HEYDE
Division of Mathematics and Statistics, CSIRO, Canberra City, Australia

This paper is concerned with the problem of estimating the parameter $\boldsymbol{\theta}$ of a stochastic process in which the asymptotic behaviour of the process is governed by a threshold theorem and depends critically on whether, say, $\boldsymbol{C}^{\mathrm{T}}\boldsymbol{\theta}<0$ or $\boldsymbol{C}^{\mathrm{T}}\boldsymbol{\theta}\geqslant 0$, $\boldsymbol{C}$ being a vector of constants. Bayesian methods have some distinct advantages in this context and allow a unified approach to estimation without the intervention of the threshold theorem. A general result on asymptotic posterior normality is provided, and its use is illustrated by application to two problems on the estimation of the traffic intensity in a queueing system.

1. Introduction

There are many stochastic processes whose intrinsic asymptotic behaviour is governed by a threshold theorem on some basic parameter(s). Suppose there is just one parameter for definiteness. Then, in this situation, the asymptotic behavior of the process is radically different depending on whether the parameter lies above or below the critical threshold level. This poses a serious problem when the parameter of the threshold theorem needs to be estimated. It is usually possible to obtain a point estimate, for example by maximum likelihood, but ordinarily one cannot provide confidence intervals without assuming that the parameter is on one side or the other of the critical value. This is rather unsatisfactory, both practically and aesthetically.

The difficulty raised by the presence of a threshold theorem can often be avoided if a Bayesian approach to the problem is adopted. This has the advantage of providing a unified procedure for all values of the parameter. There is, of course, the inevitable price to pay in terms of uncertainty over the prior distribution, but results on asymptotic posterior normality can often be invoked. We shall illustrate these points by providing a general asymptotic posterior normality result and applying it to two problems involving estimation of the traffic intensity ρ in a queueing system. Here the case $\rho<1$ corresponds to a stable system (ultimately a stationary regime) and $\rho\geqslant 1$ to a situation of prospective increasing congestion.

2. A general result on asymptotic posterior normality

In our general result we will treat the case of s parameters. Let $\{(\Omega_t, F_t)\}$ be a family of measurable spaces, t being a discrete or continuous time parameter. Suppose that

$P_t(\boldsymbol{\theta})$ is a probability measure defined on (Ω_t, F_t) and depending on the parameter $\boldsymbol{\theta}\in\Theta\subset\mathbb{R}^s$. We assume that $P_t(\boldsymbol{\theta})$ is absolutely continuous with respect to a σ-finite measure μ_t and let $p_t(\boldsymbol{\theta})$ be the density of $P_t(\boldsymbol{\theta})$ with respect to μ_t. Then, the log-likelihood function $L_t(\boldsymbol{\theta})=\log p_t(\boldsymbol{\theta})$ exists a.e. (μ_t). Let $\hat{\boldsymbol{\theta}}_t$ be a maximum likelihood estimator, namely a value of $\boldsymbol{\theta}$ maximizing $L_t(\boldsymbol{\theta})$ (uniqueness of $\boldsymbol{\theta}$ is not essential).

Let A be a real $s\times s$ matrix. The spectral norm of A is defined by

$$\|A\|^2=\sup\{\|Ax\|^2:\|x\|^2=1\}=\lambda_{\max}(A^{\mathrm{T}}A),$$

$\lambda_{\max}(B)$, $\lambda_{\min}(B)$ denoting the maximum and minimum eigenvalues of a (symmetric) matrix B and T denoting the transpose. Thus, if B is symmetric and has eigenvalues $\lambda_1,\ldots,\lambda_s$, then $\|B\|=\max_{1\leqslant k\leqslant s}|\lambda_k|$. Also, for a vector $\boldsymbol{v}$ we will use $\boldsymbol{v}>\boldsymbol{0}$ to mean that each element of $\boldsymbol{v}$ is positive.

Now suppose that $\Theta\subset\mathbb{R}^s$ and that $\boldsymbol{\theta}_0$ is an interior point of Θ. We will impose the following conditions:

(i) The prior density $\pi(\boldsymbol{\theta})$ of $\boldsymbol{\theta}$ is continuous and positive at $\boldsymbol{\theta}_0$.

(ii) $L_t(\boldsymbol{\theta})$ is a.s. twice continuously differentiable with respect to $\boldsymbol{\theta}$ in some neighbourhood $N_0\subset\Theta$ of $\boldsymbol{\theta}_0$.

(iii) $\lambda_{\min}[-L_t''(\boldsymbol{\theta}_0)]\to\infty$ a.s. as $t\to\infty$.

(iv) Given any $\boldsymbol{\delta}>\boldsymbol{0}$ such that $N_0(\boldsymbol{\delta})=[\boldsymbol{\theta}_0-\boldsymbol{\delta},\boldsymbol{\theta}_0+\boldsymbol{\delta}]\subset\Theta$, there exists $k(\boldsymbol{\delta})>0$ for which

$$\lim_{t\to\infty} P_{\boldsymbol{\theta}_0}\left\{\sup_{\boldsymbol{\theta}\in\Theta\setminus N_0(\boldsymbol{\delta})}\|-L_t''(\boldsymbol{\theta}_0)\|^{-1}[L_t(\boldsymbol{\theta})-L_t(\boldsymbol{\theta}_0)]<-k(\boldsymbol{\delta})\right\}=1.$$

(v) Given $\varepsilon>0$, there exists $\boldsymbol{\delta}=\boldsymbol{\delta}(\varepsilon)$ satisfying

$$\lim_{t\to\infty} P_{\boldsymbol{\theta}_0}\left\{\sup_{\boldsymbol{\theta}\in N_0(\boldsymbol{\delta})}\left\|\left[-L_t''(\hat{\boldsymbol{\theta}}_t)\right]^{-1/2}\left[L_t''(\hat{\boldsymbol{\theta}}_t)-L_t''(\boldsymbol{\theta})\right]\left[-L_t''(\hat{\boldsymbol{\theta}}_t)\right]^{-1/2}\right\|<\varepsilon\right\}=1.$$

Conditions (i) and (ii) require no comment while Condition (iii) essentially ensures that the process does not degenerate. Condition (iv) is little more than a simple sufficient condition for the consistency of $\hat{\boldsymbol{\theta}}_t$ for $\boldsymbol{\theta}_0$ while condition (v) is a kind of continuity condition for L_t''.

If (iii) and (v) hold, it follows that $\lambda_{\min}(-L_t''(\hat{\boldsymbol{\theta}}))\to\infty$ a.s. and hence we may define $V_t^{1/2}$ to be the nonnegative definite square root of $[-L_t''(\hat{\theta}_t)]^{-1}$, knowing that this will exist with probability tending to unity as $t\to\infty$.

Let $\boldsymbol{\Phi}$ be the measure corresponding to the multivariate normal $N(\boldsymbol{0}, I_s)$ distribution, namely,

$$\boldsymbol{\Phi}(B)=(2\pi)^{-s/2}\int_B e^{-\boldsymbol{u}^{\mathrm{T}}\boldsymbol{u}/2}\,\mathrm{d}\boldsymbol{u}.$$

Theorem. *Assume that Conditions (i)–(v) hold and that $\boldsymbol{\theta}_0$ is an interior point of Θ. Then, if $-\infty\leqslant\boldsymbol{b}\leqslant\boldsymbol{a}<\infty$, the posterior probability that $\boldsymbol{b}\leqslant V_t^{-1/2}(\boldsymbol{\theta}-\hat{\boldsymbol{\theta}}_t)\leqslant\boldsymbol{a}$ converges in $P_{\boldsymbol{\theta}_0}$ probability to $\boldsymbol{\Phi}$ ($[\boldsymbol{b},\boldsymbol{a}]$).*

This result is a multivariate extension of the theorem of Heyde and Johnstone (1979) and has been phrased to cope with continuous as well as discrete time. There is no special difficulty in the multivariate extension and the proof will be omitted. The basic methods, which also cope with continuous time, are essentially those of Walker (1969) and details are given in Johnstone (1978).

3. Applications to the traffic intensity

Our first application concerns the estimation of the traffic intensity ρ in a single server queue on the basis of observations on the queue length.

In the case of the $M/G/1$ queue there is a convenient interpretation, first noted by Kendall (1951), for the queue length process in terms of a Bienaymé–Galton–Watson branching process. It is possible to regard the first customer in the queue as the ancestor in the branching process, and the arrivals during his service period as his offspring. These in turn produce their own offspring. Then, the offspring distribution in the branching process is given by the probabilities,

$$p_j = \int_0^\infty e^{-\lambda t}(\lambda t)^j (j!)^{-1} \mathrm{d}F(t), \quad j \geqslant 0,$$

where F is the service time distribution function and λ is the rate parameter of the Poisson input stream. Furthermore, the mean of the offspring distribution is

$$\rho = \sum_{j=1}^{\infty} j p_j = \lambda \int_0^\infty t \, \mathrm{d}F(t),$$

the traffic intensity. If T has the distribution of the busy period, then $P(T<\infty)<1$ if and only if $\rho>1$. There is a critical threshold at $\rho=1$.

It is possible to provide a point estimate on ρ by using standard theory on the estimation of the mean of the offspring distribution in a branching process [e.g. Heyde (1975)]. However, confidence statements cannot avoid the criticality theorem for branching processes.

To indicate the type of results which can be obtained using the theorem of Section 2 we shall consider the special case of an $M/E_k/1$ queue where the parameter k in the Erlang distribution E_k is assumed known. Then, if

$$F(t) = \left(\beta^k/\Gamma(k)\right) \int_0^t u^{k-1} e^{-\beta u} \mathrm{d}u, \quad \beta>0,$$

we find that

$$p_j = \frac{\Gamma(j+k)}{j!\Gamma(k)} \frac{(\rho/k)^j}{(1+\rho/k)^{j+k}}, \quad j \geqslant 0,$$

where $\rho = k\lambda/\beta$ and $\{p_j\}$ has a power series distribution.

Suppose that observation begins when a single customer arrives at an unoccupied server. Let Y_0 denote the number of busy periods completed during observation. We shall need to distinguish between the cases $Y_0<1$ and $Y_0 \geqslant 1$. For either of these, however, let Z_j be the number of customers who enter the queue during the service period of the jth customer. We have

$$P(Z_1=z|Z_0=y)=\sum_{x_1+\cdots+x_y=z} p_{x_1}\cdots p_{x_y}$$

$$=\frac{(\rho/k)^z}{(\Gamma(k))^y(1+\rho/k)^{z+yk}}\sum_{x_1+\cdots+x_y=z}\frac{\Gamma(x_1+k)\cdots\Gamma(x_y+k)}{x_1!\cdots x_y!},$$

so that

$$P(Z_0=z_0,\ldots,Z_n=z_n)=\left\{\prod_{i=1}^{n}P(Z_i=z_i|Z_{i-1}=z_{i-1})\right\}P(Z_0=z_0)$$

$$=\left\{\prod_{i=1}^{n}P(Z_1=z_1|Z_0=z_0)\right\}P(Z_0=z_0)$$

$$=P(Z_0=z_0)K(z_0,\ldots,z_n)\frac{(\rho/k)^{z_1+\cdots+z_n}}{(1+\rho/k)^{z_1+\cdots+z_n+k(z_0+\cdots+z_{n-1})}}, \tag{1}$$

where $K(z_0,\ldots,z_n)$ does not depend on ρ.

In the case $Y_0<1$ let $Y_n=1+Z_1+\cdots+Z_n$ denote the number of customers who enter the queue prior to the completion of service of the nth customer. Then, by using (1) we find that the log likelihood $L_n(\rho)$ is given by

$$L_n(\rho)=\log K(Z_1,\ldots,Z_n)+(Y_n-1)\log(\rho/k)$$
$$-(Y_n-1+kY_{n-1})\log(1+\rho/k).$$

The maximum likelihood estimator of ρ is

$$\hat{\rho}=(Y_n-1)/Y_{n-1},$$

which is well known to be strongly consistent for ρ_0 [e.g. Heyde (1970)]. The conditions of the theorem are straightforward to check and we find that

$$-L_n''(\hat{\rho})=Y_{n-1}^3k(Y_n-1)^{-1}(Y_n-1+kY_{n-1})^{-1}$$

and

$$\frac{\rho-(Y_n-1)Y_{n-1}^{-1}}{(Y_n-1)^{1/2}(Y_n-1+kY_{n-1})^{1/2}Y_{n-1}^{-1}k^{-1/2}}Y_{n-1}^{1/2}\xrightarrow{p}N(0,1) \tag{2}$$

as $Y_{n-1}\to\infty$. (Note that, for fixed ρ_0, the denominator

$$(Y_n-1)^{1/2}(Y_n-1+kY_{n-1})^{1/2}Y_{n-1}^{-1}k^{-1/2}\to\rho_0^{1/2}(\rho_0+k)^{1/2}k^{-1/2}\quad\text{a.s.}$$

as $n\to\infty$.)

In the case $Y_0\geqslant 1$, let Y denote the number of customers who enter the queue during these Y_0 busy periods. With the Bienaymé–Galton–Watson process model, this is equivalent to beginning observation from a queue of length Y_0. From (1) we find that the log likelihood $L_n(\rho)$ in this case is given by

$$L_n(\rho)=K+(Y-Y_0)\log(\rho/k)-(Y-Y_0+kY)\log(1+\rho/k),$$

where K does not involve ρ. Then, the maximum likelihood estimator of ρ is

$$\hat{\rho}=(Y-Y_0)/Y,$$

and in this case the theorem yields

$$\frac{\rho-(Y-Y_0)Y^{-1}}{(Y-Y_0)^{1/2}(Y-Y_0+kY)^{1/2}Y^{-1}k^{-1/2}}Y^{1/2}\xrightarrow{p}N(0,1) \tag{3}$$

as $Y\to\infty$.

The results (2) and (3) avoid any difficulty over the magnitude of ρ and allow straightforward probabilistic assessments about ρ to be made. Similar results can also be derived from the theorem of Heyde (1979) where a Bayesian treatment of the mean of the offspring distribution in a Bienaymé–Galton–Watson process has been provided for use in the assessment of the potential severity of epidemics.

In the second application, we will deal with an $M/M/s$ system for which the queue length process is continuously observed. The queue consists of s servers with a common waiting line. Customers arrive in a Poisson stream at rate λ and the service time distribution is exponential with mean μ^{-1}. If a customer enters the system when all the servers are busy he joins the waiting line. This system is described by a process $X(t)$ in which the state represents the number of busy servers plus the length of the waiting line. It changes only through transitions from states to their nearest neighbours. Further,

$$\begin{aligned}
P(X(t+\Delta t)=i+1|X(t)=i)&=\lambda\Delta t+o(\Delta t), && i\geqslant 0,\\
P(X(t+\Delta t)=i-1|X(t)=i)&=i\mu\Delta t+o(\Delta t), && 0\leqslant i\leqslant s,\\
&=s\mu\Delta t+o(\Delta t), && i\geqslant s,\\
P(X(t+\Delta t)=j|X(t)=i)&=o(\Delta t), && |i-j|>1.
\end{aligned}$$

Here $\boldsymbol{\theta}^{\mathrm{T}}=(\lambda,\mu)$. Let $u_i(t)$ denote the number of transitions in the sample from i to $i+1$, $d_i(t)$ the number of transitions from i to $i-1$, and $\gamma_i(t)$ the total length of time spent in state i during $(0,t]$. Then, writing

$$\begin{aligned}
&q_{i,i+1}(\boldsymbol{\theta})=\lambda, \quad i\geqslant 0,\\
&q_{i,i-1}(\boldsymbol{\theta})=i\mu, \quad 0\leqslant i\leqslant s, \qquad q_i(\boldsymbol{\theta})=\lambda+i\mu, \quad 0\leqslant i\leqslant s,\\
&\qquad\quad\;\; =s\mu, \quad i\geqslant s, \qquad\qquad\qquad\;\; =\lambda+s\mu, \quad i\geqslant s,
\end{aligned}$$

the log likelihood is given by Billingsley (1961 p. 44)

$$\begin{aligned}
L_t(\boldsymbol{\theta})=&\sum_{i=0}^{\infty}u_i(t)\log q_{i,i+1}(\boldsymbol{\theta})+\sum_{i=1}^{\infty}d_i(t)\log q_{i,i-1}(\boldsymbol{\theta})\\
&-\sum_{i=0}^{\infty}\gamma_i(t)q_i(\boldsymbol{\theta}).
\end{aligned}$$

Consequently, if $u_t=\Sigma_{i=0}^{\infty}u_i(t)$ and $d_t=\Sigma_{i=1}^{\infty}d_i(t)$ are, respectively, the numbers of upward and downward transitions in the sample, then

$$L_t(\boldsymbol{\theta})=u_t\log\lambda+d_t\log\mu-t\lambda-m_t\mu+C_t, \tag{4}$$

where

$$m_t = \sum_{i=1}^{\infty} i\gamma_i(t) + \sum_{i=s+1}^{\infty} s\gamma_i(t), \quad C_t = \sum_{i=1}^{s} d_i(t)\log i + \sum_{i=s+1}^{\infty} d_i(t)\log s.$$

The maximum likelihood estimators are thus

$$\hat{\lambda}_t = u_t/t \quad \text{and} \quad \hat{\mu}_t = d_t/m_t$$

and it is easily checked that the conditions of the theorem are satisfied if $(\hat{\lambda}_t, \hat{\mu}_t)$ is consistent for (λ, μ) as $t \to \infty$.

Consistency for $\lambda \leqslant s\mu$ is covered by the theory in Billingsley (1961). Furthermore, for $\lambda > s\mu$ the queue eventually grows unboundedly with probability one and hence as $t \to \infty$,

$$\begin{aligned} m_t &= \sum_{i=1}^{s} i\gamma_i(t) + s \sum_{i=s+1}^{\infty} \gamma_i(t) \\ &\sim s \sum_{i=s+1}^{\infty} \gamma_i(t) \sim st \quad \text{a.s.} \end{aligned}$$

Also, if $P_i(t) = P(X(t) = i)$, we have

$$Eu_t = \lambda \sum_{i=0}^{\infty} \int_0^t P_i(u)\,\mathrm{d}u = \lambda t$$

and

$$\begin{aligned} Ed_t &= \sum_{i=1}^{\infty} Ed_i(t) \\ &= \sum_{i=1}^{s} i\mu \int_0^t P_i(u)\,\mathrm{d}u + \sum_{i=s+1}^{\infty} s\mu \int_0^t P_i(u)\,\mathrm{d}u. \end{aligned}$$

However, since each state is transient, $t^{-1}Ed_i(t) \to 0$ as $t \to \infty$, $0 \leqslant i \leqslant s$, and hence

$$Ed_t \sim s\mu \int_0^t \left[\sum_{i=s+1}^{\infty} P_i(u) \right] \mathrm{d}u \sim s\mu t = m_t\mu \quad \text{as } t \to \infty.$$

Then, since the input stream is Poisson with rate λ, $u_t/Eu_t \to 1$ a.s. as $t \to \infty$, while the output behaves asymptotically as a Poisson stream with rate $s\mu$ and hence $d_t/Ed_t \to 1$ a.s. as $t \to \infty$. Consistency of the maximum likelihood estimators for $\lambda > s\mu$ and hence in general follows directly.

In returning to (4), we find that

$$[-L_t''(\boldsymbol{\theta})] = \operatorname{diag}(u_t\lambda^{-2}, d_t\mu^{-2})$$

and the theorem gives

$$\begin{pmatrix} u_t^{1/2} t^{-1}(\hat{\lambda}_t - \lambda) \\ d_t^{1/2} m_t^{-1}(\hat{\mu}_t - \mu) \end{pmatrix} \xrightarrow{p} N(\mathbf{0}, I_2),$$

as $t \to \infty$ regardless of the parameter values λ, μ. This result should be contrasted with the standard maximum likelihood approach which gives a corresponding result provided $\rho = \lambda/s\mu < 1$ (Billingsley 1961). For comments on the $M/M/1$ case see also Cox (1965).

References

Billingsley, P. (1961). *Statistical Inference for Markov Processes.* Univ. of Chicago Press, Chicago, IL.

Cox, D.R. (1965). Some problems of statistical analysis connected with congestion. In *Proceedings of the Symposium on Congestion Theory*, Smith, W.L. and Wilkinson, W.E., Eds. Univ. of North Carolina Press, Chapel Hill, NC.

Heyde, C.C. (1970). Extension of a result of Seneta for the supercritical Galton–Watson process. *Ann. Math. Statist.* **41**, 739–742.

Heyde, C.C. (1975). Remarks on efficiency in estimation for branching processes. *Biometrika* **62**, 49–55.

Heyde, C.C. (1979). On assessing the potential severity of an outbreak of a rare infectious disease: a Bayesian approach. *Austral. J. Statist.*, **21**, 282–292.

Heyde, C.C. and Johnstone, I.M. (1979). On asymptotic posterior normality for stochastic processes. *J. R. Statist. Soc. B.* **41**, 184–189.

Johnstone, I.M. (1978). *Problems in Limit Theory for Martingales and Posterior Distributions from Stochastic Processes.* M.Sc. Thesis, Australian National Univ., Canberra.

Kendall, D.G. (1951). Some problems in the theory of queues. *J. R. Statist. Soc.* **13**, 151–185.

Walker, A.M. (1969). On the asymptotic behaviour of posterior distributions. *J. R. Statist. Soc. B* **31**, 80–88.

G. Kallianpur, P.R. Krishnaiah, J.K. Ghosh, eds., *Statistics and Probability: Essays in Honor of C.R. Rao*
© North-Holland Publishing Company (1982) 325–327

MINIMAX ESTIMATION IN SIMPLE RANDOM SAMPLING

J.L. HODGES, Jr. and E.L. LEHMANN*
University of California, Berkeley, CA, U.S.A.

The classical result on the sample mean as minimax estimator of the population mean in simple random sampling from a finite population is due to Aggarwal (1959). If $y_1, \ldots, y_N$ denote the population values, and

$$\bar{y} = \sum y_i / N, \qquad \tau^2 = \sum (y_i - \bar{y})^2 / N, \tag{1}$$

Aggarwal's result** states that the sample mean M is a minimax estimator of $\bar{y}$ when the loss function is squared error and the parameter space is the set of all $(y_1, \ldots, y_N)$ satisfying

$$\tau^2 \leqslant c \tag{2}$$

for some c.

With squared error as loss, some restriction on the y's is needed to give meaning to the minimax problem. However (2), although theoretically convenient, is not very natural from the point of view of applications. A more natural parameter space is

$$a \leqslant y_i \leqslant b \quad \text{for all } i \tag{3}$$

which, by a linear transformation, can be reduced to

$$0 \leqslant y_i \leqslant 1 \quad \text{for all } i. \tag{4}$$

Unfortunately, for this parameter space the minimax estimator is no longer the sample mean M.

To see this, note that

$$E(\alpha M + \beta - \bar{y})^2 = \alpha^2 \sigma_M^2 + [\beta + (\alpha - 1)\bar{y}]^2, \tag{5}$$

where

$$\sigma_M^2 = \frac{N-n}{(N-1)n} \tau^2 \tag{6}$$

is the risk of the sample mean M. We will now show that there exist α and β for which the maximum of (6) exceeds that of (5), so that the resulting estimator δ has a smaller maximum risk than M. We will show further that δ is in fact minimax.

*This paper was prepared with the support of National Science Foundation Grant MCS76-10238 and Office of Naval Research Contract No. N00014-750C-0444/NR 042-036.

**There is a gap in Aggarwal's proof. However, the result is correct; see Bickel and Lehmann, "A Minimax Property of the Sample Mean", to be published.

It is well known that τ^2 is maximum, subject to (4), when all the y's are 0 or 1, and when the proportion of 1's is $\frac{1}{2}$ for N even, and as close to $\frac{1}{2}$ as possible for N odd. This shows that

$$\max \sigma_M^2 = \begin{cases} \frac{1}{4}\dfrac{N-n}{(N-1)n}, & \text{when } N = \text{even}, \\ \frac{1}{4}\dfrac{N-n}{(N-1)n}\cdot\dfrac{N^2-1}{N^2}, & \text{when } N = \text{odd}. \end{cases} \tag{7}$$

It follows from (4) that $\tau^2 \leqslant \bar{y} - \bar{y}^2$ and hence that

$$E(\alpha M + \beta - \bar{y})^2 \leqslant \alpha^2 \frac{N-n}{(N-1)n}\bar{y}(1-\bar{y}) + [\beta + (\alpha-1)\bar{y}]^2, \tag{8}$$

with equality holding if and only if all the y's are equal to 0 or 1. The right side of (8) is a quadratic in $\bar{y}$ which reduces to the constant β^2 when

$$\alpha = \frac{1}{1+\sqrt{\dfrac{N-n}{(N-1)n}}}, \qquad \beta = \tfrac{1}{2}(1-\alpha). \tag{9}$$

The maximum risk of $\delta = \alpha M + \beta$ for these values of α and β is therefore

$$\tfrac{1}{4}(\alpha-1)^2 = \tfrac{1}{4}\frac{N-n}{(N-1)n}\left[\frac{1}{1+\sqrt{\dfrac{N-n}{(N-1)n}}}\right], \tag{10}$$

and this maximum is attained for any population of 0's and 1's.

It is easily checked that (10) is less than (7) for all $n < N$ so that M is not minimax.

On the other hand, we will now show that δ is minimax. For this purpose it is enough to prove that δ is minimax when the y's can take on only the values 0 and 1. In this case the number of one's in the sample constitutes a sufficient statistic and has a hypergeometric distribution. The population mean $\bar{y}$ is just D/N where D denotes the number of ones in the population. However, a minimax estimator for D was obtained in ref. [2] and, when divided by N agrees with that defined by (9). This completes the proof.

The minimax estimator δ is of course not unbiased. If attention is restricted to unbiased estimators, it was recently shown by Joshi (1979) as part of a stronger and more general result that the sample mean is minimax.

It is interesting to compare M and δ in terms of their relative efficiency

$$e(M,\delta) = E(\delta - \bar{y})^2 / \mathrm{Var}(M). \tag{11}$$

Using (9) and the fact that $(\alpha-1)^2 = \alpha^2(N-n)/n(N-1)$, it is easily seen that

$$e(M,\delta) = \frac{\alpha^2}{4\bar{y}(1-\bar{y})} > \frac{n}{(1+\sqrt{n})^2}\,\frac{1}{4\bar{y}(1-\bar{y})}. \tag{12}$$

The right side is just the corresponding efficiency discussed in ref. [2] for the binomial case with $\bar{y}$ in place of p. This shows that, even more strongly than in the bionomial case, for all but very small n, the risk of δ exceeds that of M for all $\bar{y}$ outside a small interval $(\frac{1}{2}-d, \frac{1}{2}+d)$; that $d \to 0$ as $n \to \infty$; and that the improvement of δ over M inside the interval becomes proportionally negligible as n becomes large. This can be summarized by saying that M is asymptotically subminimax as $n \to \infty$.

In defining the problem at the beginning of the paper no mention is made of the unit labels. However, availability of the labels does not change the minimax character of the estimator $\alpha M + \beta$ with α, β given by (9). The reason for this is the fact that the problem of estimating $\bar{y}$ is invariant under permutations of the labels. Since the group of these permutations is finite, there exists a minimax estimator which is invariant. However, an estimator is invariant under all permutations of the labels if and only if it does not depend on the labels.

References

[1] Aggarwal, O.P. (1959). Bayes and minimax procedures in sampling from finite and infinite populations. *Ann. Math. Statist.* **30**, 206–218.

[2] Hodges, Jr., J.L. and Lehmann, E.L. (1950). Some problems in minimax point estimation. *Ann. Math. Statist.* **21**, 182–197.

[3] Joshi, V.M. (1979). The best strategy for estimating the mean of a finite population. *Ann. Statist.* **7**, 531–536.

G. Kallianpur, P.R. Krishnaiah, J.K. Ghosh, eds., *Statistics and Probability: Essays in Honor of C.R. Rao*
© North-Holland Publishing Company (1982) 329–341

ANALYSES OF VARIANCE DETERMINED BY SYMMETRY AND COMBINATORIAL PROPERTIES OF ZONAL POLYNOMIALS

A.T. JAMES
The University of Adelaide, Adelaide, S.A., Australia

If G is a group of permutations of the elements of an arbitrary column vector, x, in a vector space V, we show how to analyse V into irreducible invariant subspaces V_i and calculate the symmetric idempotent matrices E_i which project orthogonally upon the V_i. The results are used to calculate *zonal polynomials*, in terms of which many multivariate distributions can be expanded, and have possible application to the inversion of matrices of normal and missing value equations arising from symmetrical experimental designs or situations.

1. Introduction

Let x be a column vector of observations on the n plots of an experimental design. The quadratic forms $x'E_\nu x$ of a decomposition of the sums of squares in an analysis of variance

$$x'x = x'E_1x + x'E_2x + \cdots \tag{1.1}$$

have idempotent matrices E_ν corresponding to orthogonal projections upon mutually orthogonal subspaces V_ν spanning the n dimensional sample space V of vectors x of observations. A group G of permutations of the plots induces a group of linear transformations of the sample space V by permutation matrices P. For example, if G contains an element (12) interchanging the first two plots, then this induces the linear transformation,

$$\begin{bmatrix} x_1 \\ x_2 \\ \vdots \\ x_n \end{bmatrix} \to \begin{bmatrix} 0 & 1 & & \\ 1 & 0 & & \\ & & 1 & \\ & & & 1 \end{bmatrix} \begin{bmatrix} x_1 \\ x_2 \\ \vdots \\ x_n \end{bmatrix}, \tag{1.2}$$

of the vectors x in the sample space.

The linear transformations (1.2) are a *representation* of G (see [1, 3, 8]). The term $x'E_\nu x$ will be invariant if and only if the subspace V_ν upon which E_ν projects is an *invariant subspace* under the representation of G, i.e. a subspace V_ν whose vectors are transformed into vectors again in V_ν by all transformations of G.

An invariant subspace, if not *irreducible* itself, is a union of *irreducible invariant subspaces*, i.e. subspaces containing no smaller invariant subspaces. Hence the terms of an invariant analysis of variance are sums of terms of what we may call an *irreducible invariant analysis of variance*, i.e. one whose terms are the squares of the lengths of orthogonal projections upon irreducible invariant subspaces.

The decomposition of a vector space V, in which a permutation group is represented, into its irreducible invariant subspaces has application to the solution of normal equation and in elucidating the structure of an experimental design. One of the most convenient and concise ways of specifying the invariant subspaces of V and the idempotents which project upon them, is to write down the decomposition of the sum of squares in an analysis of variance.

Sometimes the decomposition of V into irreducible invariant subspaces is obvious, e.g. the analysis of variance for a randomized block; if not *group characters* can be used.

2. Diallele crosses

Yates [12] has discussed this design and derived its analysis.

Suppose a brother and a sister are taken from each of, say, 4 families, and all possible crosses produced:

		female 1	2	3	4
	1	·	x	x	x
male	2	x	·	x	x
	3	x	x	·	x
	4	x	x	x	·

It is desired to analyse a set of measurements, one being taken upon an offspring of each cross; leaving out, for simplicity, the brother–sister matings.

The order in which the families are presented is quite arbitrary; but if the males are permuted their sisters must be permuted in a like manner. Hence the symmetry group of the design is the symmetric group S_4 on the 4 families.

We now want to find:

(1) which irreducible representations of S_4 occur in V,

(2) what the invariant subspaces of V in which they occur, are.

Group characters will answer question (1). The following table gives the character χ of our representation together with a table of characters of the irreducible representations of S_4 copied from Littlewood [8].

Class order χ	(1^4) 1 12	$(1^2 2)$ 6 2	$(1\,3)$ 8 0	(4) 6 0	(2^2) 3 0	Number of times representation appears
[4] (identity)	1	1	1	1	1	1
[3 1] (deviations)	3	1	0	-1	-1	2
$[2^2]$	2	0	-1	0	2	1
$[2\,1^2]$	3	-1	0	1	-1	1
$[1^4]$ (alternating)	1	-1	1	-1	1	0

Each partition of the number 4 determines a class of conjugate elements of the symmetric group S_4; e.g. the partition $(1^4)=1+1+1+1$ corresponds to the class consisting of only the identity element, the partition $(1^2 2)=2+1+1$ to the transpositions of two elements, the partition $(1\,3)=3+1$ to the cycles of 3, etc. Since, for a given representation, all group elements of a conjugate class have the same character, the character of a representation of S_4 is given by stating a value corresponding to each partition of 4.

Let us calculate the character of the representation of S_4 given by the permutations of the crosses. The identity element, comprising the class (1^4), leaves all 12 crosses invariant. Hence χ has the value 12 for this class. Any transposition, e.g. the interchange of the first 2 families, obviously leaves only 2 crosses invariant. Thus χ has the value 2 for the class $(1^2 2)$. Since all the other permutations leave no crosses invariant, χ is zero for all the remaining classes.

The values of χ are given in the third row of the table. The next rows have the character of the five irreducible representations of the symmetric group S_4, each representation being denoted by a partition of 4, written in square brackets. The last column has the number of times each irreducible representation appears in our representation. This number is obtained by using formula;

$$g^{-1} \sum_{\sigma \in G} \overline{\chi_\nu(\sigma)}\,\chi(\sigma) \tag{2.1}$$

where g is the order of the group G; e.g. the number of times the deviations representation [3 1] appears is

$$(1\times 12\times 3+6\times 2\times 1+8\times 0\times 0+6\times 0\times(-1)+3\times 0\times(-1)/24)=2.$$

Thus our representation breaks up into

$$[4]+2[3\,1]+[2\,2]+[2\,1\,1] \tag{2.2}$$

of dimensions 1, 2×3, 2, 3 respectively. The 12 dimensional vector space V, of which the vector x of observations is an element, splits up into the following irreducible invariant subspaces,

$$V=V_{[4]}+\left(V_{[3\,1]}^{[1]}\cup V_{[3\,1]}^{[2]}\right)+V_{[2\,2]}+V_{[2\,1\,1]}. \tag{2.3}$$

Since the representations [4], [3 1], [2 2], [2 1 1] are inequivalent, it follows from group representations theory that $V_{[4]}$, $V_{[2\,2]}$ and $V_{[2\,1\,1]}$ are unique mutually orthogonal subspaces of V. $V_{[3\,1]}^{[1]}$ and $V_{[3\,1]}^{[2]}$ are not unique and not necessarily orthogonal, though a pair of orthogonal ones can always be chosen. However, their union $V_{[3\,1]}^{[1]}\cup V_{[3\,1]}^{[2]}$ is unique and orthogonal to $V_{[4]}$, $V_{[2\,2]}$ and $V_{[2\,1\,1]}$.

The identity representation [4] corresponds to the correction factor, i.e. the square of the grand total divided by 12; and the two "deviations" representations [3 1], each of which are associated with the sum of squares of deviations of 4 quantities about their mean, evidently correspond to the "between rows" and "between columns" sums of squares in the analysis of variance. The representations [2 2] and [2 1 1] must correspond to interactions.

The idempotents, E_ν, of the irreducible invariant analysis of variance can be calculated as elements of an algebra, the invariant relationship algebra which we shall now investigate.

3. The invariant relationship algebra

This is the centralizer of the representation of G in V i.e. the algebra $\mathcal{B}$ of all $n \times n$ matrices B which commute with all permutation matrices $P(\sigma)$ representing the group;

$$PB=BP, \qquad P=P(\sigma), \quad \sigma \in G.$$

Since P is an orthogonal matrix

$$PBP'=B. \tag{3.1}$$

The algebra $\mathcal{B}$ is a relationship algebra in the sense of James [4] and its *primitive* idempotents i.e. idempotents which are not sums of other idempotents in $\mathcal{B}$, project upon the irreducible invariant subspaces of V and thus determine an irreducible invariant analysis of variance.

A basis for the algebra can be found as follows. The set of n^2 ordered pairs (i, j) of plots is subdivided by the permutations of the group

$$(i, j) \rightarrow (\sigma i, \sigma j), \quad \sigma \in G$$

into non-overlapping equivalence classes, two pairs in the same equivalence class being transformable from one to the other by a suitable permutation of the group while pairs in different equivalence classes are not. An equivalence class of pairs may be called an *invariant relation R*.

Define the characteristic matrix of an invariant relation as a $n \times n$ matrix of 0's and 1's whose ijth element r_{ij} is given by

$$r_{ij} = \begin{cases} 1, & \text{if } (i, j) \in R, \\ 0, & \text{otherwise.} \end{cases}$$

Theorem 1. *The characteristic matrices of the invariant relations are a basis of the invariant relationship algebra, $\mathcal{B}$.*

Proof. Eq. (3.1) implies that a matrix B belongs to $\mathcal{B}$ if and only if its elements b_{ij} are constant on all pairs (i, j) belonging to an invariant relation i.e. if and only if it is a linear combination of characteristic matrices of invariant relation.

We may assume from now on that G acts transitively upon the set S of n plots. If G does not, it is best to subdivide S into the orbits of G, i.e. the non-overlapping equivalence classes into which G divides the n plots; and study G as a transitive permutation group within each orbit.

Choose any plot i_0 as an origin and let H be the subgroup consisting of all those permutations of G which leave i_0 fixed. H is called the isotropy group at i_0. Divide the set of plots into orbits of H. Clearly, each orbit of the isotropy group H corresponds to an invariant relation.

Illustration. Diallele crosses, with k families.

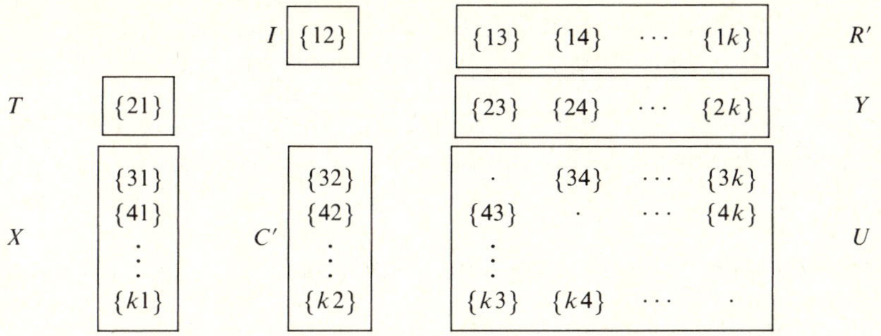

The symmetries group is now the symmetric group on the k families.

If we choose the plot $\{12\}$ as an origin of i_0, the isotropy group consists of all permutations of the last $k-2$ families. There are thus 7 orbits of the isotropy group, indicated by solid lines in the diagram together with letters to denote the corresponding invariant relations. Hence the invariant relationship algebra is 7 dimensional, and this agrees with our previous results according to a

Theorem 2. *The dimension of the invariant relationship algebra is the sum of the squares of the multiplicities of the irreducible representations present.*

For a proof, we refer the reader to a treatise on group representations. In our example, the multiplicities are $1, 2, 1, 1$. The sum of their squares is 7 which agrees with the number of invariant relations. The invariant relations have an intuitive interpretation:

Orbit of isotropy group	Symbol for invariant relation	Interpretation of invariant relation
$\{12\}$	I	Identity
$\{21\}$	T	Transpose
$\{1j\}$ $j=3,\dots,k$	R'	Same row
$\{i2\}$ $i=3,\dots,k$	C'	Same column
$\{i1\}$ $i=3,\dots,k$	X	In the column whose number is equal to the row number
$\{2j\}$ $j=3,\dots,k$	Y	In the row whose number is equal to the column number
$\{ij\}$ $i, j \geqslant 3$	U	Unrelated

The characteristic matrices of the invariant relations, for which we use the same symbols I, T, R', C', X, Y, U generate the invariant relationship algebra B. Any product of two of them is a linear combination of them.

Since the orbits of the isotropy group for the general case of k families are similar to the case of $k=4$, we surmise that the analysis of variance will be similar and, in analogy with (3.1), that the decomposition of the representation will be

$$[k]+2[k-11]+[k-22]+[k-211]. \tag{3.2}$$

Our surmise will be confirmed by verifying that these representations are actually present. To find the analysis of variance we require results about

4. Representations of transitive permutation groups

The following theory is along the lines of the work of Frame [2], see also Mackey [9]. It can also be applied to the calculation of the *zonal* functions which appear in certain multivariate distributions; see James [5].

We are still assuming that G acts transitively upon the set S of plots. If the permutation $\sigma \in G$ carries the plot i_0 into a plot i, the set of all elements of G which do so is the coset σH of the isotropy group H. Hence the plots are in one-to-one correspondence with the cosets of H

$$i \leftrightarrow \sigma H.$$

The correspondence is preserved under transformation by elements σ_1 of G:

$$\begin{array}{ccc} i & \rightarrow & \sigma_1 i \\ \updownarrow & & \updownarrow \\ \sigma H & \rightarrow & \sigma_1 \sigma H \end{array}$$

Furthermore, it follows that an orbit of H in S, which determines an invariant relation, corresponds to the union of the cosets $\sigma_1 \sigma H$ into which elements $\sigma_1 \in H$ can transform the coset σH i.e. invariant relations correspond to double cosets, $H\sigma H$. Writing the set of cosets as G/H and the double cosets as $G/H \times H$ and calling the set of invariant relations J, we can represent the isomorphisms by the diagram

$$\begin{array}{cccc} G & \rightrightarrows & S & J \\ \updownarrow & & \updownarrow & \updownarrow \\ G & \rightrightarrows & G/H & G/H \times H. \end{array}$$

The notation $G \rightrightarrows S$ is intended to indicate that G is a set of operators acting upon the elements of S.

The vector space V is, in fact, the set of functions on S. The set of all linear combinations $\Sigma_{\sigma \in G}\, a(\sigma) \cdot \sigma$ of group elements with scalar coefficients $a(\sigma)$ forms an algebra called the group algebra A. It acts as a set of linear transformations upon V. In view of the correspondence

$$S \leftrightarrow G/H,$$

the elements of V which are functions of S correspond to functions of G which are constant on the cosets of H. Likewise, from the correspondence

$$J \leftrightarrow G/H \times H$$

one sees that the elements of the invariant relationship algebra B, which are linear combinations of the characteristic matrices of invariant relations, are in 1 : 1 correspondence with the functions of G which are constant on the double cosets.

Finally, to fit all these concepts together we must introduce the *regular representation* of A. The left regular representation of the algebra A is the representation of elements $a \in A$ as linear transformations of A, regarded as a vector space, produced by left multiplication

$$a : x \to ax, \quad a \in A, x \in A. \tag{4.1}$$

The centralizer of the left regular representation i.e. the algebra of all linear transformations of A which commute with the transformations (5.1) is simply the algebra of right multiplications by elements b of A,

$$b : x \to xb, \quad b \in A$$

i.e. the right regular representation. We may depict the left and right regular representations as centralizers of each other by the following diagram:

$$A_{\text{from left}} \underset{\to}{\to} A \underset{\leftarrow}{\leftarrow} A_{\text{from right}}.$$

Let us identify a linear combination $\sum_{\sigma \in G} a(\sigma) \cdot \sigma$ of group elements with the function $a(\sigma)$ determined by its coefficients and regard either equally as an element of A. We may project A upon the subspace of functions constant on the cosets of H by postmultiplying by an idempotent $e \in A$ given by the sum of the elements in the isotropy group H divided by its order h,

$$e = h^{-1} \sum_{\sigma \in H} \sigma.$$

But the space of functions constant on the cosets is isomorphic to the space V of functions of S i.e.

$$Ae \leftrightarrow V.$$

The isomorphism may be given explicitly as

$$ae \to ae\delta_{i_0} = a\delta_{i_0} \tag{4.2}$$

where δ_{i_0} is the function in V which has the value 1 on the plot i_0 and 0 on the other plots.

Similarly, the subspace eAe consists of functions of G constant on the double cosets $H\sigma H$ and is thus isomorphic to B

$$eAe \leftrightarrow B.$$

eAe is a subalgebra of A whose right multiplication of Ae is isomorphic to the action of B upon V, i.e. eAe is the centralizer of the left regular representation of A in Ae. The results may be summarized as the

Theorem 3. *If A is the group algebra (over the field of complex numbers) of a finite transitive permutation group G of a set S, then the action of A upon the vector space V*

of complex valued functions of S is isomorphic to the left transformation of Ae by A where $e \in A$ is the idempotent obtained as the sum of the elements of the isotropy group H of G, divided by the order h of H. The action of the centralizer B of A upon V is isomorphic to the subalgebra eAe of A acting upon Ae from the right:

$$\begin{array}{ccccc} A & \rightrightarrows & V & \leftleftarrows & B \\ \downarrow & \downarrow & \downarrow & \downarrow & \downarrow \\ A_{\text{from left}} & \rightrightarrows & Ae & \leftleftarrows & eAe_{\text{from right}}. \end{array}$$

The vertical arrows signify that the algebras are isomorphic and operate isomorphically upon the isomorphic vector spaces V and Ae.

e is the unity element of the algebra eAe and can be decomposed into a sum of mutually orthogonal symmetric primitive idempotents e_ν,

$$e = e_1 + e_2 + \cdots . \tag{4.3}$$

By the isomorphism

$$eAe \leftrightarrow B \tag{4.4}$$

it follows that

$$e_\nu \leftrightarrow E_\nu$$

where E_ν are idempotent matrices which project upon the irreducible invariant subspaces of V, and from (5.3)

$$I = E_1 + E_2 + \cdots ,$$

thus giving the desired analysis of variance

$$x'x = x'E_1 x + x'E_2 x + \cdots .$$

The idempotents e_ν should be symmetric, i.e. such that the matrices E_ν are symmetric matrices.

As the matrices E_ν are clumsy to write explicitly, we obtain them as linear combinations of the characteristic matrices $I, T, R', C', \ldots$ of the invariant relations. The quadratic forms $x'x, x'Tx, \ldots$ of the latter are linear combinations of sums of squares familiar to statisticians.

5. Calculation of the idempotents

The main problem has been reduced to the calculation of primitive idempotents belonging to certain specified representations in the group ring. For the symmetric group, Alfred Young has shown us how to do this; see Rutherford [11]. An example is the calculation of the term in the analysis of variance of the dialleles corresponding to the representation $[k-22]$.

As the representation only occurs once, the idempotent must be an *orthogonal* projection and thus symmetric.

The following procedure yields an essential idempotent, i.e. a non-zero scalar multiple of an idempotent, which can then be normalized to give the true idempotent. Firstly, one writes the integers $1,2,\ldots,k$ in rows of lengths given by the numbers in the partition of k determining the representation; in this case $k-2$ and 2;

$$\begin{matrix} 3 & 4 & 5\cdots k \\ 1 & 2 & \end{matrix}. \tag{5.1}$$

The numbers may be written in any order. When the numbers are written in certain orders, the procedure may yield zero instead of the idempotent sought. If this happens, one should try again with the numbers rearranged. Elements s and a in the group algebra A are defined as:

$$s=\sum_{p\in G_1} p,$$

where p runs over the group G_1 of all permutations within the rows of the diagram (11) and

$$a=\sum \varepsilon_q q$$

where q runs over all permutations within the columns of (5.1) and $\varepsilon_q=-1$ or 1 according as q is an odd or even permutation. The essential idempotent sought is then

$$ease$$

because *as* is an essential primitive idempotent of A and hence *ease* is either zero or an essential primitive idempotent of eAe. By (4.4) it is isomorphic to a linear combination of invariant relations:

$$ease \leftrightarrow c_1 I+c_2 T+c_3 R'+c_4 C'+c_5 X+c_6 Y+c_7 U\in B. \tag{5.2}$$

We will now write a linear combination of plots to denote the function in V with values given by its coefficients. In particular, putting $i_0=\{12\}$ we shall also denote the function δ_{i_0} in V, which has the value 1 on plot $\{12\}$ and 0 on the other plots, by the symbol $\{12\}$.

The coefficients $c_1,\ldots,c_7$ in (5.2) can be found by comparing the action of B upon the function $\{12\}$ with the action of *ease* upon e. Under the isomorphism (4.2), e corresponds to $\{12\}$ and likewise according to Theorem 3, their transforms under *ease* and B, respectively, correspond.

$$\begin{array}{ccc} & \underline{ease \text{ from right}} & \\ e & \longrightarrow & eease=ease \\ \downarrow & & \downarrow \\ e\{12\}=\{12\} \xrightarrow{B} B\{12\} & & eas\{12\}=ease\{12\}. \end{array}$$

Therefore, $B\{12\}=eas\{12\}$.

A coefficient such as c_3 is thus the coefficient in $eas\{12\}$ of any plot belonging to the orbit of R', and, since e simply averages to coefficients of plots with an orbit, this is equal to the sum of the coefficients in $as\{12\}$ of those plots belonging to the orbit of R' divided by the number of plots in the orbit.

In this example, the isotropy group H, which consists of all permutations of the families $3,4,\ldots,k$, is a subgroup of G. Since all elements of a coset of H in G will have the same action upon the plot $\{12\}$ taken as origin, we need only include in s one group element from each coset of H in G, i.e. we may put $s=(()+(12))$ where the symbols () and (12) denote respectively, the identity permutation and the transposition of families 1 and 2. Thus:

$$s\{12\}=(()+(12))\{12\}=\{12\}+\{21\}$$

$$as\{12\}=(()+(13)(24)-(13)-(24))(\{12\}+\{21\})$$

$$=\{12\}+\{21\}+\{34\}+\{43\}-\{32\}-\{23\}-\{14\}-\{41\}.$$

There is only one term, i.e. $\{14\}$ in the orbit of R' and there are $k-2$ plots in the orbit; hence $c_3=-1/(k-2)$.

Calculating the other coefficients similarly we have:

$$ease \leftrightarrow I+T+\frac{2}{(k-2)(k-3)}U-\frac{1}{k-2}C'-\frac{1}{k-2}Y-\frac{1}{k-2}R'-\frac{1}{k-2}X=\lambda E_4. \tag{5.3}$$

The normalizing constant λ^{-1} can be found from the fact that the trace of an idempotent matrix is equal to the dimension of the space upon which it projects, and the dimension of the representation $[k-2\ 2]$ of S is $\frac{1}{2}k(k-3)$. Now $\operatorname{tr} I=k(k-1)$, $\operatorname{tr} T=0$, $\operatorname{tr} U=0$, etc. Hence the essential idempotent in (12) must be multiplied by the normalizing constant $\lambda^{-1}=(k-3)/2(k-1)$.

Express the quadratic forms $x'Tx$, $x'Ux$, etc. as sums of squares. Clearly:

$$x'(I+T+R'+C'+X+Y+U)x=G^2,$$

$$x'(I+T)x=\sum_{i<j} s_{ij}^2,$$

$$x'(I+R')x=\sum R_i^2,$$

$$x'(I+C')x=\sum C_j^2,$$

$$x'(2T+X+Y)x=2\sum R_i C_i$$

where $G=\Sigma_{i\neq j}x_{ij}$, $s_{ij}=x_{ij}+x_{ji}$, $R_i=\Sigma_{j(\neq i)}x_{ij}$, $C_j=\Sigma_{i(\neq j)}x_{ij}$. Substituting these relationships in (12) we obtain

$$x'E_4x=\left[\tfrac{1}{2}\sum_{i<j} s_{ij}^2-G^2/k(k-1)\right]-\left[\sum(R_i+C_i)^2/2(k-2)-2G^2/k(k-2)\right]. \tag{5.4}$$

This is the term corresponding to the representation $[k-2 2]$ and is distributed as $\sigma^2\chi^2$ on $\frac{1}{2}k(k-3)$ degrees of freedom, assuming the observations are independently normally distributed with variance σ^2.

Similarly, the term corresponding to the representation $[k-1\,1\,1]$ can be calculated as:

$$x'E_5x=\tfrac{1}{2}\sum_{i<j} d_{ij}^2-\sum_i (R_i-C_i)^2/2k, \quad d_{ij}=x_{ij}-x_{ji}, \tag{5.5}$$

on $\frac{1}{2}k(k-1)(k-2)$ degrees of freedom.

Since the term in the first square bracket of (5.4) is obviously distributed as $\sigma^2\chi^2$ on $\frac{1}{2}k(k-1)-1$ degrees of freedom, it follows that the second term must be distributed as $\sigma^2\chi^2$ on $\frac{1}{2}k(k-1)-1-\frac{1}{2}k(k-3)=k-1$ degrees of freedom. Since this term is also invariant, it must correspond to one of the representations $[k-1\;1]$ which are known to be $(k-1)$ dimensional. Also for the same reasons $\Sigma(R_i-C_i)^2/2k$ must be the other term corresponding to this representation. Hence the whole analysis of variance must be

$$x'x=G^2/k(k-1)+\left[\Sigma(R_i+C_i)^2/2(k-2)-2G^2/k(k-2)\right]+\left[\Sigma(R_i-C_i)^2/2k\right]$$
$$+x'E_4x+x'E_5x \tag{5.6}$$

where the last two terms are given by (5.4) and (5.5). These terms correspond to the representations:

$$[k], \qquad [k-1\,1], \qquad [k-1\,1], \qquad [k-2\,2], \qquad [k-2\,1\,1].$$

The purely group theoretical approach used here shows explicitly how the invariant analysis of variance follows from the group representation theory. In practice, the partitioning of the sum of squares can be expedited considerably by supplementing group theoretical methods with statistical insight and acumen. However, the group theory method above is good practice for the zonal polynomial problem.

6. Zonal polynomials

In James [6], the Shur theory relating tensor representations of the general linear group to representations of the symmetric group, was used to show that zonal polynomials correspond to primitive idempotents in a representation of the symmetric group.

The "plots" are now "doublets" and the space of "observation vectors" is the vector space of functions on the doublets. The symmetric group $\mathfrak{S}_{2f}$, operating transitively on the doublets, has a representation in the space of functions on the doublets. The symmetric idempotent matrices E_κ which project orthogonally on the irreducible subspaces correspond to the zonal polynomials.

A doublet, d, is a pairing of the integers $1,2,3,\ldots,2f-1,2f$ without regard to the order of the pairs or the order of the integers within the pairs. The set D of doublets clearly has $(2f)!/2^f f!$ elements.

The symmetric group $\mathfrak{S}_{2f}$ of all permutations of the integers $1,2,\ldots,2f$ induces a transitive group of permutations of the doublets, and the isotropy group T, which leaves the doublet $d_0=\{12\}\{34\}\cdots\{2f-1\,2f\}$ fixed, has $2^f f!$ elements.

Each invariant relation corresponds to an orbit of the isotropy group T in D and is determined by a partition $(1^{\nu_1}2^{\nu_2}\cdots)$ of f; $\nu_1+2\nu_2+3\nu_3+\cdots=f$. Under the correspondence between representations of the symmetric and linear groups discussed in James [6], this partition corresponds to the monomial $s_1^{\nu_1}s_2^{\nu_2}\cdots$ in the sums s_i of the ith powers of the latent roots of the argument matrix A.

The coefficient of the monomial $s_1^{\nu_1}s_2^{\nu_2}\cdots$ in the zonal polynomial $Z_\kappa(A)$ is the sum of the coefficients of the group elements σ in the Young essential idempotent, as for the diagram

$$\begin{array}{llllll}
1 & 2 & 3 & 4 & \cdots & 2k_1-1\,2k_1 \\
2k_1+1 & 2k_1+2 & & & \cdots & 2k_1+2k_2-1 \quad 2k_1+2k_2 \\
 & \vdots & & & & \\
 & & & & \cdots & 2k-1\,2k \quad \text{where} \quad k=\sum_i k_i,
\end{array}$$

which map d into the orbit corresponding to the partition $1^{\nu_1}2^{\nu_2}\cdots$.

Of the terms in the symmetrizer s, it is only necessary to include one element from each coset $\sigma T'$, where T' is the subgroup of T whose elements permute within the rows of the symmetry diagram. Likewise, one need only include in the alternator a, one element from each coset $T''\sigma$, where T'' is the subgroup of T which permutes within the columns. The calculation is illustrated in James [7, pp. 878–880]. The next section derives some results in the paper.

7. Orthogonality relations

Relative to a group invariant inner product, such as the ordinary sum of products of components of vectors, the subspaces in which inequivalent irreducible representations take place are orthogonal.

Now a zonal polynomial corresponds to a function on the doublets which is constant on the orbits of the isotropy group. The number of elements in an orbit is:

$$Z_{(f)\nu}=\frac{2^k k!}{\nu_1!\nu_2!\nu_3!\cdots 2^{\nu_1}4^{\nu_2}6^{\nu_3}}.$$

Since the top zonal polynomial $Z_{(f)}$ corresponds to the constant function, its coefficients are $Z_{(f)\nu}$, i.e.

$$Z_{(f)}=\sum_\nu Z_{(f)\nu}s_1^{\nu_1}s_2^{\nu_2}s_3^{\nu_3}\cdots.$$

Other zonal polynomials,

$$Z_\kappa=\sum_\nu Z_{\kappa\nu}s_1^{\nu_1}s_2^{\nu_2}\cdots,$$

$$Z_\lambda=\sum_\nu Z_{\lambda\nu}s_1^{\nu_1}s_2^{\nu_2}\cdots,$$

will correspond to functions with values $Z_{\kappa\nu}/Z_{(f)\nu}$, $Z_{\lambda\nu}/Z_{(f)\nu}$ on the orbit ν. Hence for $\kappa \neq \lambda$,

$$0=\sum_{\nu} Z_{(f)} \frac{Z_{\kappa\nu}}{Z_{(f)\nu}} \frac{Z_{\lambda\nu}}{Z_{(f)\nu}} = \sum_{\nu} \frac{Z_{\kappa\nu} Z_{\lambda\nu}}{Z_{(f)\nu}}.$$

In general we have the *orthogonality relations*,

$$\sum_{\nu} Z_{\kappa\nu} Z_{\lambda\nu} / Z_{(f)\nu} = \delta_{\kappa\lambda} N / \chi_{[2\kappa]}(1),$$

and hence,

$$\sum_{\mu} \chi_{[2\mu]}(1) Z_{\mu\kappa} Z_{\mu\lambda} = \delta_{\kappa\lambda} N Z_{(k)\kappa},$$

where $N=(2k)!/(2^k k!)$.

References

[1] Boerner, H. (1955). *Representations of Groups, With Special Consideration for the Needs of Modern Physics*. Translated from German by Murphy, P.G. et al. 1963. North-Holland, Amsterdam.

[2] Frame, J.S. (1948). Group decomposition by double coset matrices. *Bull. Am. Math. Soc.* **54**, 740–755.

[3] Higman, B. (1955). *Applied Group-Theoretic and Matrix Methods*, Oxford University Press, London.

[4] James, A.T. (1957). The relationship algebra of an experimental design. *Ann. Math. Statist.* **28**, 993–1002.

[5] James, A.T. (1961). The distribution of the latent roots of the covariance matrix. *Ann. Math. Statist.* **35**, 475–501.

[6] James, A.T. (1961). Zonal polynomials of the real positive definite symmetric matrices. *Ann. of Math.* **74**, 456–469.

[7] James, A.T. (1961). The distribution of noncentral means with known covariance. *Ann. Math. Statist.* **32**, 874–882.

[8] Littlewood, D.E. (1950). *The Theory of Group Characters*. Second Ed. Oxford University Press, London.

[9] Mackey, G.W. (1951). On induced representations of groups. *Am. J. Math.* **73**, 576–592.

[10] Murnaghan, F.D. (1952). On the multiplication of representations of the linear group. *Proc. Nat. Acad. Sci.* **38**, 738–741.

[11] Rutherford, D.E. (1948). *Substitution Analysis*. Edinburgh.

[12] Yates, F. (1947). Analysis of data from all possible reciprocal crosses between a set of parallel lines. *Heredity* **1**, 287–301.

G. Kallianpur, P.R. Krishnaiah, J.K. Ghosh, eds., *Statistics and Probability: Essays in Honor of C.R. Rao*
© North-Holland Publishing Company (1982) 343–358

ASSESSING THE PREDICTIVE INFLUENCE OF OBSERVATIONS

Wesley JOHNSON and Seymour GEISSER*
University of Minnesota, MN, U.S.A.

Introduction

Let a family of probability densities indexed by some parameter Θ, a sample of observations from this family, and prior information about θ, be given; and assume the goal is to predict a future observation from this family. Then, given a method of prediction, it may be of interest to determine those subsets of such a sample which are the most influential for the purpose of prediction. Predictive distributions are standard Bayesian inferential tools for predicting future observations; cf. Geisser (1965, 1971) and Aitchison and Dunsmore (1975). It is possible to determine predictive densities based on full, and subset deleted data sets. A subset which radically changes such critical attributes as the location and shape of the full predictive density is said to be very influential; while a subset which leaves this density relatively unchanged is said not to be influential. The discrepancy between such densities may be conveniently measured by Kullback–Leibler divergences; cf. Kullback and Leibler (1951). All possible subset deletions of fixed size may be considered, and subsets ordered from least to most influential according to the magnitudes of these divergences.

Previously the emphasis has been mainly on how subsets of observations influence the estimation of parameters; Cook (1977, 1979) and Bingham (1977). Here we redirect this focus on prediction. Our viewpoint is that for most statistical paradigms, the models used are convenient approximations to complex processes and the parameters are inevitably artifices that do not really represent observable or physical entities of the underlying process. While certainly the purpose of an analysis is well-served by an understanding and interpretation of the data, it is ultimately directed towards prediction.

Predictive densities and Kullback–Leibler divergences are reviewed and predictive influence functions defined in Section 1, and subsequently catalogued for normal paradigms. The remainder of the paper is devoted to presenting specific examples, wherein predictive influence functions are determined and studied. When explicit representations of predictive influence do not exist, approximations are made and convergence properties determined. Finally, asymptotic representations of predictive influence are derived. In all cases, the most influential subsets of a sample are characterized.

*Research supported in part by the University of Minnesota Graduate School and NIH Grant GM 25271.

In Section 2, the usual general linear regression framework assuming normal theory is considered; cf. Draper and Smith (1966). The situation where i.i.d. $N[\mu, \Theta]$ observations are collected is discussed as a special case.

1. Definitions of predictive influence

1.1. Preliminaries

The problem of prediction will be approached from the Bayesian view via predictive densities; cf. Geisser (1965, 1971) and Aitchison and Dunsmore (1975). Given a random sample from a population about which prior information is assumed, a predictive distribution which is independent of any population parameters can be derived. Inferences about future values, being based on predictive distributions or densities, will then depend on the information contained in the sample. It is of interest to determine which subsets of such a sample are most influential for purposes of prediction.

Let y be an n dimensional vector of observed values of a sample from some population, and partition y into subsets, say, $y=(y_1', y_2')'$; where y_1 has dimension n_1 and y_2 has dimension n_2; $n_1+n_2=n$. Determine the predictive density based on the full data set y, say f, and then determine the predictive density based on only y_1, say $f_{(2)}$; where the notation indicates that y_2 has been deleted. The predictive influence of the subset y_2 is then reflected in the discrepancy between the densities f and $f_{(2)}$. Suppose that a convenient measure of the discrepancy between two densities, say $I(f, f_{(2)})$, is available. Then assuming that $I(f, f_{(2)})$ will be small if f and $f_{(2)}$ are similar, and large if they are very different, it is possible to order all ${}_nC_{n_2}$ subsets of data, from least to most influential, according to the magnitudes of $I(f, f_{(2)})$. Kullback and Leibler (1951) have proposed measures, called divergences, which are useful for measuring these discrepancies. The predictive influence of a subset y_2 will be defined as Kullback–Leibler divergences between these densities.

1.2. Predictive densities

Let $\mathcal{G}$ be some arbitrary set, possibly ϕ, and let $\boldsymbol{\theta} \in R^k$ be unknown, and assume the existence of the family of probability densities $\mathcal{F}=\{f_m(\cdot|x, \boldsymbol{\theta}) | \boldsymbol{\theta} \in R^k, x \in \mathcal{G}, m = 1,2,\ldots\}$ where each density, $f_m(\cdot|x, \boldsymbol{\theta})$, has an m dimensional argument. Let $\boldsymbol{Y}$ be an n dimensional random vector with density $f_n(\cdot|x, \boldsymbol{\theta})$, and assume that $\boldsymbol{Y}=y$ has been observed. Further, assume that it is of interest to predict a future random variable which is independent of $\boldsymbol{Y}$, say Z with density $f_1(\cdot|w, \boldsymbol{\theta}), w \in \mathcal{G}$; and assume the existence of a prior probability, possibly improper, $p(\boldsymbol{\theta})\,\mathrm{d}\boldsymbol{\theta}$, for the vector $\boldsymbol{\theta}$. The posterior density, when it exists, is then defined to be $p(\boldsymbol{\theta}|\, y)= f_n(y|x, \boldsymbol{\theta})p(\boldsymbol{\theta}) / \int f_n(y|x, \boldsymbol{\theta})p(\boldsymbol{\theta})\,\mathrm{d}\boldsymbol{\theta}$. The predictive density of future Z, given w, x and y, may now be defined as $f(z|w, x, y)=\int f_1(z|w, \boldsymbol{\theta})p(\boldsymbol{\theta}|\, y)\,\mathrm{d}\boldsymbol{\theta}$, where the fact that the functional form of the density may depend on n is suppressed. Note that the density is independent of the unknown parameter $\boldsymbol{\theta}$, and consequently may be used to make inferences. As an example, cf. Geisser (1965), consider the usual linear model $Y = X\boldsymbol{\beta} + \boldsymbol{\varepsilon}$, where X is a full rank $n \times p$ matrix of observed values, $\boldsymbol{\beta}$ is a $p \times 1$ vector of unknown regression coefficients, and $\boldsymbol{\varepsilon}$ is an $n \times 1$ normally distributed vector with mean $\mathbf{0}$ and known covariance θI. Let $f(\cdot | X, \boldsymbol{\beta}, \theta) = N_n[X\beta, \theta I]$, $\mathcal{G} = \{X | X \text{ an } n \times p \text{ matrix of full rank}, n = 1, 2, \ldots\}$ and define $\mathcal{F}$ as above; where $N_n[\ ,\ ]$ defines an n dimensional multivariate normal density with given mean vector and covariance matrix. If we use the improper prior,

$$p(\boldsymbol{\beta}, \theta)\,\mathrm{d}\boldsymbol{\beta} \propto \mathrm{d}\boldsymbol{\beta}, \tag{1}$$

and assume the goal is to predict a future observation from the model $Z = \boldsymbol{w\beta} + \varepsilon^*$, where $\boldsymbol{w}$ is a $1 \times p$ vector of known values and ε^* is a $N[0, \theta]$ random variable; then, defining $v_w = \boldsymbol{w}(X'X)^{-1}\boldsymbol{w}'$, the predictive density of a future Z given an observed vector $\boldsymbol{y}$ is $N[\boldsymbol{w}\hat{\beta}, (1 + v_w)\theta]$, i.e.

$$f(z | \boldsymbol{w}, X, y) \propto \exp\left[-\tfrac{1}{2}(z - \boldsymbol{w}\hat{\boldsymbol{\beta}})^2 / \theta(1 + v_w)\right], \tag{2}$$

where $\hat{\boldsymbol{\beta}} = (X'X)^{-1}X'\boldsymbol{y}$, the usual least squares estimate of $\boldsymbol{\beta}$.

If it is assumed that θ is unknown and that the usual improper prior

$$p(\boldsymbol{\beta}, \theta)\,\mathrm{d}\boldsymbol{\beta}\,\mathrm{d}\theta \propto \theta^{-1}\,\mathrm{d}\boldsymbol{\beta}\,\mathrm{d}\theta \tag{3}$$

then the predictive density of a future, Z, is $St[n-p, \boldsymbol{w}\hat{\boldsymbol{\beta}}, (1 + v_w)s^2]$, i.e.

$$f(z | \boldsymbol{w}, X, \boldsymbol{y}) \propto \left[1 + (z - \boldsymbol{w}\hat{\boldsymbol{\beta}})^2 / (n-p)(1 + v_w)s^2\right]^{-(n-p+1)/2}, \tag{4}$$

where $s^2 = (y - X\hat{\boldsymbol{\beta}})'(y - X\hat{\boldsymbol{\beta}})/(n-p)$, the usual regression mean squared error, and $\mathrm{St}[\ ,\ ,\ ]$ denotes a Student density with specified degrees of freedom, location and dispersion.

It is not difficult to show in the above cases that the predictive densities will converge almost surely to the sampling density of the future observation, a necessary criterion for the use of the prior distribution, cf, Geisser (1971). Also, Murray (1977) has shown that the predictive densities are optimal estimates of the sampling densities in the frequency sense, among all estimators that are invariant with regard to translations and nonsingular transformations, using the Kullback–Leibler measure of divergence.

1.3. *Kullback–Leibler divergences*

Let f_1 and f_2 be generalized densities with respect to some measure ν. Let E_{f_i} denote the operator which takes expectation with respect to the density f_i, $i = 1, 2$, and define the directed divergences

$$I(f_1, f_2) = E_{f_1} \ln(f_1/f_2), \tag{5}$$

$$I(f_2, f_1) = E_{f_2} \ln(f_2/f_1), \tag{6}$$

and the divergence

$$J(f_1, f_2) = I(f_1, f_2) + I(f_2, f_1). \tag{7}$$

These expectations are all well defined and non-negative definite; Kullback (1968).

Consider the normal case and define $f_1 = N(\mu_1, \sigma_1^2)$ and $f_2 = N(\mu_2, \sigma_2^2)$, $\mu_1, \mu_2 \in R^1, \sigma_1, \sigma_2 > 0$. Then defining $\mu = (\mu_1 - \mu_2)/\sigma_2$ and $\sigma = \sigma_2/\sigma_1$, it is easy to verify that

$$2I(f_1, f_2) = \mu^2/\sigma^2 + \left[\sigma^{-2} - \ln(\sigma^{-2}) - 1\right], \tag{8}$$

$$2I(f_2, f_1) = \mu^2 + [\sigma^2 - \ln\sigma^2 - 1], \tag{9}$$

$$2J(f_1, f_2) = \mu^2(1 + \sigma^{-2}) + \left[(\sigma^2 - 1)^2/\sigma^2\right]. \tag{10}$$

It is clear that all three measures are partitionable into the sum of two components; the first reflecting a weighted difference in the means, and the second, a convex function in σ^2 with minimum at $\sigma^2 = 1$, which reflects the difference in variances. Note that if $\sigma^2 < 1$, each component of $I(f_1, f_2)$ is greater than that of $I(f_2, f_1)$, indicating the mean differences and differences in variances are weighted more heavily by $I(f_2, f_1)$ than $I(f_1, f_2)$.

Now let $f_2^* = N(\mu_2, \sigma_1^2)$ and $f_2^+ = N(\mu_1, \sigma_2^2)$. Then

$$2 \cdot I(f_2^*, f_1) = \mu^2, \qquad 2 \cdot I(f_2^+, f_1) = \sigma^2 - \ln\sigma^2 - 1. \tag{11}$$

Hence if interest lies solely in the difference in densities which have equal variances, it is sufficient to consider the first component of the divergence; or if interest rests solely on the differences in densities with equal means, it is sufficient to consider the second component.

1.4. Predictive influence

For convenience, define the predictive density of a future Z, given w, x and $\boldsymbol{y}$ as $f(z|w, x, \boldsymbol{y}) = f(z)$. Now let $\boldsymbol{y} = (\boldsymbol{y}_1', \boldsymbol{y}_2')'$ where $\boldsymbol{y}_1$ has dimension n_1, $\boldsymbol{y}_2$ has dimension n_2, $n_1 + n_2 = n$; and define the predictive density of the future, Z, given w, $x_1 \in \mathcal{G}$ and $\boldsymbol{y}_1$ as $f(z|w, x_1, \boldsymbol{y}_1) = f_{(2)}(z)$, where the notation suggests that the subset $\boldsymbol{y}_2$ has been deleted and that the reduction in $\boldsymbol{y}$ may imply a reduction in x.

As a measure of the discrepancy between the densities f and $f_{(2)}$ for all possible ${}_nC_{n_2}$ subset deletions, we will use the Kullback–Leibler divergences which are respectively $I(f, f_{(2)})$, $I(f_{(2)}, f)$ and $J(f, f_{(2)})$. For the purposes of this paper, $I(f, f_{(2)})$ will serve as the primary definition of predictive influence. However, expressions $I(f_{(2)}, f)$ and $J(f, f_{(2)})$ will be termed secondary P.I.F.'s and, in order to make a point, it will be useful to determine expressions for them in future examples.

In normal paradigms, it will usually be the case that the predictive dispersion associated with $f_{(2)}$ will be larger, when averaged over the sampling distribution of the data, than that associated with f; cf. Geisser (1971). Consequently, $I(f_{(2)}, f)$ will usually dominate $I(f, f_{(2)})$, indicating at least in normal paradigms, that the former measure gives a higher weighting to location and dispersion differences than the latter.

Suppose now that the vector $\boldsymbol{Y}$ consists of independent observations, and recall that Z is independent of $\boldsymbol{Y}$, given θ. Then prediction may be accomplished sequentially; by first conditioning $(Z, \boldsymbol{Y}_2)$ on $\boldsymbol{Y}_1 = \boldsymbol{y}_1$, and then by conditioning Z on $\boldsymbol{Y}_2 = \boldsymbol{y}_2$. Hence, the predictive influence of the subset $\boldsymbol{y}_2$ will depend on the relationship between Z and $\boldsymbol{Y}_2$, as determined by the joint predictive density for $(Z, \boldsymbol{Y}_2)$.

More formally, let $f_{(2)}(z, y_2)$ be the joint predictive density of $(Z, \boldsymbol{Y}_2)$ based on $\boldsymbol{y}_1$ only, and let $f_{(2)}(y_2)$ be the marginal predictive density of $\boldsymbol{Y}_2$. Define $f_{(2)}(z|\, y_2)= f_{(2)}(z, y_2)/f_{(2)}(y_2)$, the conditional predictive density of z given $\boldsymbol{y}_2$, based on $\boldsymbol{y}_1$ only. Then

$$f(z)=f_{(2)}(z|\, y_2). \tag{12}$$

This result is easy to show directly using Fubini's Theorem, or by noting that, given the above independence assumptions, predictive densities behave like conditional density functions.

2. Predictive influence in normal linear models

2.1. Introduction

Consider the usual general linear model defined in Subsection 1.2, $\boldsymbol{Y}=X\boldsymbol{\beta}+\boldsymbol{\varepsilon}$. Then given the prior defined by (1), when θ is assumed known, the predictive density for a single future observation will be univariate normal and P.I.F.'s can easily be derived. When θ is unknown, however, and the prior defined by (3) is assumed, the predictive density will be Student, and it will not be possible to obtain a closed form expression of predictive influence. Consequently, it will be necessary to approximate the exact P.I.F.

In this paper, P.I. will be defined as the discrepancy between univariate predictive densities. P.I. will depend upon where one is trying to predict, i.e. the future observation will be sampled from the model $z=\boldsymbol{w}\boldsymbol{\beta}+\varepsilon^*$. Hence, when predicting at $\boldsymbol{w}$, P.I. will necessarily depend on $\boldsymbol{w}$.

Cook (1977) proposed a general statistic for detecting single observations which are influential for the purpose of estimating linear functions of the regression coefficient vector $\boldsymbol{\beta}$, a special case of which is now called "Cook's distance". In what follows, it will be seen that the Cook statistic for detecting influence of a subset on the estimated regression line at $\boldsymbol{w}$, may be interpreted from the predictivist view as influence on predictive means. In a future paper, it will be shown that Cook's distance may be interpreted as the component of influence ascribable to predictive mean vectors (or point prediction) when one is trying to predict at X itself.

A basic motivation for this exposition is to point out the need to study the effect of subsets on predictive regions as well as point predictions. For purposes of prediction, a more general influence function than of Cook's influence statistics appears to be indicated; and we believe that the one proposed here eminently serves that goal.

2.2. Preliminaries

Let $\boldsymbol{Y}$ be defined as in Subsection 1.2 and let $\boldsymbol{y}$ denote a realization of the vector $\boldsymbol{Y}$. Partition $\boldsymbol{Y}$ and $\boldsymbol{y}$ in the usual way, and partition X as $(X_1', X_2')'$, where X_1 is $n_1 \times p$, and X_2 is $n_2 \times p$. Now suppose we wish to predict an observation from the model

$Z = \boldsymbol{w}\boldsymbol{\beta} + \varepsilon^*$ assuming a prior probability for $(\boldsymbol{\beta}, \theta)$. The predictive densities for this random future value given $\boldsymbol{y}$ and $\boldsymbol{y}_1$ are defined as in Subsection 2.4 as $f(z|\boldsymbol{w}, X, \boldsymbol{y}) = f(z)$, and $f(z|\boldsymbol{w}, X_1, \boldsymbol{y}_1) = f_{(2)}(z)$, respectively. P.I.F.'s are then defined as previously in Section 1. The primary definition $I(f, f_{(2)})$ will be studied in detail. However, the definition $I(f_{(2)}, f)$ will also be studied in some special situations since it lends itself readily to an interpretation of Cook's influence statistic.

2.3. *Case 1; θ known*

Assume the prior defined in (1) and define

$$S = X'X,\ S_1 = X_1'X_1,\ S_2 = X_2'X_2, \qquad \hat{\boldsymbol{\beta}} = S^{-1}X'\boldsymbol{y},\ \hat{\boldsymbol{\beta}}_{(2)} = S_1^{-1}X_1'\boldsymbol{y}_1,$$

$$\hat{\boldsymbol{y}} = X\hat{\boldsymbol{\beta}},\ \hat{\boldsymbol{y}}_{(2)} = X\hat{\boldsymbol{\beta}}_{(2)}, \qquad \hat{\boldsymbol{y}}_2 = X_2\hat{\boldsymbol{\beta}},\ \hat{\boldsymbol{y}}_{2(2)} = X_2\hat{\boldsymbol{\beta}}_{(2)}, \qquad \hat{z}_w = \boldsymbol{w}\hat{\boldsymbol{\beta}}, \qquad \hat{z}_{w(2)} = \boldsymbol{w}\hat{\boldsymbol{\beta}}_{(2)},$$

$$V_2 = X_2S^{-1}X_2', \qquad U_2 = X_2S_1^{-1}X_2', \qquad v_w = \boldsymbol{w}S^{-1}\boldsymbol{w}', \qquad u_w = \boldsymbol{w}S_1^{-1}\boldsymbol{w}',$$

$$\boldsymbol{v}_{w2} = \boldsymbol{w}S^{-1}X_2', \qquad \boldsymbol{u}_{w2} = \boldsymbol{w}S_1^{-1}X_2'.$$

These are all familiar expressions from linear models theory. For example, $\hat{\boldsymbol{\beta}}$ and $\hat{\boldsymbol{\beta}}_{(2)}$ denote the usual least squares estimates of $\boldsymbol{\beta}$ based on the full and deleted data sets respectively; $\hat{z}_w$ and $\hat{z}_{w(2)}$ denote least squares predictions at $\boldsymbol{w}$; V_2 and U_2 are proportional to sampling covariance matrices for predicted vectors at X_2, etc. Now when $n_2 = 1$ and, say, observation, i, has been deleted, the subscript (2) above will be replaced by (i), e.g. $V_2 = v_i = \boldsymbol{x}_i S^{-1}\boldsymbol{x}_i'$.

Recall now from (2) that the predictive densities f and $f_{(2)}$ are respectively $N[\hat{z}_w, (1+v_w)\theta]$ and $N(\hat{z}_{w(2)}, (1+u_w)\theta]$. If we define ξ to be the $100(1-(\alpha/2))$th percentile of the standard normal density, $0<\alpha<1$, then $1-\alpha$ predictive intervals for a future value are respectively,

$$\hat{z}_w \pm \xi\sqrt{(1+v_w)\theta} \text{ and } \hat{z}_{w(2)} \pm \xi\sqrt{(1+u_w)\theta}\,.$$

But these are precisely the same intervals that one gets via the classical tolerance interval approach where one pivots on $(Z-\hat{z}_w)/[\theta(1+v_w)]^{1/2}$ and $(Z-\hat{z}_{w(2)})/[\theta(1+u_w)]^{1/2}$, using the fact that sampling distributions are $N[0,1]$. Clearly, both interval widths and centers will be affected by influential subsets of data, regardless of whether one's approach is Bayesian or classical P.I.F.'s incorporate measures of these discrepancies.

Before going on to derive predictive influence functions, it is necessary to state and derive some useful formulas via a sequence of propositions.

Proposition 2.3.1. *The following identities are catalogued in Bingham* (1977)

$$S^{-1} = S_1^{-1} - S_1^{-1}X_2'(I+U_2)^{-1}X_2S_1^{-1}, \tag{13}$$

$$V_2 = U_2(I+U_2)^{-1}, \tag{14}$$

$$(I+U_2)^{-1} = (I-V_2), \tag{15}$$

$$(\boldsymbol{y}_2 - \hat{\boldsymbol{y}}_2) = (I+U_2)^{-1}(\boldsymbol{y}_2 - \hat{\boldsymbol{y}}_{2(2)}). \tag{16}$$

Proposition 2.3.2. *Let*

$$w=(1,w^*),\ X=(e_n,X^*),\ X_i=(e_{n_i},X_i^*)$$

for $i=1,2$ *and let* x_{ij} *be the jth row of* X_i^*, $j=1,\ldots,n_i$, $i=1,2$. *Then define*

$$\bar{X}=\sum_{i=1}^{2}\sum_{j=1}^{n_i}x_{ij}/n,\qquad \bar{X}_i=\sum_{j=1}^{n_i}x_{ij}/n_i,\qquad \tilde{X}=X^*-e_n\bar{X},\qquad \tilde{X}_i=X_i^*-e_{n_i}\bar{x}_i$$

$$S_x=\tilde{X}'\tilde{X}/n,\qquad S_x^{(i)}=\tilde{X}_i'\tilde{X}_i/n_i,\qquad J_n=e_ne_n',\qquad J_{n_i}=e_{n_i}e_{n_i}'$$

for $i=1,2$. *Then it follows from standard matrix manipulation that*

$$u_{w2}=n_1^{-1}\left[e_{n_2}'+(w^*-\bar{X}_1)S_x^{(1)-1}(X_2^*-e_{n_2}\bar{X}_1)'\right],\tag{17}$$

$$U_2=n_1^{-1}\left[J_{n_2}+(X_2^*-e_{n_2}\bar{X}_1)S_x^{(1)-1}(X_2^*-e_{n_2}\bar{X}_1)'\right],\tag{18}$$

$$u_w=n_1^{-1}\left[1+(w^*-\bar{X}_1)S_x^{(1)-1}(w^*-\bar{X}_1)'\right],\tag{19}$$

$$V_2=n^{-1}\left[J_n+(X_2^*-e_{n_2}\bar{X})S_x^{-1}(X_2^*-e_{n_2}\bar{X})'\right].\tag{20}$$

It is seen from the above that U_2 and V_2 are measures of the collective distance between where the deleted data set is observed and the center of where the non-deleted data set is observed. Similar expressions are obtained for v_{w2} and v_w.

Now define

$$T_2^2=\frac{\left[u_{w2}(I+U_2)^{-1}(y_2-\hat{y}_{2(2)})\right]^2}{u_{w2}(I+U_2)^{-1}u_{2w}\theta}=\frac{\left[v_{w2}(I-V_2)^{-1}(y_2-\hat{y}_2)\right]^2}{v_{w2}(I-V_2)^{-1}v_{2w}\theta}=t_2^2,$$

$$\eta_{w2}^2=\frac{u_{w2}(I+U_2)^{-1}u_{2w}}{1+u_w}=\frac{\eta_{w2}'^2}{1+\eta_{w2}'^2},\qquad \eta_{w2}'^2=\frac{v_{w2}(I-V_2)^{-1}v_{2w}}{1+v_w}.$$

Then

$$2I(f,f_{(2)})=T_2^2\eta_{w2}^2+\left[(1-\eta_{w2}^2)-\ln(1-\eta_{w2}^2)-1\right]\tag{21}$$

$$2I(f_{(2)},f)=T_2^2\eta_{w2}^2(1-\eta_{w2}^2)^{-1}+\left[(1-\eta_{w2}^2)^{-1}-\ln(1-\eta_{w2}^2)^{-1}-1\right].\tag{22}$$

These results are derived using simple algebra and Proposition 2.3.1. Note that by idempotence, T_2^2 has a sampling distribution which is χ_1^2, and that T_2^2 measures the distance between the vectors y_2 and $\hat{y}_{2(2)}$. The variable η_{w2}^2 indicates the relationship between where one is trying to predict and where the vector y_2 is observed, relative to where y_1 is observed. This relationship not only determines the weight a value of T_2^2 receives, and hence, the effect on point prediction, but it also solely determines the effect on predictive variances. In what follows, it will be shown that η_{w2}^2 is the multiple correlation, with respect to predictive distributions, between Z and Y_2; hence $0\leqslant\eta_{w2}^2\leqslant 1$. So values of η_{w2}^2 near one will cause the components of (21) and (22), due to the difference in predictive variances, to be unbounded. Also, the component of (22), due to the difference in predictive means, is unbounded for positive T_2^2. It is clear that $I(f_{(2)}, f) \geqslant I(f, f_{(2)})$, due to the fact that the predictive variance of $f_{(2)}$, is always larger than that of f. The fact that components of predictive influences are weighted more heavily in $I(f_{(2)}, f)$ than in $I(f, f_{(2)})$ results in different orderings. It is not clear, then, which of these definitions of P.I. is more appropriate, or even that their sum is not the best definition. A decision to this effect must ultimately rest on one's determination of the importance of the variable η_{w2}^2.

Note that when $p=1$,

$$T_2^2 = (\bar{y}_2 - \bar{y}_{(2)})^2 / \Theta(n_2^{-1} + n_1^{-1}) = (\bar{y}_2 - \bar{y})^2 / \Theta(n_2^{-1} - n^{-1}) = t_2^2,$$

$$\eta^2 = n_2 / n(n_1 + 1) = \eta'^2 (1 + \eta'^2)^{-1}, \qquad \eta'^2 = n_2/(n+1)n_1.$$

Hence, the component of P.I. due to the difference in variances is independent of the data. Subsets are ordered here according to the magnitudes of $|\bar{y}_2 - \bar{y}|$; i.e., the most influential subset will be centered distantly from the center of the combined data set. Equivalently, subsets may be ordered according to the magnitudes of $|\bar{y}_2 - \bar{y}_{(2)}|$.

Consider the case where $n_2 = 1$. Here

$$T_i^2 = (y_i - \hat{y}_{i(i)})^2 / (1 + u_i)\theta = (y_i - \hat{y}_i)^2 / (1 - v_i)\theta = t_i^2,$$

$$\eta_{wi}^2 = u_{wi}^2 / (1 + u_w)(1 + u_i), \qquad \eta_{wi}'^2 = v_{wi}^2 / (1 + v_w)(1 + v_i),$$

so that

$$2I(f_{(i)}, f) = t_i^2 \eta_{wi}'^2 + \left[(1 + \eta_{wi}'^2) - \ln(1 + \eta_{wi}'^2) - 1 \right]. \tag{23}$$

Now define the sample correlation coefficient between $\hat{Z}_w$ and $\hat{Y}_i$ as $\rho_{wi} = \text{corr}_{\text{samp.}}(\hat{Z}_w, \hat{Y}_i)$, so that $\eta_{wi}'^2 = \rho_{wi}^2 v_i (1 - v_i)^{-1} v_w (1 + v_w)^{-1}$. However, $p^{-1} t_i^2 v_i (1 - v_i)^{-1} \equiv D_i$ is just "Cook's distance", Cook (1977), for detecting the influence of the ith observation on the estimation of the regression coefficient vector; and $D_i \rho_{wi}^2 \equiv D_i(\boldsymbol{w})$ is Cook's statistic for detecting observations which are influential for the purpose of estimating the mean regression line at $\boldsymbol{w}$. Hence defining $v_i(\boldsymbol{w}) = \rho_{wi}^2 v_i (1 - v_i)^{-1}$ and $r_w = v_w (1 + v_w)^{-1}$, it follows that

$$2I(f_{(i)}, f) = D_i(\boldsymbol{w}) r_w + [(1 + v_i(\boldsymbol{w}) r_w) - \ln(1 + v_i(\boldsymbol{w})(r_w)) - 1], \tag{24}$$

thus it is seen that Cook's statistic is proportional to the first component above. Hence by (11) if one's interest rests solely on the influence on point prediction, Cook's statistic is appropriate. Further, considering the arbitrary subset deletion case, the leading component of $I(f_{(2)}, f)$ may be shown to be proportional to the appropriate generalization of Cook's statistic.

It will be useful to consider aspects of the results (21) and (22) more carefully from both classical and predictivist views; first, the predictivist view.

The prediction problem may be considered sequentially, as in Section 1. It may be initially assumed that $\boldsymbol{Y}_2$ is unobserved and that $(Z, \boldsymbol{Y}_2)$ has the joint predictive density $f_{(2)}(z, \boldsymbol{y}_2) = f(z, \boldsymbol{y}_2 | \boldsymbol{y}_1)$. By (2) this density is

$$N_{n_2+1}\left[\begin{pmatrix} \hat{z}_{w(2)} \\ \hat{\boldsymbol{y}}_{2(2)} \end{pmatrix}, \begin{pmatrix} 1 + u_w & \boldsymbol{u}_{w2} \\ \boldsymbol{u}_{2w} & I + U_2 \end{pmatrix} \theta \right].$$

Then assume $Y_2 = y_2$ is observed, and suppose it is of interest to determine whether this new information will be useful for the purpose of predicting Z. Hence conditioning on y_2, the conditional density, $f_{(2)}(z|\, y_2)$, is defined in Subsection 1.4 and is easily shown to be $N[\hat{z}_{w(2)} + u_{w2}(I+U_2)^{-1}(y_2 - \hat{y}_{2(2)}), (1+u_w - (I+U_2)^{-1}u_{2w})\theta]$. Recall from (12) that $f(z) = f_{(2)}(z|\, y_2)$, hence $I(f, f_{(2)}) = I(f_{(2)}(\cdot|\, y_2), f_{(2)})$. Now the strength of the relationship between Z and Y_2 is measured by the multiple correlation with respect to $f_{(2)}(Z, Y_2)$. But this is just η^2_{w2}. Hence $1-\eta^2_{w2}$ is the proportion of variability in Z unexplained by regression of Z on Y_2, and it is seen that the component of influence due to the difference in predictive variances will be large if the relationship between Z and Y_2 is strong and small if not. Note that observed data do not affect this relationship. Note further that $T_2^2 \alpha [E_{f_{(2)}(\cdot|\, y_2)}(Z) - E_f(Z)]^2$, i.e. T_2^2 is the standardized squared difference in predictions. Thus the component of influence due to the difference in predictive means will be large if regression on $Y_2 = y_2$ provides a significantly different prediction than one would obtain if $Y_2 = y_2$ was ignored. This component will depend on observed data through the variable T_2^2, as well as on the relationship between Z and Y_2. The problem may be similarly understood from the classical view by consideration of the sampling distribution of the residuals vector $[Z - \hat{Z}_{w(2)}, (Y_2 - \hat{Y}_{2(2)})']'$.

A canonical representation of the P.I.F. can be obtained and interpreted. Recall the existence of an orthogonal matrix, Γ, and a diagonal matrix δ_2 which satisfy $\Gamma U_2 \Gamma' = \delta_2 \equiv \text{diag.}\{\delta_i\}_{i=1}^{n_2}$; Γ and δ_2 are the matrices of eigenvectors and eigenvalues of U_2, respectively. Then by (15) it is clear that there exists $\Delta_2 = \text{diag.}\{\Delta_i\}_{i=1}^{n_2}$ such that $\Gamma V_2 \Gamma' = \Delta_2 = \delta_2(I+\delta_2)^{-1}$. Now define

$$r_{2(2)} = (I+\delta_2)^{-1/2}\Gamma(y_2 - \hat{y}_{2(2)})/\Theta \equiv (r_{1(2)}, r_{2(2)}, \ldots, r_{n_2(2)}),$$

$$\delta_{2w} = \Gamma u_{2w} \equiv (\delta_{1w}, \delta_{2w}, \ldots, \delta_{n_2 w})',$$

$$\gamma_{2w} = \left(\delta_{1w}/(1+\delta_1)^{1/2}, \ldots, \delta_{n_2 w}/(1+\delta_{n_2})^{1/2}\right)'/(1+u_w)^{1/2}$$

$$\equiv (\gamma_{1w}, \gamma_{2w}, \ldots, \gamma_{n_2 w})'.$$

Then it is easy to show that

$$2I(f, f_{(2)}) = \left\{\sum \gamma_{jw} r_{j(2)}\right\}^2 + \left[\left(1 - \sum \gamma_{jw}^2\right) - \ln\left(1 - \sum \gamma_{jw}^2\right) - 1\right]. \quad (25)$$

Observe that $\gamma_{2w} = \text{corr}_{\text{pred.}}(Z, \Gamma\gamma_2)$ and $\text{cov}_{\text{pred.}}(R_{2(2)}) = I$, where $R_{2(2)} = (I - \Delta_2)^{-1/2}\Gamma(Y_2 - y_{2(2)})/\theta$. Hence the first component of predictive influence is the square of a weighted sum of standardized canonical residuals, where the weights reflect the strength of the relationship between Z and the ith canonical variate. The second component of predictive influence depends on the sum of the above squared weights.

It will be useful in what follows to derive algebraic expressions for the variables T_2^2 and η^2_{w2}. The result will be determined by a sequence of propositions.

Proposition 2.3.3. *Assume the notation of Proposition* 2.3.2 *and define* $\tilde{S}_x = nS_x$, $\tilde{S}_x^{(i)} = n_i S_x^{(i)}$, $i=1,2$, *and* $\overline{X}_{21} = \overline{X}_2 - \overline{X}_1$. *Then*

$$\tilde{S}_x = \tilde{S}_x^{(1)} + \tilde{S}_x^{(2)} + \frac{n_1 n_2}{n} \overline{X}'_{21} \overline{X}_{21}. \tag{26}$$

Proof. The result follows by the definition of S_x and some algebra.

Proposition 2.3.4. *Define*

$$s_{xy}^{(i)} = n_i^{-1} \tilde{X}'_i y_i, \quad i=1,2, \qquad \hat{\beta}^{(1)} = S_x^{(1)-1} s_{xy}^{(1)},$$

$$\hat{\beta}^{(2)} = S_x^{(2)-1} s_{xy}^{(2)}, \qquad \hat{\beta}^{(2,1)} = \hat{\beta}^{(2)} - \hat{\beta}^{(1)}, \quad n_2 > p,$$

$$\hat{\delta} = \bar{y}_2 - \bar{y}_1 - \overline{X}_{21} \hat{\beta}^{(1)},$$

with the proviso that if $n_2 = 1$, $S_x^{(2)}$ *and* $s_{xy}^{(2)}$ *are zero matrices. Then*

$$T_2^2 = \frac{n_1 n_2 n^{-1}\Big[\big(1 + n_1 n^{-1}(w^* - \overline{X})S_x^{-1}\overline{X}'_{21}\big)\hat{\delta} + (w^* - \overline{X})S_x^{-1}\big(s_{xy}^{(2)} - S_x^{(2)}\hat{\beta}^{(1)}\big)\Big]^2}{\theta\Big[\big(1 + n_1 n^{-1}(w^* - \overline{X})S_x^{-1}\overline{X}'_{21}\big)\big(1 + \overline{X}_{21} S_x^{(1)-1}(w^* - \overline{X}_1)'\big) + (w^* - \overline{X})S_x^{-1} S_x^{(2)} S_x^{(1)-1}(w^* - \overline{X}_1)'\Big]}, \tag{27}$$

$$\eta_{w2}^2 = \frac{n^{-1} n_1^{-1} n_2\Big[\big(1 + n_1 n^{-1}(w^* - \overline{X})S_x^{-1}\overline{X}'_{21}\big)\big(1 + \overline{X}_{21} S_x^{(1)-1}(w^* - \overline{X}_1)'\big) + (w^* - \overline{X})S_x^{-1} S_x^{(2)} S_x^{(1)-1}(w^* - \overline{X}_1)'\Big]}{\Big[1 + n_1^{-1}\big(1 + (w^* - \overline{X}_1)S_x^{(1)-1}(w^* - \overline{X}_1)'\big)\Big]}. \tag{28}$$

Note that if $n_2 > p$, $s_{xy}^{(2)} = S_x^{(2)} \hat{\beta}^{(2)}$ and hence, $s_{xy}^{(2)} - S_x^{(2)}\hat{\beta}^{(1)} = S_x^{(2)}\hat{\beta}^{(2,1)}$.

Proof. By (15), (17), (20), (26) and some algebra,

$$u_{w2}(I + U_2)^{-1} = \big\{n^{-1}\big[1 + n_1(w^* - \overline{X})\tilde{S}_x^{-1}\overline{X}'_{21}\big]e'_{n_2} + (w^* - \overline{X})\tilde{S}_x^{-1}\tilde{X}'_2\big\},$$

and by further application of (17), it follows that $u_{w2}(I + U_2)^{-1} u_{w2}$ equals the numerator of (28). Hence, by applying (19), the result of (28) obtains. Now it can be shown that $\hat{y}_{2(2)} = e_{n_2}\bar{y}_1 + \tilde{X}_2\hat{\beta}^{(1)} + e_{n_2}\overline{X}_{21}\hat{\beta}^{(1)}$, from which it follows that $u_{w2}(I + U_2)^{-1}(y_2 - \hat{y}_{2(2)}) = \{n_2 n^{-1}[1 + n_1(w^* - \overline{X})\tilde{S}_x^{-1}\overline{X}'_{21}]\hat{\delta} + (w^* - \overline{X})\tilde{S}_x^{-1}(s_{xy}^{(2)} - S_x^{(2)}\hat{\beta}^{(1)})\}$; hence the result (27) obtains.

Suppose $n_2 > p$ so that it is possible to determine a least squares regression line for the deleted subset. From expression (27), it is clear that $T_2^2 = 0$ if $\hat{\delta} = 0$ and $w^* = \overline{X}$ or $\hat{\beta}^{(1)} = \hat{\beta}^{(2)}$. Now $\hat{\delta} = 0$ if and only if the deleted subset and non-deleted subset regression lines intersect at $(\bar{X}_2, \bar{y}_2)$. Hence $\hat{\delta}=0$ and $\hat{\beta}^{(1)}=\hat{\beta}^{(2)}$ imply the regression lines are the same; eliminating any difference in point predictions. Similarly predicting at the center of the full data set when the regression lines intersect at $(\bar{X}_2, \bar{y}_2)$ also implies no difference in point predictions.

An asymptotic representation predictive influence can be derived. Assume that $S_x^{(1)} \to \Sigma_x^{(1)}$, $s_{xy}^{(1)} \to \sigma_{xy}^{(1)}$, $\bar{y}_1 \to \mu_y^{(1)}$ (a.s.), $\bar{X}_1 \to \mu_x^{(1)}$ as $n \to \infty$, and define $\beta^{(1)} = \Sigma_x^{(1)}\sigma_{xy}^{(1)}$, $\tilde{\delta} = \bar{y}_2 - \mu_y^{(1)} - (\bar{X}_2 - \mu_x^{(1)})\beta^{(1)}$, $\lambda = \{[1+(w^* - \mu_x^{(1)})\Sigma_x^{(1)-1}(\bar{X}_2 - \mu_x^{(1)})']^2 + (w^* - \mu_x^{(1)})\Sigma_x^{(1)-1}S_x^{(2)}\Sigma_x^{(1)-1}(w^* - \mu_x^{(1)})'\}$. Then

$$2n^2I(f, f_{(2)}) \approx 2n^2I(f_{(2)}, f) \approx n_2 T_2^2 \lambda \quad \text{(a.s.)}$$

$$\approx \frac{n_2^2}{\theta}\Big\{\tilde{\delta}\left[1+(w^* - \mu_x^{(1)})\Sigma_x^{(1)-1}((\bar{X}_2 - \mu_x^{(1)}))'\right]$$

$$+ (w^* - \mu_x^{(1)})\Sigma_x^{(1)-1}(s_{xy}^{(2)} - S_x^{(2)}\beta^{(1)})\Big\}^2 \quad \text{(a.s.).} \tag{29}$$

These results are easily shown using (21), (22), (27), (28) and the above assumptions, after observing that $\eta_{w2}^2 = o(n^{-1})$, which implies that $n^2[(1-\eta_{w2}^2) - \ln(1-\eta_{w2}^2) - 1] = o(1)$. Now for the special case when $n_2 = 1$, the result (29) simplifies to

$$\left(y_i - \mu_y^{(1)} - (x_i - \mu_x^{(1)})\beta^{(1)}\right)^2 \Theta^{-1}\left(1+(w^* - \mu_x^{(1)})\Sigma_x^{(1)-1}(x_i^* - \mu_x^{(1)})'\right)^2. \tag{30}$$

This expression clearly implies that for a single observation to be influential, it must be observed at a considerable distance from the calculated regression line and one must be predicting at a great distance from the center.

When $p=1$, (30) reduces to $n_2(\bar{y}_2 - \mu)^2/n_2^{-1}\Theta = n_2\chi_1^2$. Hence, a subset of size n_2 is n_2 times as influential as a subset of size one for the same χ_1^2 value.

Asymptotic expressions of P.I. when $n_2 n^{-1} \to \alpha, 0<\alpha<1$, may similarly be derived, Johnson and Geisser (1979).

2.4. Case 2; exact predictive influence when θ is unknown

Let $p(\beta, \theta)$ denote the usual non-informative prior density for (β, θ), and define

$$s^2 = (y - \hat{y})'(y - \hat{y})/(n-p), \qquad s_{(2)}^2 = (y_1 - \hat{y}_{1(2)})'(y_1 - \hat{y}_{1(2)})/(n_1 - p).$$

Given the notation of Subsection 2.3, it then follows from (1.2.4) that f and $f_{(2)}$ are respectively $\mathrm{St}[n-p, \hat{z}_w, (1+v_w)s^2]$ and $\mathrm{St}[n_1 - p, \hat{z}_{w(2)}, (1+u_w)s_{(2)}^2]$.

It will be necessary to obtain some preliminary results before deriving P.I.F.'s.

Proposition 2.4.1. *Given the above notation,*

$$s^2 = \left[(n_1 - p)s_{(2)}^2 + (y_2 - \hat{y}_{2(2)})'(I+U_2)^{-1}(y_2 - \hat{y}_{2(2)})\right]/(n-p), \tag{31}$$

$$s_{(2)}^2 = \left[(n-p)s^2 - (y_2 - \hat{y}_2)'(I - V_2)^{-1}(y_2 - \hat{y}_2)\right]/(n_1 - p). \tag{32}$$

These results are given in Bingham (1977).

Proposition 2.4.2. *Let* $n_2 > 1$ *and define*

$$A_1 = (I+U_2)^{-1/2}\boldsymbol{u}_{2w}\boldsymbol{u}_{w2}(I+U_2)^{-1/2}/\boldsymbol{u}_{w2}(I+U_2)^{-1}\boldsymbol{u}_{2w}, \qquad A_2^* = I - A_1,$$
$$B_1 = (I-V_2)^{-1/2}\boldsymbol{v}_{2w}\boldsymbol{v}_{w2}(I-V_2)^{-1/2}/\boldsymbol{v}_{w2}(I-V_2)^{-1}\boldsymbol{v}_{2w}, \qquad B_2^* = I - B_1,$$
$$\boldsymbol{l}_2 = (I+U_2)^{-1/2}(\boldsymbol{y}_2 - \hat{\boldsymbol{y}}_{2(2)}), \qquad \boldsymbol{m}_2 = (I-V_2)^{-1/2}(\boldsymbol{y}_2 - \hat{\boldsymbol{y}}_2),$$
$$T_2^2 = \boldsymbol{l}_2' A_1 \boldsymbol{l}_2 / s_{(2)}^2, \qquad t_2^2 = \boldsymbol{m}_2' B_1 \boldsymbol{m}_2 / s^2,$$
$$V_2^{*2} = \boldsymbol{l}_2' A_2^* \boldsymbol{l}_2 / (n_2 - 1)s_{(2)}^2, \qquad v_2^{*2} = \boldsymbol{m}_2' B_2^* \boldsymbol{m}_2 / (n_2 - 1)s^2,$$
$$T_1^2 = \boldsymbol{l}_2'\boldsymbol{l}_2 / n_2 s_{(2)}^2, \qquad t_1^2 = \boldsymbol{m}_2'\boldsymbol{m}_2 / n_2 s^2,$$
$$C_1 = (n_1 - p)(n-p)^{-1}, \qquad C_2 = n_2(n_1 - p)^{-1}, \qquad \alpha_1 = n_2^{-1}.$$

Then

$$A_1 \text{ is idempotent of rank one}, \tag{33}$$

$$T_1^2 = \alpha_1 T_2^2 + (1-\alpha_1)V_2^{*2}, \qquad t_1^2 = \alpha_1 t_2^2 + (1-\alpha_1)v_2^{*2}, \tag{34}$$

$$t_2^2 = T_2^2/\{C_1[1 + C_2 T_1^2]\}, \qquad v_2^{*2} = V_2^{*2}/\{C_1[1 + C_2 T_1^2]\}, \tag{35}$$

$$t_1^2 = T_1^2/\{C_1[1 + C_2 T_1^2]\}. \tag{36}$$

Proof. The results (33) and (34) are easy to verify directly. The result (35) is a direct consequence of the fact that $\boldsymbol{l}_2' A_1 \boldsymbol{l}_2 = \boldsymbol{m}_2' B_1 \boldsymbol{m}_2$, $\boldsymbol{L}_2' A_2^* \boldsymbol{l}_2 = \boldsymbol{m}_2' B_2^* \boldsymbol{m}_2$ and Proposition 2.4.1. The result (36) follows directly from (34) and (35).

It is of interest to point out that the statistics T_1^2 and t_1^2 were defined in Bingham (1977), and were proposed by him as possible test statistics for testing whether $\boldsymbol{y}_2$ belongs to the assumed population. Bingham noted the relationship (2.4.5), and the fact that T_1^2 has a sampling distribution which is $F(n_2, n_1 - p)$. Then (33), (34) and Cochran's Theorem imply that $(s_{(2)}^2 T_2^2)/\theta$ and $(n_2 - 1)(s_{(2)}^2 V_2^2)/\theta$ have sampling distributions which are independent and $\chi^2(1)$ and $\chi^2(n_2 - 1)$, respectively. Further, since $\boldsymbol{y}_2 - \hat{\boldsymbol{y}}_{2(2)}$ is independent of $s_{(2)}^2$, it follows that T_2^2 and V_2^{*2} are respectively $F(1, n_1 - p)$ and hence, T_1^2 is a convex combination of "F" statistics.

Proposition 2.4.3. *Let*

$$\mu_1 = \hat{z}_w, \qquad \mu_2 = \hat{z}_{w(2)}, \qquad \mu = (\mu_2 - \mu_1)/\sigma_1, \qquad \mu' = (\mu_2 - \mu_1)/\sigma_2,$$
$$\sigma_1^2 = s^2(1 + v_w), \qquad \sigma_2^2 = s_{(2)}^2(1 + u_w), \qquad \sigma = \sigma_2/\sigma_1, \qquad \sigma' = \sigma^{-1}.$$

Then

$$\mu'^2 = T_2^2 \eta_{w2}^2, \qquad \mu^2 = t_2^2 \eta_{w2}'^2 \tag{37}$$
$$\sigma'^2 = C_1[1 + C_2 T_1^2](1 - \eta_{w2}^2),$$
$$\sigma^2 = C_1^{-1}[1 - C_1 C_2 t_1^2](1 + \eta_{w2}'^2). \tag{38}$$

Proof. The result (37) is easy to verify directly. The result (38) is a direct consequence of (31) and (32), the definitions of t_1^2 and T_1^2, and the fact that $(1+v_w)(1+u_w)^{-1}=1-\eta_{w2}^2$ and $\eta_{w2}^2=\eta_{w2}'^2/(1+\eta_{w2}'^2)$, which were indicated in Subsection 2.3.

Hence, it is seen that the difference in predictive means is very similar to the case where θ is known, the only difference is that θ is now replaced by s^2 in the definition of T_2^2. The difference from the θ known case is that the ratio of predictive variances now depends on the data as well as on η_{w2}^2. The ratio of predictive variances is a monotonic function of T_1^2, which in turn is a convex function of T_2^2 and V_2^{*2}. As in the θ known case, T_2^2 measures the distance between y_2 and $\hat{y}_{2(2)}$. The variable, V_2^2, on the other hand, will be a measure of the differences in scatter and orientation for deleted and non-deleted data sets, respectively.

It is possible to write down expressions of predictive influence, Johnson and Geisser (1979). These expressions depend upon the variables obtained in (37) and (38); however, they may not be represented explicitly.

It is seen that P.I. is a function of T_2^2, T_1^2 and η_{w2}^2. When $n_2=1, T_1^2=T_2^2=T_i^2$ so P.I. is a function of the squared studentized residual, cf. Behnken and Draper (1972), and η_{wi}^2. When $p=1$ and $n_2>p$, T_1^2 is a convex function of T_2^2 and $V_2^2=s_2^2/s_{(2)}^2$, the ratio of sample variances for deleted and non-deleted samples. When $p>1$, $n_2>p$, it will be seen that V_2^{*2} is a convex combination of the ratio of the mean squared errors for the deleted and non-deleted samples, and a function of the different slopes for deleted and non-deleted subsets.

As it is not possible to represent P.I.F.'s explicitly, it is also impossible to study the behavior of P.I.F.'s for all relevant situations. Though tedious, it is possible to show, Johnson and Geisser (1979), that $I(f, f_{(2)})$ achieves a local maximum at $V_2^{*2}=T_2^2=0$ for fixed small η_{w2}^2. If η_{w2}^2 is large, $I(f, f_{(2)})$ is decreasing in V_2^{*2} and increasing in T_2^2 in a neighborhood of zero. Further, $I(f, f_{(2)})$ is increasing in T_2^2 for all η_{w2}^2 and large T_2^2, when $V_2^{*2}=0$. A similar result holds for V_2^{*2}; and, $I(f, f_{(2)})$ is monotone and increasing in η_{w2}^2 when $T_2^2=V_2^{*2}=0$.

A more detailed analysis is deferred to the next subsection where approximate P.I.F.'s are derived and studied. It will be found that properties will be consistent and it will be inferred that exact P.I.F.'s behave in much the same way as the more tractable approximate P.I.F.'s.

Before going on to the next subsection, it will be useful to consider a further resolution of Bingham's statistic, T_1^2.

Recall, from (34), that $T_1^2=n_2^{-1}T_2^2+(1-n_2^{-1})V_2^{*2}$ and recall the definitions of Propositions 2.3.3, 2.3.4 and 2.4.2. Let $n_2>p$ and define

$$M_2=X_2(X_2'X_2)^{-1}X_2',\ y_{2(1)}=M_2y_2,\qquad s_2^2=(y_2-\hat{y}_{2(1)})'(y_2-\hat{y}_{2(1)})(n_2-p)^{-1},$$

$$A_1=(I+U_2)^{-1/2}\boldsymbol{u}_{2w}\boldsymbol{u}_{w2}(I+U_2)^{-1/2}/\boldsymbol{u}_{w2}(I+U_2)^{-1}\boldsymbol{u}_{2w},$$

$$A_2=(I+U_2)^{-1/2}(I-M_2)(I+U_2)^{-1/2},\qquad A_3=I-A_2-A_3,$$

$$l_2=(I+U_2)^{-1/2}(y_2-\hat{y}_{2(2)}),$$

$$T_2^2 = l_2' A_1 l_2 / s_{(2)}^2, \qquad V_2^2 = l_2' A_2 l_2 / (n_2 - p) s_{(2)}^2, \qquad W_2^2 = l_2' A_3 l_2 / (p-1) s_{(2)}^2,$$

$$\alpha_1' = (n_2 - p)(n_2 - 1)^{-1}, \qquad \alpha_2' = 1 - \alpha_1', \alpha_1 = n_2^{-1}, \qquad \alpha_2 = 1 - pn_2^{-1},$$

$$\alpha_3 = 1 - \alpha_1 - \alpha_2.$$

Then

$$A_1 \text{ and } A_2 \text{ are idempotent of rank one and } n_2 - p, \text{ respectively and } A_1 A_2 = 0. \tag{39}$$

$$V_2^{*2} = \alpha_1' V_2^2 + (1 - \alpha_1') W_2^2, \qquad T_1^2 = \alpha_1 T_2^2 + \alpha_2 V_2^2 + \alpha_3 W_2^2, \qquad V_2^2 = s_2^2 / s_{(2)}^2, \tag{40}$$

and if $w^* = \bar{X}_1$

$$W_2^2 = \hat{\beta}^{(2,1)\prime} \left(\tilde{S}_x^{(2)^{-1}} + \tilde{S}_x^{(1)^{-1}} \right)^{-1} \hat{\beta}^{(2,1)} / (p-1) s_{(2)}^2. \tag{41}$$

It is easy to show (39) using the fact that M_2 is idempotent. The result (40) follows directly from the fact that $A_2^* = A_2 + A_3$ and the definitions of V_2^{*2}, V_2^2 and W_2^2; and the fact that $(I - M_2) M_2 = 0$. To demonstrate the result (41) involves considerably more effort and is proved in Johnson and Geisser (1979).

Some comments are in order. The result (39) implies that the numerators of T_2^2, V_2^2 and W_2^2 have sampling distributions which are independent and proportional to $\chi^2(1)$, $\chi^2(n_2 - p)$ and $\chi^2(p-1)$, respectively. Hence, T_1^2 is a convex combination of three statistics which are distributed as $F(1, n_1 - p)$, $F(n_2 - p, n_1 - p)$ and $F(p-1, n_1 - p)$, respectively. The relation (40) implies that P.I.F.'s, as functions of V_2^2 and W_2^2, behave in the same way as they did as functions of V_2^{*2}, and that the variable V_2^2 measures the relative scatter about regression lines between deleted and non-deleted subsets. And, the relation (41) indicates that when $w^* = \bar{X}_1$, the variable W_2^2 measures the difference in slopes of regression lines relative to the sampling variability in $\hat{\beta}^{(2,1)}$, since $\mathrm{cov}_{\mathrm{samp}}(\hat{\beta}^{(2,1)}) \propto (\tilde{S}_x^{(1)-1} + \tilde{S}_x^{(2)-1})$. It seems reasonable to infer that the variable W_2^2 is measuring a similar entity but relative to a different inner product when w^* is arbitrary. Given this resolution, it follows that influential subsets will have greater scatter and will be aligned differently than non-deleted subsets.

2.5. *Approximate predictive influence*

Since exact P.I.F.'s cannot be represented explicitly, it is of interest to get good approximations. This will be accomplished by substituting appropriately scaled normal densities for Student densities in the definitions of P.I.. Johnson and Geisser (1979) show that, in large samples, exact and approximate P.I.F.'s are uniformly close on bounded intervals. They further show that pointwise convergence is of order n^{-2}, and that among all normal density approximations to a "t" density, a particular one results in optimal convergence, with respect to a Kullback–Leibler information measure.

Let f and $f_{(2)}$ be defined as in Subsection 2.4 and let μ_1, μ_2, σ_1, σ_2, μ, σ, μ' and σ' be defined as in Proposition 2.4.3. Let n, $n_1 \geqslant p+3$ and let $\tilde{f}$ and $\tilde{f}_{(2)}$ be densities which are

$$N\left[\mu_1, \sigma_1^2\left(\frac{n-p}{n-p-2}\right)\right] \quad \text{and} \quad N\left[\mu_2, \sigma_2^2\left(\frac{n_1-p}{n_1-p-2}\right)\right],$$

respectively. Variance adjustments are made so that $\mathrm{Var}_{\tilde{f}}(Z)=\mathrm{Var}_{\tilde{f}}(Z)$. Define $C_3=(n_1-p-2)(n_1-p)^{-1}$ and $C_4=(n_1-p-2)(n-p-2)^{-1}$. Then using (8) and the results of Proposition (2.4.3), it follows that

$$\hat{I}(f, f_{(2)}) \equiv I(\tilde{f}, \tilde{f}_{(2)}) = T_2^2\eta_{w2}^2(C_3) + \left[(1-\eta_{w2}^2)(C_4)(1+C_2T_1^2) - \ln\{(1-\eta_{w2}^2)(C_4)(1+C_2T_1^2)\} - 1\right]. \quad (42)$$

A similar expression may be derived for $\hat{I}(f_{(2)}, f)$. It is worth noting that the leading term in $\hat{I}(f_{(2)}, f)$ is proportional to the generalization of the Cook statistic which measures the discrepancy in mean regression lines at w^* when $n_2 \geqslant 1$ and Θ is unknown; cf. Cook (1977).

The behavior of approximate P.I.F.'s has been studied by Johnson and Geisser (1979). The results obtained are consistent with those indicated in the last section.

It is implied throughout that exact P.I.F.'s will behave in the same way as the approximate functions. Now for fixed η_{w2}^2, the most influential subsets will have large T_2^2 and V_2^{*2} values. The fact that the P.I.F. as a function of T_2^2 is J shaped appears to be a result of the difference in sample sizes for full and subset deleted data sets. Large values of T_2^2 are of greater interest than small values, and large values of η_{w2}^2 enhance the significance of a large T_2^2. As a function of V_2^{*2}, P.I. is again J shaped. Large values of V_2^{*2} are of greater interest than small values, and large values of η_{w2}^2 will tend to inhibit the significance of a large V_2^{*2}, since these variables tend to negate one another. Similarly, for fixed T_2^2 and V_2^{*2}, the P.I.F. is J shaped in η_{w2}^2; except when V_2^{*2} is small relative to T_2^2, in which case P.I. is monotone and increasing in η_{w2}^2. The reason that the variables V_2^{*2} and η_{w2}^2 negate one another becomes more apparent if one considers the different tolerance intervals obtained with full and subset deleted data sets. Let $t_{\alpha/2}(k)$ denote the $\alpha/2$th percentage point of a Student's "t" distribution with k degrees of freedom. Then $1-\alpha$ tolerance, or predictive intervals, for a future observation using full and deleted data sets are

$$\hat{z}_w \pm t_{\alpha/2}(n-p)\sqrt{1+v_w}\, s, \qquad \hat{z}_{w(2)} \pm t_{\alpha/2}(n_1-p)\sqrt{1+u_w}\, s_{(2)},$$

respectively. Now the ratio of the squared widths of these predictive intervals is proportional to

$$(1+v_w)/(1+u_w)s^2/s_{(2)}^2 \alpha(1-\eta_{w2}^2)\left[1+C_2\left(\alpha_1T_2^2+(1-\alpha_1)V_2^{*2}\right)\right].$$

Hence a large value of η_{w2}^2 implies a narrower interval, while a large value of V_2^{*2} implies a wider interval.

As a final note to this section, when $n_2=1$, the P.I.F. is always a monotone increasing function of η_{wi}^2, and a J shaped function of T_i^2. Hence the most influential single observation is observed at a point far removed from where the remaining data are observed, and at a y value which is distant from the deleted subset regression line.

Now, assume that for $T^2, V^2 \in R^+$, $T_2^2 = T^2 + o(1)$ (a.s.) and by (29), $n^2\eta_{w2}^2 = n_2\lambda + o(1)$. Defining the asymptotic P.I.F. in the same way as in section (2.3), it follows by (42), some algebra, a sequence of Taylor expressions, and the fact that $n^2 I(f, f_{(2)}) = n^2 \hat{I}(f, f_{(2)})$ a.s., that

$$\tilde{I}(f, f_{(2)}) = n_2 T_2^2 \lambda + n_2^2 \left[T_1^2 - 1\right]^2 / 2 + o(1) \quad \text{(a.s.)} \tag{43}$$

where T_1^2, T_2^2 and λ are defined as before.

Hence, asymptotically, the component reflecting different predictive variances is independent of λ, and hence, is independent of η_{w2}^2. The component reflecting different predictive means, on the other hand, depends heavily on the weighting factor λ. Observe that $1 \leqslant \lambda < \infty$, so that λ can be of considerable importance. Also note that $\tilde{I}$ is strictly increasing in T_2^2 and that $\tilde{I}$ is J shaped in V_2^{*2}, achieving a minimum at $(n_2 - T_2^2)(n_2 - 1)^{-1}$. The least influential subset will have small λ, small T_2^2 and $V_2^{*2} \approx n_2/(n_2 - 1)$. In fact, $\tilde{I}(f, f_{(2)}) = 0$ if and only if $T_2^2 = 0$ and $V_2^{*2} = n_2/(n_2 - 1)$. The most influential subset will clearly have large λ, large T_2^2 and large V_2^{*2}. In a large sample then, the above may be interpreted to mean that the most influential subsets will have large T_2^2, V_2^{*2} and η_{w2}^2, since the discordance between V_2^{*2} and η_{w2}^2 is diminished. And when $n_2 > p$, the most influential subset has large values for all four variables. Hence, the most influential subset is aligned perpendicular to the rest of the data, is more scattered about its regression line than non-deleted data, is distantly centered from the non-deleted data set regression line. Of course, if $p = 1$, asymptotic predictive influence functions are monotonic in T_2^2 and J shaped in V_2^2, indicating that the most influential subsets are those which are more distantly centered and are more widely scattered than non-deleted sets.

References

Aitchison, J. and Dunsmore, I.R. (1975). *Statistical Prediction Analysis*. Cambridge University Press, Cambridge.

Behnken, D.W. and Draper, N.R. (1972). Residuals and their variance patterns. *Technometrics* **14** 102–111.

Bingham, Christopher (1977). Some identities useful in the analysis of residuals from linear regression. U. of Minn. Technical Report No. 300. School of Statistics, University of Minnesota, MI.

Cook, R.D. (1977). Detection of influential observations in linear regression. *Technometrics* **19**, 15–18.

Cook, R.D. (1979). Influential observations in linear regression. *JASA* **74**, 169–174.

Cook, R.D and Weisberg, S. (1979). Finding influential cases in regression; a review. U. of Minn. Technical Report No. 338. School of Statistics, University of Minnesota, MI.

Draper, N.R. and Smith, H. (1966). *Applied Regression Analysis*. Wiley, New York, U.S.A.

Geisser, S. (1965). Bayesian estimation in multivariate analysis. *Ann. Math. Stat.* **36**, 150–159.

Geisser, S. (1971). The inferential use of predictive distributions. *Foundations of Statistical Inference*. Godambe and Sprott, Eds., pp. 456–466. Holt, Rinehart and Winston, Toronto.

Johnson, W., and Geisser, S. (1979). Assessing the predictive influence of observations. U. of Minn. Technical Report No. 355, School of Statistics, University of Minnesota, MI.

Kullback, S. (1968). *Information Theory and Statistics*. Peter Smith, Gloucester, MA.

Kullback, S. and Leibler, R.A. (1951). On information and sufficiency *Ann. Math. Stat.* **22**, 79–86.

Murray, Gordon D. (1977). A note on the estimation of probability density functions. *Biometrika* **64**, 150–152.

G. Kallianpur, P.R. Krishnaiah, J.K. Ghosh, eds., *Statistics and Probability: Essays in Honor of C.R. Rao*
© North-Holland Publishing Company (1982) 359–368

A CLASS OF ESTIMATORS OF A LOCATION PARAMETER IN PRESENCE OF A NUISANCE SCALE PARAMETER

A.M. KAGAN, I.A. MELAMED and A.A. ZINGER
Leningrad, U.S.S.R.

The problem of estimating a structural location parameter θ in presence of a nuisance scale parameter σ is considered on the base of an independent sample $x_1,\dots,x_n$ from a population with a distribution function $F((x-\theta)/\sigma)$. An asymptotically (when $n\to\infty$) optimal estimator of θ within the class of equivariant estimators of the form, $\bar{x}+sQ(z)$, is constructed where $\bar{x}=(x_1+\cdots+x_n)/n$, $s^2=\Sigma_1^n(x_i-\bar{x})^2/n$, $z=((x_1-\bar{x})/s,\dots,(x_n-\bar{x})/s)$ and Q runs the set of all polynomials of a given degree k. This optimal estimator depends upon F only through dimensionless characteristics of F like asymmetry and kurtosis. Its asymptotic behaviour is determined by the Fisher information contained in the space of polynomials of the degree k. Two characterization theorems are proved in this connection.

1. Introduction

Let independent identically distributed observations x_i be of the form:

$$x_i=\theta+\varepsilon_i, \quad i=1,\dots,n, \tag{1}$$

where $\theta\in R^1$ is a structural parameter and the distribution function (d.f.) $F(x/\sigma)$ of ε_i is known up to a (nuisance) parameter $\sigma\in R^1_+$. Under the assumption,

$$\int x\,\mathrm{d}F=0,$$

(which is essential only for characterization problems considered in Section 4) we shall deal with estimating θ on the base of the following dimensionless characteristics of the d.f. $F(x/\sigma)$ which eliminate σ:

$$\gamma_3=\int x^3\,\mathrm{d}F/\mu_2^{3/2},\ \gamma_4=\int x^4\,\mathrm{d}F/\mu_2^2,\dots,\gamma_{2k}=\int x^{2k}\,\mathrm{d}F/\mu_2^k \tag{2}$$

where $\mu_2=\int x^2\,\mathrm{d}F$ and $k\geqslant 1$ is an integer.

The asymmetry γ_3 and the kurtosis γ_4 of the d.f. are popular in different applications of statistical methods and that is why it seems to us useful to introduce and investigate estimators of θ which require from a priori characteristics only γ_3,γ_4 and their analogs of higher orders.

The situation considered here is intermediate between the case of estimating θ on the base of a finite number of moments of the d.f. $F(x/\sigma)$ when σ is known (Kagan, 1966; Kagan et al., 1974) and that of estimating θ in presence of a nuisance σ when the d.f. F is completely known (Kagan and Zinger, 1973).

Just as the asymptotic behaviour when $n\to\infty$ of the optimal estimator of θ, in the case of known σ and based upon the first $2k$ moments of the d.f. $F(x/\sigma)$, is determined by the Fisher information on θ contained in the space of all polynomials in x of the degree k, and that of the optimal estimator of θ in the case of unknown σ is determined by the (modified) Fisher information on θ in presence of a nuisance σ contained in x, the asymptotic behaviour of the optimal estimator of θ constructed below in terms of the characteristics (2) is determined by the Fisher information on θ in presence of a nuisance σ contained in the polynomials in x of the degree k.

The asymptotically optimal estimator (9) is a natural analog — in presence of a nuisance parameter — of the modified polynomial Pitman estimator introduced in Kagan et al. (1974). In the case of given γ_3, γ_4, the estimator (9) was investigated in Zinger and Kagan (1973); Melamed (1974) generalized their construction to the general case. Here we expound the results announced in Zinger and Kagan (1973) and Melamed (1974); we have also supplemented the characterization of a distribution by the property of minimum Fisher information on θ, in presence of a nuisance σ contained in polynomials in x of a given degree.

2. Asymptotically optimal estimators

Let:

$$\bar{x}=(x_1+\cdots+x_n)/n, \qquad s^2=\sum_1^n (x_i-\bar{x})^2/n,$$

$$z=((x_1-\bar{x})/s,\ldots,(x_n-\bar{x})/s), \qquad g_j=\left(\frac{1}{n}\right)\sum_{i=1}^{n}[(x_i-\bar{x})/s]^j, \quad j=3,\ldots,k,$$

and introduce the class of estimators of θ of the form

$$\tilde{\theta}_n=\bar{x}+sQ(z) \tag{3}$$

where $Q(z_1,\ldots,z_n)$ is a polynomial of a fixed degree k. We call (3) equivariant polynomial of the degree k estimators of θ in presence of a nuisance σ.

We shall show that on suitable choice of constants $A_2,\ldots,A_k$ the estimator

$$\bar{\theta}_n=\bar{x}+s\sum_2^k A_j g_j \tag{4}$$

will be asymptotically optimal within the class (3) in the sense that the variance of its limit (normal) distribution is less than that of any other estimator (3). Let:

$$d_{11}=\mu_2, \qquad d_{1j}=d_{j1}=\mu_2\left[\gamma_{j+1}-j\gamma_{j-1}-\tfrac{1}{2}j\gamma_3\gamma_j\right], \quad j=3,\ldots,k,$$

$$\begin{aligned} d_{ij}=d_{ji}=\mu_2\Big[&-\tfrac{1}{4}(ij-2(i+j)+4)\gamma_i\gamma_j+\tfrac{1}{4}ij\gamma_4\gamma_i\gamma_j \\ &-\tfrac{1}{2}i\gamma_i\gamma_{j+2}-\tfrac{1}{2}j\gamma_j\gamma_{i+2}+\tfrac{1}{2}ij\gamma_3\gamma_{i-1}\gamma_j+\tfrac{1}{2}ij\gamma_3\gamma_{j-1}\gamma_i \\ &+\gamma_{i+j}-i\gamma_{i-1}\gamma_{j+1}-j\gamma_{j-1}\gamma_{i+1}+ij\gamma_{i-1}\gamma_{j-1}\Big], \quad i,j=3,\ldots,k. \end{aligned}$$

$$D=\begin{Vmatrix} d_{11} & d_{13} & \cdots & d_{1k} \\ d_{31} & d_{33} & \cdots & d_{3k} \\ \vdots & & & \\ d_{k1} & d_{k3} & \cdots & d_{kk} \end{Vmatrix};$$

by D_{ij}, we will denote the cofactor of the element d_{ij} of the matrix D.

Lemma 1. *If* $\mu_{2k}=\int x^{2k}\,\mathrm{d}F<\infty$, *then the random vector*:

$$\left\{\sqrt{n}\,(\bar{x}-\theta),\sqrt{n}\,s(g_3-\gamma_3),\ldots,\sqrt{n}\,s(g_k-\gamma_k)\right\}, \tag{5}$$

is asymptotically normal, $N(0,\sigma^2 D)$, $n\to\infty$. *If in addition*,

$$\int|x|^j\,\mathrm{d}F<\infty,\quad j=1,2,\ldots, \tag{6}$$

then all the moments of the vector (5) *converge to the corresponding moments of the normal vector* $N(0,\sigma^2 D)$.

The lemma is proved by straightforward calculations similar to those in Chapter 6 of Cramér (1946).

Now let $(\xi,\eta_3,\ldots,\eta_k)$ denote a random vector with the normal distribution $N(0, D)$ and

$$E(\xi|\eta_3,\ldots,\eta_k)=\sum_3^k \hat{A}_j\eta_j.$$

Then [see Cramér (1946)],

$$\hat{A}_j=-D_{1j}/D_{11},\quad j=3,\ldots,k, \tag{7}$$

Let

$$\hat{A}_2=-\sum_3^k \hat{A}_j\gamma_j. \tag{8}$$

According to the definition of d_{ij}, the coefficients $\hat{A}_2,\ldots,\hat{A}_k$ are determined only by the characteristics (2). Consider now the estimator:

$$\hat{\theta}_n=\bar{x}-s\sum_2^k \hat{A}_j g_j. \tag{9}$$

Lemma 2. *Suppose that* F *has more than* k *growth points and* $\mu_{2k}<\infty$. *Then the random vector*

$$\left\{\sqrt{n}\,(\hat{\theta}_n-\theta),\sqrt{n}\,s(g_3-\gamma_3),\ldots,\sqrt{n}\,s(g_k-\gamma_k)\right\} \tag{10}$$

is asymptotically normal $N(0,\sigma^2\hat{D})$, $n\to\infty$ *where*

$$\hat{D}=\begin{Vmatrix} \hat{d}_{11} & 0\cdots 0 \\ 0 & \\ \vdots & D_{11} \\ 0 & \end{Vmatrix},\quad \hat{d}_{11}=(\det D)/D_{11}.$$

If in addition the condition (6) *holds, then all the moments of the vector* (10) *converge to the corresponding moments of the limit normal distribution.*

In particular, if $E_{\theta,\sigma}$ *denotes the mathematical expectation corresponding to the d.f.* $F((x-\theta)/\sigma)$, *then*:

$$E_{\theta,\sigma}\left[\sqrt{n}\left(\hat{\theta}_n-\theta\right)\right]^2=\sigma^2(\det D)/D_{11}(1+o(1)),\quad n\to\infty. \tag{11}$$

Lemma 2 is an immediate consequence of Lemma 1; the equality (11) is in fact a formula for the residual variance. The condition that F has more than k growth points guarantees the non-vanishing of both $\det D$ and D_{11}.

Let us come back to the estimators (3). Since for any $\tilde{\theta}_n$ of the form (3)

$$E_{\theta,\sigma}\left(\tilde{\theta}_n-\theta\right)^2=\sigma^2E_{0,1}\tilde{\theta}_n^2$$

there exists an optimal estimator $\bar{\theta}_n$ within the class (3):

$$E_{\theta,\sigma}\left(\bar{\theta}_n-\theta\right)^2=\min_{\tilde{\theta}_n}E_{\theta,\sigma}\left(\tilde{\theta}_n-\theta\right)^2.$$

It is easily seen that:

$$\bar{\theta}_n=\bar{x}+s\hat{Q}(z)=\bar{x}-\hat{E}_{0,1}(\bar{x}|\Lambda_k) \tag{12}$$

Here Λ_k denotes the space of all functions of the form $sQ(z)$ where Q runs the set of all polynomials of the degree k. The scalar product is defined as $(\varphi_1,\varphi_2)=E_{0,1}(\varphi_1\cdot\varphi_2)$ and $\hat{E}_{0,1}(\cdot|\Lambda_k)$ is the projection operator into Λ_k. It is clear that

$$E_{\theta,\sigma}\left(\bar{\theta}_n-\theta\right)^2\leqslant E_{\theta,\sigma}\left(\tilde{\theta}_n-\theta\right)^2.$$

The estimator $\bar{\theta}_n$ is much more complicated than $\hat{\theta}_n$; in particular the coefficients of $\bar{\theta}_n$ are not determined only by the characteristics (2). However, we shall see that the asymptotic behaviour of the estimator (12) is the same as that of the more simple estimator (9). In fact, a more stronger result holds.

Lemma 3. *Under the conditions of Lemma* 2

$$E_{\theta,\sigma}\left[\sqrt{n}\left(\bar{\theta}_n-\hat{\theta}_n\right)\right]^2=o(1),\quad n\to\infty. \tag{13}$$

Proof. We have:

$$\hat{\theta}_n-\bar{\theta}_n=\hat{E}_{0,1}(\bar{x}|\Lambda_k)-s\sum_2^k\hat{A}_jg_j=\hat{E}_{0,1}\left(\hat{\theta}_n|\Lambda_k\right). \tag{14}$$

As the right-hand side of (14) is a symmetric polynomial of z of the degree k it can be written as a polynomial in the Studentized moments $g_3,\ldots, g_k$:

$$\sqrt{n}\,(\hat{\theta}_n-\bar{\theta}_n)=s\sum B^{(n)}_{j_3\cdots j_k}\left[\sqrt{n}\,\mu_2^{1/2}(g_3-\gamma_3)\right]^{j_3}\cdots\left[\sqrt{n}\,\mu_2^{1/2}(g_k-\gamma_k)\right]^{j_k} \tag{15}$$

where the summation is extended over all non-negative integers $j_3,\ldots, j_k$ subject the condition $3j_3+\cdots+kj_k\leqslant k$. According to (14) the coefficients $B^{(n)}_{j_3\cdots j_k}$ can be found from the system of equations

$$E_{0,1}\left\{\sqrt{n}\,s\hat{\theta}_n\left[\sqrt{n}\,\mu_2^{1/2}(g_3-\gamma_3)\right]^{l_3}\cdots\left[\sqrt{n}\,\mu_2^{1/2}(g_k-\gamma_k)\right]^{l_k}\right\}=$$

$$=\sum B^{(n)}_{j_3\cdots j_k}E_{0,1}\left\{s^2\left[\sqrt{n}\,\mu_2^{1/2}(g_3-\gamma_3)\right]^{j_3+l_3}\cdots\left[\sqrt{n}\,\mu_2^{1/2}(g_k-\gamma_k)\right]^{j_k+l_k}\right\} \tag{16}$$

where $l_3,\ldots, l_k$ are non-negative integers with $3l_3+\cdots+kl_k\leqslant k$. Since $E_{0,1}(s-\mu_2^{1/2})^2\to 0$ when $n\to\infty$, we get from Lemma 2,

$$E_{0,1}\left\{s^2\left[\sqrt{n}\,\mu_2^{1/2}(g_3-\gamma_3)\right]^{j_3+l_3}\cdots\left[\sqrt{n}\,\mu_2^{1/2}(g_k-\gamma_k)\right]^{j_k+l_k}\right\}\to$$

$$\to\mu_2 E\left(\eta_3^{j_3+l_3}\cdots\eta_k^{j_k+l_k}\right) \tag{17}$$

where the (Gaussian) random variables $\eta_3,\ldots,\eta_k$ were introduced above. From Lemma 2 we also have

$$E_{0,1}\left\{\sqrt{n}\,s\hat{\theta}_n\left[\sqrt{n}\,\mu_2^{1/2}(g_3-\gamma_3)\right]^{l_3}\cdots\left[\sqrt{n}\,\mu_2^{1/2}(g_k-\gamma_k)\right]^{l_k}\right\}\to 0. \tag{18}$$

Let

$$\mathbb{C}=\left\| E\left(\eta_3^{j_3+l_3}\cdots\eta_k^{j_k+l_k}\right)\right\|$$

where $j_3,\ldots, j_k$; $l_3,\ldots, l_k$ satisfy the above conditions. Denote by M_k the linear space spanned to the random variables $\eta_3^{i_3}\cdots\eta_k^{i_k}$ where $i_3,\ldots, i_k$ run the set of all non-negative integers with $3i_3+\cdots+ki_k\leqslant k$. If the vector $(\zeta,\eta_3,\ldots,\eta_k)$ is distributed according to the normal law $N(0,\hat{D})$, then

$$0=E(\zeta|\eta_3,\ldots,\eta_k)=\hat{E}(\zeta|M_k)=\sum c_{i_3\cdots i_k}\eta_3^{i_3}\cdots\eta_k^{i_k}$$

whence

$$\sum c_{i_3\cdots i_k}E\left(\eta_3^{i_3+l_3}\cdots\eta_k^{i_k+l_k}\right)=0, \tag{19}$$

the summation being extended to all non-negative integers $i_3,\ldots, i_k$ with $3i_3+\cdots+ki_k\leqslant k$ and $l_3,\ldots, l_k$ are non-negative integers. Since the coefficients $c_{i_3\cdots i_k}$ are uniquely determined from (19) we have

$$\det\mathbb{C}\neq 0.$$

Reverting back to (16) and taking (17), (18) into consideration we get

$$B^{(n)}_{j_3\cdots j_k}\to 0,\quad n\to\infty,$$

for all non-negative integers $j_3,\ldots, j_k$ with $3j_3+\cdots+kj_k\leqslant k$. Lemma 3 is proved.

The following theorem is an immediate consequence of Lemmas 2, 3.

Theorem 1. *If the d.f. F has more than k growth points and all its moments are finite, then*

$$P_{\theta,\sigma}\left\{\sqrt{n}\,(\hat{\theta}_n-\theta)<x\right\}\to\Phi\left[x/\left(\sigma\hat{d}_{11}^{1/2}\right)\right],\quad n\to\infty$$

where $\hat{d}_{11}$ *was defined in Lemma* 2 *and* Φ *denotes the standard normal d.f. Furthermore,*

$$E_{\theta,\sigma}\left[\sqrt{n}\,(\hat{\theta}_n-\theta)\right]^2=\sigma^2\hat{d}_{11}(1+o(1)). \tag{20}$$

3. Fisher information on a location parameter contained in polynomials of a degree *k* in presence of a nuisance scale parameter

In this section, we will suppose that $F(x)$ has an absolutely continuous density function $f(x)$.

By L_f^2 we denote the Hilbert space of all functions φ with $\int\varphi^2 f\,\mathrm{d}x<\infty$ and with the scalar product:

$$(\varphi_1,\varphi_2)=\int\varphi_1\varphi_2 f\,\mathrm{d}x,\quad (\varphi,\varphi)=\|\varphi\|^2.$$

As above, the symbol $\hat{E}(\cdot|\ )$ will denote the projection operator (the mathematical expectation in wide sense). Suppose that:

$$J_1=-f'/f\in L_f^2,\qquad J_2=-(1+xf'/f)\in L_f^2 \tag{21}$$

and let

$$\hat{J}_1=J_1-\hat{E}(J_1|J_2).$$

Denote by Π_k the space of all polynomials in x of the degree k and put

$$J_1^{(k)}=\hat{E}(J_1|\Pi_k),\qquad J_2^{(k)}=\hat{E}(J_2|\Pi_k),$$
$$\hat{J}_1^{(k)}=J_1^{(k)}-\hat{E}\left(J_1^{(k)}|J_2^{(k)}\right),\qquad \hat{I}^{(k)}=\|\hat{J}^{(k)}\|^2.$$

Theorem 2. *If* $\int x^{2k}f(x)\,\mathrm{d}x<\infty$ *and the conditions* (21) *hold, then*

$$\hat{I}^{(k)}=1/\hat{d}_{11}. \tag{22}$$

The quantity $\hat{I}^{(k)}/\sigma^2$ has the meaning of the Fisher information on a (structural) parameter θ contained in the polynomials of the degree k of an observation x with the density function $(1/\sigma)f((x-\theta)/\sigma)$, in the presence of a nuisance parameter σ [see Melamed (1974)]. This fact throws additional light on the assertion of Theorem 1.

In Klebanov and Melamed (1976), the Fisher information on θ in presence of a nuisance, σ, contained in the linear space generated by $\bar{x}$ and Λ_k was calculated. It turned out to be equal to $(n\hat{I}^{(k)}/\sigma^2)(1+o(1))$, $n\to\infty$. This result jointly with Theorems 1 and 2 means that the estimator $\hat{\theta}_n$ is asymptotically efficient—and not only asymptotically optimal—within the class (3).

Proof of Theorem 2. Let:

$$\tilde{\Pi}_k = \{\varphi \in \Pi_k : (\varphi, J_2) = 0\}.$$

Lemma 4.

$$\hat{J}_1^{(k)} = \hat{E}(\hat{J}_1 | \tilde{\Pi}_k). \tag{23}$$

Proof of Lemma 4.

$$\begin{aligned}(\hat{J}_1^{(k)}, J_2) &= (J_1^{(k)}, J_2) - (\hat{E}(J_1^{(k)} | J_2^{(k)}), J_2) \\ &= (J_1^{(k)}, J_2) - (\hat{E}(J_1^{(k)} | J_2^{(k)}), J_2^{(k)}) = (J_1^{(k)}, J_2) - (J_1^{(k)}, J_2^{(k)}) \\ &= (J_1^{(k)}, J_2^{(k)}) - (J_1^{(k)}, J_2^{(k)}) = 0\end{aligned}$$

where on passing to the third and the fifth equalities we made use of the fact $J_2^{(k)} = \hat{E}(J_2 | \Pi_k)$. Hence, $\hat{J}_1^{(k)} \in \tilde{\Pi}_k$. For $\varphi \in \tilde{\Pi}_k$ we have

$$(\hat{J}_1^{(k)}, \varphi) = (J_1^{(k)}, \varphi) - (\hat{E}(J_1^{(k)} | J_2^{(k)}), \varphi) = (J_1, \varphi),$$

since $J_2^{(k)} = \hat{E}(J_2 | \Pi_k)$, $\Pi_k \supset \tilde{\Pi}_k$ and therefore $(J_2^{(k)}, \varphi) = (J_2, \varphi) = 0$.

The polynomials 1, x, $Q_j(x) = x^j/\mu_2^{j/2} - j\gamma_j x^2/(2\mu_2) - j\gamma_{j-1} x/\mu_2^{1/2} + (j/2 - 1)\gamma_j$, $j = 3, \ldots, k$ form a basis (maybe non-orthogonal) of the space $\tilde{\Pi}_k$. The straightforward calculations give that

$$\begin{aligned}&(x, x) = \mu_2 = d_{11}, \\ &(1, Q_j) = 0, \quad (x, Q_j) = d_{1j}, \quad (Q_j, Q_l) = d_{jl}, \quad j, l = 3, \ldots, k \\ &(\hat{J}_1, 1) = 0, \quad (\hat{J}_1, x) = 1, \quad (\hat{J}_1, Q_j) = 0, \quad j = 3, \ldots, k.\end{aligned}$$

Since $\hat{J}_1^{(k)} \in \tilde{\Pi}_k$ we get from the expression for the residual variance:

$$\| \hat{J}_1 - \hat{J}_1^{(k)} \|^2 = \frac{\begin{vmatrix} (\hat{J}_1, \hat{J}_1) & (\hat{J}_1, 1) & (\hat{J}_1, x) & (\hat{J}_1, Q_3) & \cdots & (\hat{J}_1, Q_k) \\ (1, \hat{J}_1) & (1, 1) & (1, x) & (1, Q_3) & \cdots & (1, Q_k) \\ \vdots & & & & & \\ (Q_k, \hat{J}_1) & (Q_k, 1) & (Q_k, x) & (Q_k, Q_3) & \cdots & (Q_k, Q_k) \end{vmatrix}}{\begin{vmatrix} (1, 1) & (1, x) & (1, Q_3) & \cdots & (1, Q_k) \\ (x, 1) & (x, x) & (x, Q_3) & \cdots & (x, Q_k) \\ \vdots & & & & \\ (Q_k, 1) & (Q_k, x) & (Q_k, Q_3) & \cdots & (Q_k, Q_k) \end{vmatrix}}$$

$$= \| \hat{J}_1 \|^2 - D_{11}/(\det D). \tag{24}$$

Since $\| \hat{J}_1 - \hat{J}_1^{(k)} \|^2 = \| \hat{J}_1 \|^2 - \| \hat{J}_1^{(k)} \|^2$ the formula (22) follows from (24) at once. Theorem 2 is proved.

4. Characterization problems

In this Section the conditions of asymptotic optimality of the sample mean, $\bar{x}$, within the class (3) will be found. This problem is closely connected with that of minimizing $\hat{I}^{(k)}$ given μ_1, μ_2.

Theorem 3. *Suppose that the d.f. $F(x)$ has more than k growth points, all its moments $\mu_1, \mu_2, \ldots$ are finite and $\mu_1=0$. The sample mean $\bar{x}$ is asymptotically optimal as an estimator of θ within the class (3) in presence of a nuisance scale parameter σ iff either $\mu_2, \ldots, \mu_{k+1}$ coincide with the corresponding moments of the normal distribution or $\mu_2, \ldots, \mu_{k+1}$ coincide with the corresponding moments of some centralized gamma distribution or $\mu_2, -\mu_3, \ldots, (-1)^{k+1}\mu_{k+1}$ coincide with the corresponding moments of some centralized gamma distribution.*

Proof. From Theorems 1 and 2 we have

$$\lim_{n\to\infty} E_{\theta,\sigma}\left[\sqrt{n}\,(\hat{\theta}_n-\theta)\right]^2=\sigma^2 G(x, Q_3, \ldots, Q_k)/G(Q_3, \ldots, Q_k) \tag{25}$$

where $G(u_1, \ldots, u_m)$ denotes the Gram determinant of the elements $u_1, \ldots, u_m$,

$$G(u_1, \ldots, u_m)=\begin{vmatrix} (u_1, u_1) & \cdots & (u_1, u_m) \\ \vdots & & \\ (u_m, u_1) & \cdots & (u_m, u_m) \end{vmatrix}.$$

As it is well known [see, for example, Akhiezer (1965)]

$$\min_{A_3, \ldots, A_k} \left\| x-\sum_3^k A_j Q_j \right\|^2 = G(x, Q_3, \ldots, Q_u)/G(Q_3, \ldots, Q_u) \leqslant \|x\|^2, \tag{26}$$

the equality holding iff:

$$(x, Q_j)=0, \quad j=3, \ldots, k. \tag{27}$$

Since

$$\lim_{n\to\infty} E_{\theta,\sigma}\left[\sqrt{n}\,(\bar{x}-\theta)\right]^2=\sigma^2\|x\|^2$$

from (25)–(27), and the explicit expressions for Q_j given in Section 3, we have the following necessary and sufficient conditions of asymptotic optimality of $\bar{x}$ within the class (3):

$$\mu_{j+1}-j\mu_3\mu_j/(2\mu_2)-j\mu_2\mu_{j-1}=0, \quad j=3, \ldots, k. \tag{28}$$

Evidently the moments of any normal distribution satisfy the conditions (28). The characteristic function of the centralized gamma distribution with parameters $\gamma>0$, $p>0$ is $(\exp -i\gamma pt)(1-i\gamma t)^{-p}$ whence we see that its moments as well as those of the centralized gamma distribution on R^1_- satisfy (28). That is why it is sufficient to show that the moments $\mu_3, \ldots, \mu_{k+1}$ are uniquely determined by (28).

If $\mu_3 = 0$, then (28) allows us to find step by step $\mu_4, \mu_5, \ldots, \mu_{k+1}$ in terms of $\mu_1 = 0, \mu_2$. In this case, the moments $\mu_2, \ldots, \mu_{k+1}$ coincide with the corresponding moments of the normal distribution.

If $\mu_3 > 0$ (resp. $\mu_3 < 0$) let us choose the centralized gamma distribution (resp. the centralized gamma distribution gamma distribution on R^1_-) with given μ_2, μ_3. Then $\mu_4, \mu_5, \ldots, \mu_{k+1}$ will be uniquely determined by the moments μ_2, μ_3 and the equations (28). As it was explained above they have to coincide with the corresponding moments of the chosen gamma distribution. Theorem 3 is proved.

Denote by $\mathfrak{T}_k$ the class of estimators (3) for a given k and by $\mathfrak{T}$ the class of estimators $\bar{x} + s\psi(z)$ where ψ is an arbitrary function (not necessarily polynomial).

Corollary. *Suppose the conditions of Theorem* 3 *are satisfied. The sample mean $\bar{x}$ is an asymptotically optimal estimator of θ within the class $\bigcup_1^\infty \mathfrak{T}_k$ iff either $F(x)$ is the d.f. of a normal distribution or $F(x)$ is the d.f. of a centralized gamma distribution or $1 - F(-x)$ is the d.f. of a centralized gamma distribution.*

The proof follows from Theorem 3 and the uniqueness of the moments problem for normal and gamma distributions. Notice that if $F(x)$ is the d.f. of either a normal or a centralized gamma distribution then $\bar{x}$ is asymptotically optimal within the class $\mathfrak{T} \supset \bigcup_1^\infty \mathfrak{T}_k$. This follows from the results of Bondesson (1973).

Now we shall find out at what distributions the minimum of the Fisher information $\hat{I}^{(k)}/\sigma^2$ introduced in Section 3 is attained given the first two moments $\mu_1 = 0, \mu_2$.

Theorem 4. *Suppose the conditions of Theorem* 2 *are satisfied. If $\mu_3 = 0$, then* min $\hat{I}^{(k)}$ *is attained at the distribution whose $(k+1)$ first moments coincide with the corresponding moments of the normal distribution $N(0, \mu_2)$. If $\mu_3 > 0$ (resp. $\mu_3 < 0$), then* min $\hat{I}^{(k)}$ *is attained at the distribution whose $(k+1)$ first moments coincide with the corresponding moments of the centralized gamma distribution (resp. the centralized gamma distribution on R^1_-) determined by the moments μ_2, μ_3.*

The proof coincides with that of Theorem 3 if to make use of the following expression for $\hat{I}^{(k)}$ obtained in Theorem 2:

$$\hat{I}^{(k)} = G(Q_3, \ldots, Q_k)/G(x, Q_3, \ldots, Q_k).$$

It is worthwhile to note that though $\hat{I}^{(k)}$ is determined by the first $2k$ moments of the d.f. F its value at the minimum point is determined only by $\mu_2, \ldots, \mu_{k+1}$ and does not depend upon $\mu_{k+2}, \ldots, \mu_{2k}$.

References

Akhiezer, N.I. (1965). *Lessons on the Approximation Theory* (in Russian). Nauka, Moscow.

Bondesson, L. (1973). Characterization of the normal and gamma distributions. *Z. Wahrscheinlichkeitstheorie und verw. Gebiete* **26** (4), 335–344.

Cramér, H. (1946). *Mathematical Methods in Statistics*. Princeton University Press, Princeton, NJ.
Kagan, A.M. (1966). On the estimation theory of location parameter. *Sankhya* **A28** (4), 335–352.
Kagan, A.M., Klebanov, L.B. and Fintushal, S.M. (1974). On asymptotic behaviour of the polynomial Pitman estimators (in Russian). *Proc. (Zapiski) Sem. LOMI* **43**, 30–39.
Kagan, A.M. and Zinger, A.A. (1973). Sample mean as an estimator of the location parameter in presence of the nuisance scale parameter. *Sankhya* **A35** (4), 447–454.
Klebanov, L.B. and Melamed, I.A. (1976). On asymptotic behaviour of some estimators of location and scale parameters (in Russian). *Problemy Peredachi Informatzii* **XII** (3), 41–56.
Melamed, I.A. (1974). On asymptotic estimation of location and scale parameters (in Russian). *Proc. Acad. Sci. Georgian SSR* **76** (3), 549–552.
Zinger, A.A. and Kagan, A.M. (1973). On using the asymmetry and kurtosis of the noise in detecting a constant signal in presence of additive noise (in Russian). In: *Abstracts of Papers of the 3rd International Symposium on Information Theory*. Vol. 1. pp. 45–47.

G. Kallianpur, P.R. Krishnaiah, J.K. Ghosh, eds., *Statistics and Probability: Essays in Honor of C.R. Rao*

A GENERALIZED CAMERON–FEYNMAN INTEGRAL*

G. KALLIANPUR
University of North Carolina, Chapel Hill, NC, U.S.A., and Nagoya University, Nagoya, Japan

A generalization of the analytic Feynman integral of Cameron and Storvick is considered which leads to integrals involving indefinite quadratic forms. The scope of application of the theory is thereby considerably widened.

1. Introduction

Among the many approaches to the Feynman integral is the one based on analytic continuation, introduced and studied extensively by Cameron and his colleagues. A succinct definition, useful for our purposes, is to be found in the papers of Cameron and Storvick [2, 3]. Its interest for probabilists lies in the role played by integration on Wiener space. It appears, however, that the definition of the analytic Feynman integral given in [3] is not general enough for applications to certain problems of quantum mechanics in which the action functional involves an indefinite quadratic form (see [1]). In their recent work, in order to handle such problems, Albeverio and Høegh–Krohn have developed a theory of Fresnel integrals (yielding corresponding Feynman path integrals) for the case of indefinite quadratic forms [1].

In this note it is shown that the analytic Feynman integral approach, appropriately generalized, also leads to integrals involving indefinite quadratic forms. The new definition which contains the one given in ref. [3] as a particular case, is based on an analytic extension to a function of two complex variables. Another feature of the approach adopted here is the use made of the abstract Wiener space model. It lends flexibility and generality to the treatment since the Wiener spaces of Cameron and Storvick, the probability space of the pinned Wiener process and the Wiener space of the (multiparameter) Yeh–Wiener process are all examples of abstract Wiener space. Analytic Feynman integrals are called Cameron–Feynman integrals in this paper.

2. The Cameron–Feynman integral

We begin with the definition of the analytic Feynman integral given by Cameron and Storvick in [3]. For the sake of clarity and greater generality, it is convenient to state it in terms of an abstract Wiener space (H, B, P). Here H is a separable Hilbert space with inner product $(\ ,\ )$ and norm $|\cdot|$. The Banach space B is the completion of

*This research was partially supported by a Grant from the National Science Foundation.

H with respect to a measurable norm $\|\cdot\|$ and P is the abstract Wiener measure on the σ-field $\mathcal{B}(B)$ of Borel sets of B. The duals of B and H will be denoted by B^* and H^*, respectively. (We shall identify H^* with H.) For details concerning abstract Wiener spaces, see the monograph by Kuo [4]. Let $F(x)$ $(x \in B)$ be a P-measurable function on B satisfying the following conditions:

For all $\lambda > 0$, the integral,

$$J(\lambda) \equiv \int_B F(\lambda^{-1/2} x)\, \mathrm{d}P(x), \tag{2.1}$$

exists and is finite. There exists a function $J^*(z)$ analytic in the half-plane $\operatorname{Re} z > 0$ such that

$$J^*(\lambda) = J(\lambda) \quad \text{for all } \lambda > 0. \tag{2.2}$$

Let q be a real parameter $(q \neq 0)$. Define

$$\mathcal{J}_1^q(F) = \lim_{\substack{z \to -\mathrm{i}q \\ \operatorname{Re} z > 0}} J^*(z) \tag{2.3}$$

as the analytic Feynman integral of F provided the limit on the right-hand side exists. We shall refer to $\mathcal{J}_1^q(F)$ as the Cameron–Feynman (CF) integral with parameter q. (The suffix 1 is used to distinguish it from the integral to be defined below.)

An important class of functions F for which $\mathcal{J}_1^q(F)$ exists in the following: Let $\mathcal{F}$ be the class of functionals F of the form

$$F(x) = \int_H \mathrm{e}^{\mathrm{i}(h,x)^\sim}\, \mathrm{d}\mu(h) \tag{2.4}$$

where μ is a complex-valued measure of bounded variation on $\mathcal{B}(H)$, the σ-field of Borel sets of H and $(h,x)^\sim$ denotes the random variable on $(B, \mathcal{B}(B), P)$ which is Gaussian with zero mean and variance $|h|^2$.

Proposition 1. *If $F \in \mathcal{F}$, then $\mathcal{J}_1^q(F)$ exists and is given by*

$$\mathcal{J}_1^q(F) = \int_H \mathrm{e}^{-(\mathrm{i}/2q)|h|^2}\, \mathrm{d}\mu(h). \tag{2.5}$$

The proof follows the main steps of [3]. Using Fubini's theorem it is easy to verify that

$$J(\lambda) = \int_H \mathrm{e}^{-|h|^2/2\lambda}\, \mathrm{d}\mu(h) \quad \text{for } \lambda > 0.$$

The function

$$J^*(z) = \int_H \mathrm{e}^{-|h|^2/2z}\, \mathrm{d}\mu(h)$$

is then analytic in $\operatorname{Re} z > 0$. The existence of the limit in (2.4) is a consequence of the Lebesgue dominated convergence theorem and (2.5) is obtained. Suppose that S is a bounded, symmetric, positive operator on H with bounded inverse. Then $[h_1, h_2] =$

$(S^{-1}h_1, h_2)$ is an inner product in H and H is complete with respect to the norm $|h|_S = +[h, h]^{1/2}$. Consider H as a Hilbert space with the new norm $|\cdot|_S$ and let (H, B, P_S) be the corresponding abstract Wiener space. The abstract Wiener measure P_S has zero mean and covariance given by the new inner product, [,] in H. Let $\mathfrak{F}_S$ be the class of functions F on B given by

$$F(x) = \int_H e^{i(h,x)^{\sim}} d\mu(h) \tag{2.6}$$

where $\mu \in \mathfrak{M}(H)$, the class of complex-valued measures of bounded variation on the Borel sets of H. From Proposition 1 it follows that the C–F integral of F exists. To emphasize the part played by the operator S we denote $\mathcal{G}_1^q(F)$ in this case by $\mathcal{G}_{1,S}^q(F)$. We then have

$$\begin{aligned}\mathcal{G}_{1,S}^q(F) &= \int_H e^{-(i/2q)|h|_S^2} d\mu(h) \\ &= \int_H e^{-(i/2q)(S^{-1}h,h)} d\mu(h).\end{aligned} \tag{2.7}$$

3. Fresnel integrals

The definition of the C–F integral in (2.5) is similar to the one given by Albeverio and Høegh–Krohn in their monograph [1]. Let f be defined on H and be of the form

$$f(y) = \int_H e^{i(h,y)} d\mu(y) \tag{3.1}$$

where $\mu \in \mathfrak{M}(H)$. The Fresnel (or Feynman) integral of f is defined in [1] by

$$I(f) = \int_H e^{-(i/2)|y|^2} d\mu(y). \tag{3.2}$$

The integral in (3.2) has the same value as that of the C–F integral in (2.5) (with $q=1$). However, the integrands in the two integrals have different domains and the Fresnel integrals are defined without reference to Wiener spaces. (This distinction is noted in [3].) There is a natural connection via the structure of abstract Wiener space between the two classes of integrands. This point will not be further pursued in this paper. As pointed out by Albeverio and Høegh–Krohn, the definitions of the Feynman integrals given above are not adequate to handle several important problems in physics, e.g. the anharmonic oscillator with n degrees of freedom. To deal with this situation, a more general theory of the Fresnel integral has been set forth in [1] which involves indefinite bilinear forms rather than positive definite forms. Only one of the definitions introduced in [1] will concern us here [1, p. 59]. Let A be a bounded symmetric operator on H with bounded inverse A^{-1}. Write $\Delta(x, y) = (x, A^{-1}y)$. For f of the form (3.1) the Fresnel integral of f with respect to the bilinear form Δ is defined to be

$$I_{\Delta}(f)=\int_H e^{-(i/2)(y,Ay)}\,d\nu(y)=\int_H e^{-(i/2)(y,A^{-1}y)}\,d\mu(y) \tag{3.3}$$

where $\nu A^{-1}=\mu$. When $A=I$, (3.3) reduces to (2.5) (with $q=1$) or (3.2). However, in general, (y, Ay) is an indefinite bilinear form. The question arises, therefore, whether the Cameron–Storvick definition of the analytic Feynman integral can be extended to include (3.3) (see the remarks in [1, p. 79]. In the next section it will be shown that such an extension is indeed possible.

4. The generalized Cameron–Feynman integral

Let the Hilbert space $H=H_1 \oplus H_2$, the orthogonal direct sum of its closed linear subspaces H_1 and H_2. Let $\|\cdot\|$ be a measurable norm on H. For $i=1,2$ denote by π_i, the orthogonal projection operator with range H_i. Then $\|h\|_i=\|\pi_i h\|$ is a measurable norm on H and hence also on H_i. Let (H, B, P), (H_i, B_i, P_i) $(i=1,2)$ be the corresponding abstract Wiener spaces. The probability space $(B, \mathfrak{B}(B), P)$ can be replaced by the product of the probability spaces $(B_i, \mathfrak{B}(B_i), P_i)$ $(i=1,2)$. For convenience of notation write $F(x)=F(x_1, x_2)$ where $x=(x_1, x_2)\in B_1 \times B_2$.

Also write $d\mu(h_1, h_2)$ for $d\mu(h)$ where $h=h_1+h_2 \in H$, $h_i \in H_i$ and $\mu \in \mathfrak{M}(H)$. We may do this with a slight abuse of notation since H and $H_1 \times H_2$ can be treated as equivalent metric spaces.

Definition of the generalized C–F integral. Let $F(x_1, x_2)$ be a measurable function on $B_1 \times B_2$ satisfying the following conditions: For all $\lambda_1>0$ and $\lambda_2>0$, the integral

$$J(\lambda_1, \lambda_2)\equiv\int_{B_1\times B_2} F(\lambda_1^{-1/2}x_1, \lambda_2^{-1/2}x_2)\,dP_1(x_1)\,dP_2(x_2) \tag{4.1}$$

exists and is finite. There exists a function $J^*(z_1, z_2)$ of the complex variables z_1 and z_2 which is analytic in the domain $D=\{(z_1, z_2)\colon \operatorname{Re} z_1>0, \operatorname{Re} z_2>0\}$ such that

$$J^*(\lambda_1, \lambda_2)=J(\lambda_1, \lambda_2) \quad \text{for all } \lambda_1>0, \lambda_2>0. \tag{4.2}$$

Let q_1, q_2 be non-zero, real parameters and let

$$\mathcal{J}_2^{q_1,q_2}(F)=\lim_{\substack{z_1\to -iq_1\\ z_2\to iq_2\\ (z_1,z_2)\in D}} J^*(z_1, z_2) \tag{4.3}$$

provided the limit exists. Then $\mathcal{J}_2^{q_1,q_2}(F)$ is defined to be the generalized Cameron–Feynman integral with parameters (q_1, q_2). If $q_1=q_2=1$ we denote the left-hand side of (4.3) by $\mathcal{J}_2(F)$. The existence of $\mathcal{J}_2(F)$ for the following class of integrands can be easily established. Let $\mathcal{F}_2$ be the class of functions F of the form

$$F(x_1,x_2)=\int_{H_1\times H_2} e^{i(h_1,x_1)^\sim+i(h_2,x_2)^\sim}\,d\mu(h_1,h_2) \tag{4.4}$$

where $\mu\in\mathfrak{M}(H)$ and $(h_1,x_1)^\sim,(h_2,x_2)^\sim$ have the same meaning as before.

Proposition 2. *If* $F\in\mathcal{F}_2$, *then* $\mathcal{G}_2^{q_1,q_2}(F)$ *exists and*

$$\mathcal{G}_2^{q_1,q_2}(F)=\int_{H_1\times H_2} e^{-(i/2q_1)|h_1|^2+(i/2q_2)|h_2|^2}\,d\mu(h_1,h_2). \tag{4.5}$$

If $q_1=q=-q_2$ ($\neq 0$), then (4.5) reduces to $\mathcal{G}_1^q(F)$, the Cameron–Storvick version of the Feynman integral. When q_1 and q_2 are of the same sign (e.g., when $q_1=q_2=1$), (4.5) is a true extension of Cameron and Storvick's integral. It includes, in particular, integrals of the type (3.3) as we shall now show.

Let A be a bounded self-adjoint operator in H with bounded inverse A^{-1}. Let m, M be the bounds of A ($m<0<M$) and $E(\cdot)$ the projection measure corresponding to the spectral representation of A. Writing $J_-=[m,0)$ and $J_+=[0,M]$ denote the projection operators $E(J_-)$ and $E(J_+)$ by E_- and E_+, respectively. Define $H_+=E_+H$, $H_-=E_-H$. Then H_+ and H_- are orthogonal subspaces and $H=H_+\oplus H_-$. For any $h\in H$, write $h=h_++h_-$ where $h_+\in H_+$ and $h_-\in H_-$. The operators $A_+=E_+A$ and $A_-=-E_-A$ are self-adjoint, positive operators on H such that $A_+H_+\subseteq H_+$ and $A_-H_-\subseteq H_-$. We then have $A=A_+-A_-$ (see [5]). Let us also define the operator $S=A_++A_-$. The operators A_+ and A_- restricted to H_+ and H_- have bounded inverses. To see this, let $f\in H_+$. Since A^{-1} exists, there is a unique $g\in H$ such that $Ag=f$. Setting $g=g_++g_-$ we obtain $f=Ag=A_+g_+-A_-g_-$. Hence $A_-g_-\in H_+\cap H_-$ so that $A_-g_-=0$. Thus $g=g_+=A_+^{-1}f$ and $|A_+^{-1}f|=|g_+|=|A^{-1}f|\leqslant\|A^{-1}\|\cdot|f|$. Similarly, $f\in H_-$ implies $A_-^{-1}f\in H_-$ and $|A_-^{-1}f|\leqslant\|A^{-1}\|\cdot|f|$. Now if $h\in H$, $h=h_++h_-$ we obtain

$$A^{-1}h=A_+^{-1}h_+-A_-^{-1}h_-,\quad h_+\in H_+,\, h_-\in H_-. \tag{4.6}$$

One proceeds in a similar manner to show that the bounded inverse S^{-1} exists, is a positive operator and that

$$S^{-1}h=A_+^{-1}h_++A_-^{-1}h_-. \tag{4.7}$$

Finally, introducing the inner product $[f,g]=(S^{-1}f,g)$ and $|f|_S^2=(S^{-1}f,f)$ in H it is readily verified that H is complete with respect to $|\cdot|_S$ and that H_+, H_- are $|\cdot|_S$-closed linear subspaces which are mutually orthogonal under the inner product [,]. We shall consider H as the Hilbert space with inner product [,]. Let $F(x_1,x_2)$ be a function given by

$$F(x_1,x_2)=\int_{H_+\times H_-} e^{i(h_+,x_1)^\sim+i(h_-,x_2)^\sim}\,d\mu(h_+,h_-)$$

where $\mu\in(\mathfrak{M})$ [see (4.4)]. Then by Proposition 2, $\mathcal{G}_2(F)$ exists and is given by (4.5) where, however, the norm $|\cdot|$ is replaced by $|\cdot|_S$. It is appropriate to write $\mathcal{G}_{2,A}(F)$

for this Cameron–Feynman integral. From (4.5), (4.6) and (4.7),

$$\begin{aligned}\mathcal{I}_{2,A}(F) &= \int_{H_+\times H_-} e^{-(i/2)|h_+|_S^2+(i/2)|h_-|_S^2}\,d\mu \\ &= \int_{H_+\times H_-} e^{-(i/2)\{(S^{-1}h_+,h_+)-(S^{-1}h_-,h_-)\}}\,d\mu \\ &= \int_{H_+\times H_-} e^{-(i/2)\{(A_+^{-1}h_+,h_+)-(A_-^{-1}h_-,h_-)\}}\,d\mu \\ &= \int_H e^{-(i/2)(A^{-1}h,h)}\,d\mu. \end{aligned} \tag{4.8}$$

The above example suggests that it is natural to define the generalized Cameron–Feynman integral starting with indefinite, inner product spaces, e.g. Krein spaces, instead of Hilbert spaces. This will be done in a later paper which will also contain details omitted here.

References

[1] Albeverio, S.A. and Høegh-Krohn, R.J. (1976). *Mathematical Theory of Feynman Path Integrals*. Lecture Notes in Mathematics, 523. Springer-Verlag, Berlin.

[2] Cameron, R.H. and Storvick, D.A. (1976). An L_2 analytic Fourier–Feynman transform. *Michigan Math. J.* **23**, 1–30.

[3] Cameron, R.H. and Storvick, D.A. (1978). Some Banach algebras of analytic Feynman integrable functionals. To appear.

[4] Kuo, H.H. (1975), *Gaussian Measures in Banach Spaces*. Lecture Notes in Mathematics, 463. Springer-Verlag, Berlin.

[5] Riesz, F. and Sz-Nagy, B. (1955). *Functional Analysis*. Ungar, New York.

(the symbol, $\approx$, henceforth will indicate that the two quantities have the same sign). Integration by parts produces

$$S(c)=-c[1-F(c)]-[1-2F(c)]\int_c^\infty[1-F(\xi)]\mathrm{d}\xi.$$

We examine

$$R(c)=\frac{S(c)}{\int_c^\infty[1-F(x)]\,\mathrm{d}x}=c\frac{Q'(c)}{Q(c)}-[1-2F(c)] \tag{12}$$

where $Q(c)=\int_c^\infty[1-F(x)]\mathrm{d}x$. Consider

$$\frac{\mathrm{d}R(c)}{\mathrm{d}c}=\frac{Q'(c)}{Q(c)}+c\left(\frac{Q'(c)}{Q(c)}\right)'+2f(c). \tag{13}$$

Let m be the mode of $f(x)$ (since the density $f(x)$ is log concave it is unimodal). For $c>m$ and $c>0$ we have

$$1-F(c)>2[1-F(c)]^2>2f(c)\int_c^\infty[1-F(x)]\,\mathrm{d}x, \tag{14}$$

the last inequality resulting from (8) in conjunction with property (v) listed at the start of the section. The foregoing relation confirms that

$$\frac{Q'(c)}{Q(c)}+2f(c)<0. \tag{15}$$

Moreover,

$$\left(\frac{Q'(c)}{Q(c)}\right)'\leqslant 0 \text{ holds for all } c \tag{16}$$

also by virtue of (8). We combine (15) and (16) to conclude that

$$\frac{\mathrm{d}R(c)}{\mathrm{d}c}<0 \quad \text{for } c>\max(0,m). \tag{17}$$

It follows in view of (17) and since $R(m)<0$ for $m\geqslant 0$ that

$$\frac{\mathrm{d}B(c)}{\mathrm{d}c}\approx R(c)<0 \quad \text{for all } c>m\geqslant 0. \tag{18}$$

Moreover, inspection of the formula (11) for c on the range $0\leqslant c\leqslant m$ reveals that each term of $R(c)$ is negative. In fact, $2F(c)-1<0$ for $c<m$ and

$$\frac{cQ'(c)}{Q(c)}=\frac{-c[1-F(c)]}{Q(c)}$$

is manifestly negative for $c>0$. It follows that

For $m>0$, then $B(c)$ is strictly decreasing for all $c>0$. (19)

We examine next the case $m<0$. Over the range $c>0$ we find as before $\mathrm{d}R(c)/\mathrm{d}c\leqslant 0$

so that $\mathrm{d}B(c)/\mathrm{d}c$ is monotone decreasing and therefore this function changes sign at most once as c traverses the positive axis. At the origin we plainly have that $\mathrm{d}B(0)/\mathrm{d}c>0$. Therefore, when $m<0$, $B(c)$ exhibits a unique mode on the positive axis. Stated formally:

For $m<0$, as c traverses 0 to ∞, $B(c)$ increases to a unique mode achieved at $c^*>0$ and strictly decreases thereafter. (20)

Implementing a completely symmetrical analysis reasoning in terms of the tails $F(c)$ rather than $1-F(c)$, we deduce:

For $m<0$, $B(c)$ strictly increases over $-\infty<c<0$ and for $m>0$ $B(c)$ achieves a unique maximum on the negative axis. (21)

The content of (20) and (21) provides the proof of Theorem 1 when $f(x)$ is continuous with mean zero.

In general, for $\mu=\int_{-\infty}^{\infty} xf(x)\,\mathrm{d}x \geqslant m=$ mode of f, by translating X which does not affect $B(c)$ we find that $B(c)$ is unimodal with its maximum located at $c^*\geqslant\mu$. When $\mu\leqslant m$, then the mode of $B(c)$ occurs at $c^*\leqslant\mu$. The proof of Theorem 1 for a general (not necessarily continuous throughout) log-concave density ensues by a straightforward approximation procedure furnishing a sequence $f_n(x)$ of continuous log concave densities convergent pointwise to $f(x)$.

When $f(x)$ is symmetric log concave, then $\mu=E[X]=$ mode $f(x)=m=0$ and the unique maximum point c^* of $B(c)$ necessarily occurs also at the origin.

3. Proof of Theorem 2

Consider first $c>0$. Observe that

$$V(c)=\frac{\int_c^{\infty} x^2 f(x)\,\mathrm{d}x}{\int_c^{\infty} f(x)\,\mathrm{d}x}-\left(\frac{\int_c^{\infty} xf(x)\,\mathrm{d}x}{\int_c^{\infty} f(x)\,\mathrm{d}x}\right)^2$$

$$=\frac{\int_0^{\infty} (x+c)^2 f(x+c)\,\mathrm{d}x}{\int_0^{\infty} f(x+c)\,\mathrm{d}x}-\left(\frac{\int_0^{\infty} (x+c)f(x+c)}{\int_0^{\infty} f(x+c)\,\mathrm{d}x}\right)^2$$

which upon expanding and simplifying becomes

$$=\frac{\int_0^{\infty} x^2 f(x+c)\,\mathrm{d}x}{\int_0^{\infty} f(x+c)\,\mathrm{d}x}-\left(\frac{\int_0^{\infty} xf(x+c)\,\mathrm{d}x}{\int_0^{\infty} f(x+c)\,\mathrm{d}x}\right)^2$$

and by obvious symmetry

$$=\frac{\int_0^\infty\int_0^\infty \tfrac{1}{2}(x^2+y^2)f(x+c)f(y+c)\,\mathrm{d}x\,\mathrm{d}y-\int_0^\infty\int_0^\infty xyf(x+c)f(y+c)\,\mathrm{d}x\,\mathrm{d}y}{\int_0^\infty\int_0^\infty f(x+c)f(y+c)\,\mathrm{d}x\,\mathrm{d}y}$$

so that

$$V(c)=\frac{\tfrac{1}{2}\int_0^\infty\int_0^\infty (x-y)^2 f(x+c)f(y+c)\,\mathrm{d}x\,\mathrm{d}y}{\int_0^\infty\int_0^\infty f(x+c)f(y+c)\,\mathrm{d}x\,\mathrm{d}y}. \tag{22}$$

By symmetry of the integral, implementing the integration over the two regions $x>y$ and $y>x$, and transforming variables, we obtain

$$V(c)=\frac{\int_0^\infty\int_y^\infty (x-y)^2 f(x+c)f(y+c)\,\mathrm{d}x\,\mathrm{d}y}{\int_0^\infty\int_y^\infty f(x+c)f(y+c)\,\mathrm{d}x\,\mathrm{d}y}.$$

An obvious change of variable $(x-y=z, \eta=y)$ yields

$$V(c)=\frac{\tfrac{1}{2}\int_0^\infty\int_0^\infty z^2 f(z+\eta+c)f(\eta+c)\,\mathrm{d}\eta\,\mathrm{d}z}{\int_0^\infty\int_0^\infty f(z+\eta+c)f(\eta+c)\,\mathrm{d}\eta\,\mathrm{d}z}. \tag{23}$$

Now let

$$M(z,c)=\int_0^\infty f(z+y+c)f(y+c)\,\mathrm{d}y=\int_0^\infty R(\xi,c)f(z+\xi)\,\mathrm{d}\xi \tag{24}$$

setting $\xi=y+c$ and identifying

$$R(\xi,c)=\begin{cases}1, & \xi>c\\ 0, & \xi<c\end{cases}.$$

The hypothesis implies that $f(z+\xi)$ is RR_2 for z, $\xi>0$, and the cumulant kernel $R(\xi,c)$ is TP_2, fact (ii) of the introduction to Section 2. By the composition formula (i) we conclude that $M(z,c)$ is RR_2 for $z>0$ and $c>0$. Thus, we have

$$V(c)=\tfrac{1}{2}\int_0^\infty z^2\tilde{M}(z,c)\,\mathrm{d}z \tag{25}$$

where

$$\tilde{M}(z,c)=\frac{M(z,c)}{\int_0^\infty M(\zeta,c)\,\mathrm{d}\zeta}$$

which is also RR_2 [see (iii)]. Since z^2 is increasing, we apply (iv) to deduce that $V(c)$ is decreasing as was to be shown.

Consider next $c<0$ and we examine the function $\tilde{V}(c)=\mathrm{Var}[X|X<c]$. By a change of variable

$$\tilde{V}(c)=\frac{\int_{-c}^{\infty} x^2 f(-x)\,\mathrm{d}x}{\int_{-c}^{\infty} f(-x)\,\mathrm{d}x}-\left(\frac{\int_{-c}^{\infty} x f(-x)\,\mathrm{d}x}{\int f(-x)\,\mathrm{d}x}\right)^2$$

Because $f(-x)$ is also log concave we deduce as before that

$$\mathrm{Var}[X|x\leqslant c] \text{ is decreasing in } -c,$$

and thus increases in c for $c\leqslant 0$. This completes the proof of Theorem 2, Part (i). The second statement is proved by similar means.

There appears to be a higher moment version of Theorem 2 as follows.

Consider a positive or a real or even random variable X following the density $f(x)$ and form,

$$V_k(c)=E\Big[(X-E(X|X>c))^k|X>c\Big], \tag{26}$$

the k order conditional moment. It is of some interest to determine natural conditions on $f(x)$ implying that $V_k(c)$ is monotone decreasing in c. This is true for the normal density. More generally, we inquire for an arbitrary increasing function $\varphi(x)$ concerning the behavior of

$$W_\varphi(c)=E[\varphi(X-E(X|X>c))|X>c]$$

as a function of c.

References

Karlin, S. (1968). *Total Positivity*. Stanford University Press, Stanford, CA.

Karlin, S. and Williams, P.T. (1981). Structured exploratory data analysis (S.E.D.A.) for determining mode of inheritance of quantitative traits. II. Simulation studies on the effect of ascertaining families through high valued probands. *Am. J. Hum. Genet.* **33**, 147–156.

Ott, J. (1979). Detection of rare major genes in lipid levels. *Hum. Genet.* **51**, 79–91.

G. Kallianpur, P.R. Krishnaiah, J.K. Ghosh, eds., *Statistics and Probability: Essays in Honor of C.R. Rao*
© North-Holland Publishing Company (1982) 383–396

ALMOST PERIODIC WEAKLY STATIONARY PROCESSES

Tatsuo KAWATA
Department of Mathematics, Keio University, Yokohama, Japan

1. Introduction

Throughout this paper, $(\Omega, \mathfrak{F}, P)$ is the underlying probability space and $X(t, \omega)$, $-\infty<t<\infty$, is supposed to be a measurable weakly stationary process with $EX(t, \omega)=0$, $-\infty<t<\infty$ and with the spectral distribution function $F(\lambda)$. Let the spectral representation of $X(t, \omega)$ be

$$X(t, \omega)=\int_{-\infty}^{\infty} e^{i\lambda t}\xi(d\lambda, \omega) \tag{1.1}$$

and the corresponding covariance function be

$$\rho(u)=\int_{-\infty}^{\infty} e^{i\lambda u} dF(\lambda), \tag{1.2}$$

where $\xi(S, \omega)$ is the spectral random measure, S being any Borel set.

We shall study in this paper the almost periodicity of sample functions of $X(t, \omega)$. Slutsky [8] was the first who investigated the problem and one of his results is that a weakly stationary process which has a uniformly almost periodic (u.a.p.) covariance function is Besicovitch almost periodic (B^2.a.p.) almost surely. Later Udagawa [10] proved that if $F(\lambda)$ is purely discontinuous (which is equivalent to the fact that the covariance function is u.a.p. as we mention again later on) with discontinuities $\{\lambda_j, j=1,2,\ldots\}$ such that

$$\inf_{j\neq k} |\lambda_j-\lambda_k|>L>0, \tag{1.3}$$

for some positive constant L, then $X(t, \omega)$ is Stepanoff almost periodic (S^2.a.p.) almost surely.

On the other hand, Paley and Wiener [7, p. 116] showed that if for a non-random function, $f(t)$, which is S^2.a.p. with Fourier exponents $\{\lambda_j\}$, (1.3) is satisfied, then there exist two positive numbers A and B such that for any set of complex numbers $a_1, a_2,\ldots, a_n$, any set of real numbers $b_1, b_2,\ldots, b_n$ and any real numbers x and y, we have

$$\int_x^{x+A}\left|\sum_{k=1}^{n} a_k f(t+b_k)\right|^2 dt\leq B\int_y^{y+A}\left|\sum_{k=1}^{n} a_k f(t+b_k)\right|^2 dt. \tag{1.4}$$

As a matter of fact, they proved that the class of $f(t)$ with the property (1.4) is equivalent to the class of S^2.a.p. functions with Fourier exponents satisfying (1.3).

The combination of above two results gives us:

Proposition 1. *If $F(\lambda)$ is purely discontinuous with discontinuities $\{\lambda_j\}$ satisfying* (1.3), *then* (1.4) *holds with $X(t, \omega)$ in place of $f(t)$ almost surely.*

Bochner [3] considered Banach-valued u.a.p. functions. although particular attention was not paid to stationarity, and discussed about the relation (1.4) for such class of functions making the meaning of (1.4) much clearer. For stationarity of u.a.p. processes or u.a.p. Banach valued functions, Bochner [4] later made investigations in connection with some sort of linear transformations.

Weakly stationary u.a.p. processes were studied again by Yu [11] who observed that for a weakly stationary process to be an $L^2(\Omega)$ valued u.a.p. function, it is necessary and sufficient that the covariance function is u.a.p.

We shall give, in Section 4, a sufficient condition under which a weakly stationary process has a version which is u.a.p. almost surely, and in Section 5, a sufficient condition under which a weakly stationary process is S^2.a.p. almost surely.

Slutsky's theorem mentioned above is true when discontinuities have no limiting point except at $\pm\infty$ and in fact the original proof seems to involve some vague points as it stands although we can improve it. We give, in Section 6, a different proof in a more general situation.

2. General remarks

We begin with the following well-known proposition.

Proposition 2. *The spectral distribution function $F(\lambda)$ is purely discontinuous, if and only if the covariance function $\rho(u)$ is u.a.p.*

Note that by definition, $F(\lambda)$ is purely discontinuous if $\sum P(\{\lambda_j\}) = F(\infty) - F(-\infty)$. The above proposition was observed by Slutsky [8] and again by Yu [11] and is easy to show.

A stochastic process which is u.a.p. as an $L^2(\Omega)$ valued function is explicitly defined as follows: A stochastic process $X(t, \omega)$ for which

$$\left\{\tau: \sup_{-\infty<t<\infty} E|X(t+\tau, \omega) - X(t, \omega)|^2 < \varepsilon\right\} \tag{2.1}$$

is relatively dense for every $\varepsilon > 0$, is u.a.p. in $L^2(\Omega)$ is called a mean u.a.p. process.

In the similar way we may define the mean S^2.a.p. property. If

$$\left\{\tau: \sup_{-\infty<t<\infty} \int_t^{t+1} E|X(t+\tau, \omega) - X(t, \omega)|^2 \, dt < \varepsilon\right\} \tag{2.2}$$

is relatively dense for every $\varepsilon > 0$, then $X(t, \omega)$ is called mean S^2.a.p.

Proposition 3. *For a weakly stationary process, the mean u.a.p. property and the mean S^2.a.p. property are equivalent to each other. For this property it is necessary and sufficient that the covariance function is u.a.p.*

From the fact that $E|X(t+u,\omega)-X(t,\omega)|^2=2[\rho(0)-\operatorname{Re}\rho(u)]$ and $|\rho(t+u)-\rho(t)|^2\leq 2\rho(0)[\rho(0)-\operatorname{Re}\rho(u)]\leq 2\rho(0)|\rho(0)-\rho(u)|\leq 2\rho(0)\sup_t|\rho(t)-\rho(t+u)|$, we readily see that $X(t,\omega)$ is mean S^2.a.p. as well as mean u.a.p., if and only if $\rho(u)$ is u.a.p.

Note that S^2.a.p. $\rho(u)$ is u.a.p. by a well-known theorem of Bochner (see ref. [2, p. 81]) that an S^2.a.p. function which is uniformly continuous on $(-\infty,\infty)$ should be u.a.p. Proposition 3 is no more than a direct consequence of this theorem of Bochner for an $L^2(\Omega)$ valued function.

We, throughout this paper, make the following assumption.

Condition 1. $\rho(u)$ is u.a.p., that is, $F(\lambda)$ is purely discontinuous. The discontinuities of $F(\lambda)$ may have limiting points. For possible limiting points μ_k, $k=0,\pm1,\pm2,\ldots$ it is assumed that

$$\inf_{k\neq l}|\mu_k-\mu_l|>1. \tag{2.3}$$

μ_k itself could be a continuity point or discontinuity point. In (2.3), 1 on the right-hand side can be replaced by a positive constant, but just for simplicity, we assume (2.3) without loss of generality. Condition 1 is not a substantial one, however, if we do not assume this condition, the situation becomes much involved and we will need the complexity of statements of theorems we are going to prove in the later sections. To avoid these sort of troubles, we consider Condition 1.

We use the following notations. First, all limiting points $\{\mu_k\}$ are enumerated to be $\mu_k<\mu_{k+1}$, $k=0,\pm1\pm2,\ldots$. We split $(-\infty,\infty)$ into $\bigcup_{n=-\infty}^{\infty}(n,n+1]$. Then the isolated discontinuities of $F(\lambda)$ in $(n,n+1]$ are denoted by $\lambda_{n,j}$ $(j=1,2,\ldots,j_n)$ when $[n,n+1]$ does not contain any μ_k. If there is a limiting point of discontinuities in $(n,n+1]$, it is denoted by μ_{k_n} and other isolated discontinuities by $\lambda_{n,j}$ with

$$\begin{aligned} &n<\lambda_{n,j}<\mu_{k_n}, && \lambda_{n,j}\uparrow\mu_{k_n}, \quad j=1,2,\ldots,\\ &\mu_{k_n}<\lambda_{n,j}\leq n+1, && \lambda_{n,j}\downarrow\mu_{k_n}, \quad j=-1,-2,\ldots. \end{aligned} \tag{2.4}$$

When μ_{k_n} happens to be a discontinuity point, μ_{k_n} is also sometimes written as $\lambda_{n,0}$. Note that each $(n,n+1]$ contains only one possible limiting point.

We also suppose that $F(\lambda)$ and $\xi(\lambda,\omega)$ (the stochastic process with orthogonal increments generated by $\xi(S,\omega)$) are left continuous.

Write

$$A(\lambda,\omega)=\xi(\lambda+0,\omega)-\xi(\lambda,\omega), \tag{2.5}$$

$$a(\lambda)=F(\lambda+0)-F(\lambda). \tag{2.6}$$

Now suppose $X(t,\omega)$ is almost periodic in some sense as a function of t, almost surely. Then by a basic fact on almost periodic functions,

$$\lim_{T\to\infty}\frac{1}{2T}\int_{-T}^{T}X(t,\omega)\,dt, \tag{2.7}$$

should exist almost surely. On the other hand, according to a result of Gaposhkin [6], in order that (2.7) exists almost surely for a weakly stationary process $X(t,\omega)$, it is necessary and sufficient that

$$\lim_{n\to\infty}\int_{|\lambda|<2^{-n}}\xi(d\lambda,\omega) \tag{2.8}$$

exists almost surely.

Let us take a spectral distribution function $F(\lambda)$ which is purely discontinuous and has discontinuities 2^{-j}, $j=1,2,\ldots$ around the origin. Then we should have that for the almost sure existence of (2.7)

$$\sum_{j=1}^{\infty} a(2^{-j})U(2^{-j},\omega) \tag{2.9}$$

should converge almost surely, where $U(2^{-j},\omega)=A(2^{-j},\omega)/a(2^{-j})$ constitutes an orthonormal sequence. Actually, (2.9) converges almost surely if $\sum a(2^{-j})\log^2 j<\infty$, by the well-known Menchoff–Rademacher theorem (e.g., Alexits [1]). On the other hand, if $a(2^{-j})$ is non-increasing and $\sum a(2^{-j})\log^2 j=\infty$, then there exists an orthonormal sequence $\{\phi_j(x)\}$, $0\leqslant x\leqslant 1$, for which $\sum a(2^{-j})\phi_j(x)$ diverges almost everywhere. This result was obtained by Tandori [9]. Therefore, we see that with probability space $\Omega=[0,1]$ ($\mathcal{F}$: class of Borel sets, P: Lebesgue measure) we can construct a $\xi(\lambda,\omega)$ for which (2.8) diverges almost surely.

This consideration indicates that some condition on the distributions of discontinuities around their limiting points will be indispensable for the sample almost periodicity of a weakly stationary process.

3. Basic lemmas

Suppose Condition 1 is satisfied. Let $(n,n+1]$ be an interval, n being an integer. Let us suppose $[n,n+1]$ contains a limiting point μ_{k_n} of discontinuities of $F(\lambda)$. Note that $[n,n+1]$ contains only one limiting point because of (2.3). Let $N_n(\alpha)$ be the number of discontinuities between n and $\mu_{k_n}-\alpha$ $(0<\alpha<\mu_{k_n}-n)$ when $\mu_{k_n}\neq n$ and $M_n(\alpha)$ be the number of discontinuities between $\mu_{k_n}+\alpha$ and $n+1$ $(0<\alpha<n+1-\mu_{k_n})$ when $\mu_{k_n}\neq n+1$.

Condition 2. Let $[n,n+1]$ contain a limiting point, μ_{k_n}, of discontinuities of $F(\lambda)$. Suppose there is a non-decreasing function $h(x)$ over $0\leqslant x<\infty$ such that $h(x)\to\infty$, as $x\to\infty$, $h(x)=$ a positive constant for $0\leqslant x\leqslant 2$ and

$$N_n(\alpha)\leqslant h(\alpha^{-1}) \quad \text{for } 0<\alpha<\mu_{k_n}-n,$$

$$M_n(\alpha)\leqslant h(\alpha^{-1}) \quad \text{for } 0<\alpha<n+1-\mu_{k_n}. \tag{3.1}$$

Suppose also that if $[n,n+1]$ does not contain any limiting point, the number of discontinuities of $F(\lambda)$ is uniformly bounded.

As a matter of fact, the second condition can be relaxed a little more and even in the first condition we may assume that $h(x)$ depends on n. But just for simplicity we place the above Condition 2 in the following discussions.

Let $g(x)$ be another non-decreasing function over $[0, \infty)$ such that

$$\int_1^\infty \frac{dx}{xg(x)} < \infty, \quad g(x)=1, \quad 0 \leqslant x \leqslant 1, \tag{3.2}$$

which is equivalent to

$$\sum_{n=1}^{\infty} \frac{1}{g(2^n)} < \infty. \tag{3.3}$$

We introduce

$$\begin{aligned} \phi(u) &= g(u^{-1})h(2^{-1}u^{-1}), \quad && u>0, \\ &=1, && u=0, \end{aligned} \tag{3.4}$$

and

$$\Phi_n = \int_n^{n+1} \phi(|\lambda - \mu_{k_n}|)\, dF(\lambda) \tag{3.5}$$

which may be finite or infinite.

We, now, give a lemma which plays a fundamental role in what follows.

Lemma 1. *Suppose the Conditions* 1 *and* 2 *are satisfied. If, for an* m, $\Phi_m < \infty$, *then*

(i)

$$\sum_{m<\lambda_{m,j} \leqslant m+1} a^{1/2}(\lambda_{m,j}) \leqslant C\Phi_m^{1/2}, \quad C = a \text{ constant}, \tag{3.6}$$

(ii)

$$\sum_{m<\lambda_{m,j} \leqslant m+1} |A(\lambda_{m,j}, \omega)| < \infty, \tag{3.7}$$

holds almost surely, and

(iii)

$$S_m(t,\omega) = \sum_{m<\lambda_{m,j} \leqslant m+1} A(\lambda_{m,j}, \omega) \exp(i\lambda_{m,j} t) \tag{3.8}$$

is a u.a.p. function of t *almost surely.*

Before proving the lemma, we give some remarks. First, (3.5) can be written by

$$\Phi_m = \left[\int_m^{\mu_{k_m}-0} + \int_{\mu_{k_m}+0}^{m+1} \right] \phi(|\lambda - \mu_{k_m}|)\, dF(\lambda) + a(\mu_{k_m})$$

when $\mu_{k_m} = \lambda_{m,0}$ is a discontinuity point of $F(\lambda)$. If not, the last term vanishes. Second, the series representing $S_m(t,\omega)$ in (3.8) originally converges in $L^2(\Omega)$ as a partial sum of the series representing $X(t,\omega)$. (iii) says that the series converges absolutely and hence uniformly almost surely from (ii), and the sum is u.a.p. almost surely. (ii) is, in turn, immediate from (i), because

$$E|A(\lambda_{m,j},\omega)|\leqslant\left[E|A(\lambda_{m,j},\omega)|^2\right]^{1/2}=a^{1/2}(\lambda_{m,j})$$

and $\Sigma E|A(\lambda_{m,j},\omega)|<\infty$ from (i), and (ii) follows.

Proof of Lemma 1. We have only to prove (i). Suppose there is an infinite number of $\lambda_{m,j}$ in (m,μ_{k_m}). Let us write

$$J_{m,n}=\left\{j:\mu_{k_m}-2^{-n}\leqslant\lambda_{m,j}<\mu_{k_m}-2^{(n+1)},\ m<\lambda_{m,j}\right\}.$$

Then for sufficiently large N,

$$\begin{aligned}\sum_{m<\lambda_{m,j}<\mu_{k_m}} a^{1/2}(\lambda_{m,j}) &= \sum_{n=0}^{\infty}\sum_{j\in J_{m,n}} a^{1/2}(\lambda_{m,j})\\
&\leqslant \sum_{n=0}^{\infty}\left[\sum_{j\in J_{m,n}} a(\lambda_{m,j})\right]^{1/2}\left[\sum_{j\in J_{m,n}} 1\right]^{1/2}\\
&\leqslant \sum_{n=0}^{\infty}\left[\sum_{j\in J_{m,n}} a(\lambda_{m,j})N_{k_m}(2^{-n-1})\right]^{1/2}\\
&\leqslant \left[\sum_{n=0}^{\infty} g^{-1}(2^n)\right]^{1/2}\\
&\quad\times\left[\sum_{n=0}^{\infty} g(2^n)\eta(n)\int_{\mu_{k_m}-2^{-n}}^{\mu_{k_m}-2^{-n-1}} h\left(2(\mu_{k_m}-\lambda)^{-1}\right)\mathrm{d}F(\lambda)\right]^{1/2},\end{aligned}$$

where $\eta(n)=1$, for $\mu_{k_m}-2^{-n}>m$, $\eta(n)=0$, for $\mu_{k_m}-2^{-n}\leqslant m$, and

$$\sum_{n=0}^{\infty} g^{-1}(2^n)\leqslant 2\int_1^{\infty}\frac{\mathrm{d}x}{xg(x)}=C^2.$$

Hence

$$\begin{aligned}\sum_{m<\lambda_{m,j}<\mu_{k_m}} a^{1/2}(\lambda_{m,j}) &\leqslant C\left[\sum_{n=0}^{\infty}\eta(n)\int_{\mu_{k_m}-2^{-n}}^{\mu_{k_m}-2^{-n-1}} g\left((\mu_{k_m}-\lambda)^{-1}\right)\right.\\
&\qquad\left.\times h\left(2(\mu_{k_m}-\lambda)^{-1}\right)\mathrm{d}F(\lambda)\right]^{1/2}\\
&\leqslant C\left[\int_m^{\mu_{k_m}-0}\phi(\mu_{k_m}-\lambda)\,\mathrm{d}F(\lambda)\right]^{1/2},\end{aligned}$$

where C's, in above and in what follows, are constants which may differ from each other.

Similarly, if there are infinitely many $\lambda_{m,j}$ in $(\mu_{k_m}, m+1]$, then we have

$$\sum_{\mu_{k_m}<\lambda_{m,j}<m+1} a^{1/2}(\lambda_{m,j})\leqslant C\left[\int_{\mu_{k_m}+0}^{m+1}\phi(\lambda-\mu_{k_m})\,\mathrm{d}F(\lambda)\right]^{1/2}.$$

We then have

$$\sum_{m<\lambda_{m,j}<m+1} a^{1/2}(\lambda_{m,j}) \leqslant C\left[\int_m^{\mu_{k_m}-0}\right]^{1/2} + C\left[\int_{\mu_{k_m}+0}^{m+1}\right]^{1/2} + a^{1/2}(\lambda_{m,0})$$

$$\leqslant C\left[\int_m^{m+1} \phi(|\lambda-\mu_{k_m}|)\,\mathrm{d}F(\lambda)\right]^{1/2}.$$

The proof of Lemma 1 is now complete.

We give another lemma which is well known.

Lemma 2. *Write, for any measurable set E and a positive integer r,*

$$\sigma_r(E) = c_r \int_E \frac{\sin^{2r}(t/2)}{(t/2)^{2r}}\,\mathrm{d}t, \tag{3.9}$$

where c_r is a constant such that $\sigma_r((-\infty,\infty))=1$. With $\sigma_r(t)=\sigma_r((-\infty,t))$,

$$\int_{-\infty}^{\infty} \mathrm{e}^{\mathrm{i}\lambda t}\,\mathrm{d}\sigma_r(t) = 0 \quad \textit{for } |\lambda| \geqslant r. \tag{3.10}$$

From this lemma, we see that, for any sequence $\{\nu_j\}$ of real numbers such that $|\nu_j - \nu_k| \geqslant r$ $(j \neq k)$, the sequence $\exp(\mathrm{i}\nu_j t)$ is an orthonormal sequence with respect to $\sigma_r(E)$.

4. Sample uniform almost periodicity

Suppose a given weakly stationary process $X(t,\omega)$ satisfies the Conditions 1 and 2. $X(t,\omega)$ is represented by

$$X(t,\omega) = \sum_{k,j} A(\lambda_{k,j},\omega)\exp(\mathrm{i}\lambda_{k,j}t), \tag{4.1}$$

in which the convergence involved in (4.1) is primarily the convergence in $L^2(\Omega)$. Let $(-\infty,\infty) = \bigcup_{n=-\infty}(n, n+1]$. When $[n, n+1]$ contains a limiting point of discontinuities of $F(\lambda)$, we defined Φ_n by (3.5). For the sake of convenience, we extend this definition to the case where $[n, n+1]$ does not contain any limiting point of discontinuities, by

$$\Phi_n = \int_n^{n+1} \mathrm{d}F(\lambda). \tag{4.2}$$

We shall prove:

Theorem 1. *The Conditions 1 and 2 are supposed to be satisfied. Let $q(x)$ be a non-decreasing function over $[0,\infty)$ such that*

$$\int_1^{\infty} \frac{\mathrm{d}x}{xq(x)} < \infty, \qquad q(x) = 1 \quad \textit{for } 0 \leqslant x \leqslant 1. \tag{4.3}$$

If

$$\sum_{n=-\infty}^{\infty} (|n|+1)q(|n|)\Phi_n < \infty, \tag{4.4}$$

then $X(t,\omega)$ *has a version which is u.a.p. almost surely.*

Proof. Suppose $[n, n+1]$ contains a limiting point of discontinuities of $F(\lambda)$. Then by Lemma 1, there is a set Ω_n with $P(\Omega_n)=1$ such that

$$S_n(t,\omega)=\sum_{n<\lambda_{n,j}\leqslant n+1} A(\lambda_{n,j},\omega)\exp(i\lambda_{m,j}t)$$

is u.a.p. and

$$\sum_{n<\lambda_{n,j}\leqslant n+1} E|A(\lambda_{n,j},\omega)| \leqslant \sum_{n<\lambda_{n,j}\leqslant n+1} a^{1/2}(\lambda_{n,j})$$

$$\leqslant C\Phi_n^{1/2}. \tag{4.5}$$

(4.5) keeps true even when $[n, n+1]$ does not contain a limiting point of discontinuities as is easily seen. Namely (4.5) holds for every interval $(n, n+1]$. Now by a well known argument, we see

$$\sum_{n=0}^{\infty} \sum_{n<\lambda_{n,j}\leqslant n+1} E|A(\lambda_{n,j},\omega)|$$

$$\leqslant C\sum_{n=0}^{\infty} \Phi_n^{1/2} = C\sum_{m=0} \sum_{n=2^m-1}^{2^{m+1}} \Phi_n^{1/2}$$

$$\leqslant C\sum_{m=0}^{\infty} 2^{m/2}\left[\sum_{n=2^m-1}^{2^{m+1}} \Phi_n\right]^{1/2} \leqslant C\sum_{m=0}^{\infty}\left[\sum_{n=2^m-1}^{2^{m+1}} (n+1)\Phi_n\right]^{1/2}$$

$$\leqslant C\left[\sum_{m=0}^{\infty} q^{-1}(2^{m-1})\right]^{1/2}\left[\sum_{m=0}^{\infty} q(2^{m-1}) \sum_{n=2^m-1}^{2^{m+1}} (n+1)\Phi_n\right]^{1/2}$$

$$\leqslant C\left[\sum_{n=0}^{\infty} (n+1)q(n)\Phi_n\right]^{1/2} < \infty.$$

Hence $\sum_{n=0}^{\infty}\sum_{n<\lambda_{n,j}\leqslant n+1}|A(\lambda_{n,j},\omega)|$ converges almost surely which implies that $\sum_{n=0}^{\infty} S_n(t,\omega)$ is uniformly and absolutely convergent almost surely. The same thing is true for $\sum_{n=-\infty}^{-1} S_n(t,\omega)$. therefore $\sum_{n=-\infty}^{\infty} S_n(t,\omega)$ is uniformly and absolutely convergent. All terms of this series are u.a.p. for $\omega \in \bigcup_{n=-\infty}^{\infty} \Omega_n$ whose probability is one and hence the sum of the series is u.a.p. almost surely. On the other hand, the series converges in $L^2(\Omega)$ to $X(t,\omega)$ and then the conclusion of the theorem follows.

Theorem 2. *Under the conditions of Theorem* 1, *the u.a.p. version of* $X(t,\omega)$ *has absolutely convergent Fourier series and the spectral measure has the property that*

$$\int_{-\infty}^{\infty} |\xi(d\lambda,\omega)| < \infty \tag{4.6}$$

almost surely.

This is included in the proof of Theorem 1.

5. Sample S^2 almost periodicity

We shall generalize the theorem of Udagawa mentioned in Section 1.

Theorem 3. *Suppose Conditions* 1 *and* 2 *are satisfied. If*

$$\sum_{n=-\infty}^{\infty} \Phi_n < \infty, \tag{5.1}$$

then $X(t,\omega)$ *is* S^2*.a.p. almost surely.*

Note that if discontinuities of $F(\lambda)$ satisfy (1.3), then (5.1) is automatically satisfied, because $\Phi_n \leqslant Ca(\lambda_n)$.

We now prove Theorem 3. As in the proof of Theorem 1, let Ω_n be a set of $\mathfrak{F}$ such that $S_n(t,\omega)$ is u.a.p. for $\omega \in \Omega_n$ and let $\Omega' = \bigcap_{n=-\infty}^{\infty} \Omega_n$, $P(\Omega')=1$.

Consider the sequence $\{S_{2n}(t,\omega),\ n=0,\pm1,\pm2,\dots\}$. Since $(2n,2n+1]$ is distant from $(2(n+1),2n+3]$ or from $(2(n-1),2n-1]$ by distance 1, by the absolute convergence of the series representing $S_n(t,\omega)$, we see from Lemma 2 that $\{S_{2n}(t,\omega)\}$ forms an orthogonal sequence with respect to $\sigma_1(E)$. For any real y, $\{S_{2n}(t+y,\omega),\ n=0,\pm1,\pm2,\dots\}$ is also an orthogonal with respect to $\sigma_1(E)$. For each $\omega\in\Omega'$ for any integers M and N,

$$\int_{-\infty}^{\infty} \left| \sum_{n=N+1}^{M} S_{2n}(t+y,\omega) \right|^2 d\sigma_1(t) =$$

$$= \int_{-\infty}^{\infty} \sum_{n=N+1}^{M} |S_{2n}(t+y,\omega)|^2 d\sigma_1(t)$$

$$\leqslant \int_{-\infty}^{\infty} \sum_{n=N+1}^{M} \left[\sum_{2n<\lambda_{2n,j}\leqslant 2n+1} |A(\lambda_{2n,j},\omega)| \right]^2 d\sigma_1(t)$$

$$= \sum_{n=N+1}^{M} \left[\sum_{2n<\lambda_{2n,j}\leqslant 2n+1} |A(\lambda_{2n,j},\omega)| \right]^2 \tag{5.2}$$

and thus

$$\sup_y \int_{-\infty}^{\infty} \left| \sum_{n=N+1}^{M} S_{2n}(t,\omega) \right|^2 d\sigma_1(t-y) \leqslant \sum_{n=N+1}^{M} \left[\sum_{2n<\lambda_{2n,j}\leqslant 2n+1} |A(\lambda_{2n,j},\omega)| \right]^2. \tag{5.3}$$

Now consider the series

$$\sum_{n=-\infty}^{\infty} \left[\sum_{2n<\lambda_{2n,j}\leqslant 2n+1} |A(\lambda_{2n,j},\omega)| \right]^2. \tag{5.4}$$

Taking expectation and using Minkowski's inequality, we have

$$\sum_{n=-\infty}^{\infty} E\left[\sum_{2n<\lambda_{2n,j}\leqslant 2n+1} |A(\lambda_{2n,j},\omega)| \right]^2 \leqslant$$

$$\leqslant \sum_{n=-\infty}^{\infty} \left\{ \sum_{2n<\lambda_{2n,j}\leqslant 2n+1} \left[E|A(\lambda_{2n,j},\omega)|^2 \right]^{1/2} \right\}^2$$

$$\leqslant C \sum_{n=-\infty}^{\infty} \Phi_{2n} < \infty$$

by Lemma 1 (i). Therefore, the series (5.4) converges almost surely. Hence

$$\lim_{M,N\to\pm\infty} \sup_y \int_{-\infty}^{\infty} \left| \sum_{n=N+1}^{M} S_{2n}(t,\omega) \right|^2 d\sigma_1(t-y) = 0 \tag{5.5}$$

almost surely.

On the other hand, the integral in (5.5) is

$$\int_{-\infty}^{\infty} \left| \sum_{n=N+1}^{M} S_{2n}(t,\omega) \right|^2 \frac{\sin^2(t-y)/2}{\pi(t-y)^2/2} dt \geqslant \frac{4}{\pi^3} \int_{y-\pi}^{y+\pi} \left| \sum_{n=N+1}^{M} S_{2n}(t,\omega) \right|^2 dt,$$

and hence,

$$\lim_{M,N\to\pm\infty} \sup_y \int_y^{y+1} \left| \sum_{n=N+1}^{M} S_{2n}(t,\omega) \right|^2 dt = 0$$

holds almost surely, that is, $\Sigma S_{2n}(t,\omega)$ converges to an S^2.a.p. function almost surely in S^2-norm: $\sup_y \int_y^{y+1} |\cdot|^2 dt$.

The same thing is true for $\Sigma S_{2n+1}(t,\omega)$ and therefore,

$$T_N(t,\omega) = \sum_{n=-N}^{N} S_n(t,\omega)$$

converges in S^2-norm to an S^2.a.p. function $X_1(t,\omega)$ almost surely, since $S_n(t,\omega)$ is u.a.p. for every n almost surely in Ω'. The convergence of $T_N(t,\omega)$ to the S^2.a.p.

function in $L^2(-T,T)$ almost surely is also included in S^2-norm convergence for every $T>0$.

While

$$\int_{-T}^{T} E|T_N(t,\omega)-X(t,\omega)|^2 \mathrm{d}t = 2T\left[\int_{N+1}^{\infty} \mathrm{d}F(\lambda)+\int_{-\infty}^{-N} \mathrm{d}F(\lambda)\right] \to 0,$$

as $N\to\infty$, for every finite fixed T, that is, $T_N(t,\omega)-X(t,\omega)$ converges in $L^2(\Omega\times(-T,T))$ to 0, which implies that for some subsequence $\{T_{N_k}(t,\omega)\}$ converges almost everywhere in $\Omega\times(-T,T)$. By Fubini argument, for almost all ω, $T_{N_k}(t,\omega)$ converges to $X(t,\omega)$ almost everywhere in $(-T,T)$. Namely in consideration of countable T, $X(t,\omega)$ is equal to $X_1(t,\omega)$ for almost all t, almost surely. Therefore, $X(t,\omega)$ is S^2.a.p. almost surely. This proves the theorem.

6. Sample B^2 almost periodicity

We shall prove Slutsky's theorem in our general situation in a quite different way.

Theorem 4. *Suppose the Conditions* 1 *and* 2 *are satisfied and* $\Phi_n<\infty$ *for every* n. *Then* $X(t,\omega)$ *is* B^2.*a.p. almost surely.*

Proof. By the same argument in deriving (5.5) and the corresponding inequalities with S_{3n}, S_{3n+1}, S_{3n+2} and with $\sigma_2(t)$ in place of $\sigma_1(t)$, we have, for $m>k$,

$$\int_{y-\pi}^{y+\pi}\left|\sum_{|n|=k}^{m} S_n(t,\omega)\right|^2 \mathrm{d}t \leqslant C\int_{-\infty}^{\infty}\sum_{p=0}^{2}\left|\sum_{k\leqslant|3n+p|\leqslant m} S_{3n+p}(t,\omega)\right|^2 \mathrm{d}\sigma_2(t-y).$$

Let A be any positive large number. Write $y=2\nu\pi$, $\nu=-N,-N+1,\ldots,N$ in this inequality, add them, where N is the smallest integer such that $(2N+1)\geqslant A$ and divide the result by $2A$. We then see that

$$\frac{1}{2A}\int_{-A}^{A}\left|\sum_{k\leqslant|n|\leqslant m} S_n(t,\omega)\right|^2 \mathrm{d}t\leqslant$$

$$\leqslant\frac{C}{2A}\sum_{\nu=-N}^{N}\int_{-\infty}^{\infty}\sum_{p=0}^{3}\left|\sum_{k\leqslant|3n+p|\leqslant m} S_{3n+p}(t,\omega)\right|^2 \mathrm{d}\sigma_2(t-2\nu\pi).$$

Since $\{S_{3n+p}(t,\omega), n=0,1,2,\ldots\}$ constitutes, for each p, an orthogonal sequence as a sequence of functions of t, with respect to $\sigma_2(E)$, in view of (5.2) with $\sigma_2(t)$ for $\sigma_1(t)$ and $S_{2n+p}(t,\omega)$ for $S_{2n}(t,\omega)$, we see that the right-hand side of the last inequality does not exceed

$$\frac{C}{2A}\sum_{\nu=-N}^{N}\int_{-\infty}^{\infty} U_{k,m}(t,\omega)\,\mathrm{d}\sigma_2(t-2\nu\pi),$$

where

$$U_{k,m}(t,\omega)=\sum_{p=0}^{3}\sum_{k\leqslant|3n+p|\leqslant m}|S_{3n+p}(t,\omega)|^2.$$

Then the above is not greater than

$$C\int_{-\infty}^{\infty}U_{k,m}(t,\omega)\frac{1}{2N}\sum_{\nu=-N}^{N}\frac{1}{(t-2\nu\pi)^4/2}\sin^4(t/2)\,dt.$$

Now it is easy to see

$$\frac{\sin^2 t}{2N}\sum_{\nu=-N}^{N}\frac{1}{(t-2\nu\pi)^4/2}\leqslant Ct^{-4},\quad \text{for } |t|>3N\pi,$$

$$\leqslant CN^{-1},\quad \text{for } |t|\leqslant 3N\pi. \tag{6.1}$$

Using this we have

$$\frac{1}{2A}\int_{-A}^{A}\left|\sum_{k\leqslant|n|\leqslant m}S_n(t,\omega)\right|^2 dt\leqslant$$

$$\leqslant C\int_{|t|>3N\pi}U_{k,m}(t,\omega)\frac{dt}{t^4}+\frac{C}{N}\int_{|t|\leqslant 3N\pi}U_{k,m}(t,\omega)\,dt. \tag{6.2}$$

Since $\Sigma_{k\leqslant|n|\leqslant m}S_n(t,\omega)$ converges to $X(t,\omega)-\Sigma_{|n|\leqslant k}S_n(t,\omega)$ as $m\to\infty$ in $L^2(\Omega\times(-A,A))$ for every $A>0$ as was mentioned in the end part of Section 5, there is a sequence m_j of integers and a set Ω^1 with $P(\Omega^1)=1$ such that $\Sigma_{k\leqslant|n|\leqslant m_j}S_n(t,\omega)$ converges to $X(t,\omega)-\Sigma_{|n|<k}S_n(t,\omega)$ almost everywhere for $\omega\in\Omega^1$. Hence for $\omega\in\Omega^1$,

$$\frac{1}{2A}\int_{-A}^{A}\left|X(t,\omega)-\sum_{|n|<k}S_n(t,\omega)\right|^2 dt\leqslant$$

$$\leqslant \liminf_{m_j\to\infty}\frac{1}{2A}\int_{-A}^{A}\left|\sum_{k\leqslant|n|\leqslant m_j}S_n(t,\omega)\right|^2 dt$$

$$\leqslant C\int_{|t|>3N\pi}U_{k,\infty}(t,\omega)\frac{dt}{t^4}+\frac{C}{N}\int_{|t|\leqslant 3N\pi}U_{k,\infty}(t,\omega)\,dt, \tag{6.3}$$

since $U_{k,m}(t,\omega)$ increases to $U_{k,\infty}(t,\omega)$ as $m\to\infty$.

Now

$$\sum_N\int_{|t|>3N\pi}E\sum_{|n|=0}^{\infty}|S_{3n+p}(t,\omega)|^2\frac{dt}{t^4}\leqslant\sum_N\int_{|t|>3N\pi}\frac{dt}{t^4}\int_{-\infty}^{\infty}dF(\lambda)=C\sum_N N^{-4}<\infty.$$

Hence,

$$\int_{|t|>3N\pi} |S_{3n+p}(t,\omega)|^2 \frac{\mathrm{d}t}{t^4} \to 0, \quad p=0,1,2,$$

for $\omega\in\Omega^2$, $P(\Omega^2)=1$. Therefore we have

$$\int_{|t|>3N\pi} U_{k,\infty}(t,\omega)\frac{\mathrm{d}t}{t^4} \to 0, \tag{6.4}$$

as $N\to\infty$, for $\omega\in\Omega^2$.

We also have

$$\limsup_{N\to\infty} \frac{1}{N}\int_{|t|\leqslant 3N\pi} \sum_{k\leqslant|3n+p|} |S_{3n+p}(t,\omega)|^2\,\mathrm{d}t=$$

$$=\limsup_{N\to\infty} \sum_{k\leqslant|r|} \frac{1}{N}\int_{|t|\leqslant 3N\pi} \left| \sum_{r<\lambda_{r,j}\leqslant r+1} A(\lambda_{r,j},\omega)e^{\mathrm{i}\lambda_{r,j}t}\right|^2 \mathrm{d}t,$$

where $r=3n+p$,

$$=\limsup_{N\to\infty}\Bigg[3\pi \sum_{k\leqslant|r|}\ \sum_{r<\lambda_{r,j}\leqslant r+1} |A(\lambda_{r,j},\omega)|^2$$

$$+\frac{1}{N}\int_{|t|\leqslant 3N} \sum_{\lambda_{r,j}\neq\lambda_{r,j'}} A(\lambda_{r,j},\omega)\overline{A(\lambda_{r,j'},\omega)}$$

$$\cdot\exp(\mathrm{i}(\lambda_{r,j}-\lambda_{r,j'})t)\,\mathrm{d}t\Bigg].$$

By keeping Lemma 1 (ii) in mind, it is seen that the superior limit as N of the integral of the above is zero because

$$\lim_{N\to\infty}\frac{1}{N}\int_{|t|<3N\pi} e^{\mathrm{i}\lambda t}\,\mathrm{d}t=0 \quad \text{for } \lambda\neq 0.$$

Therefore,

$$\limsup_{N\to\infty}\frac{1}{N}\int_{|t|\leqslant 3N\pi} U_{k,\infty}(t,\omega)\,\mathrm{d}t=$$

$$=3\pi\sum_{p=0}^{2}\ \sum_{k\leqslant|r|}\ \sum_{r\leqslant\lambda_{r,j}\leqslant r+1} |A(\lambda_{r,j},\omega)|^2$$

$$=3\pi\sum_{k<|n|}\ \sum_{n<\lambda_{n,j}\leqslant n+1} |A(\lambda_{n,j},\omega)|^2.$$

Combining this with (6.3) and (6.4), we have for $\omega\in\Omega^1\cap\Omega^2$,

$$\limsup_{A\to\infty}\frac{1}{2A}\int_{-A}^{A}\left|X(t,\omega)-\sum_{|n|<k} S_n(t,\omega)\right|^2\mathrm{d}t \leqslant 3\pi\sum_{k<|n|}\ \sum_{n<\lambda_{n,j}\leqslant n+1} |A(\lambda_{n,j},\omega)|^2.$$

The expectation of the last term is $\int_{|\lambda|\geqslant k} \mathrm{d}F(\lambda)$. Hence choosing $k=k_l$ so that

$$\sum_{l=1}^{\infty} \int_{|\lambda|>k_l} \mathrm{d}F(\lambda)<\infty$$

which is obviously possible, we have

$$\sum_{l=1}^{\infty} E \sum_{k_l<|n|} \sum_{n<\lambda_{n,j}\leqslant n+1} |A(\lambda_{n,j}, \omega)|^2 < \infty .$$

Hence for $\omega\in\Omega^3$ for some Ω^3 with $P(\Omega^3)=1$,

$$\lim_{l\to\infty} \sum_{k_l<|n|} \sum_{n<\lambda_{n,j}\leqslant n+1} |A(\lambda_{n,j}, \omega)|^2 = 0.$$

Therefore, for $\omega\in\Omega^1\Omega^2\Omega^3$, the probability of this set being one,

$$\lim_{l\to\infty} \limsup_{A\to\infty} \frac{1}{2A} \int_{-A}^{A} \left| X(t, \omega) - \sum_{|n|\leqslant k_l} S_n(t, \omega)\right|^2 \mathrm{d}t = 0.$$

Since $S_n(t, \omega)$ is u.a.p. almost surely, $X(t, \omega)$ is B^2.a.p. almost surely. This completes the proof.

References

[1] Alexits, G. (1961). *Convergence Problems of Orthogonal Series.* (English Translation). Pergamon Press, New York–Oxford–London–Paris.

[2] Besicovitch, A.S. (1932). *Almost Periodic Functions.* Cambridge University Press, London.

[3] Bochner, S. (1936). On general Fourier series with gaps. *Prace Math.-Fizy.* **43**, 63–79.

[4] Bochner, S. (1954). Stationarity, boundedness, almost periodicity of random valued functions. *Proc. Third Berkeley Symp. Math. Statist. and Prob.* **2**, 7–27.

[5] Bohr, H. (1932). *Fastperiodische Funktionen.* Springer, Berlin.

[6] Gaposhkin, V. F. (1977). Criteria for the strong law of large numbers for some classes of second order stationary processes and homogeneous random fields. *Theory Prob. Appl.* **22**, 286–310.

[7] Paley, R.E.A.C. and Wiener, N. (1934). Fourier transforms in the complex domain. *Amer. Math. Soc. Colloq. Publ.* **19**.

[8] Slutsky, E. (1938). Sur les fonction aléatoires presque périodiques et sur la decomposition des fonctions aléatoires stationaires en composantes. *Actualites Sci. Ind.* No. **738**, 33–55.

[9] Tandori, K. (1957). Uber die orthogonal Funktionen 1. *Acta Sci. Math.* **18**, 57–130.

[10] Udagawa, M. (1953). Asymptotic properties of distributions of some functionals of random variables. *Rep. Statist. Appl. Res. JUSE.* **2**, 1–66.

[11] Yu, H. (1973). Almost periodic weakly stationary process. Dissertation. Korean University, Korea.

G. Kallianpur, P.R. Krishnaiah, J.K. Ghosh, eds., *Statistics and Probability: Essays in Honor of C.R. Rao*
© North-Holland Publishing Company (1982) 397–410

CLASSIFICATORY DATA STRUCTURES AND ASSOCIATED LINEAR MODELS

Oscar KEMPTHORNE
Statistical Laboratory, Iowa State University, Ames, IA, U.S.A.

C. R. Rao has been a consistent contributor to the theory and applications of statistics since the 1940s. The breadth of his contributions to theory of statistics has been great, and he has contributed strongly to statistical methods also, particularly with respect to linear models and multivariate analysis. To contribute to a volume honoring C. R. Rao is a pleasure also because of valuable personal association over the decades.

1. Introduction

Linear model theory was developed, I suggest, first in reaction to the design and analysis of experiments. I shall not attempt to support this view and shall merely point to the work of Fisher of the 1920s and Yates of the 1930s. Unfortunately, though I suppose it is inevitable, the work of "founding fathers" gets lost in the "mist of time". Increasingly, we see teaching texts that give no cognizance to this history; and I surmise that this happens partially because the writers have either not been taught or have not read that history. We are exposed increasingly to writers who suggest that the origins of the subject lie in finite-dimensional vector spaces and projections. This is, I suggest, rather appalling for the very simple reason that projections come from least squares (and its generalizations) and then from normal equations. One needs, indeed, only to look at, say, Fourier analysis to see that the basic idea in the whole area is least squares approximation, *leading to normal equations*. Furthermore, normal equations are indeed simple to get. We can teach them to individuals who know only a little partial differentiation and a little bit of the theory of linear equations. It is true, of course, that we can imbed the whole matter in *linear* space theory and functional analysis, so that if one has learned this fine body of mathematical thought, then one can wrap up some of the area of linear models as an application of such analysis. Also, one can envisage that area of mathematical development contributing to linear model theory. However, to embed linear model theory in linear functional analysis, and then to take the view that the way to teach linear model theory is first to teach linear functional analysis is to commit errors of two kinds. On the other hand, one should not use a "sledgehammer to crack a nut"; when one does, one smashes the nut to smithereens and can lose it. On the other hand, this subject, like any other branch of applied mathematics, gets its life from applications. Functional analysis grew out, partially, from problems of mathematical physics. Linear model theory has grown out of problems of data analysis and interpretation. If one follows the approach of one field in developing the other, one misses many of the significant problems: the problems that give a field its life. Furthermore, the student has a finite time and cannot often study advanced areas by which what he needs becomes elementary.

Another aspect that I regard as highly significant and one to which a considerable portion of this paper will be addressed is that models that are not of full rank are essentially intrinsic to a class of situations that may be characterized as classificatory data structures. So in linear models as used in statistics, one uses intrinsically what is an absurdity in linear functional analysis, namely that the functions we use, which are finite sequences or finite vectors, or functions on $\{1,2,\ldots,n\}$ are *not* independent. I use the word "absurd" because I ask: Who would use a function space of non-independent elements?

Another curious aspect of the "column-space" or "manifold" approach, or the so-called "coordinate-free approach" is that it looks at one aspect but does not look at all at another aspect. In a proper discussion of the model: $y \doteq X\beta$ we have to be concerned with both the column space and the row space of X. It is relevant to note that mathematicians prefer to talk about transformation. A proper attention to linear models requires what may be called the column kernel and the row kernel of a matrix.

The main aspect I shall discuss is the case in which the model matrix X (*not* the design matrix) is derived by superimposing a classificatory structure on the data. It is inherent, I shall suggest, in such data structures that one considers models that are not of full rank, even though given a sample linear model $y \doteq X\beta$, it is trivial to get a full rank column (or row) reparametrization to say $y \doteq Z\gamma$, or even an orthogonal column reparametrization in which $Z'Z$ is diagonal or even a normalized orthogonal reparametrization in which $Z'Z = I$. The formulation of ideas of linear models for classificatory data structures has many interesting features. Many of these features are well known, in the sense that *practising* statisticians have used the ideation for some decades; but a formal description of what is happening is not in the literature, I believe.

The ideation I shall describe was developed at Iowa State University in collaboration over time, with Wilk (1955), Zyskind (1958, 1962), Throckmorton (1961) and Carney (1967) with myself as the "permanent" continuing member of the effort. The ideation I give is related, clearly, to that of Nelder (1965a,b, 1977) that came after these writings I have cited.

2. Approximative versus stochastic models

An idea that I have been expositing in my courses for more than a decade is that it is extremely useful, *epistemologically*, to make a distinction between the following:

(a) *An approximative linear model*:

$$y \doteq X\beta$$

in which we have, say, y being $n\times 1$, X being $n\times p$ and β being $p\times 1$. Then our problem, in the approximative mode, is to determine a vector Xb such that

$$\| y - Xb \| = \min_{\beta} \| y - X\beta \|.$$

I am using slightly sophisticated terminology here *solely* for reasons of economy. Common norms of approximation are:

$$l_1^{(n)}: \sum_{i=1}^{n} |y_i - (X\beta)_i|,$$

$$l_2^{(n)}: \left\{ \sum_{i=1}^{n} [y_i - (X\beta)_i]^2 \right\}^{1/2}$$

and

$$l_\infty^{(n)}: \max_{1 \leq i \leq n} |y_i - (X\beta)_i|.$$

Fitting a linear model should be regarded, I believe, as a purely numerical or approximative affair. Viewed in this way, analysis of variance is a characterization of the accuracy of an approximation, and as a characterization of how a broader approximation does better than a narrower approximation (as in comparing the approximative models $y \doteq X_1\beta_1$ and $y \doteq X_1\beta_1 + X_2\beta_2$).

In contrast with the approximative model, we have what I like to call:

(b) *A stochastic linear model*

$$y = X\beta + e,$$

with y and e both random vectors. Then the discourse makes contact with the general area of mathematical statistics.

In the linear approximative model context, I consider it preferable to use an idea of identifiability in which $\lambda'\beta$ is (linearly) identifiable in a linear approximative, $y \doteq X\beta$ model if $\lambda'\beta = a'X\beta$ for all specified β for some vector a.

In the stochastic linear model, the standard idea is of estimability (again, linear) and if $\mathcal{E}(e) = \varphi$, (the null vector), $\lambda'\beta$ is (linearly) estimable if $\lambda'\beta = a'X\beta$ for all specified β and for some vector a.

I shall talk entirely about linear approximative models. The passage to linear stochastic models is trivial if Gauss–Markoff assumptions, $\mathcal{E}(e) = \varphi$, $\mathcal{E}(ee') = \sigma^2 I$, hold and distribution theory of analysis of variance with the additional assumption of normality goes rather easily.

3. Full rank reparametrization

Given $y \doteq X\beta$, least squares approximation gives the Normal Equations (NE), $X'Xb = X'y$, which are consistent, and have a unique solution for Xb which gives a global minimum for $(y - X\beta)'(y - X\beta)$. One can approach this situation in many ways that I shall not enumerate. One way is to assert that the model vector, $X\beta$, is a vector in the column space of X, and so one may obtain a full rank parametrization,

$X\beta = Z\gamma$ with Z of full column rank. Then $Z'Z$ is invertible and:

$$\hat{X\beta} = \hat{Z\gamma} = (Z'Z)^{-1}Z'y.$$

The great bulk of theorems can then be proved easily.

So one may ask: Why all the writing about non-full rank models?

A thesis of the present paper is that it is both natural and informative to use non-full rank models in a widely occurring set of circumstances. Concurrent with this thesis is the view that what is called the "coordinate-free approach" is only partially successful and fails with respect to some critical needs of scientists who need to use linear models in their scientific activities.

With the linear model $y \doteq X\beta$ and the use of the $l_2^{(n)}$ norm as a measure of badness of fit, we are led to least squares and the normal equations, $X'Xb = X'y$. Then by consistency for all $y \in R^n$, $\exists B \ni X'XB = X'$ and XB is unique, symmetric and idempotent. These results are very easily proved. One needs only the elementary theory of linear equations. Then we can define $P_X = XB$, and it is the matrix of the orthogonal projection onto $\mathcal{C}(X)$, the column space of X. Also

$$\text{Fit}(X\beta) = P_X y,$$

and $P_X y$ and $(I - P_X)y$ are orthogonal vectors in R^n. It is easy to see, then, that some of the mathematics is easily derivable from the use of P_X. One may call this *a* coordinate-free approach if one wishes. One can easily see that

$$P_X = X(X'X)^{-}X',$$

for any generalized inverse of $X'X$. Also one notes that one will rarely see $(X'X)^{-}$ except in expressions that are invariant with respect to choice of generalized inverse. So the argumentation about the merits of matrices and coordinate free approach is really an argument about trivia. The way I teach *some* aspects is by P-matrices because of two simple properties:

(1) $P = P' = P^2$;

(2) Suppose we have 2 matrices each of n rows, X and Z. Then obviously we can define P_X and P_Z, and then $P_Z - P_X$ is a projection matrix iff $\mathcal{C}(X) \subset \mathcal{C}(Z)$ and if this is the case,

$$P_X y \quad \text{and} \quad (P_Z - P_X)y,$$

are orthogonal vectors. So with subspaces $S_1 \subset S_2 \subset \cdots S_k$, and associated projection matrices $P_1, P_2, \ldots, P_k$, then

$$I = P_1 + (P_2 - P_1) + \cdots + (P_k - P_{k-1}) + (I - P_k).$$

The whole of the elementary algebra and statistical theory then "falls out" very easily.

All this is, I suppose, coordinate-free but it is coordinate-free only given the projection matrices. The only way to get a projection matrix onto S_1 is to coordinatize S_1, e.g. by stating that S_1 is $\{X\beta : \beta \in R^p\}$ and then obtain by one route or another the least squares fit of $y \doteq X\beta$.

I conclude, then, that the presentations of the coordinate-free approach do not address the basic issues of examining linear models. Whether one uses P_X or

$X(X'X)^{-}X'$ is largely a matter of taste. However, there is a clear advantage to the former typographically and mnemonically.

Regarding the whole matter as entirely one of column spaces has, also, a grave defect. The data analyst who uses the linear model, $y \doteq X\beta$ or $y = X\beta + e$ does so *not* because of a wish to determine the fit or estimator of $X\beta$. Rather the aim is to determine identifiable or estimable functions. As soon as one contemplates this, one sees that one must also be involved in the row space of X. It is true, of course that a model can be expressed as one of full rank on a full set of identifiable or estimable functions [cf., Kempthorne (1951, p. 78)], and given this use of proper inverses is entirely valid. Furthermore, as is obvious in general, the modeller or data analyst thinks in terms of a particular modelling process with a particular parametrization. So the start is always with $X\beta$, and β to be determined as far as possible. And then we know that a function $\lambda'\beta$ can be examined if and only if λ' belongs to the row space of X, i.e.

$$\lambda' \in \mathcal{R}(X) = \mathcal{R}(X'X).$$

So I express my view that the matrix X in the model, $y \doteq X\beta$ or $y = X\beta + e$, is critical. It cannot be replaced by its associated column space. The absence of discussion of $\mathcal{R}(X)$ in presentations of linear model theory is, I believe, a grave error.

4. The "geometry" of the two-part ordered linear model

We are exposed, frequently, to the idea that the way to present linear model theory and analysis of variance is by geometry, by a geometrical picture. I must express my opinion that this is largely illusory, for the simple reason that one can represent only a two-dimensional situation accurately by a two-dimensional picture, and a three-dimensional situation by a two-dimensional picture only by what I term "sleight-of-eye". This is not to imply, of course, that such pictures may be suggestive.

A picture was presented by the late Henry Scheffé (1959, p. 43). If one understands it, then it is useful. I prefer the following picture, which is worth recording, I think. Consider the model:

$$y \doteq X_1\beta_1 + X_2\beta_2.$$

Then we have $P_1 y$, where P_1 is the projector onto $\mathcal{C}(X_1)$, and $P_{12}y$, where P_{12} is the projector onto $\mathcal{C}(X_1 X_2)$. Then consider the following picture (Fig. 1).

This rectangular parallelepiped represents only part of R^n, of course. It is imbedded in R^n. We will represent vectors by two letters *in order*.

Then the rectangle OABC represents (part of) $\mathcal{C}(X_1 X_2)$. OA represents (part of) $C(X_1)$; OF$=y$; OA$=P_1y$; AF$=(I-P_1)y$; OB$=P_{12}y$; AB$=(P_{12}-P_1)y$ and BF$=(I-P_{12})y$.

We "see" the derived perpendicularity properties. It is, we suggest, absurd to claim that this "gives" the analysis of variance in a form that the student can understand.

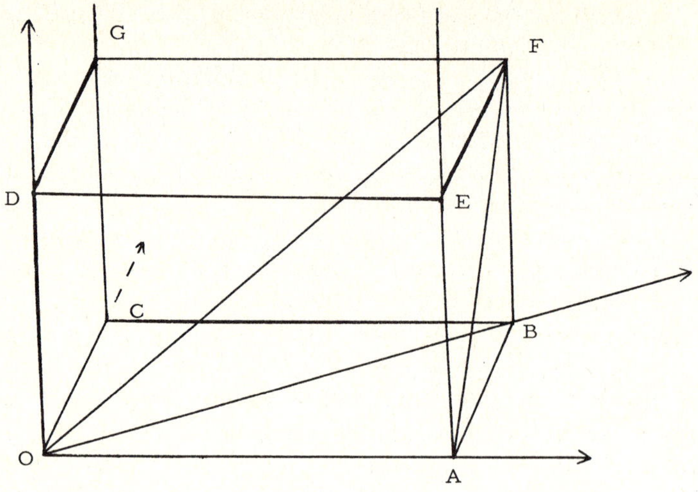

Fig. 1. The "geometry" of an ordered two-part model.

When the student understands projections and their properties the picture makes sense.

It is only by understanding the basic algebra, e.g. that $P_1' = P_1$, $P_1(P_{12} - P_1) = \varphi$, and so on, that the student may be expected to understand the picture.

5. Classificatory data structures

We start from a population of individuals that we will index by the variable, $u = 1, 2, \ldots, n$. In certain situations, that we will not discuss here, it is useful to envisage n as being countably infinite.

Then we define a factor as a partition of the population. To each subset of the partition we attach a label, and we call the separate such labels the levels of the factor or partition. If we name our factor by A we have that the level of individual u with respect to factor A is a function from $\{u\}$ to $\{\text{levels of } A\}$ that we can denote by $x_A(u)$.

Now suppose we have a second factor or partition B. Then it is obvious that the partition given by B can be a refinement of the partition given by A. If this happens, we say that factor B is nested by factor A. This is a definition that is natural, useful and extremely easy to understand.

If factor B is not nested by factor A, then we say that B is crossed with A. We do not wish to imply thereby that there is complete crossing, so that there are individuals with level i of factor A and level j of factor B.

A mode of thinking that I find useful is to note that if A represents a partition and B represents a partition then we can consider the product partition which I will denote by $A \otimes B$. Then we see that if B is nested by A, it is the case that the product partition $A \otimes B$ is the same partition as partition B.

A useful representation of nesting of B by A is given by the formal mathematical statement:

$$B \text{ is nested by } A \quad \text{iff}$$
$$x_B(u)=x_B(u') \Rightarrow x_A(u)=x_A(u').$$

In words, B is nested by A if two individuals having the same levels of B necessarily have the same levels of A.

It is of interest, perhaps, to note that if B is nested by A, then the subsets formed by B are atomic elements of a field and then the field generated by the elements formed by A is a subfield. From this basis, one can, I believe, make a correspondence of the whole σ-algebra panorama to analysis of variance and to projections. I shall not pursue this here.

Let us now turn to the case of 3 factors A, B and C. I shall use the words factor and partition interchangeably. The applied scientist likes, naturally, the term "factor". The theoretical statistician will find the word "partition" of considerable use.

What are the possibilities with 3 factors? We first note that if C is nested by B and B is nested by A, then C is nested by A. This is trivial from the formal definition I give above. So we can think about possibilities merely by thinking about 3×3 matrices, in which nesting is indicated. Consider, for example, the table:

	A	B	C
A	1	0	1
B	0	1	0
C	0	0	1

Here our formalism is as follows:

$$A \text{ nests } C \quad \text{or} \quad C \text{ is nested by } A.$$

It is obvious that any factor nests itself. A realization of this structure arises in looking at humans of the USA:

$$A=\text{states}, \qquad B=\text{sex}, \qquad C=\text{counties}.$$

If we label our factors as $F_1, F_2, \ldots, F_m$, then the nesting structure is given by a $m \times m$ table or matrix $S=(S_{ij})$(S for structure) in which

$$S_{ii}=1, \quad \forall i,$$
$$S_{ij}=1 \Rightarrow S_{ji}=0,$$
$$S_{ij}=1,\ S_{jk}=1 \Rightarrow S_{ik}=1.$$

Here then are the possible structure matrices with 3 factors:

$$\begin{pmatrix} 1 & 1 & 1 \\ 0 & 1 & 1 \\ 0 & 0 & 1 \end{pmatrix}, \begin{pmatrix} 1 & 1 & 1 \\ 0 & 1 & 0 \\ 0 & 0 & 1 \end{pmatrix}, \begin{pmatrix} 1 & 1 & 0 \\ 0 & 1 & 0 \\ 0 & 0 & 1 \end{pmatrix}, \begin{pmatrix} 1 & 0 & 1 \\ 0 & 1 & 1 \\ 0 & 0 & 1 \end{pmatrix}, \begin{pmatrix} 1 & 0 & 0 \\ 0 & 1 & 0 \\ 0 & 0 & 1 \end{pmatrix}.$$

A mode of representing data structures was given in a Ph.D. Thesis by Throckmorton (1961) which is extremely useful. It was presented by Kempthorne and Folks (1971, Section 16.11).

Here are structure diagrams for the 5 cases:

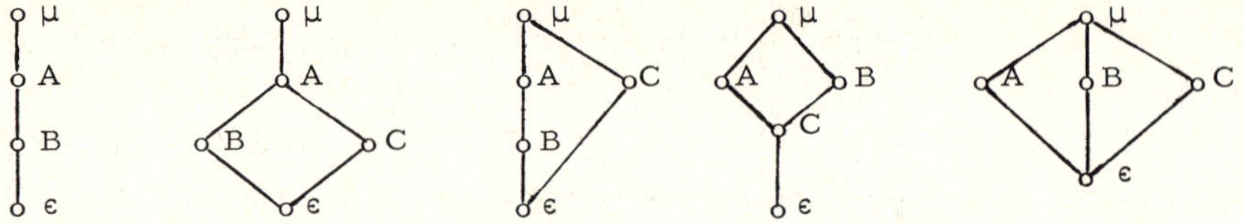

The diagrams are purposely drawn with a "top" denoted by "μ", which tells us that all our individuals belong to a trivial factor with one level, this arising because they all belong to the set of individuals under consideration. Also, they are drawn with a "bottom" denoted by "ε", which is a factor or partition in which every individual comprises a different subset induced by the partition. The choice of names, μ and ε, is obviously motivated by the sorts of linear model we shall contemplate.

By an error of writing a realization of the second structure was not given properly. Here is one:

A is locations,
B is rows within locations,
C is columns within locations;

there is no correspondence envisaged between rows (columns) between locations.

The next step is to consider the case of 4 factors, say *A*, *B*, *C* and *D*. Throckmorton (1961) enumerated all the possibilities of which there are 15 (a curious number). I shall not enumerate them all here. They are given in Kempthorne et al. (1961, p. 88). I shall merely give 3:

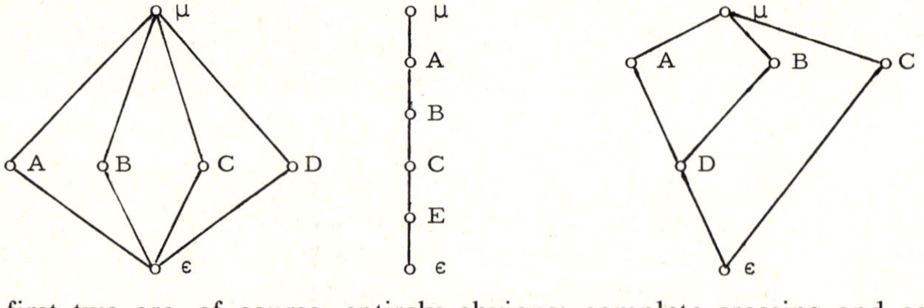

The first two are, of course, entirely obvious; complete crossing and complete nesting. I leave, to the reader, the task of formulating a data problem in which the third structure will be appropriate.

Throckmorton also enumerated in his thesis all the possible 5-factor structures, finding that there were 63 in all. I give only one, a rather "weird" one:

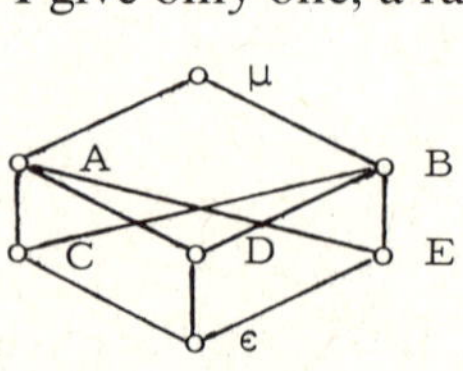

Throckmorton (1961) also gave a process for incorporating randomization, which is random association of factors, into the structure diagrams.

6. Renaming of factors and renaming of individuals

Suppose we have factors A and B. Then we have partitions A, B, and, of course, the product partition

$$A \otimes B.$$

If B is nested by A then the partition $A \otimes B$ is the same as the partition B. So we will agree to rename the B partition with the name AB. Take:

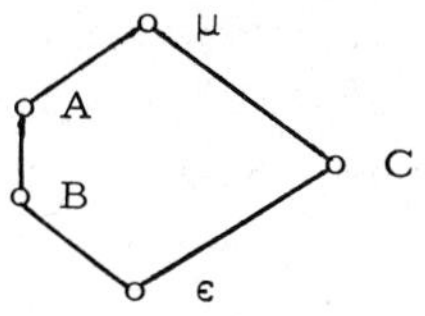

then a list of the partitions and product partitions is μ, A, B, $A \otimes B, C, A \otimes C, B \otimes C, A \otimes B \otimes C$. We rename, taking account of the nesting, as μ, A, AB, C, AC, ABC.

How is this useful? It is obvious. Suppose we have a balanced population, the number of levels of each nested partition within a level of a nestor partition is constant and in which the numbers of individuals in the subsets given by a partition are all equal. Then we have a single ANOVA:

Source
μ
A
AB
C
AC
ABC
Residual.

How can we see this? Let us label individuals by the levels of factors with a certain rule. We may label by levels of crossed factors and by levels of nested factors within levels of factors or partitions that nest each nested factor. So in this case we shall index by i level of A, j level of B within level of A and k level of C and finally by l= index of individual within the ultimate cells. Then we have:

$$\begin{aligned} y_{ijkl} = y_{....} &+ (y_{i...} - y_{....}) \\ &+ (y_{ij..} - y_{i...}) \\ &+ (y_{..k.} - y_{....}) \\ &+ (y_{i\cdot k\cdot} - y_{i...} - y_{..k.} + y_{....}) \\ &+ (y_{ijk\cdot} - y_{ij..} - y_{i\cdot k\cdot} + y_{i...}) \\ &+ (y_{ijkl} - y_{ijk\cdot}). \end{aligned}$$

General theory of this way of thinking was written out extensively by Zyskind (1958, 1962).

7. Linear classificatory models

The idea is very simple. We have factors or partitions. We have product partitions. Every partition, simple or product, separates our individuals into disjoint subsets. So we envisage a contribution in the model from every partition and every product partition.

Suppose we have the structure:

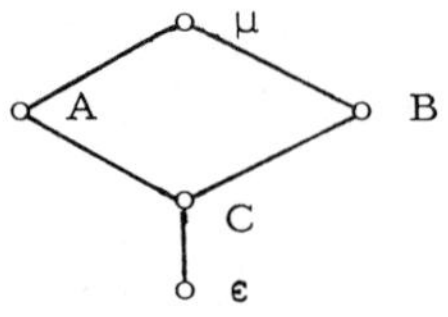

Then the partitions, using our renaming of partitions, are μ, A, B, AB, ABC.

Suppose now that we use conventional labelling, then our observations are y_{ijkl}, with i, j, k labelling levels of factors A, B and C. The associated full linear model is then

$$y_{ijkl} \doteq \mu + a_i + b_j + ab_{ij} + abc_{ijk}.$$

We can write this matricially as

$$y \doteq \mathcal{I}\mu + X_a a + X_b b + X_{ab} ab + X_{abc} abc.$$

From this way of thinking, I get to the idea of a well-formulated model.

A model is well formulated if given that it contains a contribution associated with partition π, it contains terms associated with all partitions that nest that partition π.

Consider a simple structure:

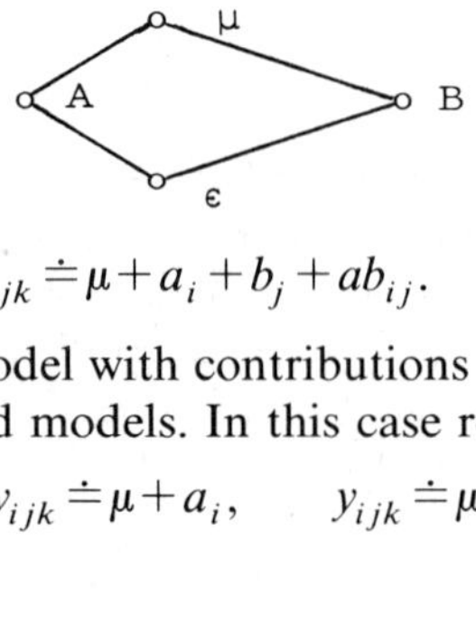

$$y_{ijk} \doteq \mu + a_i + b_j + ab_{ij}.$$

Then from a full model, i.e. a model with contributions from every partition, we can obtain what we may call reduced models. In this case reduced models are:

$$y_{ijk} \doteq \mu + a_i + b_j, \qquad y_{ijk} \doteq \mu + a_i, \qquad y_{ijk} \doteq \mu + b_j, \qquad y_{ijk} \doteq \mu.$$

What about

$$y_{ijk} \doteq \mu + a_i + ab_{ij}?$$

With the way I think about this area and the idea of well-formulated models, this is not well formulated. Why? Because the partition $A \otimes B$ is entertained as potentially contributing effects. But the $A \otimes B$ partition is nested by the A partition and the B partition. So if a model contains ab terms, it must contain "a" terms and "b" terms.

We can apply the ideation to matricial representation very handily. Suppose we label the levels of our factors; and we label levels of nested factors within levels of factors or partitions that nest a factor, and then suppose we index individuals by the levels of partitions within which they lie. Then we achieve a very simple matricial form for the overall model

$$y \doteq \mathcal{I}\mu + X_a a + X_b b + X_{ab} ab$$
$$= \mathcal{I}\mu + X_a a + X_b b + X_a \odot X_b.$$

Here we have the property, in which u labels individuals and i, j the columns of X_a and X_b, and we label columns of X_{ab} lexicographically

$$(X_a \odot X_b)_{u,ij} = (X_a)_{u,i} \cdot (X_b)_{u \odot j}.$$

Why do I call

$$y_{ijk} \doteq \mu + a_i + ab_{ij}$$

not well formulated?

From one viewpoint, define numbers $\{w_j\}, \Sigma w_j = 1$, and then define

$$b_j = \sum_j w_j ab_{ij}, \qquad ab^*_{ij} = ab_{ij} - b_j,$$

and our model is

$$y \doteq \mathcal{I}\mu + a_i + b_j + ab^*_{ij},$$

and we see that mere reparametrization brings in B effects.

From another, equivalent viewpoint if we write

$$y \doteq \mathcal{I}\mu + X_a a + X_{ab} ab,$$

then $\mathcal{C}(X_b) \subset \mathcal{C}(X_{ab})$. So even if one thinks one is not including effects of the B factor, one is doing so.

8. The adjoining of conditions on parameters

Let me give a simple example

$$y_{ij} \doteq \mu + a_i.$$

This is non-full rank. Now let us adjoin conditions on parameters,

$$\sum a_i = 0.$$

We then have 2 candidate models

I	II
$y_{ij} \doteq \mu + a_i$	$y_{ij} \doteq \mu + a_i$ $\sum a_i = 0.$

Are these models the same? *Obviously not*. Are the models essentially equivalent? *What should one mean by essentially equivalent*? Is this a matter of column spaces? NO. It is a matter of row spaces.

Let us consider:

I	II
$y \doteq X\beta$	$y \doteq X\beta$
	$C\beta = c$
$\lambda'\beta$ is estimable	$\lambda'\beta$ is estimable
$\lambda' \in \mathcal{R}(X)$	$\lambda' \in \mathcal{R}\begin{pmatrix} X \\ C \end{pmatrix}$.

Suppose $\mathcal{R}(C) \cap \mathcal{R}(X) = \{\varphi\}$. Then I shall say that these models are essentially equivalent. They are equivalent datawise. In Case I, the NE are $X'Xb = X'y$ giving $\lambda'\beta = \rho' X' y$, where $X'X\rho = \lambda$. In Case II, the NE are

$$X'Xb + C'm = X'y,$$
$$Cb = c,$$

and under the conditions these are equivalent to

$$X'Xb = X'y,$$
$$Cb = c,$$

so the only addition is that provided by the model formulation and *not* by the data.

The argument here is related, of course, to the impositions of conditions on solutions of the NE for Model I, as done early by Yates (1935). If adjoining $Cb = c$ retains consistency of the NE, then $\rho' C\beta$ is not identifiable or estimable unless $\rho' C = \text{null}$.

I shall not go into details of the elementary mathematics of Model II, except to note that if an inverse of

$$\begin{pmatrix} X'X & C' \\ C & \phi \end{pmatrix} \text{ is } \begin{pmatrix} D & E \\ E' & F \end{pmatrix},$$

then

$$X\hat{\beta} = XDX'y + XEc$$

and XDX' is a symmetric projection matrix. The mathematics here is very similar to that of Rao (1971) with his general theory, and naturally so.

As a final remark, I suggest

Definition The models

$$\text{I:} \quad y \doteq X_1\theta_1$$
$$C_1\theta_1 = \varphi,$$

and

$$\text{II:} \quad y \doteq X_2\theta_2$$
$$C_2\theta_2 = \varphi$$

are essentially equivalent if

$$\mathcal{C}[X_1(I - C_1^- C_1)] = \mathcal{C}[X_2(I - C_2^- C_2)].$$

This type of thinking is useful, I believe, because it helps to "make sense" of the many different ways of writing models. For example, we see

$$\text{I:} \quad y_{ijk} \doteq \mu + a_i + b_j$$

$$\text{II:} \quad y_{ijk} \doteq \mu + a_i + b_j$$

$$\sum a_i = 0, \qquad \sum b_j = 0$$

$$\text{III:} \quad y_{ijk} \doteq \tau_{ij}$$

$$\tau_{ij} - \tau_{i'j} - \tau_{ij'} + \tau_{i'j'} = 0, \quad \forall i, i', j, j'.$$

It seems clear that I or II are preferable for most, if not all purposes.

9. Conditioning in a multipart ordered model

It is useful, perhaps, to give the following. Consider

$$\text{I:} \quad y \doteq X_1\beta_1 + X_2\beta_2 + \cdots + X_k\beta_k$$

and

$$\text{II:} \quad y \doteq X_1\beta_1 + X_2\beta_2 + \cdots + X_k\beta_k$$

$$C_1\beta_1 = \varphi$$

$$C_2\beta_2 = \varphi$$

$$\cdots\cdot$$

$$C_k\beta_k = \varphi.$$

Then I suggest that we define these as essentially equivalent if

$$\mathcal{R}(X_1) \cap \mathcal{R}(C_1) = \{\varphi\},$$

$$\mathcal{R}(X_2) \cap \mathcal{R}(C_2) = \{\varphi\},$$

and so on.

10. Fixed and random models

The mode of presentation of classificatory linear models that I give above leads naturally to a way of addressing models with random factors, in that a random factor is one with an infinite number of levels, of which we observe a finite number. This point of view has been pursued extensively by my coworkers, starting from Wilk (1955) and Wilk and Kempthorne (1955).

This viewpoint is clearly different from that of Nelder (1977), but I shall not discuss this here.

References

Carney, E.J. (1967) Computation of variances and covariances of variance component estimates. Ph.D. Thesis.

Kempthorne, O. (1952). *The Design and Analysis of Experiments*. Wiley, New York. Reprinted by Krieger, 1973.

Kempthorne, O. and J.L. Folks (1971). *Probability, Statistics, and Data Analysis*. Iowa State University Press.

Nelder, J.A. (1965*a*). The analysis of randomized experiments with orthogonal block structure. I. Block structure and the null analysis of variance. *Proc. R. Soc. A* **283**, 147–162.

Nelder, J.A. (1965*b*). The analysis of randomized experiments with orthogonal block structure. II. Treatment structure and the general analysis of variance. *Proc. R. Soc. A* **283**, 163–178.

Nelder, J.A. (1977). A reformulation of linear models. *J. R. Statist. Soc. A* **140**, 48–63.

Rao, C.R. (1971). Unified theory of linear estimation. *Sankhyā A* **33**, 371–394.

Scheffé, H. (1959). *The Analysis of Variance*. Wiley, New York. p. 43.

Throckmorton, T.N. (1961). Analysis of Variance Procedures. Aeronautical Research Laboratory, Office of Aerospace Research, United States Air Force. Wright-Patterson Air Force Base, Ohio. 1961. (With O. Kempthorne, G. Zyskind, S. Addelman, R.F. White.)

Throckmorton, T.N. (1961). Structures of Classificatory Data. Ph.D. Thesis.

Wilk, M.B. (1955). Linear Models and Randomized Experiments. Ph.D. Thesis.

Wilk, M.B. and O. Kempthorne (1955). Fixed, mixed, and random models. *JASA* **50**, 272, 1144–1167.

Zyskind, G. (1958). Error Structures in Experimental Designs. Ph.D. Thesis.

Zyskind, G. (1962). On structure, relation, Σ, and expectation of mean squares. *Sankhyā 24* 115–148.

G. Kallianpur, P.R. Krishnaiah, J.K. Ghosh, eds., *Statistics and Probability: Essays in Honor of C.R. Rao*
© North-Holland Publishing Company (1982) 411–417

A THEOREM ON QUADRATIC FORMS FOR NORMAL VARIABLES

C.G. KHATRI
Gujarat University, Ahmedabad, India

Let x be distributed as $N(\mu, I)$ and let $q=x'Ax$ be a quadratic form such that $q=\Sigma_{i=1}^{k} q_i$ where $q_i=x'A_ix$ $(i=1,2,\dots,k)$. In this connection, a natural extension of the result of Graybill and Marsaglia (1957) is given for the quadratic forms to be distributed as linear functions of independent chi-square variates. The results are obtained by modifying some of the conditions given by Khatri (1977). Some results are mentioned when the covariance matrix of x is V, and the vector x is replaced by the matrix X.

1. Introduction

When x is distributed as $N(\mu, I_n)$ and a quadratic form $q=x'Ax$ is decomposed into k quadratic forms, $q_i=x'A_ix$ $(i=1,2,\dots,k)$ such that $q=\Sigma_{i=1}^{k} q_i$. Then, Graybill and Marsaglia (1957) established the following:

Theorem 1. *Consider the conditions*:

(a) q_i *are distributed as chi-squares (central or non-central) for all* $i=1,2,\dots,k$;
(b) q_i *and* $q_{i'}$ *are independently distributed for all* $i\neq i'$, $i, i'=1,2,\dots,k$;
(c) q *is distributed as chi-square (central or non-central) and*
(d) Rank $A=\Sigma_{i=1}^{k}$ Rank A_i.
Then:
(i) (c) *and* (d) *imply all conditions and*
(ii) any two of (a), (b) *and* (c) *imply all conditions.*

The above result was extended by Khatri (1977) and is given by:

Theorem 2. *Let* $\lambda_1, \lambda_2, \dots, \lambda_m$ *be distinct non-zero real numbers and* $m>1$. *Consider the following conditions*:

(a) q_i *is distributed as* $\Sigma_{j=1}^{m} \lambda_j \chi_{ij}^2(f_{ij}, v_{ij})$, *for all* $i=1,2,\dots,k$, *where* χ_{ij}^2*'s are independent non-central chi-square variates for* $j=1,2,\dots,m$ *(some of* f_{ij}*'s may be zero).*

(b) q_i *and* $q_{i'}$ *are independently distributed for* $i\neq i'$, $i, i'=1,2,\dots,k$.

(c) q *is distributed as* $\Sigma_{j=1}^{m} \lambda_j \chi_j^2(f_j, v_j)$ *where* χ_j^2*'s are independent non-central chi-square variates.*

(d) Rank $A=\Sigma_{i=1}^{k}$ Rank A_i.

(e) Rank $A=\Sigma_{i=1}^{k}\Sigma_{j=1}^{m}$ Rank(A_iE_j) *where* $A=\Sigma_{j=1}^{m}\lambda_jE_j$, $E_j^2=E_j$ *and* $E_jE_{j'}=0$ *for* $j\neq j'$, $j, j'=1,2,\dots,m$.

Then:

(i) (a) *and* (b) *imply all conditions*;

(ii) (a), (c) *and* (d) *imply all conditions, either for* $m=2$ *if* λ_j*'s are of the same sign or for* $m=2$ *and* $m=3$ *if* λ_j*'s are of different signs*;

(iii) (b) *and* (c) *imply all conditions and*

(iv) (c), (d) *and* (e) *imply all conditions.*

If we compare Theorem 2 with Theorem 1 for $m=1$, then the results (i), (iii) and (iv) are the same as Theorem 1, but the result (ii) is not appropriate. Hence, in order to get the same result as Theorem 1 for $m=1$, we modify the two conditions (d) and (e) by the following conditions, (d′) and (e′), namely:

(d′) Rank $A=\Sigma_{i=1}^{k}$ Rank $A_i=\Sigma_{i=1}^{k}\Sigma_{j=1}^{m}$ Rank(A_iE_j) where $A=\Sigma_{j=1}^{m}\lambda_jE_j$ with $E_j^2=E_j$ and $E_jE_{j'}=0$ for $j\neq j'$, $j,j'=1,2,\ldots,m$; and

(e′) any $(m-1)$ relations from the m relations $f_j=\Sigma_{i=1}^{k}f_{ij}$ for $j=1,2,\ldots,m$ are valid.

Then, we get the following results:

(i) (a) and (b) $\Rightarrow$ all conditions;

(ii) (a), (c) and (e′) $\Rightarrow$ all conditions;

(iii) (b) and (c) $\Rightarrow$ all conditions and

(iv) (c) and (d′) $\Rightarrow$ all conditions.

Notice that when $m=1$, the result (ii) is the same as corresponding result of Theorem 1 and all other results are the same as Theorem 2. Thus, the modified result (called Theorem 3) is the real extension of Theorem 1.

Some extensions of the above result are given when the covariance matrix of x is V and the vector x is replaced by the matrix X.

2. Proof of Theorem 3

Since the results of Theorem 3, namely, (i), (iii) and (iv) are established by Khatri (1977), we shall only establish the result (ii), namely, (a), (c) and (e′) $\Rightarrow$ (b) and (d′). In the proof, we require the following lemmas:

Lemma 1. *Let* $B=(b_{ij})$ *be* $n\times n$ *positive definite matrix. Then* $|B|=\Pi_{i=1}^{n}b_{ii}$ *iff* $B=\mathrm{diag}(b_{11},b_{22},\ldots,b_{nn})$.

Proof. Let

$$B=\begin{pmatrix} b_{11} & b'_{(n)} \\ b_{(n)} & B_{(11)} \end{pmatrix},\qquad B_{(11)}=\begin{pmatrix} b_{22} & b'_{(n-1)} \\ b_{(n-1)} & B_{(22)} \end{pmatrix},\ldots,$$

$$B_{(n-2,n-2)}=\begin{pmatrix} b_{n-1,n-1} & b_{(2)} \\ b_{(2)} & B_{(n-1,n-1)} \end{pmatrix},\qquad B_{(n-1,n-1)}=b_{nn}\quad\text{with } b_{(2)}=b_{n-1,n}.$$

Then, it is easy to see that:

$$|B| \leqslant b_{11}|B_{(11)}| \leqslant b_{11}b_{22}|B_{(22)}| \leqslant \cdots \leqslant b_{11}b_{22}\cdots b_{nn}$$

where the first equality holds iff $b_{(n)}=0$, the second equality holds iff $b_{(n-1)}=0$ and so on. This proves that $|B|=\prod_{i=1}^{n} b_{ii}$ iff $B=\text{diag}(b_{11},\dots, b_{nn})$.

Lemma 2. *Let $A_1, A_2,\dots,A_k$ and $A=\sum_{i=1}^{k} A_i$ be symmetric (or Hermitian if the elements are complex) matrices. Suppose the following conditions hold*:

(a) *Non-zero distinct eigenvalues λ_j's of A_i are repeated $f_{i1}, f_{i2},\dots,f_{im}$ times respectively and some of the f_{ij}'s may be zero, for $i=1,2,\dots,k$;*

(c) *Non-zero distinct eigenvalues λ_j's of A are repeated $f_1, f_2,\dots,f_m$ times, respectively and*

(e′) *any $(m-1)$ relations from $f_j=\sum_{i=1}^{k} f_{ij}$ $(j=1,2,\dots,m)$ hold.*

Then, we have:

(b) *$A_iA_{i'}=0$ for $i\neq i'$, $i, i'=1,2,\dots,k$ and*

(d′) Rank $A=\sum_{i=1}^{k}$ Rank $A_i=\sum_{i=1}^{k}\sum_{j=1}^{m}$ Rank(A_iE_j) *where $A=\sum_{j=1}^{m}\lambda_jE_j$, $E_j=E_{j'}$ and $E_jE_{j'}=0$ for $j\neq j'$, $j, j'=1,2,\dots,m$.*

Proof. On account of (a) and (c), the spectral decompositions of $A, A_1,\dots, A_k$ can be written as:

$$A=\sum_{j=1}^{m}\lambda_jE_j, \qquad A_i=\sum_{j=1}^{m}\lambda_jE_{ij}$$

where $E_j^2=E_j$, $E_{ij}^2=E_{ij}$, $\operatorname{tr}E_{ij}=f_{ij}$, $\operatorname{tr}E_j=f_j$, $E_{ij}E_{ij'}=0$ and $E_jE_{j'}=0$ for $j\neq j'$, $j, j'=1,2,\dots, m$ and $i=1,2,\dots, k$.

Since $A=\sum_{i=1}^{k} A_i$, we have $\operatorname{tr}A=\sum_{i=1}^{k}\operatorname{tr}A_i$ and hence:

$$\sum_{j=1}^{m}\lambda_jf_j=\sum_{i=1}^{k}\sum_{j=1}^{m}\lambda_jf_{ij}=\sum_{j=1}^{m}\lambda_j\left(\sum_{i=1}^{k}f_{ij}\right).$$

Now, using (e′), we get $f_j=\sum_{i=1}^{k} f_{ij}$ for all $j=1,2,\dots, m$. Further, using spectral decomposition to $A=\sum_{i=1}^{k} A_i$, we get $A=\sum_{j=1}^{m}\lambda_jE_j=\sum_{j=1}^{m}\lambda_jP_j$ with $P_j=\sum_{i=1}^{k} E_{ij}$. Observe that:

$$\sum_{j=1}^{m} f_j=\text{Rank } A\leqslant \sum_{j=1}^{m}\text{Rank } P_j\leqslant \sum_{j=1}^{m}\sum_{i=1}^{k}\text{Rank } E_{ij}=\sum_{j=1}^{m}\sum_{i=1}^{k}f_{ij}=\sum_{j=1}^{m}f_j.$$

Hence,

$$\text{Rank } P_j=f_j=\sum_{i=1}^{k}\text{Rank } E_{ij}=\sum_{i=1}^{k}f_{ij}.$$

Since E_{ij}'s are symmetric idempotent matrices, we can write $E_{ij}=F_{ij}F'_{ij}$ where $F'_{ij}F_{ij}=I_{f_{ij}}$ for $i=1,2,\dots, k$ and $j=1,2,\dots, m$.

Let $F_j=(F_{1j}, F_{2j},\dots, F_{kj})$, omitting F_{ij}'s which are null matrices on account of f_{ij}'s being zero. Then,

$$P_j=F_jF'_j, \quad \text{Rank } F_j=f_j(>0),$$

and the diagonal elements of $F_j'F_j$ are all unities. Define

$$F=(F_1, F_2, \ldots, F_m),$$

and

$$E_j=T_jT_j', \quad T=(T_1, T_2, \ldots, T_m).$$

Then,

$$T'T=I_r, \quad r=\sum_{j=1}^{m} f_j,$$

and the diagonal elements of $F'F$ are unities. Further, if $D=\text{diag}(\lambda_1 I_{f_1}, \ldots, \lambda_m I_{f_m})$, then

$$\sum_{j=1}^{m} \lambda_j E_j=\sum_{j=1}^{m} \lambda_j P_j, \qquad TDT'=FDF',$$

and $R=T'F$ is non-singular. From this, we get:

$$D=RDR' \quad \text{and} \quad (F'F)^{-1}R'DR(F'F)^{-1}=D.$$

Taking determinants, we get $|F'F|=1$ and by Lemma 1, we get $F'F=I_r$. Thus, $\sum_{j=1}^{m} \lambda_j P_j$ is a spectral representation of A and since the spectral representation is unique, we must have,

$$E_j=P_j=\sum_{i=1}^{k} E_{ij}, \qquad E_{ij}E_{i'j'}=0 \quad \text{for } i\neq i' \text{ or } j\neq j', \text{ or both},$$

which proves (b) and (d′). This proves Lemma 2.

The following Lemma 3 is the generalization of Good (1969).

Lemma 3. (a) *Let A be an $n\times n$ symmetric (or Hermitian for complex elements) matrix and let $w_1\neq w_2\neq \cdots \neq w_m\neq 0$ be real numbers such that* $\text{tr}\, A^j=\sum_{i=1}^{m} w_i^j f_i$ *for $j=1,2,\ldots,2m$, where $f_1,\ldots, f_m$ are some positive integers. Further, if*

$$\text{Rank}\left[I-\prod_{j=1}^{m}(I-A/w_j)\right]=\sum_{j=1}^{m} f_j,$$

then, $w_1, w_2, \ldots, w_m$ are the distinct non-zero eigenvalues of A, repeated $f_1,\ldots, f_m$ times, respectively.

(b) *Let A be an $n\times n$ symmetric (or Hermitian for complex elements) matrix and let $w_1\neq w_2\neq \cdots \neq w_m\neq 0$ be real numbers such that* $\text{tr}\, A^j=\sum_{i=1}^{m} w_i^j f_i$ *for $i=1,2,\ldots, nm$ where $f_1,\ldots, f_m$ are some real numbers. Then, $f_1,\ldots, f_m$ are positive integers and $w_1, w_2, \ldots, w_m$ are the distinct non-zero eigenvalues of A, repeated $f_1,\ldots, f_m$ times, respectively.*

Proof. (a) Let $R=I-\prod_{j=1}^{m}(I-A/w_j)$. Then $R=\sum_{j=1}^{m}(-1)^{j+1}c_jA^j$ where $c_j=\sum_{i_1>i_2>\cdots>i_j>1}^{m}(w_{i_1}w_{i_2},\ldots,w_{i_j})$ for $j=1,2,\ldots,m$. Then, it is easy to verify that:

$$\operatorname{tr} R=\operatorname{tr} R^2=\sum_{j=1}^{m} f_j,$$

while $\sum_{j=1}^{m} f_j=\text{Rank } R$. Then, by Shanbhag (1968),

$$R^2=R.$$

This shows that the distinct eigenvalues of R are 0 and 1. Hence, if w is any non-zero eigenvalue of A, then

$$1-\prod_{j=1}^{m}(1-w/w_j)=1 \text{ or } =0$$

or

$$\prod_{j=1}^{m}(1-w/w_j)=0 \text{ or } =1.$$

This gives $w=w_1,w_2,\ldots,w_m,\ w_1',w_2',\ldots,w_\alpha'$ where $w_1',w_2',\ldots,w_\alpha'$ $(\alpha<m-1)$ are the distinct non-zero solutions of

$$\prod_{i=1}^{m}(1-w/w_i)=1.$$

Let us suppose that the above non-zero eigenvalues of A are respectively repeated $r_1,r_2,\ldots,r_m,\ s_1,s_2,\ldots,s_\alpha$ times. Here, some of the r_i's or s_i's may be zero if the corresponding w_i's are not the eigenvalues of A. Then,

$$\sum_{i=1}^{m} w_i^j f_i=\operatorname{tr} A^j=\sum^{m} w_i^j r_i+\sum_{i=1}^{\alpha} w_i'^j s_i \quad \text{for } j=1,2,\ldots,2m$$

or

$$\sum_{i=1}^{m} w_i^j(r_i-f_i)+\sum_{i=1}^{\alpha} w_i'^j s_i=0 \quad \text{for } j=1,2,\ldots,2m.$$

Since $w_1,w_2,\ldots,w_m,\ w_1',\ldots,w_\alpha'$ are all distinct and non-zero, the above equations give the unique solution iff $r_i=f_i$ for $i=1,2,\ldots,m$ and $s_i=0$ for $i=1,2,\ldots,\alpha$. This proves the first part of Lemma 3.

(b) Let us define:

$$R=I-\prod_{j=1}^{m}(I-A/w_j)=\sum_{j=1}^{m}(-1)^{j+1}c_jA^j$$

where c_j is the same as defined in (a). Since $\operatorname{tr} A^j=\sum_{i=1}^{m} w_i^j f_i$ for $j=1,2,\ldots,nm$, we can verify (after some simplifications) that:

$$\operatorname{tr} R=\operatorname{tr} R^2=\cdots=\operatorname{tr} R^n=\sum_{i=1}^{m} f_i.$$

Then, using Good (1969), $\sum_{i=1}^{m} f_i$ is a positive integer and $R^2=R$. Then, using the arguments given in the proof of case (a), we get the required result. This proves Lemma 3.

The referee has raised the question of justifying the results of Shanbhag (1968) and Good (1969) for complex matrix R. It may be noted that the eigenvalues of Hermitian matrix R are real and the above conditions of Shanbhag (1968) and Good (1969) show that the non-zero eigenvalues of R are unity. Hence R becomes an idempotent matrix. This justifies their results to Hermitian matrices.

Lemma 4. *Let x be $N(\mu, I)$ and let $q=x'Ax$. Then for distinct $\lambda_1,\ldots,\lambda_m$, q is distributed as $\Sigma_{j=1}^m \lambda_j \chi_j^2(f_j, v_j)$ iff $\lambda_1,\ldots,\lambda_m$ are the distinct eigenvalues of A, repeated $f_1,\ldots,f_m$ times or the structural representation of A is*

$$A=\sum_{j=1}^{m} \lambda_j E_j, \qquad E_j^2=E_j, \qquad E_j E_{j'}=0 \quad \text{and} \quad \operatorname{tr} E_j=f_j$$

for all $j\neq j'$, $j, j'=1,2,\ldots,m$ [see Khatri (1977, 1980)].

Now, the proof of Theorem 3 becomes complete by using Lemmas 4 and 2. Lemma 3 gives an alternative representation of conditions (a) and (c). The following is the modified Theorem 3 when the covariance matrix of x is V.

Theorem 4. *Let x be distributed as $N(\mu, V)$ and assume that $\mu=Vd$ for some vector d. Let $q=x'Ax$ and $q_i=x'A_ix$ $(i=1,2,\ldots,k)$ be quadratic forms such that $q=\Sigma_{i=1}^k q_i$. Consider the conditions (keep* (a), (b) *and* (c) *the same conditions as given in Theorem* 2):

(d) $\operatorname{Rank}(VAV)=\Sigma_{i=1}^k \operatorname{Rank}(VA_iV)=\Sigma_{i=1}^k \Sigma_{j=1}^m \operatorname{Rank}(VA_iE_j)$ *where* $VAV=\Sigma_{j=1}^m \lambda_j E_j$, $E_jV^-E_j=E_j$ *and* $E_jV^-E_{j'}=0$ *for* $j\neq j'$ *and*

(e) *any* $(m-1)$ *relations from the m relations* $f_j=\Sigma_{i=1}^k f_{ij}$ $(j=1,2,\ldots,m)$ *are true. Then,*

(i) (a) *and* (b) $\Rightarrow$ *all conditions*;
(ii) (a), (c) *and* (e) $\Rightarrow$ *all conditions*;
(iii) (b) *and* (c) $\Rightarrow$ *all conditions and*
(iv) (c) *and* (d) $\Rightarrow$ *all conditions.*

A $p\times n$ matrix X is said to be distributed as $N_{p,n}(\mu, V_1, V_2)$ where μ, V_1 and V_2 are respectively $p\times n$, $p\times p$ and $n\times n$ matrices, iff $x_*=(x_1', x_2',\ldots,x_n')'$ is distributed as $N_{pn}(\mu_*, V)$ where $x_1, x_2,\ldots,x_n$ are the column vectors of X, $\mu_*=(\mu_1',\ldots,\mu_n')'$, μ_i is the ith column of μ and $V=V_2\otimes V_1$ is the Kronecker product of V_2 and V_1. With this notation, we can rewrite Theorem 4 as:

Theorem 5. *Let X be distributed as $N_{p,n}(\mu, V_1, V)$ and $\mu=TV$ for some matrix T. Let $Q=XAX'$, $Q_i=XA_iX'$ $(i=1,2,\ldots,k)$ be such that $Q=\Sigma_{i=1}^k Q_i$. Then, consider the following statements*:

(a) *Q_i is distributed $\Sigma_{j=1}^m \lambda_j W_{ij}(f_{ij}, V_i, \Omega_{ij})$ where $W_{ij}(f_{ij}, V_1, \Omega_{ij})$ are independent non-central Wishart distribution with degrees of freedom f_{ij}, non-central parameter Ω_{ij}*

and scale factor V_1. *Here* f_{ij} *may be zero or may be less than* p;

(b) Q_i *and* $Q_{i'}$ *are independently distributed for all* $i \neq i'$;

(c) Q *is distributed as* $\Sigma_{j=1}^{m} \lambda_j W_j(f_j, V_1, \Omega_j)$;

(d) *the same as condition* (d) *of Theorem* 4 *and*

(e) *the same as condition* (e) *of Theorem* 4.

Then,

(i) (a) *and* (b) $\Rightarrow$ *all conditions*;

(ii) (a), (c) *and* (e) $\Rightarrow$ *all conditions*;

(iii) (b) *and* (c) $\Rightarrow$ *all conditions and*

(iv) (c) *and* (d) $\Rightarrow$ *all conditions.*

References

Good, I.J. (1969). Conditions for a quadratic form to have a Chi-squared distribution. *Biometrika* **56**, 215–216. Corrigenda (1970). *Biometrika* **57**, 225.

Graybill, F.A. and Marsaglia, G. (1957). Idempotent matrices and quadratic forms in the general linear hypothesis. *Ann. Math. Statist.* **28**, 678–686.

Khatri, C.G. (1977). Quadratic forms and extension of Cochran's theorem to normal vector variables. In: Krishnaiah, P.R., Ed., *Multivariate Analysis*-IV. North-Holland, New York, pp. 79–94.

Khatri, C.G. (1980) Quadratic forms. In: Krishnaiah, P.R., Ed., *Handbook of Statistics*. North-Holland, New York, Chapter 8.

Shanbhag, D.N. (1968). Some remarks concerning Khatri's result on quadratic forms. *Biometrika* **55**, 593–595.

G. Kallianpur, P.R. Krishnaiah, J.K. Ghosh, eds., *Statistics and Probability: Essays in Honor of C.R. Rao*

OPTIMUM RATES FOR NON-PARAMETRIC DENSITY AND REGRESSION ESTIMATES, UNDER ORDER RESTRICTIONS*

J. KIEFER (deceased 1981)
University of California, Berkeley, CA, U.S.A.

1. Introduction

Among the many contributions for which Professor Rao will be remembered is his work on the information inequality for estimators in regular parametric (finite-dimensional space of distributions) problems. The lower bound on squared error obtained by that method is useful in certain non-parametric problems as well, by reduction to an appropriate finite-dimensional problem. Čenčov (1962) was the first to announce a lower bound result for a case of the global density estimation problem in which the "best rate" is obtained, and he mentioned the Cramér–Rao inequality as part of the method. Farrell (1967, 1980) pioneered the use of the inequality in the local density estimation problem, and also (1972) the use of the Bayes method to obtain general results on rates in these problems. [The latter method has not always been termed such, but is the principal method used in the lower bound results mentioned herein, except for Farrell (1967, 1980), who alone makes essential use of the inequality in his calculations. Čenčov, Hasminskii, and Samarov make use of a Hajek-type asymptotic efficiency result, which can be proved by asymptotic Bayesian methods; Bretagnolle and Huber (1979) use Kullback–Leibler information in an inequality on risk, but then use the device of Farrell (1972) of estimating a maximum risk from an average.] It is to be emphasized that these lower bounds apply to *all* estimators, not merely usual ones of kernel types, etc. Thus, results on the properties of good kernel estimators are referred to later in this section not because they yield lower bounds for all estimators, but rather because they justify that such bounds give an attainable rate.

Some problems which have been attacked by one or both of these methods are those of density estimation, failure rate estimation, spectral density estimation, and regression function estimation. Moreover, both the local and global performance of estimators have been considered. In principle, at least for squared error loss, there are thus 16 possible combinations of method, problem, performance, not all of which have been carried out. The Bayes method has seemed easiest to apply, both in terms of details of the proof and also in the greater generality of the loss functions it treats; but the information inequality method has thus far yielded the most explicit results in the form of a close-to-best constant [C in (1.2) below, or its analogue for local estimation] that multiplies the power of the sample size, n, shown to be

*Research sponsored by NSF Grant MCS 78-25301 and ONR Grant N00014-75-C-0444.

optimum in these investigations, in Farrell (1980); and Farrell has used Wolfowitz's extension of the information inequality to consider also sequential procedures. These advantages of the two methods may change as more precise (and perhaps sequential) Bayes estimates are used, or as the extensions of Barankin and Rao of the information inequality for other loss functions, are invoked. In any event, it seems worth while to give both developments here, even though the result of Section 4 can also be obtained by the method of Section 2 if one keeps track of the constants.

The various investigations differ in the classes of functions (regularity conditions) imposed, but typically are of the following nature, described here for the problem of local density estimation. The rv's $X_1, X_2, \ldots, X_n$ are iid with common density, f on R^d, assumed to belong to some class $\mathcal{F}$. For global performance, one estimates all of f; for local performance, one estimates $f(b)$ for some specified point b, usually represented as 0. (More generally, estimation of Tf has been considered, where T is a differential operator.) General loss functions have been considered, but simplest are zero–one and squared error. For example, for zero–one loss and local performance, one shows for an appropriate p (depending on d and $\mathcal{F}$) that

$$\lim_{n\to\infty} \sup_{f\in\mathcal{F}} P_f\{|t_n - f(0)| > cn^{-p}\}$$

approaches 1 as $c \to 0$, for every sequence of estimators $t_n = t_n(X_1, \ldots, X_n)$; particular t_n's are then shown to yield 0 as the limit of (1.1) when $c \to \infty$; thus, n^{-p} is the best "rate in probability" possible for t_n, a rough asymptotic minimax result. Similarly, for squared integrated error and global performance, with $t_n(x) = t_n(x; X_1, \ldots, X_n)$ estimating $f(x)$, one shows that

$$\lim_{n\to\infty} n^{2p} \sup_{f\in\mathcal{F}} E_f \int_B [t_n(x) - f(x)]^2 \, dx \geq C, \tag{1.2}$$

where B has positive measure and C is a positive constant depending on B and $\mathcal{F}$; particular t_n's are shown to make the left side of (1.2) finite. [Usually $B = R^d$ in the literature, but we must modify this in Section 3. For the local analogue of (1.2), at the point x, omit the integration in (1.2).] While the assumptions of different authors vary slightly, a typical formulation for local density estimation is that, if k^{th} derivatives of f are assumed bounded uniformly in $\mathcal{F}$ [this can be weakened to a Lipschitz condition on the $(k-1)$st derivative], then $p = k/(2k+d)$ in (1.1) and (1.2). (If Tf is estimated where T is a differential operator of order $s < k$, the corresponding result is $p = (k-s)/(2k+d)$.) The analogue of (1.2) for the maximum deviation of t_n from f on an interval B is, for the same p,

$$\lim_{n\to\infty} (n/\log n)^{2p} \sup_{f\in\mathcal{F}} E_f \sup_{x\in B} [t_n(x) - f(x)]^2 \geq C. \tag{1.3}$$

In some non-parametric settings, particularly ones arising in life testing or reliability theory, f is restricted by the context by an *ordering condition* of a different nature from the *regularity condition* indicated above. Examples of such orderings are f decreasing, (and with $d=1$) f with decreasing failure rate (DFR), f "star-shaped", or f with increasing failure rate on average (IFRA). The problem of estimating the df F corresponding to f, or f itself, or the failure rate $r_f = f/(1-F)$, has been considered frequently; see Barlow et al. (1972). One may wonder, if one imposes such an additional restriction, can one increase the value of p in (1.1) and (1.2)? For example, if $\mathcal{F}$ consists of those f that are decreasing on $(-1,\infty)$ and for which $-f' \leqslant M < \infty$ (as in the $k=1$ condition above), is there a sequence t_n that makes use of the monotonicity of f to achieve some $p > \frac{1}{3}$ in (1.1), or (1.2) for B a compact subinterval of $(-1,\infty)$? The present paper answers "no" to this question, although this does not prove that the best constant C in (1.2) might not be improved by imposing the additional restriction.

In view of the variety of problems, losses, performance characteristics, and regularity and ordering conditions of $\mathcal{F}$ that have been (or can be) considered, no attempt is made herein to treat all cases. Rather, three examples are treated in detail to indicate the type of argument needed to modify existing proofs when an ordering condition is imposed. These are the treatments of Farrell (1967), Stone (1980) and Samarov (1978) for the density estimation problem. Modifications, imposing an order restriction on the class of functions to be estimated, can be made in such works as those of Farrell (1972), Meyer (1977), Bretagnolle and Huber (1979), Hasminskii (1979) and Wahba (1975) for this problem, of those of Stone (1980, 1981) and Ibragimov and Hasminskii (1980) for the regression problem, and of those of Samarov (1977) and Farrell (1980) for the spectral density problem (perhaps the least natural of these on which to impose an order restriction), with varying amounts of difficulty; for example, the choice of $\mathcal{F}$ in Bretagnolle and Huber (1979) (in the explicit form of f_0 of the next paragraph) is such that more alterations are required in the proof to handle (say) decreasing densities than in the presentation of Samarov (1978), although one eventually obtains a more explicit C in (1.2) from the former; the regularity restrictions of $\mathcal{F}$ differ in the two works. The lower bound rate for estimating f, where $F < 1-\varepsilon$, yields the same rate for estimating r_f. We remark that, for some of the classes $\mathcal{F}$ considered by Wahba and by Meyer, the order obtained in the lower bound is not quite attained by any known sequence t_n. In the three treatments considered herein, the order given by the bound is attainable.

The idea of the modification of the Bayes method needed to handle order restrictions, is fairly obvious. The lower bound is typically obtained in the treatments without order restrictions by considering a subfamily of densities of the form $f_0 + q_n$, where f_0 is fixed and q_n is restricted to a sufficiently large finite-dimensional family. By considering $f_0^* + \varepsilon q_n$ where ε is sufficiently small and f_0^* satisfies the order property *strictly* enough (e.g., for decreasing densities, f_0^* and q_n on R^1 with compact support and $(f_0^*)' < -L < 0$) one obtains a family satisfying the order restrictions for which the derivation of the lower bound can be carried through with appropriate modifications. This can be formalized in general in terms of $\mathcal{F}$ (with order restrictions) containing an appropriate ball, but the resulting condition must still be verified in each case in the manner indicated above (and which is used in Sections 2 and 3). This is more difficult for some order restrictions than for others.

The information inequality method (Section 4) requires more effort for modification, since one must go through all the information inequality calculations for the parametric family, satisfying the order restriction, that replaces the family used by Farrell.

We do not discuss in detail here the existence of sequences of estimators satisfying the order restriction and which achieve the optimum rate. (Of course, the rate is achieved by estimators in the literature that take on values f not necessarily satisfying the order restriction, but it seems unsatisfactory to use such estimators.) Some estimators satisfying order restrictions are in the literature (e.g., Barlow et al. (1972), Section 5.4 for local estimation of r_f when $k=2$); it can be verified that others can be obtained by isotonizing standard estimators for problems without restrictions. Some of the many papers that study estimators for the latter problems are listed in the references to the present paper. The papers of Rosenblatt (1971), Wahba (1975), Walter and Blum (1979) and Wegman (1972a, 1972b) contain additional references and summaries of many results. We omit reference to the work of the author and Wolfowitz, and of Millar, on the basically different problem of asymptotically optimum estimation of a d f under order restrictions.

A word is in order regarding estimation of f at or near an endpoint of the support of f, say the left end point x_f in the DFR case. Most of the literature does not handle this case, but Sacks and Ylvisaker (1978) give asymmetric kernels that attain the desired rate for estimation at known x_f or at $x_f+n^{-2/(2k+d)}t$ for $t>0$. Since x_f can be estimated to within order n^{-1} in probability (with corresponding results on the moments of the discrepancy), the S–Y estimator can be used to estimate f at known or unknown x_f in the setting of estimation at a point (Section 2). The family of estimators at $x_f+n^{-2/(2k+d)}t$ seem suitable for the global estimation problem of Section 3 where (1.2) is used with $B=R^1$. When $B=R^1$ (rather than $B=I_1$ of Section 3), (1.3) is no longer valid with x_f unknown, since an error of order 1 is made in estimating f, just to the right or left of x_f.

2. Local density estimation*

For brevity we treat only estimation of $f(0)$, although estimation of $(Tf)\,(0)$ (see Section 1) is handled similarly; also, we describe regularity on f in terms of derivatives, although this can be replaced by Stone's Lipschitz condition. Although the desired conclusion for decreasing densities is a consequence of that for DFR considered in the next two paragraphs, we first consider the almost trivial application of our method to the former case, in order to illustrate the method when the calculations are simplest. Thus, we first consider the problem of estimating $f(0)$ for suitably regular densities f for which f is non-increasing in all d coordinate variables, on its support D_f; let I be a fixed rectangle of R^d with $0\in\operatorname{int} I$, and for k a positive integer let $\mathcal{F}_{D,k,M}$ consist of all such f for which $I\subset\operatorname{int} D_f$ and for which all kth derivatives are bounded in absolute value on $\operatorname{int} D_f$ by $2M<\infty$. For convenience rescale so that $I\subset(-1,1)^d$. Let $f_0(x_1,\ldots,x_d)=h[1-d^{-1}\Sigma_1^d x_i]$ on $D_{f_0}=[-1,1]^d$ where $h=\min(1,M)$ (in case $k=1$). Let $x_0=(\frac{1}{2},\ldots,\frac{1}{2})$. We let g_n be the functions

*Stone (1980).

defined by Stone (1980, Section 3). These are infinitely differentiable, have compact support that shrinks to $\{0, x_0\}$ as $n \to \infty$, have kth derivatives bounded in absolute value by M, have $\lim_n \max_x g_n(x)=0$, and have $\int f_0 g_n \mathrm{d}x=0$. Clearly $f_0(1+\varepsilon g_n)$ has negative first derivatives for n large and ε sufficiently small and positive, and is in $\mathscr{F}_{D,k,M}$. Stone's Bayes argument applies to f_0, $f_0(1+\varepsilon g_n)$ and thus yields $p=k/(2k+d)$ in (1.1) and in the local analogue of (1.2). [Stone's lower bound argument is valid although f_0 here is not continuous. The reason for assuming $D_f \supset I$ is that Stone's or other standard estimators, or modifications of them to make them non-increasing, will then achieve the rate p uniformly on $\mathscr{F}$; if we only assume $0 \in \operatorname{int} D_f$, the proof of uniformity fails because f can have a discontinuity arbitrarily near 0.]

Now suppose $d=1$. The failure rate of f with df F is defined to be $r_f=f/(1-F)$ on int D_f. If f is DFR (r_f non-decreasing), f is decreasing on int D_f and $D_f=[b,+\infty)$ for some finite b. Let $\mathscr{F}_{\mathrm{DFR},k,M}$ be the subset of $\mathscr{F}_{D,k,M}$ with DFR. Now let $f_0(x)=L(x+2)^{-(L+1)}$ on $D_{f_0}=[-1,\infty)$ where $L>0$ is small enough that $|f_0^{(k)}|\leqslant M$ on $(-1,\infty)$. Then $r_{f_0}(x)=L(x+2)^{-1}$ and $r'_{f_0}(x)=-L(x+2)^{-2}$. For g_n as above, defined w.r.t. this f_0 and with support taken to be in $[-\frac{1}{2},1]$, the failure rate of $f_0(1+\varepsilon g_n)$ is $f_0(1+\varepsilon g_n)/[1-F_0-\varepsilon\int_{-\infty}^{x} f_0 g_n]=f_0(1+\varepsilon g_n)/U_n$ (say), with derivative

$$U_n^{-2}\{U_n[f_0'+\varepsilon(f_0 g_n)']+[f_0(1+\varepsilon g_n)]^2\}. \tag{2.1}$$

For $x>1$, $f_0 g_n$ vanishes; and for ε small, (2.1) on $(-1,1]$ is uniformly within $L/10$ of r'_{f_0}, while $r'_{f_0}\leqslant -L/9$ on $(-1,1]$. Thus, for ε sufficiently small, $f_0(1+\varepsilon g_n)$ is in $\mathscr{F}_{\mathrm{DRF},k,M}$. Again applying Stone's result, we conclude that $p=k/(2k+1)$ for $\mathscr{F}_{\mathrm{DFR},k,M}$.

A natural generalization of DFR to $d>1$ is obtained by defining $G_f(x)=\int_{H_x} f(u)\mathrm{d}u$ where $H_x=\{u: u_i \geqslant x_i \forall i\}$, $r_f=f/G_f$, and f to be DFR if r_f is non-increasing on int D_f. Taking $f_0(x)=L^d\prod_1^d(x_i+2)^{-(L+1)}$ on $D_{f_0}=[-1,\infty)^d$, one sees that a development like that above yields $p=k/(2k+d)$ for $\mathscr{F}_{\mathrm{DFR},k,M}$, the subset of DFR densities of $\mathscr{F}_{D,k,M}$.

Such classes as densities increasing on D_f, or with increasing failure rate (IFR) or IFR on average or star-shapedness can be treated by the same method. The same derivation of optimum rate holds for estimation of f at an endpoint (known or unknown) of D_f; see also the end of Section 1.

3. Global density estimation*

For $d=1$, let I_1 and I_2 be fixed intervals with $I_1 \subset \operatorname{int} I_2$ and I_1 of positive (perhaps infinite) length. Let $\mathscr{F}_{\mathrm{DFR},k,M}$ consist of densities f for which $D_f \supset I_2$, f has DFR on int D_f, and the kth derivative satisfies $|f^{(k)}|\leqslant M<\infty$ on I_2. (Samarov considers a more general Lipschitz condition, as well as an alternative integral-Lipschitz condition; for brevity we omit consideration of these, but the treatment that follows works for these regularity conditions.) The reason for the formulation involving I_1 and I_2 is that we want to show the optimum rate has nothing to do with the discontinuity at

*Samarov (1978).

the left end of D_f, and conventional estimators suffice for estimation on I_1. Thus, to obtain a rate attainable by a known method of estimation (suitable kernel with compact support), or by a suitable DFR modification thereof, we consider (1.2) with $B=I_1$. This causes only obvious modifications in Samarov's proof for $B=R^1$, which we do not discuss. The lower bound of course also holds with $B=R^1$; see the end of Section 1 regarding attainment of the rate in that case.

We verify that $p=k/(2k+1)$ in (1.2) for $\mathscr{F}_{\mathrm{DFR},k,M}$, and this order is consequently also optimum for non-increasing densities satisfying the same regularity restrictions. Take $I_2=[0,\infty)$, $I_1\subset\operatorname{int} I_2$, and $f_0(x)=L(x+2)^{-(L+1)}$ on $D_f=[-1,\infty)$ as in Section 2. Let ϕ be a bounded function with support $[-1,1]$, bounded kth derivative, and $\int_{-1}^{1}\phi(x)\mathrm{d}x=0$. Let $[a,b]$ be a non-degenerate interval interior to I_1, and, with m the integral part of $n^{1/(2k+1)}(b-a)/2$, let $x_{n,j}=a+n^{-1/(2k+1)}(2j-1)$ and $\phi_{n,j}(x)=n^{-k/(2k+1)}\phi(n^{1/(2k+1)}(x-x_{n,j}))$ for $1\leqslant j\leqslant m$. Following Samarov, we let $\theta=(\theta_1,\ldots,\theta_m)$ and $f_\theta=f_0+\varepsilon\Sigma_{j=1}^{m}\theta_j\phi_{n,j}$ where $\theta\in\Omega_n=\{\theta:\max_{1\leqslant i\leqslant m}|\theta_i|\leqslant 1\}$. The functions $\Sigma_j\theta_j\phi_{n,j}$ have the role of the f_0g_n of Section 2, and it is easily seen that the $\phi_{n,j}$ have disjoint supports in $[a,b]$ and that for ε sufficiently small (by an argument like that of Section 2) $f_\theta\in\mathscr{F}_{\mathrm{DFR},k,M}\ \forall\theta\in\Omega_n$. Samarov's argument then yields the desired result.

We remark that for $d>1$ Samarov's argument can be extended to yield $p=k/(2k+d)$ and the f_0 of Section 2 yields the desired result in this case. Indeed, Bretagnolle and Huber have sketched the modification to the case $d>1$ of their development for $d=1$, which shares some elements (a similar class of f_θ's, but with each $\theta_i=\pm 1$) with Samarov's development.

The development for DFR above applies also to the proof of (1.3) by Hasminskii (1979b). In fact, the latter works with essentially the above f_θ's but with a simpler subset of the above Ω_n.

4. Information inequality development*

The author is indebted to Roger Farrell for pointing out that differentiation under the integral sign is not justified for the particular family considered in his 1967 paper. However, his basic outline of *method* is correct and can be applied to more regular families to obtain his main conclusion on lack of uniform consistency. The DFR family $\{g_{\gamma,\theta}\}$ defined below is such a family, and it yields the desired conclusion (of our being unable to estimate $g_{\gamma,\theta}(0)$ with uniformly small risk, for *any* sample size). This family has support depending on γ (*not* the differentiation parameter); we spend additional time to define $\{f_{\gamma,\theta}\}$ with support $[-\frac{1}{4},\infty)$ for all members of the family because the conclusion is then more satisfactory.

In all that follows, $\Omega=\{(\gamma,\theta):0<\gamma<\frac{1}{8},\ 3\gamma<\theta<4\gamma\}$. The development will take γ as known but arbitrary in the interval $0<\gamma<\frac{1}{8}$, and the information inequality will be used w.r.t. θ. We first define densities $g_{\gamma,\theta}$ with support $[-\gamma,\infty)$ and then densities $h_{\gamma,\theta}$ with support $[-\frac{1}{4},-\gamma]$. These are defined so that $g_{\gamma,\theta}(-\gamma)=h_{\gamma,\theta}(-\gamma)$.

*Farrell (1967, 1980).

Let $h^* = h$ except that $h^*(-\gamma)=0$. Since the densities g and h are each continuous on their support, are bounded uniformly on Ω, and are piecewise continuously differentiable, it follows that $f_{\gamma,\theta}=\frac{1}{2}(g_{\gamma,\theta}+h^*_{\gamma,\theta})$ is also bounded uniformly on Ω, and is continuous and piecewise continuously differentiable on $[-\frac{1}{4},\infty)$. As Farrell (1967, p. 472) points out, the result for $\{f_{\gamma,\theta}\}$ then implies it for continuously differentiable f. Making a scale transformation (to alter the bound on f), we obtain the

Theorem. *For $M>0$, there is a value $b<0$ such that, if $\mathcal{F}$ consists of the DFR densities bounded by M and with continuous derivative on (b,∞), then for some $c>0$*

$$\inf_{t_n}\sup_{f\in\mathcal{F}} E_f(t_n-f(0))^2>c, \tag{4.1}$$

for every $n>0$.

In fact, it is not difficult to modify the definition of $f_{\gamma,\theta}$ to allow b, in the theorem, to be specified, $-\infty<b<0$; we omit the extra arithmetic. Also, the corresponding sequential statement (below (4.1), "for every stopping variable n with $\sup_f En<M'$ where $M'<\infty$" can be verified as in Farrell's development. We have not bothered herein to obtain a sharp constant for $\{f_{\gamma,\theta}\}$, with which to replace the right side of (4.1).

Proof of theorem. We define

$$g_{\gamma,\theta}(x)=\gamma(\theta-\gamma)^{\gamma}(x+\theta)^{-\gamma-1}, \quad \text{on } D_g=[-\gamma,\infty). \tag{4.2}$$

Then $r_g(x)=\gamma(x+\theta)^{-1}$ on D_g, so g is DFR. A direct calculation yields the Fisher information

$$I_g=-E_{\gamma,\theta}\partial^2\log g_{\gamma,\theta}(X)/\partial\theta^2=\gamma/(\gamma+2)(\theta-\gamma)^2. \tag{4.3}$$

We write $J=g_{\gamma,\theta}(-\gamma)=\gamma/(\theta-\gamma)$.

We define

$$h_{\gamma,\theta}(x)=\begin{cases}\alpha(x+\beta)^{-2}, & \text{if } -\frac{1}{4}\leqslant x\leqslant -2\gamma,\\ A-B(x/\gamma)+C(x/\gamma)^2, & \text{if } -2\gamma\leqslant x\leqslant -\gamma,\end{cases} \tag{4.4}$$

where

$$\beta=\tfrac{1}{4}+\frac{(\frac{1}{4}-2\gamma)^2}{\frac{1}{4}+\frac{17}{12}\gamma}, \qquad \alpha=2(\beta-2\gamma)^2;$$
$$A=-4+8J, \qquad B=5-10J, \qquad C=3J-1. \tag{4.5}$$

It can be verified that the definitions (4.5) make (4.4) a density, make the two lines of (4.4) have the common value 2 at $x=-2\gamma$, and make $h_{\gamma,\theta}(-\gamma)=J=g_{\gamma,\theta}(-\gamma)$. Note that $h_{\gamma,\theta}(x)$ does not depend on θ for $-\frac{1}{4}\leqslant x\leqslant -2\gamma$. On $-2\gamma\leqslant x\leqslant -\gamma$, $h_{\gamma,\theta}(x)=P_1(x/\gamma)+P_2(x/\gamma)J$, where the P_i's are quadratic polynomials with constant coefficients. Thus, on $(-2\gamma,-\gamma)$, we have $\partial h_{\gamma,\theta}(x)/\partial\theta=-P_2(x/\gamma)\gamma/(\theta-\gamma)^2$. Since $\frac{1}{3}<J<\frac{1}{2}$ on Ω, we have $B>0, C>0$, and hence $h_{\gamma,\theta}$ is decreasing on $[-2\gamma, -\gamma]$ where it has minimum $h_{\gamma,\theta}(-\gamma)=J>\frac{1}{3}$. Consequently, if $\max_{1\leqslant u\leqslant 2}|P_2(u)|=K$, we have

$$I_h=\int_{-2\gamma}^{-\gamma}\{[\partial h_{\gamma,\theta}(x)/\partial\theta]^2/h_{\gamma,\theta}(x)\}\,dx\leqslant 3\gamma K^2\left[\gamma/(\theta-\gamma)^2\right]^2<K^2/\gamma. \quad (4.6)$$

Defining $f=(g+h^*)/2$ as described earlier, we have $I_f=(I_g+I_h)/2$ since the supports of g and h^* are disjoint. From (4.3) and (4.6), $I_f<K'/\gamma$ on Ω where $K'=\frac{1}{2}(K^2+1)$. We define

$$\begin{aligned} Q=\partial f_{\gamma,\theta}(0)/\partial\theta&=\gamma(\theta-\gamma)^{\gamma-1}\theta^{-\gamma-2}[\gamma\theta-(\gamma+1)(\theta-\gamma)]\\ &<-K''/\gamma, \end{aligned} \quad (4.7)$$

where $K''>0$. The information inequality for an estimator t_n of $f(0)$ with bias $b(\gamma,\theta)=E_{\theta,\gamma}t_n-f_{\gamma,\theta}(0)$ is

$$E_{\theta,\gamma}\left(t_n-f_{\gamma,\theta}(0)\right)^2\geqslant b^2+(Q+\partial b/\partial\theta)^2/nI_f. \quad (4.8)$$

Following Farrell, if there is a θ in $(3\gamma,4\gamma)$ for which $\partial b/\partial\theta\leqslant -Q/2$ (positive), we see that the right side of (4.1) is $\geqslant Q^2/4nI_f$ at that θ. If there is no such θ, then b is increasing in θ on $(3\gamma,4\gamma)$, with derivative $>-Q/2$, and hence at some θ in that interval we have $|b|>(\gamma/2)(-Q/2)$ and the right side of (4.1) is $\geqslant Q^2\gamma^2/16$ at that θ. Thus, since $I_f<K'/\gamma$ and $Q<-K''/\gamma$, we obtain

$$\begin{aligned} \max_{3\gamma<\theta<4\gamma} E_{\theta,\gamma}(t_n-f_{\gamma,\theta}(0))^2&\geqslant\\ &\geqslant\min(Q^2/4nI_f,\, Q^2\gamma^2/16)\\ &>(K'')^2\min(1/4K'n\gamma,\tfrac{1}{16}). \end{aligned} \quad (4.9)$$

Taking the supremum over $0<\gamma<\frac{1}{8}$ yields the right side of (4.1), but we must still show f is DFR.

We write F, H for the df's of f, h. Since $f=g/2$ on $[-\gamma,\infty)$, $r_f=r_g$ there and thus f has DFR on $[-\gamma,\infty)$. On $[-\frac{1}{4},-\gamma]$, $r_f=h/2[1-H/2]$, so $(2-H)^2r_f'=(2-H)h'+h^2\leqslant h'+h^2$ since $h'<0$. We shall show $h'+h^2<0$ on $[-\frac{1}{4},-\gamma]$.

On $[-\frac{1}{4},-2\gamma]$ we have, for $0<\gamma<\frac{1}{8}$,

$$\begin{aligned} \alpha^{-1}(x+\beta)^4\left[h'(x)+h^2(x)\right]&=-2(x+\beta)+\alpha\\ &\leqslant\tfrac{1}{2}-2\beta+\alpha\\ &=2(\tfrac{1}{4}-2\gamma)^2(\tfrac{1}{2}+\tfrac{17}{6}\gamma)^{-2}(-8\gamma+\tfrac{49}{36}\gamma^2)<0. \end{aligned} \quad (4.10)$$

On $[-2\gamma,-\gamma]$ we have

$$\begin{aligned} h'+h^2&=-B/\gamma+2Cx/\gamma^2+\left[A-B(x/\gamma)+C(x/\gamma)^2\right]^2\\ &<-(B+2C)/\gamma+(A+2B+4C)^2\\ &\leqslant-(3-4J)/\gamma+4<-1/\gamma+4<0. \end{aligned} \quad (4.11)$$

This completes the proof.

Farrell (1980) uses the information inequality method to obtain the lower bound [local analogue of (1.2)] when $k=2$, with a determination of a C that is close to best possible. One can define a family of DFR densities to obtain the right rate in that case using the present method, but without invoking Farrell's technique to obtain a good constant. The derivation is tedious compared to that of Section 2. However, for decreasing (rather than DFR) densities there may be hope that a reasonable C can be obtained by Farrell's method.

Acknowledgements

The author is grateful to Jerry Sacks for pointing out the applicability of the S–Y estimators in Section 1.

References

[1] Alekseev, V. (1974). On uniform convergence of spectral density estimates of a Gaussian stationary random process. *Teor. Veroyatnost. i Primenen.* **19**, 198–206.

[2] Barlow, R.E., Bartholomew, D.J., Bremner, J.M. and Brunk, H.D. (1972). *Statistical Inference Under Order Restrictions.* Wiley, London.

[3] Bickel, R.J. and Rosenblatt, M. (1973). On some global measures of the deviations of density function estimates. *Ann. Statist.* **1**, 1071–1095.

[4] Bretagnolle, J. and Huber, C. (1979). Estimation des densités: risque minimax. *Z. Wahrsch.* **47**, 119–137.

[5] Cencov, N.N. (1962). Evaluation of an unknown distribution density from observations. (Translation) *Soviet Math.* **3**, 1559–1562.

[6] Farrell, R.H. (1967). On the lack of a uniformly consistent sequence of estimators of a density function in certain cases. *Ann. Math. Statist.* **38**, 471–474.

[7] Farrell, R.H. (1972). On best obtainable asymptotic rates of convergence in estimation of a density function at a point. *Ann. Math. Statist.* **43**, 170–180.

[8] Farrell, R.H. (1979). Asymptotic lower bounds for the risk of estimators of the value of a spectral density function. *Z. Wahrsch.* **49**, 221–234.

[9] Farrell, R.H. (1980). On the efficiency of density function estimators. (To appear.)

[10] Hasminskii, R.Z. (1979a). Lower bound for the risk of nonparametric estimates of the mode. In *Contribution to Statistics* (Hajek Memorial Volume) Jurečkova, J. Ed., pp. 91–97. Academia, Prague.

[11] Hasminskii, R.Z. (1979b). A lower bound for risks of nonparametric estimates of density in the uniform metric. *Teor. Veroyathost. i Primenen.* **24**, 824–828.

[12] Hasminskii, R.Z. and Samarov, A. (1978). On the quality of some non-parametric estimates in uniform metric. (To appear.)

[12a] Ibragimov, I.A. and Hasminskii, R.Z. (1980). On nonparametric estimation of regression. *Soviet Math. Dokl.* **21**, 810–814.

[13] Leadbetter, M.R. and Watson, G.S. (1963). On the estimation of a probability density, I. *Ann. Math. Statist.* **34**, 480–491.

[14] Meyer, T.G. (1977). Bounds for estimation of density functions and their derivatives. *Ann. Statist.* **5**, 136–142.

[15] Nadaraya, E.A. (1965). On nonparametric estimates of regression curves. *Theor. Prob. Applns.* **10**, 186–190.

[16] Nadaraya, E.A. (1970). Remark on nonparametric estimates of density functions and regression curves. *Theor. Prob. Applns.* **15**, 134–37.

[17] Nadaraya, E.A. (1974). On the integral mean square error of some nonparametric estimates for the density function. *Theor. Prob. Applns.* **19**, 133–141.

[18] Parzen, E. (1962). On the estimation of a probability density function and the mode. *Ann. Math. Statist.* **33**, 1065–1076.
[19] Pickands. J. (1969). Efficient estimation of a probability density. *Ann. Math. Statist.* **40**, 854–864.
[20] Revesz, P. (1972). On empirical density function. *Periodica Math. Hungar.* **2**, 85–110.
[21] Rosenblatt, M. (1956). Remarks on some nonparametric estimates of a density function. *Ann. Math. Statist.* **27**, 832–837.
[22] Rosenblatt, M. (1971). Curve estimates. *Ann. Math. Statist.* **42**, 1801–1823.
[23] Sacks, J. and Ylvisaker, D. (1978). Asymptotically optimum kernels for density estimation at a point. (To appear.)
[24] Samarov, A. (1977). Lower bound for risk of spectral density estimates. *Problems of Info. Theor. Transm.* **13**, 48–51.
[25] Samarov, A. (1978). Lower bound for integral risk of density estimates. In *Problems of Construction of Systems for Information Transmission*. Bloch, E. Ed., (To appear.)
[26] Schuster, E.F. (1969). Estimation of a probability density function and its derivatives. *Ann. Math. Statist.* **40**, 1187–1195.
[27] Schwartz, S.C. (1967). Estimation of a probability density by an orthogonal series. *Ann. Math. Statist.* **38**, 1261–1265.
[28] Stone, C.J. (1980). Optimal rates of convergence for nonparametric estimators. (To appear in *Ann. Math. Statist.*)
[28a] Stone, C.J. (1981). Optimal global rates of convergence for nonparametric regression. Preprint.
[29] Van Ryzin, J. (1973). A histogram method of density estimation. *Comm. Statist.* **2**, 493–506.
[30] Wahba, G. (1975). Optimal convergence properties of variable knot, kernel, and orthogonal series methods for density estimation. *Ann. Math. Statist.* **3**, 15–29.
[31] Walter, G. and Blum, J. (1979). Probability density estimation using delta sequences. *Ann. Statist.* **7**, 328–340.
[32] Wegman, E.J. (1972a). Nonparametric probability density estimation: I. A summary of available methods. *Technometrics.* **14**, 533–546.
[33] Wegman, E.J. (1972b). Nonparametric probability density estimation: II. A comparison of density estimation methods. *J. Statist. Comput. Simul.* **1**, 225–245.
[34] Weiss, L. and Wolfowitz, J. (1967). Estimation of a density at a point. *Z. Wahrsch.* **7**, 327–335.
[35] Winter, B.B. (1975). Rate of strong consistency of two nonparametric density estimators. *Ann. Statist.* **3**, 759–766.
[36] Woodroofe, M.B. (1970a). On the maximum deviation of the sample density. *Ann. Math. Statist.* **38**, 475–481.
[37] Woodroofe, M.B. (1967b). The maximum deviation of sample spectral densities. *Ann. Math. Statist.* **38**, 1558–1569.
[38] Woodroofe, M.B. (1970). On choosing a delta-sequence. *Ann. Math. Statist.* **41**, 1665–1671.
[39] Woodroofe, M.B. and Van Ness, J.W. (1967). The maximum deviation of sample spectral densities. *Ann. Math. Statist.* **38**, 1558–1569.

G. Kallianpur, P.R. Krishnaiah, J.K. Ghosh, eds., *Statistics and Probability: Essays in Honor of C.R. Rao*

TWO "BEST" UNBIASED ESTIMATORS OF NORMAL INTEGRAL MEAN

Y. KOJIMA
Tōa Boshoku Co. Ltd., Osaka, Japan

H. MORIMOTO
Osaka City University, Osaka, Japan

K. TAKEUCHI
Tokyo University, Tokyo, Japan

The locally best unbiased estimator (LBUE) for integer-mean of normal distribution with known variance is given and compared with the best invariant estimator (wrt. the discrete additive group of integers), which is also unbiased, and a few other unbiased estimators known so far. It is proved that the LBUE, while minimizing the variance at one parameter point, has an infinite variance at all other points. A general method for arriving at the locally best unbiased estimator of an integral parameter as a limit of such an estimator for a case of finite parameter space is given.

1. Introduction

Suppose that we observe a sample from a normal distribution with unknown integer-valued mean m and known variance. The problem of unbiased estimation of m is a typical instance of an incomplete sufficient statistic and infinity of unbiased estimators based on it. Thus Hammersley (1950), who first treated the problem, remarked that "the nearest integer to the sample mean $\bar{x}$" is unbiased as well as it is MLE. Stein pointed out in the discussion of the same paper that "the nearest even (odd) integer to $\bar{x}$" is also unbiased. Later, Khan (1973) proved inadmissibility w.r.t. the mean square error of Hammersley's estimator, by showing that it is dominated by a certain convex combination of itself and $\bar{x}$. Unni (1977) characterized the totality of the functions of m which have UMVUE and pointed out the non-existence of UMVUE for m itself.

In Section 2 of the present paper, we will give the LBUE (locally best unbiased estimator) of m, by the method of Stein (1950). It turns out to be an extremely strange-looking function of $\bar{x}$ whose variance at a specified value of m, say $m=a$, is nearly $0.95V(\bar{x})$ when $V(\bar{x})$ is around 1 and much smaller when the latter is smaller, but is infinity at all other values of m, i.e. $m \neq a$. Thus the non-existence of UMVUE for m can be seen here again, somewhat more strikingly.

The constructive method by which we arrived at our estimator is described in Section 3, while in Section 2 the same estimator is just put forward and is proved to be LBUE. The method consists in starting with a finite parameter space $\{-k, -k+1, \ldots, -1, 0, +1, \ldots, k-1, k\}$, for which case it is much easier to find out the LBUE, and obtaining the LBUE for the original problem as its limit when k goes to infinity. The limiting process summarized in Theorem 3 may be applicable to a fairly wide class of similar problems*.

In Section 4, a more gentle unbiased estimator is displayed so that it contrasts with the wild behaviour of LBUE. It is the best invariant estimator wrt. the discrete additive group of integers. It was given by Ghosh and Subramanyam (1975), but its performance as a small-sample unbiased estimator has not much been examined so far. We make use of Fourier expansion to get an expression convenient for numerical calculation, which reveals its very moderate behaviour as a function of $\bar{x}$. Its variance, too, is much smaller than $V(\bar{x})$, especially when the latter is smaller than 0.2, say.

A few tables are attached to give the numerical values of these two estimators as functions of $\bar{x}$ and to compare the variances of various estimators.

2. Locally best unbiased estimator

Let $x=(x_1,\ldots,x_n)$ be the observed value from a density $p(x,m)$, m be the parameter to be estimated and X and M denote the sample space and the parameter space, respectively. One of the results of Stein (1950) on LBUE can be summarized as follows:

Fix a value a in M and define

$$A(u,v)=\int_X \frac{p(x,u)p(x,v)}{p(x,a)}\,\mathrm{d}x \tag{1}$$

for each u and v in M. The integration is w.r.t. the Lebesgue measure. Suppose that there exists a (signed) measure $\mathrm{d}s(u)$ on M which satisfies

$$\int_M A(u,v)\,\mathrm{d}s(u)=v \tag{2}$$

for all v in M. Then LBUE at $m=a$ is given by

$$f(x)=\int_M \frac{p(x,u)}{p(x,a)}\,\mathrm{d}s(u), \tag{3}$$

and the minimized variance of $f(x)$ at $m=a$ is

$$V(f(x))=\int_M u\,\mathrm{d}s(u)-a^2, \tag{4}$$

provided the changes of order of integrations involved in (5), (6) and (7) below are justified.

*Part of the results of this paper, namely, the construction of the LBUE by the method of Section 3 and resulting non-existence of UMVUE, had been presented at the Third Soviet–Japan Symposium on Probability Theory in Tashkent, August–September, 1975.

In fact, the unbiasedness of $f(x)$ would follow thus:

$$\int f(x)\,p(x,m)\,\mathrm{d}x=\int\left[\int\frac{p(x,u)}{p(x,a)}\,\mathrm{d}s(u)\right]p(x,m)\,\mathrm{d}x$$
$$=\int\left[\int\frac{p(x,u)}{p(x,a)}p(x,m)\,\mathrm{d}x\right]\mathrm{d}s(u)$$
$$=\int A(u,m)\,\mathrm{d}s(u)=m. \tag{5}$$

Now, let $h(x)$ be an unbiased estimator of 0 with a finite variance at $m=a$. Then we would have

$$\int f(x)h(x)p(x,a)\,\mathrm{d}x=\int\left(\int\frac{p(x,u)}{p(x,a)}\,\mathrm{d}s(u)\right)h(x)\,p(x,a)\,\mathrm{d}x$$
$$=\int\left(\int h(x)p(x,u)\,\mathrm{d}x\right)\mathrm{d}s(u)$$
$$=0, \tag{6}$$

which is the well-known condition for $f(x)$ to be LBUE at $m=a$ first proved by C. R. Rao (1947).

Further, the variance of $f(x)$ at $m=a$ would be calculated as

$$\int f(x)^2\,p(x,a)\,\mathrm{d}x-a^2=\int\int\left[\int\frac{p(x,u)p(x,v)}{p(x,a)}\,\mathrm{d}x\right]\mathrm{d}s(v)\,\mathrm{d}s(u)-a^2$$
$$=\int u\,\mathrm{d}s(u)-a^2. \tag{7}$$

Let us now apply the foregoing result to the estimation of a normal integral mean. Suppose that we have a sample $x=(x_1,\ldots,x_n)$ from $N(m,nd^2)$. Since $\bar{x}=\Sigma x_i/n$ is a sufficient statistic we will take $\bar{x}$ in place of x and assume that $a=0$. In this context $X=R$, $p(\bar{x},m)$ is the density of $N(m,d^2)$, M is the set of all integers and $\mathrm{d}s(u)$ should be a discrete measure putting a mass $s(u)$ on each integer u in M. $A(u,v)$ is calculated as

$$A(u,v)=\int\frac{1}{\sqrt{2\pi}}\exp\left[\left((\bar{x}-u)^2+(\bar{x}-v)^2-\bar{x}^2\right)/-2d^2\right]\mathrm{d}\bar{x}=\exp(uv/d^2),$$

and equation (2) reduces to an infinite system of linear equations:

$$\sum_{u=-\infty}^{+\infty} s(u)\exp\frac{uv}{d^2}=v,\quad v=0,\pm1,\pm2,\ldots. \tag{8}$$

A solution of these equations, as is verified later, is given by:

$$s(u)=(-1)^{u+1}[\exp(u(u-1)/2d^2)\cdot(\exp(u/d^2)-1)]^{-1},\qquad u>0,$$
$$=0,\qquad u=0,$$
$$=-s(-u),\qquad u<0. \tag{9}$$

Hence, $f(x)$ in (3) takes the form given in (10) below.

Theorem 1. *The LBUE at $m=0$ is*

$$f(\bar{x})=\sum_{u=1}^{\infty}(-1)^{u+1}\left[\exp(u(u-1)/2d^2)\left(\exp(u/d^2)-1\right)\exp(u^2/2d^2)\right]^{-1}$$
$$\times\left[\exp(u\bar{x}/d^2)-\exp(-u\bar{x}/d^2)\right]. \tag{10}$$

Its local minimum variance at $m=0$ is

$$V(f)=2\sum_{u=1}^{\infty}(-1)^{u+1}u\left[\exp(-u(u-1)/2d^2)/\left(\exp(u/d^2)-1\right)\right. \tag{11}$$

Proof. For simplicity of printing let $d=1$. For other values of d we only need to replace all e in the proof with $\exp(1/d^2)$.

(i) $s(u)$ is a solution of (8).

Substitute (9) in (8). Then

LHS of (8) =

$$=\sum_{u=1}^{\infty}(-1)^{u+1}e^{-u(u-1)/2}(e^{uv}-e^{-uv})/(e^u-1)$$

$$=\sum_{u=1}^{\infty}(-1)^{u+1}e^{-u(u-1)/2}[e^{-vu}+e^{(-v+1)u}+\cdots+e^{(v-1)u}]$$

$$=\sum_{p=0}^{v-1}\left[\sum_{u=1}^{\infty}(-1)^{u+1}e^{-u(u-1)/2}e^{u(-v+p)}+\sum_{u=1}^{\infty}(-1)^{u+1}e^{-u(u-1)/2}e^{u(v-1-p)}\right],$$

as the series converges absolutely. The formula in [ ] is equal to

$$\sum_{u=1}^{\infty}(-1)^{u+1}\exp\left[\tfrac{1}{2}[-u(u+(2v-2p-1))]\right]$$
$$+\sum_{u=1}^{\infty}(-1)^{u+1}\exp\left[\tfrac{1}{2}[-u(u-(2v-2p-1))]\right].$$

The terms for $u=1,2,\ldots$ in the first series and those for $u=(2v-2p-1)+1,(2v-2p-1)+2,\ldots$ in the second series, respectively, cancel each other and the first $(2v-2p-1)$ terms of the second series remain. Among them the terms for $u=j$ and $u=(2v-2p-1)-j$, $j=1,2,\ldots,v-p-1$, respectively, cancel each other and finally all but the term for $u=2v-2p-1$, which is equal to 1, disappear. This happens for $p=1,2,\ldots,v$. Hence LHS of (8) is equal to v.

(ii) $f(\bar{x})$ is unbiased.

Note that $f(\bar{x})$ is integrable w.r.t. $N(m,1)$ for all m in M. In fact,

$$|f(\bar{x})|\leqslant\sum_{u=1}^{\infty}e^{-u^2+u/2}(e^{u\bar{x}}+e^{-u\bar{x}}),$$

and

$$\frac{1}{\sqrt{2\pi}}\int e^{u\bar{x}}e^{-(\bar{x}-m)^2/2}\,d\bar{x}=e^{u^2/2}\cdot e^{mu}.$$

Hence,

$$\int |f(\bar{x})|p(\bar{x},0)\,d\bar{x}\leqslant \sum_{u=1}^{\infty} e^{u^2/2}(e^{mu}+e^{-mu})e^{-u^2+u/2}$$

$$=\sum_{u=1}^{\infty}\left(e^{-u^2/2+u/2+mu}+e^{-u^2/2+u/2-mu}\right)<\infty.$$

Hence the change of order of integration in (5) is justified.

(iii) $f(\bar{x})$ is LBUE at $m=0$.

Notice that $f(\bar{x})$ is square-integrable w.r.t. $N(0,1)$. In fact,

$$|f(\bar{x})^2|=\Bigg|\sum_u\sum_v\frac{1}{(e^u-1)(e^v-1)}e^{-u^2-v^2+u/2+v/2}$$

$$\times\left(e^{(u+v)\bar{x}}-e^{(u-v)\bar{x}}-e^{(v-u)\bar{x}}+e^{-(u+v)\bar{x}}\right)\Bigg|$$

$$\leqslant\sum_u\sum_v\frac{4}{(e^u-1)(e^v-1)}e^{-u^2-v^2+u/2+v/2}\left(e^{(u+v)\bar{x}}+e^{-(u+v)\bar{x}}\right). \tag{12}$$

Now

$$\frac{1}{\sqrt{2\pi}}\int\frac{1}{(e^u-1)(e^v-1)}e^{-u^2-v^2+u/2+v/2}e^{(u+v)\bar{x}}e^{-\bar{x}^2/2}\,d\bar{x}=$$

$$=\frac{1}{(e^u-1)(e^v-1)}e^{-u^2-v^2+u/2+v/2}e^{(u+v)^2/2}\leqslant\frac{e^{(u+v)/2}}{(e^u-1)(e^v-1)}.$$

The sum of the last term for u, $v=1,2,\ldots$ converges.

Thus the change of order of integration in (6) [and also in (7)] is justified.

(iv) The variance of $f(x)$ at $m=0$.

The result is straightforward, as the equality (7) has been justified above.

Table 1 shows the values of $f(\bar{x})$ for some values of $\bar{x}$. Notice that $f(\bar{x})$ is much smaller than $\bar{x}$, indeed very close to 0 when $\bar{x}$ is not far from 0, and in some cases even when $\bar{x}$ goes beyond 1. This difference has to be balanced by divergent oscillations of $f(\bar{x})$ for larger values of $\bar{x}$, to maintain its unbiasedness.

Values for $\bar{x}=0, 0.05,\ldots,0.5$ are to be compared with the corresponding values of $g(\bar{x})$, the best invariant estimator given in Section 4 and Tables 4 and 5. The values of $f(\bar{x})$ around its apparent maxima and minima, and also some intermediate values are included to give rough idea of the functional behaviour of $f(\bar{x})$. All other entries are omitted, as we do not know if anyone wants them for any purpose of statistical practice.

Table 1
Values of $f(\bar{x})$ against $\bar{x}$

$\bar{x}=$	$d^2=0.2$	$d^2=0.5$	$d^2=1$	$d^2=2$
0.00	0.00000E00	0.00000E00	0.00000E00	0.00000E00
0.05	2.81327E-04	1.15165E-02	3.37610E-02	4.80269E-02
0.10	5.80330E-04	2.31477E-02	6.75949E-02	9.60642E-02
0.15	9.15792E-04	3.50094E-02	1.01575E-01	1.44122E-02
0.20	1.30879E-03	4.72198E-02	1.35773E-01	1.92212E-01
0.25	1.78401E-03	5.99002E-02	1.70264E-01	2.40343E-01
0.30	2.37132E-03	7.31769E-02	2.05120E-01	2.88526E-01
0.35	3.10761E-03	8.71818E-02	2.40416E-01	3.36770E-01
0.40	4.03913E-03	1.02054E-01	2.76226E-01	3.85087E-01
0.45	5.22442E-03	1.17941E-01	3.12625E-01	4.33484E-01
0.50	6.73794E-03	1.35000E-01	3.49687E-01	4.81973E-01
0.60	1.11566E-02	1.73323E-01	4.26106E-01	5.79258E-01
0.70	1.84231E-02	2.18540E-01	5.06094E-01	6.77012E-01
0.80	3.03920E-02	2.72436E-01	5.90272E-01	7.75300E-01
0.90	5.01185E-02	3.37127E-01	6.79268E-01	8.74178E-01
1.00	8.26378E-02	4.15143E-01	7.73716E-01	9.73696E-01
1.10	1.36250E-01	5.09511E-01	8.74251E-01	1.07390E00
1.20	2.24640E-01	6.23869E-01	9.81506E-01	1.17481E00
1.30	3.70368E-01	7.62579E-01	1.09610E00	1.27646E00
1.40	6.10628E-01	9.30872E-01	1.21864E00	1.37885E00
1.50	1.00674E00	1.13500E00	1.34969E00	1.48197E00
1.60	1.65978E00	1.38240E00	1.48975E00	1.58582E00
1.70	2.73639E00	1.68190E00	1.63925E00	1.69035E00
1.80	4.51118E00	2.04383E00	1.79851E00	1.79552E00
1.90	7.43670E00	2.48025E00	1.96770E00	1.90125E00
2.00	1.22584E01	3.00495E00	2.14680E00	2.00747E00
2.50	1.48420E02	7.52901E00	3.16687E00	2.54142E00
3.00	1.67189E03	1.57521E01	4.16093E00	3.06316E00
3.10	2.59774E03	1.72343E01	4.29314E00	3.16340E00
3.20	3.85144E03	1.80681E01	4.37763E00	3.26163E00
3.30	5.17701E03	1.76381E01	4.39832E00	3.35755E00
3.40	5.34713E03	1.49666E01	4.33579E00	3.45090E00
3.50	1.49420E02	8.52901E00	4.16687E00	3.54142E00
4.00	-2.99720E06	-2.19196E02	6.15888E-01	3.94527E00
4.50	-4.78627E08	-2.16657E03	-1.12467E01	4.27501E00
5.00	-6.60576E10	-1.31005E04	-3.19677E01	4.62482E00
5.10	-1.69183E11	-1.72437E04	-3.49649E01	4.71755E00
5.20	-4.13249E11	-2.13438E04	-3.59870E01	4.82484E00
5.30	-9.14104E11	-2.36432E04	-3.36519E01	4.95059E00
5.40	-1.54681E12	-2.01065E04	-2.59904E01	5.09909E00
5.50	-4.78627E08	-2.16557E03	-1.02467E01	5.27501E00
6.00	1.77248E16	1.89881E06	3.78164E02	6.74329E00
6.50	3.44635E19	4.79296E07	2.42675E03	9.52055E00
7.00	5.79430E22	7.73218E08	8.59680E03	1.27559E01
7.10	2.44671E23	1.23808E09	1.00627E04	1.30958E01
7.20	9.85325E23	1.86106E09	1.10766E04	1.31662E01
7.30	3.59340E24	2.49205E09	1.09902E04	1.28559E01

Table 1
Values of $f(\bar{x})$ against $\bar{x}$

$\bar{x}=$	$d^2=0.2$	$d^2=0.5$	$d^2=1$	$d^2=2$
7.40	1.00248E25	2.51465E09	8.70591E03	1.20288E01
7.50	3.44635E19	4.79296E07	2.42775E03	1.05206E01
8.00	¯2.30757E30	¯8.45081E11	¯2.63533E05	¯1.50898E01
8.50	¯5.46596E34	¯5.76429E13	¯2.66897E06	¯9.59572E01
9.00	¯1.11955E39	¯2.52230E15	¯1.51214E07	¯2.27232E02
9.10	¯7.79421E39	¯4.93031E15	¯1.93828E07	¯2.44162E02
9.20	¯5.17508E40	¯9.04497E15	¯2.32651E07	¯2.48352E02
9.30	¯3.11164E41	¯1.47720E16	¯2.49077E07	¯2.32209E02
9.40	¯1.43122E42	¯1.81308E16	¯2.04917E07	¯1.85544E02
9.50	¯5.46596E34	¯5.76429E13	¯2.66896E06	¯9.49572E01
10.00	6.61714E48[a)]	2.04290E19	1.32234E09	1.79601E03

[a)] 6.61714E48, e.g. denotes 6.61714×10^{48}.

Table 2
Variances of various estimators at $m=0$.

Estimator	$V(\bar{x})=0.05$	0.1	0.2	0.5	1.0	2.0
LBUE ($f(\bar{x})$)	0.412×10^{-8}	0.908×10^{-4}	0.0136	0.303	0.949	1.999
Best invariant estimator ($g(\bar{x})$)[a)]	0.0165	0.0845	0.199	0.500	1.000	2.000
Nearest integer to $\bar{x}$	0.0253	0.114	0.266	0.544	1.083	2.084
Nearest even number to $\bar{x}$	0.312×10^{-4}	0.00626	0.101	0.629	1.302	2.333
Nearest odd number to $\bar{x}$	1.00000	1.00000	1.00006	1.037	1.365	2.333

[a)] For the best invariant estimator, see also Tables 4 and 5.

Table 2 shows the variances of $f(\bar{x})$ and $g(\bar{x})$, the best invariant estimator for a few values of d^2, together with those of three other estimators mentioned in Section 1.

In view of the tremendous fluctuation of $f(\bar{x})$, it would not be surprising to have the following

Theorem 2. *LBUE at $m=0$ has infinite variance at all $m\neq0$.*

Proof. Look at RHS of (12) and pick out a term containing $e^{\pm(u+v)\bar{x}}$ with $u=v$. Its integral w.r.t. $N(m,1)$ is

$$\frac{1}{\sqrt{2\pi}}\int\frac{1}{(e^u-1)^2}e^{-2u^2+u}e^{\pm2u\bar{x}}e^{-(\bar{x}-m)^2/2}d\bar{x}=\frac{1}{(e^u-1)^2}e^{-2u^2+u}e^{2u^2}e^{\pm2um}$$

$$=\frac{1}{(e^u-1)^2}e^{u(2m+1)}.$$

The sum of the last term over $u=1,2,\ldots$ diverges provided $m\neq 0$. A fortiori, the square-integral of $f(\bar{x})$ diverges.

This reminds us of a result by Ghosh and Subramanyam (1975) which says if an estimator attains asymptotically an analogue of Cramér–Rao bound at one parameter point then its asymptotic variance is infinity at all other points.

It is further conjectured that there is no estimator which minimizes the variance at $m=0$ in the class of unbiased estimators whose variances are finite for *all* m in M.

3. Construction of LBUE as a limit of estimators for finite parameter spaces

When we try to construct an LBUE by the method of Stein, as is seen in the last section, a crucial point lies in how we get hold of $s(u)$ by solving the equation (2). It is an infinite system of linear equations in the case of discrete parameters, and a standard technique would be to solve a finite system which is a part of the infinite system and to obtain the desired solution as a limit of the solutions of such finite systems. The following theorem is designed to give a description of such a technique generally applicable to discrete parameters.

Theorem 3. *Suppose that we have an observation x from a density $p(x,m)$ and assume the following conditions.*

(i) *The parameter space M is the limit of an increasing sequence of its subsets $M(k)$, $k=1,2,\ldots$.*

(ii) *A point a belongs to $M(k)$ for all k.*

(iii) *For each k, there is a function $f(x;k)$ which minimizes the variance at $m=a$ in the class of those functions which are unbiased for m in $M(k)$.*

(iv) *There exists a limit $f(x)=\lim_{k\to\infty} f(x;k)$ for almost all x w.r.t. $p(x,m)$ for all m in M.*

(v) *$f(x)$ is unbiased for m in M. (In other words, limit and integration can be interchanged thus:* $\lim_{k\to\infty}\int f(x;k)p(x,m)\,\mathrm{d}x=\int \lim_{k\to\infty} f(x;k)p(x,m)\,\mathrm{d}x$. *Notice that for each m in M, $f(x;k)$ is unbiased for sufficiently large k.)*

Then $f(x)$ is LBUE at $m=a$ and $V(f(x))=\lim_{k\to\infty}V(f(x;k))$ at $m=a$.

Proof. Take an unbiased estimator $h(x)$ for m in M. As $h(x)$ is unbiased for m in each of $M(k)$, we have, by (iii),

$$\int f^2(x:k)p(x,a)\,\mathrm{d}x\leqslant\int h^2(x)p(x,a)\,\mathrm{d}x, \tag{13}$$

for each k. In particular,

$$\int f^2(x;k)p(x,a)\,\mathrm{d}x\leqslant\int f^2(x)p(x,a)\,\mathrm{d}x. \tag{14}$$

On the other hand, it follows from Fatou's lemma and (13) that

$$\int f^2(x)p(x,a)\,\mathrm{d}x\leqslant\liminf_{k\to\infty}\int f^2(x;k)p(x,a)\,\mathrm{d}x\leqslant\int h^2(x)p(x,a)\,\mathrm{d}x.$$

This means that $f(x)$ is LBUE at $m=a$, and implies, together with (14), that

$$\int f^2(x)p(x,a)\,dx = \lim_{k\to\infty}\int f(x;k)^2 p(x,a)\,dx.$$

For the case of normal integral mean we start with a finite parameter space $M(k)=\{-k,-k+1,\ldots,-1,0,+1,\ldots,k-1,k\}$ and solve the finite system of equations

$$\sum_{u=-k}^{k} s(u;k)\exp(uv/d^2)=v,\quad v=0,\pm1,\ldots,\pm k,$$

to get a measure $s(u;k)$ on $M(k)$. It turns out

$$\begin{aligned} s(u;k)&=s(u)\prod_{p=k+1}^{p=k+u}\frac{e^{p/d^2}-e^{u/d^2}}{e^{p/d^2}-1}, && u=1,2,\ldots,k,\\ &=0, && u=0,\\ &=-s(-u;k), && u=-1,-2,\ldots,-k, \end{aligned} \tag{15}$$

where $s(u)$ is same as in (9). The process to get (15) from the Cramer's formula is extremely cumbersome and is omitted, as we have established our result in the last section. On the other hand, the remaining steps are much easier. In fact, it follows immediately that

$$f(\bar{x};k)=\sum_{u=-k}^{k} s(u;k)e^{-u^2/2d^2}\left(e^{u\bar{x}/d^2}-e^{-u\bar{x}/d^2}\right)$$

satisfies the condition (iii) of Theorem 3 with $\bar{x}$ in place of x, as the equalities (5), (6) and (7) follow trivially from the finiteness of $M(k)$. It is easy to see that $f(\bar{x};k)$ converges to $f(\bar{x})$. Further,

$$|f(\bar{x};k)|\leqslant K\left(e^{(\bar{x}+1/2)^2/4}+e^{(\bar{x}-1/2)^2/4}\right).$$

Because, as $|s(u;k)|\leqslant|s(u)|$ and $|e^{u\bar{x}/d^2}-e^{-u\bar{x}/d^2}|\leqslant e^{u\bar{x}/d^2}$ for $\bar{x}\geqslant 0$,

$$\begin{aligned} f(\bar{x};k)&=\sum_{u=1}^{\infty}\frac{1}{e^{u/d^2}-1}\,e^{-u(u-1)/d^2}e^{-u^2/2}e^{u\bar{x}/d^2}\\ &=\left[\sum_{u=1}^{\infty}\frac{1}{e^{u/d^2}-1}\,e^{-(u-(2\bar{x}+1)/4)^2}\right]e^{(\bar{x}+1/2)^2/4}\\ &\leqslant Ke^{(\bar{x}+1/2)^2/4}, \end{aligned}$$

for $\bar{x}\geqslant 0$. Similarly we have,

$$|f(\bar{x};k)|\leqslant Ke^{(\bar{x}-1/2)^2/4},$$

for $\bar{x}\leqslant 0$. Thus (v) is satisfied and hence $f(\bar{x})$ is LBUE.

This method has been successfully applied to the estimation of an integral mean of Poisson distribution and is expected to have similar applications. As they do not fall in the scope of the present paper, the results will be published elsewhere.

4. Best invariant (and unbiased) estimator

Suppose that we have n independent observations $x_1,\ldots,x_n$ from a density $p(x,m) = p(x-m)$ with an integral parameter m. An estimator $h(x_1,\ldots,x_n)$ is said to be invariant w.r.t. the discrete additive group I if $h(x_1+m,\ldots,x_n+m)=h(x_1,\ldots,x_n)+m$ for all m in M, the set of all the integers. Then the best invariant estimator, which minimizes the mean square error among all invariant estimators, is given by

$$g(x_1,\ldots,x_n)=\frac{\sum\limits_{m=-\infty}^{\infty} m\prod p(x_i-m)}{\sum\limits_{m=-\infty}^{\infty} \prod p(x_i-m)}. \tag{16}$$

In fact, if we put

$$D=(x_1-[x_1],\ldots,x_n-[x_1]),$$

where $[x_1]$ denotes the integer part of x_1, then D is the maximal invariant and it is easily seen that g can be expressed as

$$g(x_1,\ldots,x_n)=x_1-E_0(x_1|D),$$

from which follows conditional unbiasedness:

$$E_m(g|D)=E_m(x_1|D)-E_0(x_1|D)=m.$$

Let h be any invariant estimator. Then $g-h$ is a function of D, and it is shown that

$$E(h-m)^2=E(g-m)^2+E(h-g)^2 \geqslant E(g-m)^2, \tag{17}$$

as is to be proved.

When $x_1,\ldots,x_n$ are i.i.d. normal with mean m and variance nd^2, we get

$$g(x_1,\ldots,x_n)=\frac{\sum\limits_m m\exp\left[-(\bar{x}-m)^2/2d^2\right]}{\sum\limits_m \exp\left[-(\bar{x}-m)^2/2d^2\right]}$$

$$=\bar{x}-\frac{t'(\bar{x})}{t(\bar{x})}d^2, \tag{18}$$

where $\bar{x}=\Sigma x_i/n$ and $t(\bar{x})$ is defined as

$$t(\bar{x})=\sum_{m=\infty}^{\infty}\frac{1}{\sqrt{2\pi}\,d}\mathrm{e}^{-(\bar{x}-m)^2/2d^2}. \tag{19}$$

The foregoing result was given by Ghosh and Subramanyam (1975).

Since t is a continuous, even and periodic function, it can be expanded into a Fourier series as

$$t(\bar{x})=\frac{a_0}{2}+\sum_{k=1}^{\infty} a_k\cos 2k\pi\bar{x} \tag{20}$$

and it also holds that

$$t'(\bar{x}) = -\sum_{k=0}^{\infty} 2k\pi a_k \sin 2k\pi\bar{x}.$$

The coefficient a_k is given by

$$\begin{aligned}a_k &= 2\int_0^1 \cos 2k\pi\bar{x}\, t(\bar{x})\,\mathrm{d}\bar{x}\\ &= 2\int_0^1 \sum_{m=-\infty}^{\infty} \cos 2k\pi\bar{x}\frac{1}{\sqrt{2\pi}\,d}\,\mathrm{e}^{-(\bar{x}-m)^2/2d^2}\,\mathrm{d}\bar{x}.\\ &= 2\int_{-\infty}^{\infty} \cos 2k\pi\bar{x}\cdot\frac{1}{\sqrt{2\pi}\,d}\,\mathrm{e}^{-\bar{x}^2/2d^2}\,\mathrm{d}\bar{x}\\ &= 2E(\cos 2k\pi\bar{x})\\ &= 2\mathrm{Re}[E(\exp 2k\pi\mathrm{i}\bar{x})] = 2\mathrm{e}^{-2k^2\pi^2 d^2}\end{aligned} \tag{21}$$

Summarizing the foregoing paragraphs we give the following

Theorem 4. *The best invariant estimator of the integer mean of normal distribution is given by* (18), *where* $t(\bar{x})$ *is given by* (19), *or* (20) *and* (21). *It is also unbiased with the variance*:

$$V(g(\bar{x})) = V(\bar{x}) - E\left(\frac{t'(\bar{x})}{t(\bar{x})}d^2\right)^2. \tag{22}$$

Proof. Unbiasedness follows from the fact that $t'(\bar{x})/t(\bar{x})$ is an odd function. Further, from (17) and the invariance of $\bar{x}$ it follows that

$$E(\bar{x}-m)^2 = E(g(\bar{x})-m)^2 + E(\bar{x}-g(\bar{x}))^2,$$

and hence, (22).

Note that the coefficient a_k converges very quickly to 0 as k increases, so $g(\bar{x})$ is very moderately behaved in contrast with the wild behaviour of the LBUE. In fact, it is continuous, with $g(\bar{x}) = \bar{x}$ for $\bar{x} = 0, \pm 0.5, \pm 1, \ldots,$ as is seen from (18). Unlike LBUE, it is monotone increasing in $\bar{x}$ and satisfies $[\bar{x}] \leqslant g(\bar{x}) \leqslant [\bar{x}]+1$. Between $\bar{x}=0$ and 0.5, it is smaller than $\bar{x}$, but is not so close to 0 as LBUE. For $0.5 < \bar{x} < 1$, it is larger than $\bar{x}$, as its graph is symmetric wrt. the point (0.5,0.5).

The coefficients a_k in the expression

$$g(\bar{x}) = \bar{x} + \frac{\sum_k 2k\pi a_k \sin 2\pi\bar{x}}{1+\sum_k a_k \cos 2\pi\bar{x}}$$

are very small unless d^2 is small. For some values of d^2 they are given in Table 3. The estimator is, therefore, easily calculated when $d^2 \geqslant 0.1$, say. When $d^2 \geqslant 1$, it is virtually equal to $\bar{x}$. This suggests why the variance of an estimator, e.g. the LBUE,

Table 3
Values of a_k for small values of d^2

d^2	a_1	a_2	a_3	a_4
1	5.35×10^{-9}	—	—	—
0.5	1.0345×10^{-4}	—	—	—
0.2	0.038593	2.77×10^{-7}	—	—
0.1	0.277822	0.000745	3.85×10^{-8}	—
0.05	0.745412	0.038593	0.000278	1.39×10^{-7}
0.02	1.347651	0.206153	0.028637	0.001806

can be made significantly smaller than that of $\bar{x}$ at $m=0$, only at the cost of very large variance at other parameter values. See Table 4.

When d^2 is not very small, we have approximately,

$$E(\bar{x}-g(\bar{x}))^2 \doteqdot a_1^2 d^4 E(\cos^2 2\pi\bar{x})$$
$$= \frac{a_1^2 d^4}{2} E(\cos 4\pi\bar{x}+1)$$
$$\doteqdot \frac{a_1^2 d^4}{2} = 2d^4 e^{-2\pi^2 d^2},$$

which indicates the magnitude of the gain in terms of the variance of $g(\bar{x})$ over $\bar{x}$.

On the other hand, when d^2 is small, $g(\bar{x})$ can be calculated from (18) itself. For each value of $\bar{x}$, it seems enough to take four terms each in the numerator and the denominator, corresponding to those values of m which are nearer to $\bar{x}$. When $d^2 \leqq 0.05$, only two terms are enough and we have approximately

$$g(\bar{x}) \doteqdot \frac{1}{1+\exp(1-2\bar{x})/2d^2}, \quad 0<\bar{x}<0.5.$$

Furthermore, when d^2 is even smaller than 0.01, $g(\bar{x})$ is very close to the Hammersley's estimator, the nearest integer to $\bar{x}$. See Table 5.

Table 4
Values of $g(\bar{x})$ against $\bar{x}$ and its variance against d^2, (i).

$x=$	$d^2=0.5$	0.2	0.1
0	0	0	0
0.05	0.049900	0.035544	0.006917
0.1	0.099809	0.072357	0.015514
0.15	0.149737	0.111635	0.027812
0.2	0.199691	0.154420	0.046517
0.25	0.249675	0.201503	0.075309
0.3	0.299691	0.253320	0.118874
0.35	0.349737	0.309855	0.182232
0.4	0.399809	0.370576	0.268835
0.45	0.449900	0.434443	0.377493
0.5	0.5	0.5	0.5
$V(g(\bar{x}))$ at $m=0$	$0.5-5.28\times10^{-8}$	0.19882	0.08447

Table 5
Values of $g(\bar{x})$ against $\bar{x}$ and its variance against d^2, (ii)

$\bar{x}=$	$d^2=0.05$	0.02	0.01
0	0	0	0
0.05	0.000107	0.000000	0.000000
0.1	0.000329	0.000000	0.000000
0.15	0.000909	0.000000	0.000000
0.2	0.002472	0.000000	0.000000
0.25	0.006693	0.000004	0.000000
0.3	0.017986	0.000045	0.000000
0.35	0.047426	0.000553	0.000000
0.4	0.119203	0.006693	0.000045
0.45	0.268941	0.075858	0.006693
0.5	0.5	0.5	0.5
$V(g(\bar{x}))$	0.01653	0.0000383	1.11×10^{-2}

$g(\bar{x})=2\bar{x}-g(\bar{x}-0.5)$, for $\bar{x}>0.5$.

Acknowledgements

The authors are grateful to A. Kagan who made useful comments when some of the results were announced in the Third Japan–Soviet Symposium on Probability Theory at Tashkent in 1975. Thanks are due to J.K. Ghosh, M. Sibuya and T. Suzuki for their valuable cooperation and suggestions.

Added in proof

Recently the authors' attention was drawn by J.K. Ghosh to Ghosh, M. and G. Meeden (1978) which proves admissibility of Hammersley's estimator in the class of all the integer-valued estimators. The same estimator was also shown by Ghosh, J.K. and K. Subramanyam (1975) to be best in the class of all the integer-valued invariant estimators.

References

[1] Ghosh, J.K. and Subramanyam (1975). Inference about separated families in large samples. *Sankhyā Ser. A* **37**, 502–513.
[2] Ghosh, M. and G. Meeden (1978). Admissibility of the MLE of the normal integer mean. *Sankhyā Ser. B* **40**, 1–10.
[3] Hammersley, J. (1950). On estimating restricted parameters. *J. R. Statist. Soc. Ser. B.* **12**, 192–240.
[4] Khan, R. (1973). On some properties of Hammersley's estimator of an integer mean. *Ann. of Statist.* **1**, 756–762.
[5] Rao, C. R. (1947). Minimum variance and the estimation of several parameters. *Proc. Cambridge Philos. Soc.* **43**, 81–91.
[6] Stein, C. (1950). Unbiased estimates of minimum variance. *Ann. Math. Statist.* **21**, 405–415.
[7] Unni, K. (1977). The theory of estimation in algebraic and analytic exponential families with applications to variance components models. Ph.D. Thesis. Indian Statistical Institute.

G. Kallianpur, P.R. Krishnaiah, J.K. Ghosh, eds., *Statistics and Probability: Essays in Honor of C.R. Rao*

STABLE AND SEMISTABLE PROBABILITY MEASURES ON A HILBERT SPACE*

R.G. LAHA

Bowling Green State University, Bowling Green, OH, U.S.A.

V.K. ROHATGI

Bowling Green State University, Bowling Green, OH and Ohio State University, Columbus, OH, U.S.A.

This paper surveys some recent developments in the theory of stable and semistable probability measures on a real separable Hilbert space. An exposition of some recent work on operator-stable and operator-semistable measures is also included.

1. Introduction

Let $\mathcal{H}$ be a real separable Hilbert space with inner product $\langle \cdot, \cdot \rangle$ and norm $\|\cdot\|$. Let $\mathcal{B}$ denote the σ-field generated by the class all open subsets of $\mathcal{H}$. Let $\mathcal{P}$ denote the class of all probability measures on $\mathcal{B}$. For every $P \in \mathcal{P}$ the Fourier transform $\hat{P}$ of P is a complex-valued function defined on $\mathcal{H}$ by the formula

$$\hat{P}(y) = \int_{\mathcal{H}} \exp(\mathrm{i}\langle x, y \rangle)\,\mathrm{d}P(x), \quad y \in \mathcal{H}. \tag{1.1}$$

Let $P_1, P_2 \in \mathcal{P}$. Then the convolution of P_1 and P_2 is the probability measure $P_1 * P_2$ on $\mathcal{B}$ defined by

$$P_1 * P_2(E) = \int_{\mathcal{H}} P_1(E-y)\,\mathrm{d}P_2(y) = \int_{\mathcal{H}} P_2(E-y)\,\mathrm{d}P_1(y) \tag{1.2}$$

for $E \in \mathcal{B}$. For $n \geqslant 1$ let P^{*n} denote the n-fold convolution of $P \in \mathcal{P}$. For any $P \in \mathcal{P}$ and any real number $a \neq 0$ we define the probability measure $T_a P$ on $\mathcal{B}$ by setting

$$T_a P(E) = P(a^{-1}E), \quad E \in \mathcal{B}. \tag{1.3}$$

For $a=0$ we set $T_a P = \delta_0$ where δ_x is the probability measure degenerate at $x \in \mathcal{H}$. Clearly $\widehat{T_a P}(y) = \hat{P}(ay)$, $y \in \mathcal{H}$ and for all real a. Let $P_n \in \mathcal{P}$, $n \geqslant 1$. We say that P_n

*Research supported by the National Science Foundation Grant #MCS78-01338. Invited paper read at IMS Meetings, Charleston, South Carolina, March 1980.

converges weakly to a probability measure $P \in \mathcal{P}$ if and only if

$$\lim_{n\to\infty} \int_{\mathcal{H}} f \,\mathrm{d}P_n = \int_{\mathcal{H}} f \,\mathrm{d}P \tag{1.4}$$

for every bounded continuous real-valued function f, on $\mathcal{H}$. In this case we write $P_n \Rightarrow P$. A sequence $\{P_n\}$, $P_n \in \mathcal{P}$, $n \geq 1$, is said to be tight if for every $\varepsilon > 0$ there exists a compact set $K_\varepsilon \in \mathcal{B}$ such that $P_n(K_\varepsilon) > 1 - \varepsilon$ for every $n \geq 1$.

The following elementary properties of the Fourier transform $\hat{P}$ of $P \in \mathcal{P}$ are easy to verify. See, for example, Laha and Rohatgi [15, pp. 462–464],

(i) $\hat{P}(0) = 1$.
(ii) $|\hat{P}(y)| \leq 1$, $y \in \mathcal{H}$.
(iii) $\hat{P}(-y) = \overline{\hat{P}(y)}$, $y \in \mathcal{H}$.
(iv) $\hat{P}$ is uniformly continuous with respect to the usual norm topology on $\mathcal{H}$.
(v) Let $P_1, P_2 \in \mathcal{P}$. Then $\hat{P}_1 = \hat{P}_2$ on $\mathcal{H} \Leftrightarrow P_1 = P_2$ on $\mathcal{B}$.
(vi) Let $P_1, P_2 \in \mathcal{P}$. Then $\widehat{P_1 * P_2} = \hat{P}_1 \cdot \hat{P}_2$.
(vii) Let $P_n \in \mathcal{P}$, $n \geq 1$, and $P \in \mathcal{P}$ be such that $P_n \Rightarrow P$. Then $\hat{P}_n \to \hat{P}$ on $\mathcal{H}$. Conversely, let $P_n \in \mathcal{P}$, $n \geq 1$, be a tight sequence such that $\hat{P}_n \to \theta$ where θ is a complex-valued function on $\mathcal{H}$. Then there exists a $P \in \mathcal{P}$ such that $P_n \Rightarrow P$ and, moreover, $\theta = \hat{P}$.

Let $P \in \mathcal{P}$. Then P is said to be infinitely divisible (i.d.) if for every positive integer $n \geq 1$, there exists a $P_n \in \mathcal{P}$ such that

$$P = P_n^{*n}.$$

Equivalently $P \in \mathcal{P}$ is i.d. if its Fourier transform $\hat{P}$ satisfies the following condition: For every $n \geq 1$ there exists a Fourier transform $\hat{P}_n$ of a probability measure $P_n \in \mathcal{P}$ such that $\hat{P} = (\hat{P}_n)^n$. If P is i.d. we say that $\hat{P}$ is i.d. The following representation of an i.d. Fourier transform $\hat{P}$ of $P \in \mathcal{P}$ was obtained by Varadhan [24].

Theorem 1. *Let ϕ be a complex-valued function on $\mathcal{H}$. Then ϕ is the Fourier transform of an i.d. probability measure $P \in \mathcal{P}$ if and only if*

$$\ln \phi(y) = \mathrm{i}\langle a, y\rangle - \tfrac{1}{2}\langle Sy, y\rangle + \int_{\mathcal{H}} \left\{ \exp \mathrm{i}\langle x, y\rangle - 1 - \frac{\mathrm{i}\langle x, y\rangle}{1 + \|x\|^2} \right\} \mathrm{d}M(x) \tag{1.5}$$

for $y \in \mathcal{H}$, where M is a σ-finite measure on $\mathcal{B}$ with finite mass outside every neighborhood of $x = 0 \in \mathcal{H}$ and satisfying

$$\int_{\|x\| \leq 1} \|x\|^2 \,\mathrm{d}M(x) < \infty,$$

$a \in \mathcal{H}$ is a fixed vector and S is an S-operator on $\mathcal{H}$. In this case a, S and M are uniquely determined by ϕ. Here M is called the Lévy spectral measure associated with ϕ.

Next we define a Gaussian measure on $\mathcal{B}$. Let ϕ be a complex-valued function on $\mathcal{H}$ defined by

$$\phi(y) = \exp\{\mathrm{i}\langle a, y\rangle - \tfrac{1}{2}\langle Sy, y\rangle\}, \quad y \in \mathcal{H} \tag{1.6}$$

where $a \in \mathcal{H}$ and S is an S-operator on $\mathcal{H}$. Then ϕ is the Fourier transform of a probability measure P on $\mathcal{B}$ satisfying the condition $\int_{\mathcal{H}} \|x\|^2 \mathrm{d}P(x) < \infty$. Here the measure P is called a Gaussian measure on $\mathcal{B}$ with mean vector a and covariance operator S. We note that a Gaussian measure is i.d.

As in the finite dimensional case the class of i.d. probability measures on $(\mathcal{H}, \mathcal{B})$ plays a basic role in the study of the Central Limit Problem on $\mathcal{H}$. More precisely we consider a sequence $\{X_{nk}, 1 \leq k \leq k_n, n \geq 1\}$ of $\mathcal{H}$-valued, row-wise, independent, random variables defined on some probability space (Ω, ζ, P) satisfying the following (u.a.n.) condition: For every $\varepsilon > 0$

$$\lim_{n\to\infty} \max_{1 \leq k \leq k_n} P\{\|X_{nk}\| \geq \varepsilon\} = 0.$$

Set $S_n = \sum_{k=1}^{k_n} X_{nk}$. It is well known (Laha and Rohatgi [15, p. 519], or Parthasarathy [19, p. 199]) that the class of limit distributions of the sequence $\{S_n\}$ coincides with the class of all i.d. probability measures on $\mathcal{H}$.

In this paper we consider some special subclasses of the class of all i.d. probability measures on $(\mathcal{H}, \mathcal{B})$. In particular, we present a brief exposition of some recent developments in the theory of stable and semistable probability measures on $(\mathcal{H}, \mathcal{B})$ (Section 2). We also survey some recent results on operator stable and operator semistable probability measures (Section 3). In Section 4 we state some important unsolved problems.

2. Stable and semistable probability measures

Let $P \in \mathcal{P}$. Then P is said to be stable if for every $a > 0$ and $b > 0$ there exists $c = c(a, b) > 0$ and an $x \in \mathcal{H}$ such that the relation

$$T_a P * T_b P = T_c P * \delta_x \tag{2.1}$$

holds. We note that $P \in \mathcal{P}$ is stable if and only if for every $n \geq 1$ there exist an $a_n > 0$ and an element $x_n \in \mathcal{H}$ such that

$$P^{*n} = T_{a_n} P * \delta_{x_n}. \tag{2.2}$$

Equivalently, P is stable if and only if

$$[\hat{P}(y)]^n = \hat{P}(a_n y) \exp\{\mathrm{i}\langle x_n, y\rangle\}, \quad y \in \mathcal{H}, n \geq 1. \tag{2.3}$$

It follows easily that a stable probability measure on $\mathcal{H}$ is i.d. In particular every Gaussian measure on $(\mathcal{H}, \mathcal{B})$ is stable.

The following representation of the Fourier transform of a stable probability measure on $(\mathcal{H}, \mathcal{B})$ is due to Kuelbs [11].

Theorem 2.1. *Let $P \in \mathcal{P}$. Then P is stable if and only if*

(i) *P is Gaussian, and*

(ii) *there exists a constant α, $0 < \alpha < 2$, an element $a \in \mathcal{H}$, and a finite measure ν concentrated on the boundary of the unit sphere $S = \{y \in \mathcal{H}: \|y\| = 1\}$ such that for*

$y \in \mathcal{H}$

$$\ln \hat{P}(y) = \mathrm{i}\langle a, y\rangle - \int_S |\langle x, y\rangle|^\alpha \mathrm{d}\nu(x) + \mathrm{i}C(\alpha, y) \tag{2.4}$$

where

$$C(\alpha, y) = \begin{cases} \tan\frac{1}{2}\pi\alpha \int_S \langle x, y\rangle |\langle x, y\rangle|^{\alpha-1} \mathrm{d}\nu(x) & \textit{for } \alpha \neq 1, \\ (2/\pi)\int_S \langle x, y\rangle \ln|\langle x, y\rangle| \mathrm{d}\nu(x) & \textit{for } \alpha = 1. \end{cases} \tag{2.5}$$

In this case α $(0<\alpha<2)$ *is called the index or type of the stable measure* P. *If* P *is Gaussian we say that* P *has index* 2.

This result extends the work of Lévy [18] for the one-dimensional case and Rvačeva [21] for the finite dimensional Euclidean space. Recently, Jurek and Urbanik [3] extended Theorem 2.1 to a real separable Banach-space.

Next we define and characterize the domain of attraction of a stable probability measure on $\mathcal{B}$. Let $P \in \mathcal{P}$. We say that P belongs to the domain of attraction of $\mu \in \mathcal{P}$ if sequences $a_n > 0$ and $x_n \in \mathcal{H}$ exist such that

$$T_{a_n} P^{*n} * \delta_{x_n} \Rightarrow \mu.$$

It is known [14] that stable probability measures and only stable probability measures have nonempty domains of attractions. The domain of attraction of a Gaussian measure was characterized independently by Klowsowka [4,5], and Kuelbs and Mandrekar [15].

Theorem 2.2 (Kuelbs and Mandrekar [15]). *A probability measure* $P \in \mathcal{P}$ *belongs to the domain of attraction of a non-degenerate Gaussian measure* $\mu \in \mathcal{P}$ *with mean vector* $0 \in \mathcal{H}$ *and covariance operator* S *on* $\mathcal{H}$ *if and only if*:

(i)

$$\lim_{\lambda\to\infty} \frac{\lambda^2 \int_{\|y\| \geq \lambda} \mathrm{d}P(y)}{\int_{\|y\|<\lambda} \|y\|^2 \mathrm{d}P(y)} = 0.$$

(ii)

$$\lim_{\lambda\to\infty} \frac{\int_{\|y\|<\lambda} \langle x, y\rangle^2 \mathrm{d}P(y)}{\int_{\|y\|<\lambda} \langle z, y\rangle^2 \mathrm{d}P(y)} = \frac{\langle Sx, x\rangle}{\langle Sz, z\rangle}$$

for $z \neq 0$ *and* $y, z \in \mathcal{H}$ *provided* $\int_{\mathcal{H}} \|y\|^2 \mathrm{d}P(y) = \infty$, *or*

(ii)′

$$\lim_{\lambda\to\infty} \frac{\int_{\|y\|<\lambda} \langle y-a, x\rangle^2 \mathrm{d}P(y)}{\int_{\|y\|<\lambda} \langle y-a, z\rangle^2 \mathrm{d}P(y)} = \frac{\langle Sx, x\rangle}{\langle Sz, z\rangle}$$

for $z \neq 0$, $y, z \in \mathcal{H}$ *provided* $\int_{\mathcal{H}} \|y\|^2 \mathrm{d}P(y) < \infty$, *where* $a \in \mathcal{H}$ *is determined uniquely by* $\langle a, x\rangle = \int_{\mathcal{H}} \langle x, y\rangle \, \mathrm{d}P(y)$, $x \in \mathcal{H}$.

(iii)

$$\alpha_m = \limsup_{\lambda\to\infty} \frac{\int_{\|y\|<\lambda} \|\pi_m y\|^2 \mathrm{d}P(y)}{\int_{\|y\|<\lambda} \|y\|^2 \mathrm{d}P(y)} > 0$$

for $m \geqslant 1$ *and*

$$\lim_{m\to\infty} \alpha_m = 0$$

where $\pi_m y = \sum_{\nu=m}^{\infty} \langle y, e_\nu\rangle e_\nu$ *for some orthonormal basis* $e_\nu \in \mathcal{H}$ *and* $y \in \mathcal{H}$.

The domain of attraction of a non-Gaussian stable probability measure was characterized by Klosowska [6], and Kuelbs and Mandrekar [15]. For simplicity in presentation we consider only the symmetric case.

Theorem 2.3 (Kuelbs and Mandrekar [15]). *Let* $P \in \mathcal{P}$. *Then* P *belongs to the domain of attraction of a nondegenerate symmetric stable probability measure* μ *of type* α $(0 < \alpha < 2)$ *if and only if the following two conditions are satisfied*:

(i)

$$\lim_{\lambda\to\infty} \frac{\int_{\|y\|\geqslant\lambda} \mathrm{d}P(y)}{\int_{\|\pi_m y\|\geqslant k\lambda} \mathrm{d}P(y)} = \frac{c_1 k^\alpha}{c_m}$$

where for every $m \geqslant 1$, $c_m > 0$, $\lim_{m\to\infty} c_m = 0$ *and* π_m *is as defined in Theorem* 2.2.

(ii)

$$\lim_{\lambda\to\infty} \frac{P\{y \in \mathcal{H}: \|y\| \geqslant \lambda, y/\|y\| \in A\}}{P\{y \in \mathcal{H}: \|y\| \geqslant \lambda, y/\|y\| \in A^*\}} = \frac{\nu(A)}{\nu(A^*)}$$

for all continuity sets $A, A^* \in \mathcal{B}(S)$, *the Borel* σ-*field of subsets of* S, *with* $\nu(A^*) > 0$.

The domains of attractions of stable measures on a finite dimensional Euclidean space were studied earlier by Rvačeva [21].

Let $P\in\mathcal{P}$. Suppose a sequence $a_n>0$, and a sequence of elements $x_n\in\mathcal{H}$ exist such that

$$T_{a_n}P^{*n}*\delta_{x_n}\Rightarrow\mu \tag{2.6}$$

where $\mu\in\mathcal{P}$. Then we know that μ is stable and P belongs to the domain of attraction of μ. Suppose, however, that the parameter sequence $\{n\}$ in (2.6) does not run through all natural numbers but through some infinite subsequence $\{n_j\}$, $n_j\to\infty$ as $j\to\infty$. Then we say that P belongs to the domain of partial attraction of μ. It is known [2] that only infinitely divisible probability measures may have domain of partial attraction. Moreover, every i.d. probability measure on $\mathcal{B}$ has a nonempty domain of partial attraction [1]. We now consider the special class $\mathfrak{A}$ of limiting measures μ of the sequence $\{T_{a_{n_j}}P^{*n_j}*\delta_{x_{n_j}},\ j\geqslant 1\}$ when $\{n_j\}$ satisfies the additional condition

$$\begin{cases} 1\leqslant n_j<n_{j+1} & \text{for all } j\geqslant 1,\\ \lim\limits_{j\to\infty}\dfrac{n_{j+1}}{n_j}=r, & 1\leqslant r<\infty. \end{cases} \tag{2.7}$$

Following Kruglov [7], we call members of the class $\mathfrak{A}$ semistable. It is clear that $\mathfrak{A}$ is a subclass of the class of all i.d. probability measures on $\mathcal{B}$ and contains the class of all stable probability measures on $\mathcal{B}$. We give below two results characterizing the Fourier transform of a semistable probability measure on $\mathcal{B}$ which are due to Laha and Rohatgi [16]. Kumar [13] also obtained essentially similar results by a somewhat different approach.

Theorem 2.4. *A complex-valued function ϕ on $\mathcal{H}$ is the Fourier transform of a semistable probability measure on $\mathcal{B}$ if and only if ϕ satisfies the functional equation*

$$\phi(y)=\phi^r(ay)\exp(\mathrm{i}\langle x,y\rangle),\quad y\in\mathcal{H} \tag{2.8}$$

where $x\in\mathcal{H}$, $0<a<1$ and $r\geqslant 1$.

Theorem 2.5. *A complex-valued function ϕ on $\mathcal{H}$ is the Fourier transform of a semistable probability measure on $\mathcal{B}$ if and only if ϕ satisfies one of the following conditions*:

(i) *ϕ is the Fourier transform of a stable probability measure on $\mathcal{B}$ with index α, $0<\alpha<2$.*

(ii) *ϕ is the Fourier transform of a Gaussian probability measure on $\mathcal{B}$.*

(iii) *The Lévy spectral measure μ associated with ϕ in the representation* (1.5) *satisfies the equation*

$$M(b^{-1}E)=b^{\rho}M(E)$$

where $0<\rho<2$, $0<b<1$ and $E\in\mathcal{B}$, $0\notin E$.

The Fourier transforms of probability measures on $\mathbb{R}$ satisfying the functional equation of the form (2.8) were first considered by Lévy [18]. Kruglov [7], however, made a systematic investigation of class $\mathfrak{A}$ on $\mathbb{R}$. In particular, he obtained an analog of the Lévy–Khintchin representation of the Fourier transform of a semistable probability measure.

3. Operator-stable and operator-semistable probability measures

Let $\mathcal{H}$ be a real separable Hilbert space and let $\mathcal{B}$ be the Borel σ-field of subsets of $\mathcal{H}$. Let $\mathcal{L}=\mathcal{L}(\mathcal{H})$ denote the algebra of all bounded linear operators on $\mathcal{H}$ with the usual norm topology. Let $\mathcal{G}=\mathcal{G}(\mathcal{H})$ denote the multiplicative group of all invertible operators in $\mathcal{L}$. Let $\mathcal{P}=\mathcal{P}(\mathcal{H})$ be the class of all probability measures on $\mathcal{B}$.

Let $P\in\mathcal{P}$. Then a closed subset $S(P)\subseteq\mathcal{H}$ is said to be the support of P if the complement of $S(P)$ in $\mathcal{H}$ has P-measure zero and, moreover, for any $x\in S(P)$ and any neighborhood V_x of x, $P(V_x)>0$. We say that $P\in\mathcal{P}$ is full if its support is not contained in any proper closed subspace of $\mathcal{H}$.

Let $A\in\mathcal{G}$ and $P\in\mathcal{P}$. We define the measure AP on $\mathcal{B}$ by setting

$$AP(E)=P(A^{-1}E), \quad E\in\mathcal{B}.$$

Then $AP\in\mathcal{P}$ and, moreover,

$$\widehat{AP}(y)=\hat{P}(A^*y), \quad y\in\mathcal{H},$$

where A^* is the adjoint of A. We note that for $A\in\mathcal{G}$ and $P_1, P_2\in\mathcal{P}$

$$A(P_1*P_2)=AP_1*AP_2$$

and if $A_n\in\mathcal{G}$, $n\geq 1$, $A\in\mathcal{G}$, P_n, $P\in\mathcal{P}$ such that $A_n\to A$ and $P_n\Rightarrow P$, then $A_nP_n\Rightarrow AP$.

Let $P\in\mathcal{P}$. Then P is said to be operator stable, if sequences $A_n\in\mathcal{G}$, $x_n\in\mathcal{H}$ and a measure $Q\in\mathcal{P}$ exist such that

$$A_nQ^{*n}*\delta_{x_n}\Rightarrow P. \tag{3.1}$$

The class of operator-stable measures on a finite-dimensional Euclidean space has been investigated by Sharpe [23]. Let $\mathcal{H}=\mathcal{V}$ be a finite dimensional Euclidean space and let End $\mathcal{V}$ and Aut $\mathcal{V}$, denote, respectively, the algebra of all endomorphisms and the group of all automorphisms of the space $\mathcal{V}$. Clearly $\mathcal{L}=\text{End}\,\mathcal{V}$ and $\mathcal{G}=\text{Aut}\,\mathcal{V}$. The principal results of Sharpe [23] can now be stated in the following theorem.

Theorem 3.1.

(i) *Let $P\in\mathcal{P}$ be a full measure. Then P is operator-stable if and only if for every $n\geq 1$ there exists a $B_n\in\text{Aut}\,\mathcal{V}$ and an $x_n\in\mathcal{V}$ such that the relation*

$$P^{*n}=B_nP*\delta_{x_n} \tag{3.2}$$

holds. In particular, every full operator-stable probability measure on $\mathcal{B}$ is i.d.

(ii) *A full probability measure P on $(\mathcal{V},\mathcal{B})$ is operator-stable if and only if there exists a $B\in\text{Aut}\,\mathcal{V}$ and an $x_t\in\mathcal{V}$ such that the relation*

$$P^t=t^BP*\delta_{x_t} \tag{3.3}$$

holds for all $t>0$. Here P^t is the (unique) probability measure whose Fourier transform is $\hat{P}^t$, $t^B=\exp(\ln tB)$, $t>0$, and $B\in\text{Aut}\,\mathcal{V}$ is called the exponent of P.

Moreover, in this case the spectrum of B is contained in the half-plane $\operatorname{Re} z \geq \frac{1}{2}$, *and all the eigenvalues of B on the line* $\operatorname{Re} z = \frac{1}{2}$ *are simple.*

(iii) *Let* $P \in \mathcal{P}$ *be a full operator-stable measure. Then P can be written as the convolution* $P = P_1 * P_2$ *of measures* P_1 *and* P_2 *concentrated respectively on subspaces* $\mathcal{V}_1$ *and* $\mathcal{V}_2$ *of* $\mathcal{V}$ *each of which is invariant with respect to B such that* $\mathcal{V} = \mathcal{V}_1 \oplus \mathcal{V}_2$. *Here* P_1 *is a full Gaussian measure on* $\mathcal{V}_1$ *and* P_2 *is a full operator-stable measure on* $\mathcal{V}_2$ *having no Gaussian component.*

Kucharczak [9] showed that every (not necessarily full) operator-stable measure admits the representation (3.2). In a separate paper [8], he obtained the following representation of the Fourier transform of a full operator-stable probability measure on $(\mathcal{V}, \mathcal{B})$ using the method of extreme points.

Theorem 3.2. *A complex-valued function* ϕ *on* $\mathcal{V}$ *is the Fourier transform of a full operator-stable probability measure on* $\mathcal{B}$ *if and only if*

$$\ln \phi(y) = \mathrm{i}\langle a, y\rangle - \tfrac{1}{2}\langle Sy, y\rangle + \int_{\mathcal{V}-\{0\}} \int_0^\infty \left\{ \exp\left[\mathrm{i}\langle t^B x, y\rangle\right] - 1 - \frac{\mathrm{i}\langle t^B x, y\rangle}{1 + \|t^B x\|^2} \frac{1}{t^2} \mathrm{d}t\, \mathrm{d}\mu(x) \right. \tag{3.4}$$

where $a \in \mathcal{V}$, $B \in \operatorname{Aut} \mathcal{V}$ *such that all the eigenvalues of B are in the half-plane* $\operatorname{Re} z \geq \frac{1}{2}$, *S is a non-negative symmetric operator on* $\mathcal{V}$ *which vanishes on the (B-invariant) subspace* $\mathcal{V}_0 \subset \mathcal{V}$ *spanned by those eigenvectors which are associated with the eigenvalues of B lying in the halfplane* $\operatorname{Re} z > \frac{1}{2}$, *and* μ *is a finite measure on* $\mathcal{B}$ *concentrated on* $\mathcal{V}_0$.

Parthasarathy and Schmidt [20] considered a more general notion of operator-stability. We identify the group $\operatorname{Aut} \mathcal{V}$ with the general linear group $\mathrm{GL}(d, \mathbb{R})$ where $d = \dim_{\mathbb{R}} \mathcal{V}$. Let $G \subseteq \mathrm{GL}(d, \mathbb{R})$ be a fixed subgroup. Then $P \in \mathcal{P}$ is said to be operator stable with respect to G if for every $A_1, A_2 \in G$ there exist an $A_3 = A_3(A_1, A_2) \in G$ and an $x \in \mathcal{V}$ such that

$$A_1 P * A_2 P = A_3 P * \delta_x. \tag{3.5}$$

It can be easily verified from (3.3) in Theorem 3.1 that the class of operator-stable measures on $(\mathcal{V}, \mathcal{B})$ studied by Sharpe coincides with the class of probability measures which are operator-stable with respect to the one parameter subgroup $G = \{t^B, t > 0\}$ of $\mathrm{GL}(d, \mathbb{R})$ generated by a fixed $B \in \mathrm{GL}(d, \mathbb{R})$. It is shown in [20] that every probability measure which is operator stable with respect to a fixed subgroup $G \subseteq \mathrm{GL}(d, \mathbb{R})$ is i.d. Moreover, the class of probability measures on $\mathcal{B}$ which are operator-stable with respect to the full group $\mathrm{GL}(d, \mathbb{R})$ coincides with the class of all non-singular Gaussian and the degenerate probability measures.

Schmidt [22] has made a detailed investigation of the class of probability measures which are operator-stable with respect to the subgroups of all invertible diagonal matrices and invertible lower triangular matrices of $\mathrm{GL}(d, \mathbb{R})$.

We now return to the case when $\mathcal{H}$ is a real separable Hilbert space and use the notation introduced at the beginning of this section. Very little work appears to have been done on operator-stable probability measures in this case. The following characterization of an operator-stable probability measure P on $(\mathcal{H}, \mathcal{B})$ is due to Laha and Rohatgi [17].

Theorem 3.3. *Let $P \in \mathcal{P}$ be a full measure. Then P is operator stable if and only if for every $n \geq 1$ the relation*

$$P = B_n P^{*n} * \delta_{y_n} \tag{3.6}$$

holds for $B_n \in \mathcal{G}$ and $y_n \in \mathcal{H}$.

It follows as an immediate consequence of (3.6) in Theorem 3.3 that every full operator-stable probability measure P on $(\mathcal{H}, \mathcal{B})$ is i.d. A similar characterization of operator-stable probability measures on Banach spaces admitting a homogeneous decomposition has been obtained by Kucharczak and Urbanik [10].

Next we consider operator-semistable probability measures on $(\mathcal{H}, \mathcal{B})$. Let $P \in \mathcal{P}$. We say that P is operator-semistable if there exist a measure $Q \in \mathcal{P}$, and sequences $A_n \in \mathcal{G}$, $x_n \in \mathcal{H}$ and a sequence of positive integers k_n, $k_n \to \infty$ as $n \to \infty$, with $k_{n+1}/k_n \to r$ for $r \geq 1$ such that

$$A_n Q^{*k_n} * \delta_{x_n} \Rightarrow P. \tag{3.7}$$

If $\mathcal{H}$ is infinite-dimensional we assume that the sequence $\{A_n\}$, $A_n \in \mathcal{G}$, satisfies the following condition:

$$\begin{cases} \text{The semigroup,} \\ \text{Sem}\big(\{A_m A_n^{-1}, n=1,2,\dots, m; m=1,2,\dots\}\big), \text{ is compact} \\ \text{in the norm topology of } \mathcal{L}. \end{cases} \tag{3.8}$$

We note, however, that for full probability measures on finite-dimensional Euclidean spaces the compactness condition (3.8) may be omitted (see, for example, [2]).

The following characterization of an operator-semistable probability measure on $(\mathcal{H}, \mathcal{B})$ is due to Laha and Rohatgi [17].

Theorem 3.4. *Let $P \in \mathcal{P}$ be a full measure. Then P is operator-semistable if and only if*

(i) *P is i.d., and*

(ii) *there exist a c, $0 < c < 1$, an element $x \in \mathcal{H}$, and an operator $A \in \mathcal{G}$, such that the relation*

$$P^c = AP * \delta_x \tag{3.9}$$

holds. (Here P^c denotes the probability measure whose Fourier transform is $\hat{P}^c$.)

In the case when $\mathcal{H} = \mathcal{V}$ is a finite-dimensional Euclidean space, Jajte [2] has obtained a more complete description of the class of operator-semistable measures. In this case he has additionally shown that the spectrum of A occurring in (3.9) is contained in $\{|z|^2 \leq c\}$ and the eigenvalues of A on $\{|z|^2 = c\}$ are simple. Moreover, the measure P can be written as a convolution $P = P_1 * P_2$ of two measures P_1 and P_2 concentrated, respectively, on subspaces $\mathcal{V}_1$ and $\mathcal{V}_2$ of $\mathcal{V}$ (each of which is invariant with respect to A) such that $\mathcal{V}=\mathcal{V}_1\oplus\mathcal{V}_2$. In this case P_1 is a full operator-semistable measure without any Gaussian component and P_2 is a full Gaussian measure.

4. Concluding remarks

In the following we briefly describe some important problems which appear to be as yet unsolved. It will be of considerable interest to obtain an analog of the Lévy–Khintchine representation of the Fourier transform of a semistable probability measure on $(\mathcal{H}, \mathcal{B})$. It may then be possible to describe the domain of attraction of semistable probability measures on $(\mathcal{H}, \mathcal{B})$. For the case of the real line this has been done by Kurglov [7].

For the case of probability measures on $(\mathcal{V}, \mathcal{B})$ the major problem is to determine the class of probability measures which are operator-stable with respect to any fixed subgroup G of $GL(d, \mathbb{R})$ in the sense of Parthasarathy and Schmidt [20]. Once this has been accomplished one can proceed to study the domains of attraction of these probability measures and related problems.

For the case of operator-stable and operator-semistable probability measures on $(\mathcal{H}, \mathcal{B})$ where $\mathcal{H}$ is infinite-dimensional it is of interest to obtain an explicit representation of their Fourier transforms. Once this has been achieved one can investigate the domains of attractions of these probability measures. It would also be of interest to obtain a precise description of the structure of the spectrum of the operator A in (3.9).

References

[1] Baránska, J. (1973). Domains of partial attraction for infinitely divisible distributions in a Hilbert space. *Colloq. Math.* **38**, 317–322.

[2] Jajte, R. (1977). Semistable probability measures on R^N. *Studia Math.* **66**, 29–39.

[3] Jurek, Z. and Urbanik, K. (1978). Remarks on stable measures on Banach spaces. *Colloq. Math.* **38**, 269–276.

[4] Klosowska, M. (1972). The domain of attraction of a normal distribution in a Hilbert space. *Studia Math.* **43**, 195–208.

[5] Klosowska, M. (1973). A characterization of the domain of attraction of a normal distribution in a Hilbert space. *Colloq. Math.* **28**, 323–327.

[6] Klosowska, M. (1974). The domain of attraction of a non-Gaussian stable distribution in a Hilbert space. *Colloq. Math.* **32**, 127–136.

[7] Kruglov, V.M. (1972). On the extension of the class of stable distributions. *Theor. Probability Appl.* **17**, 685–694.

[8] Kucharczak, J. (1975). Remarks on operator-stable measures. *Colloq. Math.* **34**, 109–119.

[9] Kucharczak, J. (1975). On operator-stable probability measures. *Bull. Acad. Polon. Sci.* **23**, 571–576.

[10] Kucharczak, J. and Urbanik, K. (1977). Operator-stable measures on some Banach spaces. *Bull. Acad. Polon. Sci.* **25**, 585–588.

[11] Kuelbs, J. (1973). A representation theorem for symmetric stable processes and stable measures on *H. Z. Wahrscheinlichkeitstheorie und Verw. Gebiete* **20**, 259–271.

[12] Kuelbs, J. and Mandrekar, V. (1974). Domains of attraction of stable measures on a Hilbert space. *Studia Math.* **50**, 149–162.

[13] Kumar, A. (1976). Semistable probability measures on a Hilbert space. *J. Multivar. Anal.* **6**, 309–318.
[14] Kumar, A. and Mandrekar, V. (1972). Stable probability measures on Banach spaces. *Studia Math.* **42**, 133–144.
[15] Laha, R.G. and Rohatgi, V.K. (1979). *Probability Theory*. Wiley, New York.
[16] Laha, R.G. and Rohatgi, V.K. (1980). Semistable measures on a Hilbert space. *J. Multivar. Anal.* **10**, 88–94.
[17] Laha, R.G. and Rohatgi, V.K. (1980). Operator semi-stable probability measures on a Hilbert space. *Bull Austral. Math. Soc.* **22**, 397–406, 479–480.
[18] Lévy, P. (1954). *Theorie de l'Addition des Variables Aléatoires*. Gauthier-Villars, Paris.
[19] Parthasarathy, K.R. (1967). *Probability Measures on Metric Spaces*. Academic Press, New York.
[20] Parthasarathy, K.R. and Schmidt, K. (1975). Stable positive definite functions. *Trans. Am. Math. Soc.* **203**, 161–174.
[21] Rvačeva, E.L. (1962). On domains of attraction of multidimensional distributions. In: *Selected Translations in Mathematics and Statistics*. **2**. American Math. Soc., Providence, RI, pp. 183–205.
[22] Schmidt, K. (1975). Stable probability measures on R^ν. *Z. Wahrscheinlichkeitstheorie und Verw. Gebiete* **33**, 19–31.
[23] Sharpe, M. (1969). Operator-stable probability distributions on vector spaces. *Trans. Am. Math. Soc.* **136**, 51–65.
[24] Varadhan, S.R.S. (1962). Limit theorems for sums of independent random variables with values in a Hilbert space. *Sankhȳa* **24**, 213–238.

Note added in proof. Since this paper was completed a considerable amount of new material has become available. A general treatment of the central limit theory for Banach space valued random variables is given in the monograph "*The Central Limit Theorem for Real and Banach Valued Random Variables*," by A. Araujo and E. Ginê, Wiley, New York, 1979. Priority for Theorems 2.4 and 2.5 goes to V.M. Kruglov (On a class of limit distributions in a Hilbert space, *Lietuvos Mat. Rinkinys* **12**(1972) 85–88 (in Russian)). For a characterization of the domain of attraction in a separable Banach space we refer to the survey paper of V. Mandrekar (Domain of attraction problem on Banach spaces, 1980-preprint). The domain of attraction of semistable measures on $\mathbb{R}_n$ has been investigated by E. Hensz and M. Klosowska (On the semi-attraction to the stable measures, *Trans. 8th Prague Conference on Information Theory*, *Vol. C*, 135–142, Reidel, 1979). Full operator-stable probability measures on a Banach space have been characterized by W. Krakowiak (Operator stable probability measures on Banach spaces, *Colloquium Math.* **41**(1979) 313–326). He also obtains a representation of the characteristic functional of an operator stable measure. See also Z.J. Jurek, Convergence of types, self decomposability and stability of measures on linear spaces, Lecture Notes in Mathematics: Proc. IIIrd Conference on Probablity on Banach Spaces, 1981.

G. Kallianpur, P.R. Krishnaiah, J.K. Ghosh, eds., *Statistics and Probability: Essays in Honor of C.R. Rao*
© North-Holland Publishing Company (1982) 455–463

LIMIT THEOREMS FOR EMPIRICAL MEASURES AND POISSONIZATION

L. LE CAM
University of California, Berkeley, CA, U.S.A.

1. Introduction

Let $X_{n,j}$; $j=1,2,\ldots,n$ be independent random variables taking their values in a measurable space $(\Omega, \mathscr{Q})$. For each $A \in \mathscr{Q}$ let $N_n(A)=\Sigma_j I_A(X_{n,j})$ and let $\mu_n(A)= n^{-1}N_n(A)$. This is the empirical measure defined by the $X_{n,j}$. The literature contains many theorems concerning the limiting behavior of μ_n when the $X_{n,j}$ are not only independent but also identically distributed. The intent of the present paper is to show that, in a large class of such theorems, one can delete the assumption that the distributions are identical.

The proof is carried out by replacing $X_{n,j}$ by a Poisson number ν_j of independent variables $X_{n,j,k}$ all distributed with the same law as $X_{n,j}$.

The technique of proof requires the measurability of certain operations. Here we have opted for a simplified system using weakly defined measures. The relevant definitions are given in Section 2 below. In some theorems this is essentially the same thing as working with equivalence classes of functions instead of the functions themselves, as done by Pollard in a paper to be published. An alternative would be to use conditions of the Souslin type, as done by Dudley in [5]. The essential features of our argument would be unchanged.

Section 3 states some of the limit theorems obtainable by our procedure. Section 4 illustrates it further by giving a proof of an inequality relative to the tails of the distributions of the empirical cumulatives in finite dimensional spaces. The technique cannot possibly give the best attainable results. However, it is simple and gives results without elaborate computations.

2. Weakly defined measures

In the present section, we recall certain results relative to symmetrization operations and to the replacement of the distribution of a sum by its natural accompanying infinitely divisible distribution. This is done in a context using weakly defined Radon measures on Banach spaces. For the applications to empirical measures the Banach space will be the space, $\mathfrak{X}$, of bounded numerical functions on a class $\mathbb{S} \subset \mathscr{Q}$ with its supremum norm. Typically $\mathfrak{X}$ is not separable and the addition operation is not measurable, hence the necessity for some precautions. A possibility is as follows.

Let $\mathcal{X}$ be a Banach space with a norm noted $\|x\|$ and with a dual $\mathcal{Y}$. For each finite set $F \subset \mathcal{Y}$ let T_F be the map from $\mathcal{X}$ to $\mathbb{R}^F$ defined by $T_F(x) = \{\langle y, x\rangle; y \in F\}$. Give $\mathbb{R}^F$ its maximum coordinate norm and let Λ_F be the space of bounded numerical functions defined on F and satisfying there a Lipschitz condition. Let H be the space of functions which are defined on $\mathcal{X}$ by expressions of the type $\gamma = f \circ T_F$ for a finite set $F \subset \mathcal{Y}$ and for an $f \in \Lambda_F$. This set H is a vector lattice which contains the constant functions. Let B denote the unit ball of $\mathcal{Y}$. For $F \subset B$, let H_F be the subset of H formed by functions $\gamma = f \circ T_f$ where on $\mathbb{R}^F$ the function f satisfies the condition $|f| \leq 1$ and $|f(u_1) - f(u_2)| \leq |u_1 - u_2|$.

If μ is a linear functional defined on H, give it a norm by the expression

$$\|\mu\|_B = \sup_{F \subset B} \sup_{\gamma \in H_F} |\langle \mu, \gamma \rangle|$$

in which F runs through all finite subsets of the ball B. This amounts to the same thing as taking the images $T_F\mu$ and using for them the dual Lipschitz norm used by Dudley. Then one takes a supremum over $F \subset B$.

Let $\mathcal{Y}^*$ be the algebraic dual of $\mathcal{Y}$. Give it the weak topology $w(\mathcal{Y}^*, \mathcal{Y})$ induced by $\mathcal{Y}$. Note that the elements of H are uniformly continuous for $w(\mathcal{Y}^*, \mathcal{Y})$. Therefore, they have uniquely defined uniformly continuous extensions to the entire $\mathcal{Y}^*$. In particular they extend to the second dual $\mathcal{Z} \subset \mathcal{Y}^*$ of the space $\mathcal{X}$. The extension to $\mathcal{Z}$ of $\gamma \in H$ will be called $\hat{\gamma}$. The space H extends to $\hat{H}$. The spaces of linear functionals on H and $\hat{H}$ are canonically isomorphic. Two classes of linear functionals will be used below.

Definition. Let μ be a positive linear functional on $\hat{H}$. It is a weak Radon expectation if $\langle \hat{\mu}, 1 \rangle = 1$ and if it is the restriction to $\hat{H}$ of a measure which is a Radon measure on $\mathcal{Z}$ for the $w(\mathcal{Z}, \mathcal{Y})$ topology.

A weak Radon expectation is strong if it is the restriction to $\hat{H}$ of a measure which is a Radon measure for the strong topology of $\mathcal{Z}$.

A weak Radon expectation on H has for parent a Radon measure $\bar{\mu}$ entirely determined by the values of μ on $\hat{H}$. All $w(\mathcal{Z}, \mathcal{Y})$ closed subsets of $\mathcal{Z}$ are in the domain of $\bar{\mu}$. The open balls $\{z; \|z - z_0\| < b\}$ are also in the domain of $\bar{\mu}$. The addition map from $\mathcal{Z} \times \mathcal{Z}$ is continuous for the weak topology $w(\mathcal{Z}, \mathcal{Y})$. Therefore, there is no difficulty in the definition of convolution of weak Radon expectation. Equivalently there is no difficulty in the definition of addition for the corresponding random elements with values in $\mathcal{Z}$.

The following lemma gives a characterization of weak Radon expectations. Let P be a positive linear functional defined on H and such that $P1 = 1$. Let V be a convex symmetric subset of $\mathcal{Y}$ and let $K = V^\circ$ be its polar in $\mathcal{Y}^*$.

Lemma 1. *For each $\varepsilon \in [0, 1]$, the following statements are equivalent*:

(i) *For $\gamma \in H$ such that $0 \leq \gamma \leq 1$ and such that γ vanishes on K one has $P\gamma \leq \varepsilon$.*

(ii) *There is a probability measure P_1 which is a Radon measure on K for $w(\mathcal{Y}^*, \mathcal{Y})$ and a positive linear functional P_2 or H such that P is the restriction to H of $(1-\varepsilon)P_1 + \varepsilon P_2$.*

Another equivalent form of the above conditions can be stated using the "process" $y \to \mathfrak{Z}(y)$ whose distribution is P on $\mathfrak{Y}^*$. It means that for every finite set $F \subset V$ one has

$$P\{\sup[|Z(y)|;\, y \in F] > 1\} \leqslant \varepsilon.$$

The ideas of the proof of Lemma 1 go back to Mourier [10]. More general results have been given by de Acosta [4].

In the sequel we shall use the space of weak Radon expectations and define the topology of that space by the norm $\|\mu - \nu\|_B$ defined above. Convergence for this norm is related to the more commonly used Prohorov distance or the distance used by Dudley for measures defined on the σ-field of $\mathfrak{Z}$ generated by its balls.

The two are equivalent in $C[0,1]$ as shown by Bartoszynski [2]. A description of the situation in uniform spaces was given by Caby [3].

One easy result is as follows:

Let μ and ν be two weak Radon expectations on H. Assume that $\|\mu - \nu\|_B \leqslant \varepsilon \leqslant \frac{1}{4}$. Let K be a $w(\mathfrak{Z}, \mathfrak{Y})$ closed convex symmetric subset of $\mathfrak{Z}$. Let K'_ε be the set $\{z\colon z \in \mathfrak{Z}, \inf[\|z - t\|;\, t \in K] \leqslant 2\sqrt{\varepsilon}\}$. Then $\bar{\nu}(K'_\varepsilon) \geqslant \bar{\mu}(K) - \sqrt{\varepsilon}$. The proof of this statement mimics the usual proof of equivalence of the Prohorov distance and the dual Lipschitz distance. We shall also need a result concerning "tightness" as follows. To state it, for any set $K \subset \mathfrak{Z}$ let $K^\varepsilon = \{z;\, [\inf\|z - t\|;\, t \in K] \leqslant \varepsilon\}$.

Lemma 2. *Let $\mathfrak{S}$ be the space of all weak Radon expectations on H. Let Φ be a filter on $\mathfrak{S}$. Assume that along Φ and for each $h \in H$ the numbers νh converge to a limit ϕh. Then the following statements are equivalent*:

(i) *ϕ is a strong Radon expectation on H and $\|\nu - \phi\|_B$ converges to zero along Φ.*

(ii) *For each $\varepsilon > 0$, there is a strongly compact set $K \subset \mathfrak{Z}$ and a set $S \in \Phi$ such that $\bar{\nu}(K^\varepsilon) \geqslant 1 - \varepsilon$ for all $\nu \in S$.*

(iii) *For each $\varepsilon > 0$, there is a strongly compact set $K \subset \mathfrak{Z}$ such that for each $\alpha > 0$ there is a set $S_\alpha \in \Phi$ for which $\bar{\nu}(K^\alpha) \geqslant 1 - \varepsilon$ for all $\nu \in S_\alpha$.*

The proof resembles that of the usual Prohorov theorem and will be skipped. The result is, however, essential for the proof of the other results stated below, and these results are essential to the Poissonization approach. To express them, let us say that a sequence $\{P_\nu\}$, $P_\nu \in \mathfrak{S}$ is *shift compact with strong limits* if there are constants $a_\nu \in \mathfrak{Z}$ with the following property. Let P'_ν be P_ν shifted by a_ν. That is if $P_\nu = \mathfrak{L}(X_\nu)$ then $P'_\nu = \mathfrak{L}(X_\nu + a_\nu)$. Then every subsequence of $\{P'_\nu\}$ contains a further subsequence which converges for the norm of $\mathfrak{S}$ to a limit which is a strong Radon expectation on H.

The sequence $\{P_\nu\}$ will be called *relatively compact with strong limits* if it possesses the property just described, with constants $a_\nu = 0$.

If a strong Radon expectation on H has its support in $\mathfrak{X}$ itself, call it a strong Radon expectation on $\mathfrak{X}$.

Theorem 1. *Let $\{P_\nu\}$ and $\{Q_\nu\}$ be two sequences of weak Radon expectations on H. Assume that the convolution $P_\nu Q_\nu$ converges for the norm of $\mathfrak{S}$ to a limit ϕ which is* *a strong Radon expectation on $\mathcal{X}$. Then the sequence $\{P_\nu\}$ is shift compact with strong limits in $\mathcal{X}$.*

The next result refers to the Poissonization approach. If P is any weak Radon expectation on H one can define its Poisson exponential $\exp\{P-I\}$ by the convolution series $e^{-1}\Sigma_n P^n/n!$. If $P=\mathfrak{L}(X)$, this means that $\exp(P-I)$ is the distribution of $\Sigma_{j\leq\nu} X_j$ where the X_j are independent, $\mathfrak{L}(X_j)=\mathfrak{L}(X)$ and ν is a Poisson variable independent of the X_j with $E\nu=1$.

Theorem 2. *Let $\{P_{\nu,j};\ j=1,2,\ldots,k_\nu;\ \nu=1,2,\ldots\}$ be a double array of weak Radon expectations. Let Q_ν be the Poisson exponential $Q_\nu=\exp\{\Sigma_j(P_{\nu,j}-I)\}$ and let P_ν be the convolution $P_\nu=\Pi_j P_{\nu,j}$.*

Assume that, as ν tends to infinity, the sequence $\{Q_\nu\}$ converges for the norm of $\mathcal{E}$ to a limit Q which is a strong Radon expectation. Then the P_ν are shift compact with strong limits.

The preceding Theorems 1 and 2 will not be proved here. The proof of Theorem 1 parallels that of [11, Theorem 2.2, Chapter III]. The proof of Theorem 2 parallels that of [8, Théorème 3].

For added information for the infinitesimal case see Araujo [1].

Note, however, that here as in Le Cam [8] convergence of P_ν does not imply shift compactness of Q_ν even if the $P_{\nu,j}$ are symmetric. Note also that the result is stated only for limits which are *strong* Radon expectations. The reason for this is as follows. Take for $\mathcal{Z}$ the Banach space of bounded sequences with the uniform norm. If P is a weak Radon expectation on H, let P^s be its symmetrized, convolution of P with its symmetric P' such that $P'(A)=P(-A)$. One can find a family $\{P_\xi\}$ of weak Radon expectations on H with the following properties:

(i) ξ runs through a set Ξ which has the cardinality of the continuum.

(ii) P_ξ^s is independent of ξ.

(iii) Let $Z(\xi)$ have distribution P_ξ. Then for any two distinct points ξ and η of Ξ and any two elements a and b of $\mathcal{Z}$ one has

$$\|\mathfrak{L}[a+Z(\xi)]-\mathfrak{L}[b+Z(\eta)]\|_B\geq\tfrac{1}{4}.$$

Here Z is the second dual of the space c_0 of sequences which tend to zero. Thus it is allowable in one framework. To construct the $Z(\xi)$ proceed as follows. Let u_k be the sequence $u_k\in\mathcal{Z}$ whose entries are all zero except the kth one which is equal to unity. Let Ξ be the space of sequences $\xi=\{\xi_k;\ k=1,2,\ldots\}$ such that $\xi_k^2=1$. Let X_k be a sequence of independent identically distributed real valued variables such that $\Pr[X_k=-1]=\frac{1}{4}$ and $\Pr[X_k=1]=\frac{3}{4}$. Let $\mathcal{Z}(\xi)=\Sigma_k \xi_k X_k u_k$. The distributions $P_\xi=\mathfrak{L}[Z(\xi)]$ satisfy the requirements expressed above. We presume, but do not know, that similar examples can be constructed for any non-reflexive space $\mathcal{X}$. For reflexive spaces such examples are precluded by Theorem 1 since all weak Radon expectations are strong.

3. Limit theorems for empirical measures

Let Ω be a set carrying a σ-field $\mathcal{A}$. Let $\mathcal{S}\subset\mathcal{A}$ be a certain class of sets. We shall proceed below using for the spaces $\mathfrak{X}$ of Section 2 the space $\mathfrak{X}=\mathcal{B}(\mathcal{S})$ of bounded numerical functions on $\mathcal{S}$ with the uniform norm $\|f\|=\sup\{|f(s)|;\ S\in\mathcal{S}\}$. (One can identify $\mathfrak{X}$ with the second dual of the closure of the space of functions with finite support on $\mathcal{S}$).

Now for integer n, let $X_{n,j}$; $j=1,2,\dots,n$ be a sequence of independent random variables with values in $(\Omega,\mathcal{A})$. Let $P_{n,j}$ be the distribution of $X_{n,j}$. Let $M_n=\Sigma_{j=1}^n P_{n,j}$ and let $\bar{P}_n=n^{-1}M_n$. The $X_{n,j}$ define an empirical measure by the relation $\nu_n(A)=(1/n)\Sigma_{j=1}^n I_A(X_{n,j})$. Let Z_n be the stochastic process defined on $\mathcal{S}$ by:

$$Z_n(S)=\sqrt{n}\left[\mu_n(S)-\bar{P}_n(S)\right].$$

It will be called the empirical process of the $X_{n,j}$.

It is a second order process with $EZ_n(S)=0$ and covariances,

$$EZ_n(A)Z_n(B)=\frac{1}{n}\sum_{j=1}^{n}\left[P_{n,j}(A\cap B)-P_{n,j}(A)P_{n,j}(B)\right].$$

$$=\bar{P}_n(A\cap B)-\frac{1}{n}\sum_{j=1}^{n}P_{n,j}(A)P_{n,j}(B).$$

To such a system, one can associate other processes Z_n^* as follows. Let $X_{n,j}^*$ be independent random variables all distributed according to the average distribution $\bar{P}_n$. The Z_n^* have zero expectations and covariances:

$$EZ_n^*(A)Z_n^*(B)=\bar{P}_n(A\cap B)-\bar{P}_n(A)\bar{P}_n(B).$$

Consider also the Gaussian process Y_n which has zero expectation and the same covariance structure as Z_n. Each of these processes defines an expectation on the space H attached to $\mathfrak{X}=\mathcal{B}(\mathcal{A})$. They will be denoted by symbols such as $\mathcal{L}(Z_n)$, $\mathcal{L}(Y_n)$ etc. Note that both $\mathcal{L}(Z_n)$ and $\mathcal{L}(Z_n^*)$ are always weak Radon expectations. However, the same need not be true of $\mathcal{L}(Y_n)$.

One result is as follows:

Theorem 3. *Assume that there is a strong Radon expectation G such that $\|\mathcal{L}(Z_n^*)-G\|_B$ tends to zero. Then the $\mathcal{L}(Z_n)$ are relatively compact with strong limits.*

Proof. For each pair (n,j) let $X_{n,j,k}$; $k=1,2,\dots$ be independent variables with common distribution $\mathcal{L}(X_{n,j})$. Let ν_j be a Poisson variable with $E\nu_j=1$. Take all the $X_{n,j,k}$ and ν_j mutually independent.

The process Z_n can be written in the form

$$Z_n(A)=\frac{1}{\sqrt{n}}\sum_j\left[I_A(X_{n,j})-P_{n,j}(A)\right]$$

$$=\frac{1}{\sqrt{n}}\sum_j\left[I_A(X_{n,j})-\bar{P}_n(A)\right].$$

Consider also the process W_n defined by:

$$W_n(A)=\frac{1}{\sqrt{n}}\sum_j\sum_{k\leq\nu_j}\left[I_A(X_{n,j,k})-\bar{P}_n(A)\right].$$

Let $P'_{n,j}$ be the distribution of a process,

$$\{I_S(X_{n,j})-\bar{P}_n(S);\ S\in\mathbb{S}\}$$

on the space $\mathfrak{Z}$. The distribution of W_n depends only on the sum $\Sigma_j P'_{n,j}=n\bar{P}'_n$ where $\bar{P}'_n$ is the distribution of a process where $X_{n,j}$ is replaced by $X^*_{n,j}$. Thus W_n has the same distribution as W^*_n defined by

$$W^*_n(A)=\frac{1}{\sqrt{n}}\sum_{j=k}^{N_n}\left[I_A(X^*_{n,j})-\bar{P}_n(A)\right],$$

with $N_n=\Sigma_j\nu_j$. Now $(n/N_n)^{1/2}W^*_n$ is the empirical process associated with a Poisson number N_n of the identically distributed $X^*_{n,j}$. Its distribution being an average of distributions which converge also converges to the same limit. Since N_n/n tends to one, so does the distribution of W^*_n, thus $\mathfrak{L}(W_n)$ converges. According to Theorem 2, this implies that the sequence $\mathfrak{L}(Z_n)$ is shift compact with strong limits. However, in the present case the distributions $\mathfrak{L}[Z_n(A)]$ are approximable by centered Gaussian distributions and this uniformly for $A\in\mathbb{S}$. Thus no recentering is needed. The result follows.

The preceding Theorem 3 can be applied in particular to cases where $\bar{P}_n$ is a certain P independent of n and where the process of distribution G admit uniformly continuous sample paths on $\mathbb{S}$ metrized by the distance $d(A,B)=P[A\Delta B]$, at least if one assumes $\mathbb{S}$ precompact for this distance.

For instance, the theorem is applicable to cases where Ω is Euclidean and $\mathbb{S}$ is one of the usual classes: half-spaces, octants in a given coordinate system, etc. It applies also to unstructured spaces Ω when $\mathbb{S}$ is a Vapnik–Červonenkis class, as in Dudley [5].

Note, however, that in all such cases the passage to the weak Radon measures replaces items such as $\sup\{|Z_n(S)|;\ S\in\mathbb{S}\}$ by the corresponding supremum of equivalence classes. This is immaterial for many classes and for instance for the classes which satisfy an entropy with inclusion condition as in [5].

For other classes more specific measurability conditions seem to be needed. Under the Souslin type conditions used in [5] one can prove theorems similar to Theorems 1 and 2 and therefore Theorem 3. However, the proofs are more complex in their measurability aspects.

4. An inequality for empirical measures

Consider independent variables $X_{n,j}$ as in Section 3, but suppose that the set Ω is a Euclidean space and that the class $\mathbb{S}$ is the class of all sets of the form $\{x\colon x\leq y\}$ where $x\leq y$ means that no coordinate of x exceeds the corresponding coordinate of

y. The conclusions of Section 3 are applicable to this case. However, letting $F_n(y)=\mu_n\{x: x\leq y\}$ and $\bar{F}_n(y)=\bar{P}_n\{x; x\leq y\}$ one would like to have bounds for the tail probabilities of the variable $V_n=\sqrt{n}\sup|F_n(y)-\bar{F}_n(y)|=\sup\{|Z_n(S)|; S\in\mathcal{S}\}$ in the notation of Section 3.

Such bounds have been given by Kiefer and Wolfowitz [7] in the case of identically distributed variables. We shall give here a weaker bound valid for independent variables whose distributions vary arbitrarily with the index j.

Theorem 4. *Let Ω be Euclidean with dimension k. Let $\mathcal{S}$ be the class of sets of the form $S=\{x: x\leq y\}$ for $y\in\Omega$. Let $X_{n,j}$ be independent with $\mathcal{L}(X_{n,j})=P_{n,j}$. Let Z_n be the corresponding empirical process $Z_n(A)=(1/\sqrt{n})\Sigma[I_A(X_{n,j})-P_{n,j}(A)]$.*

Then, for every $r\geq 0$ and for $\|Z_n\|=\sup\{|Z_n(S)|; S\in\mathcal{S}\}$ one has

$$\Pr\left\{\|Z_n\|\geq r+\frac{1}{\sqrt{n}}\right\}\leq 2^{k+6}\exp\left\{-\frac{r^2}{48}\right\}.$$

Proof. According to a result of Jogdeo and Samuels the median $m_s(s)$ of any $Z_n(S)$ is such that $|m_n|\leq 1/\sqrt{n}$. Thus it will be sufficient to obtain bounds for the probability that $\|Z_n-m_n\|\geq r$.

Let $U_{n,j}(S)=(1/\sqrt{n})I_S(X_{n,j})$. Then Z_n is a centered version of $\Sigma_j U_{n,j}$.

Consider also another set of processes $U'_{n,j}$, independent replicates of the $U_{n,j}$. Let Z_n^s be the symmetrized process $Z_n^s=\Sigma_j(U_{n,j}-U'_{n,j})$.

According to Paul Lévy's symmetrization inequalities one has,

$$\Pr\{\|Z_n-m_n\|\geq r\}\leq 2\Pr\{\|Z_n^s\|\geq r\}.$$

Take independent random variables ξ_j with $P[\xi_j=1]=P[\xi_j=0]=\frac{1}{2}$ and consider instead of Z_n the sum $S_n=\Sigma_j\xi_j(U_{n,j}-U'_{n,j})$ and a Poissonized sum $T_n=\Sigma_j\Sigma_{k\leq\nu_j}[U_{n,j,k}-U'_{n,j,k}]$ where ν_j is a Poisson variable with $E\nu_j=\log 2$. The argument of [8, Théorème 3] shows that:

$$\Pr\{\|S_n\|\geq x\}\leq 2\Pr\{\|T_n\|\geq x\}.$$

Also S_n and $S'_n=\Sigma(1-\xi_j)(U_{n,j}-U'_{n,j})$ are identically distributed and $Z_n^s=S_n+S'_n$. This gives

$$\Pr\{\|Z_n\|\geq x\}\leq 2\Pr\{\|S_n\|\geq\tfrac{1}{2}x\}\leq 4\Pr\{\|T_n\|\geq\tfrac{1}{2}x\}.$$

Consider also the sum

$$V_n=\sum_j\sum_{k\leq\nu_j}U_{n,j,k}-\sum_j\sum_{k\leq\nu'_j}U_{n,j,k}$$

where the ν_j are independent replicates of the ν_j. Suppose that $\Pr\{\|V_n\|\geq x\}\leq f(x)$. Then there is some function a_n such that:

$$\Pr\left\{\left\|\left(\sum_j\sum_{k\leq\nu_j}U_{n,j,k}\right)-a_n\right\|\geq x\right\}\leq f(x).$$

This yields $\Pr\{\|T_n\|\geq x\}\leq 2f(\frac{1}{2}x)$.

Putting all these relations together, we obtain

$$\Pr\left\{\|Z_n\| \geqslant x + \frac{1}{\sqrt{n}}\right\} \leqslant 8f(\tfrac{1}{4}x).$$

It remains to evaluate the function f. For this consider the process V_n. It is a symmetrized version of a Poisson process with base $M_n = \Sigma_j P_{n,j}$ and step size $n^{-1/2}$. It is a process where disjoint sets yield independent variables. It follows then from the symmetry argument of Lévy that

$$\Pr\{\|V_n\| \geqslant x\} \leqslant 2^k \Pr\{|V_n(\Omega)| \geqslant x\}.$$

This is a standard result for $k=1$. To obtain it for $k>1$ one takes a finite subset of $\mathbb{R}^k$ and reorder it lexicographically. One can then apply the usual Lévy inequalities to the last coordinate. This permits to replace that last coordinate by $(+\infty)$ and one proceeds downwards.

Thus we are led to evaluate a quantity such as $\Pr\{|V_n(\Omega)| \geqslant x\}$. The distribution of $V_n(\Omega)$ is the same as that of $N - N'/\sqrt{n}$ where N and N' are independent Poisson variables with $EN = EN' = n\log 2$. Note also that the number r entering in the statement of the result can be taken smaller than $\sqrt{n}$ since $\|Z_n\| \leqslant \sqrt{n}$. Thus we need only evaluate $\Pr\{|N-N'| \geqslant \frac{1}{4}x\sqrt{n}\}$ for $x \leqslant \sqrt{n}$. Taking this into account, a straightforward computation using the function $E\exp\{t(N-N')\} = \exp\{[2n\log 2][\operatorname{Ch} t - 1]\}$ will give the result as stated.

Remark 1. The tails of the distribution of $N - N'$ are not decreasing in the square exponential manner described here. The restriction $x \leqslant \sqrt{n}$ is essential in the preceding argument.

Remark 2. The result given here can be compared with results of Van Zuijlen in [12], obtained by a different method.

Remark 3. The method used here cannot possibly give the best achievable bounds. Bickel and Van Zwet have conjectured that the identically distributed case would be the most dispersed case, as in the results of Hoeffding [6]. If so, the exponent $\frac{1}{48}r^2$ should be replaceable by $(2-\varepsilon)r^2$, $\varepsilon > 0$. For the case of $k=1$, Bretagnolle has shown us an argument which indeed yields the $(2-\varepsilon)r^2$ result. The argument may be extendable to higher dimensions.

5. Concluding remarks

The subject treated here arose, of course, in a statistical context. In many cases, one would propose to estimate a parameter by a minimum distance method using the empirical cumulative. This may be done occasionally for independent identically distributed observations $X^*_{n,j}$. However, if one wants to obtain robust estimates (see, for instance [9]), one needs to consider small departures from the distributions so that the variables observed are no longer $X^*_{n,j}$ but $X_{n,j}$ with $\mathcal{L}(X_{n,j})$ close to $\mathcal{L}(X^*_{n,j})$.

It is usual in the literature to assume that the $X_{n,j}$ are still independent identically distributed. The inequality given in Section 4 originated through a question of Wang who wanted a result where independence was maintained but $\mathfrak{L}(X_{n,j})$ was allowed to vary with j.

References

[1] Araujo, A. (1975). On infinitely divisible laws in $C[0,1]$. *Proc. Am. Math. Soc.* **51**, 179–185.

[2] Bartoszynski, R. (1961). A characterization of the weak convergence of measures. *Ann. Math. Statist.* **33**, 561–576.

[3] Caby, E.C. (1976) Convergence of measures on uniform spaces. Thesis, Univ. of California, Berkeley, CA.

[4] De Acosta, A. (1971). On the concentration and extension of cylinder measures. *Trans. Am. Math. Soc.* **160**, 217–228.

[5] Dudley, R.M. (1978). Central limit theorems for empirical measures. *Ann. Probability* **6**, 899–929.

[6] Hoeffding, W. (1955). The extrema of the expected value of a function of independent random variables. *Ann. Math. Statist.* **26**, 268–275.

[7] Kiefer, J. and Wolfowitz, J. (1958). On the deviations of the empiric distribution function of vector chance variables. *Trans. Am. Math. Soc.* **87**, 173–186.

[8] Le Cam, L. (1970). Remarques sur le théorème limite central dans les espaces localement convexes. In: *Les probabilités sur les structures algébriques*. C.N.R.S., Paris, pp. 233–249.

[9] Millar, P.W. (1979). Robust estimation via minimum distance methods. Submitted for publication.

[10] Mourier, E. (1954). Eléments aléatoires dans un espace de Banach. Thèse. Faculté des Sciences, Paris.

[11] Parthasarathy, K.R. (1967). *Probability Measures on Metric Spaces*. Academic Press, New York, 276 pp.

[12] Van Zuijlen, M.C.A. (1978). Properties of the empirical distribution for independent nonidentically distributed random variables. *Ann. Probability* **6**, 250–266.

G. Kallianpur, P.R. Krishnaiah, J.K. Ghosh, eds., *Statistics and Probability: Essays in Honor of C.R. Rao*

ON SOME RECENT ADVANCES IN THE THEORY OF UNIVARIATE CHARACTERISTIC FUNCTIONS AND ON THE DEVELOPMENTS WHICH LED TO THEM

Eugene LUKACS
Professor Emeritus, Catholic University, University Professor (rtd), Bowling Green State University

Introduction

The theory of characteristic functions* has shown a considerable growth during the last ten years. In this paper we describe some of these results. However, we also include some older work directly connected to recent developments. In this survey we do not discuss characterization problems nor the literature on stability theorems. These omissions are justified by the fact that characterization problems are extensively covered in ref. [21], while the present author wrote two surveys on stability theorems [33, 35]. A number of mathematicians studied generalizations of c.f. such as Fourier transforms of functions of bounded variation and ridge functions. These generalizations as well as the theory of metrics in the space of distribution functions will not be considered in this paper.

1. Some recently investigated properties

One of the important characteristics of d.f. are the moments. The existence of moments of a d.f. is closely connected to the differentiability of the corresponding c.f. and was studied for moments of integer order a long time ago. Recently the connection between fractional derivatives of c.f. and the existence of moments of non-integer order was investigated by Wolfe [58] and Laue [29].

Let λ be a positive real number, not necessarily an integer, and let $F(x)$ be a d.f. We say that $F(x)$ has a moment of order λ if the integral $\beta_\lambda = \int_{-\infty}^{\infty} |x|^\lambda \,\mathrm{d}F(x)$ exists and is finite.

Following Marchaud [36], we define fractional derivatives in the following way. Let k be a positive integer and $0<\lambda<1$. The derivative of a function $f(t)$ of order $k+\lambda$ is defined by

$$D^{k+\lambda}f = \frac{\mathrm{d}^{k+\lambda}}{\mathrm{d}t^{k+\lambda}} f(t) = \frac{\lambda}{\Gamma(1-\lambda)} \int_{-\infty}^{t} \frac{f^{(k)}(t) - f^{(k)}(u)}{(t-u)^{1+\lambda}} \,\mathrm{d}u. \tag{1.1}$$

Here $f^{(k)}(t)$ is the kth derivative of $f(t)$.

*We will use the abbreviations c.f. for characteristic function(s), d.f. for distribution function(s), d. for distribution(s) and a.c.f. for analytic characteristic function(s), also i.d. for infinitely divisible.

Theorem 1.1. *Let $F(x)$ be a d.f. with c.f. $f(t)$ and let $k=2n$ be an even integer and $0<\lambda<1$. The absolute moment $\beta_{2n+\lambda}$ of $F(x)$ exists if, and only if:*
(i) $\beta_{2n}<\infty$ *and*
(ii) $\operatorname{Re}[D^{2n+\lambda}f(t)|_{t=0}]$ *exists.*
Then

$$\beta_{2n+\lambda}=\frac{1}{\cos(\lambda\pi/2)}\operatorname{Re}\left[(-1)^{n}D^{2n+\lambda}f(t)|_{t=0}\right].$$

Theorem 1.2. *Let $F(x)$ be a d.f. with c.f. $f(t)$ and let $k=2n+1$ be an odd integer and $0<\lambda<1$. The absolute moment $\beta_{2n+1+\lambda}$ of $F(x)$ exists if, and only if:*
(i) $\beta_{2n+1}<\infty$;
(ii) $\operatorname{Re}[D^{2n+1+\lambda}f(t)|_{t=0}]$ *exists and*
(iii) $\lim_{t\to 0}\dfrac{1-\operatorname{Re}h_{2n}(t)}{t^{1+\lambda}}$ *exists.*
Then

$$\beta_{2n+1+\lambda}=\frac{1}{\sin(\pi\lambda/2)}\operatorname{Re}\left[(-1)^{n+1}D^{2n+1+\lambda}f(t)|_{t=0}\right].$$

Here $h_{2n}(t)=(1/\beta_{2n})\int_{-\infty}^{\infty}e^{itx}x^{2n}\,dF(x)$.

2. Closeness of distribution functions

In view of the continuity theorem one can expect that two d.f. whose c.f. do not differ much are also close to each other in some sense. In order to obtain such results one has to define the closeness of d.f. This is done by defining a distance between d.f. so that the set of d.f. becomes a metric space. A systematic study of metrics in the space of d.f. was recently made by Senatov [47] and Zolotarev [61]. Here we will use only two definitions for the distance between two d.f. $F(x)$ and $G(x)$.

$$\rho(F,G)=\sup_{x}|F(x)-G(x)|, \tag{2.1}$$

$$L(F,G)=\inf\left[h\colon F(x-h)-h\leqslant G(x)\leqslant F(x+h)+h \text{ for all } x\right]. \tag{2.2}$$

The metric (2.1) is called the uniform (sometimes the Kolmogorov) metric while (2.2) is called the Lévy metric.

The following estimates exist for the uniform metric.

Theorem 2.1. *Let A, T and ε be arbitrary positive constants and let $F(x)$ and $G(x)$ be d.f. Let $f(t)$ and $g(t)$ be the c.f. of $F(x)$ and $G(x)$, respectively. Assume that:*
(i) $F(-\infty)=G(-\infty)$, $\quad F(+\infty)=G(+\infty)$;
(ii) $\int_{-\infty}^{\infty}|F(x)-G(x)|\,dx<\infty$;
(iii) $\int_{-T}^{T}\left|\dfrac{f(t)-g(t)}{t}\right|dt=\varepsilon$ *and*
(iv) $G'(x)$ *exists for all x and* $|G'(x)|\leqslant A$.

Then to every k there corresponds a finite positive c(k), depending only on k, such that:

$$|F(x)-G(x)|\leqslant k\frac{\varepsilon}{2\pi}+c(k)\frac{A}{T}.$$

Theorem 2.1 supposes that $G(x)$ is absolutely continuous, the next theorem applies to purely discrete distributions.

Theorem 2.2. *Let A, T and ε be arbitrary positive constants and let F(x) be a purely discrete d.f. and G(x) be a d.f. Suppose that the Fourier–Stieltjes transform of F(x) and G(x) satisfies the conditions* (i), (ii) *and* (iii) *of Theorem* 2.1. *Assume further*, (iv), *that the functions F(x) and G(x) have discontinuity points only at the points* x_v $(v=0,\pm 1,\pm 2,\ldots;\ x_{v+1}>x_v)$ *and that there exists a constant* $L>0$ *such that* $\inf(x_{v+1}-x_v)\geqslant L$

(v) $|G'(x)|\leqslant A$ for $x\neq x_v$ $\quad (v=0,\pm 1,2,\ldots)$.

Then to every number $k>1$, *two finite positive numbers correspond*, $c_1(k)$ *and* $c_2(k)$, *depending only on k, such that*

$$|F(x)-G(x)|\leqslant k\frac{\varepsilon}{2\pi}+c_1(k)\frac{A}{T}$$

provided that $TL\geqslant c_2(k)$.

We note that these theorems, due to Esseen, do not cover all possibilities since they impose various restrictions on $F(x)$ and $G(x)$.

Zolotarev [60] considered the Lévy distance as a measure of the closeness of d.f. He did not need any restrictive conditions on the d.f.

Theorem 2.3. *Let F(x) and G(x) be two d.f. with c.f. f(t) and g(t) respectively. Then*

$$L(F,G)\leqslant\frac{2}{\pi}\int_0^T|f(t)-g(t)|\frac{dt}{t}+2e\frac{\log T}{T}$$

provided that $TL\geqslant 1.3$.

3. Unimodality

A d.f. $F(x)$ is said to be unimodal if at least one value, a, exists such that $F(x)$ is convex for $x<a$ but concave for $x>a$. The point a is called the vertex of $F(x)$; mostly we will assume that $a=0$.

Theorem 3.1. *A d.f. F(x) is unimodal with vertex* $x=0$ *if, and only if its characteristic function f(t) can be represented as*

$$f(t)=\frac{1}{t}\int_0^t g(u)\,du \quad (-\infty<t<\infty)$$

where g(u) is a characteristic function.

This theorem is due to Khinchine [23].

Laha [27] gave a necessary and sufficient condition for the unimodality of symmetric and absolutely continuous d.f.

The next theorem deals with a sufficient condition for unimodality. It is somewhat similar to the famous theorem of Pólya.

Theorem 3.2. *Let $f(t)$ be a real valued, continuous function such that $f(0)=1$, $f(-t)=f(t)$, $\lim_{t\to\infty} f(t)=0$ and suppose that $-f'(t)$ is convex for $t>0$. Assume further that $-f'''(t)$ exists and is non-negative for $t\geqslant 0$ and that*

$$\int_0^\infty t|f(t)|\,dt<\infty \quad \textit{while} \quad \lim_{t\to\infty} t^2 f(t)=\lim_{t\to\infty} t^3 f(t)=\lim_{t\to\infty} t^4 f(t)=0.$$

Then $f(t)$ is the characteristic function of a unimodal distribution.

This theorem and other similar results are due to Askey [2]. Medgyessy [39] and Dharmadikari and Jogdeo [9] studied unimodal discrete distributions.

Chung [6] noted that the convolution of two unimodal d. is not necessarily unimodal. It seemed, therefore, desirable to find conditions which assure that the convolution of two unimodal d. be unimodal.

Ibragimov [17] calls a unimodal d.f. strongly unimodal if its convolution with any unimodal d.f. is unimodal. He obtained the following result.

Theorem 3.3. *A non-degenerate unimodal d.f. $F(x)$ is strongly unimodal if, and only if, $F(x)$ is continuous and if* $\log F'(x)$ *is concave on the set E of points on which neither the right-hand, nor the left-hand derivative of $F(x)$ vanishes.*

Yamazato [59] showed that the convolution of two unimodal d.f. is unimodal if the "factors" $G(x)$ and $H(x)$ satisfy certain conditions. Yamazato's result is very useful in the study of the unimodality of stable distributions and of the *L*-class (see Section 8).

4. Tables

Extensive tables of c.f. were prepared by Oberhettinger [40]. These are restricted to c.f. of absolutely continuous d.f. and also contain very cleverly arranged tables of inverses which permit one to find the d.f. that belongs to a given c.f. The table [40] contains over 1000 entries and also presents a brief table of the transforms of univariate discrete d. and of some multivariate density functions.

There are tables for special classes of d.f. Here we mention only Bolshev et al. [5] Mandelbrot and Zarafalla [3] and Holt and Crow [16] which are restricted to stable *d*.

5. Analytic and entire characteristic functions

A c.f. $f(t)$ is said to be an analytic characteristic function (a.c.f.) if a function $A(z)$ of the complex variable $z=t+iy$ (t, y, real) exists which is regular in $|z|<\rho,(\rho>0)$ and if a constant Δ exists such that $A(t)=f(t)$ for $|t|<\Delta$.

An a.c.f. is always regular in a horizontal strip which contains the real axis. If this strip is the whole plane, we are then speaking about an entire c.f.

Many a.c.f. can be continued beyond their strip of regularity and can have a natural boundary.

Theorem 5.1. *Let G be a domain in the complex z-plane which contains the strip $-\alpha<\operatorname{Im} z<\beta$ and which is symmetric with respect to the imaginary axis. Suppose further that the points $i\beta$ and $-i\alpha$ belong to the boundary ∂G of G. Then there exists an a.c.f. whose natural domain of analyticity is G.*

A proof of this theorem can be found in Cuppens and Lukacs [8].

Characteristics describing the growth of entire characteristic functions were studied by many authors. Here we mention only Ramachandran [44] who generalized considerably the results obtained earlier by other authors.

Let $f(z)$ be an entire c.f., we put $M(r,f)=\sup_{|z|\leq r}|f(z)|$ and write $\lambda=\limsup_{r\to\infty}[\log\log\log M(r,f)/\log r]$. Ramachandran [44] assumed that a d.f. $F(x)$ has the property that

$$T(x)\leq L\exp\left[-\lambda x(\log x)^{\delta}\right]$$

and showed that the c.f. of $F(x)$ is an entire c.f. such that $\lambda\leq 1/\delta$. Here L,λ,δ are positive constants while $T(x)=1-F(x)+F(-x)$. In the same paper Ramachandran also obtained the following result.

Theorem 5.2. *The d.f. $F(x)$ has an entire c.f. $f(z)$ such that*

$$\limsup\frac{\log\log\log M(r,f)}{\log r}=\alpha,$$

if, and only if,

(i) $T(x)>0$ *for all* $x>0$,

(ii) $$\liminf_{x\to\infty}\frac{\log\left[x^{-1}\log[T(x)]^{-1}\right]}{\log\log x}=1/\alpha.$$

The study of the growth of entire functions can be refined by introducing proximate orders and their types.

A real valued function $\rho(r)$ is called a proximate order if it is defined for all $r>0$ and if it admits a left and right derivative everywhere, and if it satisfies the following conditions

(i) $\lim_{r\to\infty}\rho(r)=\rho$, $(0<\rho<\infty)$.

(ii) $\rho'(r)$ exists for sufficiently large r.

(iii) $\lim_{r\to\infty}\rho'(r)r\log r=0$.

If $\limsup_{r\to\infty} \log M(r)/r^{\rho(r)} = \tau$ $(0<\tau<\infty)$ then τ is called the type of $f(z)$ with respect to the proximate order $\rho(r)$.

The connection between proximate orders and their types and the behaviour of the corresponding distribution functions was studied by several authors. Here we mention Rossberg [46a, b, c] and Dewess and Riedel [8a].

6. Infinitely divisible characteristic functions

The theory of infinitely divisible (i.d.) c.f. is one of the most important parts of the theory of c.f. and had been developed a long time ago. This is due to the crucial role which i.d.d.f. play in the theory of limit distributions. Properties of i.d.c.f. and i.d.d.f. which are needed in this connection can be found in many books on probability theory and are, therefore, not treated here. We restrict ourselves to the discussion of some recent work which is in general not connected to the theory of limit distributions and which is of intrinsic mathematical interest.

The tail behaviour of i.d.d.f. was studied by Steutel [51] and by Horn [16]. Here we mention one of these results, due to Horn.

Theorem 6.1. *Let* $F(x)$ *be an i.d.d. such that its tail* $T(x)=1-F(x)+F(-x)=O\{\exp[-xM(x)]\}$ *as* $x\to\infty$. *Here* $M(x)$ *is a nonnegative, measurable function. The d.f.* $F(x)$ *is normal (possibly degenerate) if, and only if,* $M(x)/\log x\to\infty$ *as* $x\to\infty$ *and if* $M(x)$ *is continuous and strictly monotone for large* x.

The connection between the existence of moments of an i.d.d.f. and the existence of moments of the corresponding spectral function was also investigated. It can be shown that the moment of order $(2k)$ of an i.d.d.f. exists if, and only if, the moment of order $(2k)$ of its spectral function exists.

The unimodality of the spectral function in the Lévy canonical representation was studied by Alf and O'Connor [1] and continued by O'Connor [41].

Conditions for the infinite divisibility of certain discrete d. are also known. We give an example.

Theorem 6.2. *Let* p_n *be a probability distribution on the non-negative integers with* $p_0>0$. *A necessary and sufficient condition for the infinite divisibility of this distribution is that*

$$np_n = \sum_{j=0}^{n-1} p_j q_{n-j-1},$$

where $q_j\geqslant 0$ $(j=1,2,\ldots)$ *and* $\sum_{j=0}^{n-1} p_j q_{n-j-1}<\infty$.

This result is due to Steutel [49]. He also showed (Steutel [50]) that all completely monotone densities are i.d.

It is often difficult to decide whether or not a given density is i.d. A number of authors treated problems of this kind. The infinite divisibility of Student's *t*-

distribution was shown by Grosswald [14]; Thorin [52] showed that the lognormal d. is i.d. and also (Thorin [53]) that the Pareto d. is i.d. Lewis [30] showed that the von Mises d. is i.d. for certain values of the parameter. Barndorff–Nielsen and Halgreen [3] studied the infinite divisibility of the hyperbolic and the generalized inverse Gaussian d.

Stable distributions are defined by a functional equation. A d.f. $F(x)$ is said to be stable if to every pair of positive numbers b_1 and b_2 and real numbers c_1 and c_2 corresponds a $b>0$ and a real c such that

$$F\left(\frac{x-c_1}{b_1}\right)*F\left(\frac{x-c_2}{b_2}\right)=F\left(\frac{x-c}{b}\right). \tag{6.1}$$

The c.f. $f(t)$ of a stable d.f. must therefore satisfy the equation

$$f(b_1t)f(b_2t)=f(bt)\exp(\mathrm{i}\gamma t), \tag{6.2}$$

where $\gamma=c-c_1-c_2$. Iterating eq. (6.2), one obtains the functional equation

$$[f(t)]^n=f(b_nt)\exp(\mathrm{i}\gamma' t) \tag{6.3}$$

where $b_n>0$ and γ' is real. Equation (6.3) shows that all stable d. are i.d. Equation (6.2) proves that a stable c.f. necessarily has the form

$$\log f(t)=\mathrm{i}at-c|t|^{\alpha}\left\{1+\mathrm{i}\beta\frac{t}{|t|}\omega(|t|,\alpha)\right\} \tag{6.4}$$

where

$$\omega(|t|,\alpha)=\begin{cases}\tan \pi\alpha/2, & \text{if } \alpha\neq 1,\\ \dfrac{2}{\pi}\log|t|, & \text{if } \alpha=1,\end{cases} \tag{6.4a}$$

and where $c\geqslant 0$, $|\beta|\leqslant 1$, $0<\alpha<2$, a real. For $\beta=0$ stable d.f. are symmetric, it is also possible to show that a one-sided, stable d.f. exists.

Bochner [4] considered equation (6.3) but did not assume that $f(t)$ is a c.f. (i.e. positive definite), but only that it is continuous in $0\leqslant x<\infty$ and that $f(0)=1$. He determined all functions having this property.

It is easily seen that all stable d. are absolutely continuous. Properties of their frequency functions were investigated by a number of authors (see Lukacs [34]).

It has been long known that all symmetric stable d. are unimodal. It was believed that all stable d., even if not symmetric, are unimodal. However, proof of the conjecture of the unimodality of all stable d. has a very strange history. Gnedenko and Kolmogorov gave a proof of the unimodality of stable d. in their book [12]. This proof depended on a theorem of Lapin which stated that the convolution of two unimodal d. with mode 0 is unimodal. However, Chung [12, 7] proved that Lapin's theorem was incorrect by constructing a counterexample to Lapin's assertion. In their book, Ibragimov and Linnik [18] presented a proof for the unimodality of stable d., but Kanter [22] showed that this proof contained a gap and it is, therefore, also invalid. More recently Yamazato [41] studied a class (the so-called L-class) which contains all stable d. and he succeeded in proving its unimodality. Therefore, this implies that all stable d. are unimodal. We will discuss the L-class later.

The concept of stable d.f. (or c.f.) admits several generalizations. These wider classes of c.f. are, like the stable d.f., defined by means of functional equations which their c.f. satisfy.

Already Lévy [29] introduced the first extension by defining semistable d. as d.f. whose c.f. satisfy, for all t, the equation

$$f(t)=[f(\beta t)]^{\gamma} \tag{6.5}$$

where $0<|\beta|<1$, $\gamma>0$. The solutions of (6.5) admit the canonical representation (see Kagan et al. [21])

$$\begin{aligned}\log f(t)=\mathrm{i}at+&\int_{-\infty}^{\infty}(\mathrm{e}^{\mathrm{i}tx}-1-\frac{\mathrm{i}tx}{1+x^2})\,\mathrm{d}M(x)\\ +&\int_{-\infty}^{\infty}(\mathrm{e}^{\mathrm{i}tx}-1+\frac{\mathrm{i}tx}{1+x^2})\mathrm{d}N(x),\end{aligned} \tag{6.6}$$

where $M(x)$ and $N(x)$ have the form

$$M(x)=\xi_1(\log|x|)|x|^{-\lambda}, \qquad N(x)=\xi_2(\log x)x^{-\lambda}.$$

Here ξ_1 and ξ_2 are non-negative functions defined on R_1 and they are periodic with period$-\log\beta$. The constant λ is the unique real solution of the equation $\gamma|\beta|^{\lambda}=1$.

Theorem 6.3. *All semistable distributions are infinitely divisible. This follows immediately from formula* (6.6). *Ramachandran and Rao* [45] *studied a more far-reaching generalization of stable d.f.*

A c.f. is said to belong to a generalized stable d. if it is non-vanishing and if it satisfies the functional equation

$$\prod_{j=1}^{s}[f(c_jt)]^{\gamma_j}=\prod_{j=s+1}^{s+k}[f(c_jt)]^{\gamma_j} \tag{6.7}$$

for all t. Here $\gamma_j>0$ and $0<|c_j|\leqslant 1$ for $1\leqslant j\leqslant s+k$. Two special cases are

$$f(t)=\prod_{j=1}^{r}[f(c_jt)]^{\gamma_j}, \tag{6.7a}$$

with $\gamma_j>0$ and $0<|c_j|<1$ and

$$f(t)=\prod_{j=1}^{r}[f(c_jt)]^{\gamma_j}, \tag{6.7b}$$

with $\gamma_j>0$ and $0<c_j<1$. For $r=1$, equation (6.7b) becomes the equation defining semistable c.f.

Let λ be the unique real solution of the equation

$$\sum_j \gamma_j c_j^{\lambda}=1.$$

Ramachandran and Rao obtained a number of interesting results. Here we formulate only part of their results.

Theorem 6.4. *Suppose that either* $0<\lambda<1$ *or* $1<\lambda<2$, *then the non-degenerate solutions* of* (1.7b) *belong to absolutely continuous, infinitely divisible distributions. They have moments of all positive orders inferior to* λ.

Theorem 6.5. *Let* $f(t)$ *be the characteristic function of a generalized stable distribution* [*i.e. characteristic function satisfying* (6.7)] *then it is infinitely divisible. If* λ *is the unique real solution of* $\Sigma_j \gamma_j |c_j|^\lambda = 1$, *then the corresponding distribution function has moments of all positive orders inferior to* λ *but no moments of order* $\geqslant\lambda$. *If* $\lambda=2$ *then the distribution is normal.*

A d.f. $G(x)$ is said to belong to the domain of attraction of a d.f. $F(x)$ if for some sequences of constants $\{A_n\}$ and $\{B_n\}$, $\operatorname{Lim}_{n\to\infty} G^{*n}(B_n x + A_n) = F(x)$.

Shimizu [48] also gave a necessary and sufficient condition which assures that a d.f. $G(x)$, belongs to the domain of attraction of a semistable d.f., $F(x)$.

Lévy [29] considered a rather wide class of c.f. which were also defined by a functional equation. This is the class of self-decomposable c.f. (or class L).

A c.f. $f(t)$ is said to be self-decomposable (belong to the class L) if for every $c\in(0,1)$ the relation

$$f(t)=f(ct)f_c(t) \tag{6.8}$$

holds where $f_c(t)$ is some c.f.

Theorem 6.6. *All self-decomposable characteristic functions, as well as the functions* $f_c(t)$, *are infinitely divisible. All stable characteristic functions belong to the L-class.*

Since the functions of the L-class are i.d., they are determined by their spectral functions. One has the following result.

Theorem 6.7. *An infinitely divisible characteristic function belongs to the L-class if, and only if, the functions* $M(u)$ *and* $N(u)$, *in their canonical representations, have left- and right-hand derivatives everywhere and if the function* $uM'(u)$ *is non-increasing for* $u<0$ *while* $uN'(u)$ *is non-decreasing for* $u>0$. *Here* $M'(u)$ *and* $N'(u)$ *denote either the right or left derivatives, possibly different ones at different points.*

Urbanik [54, 55] gave a representation for c.f. of the class L.

Theorem 6.8. *Let* $f(t)$ *be a self-decomposable characteristic function* (*i.e.* $f(t)\in L$). *Then*

$$\log f(t) = \mathrm{i}at + \int_{-\infty}^{\infty}\left[\int_0^u \frac{\exp(\mathrm{i}v-1)}{v}\,\mathrm{d}v - \mathrm{i}t\arctan u\right]\frac{\mathrm{d}Q(u)}{\log(1+u^2)}.$$

*We omit the discussion of cases where (6.7b) yields degenerate c.f. as solutions.

Here a is a real constant, Q is a finite Borel measure on R_1, the integrand is defined for $u=0$ by continuity to be $-t^2/4$. Moreover, a and Q are uniquely determined by $f(t)$.

It is also possible to define a sequence of classes of c.f. in the following way: A c.f. is said to belong to the class L_m if for every $a\in(0,1)$, there exists a c.f. $f_a(t)$, $f_a(t)\in L_{m-1}$, such that

$$f(t)=f(at)f_a(t). \tag{6.9}$$

The class L_{-1} is, by definition, the class of all c.f.; the class $L_0=L$ is the class of self-decomposable c.f.

Urbanik [55] and also Kumar and Schreiber [26] studied the classes L_m and obtained representations similar to the representation of c.f. from L. The classes L_m form a decreasing sequence of sets. These are closed under shifts, changes of scale, convolutions and weak convergences.

In recent years, several papers dealt with the question of whether the distributions of class L were unimodal. Already Wintner [56] had shown that every symmetric L-function is unimodal. Several authors tried subsequently to solve the general problem of the unimodality of the L-class. The following partial result is due to Wolfe [57].

Theorem 6.9. *Suppose that $F(x)$ is a distribution function of the L-class and that the support of its spectrum in the Lévy canonical representation is either the half line $(-\infty,0)$ or the half line $(0,\infty)$, then $F(x)$ is unimodal.*

Using Theorem 3.2, Wolfe also showed that every d. of the class L is the convolution of two unimodal L-distributions.

The problem of the unimodality of the L-class was solved by Yamazato [59] who obtained the following result.

Theorem 6.10. *All self-decomposable distributions are unimodal.*

This theorem is very significant for the following reason. The set of stable c.f. is a proper subset of the set of self-decomposable c.f. Theorem 6.10 implies therefore the unimodality of all stable d.f. Thus this old problem has been finally solved.

7. Decomposition of characteristic functions

This area was surveyed systematically by Livsic et al. [32]. This survey was supplemented and brought up to date by Ostrovskii [43]. The general decomposition theorems, owing to Khinchine and to Cramér, are well known and are by now classical results which can be found in textbooks and monographs and will, therefore, not be discussed here. Cramér's result shows that infinitely divisible distributions can have indecomposable factors.

Let $F(x)$ be a d. which has indecomposable factors; we denote the set of all indecomposable factors of $F(x)$ by $N(F)$ and the set of all indecomposable factors whose spectrum is the set S by $N_S(F)$. Ilinskii [19, 20] studied the set $N_S(F)$ for certain i.d.d.f., such as the geometric and exponential distributions and also for a few other d. Kudina [24] studied the spectra of certain i.d.d.f., she also showed [25] that the set of indecomposable c.f. is dense, in the sense of weak convergence, in the set of all c.f. The existence of indecomposable factors of i.d.d. was also studied by Mase [38].

The study of the class I_0 of i.d.c.f. which have only i.d. factors is one of the most important problems in the arithmetic of d.f. In this connection the class $\mathfrak{L}$ of d.f. with finite or denumerable Poisson spectrum which satisfy certain conditions is important. It was shown by Goldberg and Ostrovskii [13] that c.f. exist which belong to I_0 but not to $\mathfrak{L}$.

Fryntov and Cistjakov [11] obtained a necessary and sufficient condition which assures that a lattice d. belongs to I_0.

It is known that convolutions of three or more Poisson type d. can have an indecomposable factor. Cuppens [7] derived sufficient conditions which assure that a finite convolution of Poisson type d. belongs to I_0. Denumerable convolutions of Poisson type d. were studied by Fryntov [10].

In analogy to the class I_0 one can also define a class I_0^α of all i.d.c.f. (or i.d.d.f.) which have only i.d. α-components. Clearly, $I_0^\alpha \subset I_0$. Sufficient conditions for membership in I_0^α were given by Linnik and Ostrovskii [31] and also by Ostrovskii [42].

References

[1] Alf, C. and O'Connor, T. (1977). Unimodality of the Lévy spectral function. *Pacific J. Math.* **69**, 285–290.

[2] Askey, R. (1975). Some characteristic functions of unimodal distribution. *J. Math. Anal. Appl.* **50**, 465–469.

[3] Barndorff-Nielsen, O. and Halgreen, C. (1977). Infinite divisibility of the hyperbolic and generalized inverse Gaussian distributions. *Z. Wahrschtheorie verw. Geb.* **38**, 309–311.

[4] Bochner, S. (1974/75). A formal approach to stochastic stability. *Z. Wahrschtheorie verw. Geb.* **31**, 187–198.

[5] Bolshev, L.N., Zolotarev, V.M., Kedrova, E.S. and Rubinskaya, M.A. (1968). Tables of cumulative distributions of one sided stable distributions. *Theory Probab. Appl.* **15**, 299–309.

[6] Chung, K.L. (1953). Sur les lois de probabilité unimodales. *C.R. Acad. Sci. Paris* **236**, 583–584.

[7] Cuppens, R., Sur un théorème de Paul Lévy. *Publ. Inst. Stat. Univ. Paris* **XVII**, 1–6.

[8] Cuppens, R. and Lukacs, E. (1970). On the domains of definition of analytic characteristic functions. *Ann. Math. Stat.* **41**, 1096–1101.

[8a] Dewess, M. and Riedel, M. (1977). The connection between the proximate order of an entire characteristic function and the corresponding distribution function. *Czechoslov. Math. J.* **5**(27) 102, 173–185.

[9] Dharmadikari, S.W. and Jogdeo, C. (1974). On the characterization of the unimodality of discrete distributions. *Z. Wahrschtheorie verw. Geb.* **30**, 203–206.

[10] Fryntov, A.E. (1968). On the factorization of compositions of a countable number of Poisson laws. *Mat. Sbornik* 99(141), 176–191. [Engl. Transl. *Math. USSR Sbornik* **28**, 153–167.]

[11] Fryntov, A.E. and Cistjakov, G.P. (1977). On the membership of lattice probability laws in the class I_0. *Izvestija A.N. USSR* **41**, 462–475. [Engl. Transl. *Math. USSR Izvestija* **41**, 441–452].

[12] Gnedenko, B.V. and Kolmogorov, A.N. (1954). *Limit Distributions for Sums of Independent Random Variables.* Addison Wesley, Cambridge, MA (See also appendix 2 of Chung's English translation.)

[13] Goldberg, A.A. and Ostrovskii, I.V. (1970). An application of a theorem of W.K. Hayman to a problem in the theory of decomposition of probability laws. *Selected Transl: Math. Stat. Probabil.* 147–151. (Russian original 1967.)
[14] Grosswald, E. (1976). The Student *t*-distribution of any degree of freedom is infinitely divisible. *Z. Wahrschtheorie verw. Geb.* **36**, 103–109.
[15] Holt, D. and Crow, E.L. (1973). *Tables and Graphs of Stable Probability Density Functions*. NBS, Boulder, CO.
[16] Horn, R.A. (1972). A necessary and sufficient condition for an infinitely divisible distribution to be normal or degenerate. *Z. Wahrschtheorie verw. Geb.* **21**, 273–276.
[17] Ibragimov, I.A. (1953). On the composition of unimodal distributions. *Theory Probabil. Appl.* **1**, 255–260.
[18] Ibragimov, I.A. and Linnik Yu.V. (1971). *Independent and Stationary Sequences of Random Variables*. Wolters-Noordhoff, Groningen, The Netherlands. (Russian original, Nauka Moscow 1965).
[19] Ilinskii, A.I. (1974). The indecomposable components of certain infinitely divisible laws. *Dokl. Akad. Nauk USSR* **215**, 529–531. [English Transl. Soviet Math. Dokl. 15, 542–545].
[20] Ilinskii A.I. (1977). O nezrazlozemih komponentah bezgranicno delemih zakonov. *Teor. funkts. funktsionaln. analis prilozenija vipusk* **27**, Harkov 61–71.
[21] Kagan, A.M., Linnik, Yu.V. and Rao, C.R. (1973). *Characterization Problems in Mathematical Statistics*. Wiley, New York. [Russian original Nauka, Moscow 1972].
[22] Kanter, M. (1976). On the unimodality of stable distributions. *Ann. Prob.* **4**, 1006–1008.
[23] Khinchine, A.Ya. (1938). On unimodal distributions. *Izv. Naucno Issled. Inst. Mat. Meh. Tomskgos. Univ.* **2**, 1–7.
[24] Kudina, L.S. (1972). Indecomposable laws with a preassigned spectrum. *Teoria funkts. funktsionaln. analis. prilozenija* **16**, 206–212.
[25] Kudina, L.S. (1974). The closure of the set of indecomposable distributions with a fixed spectrum. *Teoria funkts. funktsionaln. analis prilozenija* **16**, 206–212.
[26] Kumar, A. and Schreiber, B.M. (1978). Characterization of subclasses of class *L* probability distributions. *Ann. Prob.* **6**, 179–293.
[27] Laha, R.G. (1961). On a class of unimodal distributions. *Proc. Amer. Math. Soc.* **12**, 181–184.
[28] Laue, G. Remarks on relations between fractional moments and fractional derivatives. Accepted for publication in the *J. Appl. Probabil.*
[29] Lévy, P. (1937). *Théorie de l'Addition des Variables Aléatoires*. Gauthier-Villars, Paris.
[30] Lewis, T. (1976). On the infinite divisibility of the von Mises distribution. *J. Australian Math. Soc. Ser. A* **22**, 332–342.
[31] Linnik, Yu.V. and Ostrovskii, I.V. (1972). *Decomposition of Random Variables and Random Vectors*. American Math. Soc. Providence, R.I. USA. [Russian original Nauka, Moscow, 1972].
[32] Livsic, L.Z., Ostrovskii I.V. and Cistjakov, G.P. (1975) Aritmetika verojatnostih zakonov. *Itogi Nauki. Ser. Verojatn. Mat. Stat. Teoretickaja Kibernetika* **12**, 5–42.
[33] Lukacs, E. (1978). Stability theorems for characterizations. *Bull. Intern. Stat. Inst.* **47**(2), 366–381.
[34] Lukacs, E. (1970). *Characteristic Functions*. Charles Giffin & Co., London.
[35] Lukacs, E. (1977). Stability theorems. *Adv. Appl. Prob.* **9**, 336–363.
[36] Marchaud, A. (1972). Sur les dérivées et sur les differences des fonctions de variables réelles. *J. Math. Pures Appl. IX Ser.* **6**, 337–452.
[37] Mandelbrot, B. and Zarafalla, R. (1959). Five Place Tables of Certain Stable Distributions. IBM research report RC 421, IBM Research Center, Yorktown, New York.
[38] Mase, S. (1975). Decomposition of infinitely devisible characteristic functions with absolutely continuous Poisson spectral meas. *Ann. Stat. Math.* **27**, 289–298.
[39] Medgyessy, P. (1972). On the unimodality of discrete distributions. *Periodica Math. Hung.* **2**, 245–257.
[40] Oberhettinger, F. (1973). *Fourier Transforms of Distributions and Their Inverses*. Academic Press, New York.
[41] O'Connor, T.A. (1979). Infinitely divisible distributions with unimodal Lévy spectral function. *Ann. Prob.* **7**, 494–499.

[42] Ostrovskii, I.V. (1970). On some classes of infinitely divisible laws. *Izvestija A.N. USSR Ser. Mat.* **34**, [English transl. *Math. USSR Izvestija* **41**, 931–952].

[43] Ostrovskii, I.V. (1977). The arithmetic of probability distributions. *J. Multivariate Anal.* **7**, 475–490.

[44] Ramachandran, B. (1962). On the order and type of entire characteristic functions. *Ann. Math. Stat.* **33**, 1238–1255.

[45] Ramachandran, B. and Rao, C.R. (1968). Some results on characteristic functions and characterizations of the normal and generalized stable laws. *Sankhya A* **30**, 125–140.

[46] Raikov, D.A. (1938). On the decomposition of Gauss and Poisson laws. *Izvestija Akad. Nauk USSR, Ser. Mat.* **2**, 91–124.

[46a] Rossberg, H.J. (1966). Wachstum und Nullstllenverteilung ganzer charakteristischer Funktionen. Monatsber. Deutsch. Akad. Wiss. Berlin 8, 275–287.

[46b] Rossberg, H.J. (1967). Der Zusammenhang zwischen einer ganzen Funktion einer verfeinerten Ordnung und ihrer Verteilungsfunktion. *Czechoslov. Math. J.* **17** (92), 317–333.

[46c] Rossberg, H.J. (1968). Eigenschaften der charakteristischen Funktionen einseitig beschrankter Verteilungsfunktionen und ihre Anwendung auf ein Charakterisierunsproblem. *Math. Nachr.* **37**, 37–57.

[47] Senatov, V.V. (1971). On some properties of metrics in the set of distribution functions. *Mat. Sbornik* **102** (144), 425–434.

[48] Shimizu, R. (1970). On the domain of attraction of semistable distributions. *Ann. Inst. Stat. Math.* **22**, 245–255.

[49] Steutel, F.W. (1969). Note on complete monotone densities. *Ann. Math. Stat.* **10**, 1130–1131.

[50] Steutel, F.W. (1971). On the zeros of infinitely divisible densities. *Ann. Math. Stat.* **42**, 812–815.

[51] Steutel, F.W. (1974). On the tails of infinitely divisible distributions. *Z. Wahrschtheorie verw. Geb.* **28**, 273–276.

[52] Thorin, O. (1977). On the infinite divisibility of the log-normal distribution. *Skand. Aktuarietidskr.* 121–148.

[53] Thorin, O. (1977). On the infinite divisibility of the Pareto distribution. *Skand. Aktuarietidskr.* 31–40.

[54] Urbanik, K. (1968). A representation of self-decomposable distributions. *Bull. Acad. Polon. Sci. Section Sci. Mat. Astr. Physics* **XVI**, 209–214.

[55] Urbanik, K. (1973). Limit laws for sequences of normed sums satisfying some stability conditions. *Multivariate Analysis* III (P.R. Krishnaiah, Ed.). pp. 225–237 Academic Press, New York.

[56] Wintner, A. (1956). Cauchy's stable distributions and an "explicit formula" of Mellin. *Amer. J. Math.* **78**, 819–861.

[57] Wolfe, S.J. (1971). On the unimodality of *L*-functions. *Ann. Math. Stat.* **42**, 912–918.

[58] Wolfe, S.J. (1975). On moments of probability distributions in fractional calculus and applications. Springer lecture notes 457, pp. 306–317. Springer Verlag, Berlin.

[59] Yamazato, M. (1978). Unimodality of infinitely divisible distributions of class *L*. *Ann. Prob.* **6**, 523–531.

[60] Zolotarev, V.M. (1971). Estimates of the difference between distributions in the Lévy metric. *Proc. Steklov Inst. Math.* **112**, 232–240.

[61] Zolotarev, V.M., Metric distances in spaces of random variables and distributions. *Math. Sbornik* **101** (143), 416–454. (To be translated).

G. Kallianpur, P.R. Krishnaiah, J.K. Ghosh, eds., *Statistics and Probability: Essays in Honor of C.R. Rao*

ON WEIGHTED DISTRIBUTIONS

M. MAHFOUD* and G. P. PATIL

Department of Statistics, The Pennsylvania State University, University Park, PA 16802, U.S.A.

1. Introduction and summary

The concept of weighted distributions can be traced to the study of "the effect of methods of ascertainment upon estimation of frequencies" by Fisher (1934). It was Rao (1965), however, who saw the need for a unifying concept and formulated it in his remarkable paper presented at the First International Symposium on Classical and Contagious Discrete Distributions held at McGill University in 1963. He identified various situations that can be modeled by what he called weighted distributions. These situations refer to instances where the recorded observations cannot be considered as a random sample from the original distribution. This may occur because of non-observability of some events or a damage caused to original observation resulting in a reduced value, or adoption of a sampling procedure which gives unequal chances to the units in the original. Rao's paper has stimulated considerable research on damage models, characterization of discrete distributions, and sampling mechanisms generating a wide variety of weighted distributions.

In all these cases, weighted distributions are justified because of the built-in probability sampling at some stage of data collection. But they arise from distinctly different recording mechanisms also; see for example, Patil and Rao (1976, 1978). Furthermore, the concept of weighted distributions can be extended to accommodate moment distributions known in economics (Hart, 1975) and mass-size distributions used in small particle physics and sedimentology (Herdan, 1960). For a comprehensive survey of examples of weighted distributions and how they arise in a wide range of scientific areas, see Patil and Rao (1976). Situations leading to discrete weighted distributions include the analysis of family data, the aerial survey involving visibility bias in wildlife ecology, and line transect sampling. Examples of continuous weighted distributions refer to cell kinetics, early disease detection, heart transplant statistics, etc.

In this paper, the properties of weighted distributions are studied in comparison with those of the original distributions. We examine how some parameters of the weighted distribution relate to those of the original distributions. As most often the interest is to make statements about the parameters or the form of the original distribution, several questions can be raised as to how the weight function affects the original distribution. Some relationships between the parameters of the two distribu-

*Present address: Institut National de la Statistique et de l'Economie Appliquee, Rabat, Morocco.

tions prove to be characteristics of specific pdf's such as log-normal, gamma and Poisson. These characterizations are examined and also the effect of size-biased sampling on the mixtures of specific distributions is investigated.

Further, we study bivariate weighted distributions with different wf's. The wf $w(x, y)=x^{\alpha}$ is considered. The effects of $w(x, y)=x$ on the marginal and conditional variances of X are examined. The trinomial, the negative trinomial and the double Poisson distributions are characterized as the only bivariate SSPSD's for which $\eta^2(X|Y)=\eta^2(X^w|Y^w)$ where $w(x, y)=x$. Bivariate weighted distributions with wf $w(x, y)=\max(x, y)$ are also studied.

2. Notation and terminology

In this section, we introduce the notation and terminology to be used in the paper. The statistical concepts are recorded as definitions. Table 1 provides a list of needed univariate distributions and related notation.

Consider a natural mechanism generating a random variable (rv) X with probability (density) function (pdf/pf) $f(x)$. Let $w(x)$ be a nonnegative weight function (wf) and assume that X is such that $E[w(X)]$ exists. Denote a new pdf by:

$$f^w(x)=\frac{w(x)f(x)}{E[w(X)]},$$

and denote by X^w the rv whose pdf is $f^w(x)$. X and X^w are referred to as original and weighted rv's and their respective distributions are called original and weighted distributions.

When sampling a population where the rv $X \sim f(x)$, the recorded observation X^w turns out to be a weighted version of X with wf $w(x)=x^{\alpha}$, we say that X^w is size-biased of order α (sb(α)). Such a selection procedure is called size-biased sampling of order α (sbs(α)).

Table 1
A list of univariate distributions

X	pf/pdf
Poisson (θ)	$e^{-\theta}\theta^x/x!$, $x=1,2,\ldots$, $\theta>0$
Binomial (n,θ)	$\binom{n}{k}\theta^x(1-\theta)^{n-x}$, $x=0,1,\ldots,n$, $0<\theta<1$
Negative-binomial (k,θ)	$\binom{k+x-1}{x}(1-\theta)^x\theta^k$, $x=0,1,\ldots$, $0<\theta<1$
Geometric (θ)	$(1-\theta)^x\theta$, $x=0,1,\ldots$, $0<\theta<1$
Log-series (θ)	$\theta^x/[-x\ln(1-\theta)]$, $x=1,2,\ldots$, $0<\theta<1$
Uniform (0, θ)	θ^{-1}, $\theta>0$, $0\leqslant x<\theta$
Beta (a,b)	$(\Gamma(a+b)/\Gamma(a)\Gamma(b))x^{a-1}(1-x)^{b-1}$, $0<x<1$, $a>0$, $b>0$
Gamma (k,θ)	$(1/\Gamma(k))(x^{k-1}/\theta^k)e^{-x/\theta}$, $x>0$, $\theta>0$, $k>0$
Pareto (θ_0,λ)	$(\lambda/x)(\theta_0/x)^{\lambda}$, $x>\theta_0$, $\theta_0>0$, $\lambda>0$
Log-normal (μ,σ)	$\left(1/\sqrt{2\pi}\cdot x\cdot\sigma\right)^{-1}\exp[-1/2((\ln x-\mu)/\sigma)^2]$, $x>0$, $\sigma>0$, $-\infty<\mu<+\infty$

Because of the wide use of the wf $w(x)=x$, its corresponding weighted rv X^w and weighted distribution $f^w(x)$ are denoted by X^* and $f^*(x)$. Hence we have

$$f^*(x)=\frac{xf(x)}{E(X)}, \quad x\geqslant 0.$$

Throughout this study we will make use of the following notation:
(a) $x^{(r)}=x(x-1)\cdots(x-r+1)$.
(b) $\mu'_r=E(X^r)$, $\mu_{(r)}=E(X^{(r)})$, $\mu=E(X)$, $\mu^*=E(X^*)$.
(c) $\mu_r=E(X-\mu)^2$, $\mu_2=V(X)$.
(d) The harmonic mean of X: $H(X)=[E(X^{-1})]^{-1}$.
(e) Coefficient of variation of X: $c=\mu_2^{1/2}/\mu$.
(f) Coefficient of skewness: $\gamma_1=\mu_3/\mu_2^{3/2}$. If γ_1 is negative, we say that the distribution of X is negatively skewed.
(g) Coefficient of kurtosis: $\gamma_2=\mu_4/\mu_2^2$.
(h) The cumulative distribution function (cdf) of X is $F(x)=\int_{-\infty}^{x} f(t)\,dt$. $\bar{F}(x)=1-F(x)$.

Definition 1.1. Consider the rv's $X\sim f_X(x)$, $Y\sim f_Y(y)$, $Z\sim g_Z(z)$ and a real number, p, such that $0<p<1$, then g_Z is a mixture of f_X and f_Y if

$$g_Z(x)=pf_X(x)+(1-p)f_Y(x).$$

Definition 1.2. Consider the rv's $X\sim f_X(x)$, $Y\sim f_Y(y)$, $Z\sim g_Z(z)$ and a real number $b>1$, then if

$$g_Z(x)=bf_X(x)-(b-1)f_Y(x),$$

we say that g_Z is a negative mixture of f_X with f_Y.

Observe that g_Z is a negative mixture of f_X with f_Y whenever f_X is a mixture of f_y and g_Z.

3. Univariate weighted distributions

3.1. *Properties of the weighted distributions*

Some properties of weighted distributions, which will be used later, are studied. Since these properties are straightforward to establish, their proofs are omitted. Let a non-negative rv $X\sim f(x)$, for which the first three moments exist and subject to sbs. So the rv $X^*\sim f^*(x)$ and some of its properties are given below.

Property 1. $\mu^*-\mu=\mu_2/\mu$.

Property 2. $\mu_2^*-\mu_2=(\mu_2/\mu)^2(\gamma_1/c-1)$.

Remark. While the mean of X^* is always greater the mean of X, the variance of X^* is greater than the variance of X if γ_1 exceeds c. Also for all negatively skewed original distributions, μ_2^* is less than μ_2.

Property 3. $(\mu^*_{-1})^{-1} = \mu$.

The harmonic mean of X^* is equal to the mean of X.

Property 4. $\mu_2 = (\mu^*_{-1})^{-1}[\mu^* - (\mu'^*_{-1})^{-1}]$.

From Properties 1 and 4, the variance to mean ratio of X is given by the difference between the arithmetic mean and the harmonic mean of X^*.

3.2. *Effects of the weight function on the original distribution*

The effects of the wf on the original distribution are discussed for commonly used forms of $w(\cdot)$. We also study the effect of sbs on the hazard rate function.

Definition. Let the rv $X \sim f(x; \theta)$, then we say that $f(x; \theta)$ is form invariant under the wf $w(x) = x^\alpha$, if $X^w \sim f(x; \eta)$, where η is the new parameter depending on θ and α.

First, consider a rv $X \sim \text{Gamma}(k, \theta)$. The size-biased rv X^* is also distributed as $\text{Gamma}(k+1, \theta)$. However, if the rv $X \sim \text{LSD}(\theta)$, then $X^* \sim \text{Geometric}(\theta)$. In this case the rv X^* is not distributed as another log-series distribution. In connection with this result, Rao (1965) noticed that the geometric distribution is sometimes found to provide a good fit to an observed distribution of family size, but this may be due to a sampling procedure giving large probability of selection to families with large sizes and, it may well be that the original distribution of the family size is logarithmic. In the light of these two examples, the question arises as to under which conditions the original distribution is form invariant under sbs(α). Under some regularity conditions, Patil and Ord (1975) prove the following result.

Theorem 3.1. *Let the rv $X \sim f(x, \theta)$ such that μ'_α exists and the wf be $w(x) = x^\alpha$. If X satisfies the regularity conditions of continuity and expectation relative to α, then X subject to the wf $w(x) = x^\alpha$ is form invariant if and only if its pdf is given by*

$$f(x; \theta) = a(x) x^\theta / m(\theta) = \exp[\theta \ln x + A(x) - B(\theta)]$$

where $a(x) = \exp[A(x)]$ and $m(\theta) = \exp[B(\theta)]$.

Table 2 gives some well-known distributions which belong to the log–exponential family of distributions. It also defines the parameter θ and the range for α.

Sampling components which are operating in a system at a fixed time and then observing the life lengths of these components leads to a size-biased sample (Blumenthal, 1967; Cox, 1969; Schaeffer, 1972). As in life length studies the concept

Table 2
Some members of the log-exponential family of distributions

Distribution	pdf	θ	Range for α
Log-normal	$(2\pi\sigma^2)^{-1/2}\exp[-(1/2\sigma^2)(\ln x-\mu)^2-\ln x]$	$(\mu/\sigma^2)-1$	$\alpha>0$
Pareto	$(\lambda/x)(\theta_0/x)^\lambda$, $x>\theta_0$, $\theta_0>0$, $\lambda>0$	$-\lambda$	$\alpha<\lambda$
Gamma	$x^{k-1}e^{-x}/\Gamma(k)$, $x>0$, $k>0$	$k-1$	$\alpha>-k$
Beta 1st kind	$x^{a-1}(1-x)^{b-1}/B(a,b)$, $0<x<1$, $a>0$, $b>0$	$a-1$	$\alpha>-a$
Beta 2nd kind	$x^{a-1}(1+x)^b/B(a,b-a)$, $x>0$, $a>0$, $b>0$	$a-1$	$\alpha<b-a$
Pearson type V	$x^{-k-1}\exp(-x^{-1})/\Gamma(k)$	$-k-1$	$\alpha<k$

of hazard rate function, $h(x)$, is an alternative way to describe the life distribution of the components, we propose to examine the effect of sbs on $h(x)$.

Theorem 3.2. *Consider a non-negative real-valued rv X subject to sbs and such that $E(X)$ exists. If $h(x)$ is the hazard rate function of X, then*

$$h^*(x)\leqslant h(x),\quad x\geqslant 0.$$

Proof. By definition we have

$$h(x)=f(x)\Big/\int_x^\infty f(t)\,dt,$$

$$h^*(x)=xf(x)\Big/\int_x^\infty tf(t)\,dt.$$

Consider the ratio $h^*(x)/h(x)$ and note that

$$\left[\int_x^\infty tf(t)\,dt\Big/\int_x^\infty f(t)\,dt\right]>x.$$

Remark. On intuitive grounds, one can expect such a result since the sampling mechanism tends to select an unduly low proportion of components which become defective at early age.

3.3. Characterizations based on weighted distributions

Some relationships between parameters of the original distribution and those of the weighted distribution turn out to be characteristic properties of specific distributions. Thus we characterize log-normal, gamma and Poisson distributions.

The characterization of the log-normal distribution came about by connecting an empirical remark made by Krumbein and Pettijohn (1938) with the fact that this distribution provides a good fit to the observed particles sizes. They noticed from experimental data that when plotting, separately, the logarithm of the particle sizes, X, versus their frequencies, and versus their total weights, X^w, the variability in $\ln X$ and $\ln X^w$ appears to be the same from the two histograms.

Theorem 3.3. *Consider a non-negative real-valued rv* $X \sim f(x)$ *and such that* $E(X^\alpha) = \nu_\alpha$ *exists for* $\alpha \geqslant 0$. *Let* X^w *be the weighted form of* X *with wf* $w(x) = x^\alpha$. *If* $Y = \ln X$ *and* $Z_\alpha = \ln X^w$, *then* X *is log-normally distributed if and only if* $V(Y) = V(Z_\alpha)$ *for all* $\alpha > 0$.

Proof. If $X \sim \mathrm{LN}(\mu, \sigma^2)$, then $X^w \sim \mathrm{LN}(\mu + \alpha\sigma^2, \sigma^2)$ and therefore $V(Y) = V(Z)$.

Conversely, if $X \sim f(x)$ and ν_α exists for $\alpha \geqslant 0$, then with $Y = \ln X$ and $Z = \ln X^w$ we have

$$Y \sim g(y) = e^y f(e^y)$$

and

$$Z \sim g^\nu(z) = e^{\alpha z} g(z) / M_Y(\alpha), \tag{3.1}$$

where $M_Y(\alpha)$ is the mgf of Y. From (3.1) it can be shown (Patil and Shorrock, 1965) that

$$V(Z) = \frac{d}{d\alpha}[M'_Y(\alpha)/M_Y(\alpha)]$$

where $M'_Y(\alpha) = (d/d\alpha)M_Y(\alpha)$. Now let $V(Y) = \sigma^2$ and assume that $V(Z) = \sigma^2$, that is

$$\frac{d}{d\alpha}[M'_Y(\alpha)/M_Y(\alpha)] = \sigma^2. \tag{3.2}$$

With $M_Y(0) = 1$ and $M'_Y(0) = E(Y) = \mu$, the solution for (3.2) is given by

$$M_Y(\alpha) = e^{\mu\alpha + \sigma^2\alpha^2/2},$$

which implies that $X \sim \mathrm{LN}(\mu, \sigma^2)$.

The next theorem gives a characterization of the Poisson distribution based on the equality of the variances of the rv's X and X^*.

Theorem 3.4. *Consider a non-negative real-valued rv* $X \sim f(x; \theta)$ *where*

$$f(x; \theta) = a(x) e^{\theta x} / M(\theta), \tag{3.3}$$

and subject to sbs, then X *has Poisson distribution if and only if* $V(X) = V(X^*)$.

Theorem 3.5. *Consider a non-negative real-valued rv* X *with pdf*

$$f(x; \theta) = \frac{a(x) e^{\theta x}}{M(\theta)} \tag{3.4}$$

then the rv X *has gamma distribution if and only if the coefficient of variation of* X *does not depend on* θ.

Proof. If $X \sim \mathrm{Gamma}(k, \theta)$ with k known, then the coefficient of variation of X is $c = k^{-1/2}$, which does not depend on θ. Conversely, suppose that the pdf of X is given by (3.4) and that

$$\frac{\mu_2}{\mu^2}=a, \quad a>0 \tag{3.5}$$

and does not depend on θ. (3.5) can be rewritten as

$$\frac{M''(\theta)}{M'(\theta)}=(1+a)\frac{M'(\theta)}{M(\theta)}$$

implying that

$$M'(\theta)=b[M(\theta)]^{(a+1)}, \quad b>0. \tag{3.6}$$

The solution of (3.6) is given by

$$\frac{1}{-a[M(\theta)]^{a}}=b\theta+d.$$

With $M(0)=1$ we have $d=-a^{-1}$ and therefore $M(\theta)$ can be written as

$$M(\theta)=(1-ab\theta)^{-1/a}$$

which implies that X has gamma distribution.

Warren (1974) observed that, under sbs, the relationship between the coefficient of skewness, γ_1, and the coefficient of variation, c, is the same for the rv's X and X^*, when the rv X has gamma distribution. Also if the original distribution is log-normal, γ_1, c and their relationship are invariant under sbs. In the following theorems we generalize the above results to the case of sbs(α) and including its implications on the coefficients of kurtosis, γ_2.

Theorem 3.6. *Consider a rv X subject to sbs(α). If X has a gamma distribution, then the following relationships*

$$\gamma_1=2c, \tag{3.7}$$

$$\gamma_2=3(1+2c^2) \tag{3.8}$$

hold for X and their form is invariant for X^w.

Proof. If $X\sim\text{Gamma}(k,\theta)$, then $X^w\sim\text{Gamma}(k+\alpha,\theta)$. Without loss of generality, the parameter θ can be taken to be one. Using the fact that $E(X^r)=\Gamma(k+r)/\Gamma(k)$, the relationships (3.7) and (3.8) hold for X. Also for X^w we have

$$\gamma_1^w=2c^w$$

and

$$\gamma_2^w=3\left[1+2(c^w)^2\right].$$

Theorem 3.7. *Consider a rv X subject to sbs(α). If X has a log-normal distribution, then $c^w=c$, $\gamma_1^w=\gamma_1$ and $\gamma_2^w=\gamma_2$.*

Proof. By observing that if $X \sim \mathrm{LN}(\mu, \sigma^2)$, $X^w \sim \mathrm{LN}(\mu+\alpha\sigma^2, \sigma^2)$ and that $E(X^r) = \exp(\mu r + \frac{1}{2}\sigma^2 r^2)$, the theorem is readily proved.

As a consequence of the above result, for the log-normal distribution the relationship $\gamma_1 = c(c^2+3)$ hold for the rv X as well as the rv X^w with $w(x) = x^\alpha$.

4. Multivariate weighted distributions

4.1. Introduction

Let (X, Y) be a pair of non-negative rv's with a joint pdf $f(x, y)$ and let $w(x, y)$ be a non-negative wf. The joint pdf of (X, Y) is such that $E[w(X, Y)]$ exists. The weighted form of $f(x, y)$ is

$$f^w(x, y) = \frac{w(x, y)f(x, y)}{E[w(X, Y)]}.$$

The pair of rv's whose joint pdf is $f^w(x, y)$ is denoted by $(X, Y)^w$.

The extension to s-variate weighted distributions is straightforward.

4.2. *Bivariate weighted distributions with wf* $w(x, y) = x^\alpha$

Let (X, Y) be a pair of non-negative rv's with joint pdf $f(x, y)$ such that $E(X^\alpha)$ exists. With wf $w(x, y) = x^\alpha$, the joint pdf of $(X, Y)^w$ is

$$f^w(x, y) = \frac{x^\alpha f(x, y)}{E(X^\alpha)}. \tag{4.1}$$

From (4.1) we obtain the following marginal and conditional distributions

$$f_X^w(x) = \frac{x^\alpha f_X(x)}{E(X^\alpha)}, \tag{4.2}$$

$$f_Y^w(y) = \frac{E(X^\alpha | y) f_Y(y)}{E[E(X^\alpha | y)]}, \tag{4.3}$$

$$f_X^w(x|y) = \frac{x^\alpha f_X(x|y)}{E(X^\alpha | y)}, \tag{4.4}$$

$$f_Y^w(y|x) = f_Y(y|x). \tag{4.5}$$

When the wf $w(x, y) = x$, the marginal distributions of X^w and Y^w are given by (4.2) and (4.3) with $\alpha = 1$. So the marginal distribution of Y^w is the weighted marginal distribution of Y with wf being the regression of X on Y, i.e. $w(y) = E(X | y)$. Furthermore, if the regression of X on Y is linear, i.e $E(X | y) = ay + b$, then the marginal distribution of Y^w can be written as

$$f_Y^w(y)=a\frac{E(Y)}{E(X)}\cdot f_Y^*(y)+\left[1-a\frac{E(Y)}{E(X)}\right]f_Y(y).$$

We distinguish three cases: (i) $aE(Y)/E(X)<0$, f_Y^w is a negative mixture of f_Y with f_Y^* (Patil et al. 1975); (ii) $0<aE(Y)/E(X)<1$, f_Y^w is a mixture of f_Y^* and f_Y and (iii) $aE(Y)/E(X)>1$, f_Y^w is a negative mixture of f_Y^* with f_Y.

For widely used bivariate distributions, the regression of X on Y is linear. Some of these bivariate distributions are multinomial, negative multinomial, hypergeometric, logarithmic, Poisson, Dirichlet, gamma, Pareto and F-distribution (Mardia, 1970).

The effect of the wf $w(x, y)=x$ on the regression of X on Y is given by

$$E(X^w|Y^w=y)-E(X|Y=y)=V(X|Y=y)/E(X|Y=y).$$

In the following theorem we show how the correlation ratio and the regression of X on Y can be obtained from the bivariate weighted distribution.

Theorem 4.1. *Let (X, Y) be a pair of non-negative rv's with joint pdf $f(x, y)$ such that $E(X)$ and $E(X|y)$ exist for all $y>0$. Consider the wf $w(x, y)=x$, then we have*

$$E(X|Y=y)=H(X^w|Y^w=y) \tag{4.6}$$

and

$$\eta^2(X|Y)=\frac{E[H(X^w|Y^w)]-H(X^w)}{E(X^w)-H(X^w)}. \tag{4.7}$$

Proof. (4.6) is trivial. To establish (4.7) we have

$$V(X)=H(X^w)[E(X^w)-H(X^w)],$$

and

$$H(X^w|Y^w=y)=E(X|Y=y),$$

therefore,

$$\begin{aligned}V[E(X|Y)]&=H[E(X|Y)]\cdot\{E[E(X|Y)]-H[E(X|Y)]\}\\&=H[H(X^w|Y^w)]\cdot\{E[H(X^w|Y^w)]-H[H(X^w|Y^w)]\}\\&=H(X^w)\cdot\{E[H(X^w|Y^w)]-H(X^w)\}.\end{aligned}$$

Hence

$$\eta^2(X|Y)=\frac{V[E(X|Y)]}{V(X)}=\frac{E[H(X^w|Y^w)]-H(X^w)}{E(X^w)-H(X^w)}.$$

Table 3 gives the effect of the wf $w(x, y)=x$ on the correlation ratio of X on Y. The bivariate distributions considered are multinomial, negative multinomial, logarithmic and Dirichlet.

Table 3
Comparisons between $\eta^2(X|Y)$ and $\eta^2(X^w|Y^w)$ when $w(x, y)=x$.

Distribution	$\eta^2(X\|Y)$	$\eta^2(X^w\|Y^w)$	$\eta^2(X\|Y)/\eta^2(X^w\|Y^w)$
$BM(n,\theta_1,\theta_2)$	$\theta_1\theta_2/(1-\theta_1)(1-\theta_2)$	$\theta_1\theta_2/(1-\theta_1)(1-\theta_2)$	1
$BNM(k,\theta_1,\theta_2)$	$\theta_1\theta_2/(1-\theta_1)(1-\theta_2)$	$\theta_1\theta_2/(1-\theta_1)(1-\theta_2)$	1
$BLSD(\theta_1,\theta_2)$	$V[E(X\|Y)]/V(X)$*	$\theta_1\theta_2/(1-\theta_1)(1-\theta_2)$	$\eta^2(X\|Y)/\eta^2(X^w\|Y^w)$
Dirichlet (l,m,n)	$lm/(m+n)(l+n)$	$(l+1)(l+1+n)(m+n)$	$\dfrac{l(l+1+n)}{(l+1)(l+n)}$

4.3. Bivariate weighted SSPSD's with wf w(x, y)=x

Patil (1968) has shown that some probability models arising from sampling with replacement from population with multiple characteristics can be expressed in the form of what he has called "Sum-Symmetric Power Series Distributions" (SSPSD's).

Definition. A pair of non-negative integer valued rv's (X, Y) with joint pf:

$$p(x, y)=a(x+y)\binom{x+y}{x}\theta_1^x\theta_2^y/f(\theta_1+\theta_2),$$

where $\theta_1, \theta_2>0$, and

$$f(\theta_1+\theta_2)=\sum_{x,y} a(x+y)\binom{x+y}{x}\theta_1^x\theta_2^y,$$

is said to have a bivariate SSPSD$(\theta_1, \theta_2, f(\theta_1+\theta_2))$.

$$*V(X)=(\theta_1\gamma_{12}-\theta_1+\theta_3\gamma_{12})\theta_1/[\theta_3\gamma_{12}]^2,$$

$$V[E(X|Y)]=\left[\frac{\theta_1}{(1-\theta_1)\gamma_1}-\frac{\theta_1}{\theta_3\gamma_{12}}\right]\gamma_2/\gamma_{12}+$$

$$+\left(\frac{\theta_1}{1-\theta_1}\right)^2\left(1-\frac{\gamma_1}{\gamma_{12}}\right)\frac{\theta_2[(1-\theta_1)\gamma_{12}+(1-\theta_1)\gamma_1-\theta_2]}{\theta_3^2(\gamma_{12}+\gamma_1)^2}$$

where

$$\theta_3=1-\theta_1-\theta_2,\qquad \gamma_i=-\ln(1-\theta_i),\quad i=1,2,$$

$$\gamma_{12}=-\ln(1-\theta_1-\theta_2).$$

The following theorem characterizes double Poisson, bivariate multinomial, and bivariate negative binomial.

Theorem 4.2. *Let $(X, Y)\sim$ SSPSD$(\theta_1, \theta_2, f(\theta_1+\theta_2))$ and let the wf be $w(x, y)=x$, then $\eta^2(X|S)=\eta^2(X^w|S^w)$ if and only if (X, Y) is distributed as double Poisson, bivariate multinomial, or bivariate negative multinomial.*

4.4. Bivariate weighted distribution with wf $w(x, y)=\max(x, y)$

A situation where the wf $w(x, y)=\max(x, y)$ arises is when the recorded observation is obtained from a renewal process sampled at a given time t. Let $(X, Y)\sim f(x, y)$ be the lifetimes of two components in parallel and forming a kit. The lifetime of a kit is $Z=\max(X, Y)$. Consider a renewal system of kits. If at time t, one records the lifetime, Z^*, of the working kit, the pdf of Z^* is the weighted pdf of Z with wf $z=\max(x, y)$ (Cox, 1962). If the lifetimes of both components in the sampled kit are recorded, their joint pdf is the weighted form of $f(x, y)$ with wf $w(x, y)=\max(x, y)$.

In general, let (X, Y) be a pair of non-negative rv's with joint pdf $f(x, y)$ such that $E(\max(X, Y))$ exists, and let $(X, Y)^w$ be distributed as $f^w(x, y)$, where

$$f^w(x, y)=\frac{\max(x, y)f(x, y)}{E[\max(X, Y)]}. \tag{4.8}$$

The marginal pdf of Y^w is proportional to

$$\int_0^\infty \max(x, y)f(x, y)\,\mathrm{d}x=\int_0^y yf(x, y)\,\mathrm{d}x+\int_y^\infty xf(x, y)\,\mathrm{d}x$$

$$=F_{X|y}(y)\cdot yf_Y(y)+\bar{\mu}_{X|y}(y)\cdot\bar{F}_{X|y}(y)\cdot f_Y(y)$$

where $F_{X|y}(y)=\int_0^y f_X(x|y)\,\mathrm{d}x$, $\bar{F}_{X|y}(y)=1-F_{X|y}(y)$ and $\bar{\mu}_{X|y}(y)=\int_y^\infty xf_X(x|y)\,\mathrm{d}x/\bar{F}_{X|y}(y)$. So the marginal pdf of Y^w is

$$f_Y^w(y)=\frac{\left[yF_{X|y}(y)+\bar{\mu}_{X|y}(y)\bar{F}_{X|y}(y)\right]f_Y(y)}{E\left[YF_{X|Y}(Y)+\bar{\mu}_{X|Y}(Y)\bar{F}_{X|Y}(Y)\right]}. \tag{4.9}$$

Assuming that X and Y are independent, the following theorems examine the effects of the wf $w(x, y)=\max(x, y)$ on the independence of X^w and Y^w and the regression of Y^w on X^w.

Theorem 4.3. *Let (X, Y) be a pair of non-negative independent rv's with joint pdf $f(x, y)=f_X(x)f_Y(y)$; with wf $w(x, y)=\max(x, y)$, the rv's $(X, Y)^w$ with joint pdf given by* (4.8) *are dependent.*

Proof. Using the independence of X and Y, the conditional pdf of $Y^w|X^w=x$ can be written as

$$f_Y^w(y|x)=\begin{cases} u(x, y)xf_Y^w(y), & x\geq y,\\ u(x, y)yf_Y^w(y), & x<y\end{cases} \tag{4.10}$$

where

$$u(x, y)=E[\max(X, Y)]/\left[yF_X(y)+\bar{\mu}_X(y)\bar{F}_X(y)\right]\left[xF_Y(x)+\bar{\mu}_Y(x)\bar{F}_Y(x)\right],$$

hence X^w and Y^w are dependent.

Lemma. *Let (X, Y) be a pair of non-negative independent rv's with joint pdf $f(x, y)=f_X(x)f_Y(y)$. With the wf $w(x, y)=\max(x, y)$, the regression of Y^w on X^w is given by*

$$E(Y^w|X^w=x)=\frac{xF_Y(s)\mu_Y(x)+\bar{F}_Y(x)\bar{\mu}'_{2Y}(x)}{sF_Y(x)+\bar{F}_Y(x)\bar{\mu}_Y(x)} \tag{4.11}$$

where $\mu_Y(x)=\int_0^x yf_Y(y)\,\mathrm{d}y/F_Y(x)$ and $\bar{\mu}'_{2Y}(x)=\int_x^\infty y^2f_Y(y)\,\mathrm{d}y/\bar{F}_Y(x)$.

Proof. Using (4.10), we have

$$E[Y^w|X^w=x]=\int_0^x u(x, y)xf_Y^w(y)\,\mathrm{d}y+\int_x^\infty u(x, y)yf_Y^w(y)\,\mathrm{d}y$$

$$=\int_0^x \frac{x\cdot yf_Y(y)\,\mathrm{d}y}{xF_Y(x)+\bar{\mu}_Y(x)\bar{F}(x)}+\int_x^\infty \frac{y^2f_Y(y)\,\mathrm{d}y}{xF_Y(x)+\bar{\mu}_Y(x)\bar{F}_Y(x)}$$

$$=\frac{x\mu_Y(x)\cdot F_Y(x)+\bar{\mu}'_{2Y}\bar{F}_Y(x)}{xF_Y(x)+\bar{\mu}_Y(x)\bar{F}_Y(x)}.$$

Four properties of the regression of Y^w on X^w are given as Theorem 4.4.

Theorem 4.4. *Let (X, Y) be a pair of non-negative independent rv's with joint pdf $f(x, y)=f_X(x)f_Y(y)$ and such that $E[\max(X, Y)]$ exists, then, with the wf $w(x, y)=\max(x, y)$, the regression of Y^w on X^w is such that*

(i) *$E(Y^w|X^w=x)$ is a decreasing function of x,*
(ii) *$E(Y^w|X^w=0)=E(Y)+V(Y)/E(X)$,*
(iii) *$E(Y^w|X^w=x)>E(Y)$,*
(iv) *$\lim E(Y^w|X^w=x)=E(Y)$ as $x\to\infty$.*

Proof. Part (i) follows by differentiating (4.11) with respect to x. We, then, obtain the derivative as

$$\frac{\int_0^x yf_Y(y)\,\mathrm{d}y\int_x^\infty yf_Y(y)\,\mathrm{d}y-F_Y(x)\int_x^\infty y^2f_Y(y)\,\mathrm{d}y}{\left[xF_Y(x)+\bar{F}_Y(x)\bar{\mu}_Y(x)\right]^2}$$

which is negative since

$$\int_0^x yf_Y(y)\,\mathrm{d}y/F_Y(x)<\int_x^\infty yf_Y^*(y)\,\mathrm{d}y/\bar{F}_Y^*(x).$$

Parts (ii), (iii) and (iv) are straightforward.

As an example, let

$$(X, Y)\sim f(x, y)=\lambda_1 e^{-\lambda_1 x}\cdot\lambda_2 e^{-\lambda_2 y},\quad x, y>0,$$

then, with $w(x, y)=\max(x, y)$, we have

$$f^{w}_{X,Y}(x, y)=\begin{cases}\lambda_1 xe^{-\lambda_1 x}\lambda_2 e^{-\lambda_2 y}/E[\max(X,Y)], & x\geqslant y,\\ \lambda_1 e^{-\lambda_1 x}\lambda_2 ye^{-\lambda_2 y}/E[\max(X,Y)], & x<y,\end{cases} \tag{4.12}$$

where $E[\max(X,Y)]=\lambda_1^{-1}+\lambda_2^{-1}+\lambda^{-1}$, $\lambda=\lambda_1+\lambda_2$. The marginal pdf of X^w is

$$f^{w}_{X}(x)=\frac{\lambda_1^{-1}}{\lambda_1(\lambda_2\lambda)^{-1}+\lambda_1^{-1}}\lambda_1^2 xe^{-\lambda_1 x}+\frac{\lambda_1(\lambda_2\lambda)^{-1}}{\lambda_1(\lambda_2\lambda)^{-1}+\lambda_1^{-1}}\lambda e^{-\lambda x}$$

which is a mixture of Gamma$(2,\lambda_1^{-1})$ and Gamma$(1,\lambda^{-1})$. Using part (ii) of Theorem 4.4, we have

$$E(Y^w|X^w=0)-E(Y)=E(Y).$$

The moment generating function of $(X,Y)^w$ is

$$M_{(X,Y)^w}(s,t)=\frac{1}{\lambda_1^2+\lambda_2\lambda}\left\{\frac{\lambda_2^2}{(1-\lambda_1^{-1}s)^2[1-\lambda^{-1}(x+t)]}+\frac{\lambda_1\lambda}{(1-\lambda_1^{-1}s)[1-\lambda_2^{-1}t]^2}\right\}$$

which can be written as

$$M_{(X,Y)^w}(s,t)=\frac{1}{\lambda_1^2+\lambda_2\lambda}\left\{\lambda_2^2\left[1+2\frac{s}{\lambda_1}+3\left(\frac{s}{\lambda_1}\right)^2+\cdots\right]\left[1+\frac{s+t}{\lambda}+\frac{(s+t)^2}{\lambda}+\cdots\right]\right.$$
$$\left.+\lambda_1\lambda\left[1+\frac{s}{\lambda_1}+\left(\frac{s}{\lambda_1}\right)^2+\cdots\right]\left[1+2\frac{t}{\lambda_2}+3\left(\frac{t}{\lambda_2}\right)^2+\cdots\right]\right\}.$$

Hence,

$$\mathrm{Cov}(X^w,Y^w)=-\lambda_1\lambda_2(\lambda^2+\lambda_1\lambda+\lambda_2^2)/\lambda^2(\lambda_2^2+\lambda_1\lambda)^2,$$

which implies that X^w and Y^w are negatively correlated.

References

Blumenthal, S. (1967). Proportional sampling in life studies. *Technometrics* **9**, 205–218.

Cook, R.D. and Martin, F.B. (1974). A model for quadrat sampling with visibility bias. *J. Am. Statist. Assoc.* **69**, 345–349.

Cox, D.R. (1962). *Renewal Theory*. Frome and Long; Butler & Tanner, Ltd.

Cox, D.R. (1969). Some sampling problems in technology. In: Johnson, N.L., Ed. *New Developments in Survey Sampling*. Wiley-Interscience, New York, pp. 506–527.

Fisher, R.A. (1934). The effects of methods of ascertainment upon the estimation of frequencies. *Ann. of Eugenics* **6**, 13–25.

Gates, C.E. (1979). Line transect and related issues. In: Cormack, R.M., Patil, G.P. and Robson, D.S., Eds., *Sampling Biological Populations*. The International Co-operative Publishing House, Fairland, MD.

Haldane, J.B.S. (1938). The estimation of the frequency of recessive conditions in man. *Ann. of Eugenics* **7**, 255–262.oy

Hart, P.E. (1975). Moment distributions in economics: an exposition. *J. Roy. Statist. Soc. A* **138**, 423–434.

Herdan, G. (1960). *Small Particle Statistics*. Elsevier, Amsterdam.
Krumbein, W.C. and Pettijohn, F.J. (1938). *Manual of Sedimentary Petrography*. New York.
Mardia, K.V. (1970). *Families of Bivariate Distributions*. Charles Griffin & Co., Ltd., London.
Ord, K. (1975). Statistical models for personal income distributions. In: Palit, G.P., Kotz, S. and Ord, K., Eds., *Statistical Distributions in Scientific Work*, Vol. 2. pp. 151–158.
Packhman, G.H. (1955). Volume, weight and number frequency analysis of sediments from thin-section data. *J. Geology* **63**, 50–58.
Patil, G.P. (1968). On sampling with replacement from populations with multiple characters. *Sankhyā B* **30**, 355–366.
Patil, G.P., Boswell, M.T. and Friday, D.S. (1975). Chance mechanisms in computer generation of random variables. In: Patil, G.P., Kotz, S. and Ord, K., Eds., *Statistical Distributions in Scientific Work*, Vol. 2. pp. 37–50.
Patil, G.P. and Rao, C.R. (1976). Weighted distributions: a survey of their applications. Proceedings of the Conference of Statistics and Applications, Dayton, OH, pp. 1–40.
Patil, G.P. and Rao, C.R. (1978). Weighted distributions and size biased sampling with applications to wildlife populations and human families. *Biometrics* **34**, 179–189.
Schaeffer, R.L. (1972). Size biased sampling. *Technometrics* **14**, 634–644.

G. Kallianpur, P.R. Krishnaiah, J.K. Ghosh, eds., *Statistics and Probability: Essays in Honor of C.R. Rao*

SOME RESULTS ON COCYCLES AND SPECTRA OF ERGODIC FLOWS

Joseph MATHEW and M.G. NADKARNI
Indian Statistical Institute, Calcutta, and University of Bombay, Bombay, India

Introduction

By a flow on a finite measure space $(\Omega, \mathcal{Q}, \mu)$ we mean a group of invertible non-singular transformations (i.e., automorphisms) τ_t, $t \in R$, on Ω indexed by the real line such that for any $A \in \mathcal{Q}$, $\mu(\tau_t A)$ is continuous in t; the measure space being assumed isomorphic to unit interval Borel σ-algebra and Lebesgue measure (except that total measure need not be one). Two flows τ_t, $t \in R$, and σ_t, $t \in R$, on two measure spaces $(\Omega_1, \mathcal{Q}_1, \mu_1)$, $(\Omega_2, \mathcal{Q}_2, \mu_2)$ are said to be isomorphic if there is a one–one onto Borel map $\phi: \Omega_1 \to \Omega_2$ such that ϕ and ϕ^{-1} preserve null sets and $\phi \tau_t \phi^{-1} = \sigma_t$ for all t. A flow τ_t, $t \in R$, is said to be ergodic if $A \in \mathcal{Q}$ and $\tau_t A = A$ for all t, then $\mu(A) = 0$ or $\mu(\Omega - A) = 0$.

The purpose of this paper is to give some results on cocycles and spectra of ergodic flows by using the notion of flow built under a function. In Section 1 we explain and obtain some facts about flow built under a function. Our flows built under a function are non-singular in accordance with Krengel [9] and Dani [5], rather than measure preserving with which Ambrose [1] and Ambrose and Kakutani [2] were concerned. In Section 2 we describe cocycles on a flow built under a function in terms of cocycles on the base transformation. This description is used in Section 3 to give multiplicative integral representations of cocycles on any ergodic flow, thus generalising a result of Helson [7] and Carlson [4]. In Section 4 we prove a lemma connecting a Radon–Nikodym derivative on the flow with one on the base space and use it in Section 5 to describe completely the spectral measure of a unitary group associated with a well built flow in terms of a unitary group associated with the base transformation. In Section 6 we prove a result which implies that spectral measure of a unitary group associated with any ergodic flow gives non-zero spectral measure to every non-empty open set.

1. Construction of a flow built under a function

Let $(X, \mathcal{B}, m)$ be a measure space isomorphic to the unit interval, Borel σ-algebra and Lebesgue measure, except that $m(X)$ although finite need not be one. Let $S: X \to X$ be an ergodic automorphism and let F be a Borel function on X non-negative, bounded away from zero and such that $\int_X F \, \mathrm{d}m = 1$. Give R the Borel

σ-algebra and Lebesgue measure. Give $X\times R$ the product σ-algebra and product measure. Restrict these to the portion under the graph of F, i.e., to the set $Y=\{(x,u)\colon 0\leqslant u<F(x)\}$. Let $(Y,\mathcal{C},n)$ be this new measure space. Define on Y a flow T_t, $t\in R$, as follows: Each point (x,u) in Y moves straight up at unit speed until it hits $(x,F(x))$. It then goes to $(Sx,0)$ and continues to move up at unit speed and so on up to time t. The point thus reached at time $t>0$ is defined to be $T_t(x,u)$. For $t<0$, $T_t(x,u)$ is defined to be the point (y,v) such that $T_{|t|}(y,v=)(x,u)$. This pictorial description of T_t, $t\in R$, is slightly ambiguous since the point $(x,F(x))$ is not in Y and the meaning of $T_t(x,u)$ hitting $(x,F(x))$ is not clear. T_t, $t\in R$, is therefore described more accurately but cumbersomely as follows:

$$T_t(x,u)=\begin{cases}\left(S^n x,\, u+t-\sum_{k=0}^{n-1} F(S^k x)\right)\\ \quad \text{if } \sum_{k=0}^{n-1} F(S^k x)\leqslant u+t<\sum_{k=0}^{n} F(S^k x),\quad n=0,1,2,\ldots,\\ \left(S^{-n}x,\, u+t+\sum_{k=1}^{n} F(S^{-k}x)\right)\cdot\\ \quad \text{if } -\sum_{k=1}^{n} F(S^{-k}x)\leqslant u+t<-\sum_{k=1}^{n-1} F(S^{-k}x),\quad n=1,2,\ldots.\end{cases}$$

In the above formula it is understood that $\Sigma_{k=0}^{-1}=0$. We call T_t, $t\in R$, a flow built under the function F on the transformation S. X is called the base space and S the base transformation. F is called the ceiling function. It can be shown that T_t, $t\in R$, is indeed a non-singular flow on $(Y,\mathcal{C},n)$ which is measure preserving whenever S is measure preserving. When $F(x)=1$ for all $x\in X$, T_t, $t\in R$, is called a well-built flow. We note that $T_{F(x)}(x,0)=(Sx,0)$, a relation which will be useful in the sequel.

Notation. For each $t\in R$ and $x\in X$ let

$$[t]_x=\text{the integer } n \text{ such that}\begin{cases}\sum_{k=0}^{n-1} F(S^k x)\leqslant t<\sum_{k=0}^{n} F(S^k x) & \text{if } t\geqslant 0,\\ -\sum_{k=1}^{-n} F(S^{-k}x)\leqslant t<-\sum_{k=1}^{-(n+1)} F(S^{-k}x) & \text{if } t<0,\end{cases}$$

$$\langle t\rangle_x=\begin{cases} t-\sum_{k=0}^{[t]_x-1} F(S^k x) & \text{if } [t]_x\geqslant 0,\\ t+\sum_{k=1}^{-[t]_x} F(S^{-k}x) & \text{if } [t]_x<0.\end{cases}$$

Remark. One can visualise translation on the real line as a flow built under the constant function 1 with base space X the integers and S as addition by one. If this is done, $[t]_x$ is independent of x and equals the integral part of t. For the integral part $[t]$ of t, the following relation is easily proved:

$$[t_1+t_2+t_3]=[t_1+t_2]+[t_3+\langle t_1+t_2\rangle],$$

where $\langle t\rangle$ means the fractional part of t. The next lemma is a generalization of this fact.

Lemma 1.

$$[t_1+t_2+t_3]_x=[t_1+t_2]_x+[t_3+\langle t_1+t_2\rangle_x]_{S^{[t_1+t_2]_x}x}.$$

Proof. Let

$$n_1=[t_1+t_2]_x,$$
$$n_2=[t_3+\langle t_1+t_2\rangle_x]_{S^{[t_1+t_2]_x}x}$$

and

$$n=[t_1+t_2+t_3]_x.$$

Consider the case when $n_1, n_2\geqslant 0$. By definition,

$$\langle t_1+t_2\rangle_x=t_1+t_2-\sum_{k=0}^{n_1-1} F(S^kx),$$

and

$$\sum_{k=0}^{n_2-1} F\big(S^k(S^{n_1}x)\big)\leqslant t_3+\langle t_1+t_2\rangle_x<\sum_{k=0}^{n_2} F\big(S^k(S^{n_1}x)\big),$$

i.e.

$$\sum_{k=n_1}^{n_1+n_2-1} F(S^kx)\leqslant t_3+t_1+t_2-\sum_{k=0}^{n_1-1} F(S^kx)<\sum_{k=n_1}^{n_1+n_2} F(S^kx).$$

So,

$$\sum_{k=0}^{n_1+n_2-1} F(S^kx)\leqslant t_1+t_2+t_3<\sum_{k=0}^{n_1+n_2} F(S^kx),$$

i.e. $n=[t_1+t_2+t_3]_x=n_1+n_2$. Similarly we can consider the remaining cases and verify Lemma 1.

Remark. It is a consequence of the work of Krengel [9] and Dani [5] that every ergodic flow is isomorphic to a flow built under a function. For a measure preserving ergodic flow the result is due to Ambrose [1] and Ambrose and Kakutani [2].

2. Cocycles on a flow built under a function in terms of cocycles on the base transformation

In this section we show that every cocycle on the base space gives rise to a cocycle on the flow and that every cocycle on the flow is cohomologous to one which arises in this manner.

Let $\mathcal{U}$ denote the group of unitary operators on a complex separable Hilbert space $\mathcal{H}$ and let $\mathcal{U}$ be equipped with its Borel σ-algebra. Let Z denote the group of integers. A (Z, X) cocycle is a measurable function on $Z \times X$ taking values in $\mathcal{U}$ and satisfying

$$A(0, x)=I=\text{identity operator},$$
$$A(m+n, x)=A(m, x)A(n, S^m x).$$

Similarly, by an (R, Y) cocycle is meant a measurable function A on $R \times Y$ with values in $\mathcal{U}$ such that

$$\begin{cases} A(0, y)=I=\text{identity}, \\ A(t+s, y)=A(t, y)A(s, T_t y) \end{cases} \tag{1}$$

for all $t, s \in R$ and $y \in Y$. Two (Z, X) cocycles A and B are said to be cohomologous or cohomologous (ρ) if there is a $\mathcal{U}$ valued measurable function ρ such that

$$B(m, x)=\rho(x)A(m, x)(\rho(S^m x))^{-1}$$

for all integers m and for all $x \in X$. In case A and B are (R, Y) cocycles and

$$B(t, y)=\rho(y)A(t, y)(\rho(T_t y))^{-1}$$

for all $t \in R$, $y \in Y$, then we say again that A and B are cohomologous or cohomologous (ρ), where ρ now is a function on Y, $\mathcal{U}$ valued and measurable. For a measurable $\mathcal{U}$-valued function β put

$$A_\beta(m, x)=\begin{cases} \beta(x)\beta(Sx)\cdots\beta(S^{m-1}x), & m>0, \\ I, & m=0, \\ (\beta(S^{-1}x))^{-1}\cdots(\beta(S^m x))^{-1}, & m<0. \end{cases} \tag{2}$$

Then A_β is a (Z, X) cocycle. Conversely every (Z, X) cocycle is equal to A_β with $\beta(x)=A(1, x)$. The main result of this section is the following:

Theorem 1. *Let A be a (Z, X) cocycle. Define $\bar{A}$ on $R \times Y$ as follows:*

$$\bar{A}(t,(x, u))=A([t+u]_x, x).$$

Then $\bar{A}$ is an (R, Y) cocycle. Two (Z, X) cocycles A_1 and A_2 are cohomologous if and only if $\bar{A}_1$ and $\bar{A}_2$ are cohomologous. Every (R, Y) cocycle is cohomologous to $\bar{A}$ for some (Z, X) cocycle A.

Proof. We first prove that $\bar{A}$ is jointly measurable on $R \times Y$. Since A is measurable, it is enough to prove that the map $(t,(x, u)) \to ([t+u]_x, x)$ is a measurable map from $R \times Y$ into $Z \times X$. Since $(t,(x, u)) \to x$ is measurable, it is enough to show that $(t,(x, u)) \to [t+u]_x$ is measurable. Let n_0 be an integer and consider $E=\{(t,(x, u)): [t+u]_x=n_0\}$. Now $\phi:(t,(x, u)) \to (t+u, x)$ is measurable from $R \times Y$ into $R \times X$.

Next the set

$$V=\{(t,x)\colon [t]_x=n_0\}=\left\{(t,x)\colon \sum_{k=0}^{n_0-1} F(S^k x)\leqslant t< \sum_{k=0}^{n_0} F(S^k x)\right\}$$

is measurable in $R\times X$. Hence $E=\{(t,(x,u))\colon [t+u]_x=n_0\}=\phi^{-1}(V)$ is a measurable subset of $R\times Y$. Hence $\bar{A}$ is jointly measurable. Next we verify that $\bar{A}$ satisfies the cocycle identity.

$$\begin{aligned}\bar{A}(t_1+t_2,(x,u))&=A([t_1+t_2+u]_x,x)\\&=A([t_1+u]_x+[t_2+\langle t_1+u\rangle_x]_{S^{[t_1+t_2]_x}x},x)\\&=A([t_1+u]_x,x)A\left([t_2+\langle t_1+u\rangle_x]_{S^{[t_1+t_2]_x}x},S^{[t_1+u]_x}x\right)\\&=\bar{A}(t_1,(x,u))\cdot\bar{A}\left(t_2,(S^{[t_1+u]_x}x,\langle t_1+u\rangle_x)\right)\\&=\bar{A}(t_1,(x,u))\cdot\bar{A}\left(t_2,T_{t_1}(x,u)\right).\end{aligned}$$

which shows that $\bar{A}$ is an (R,Y) cocycle. It can be verified that if two (Z,X) cocycles A_1 and A_2 are cohomologous (ρ_0), then $\bar{A}_1$ and $\bar{A}_2$ are cohomologous ρ, where $\rho(x,u)=\rho_0(x)$ and that if $\bar{A}_1$ and $\bar{A}_2$ are cohomologous (ρ), then A_1 and A_2 are cohomologous ρ_0 where $\rho_0(x)=\rho(x,0)$. We finally show that every (R,Y) cocycle is cohomologous to $\bar{A}$ for some (Z,X) cocycle A. Let B be an (R,Y) cocycle. Define $\beta(x)=B(F(x),(x,0))$ and let A denote the cocycle A_β given by β according to (2). Define ρ on Y by $\rho(x,u)=B(u,(x,0))$. ρ is measurable on Y. For $t\in R$, $(x,u)\in Y$,

$$\begin{aligned}&\rho(x,u)B(t,(x,u))\rho(T_t(x,u))^{-1}=\\&\quad=B(u,(x,0))B(t,(x,u))\left(B(\langle t+u\rangle_x,(S^{[t+u]_x}x,0))\right)^{-1}\\&\quad=B(u+t,(x,0))\left(B(\langle t+u\rangle_x,(S^{[t+u]_x}x,0))\right)^{-1}\\&\quad=B\left(F(x)+\cdots+F(S^{[t+u]_x}x),(x,0)\right)\\&\quad=A([t+u]_x,x)=\bar{A}(t,(x,u)).\end{aligned}$$

Thus B and $\bar{A}$ are cohomologous.

Remark 1. Proof of Theorem 1 is much simpler for well-built flows since $[t]_x$ is independent of x. By using this, one can see Theorem 1, in general, as follows: Since two flows built under functions on the same base space are weakly equivalent and since cohomology classes of cocycles are invariant under weak equivalence and since cohomology classes of a well built flow correspond one–one to the cohomology classes of the base transformation, it follows that the cohomology classes of cocycles for any flow built under a function correspond one–one to the cohomology classes of the base transformation. This, however, does not tell us how the cocycles correspond and we will need in the sequel our concrete method of associating a cocycle on the flow with one on the base space.

Remark 2. In the usual definition of a cocycle, the identity (1) is required to hold a.e. (t, s, y) rather than everywhere as we have done. However, it can be shown that if (1) holds a.e. (t, s, y), then by modifying A on a null set we can obtain strict equality. This has been shown by Gamelin [6, p. 187] in a special case and generally by Mathew [11, Ch. II, pp. 40–43]. Also in a special case Theorem 1 is due to Gamelin [6, p. 184].

3. Multiplicative integral representation for cocycles

We first define a multiplicative integral in the form we need and state some of its properties. Let H be a function on R taking values in $B(\mathcal{H})$ the bounded operators on the Hilbert space $\mathcal{H}$. Let H be Riemann integrable on every closed bounded interval $[a, b]$ in the sense that

$$\lim_{\max \Delta t_j \to 0} \sum_{j=0}^{n-1} H(\tau_j)(t_{j+1}-t_j)$$

exists where $a=t_0<t_1<\cdots<t_n=b$, $\Delta t_j=t_{j+1}-t_j$. $\tau_j \in [t_j, t_{j+1}]$, the limit being independent of choice of $t_0,\ldots,t_n$, and of the choice of τ_j's. The multiplicative Riemann integral of H is defined to be the limit:

$$\lim_{\max \Delta t_j \to 0} \prod_{j=0}^{n-1} e^{H(\tau_j)(t_{j+1}-t_j)}$$

and denoted by

$$\overset{\curvearrowright}{\int_a^b} e^{H(t)}\,dt.$$

It can be shown that when H is Riemann integrable in the above sense, the multiplicative Riemann integral of H exists and is independent of $t_0, t_1,\ldots, t_n$ and of the choice of τ_j's. For $b<a$, we define

$$\overset{\curvearrowright}{\int_a^b} e^{H(t)}\,dt-\left(\overset{\curvearrowright}{\int_b^a} e^{H(t)}\,dt\right)^{-1}.$$

Further one can establish for real a, b, c,

$$\overset{\curvearrowright}{\int_a^b} e^{H(t)}\,dt=\overset{\curvearrowright}{\int_a^c} e^{H(t)}\,dt\cdot\overset{\curvearrowright}{\int_c^b} e^{H(t)}\,dt, \tag{3}$$

$$\overset{\curvearrowright}{\int_a^b} e^{H(s+t)}\,dt=\overset{\curvearrowright}{\int_{a+s}^{b+s}} e^{H(t)}\,dt. \tag{4}$$

We refer the reader to papers of Masani [10] and Potapov [13, p. 229] for a general and detailed treatment of multiplicative Riemann integral.

Let H be a bounded Hermitian operator valued measurable function on Y such that for every $y \in Y$, $H(T_t y)$ as a function in t is Riemann integrable on every bounded closed interval. Put

$$B(t, y) = \overset{\curvearrowright}{\int_0^t} e^{iH(T_u y)} du. \tag{5}$$

Then B is a cocycle as can be readily verified using (3) and (4).

Theorem 2. *Every (R, Y) cocycle is cohomologous to one of the form* (5) *where H is such that for each y, $H(T_u y)$ as a function in u takes finite or countable number of distinct values.*

Proof. Let C be an (R, Y) cocycle which, by Theorem 1, may be assumed to be of the form $\bar{A}$ for some (Z, X) cocycle A. Let $A(1, x) = e^{iG(x)}$, where G is a bounded measurable Hermitian operator valued function on X, spectrum of $G(x)$, for each x, being in the interval $[0, 2\pi]$. Define H on Y by $H(x, u) = G(x)/F(x)$ where F is the ceiling function. For fixed (x, u), $H(T_t(x, u))$ as a function of t, is constant on each of the intervals

$$\left[\left(\sum_{j=0}^{k-1} F(S^j x)\right) - u, \quad \left(\sum_{j=0}^{k} F(S^j x)\right) - u\right), \qquad k = 1, 2, 3, \ldots,$$

with value $G(S^k x)/F(S^k x)$ on kth interval. The multiplicative integral of $H(T_t(x, u))$ on the kth interval is then equal to $e^{iG(S^k x)}$. $H(T_t(x, u))$ is equal to $G(x)/F(x)$ on the interval $[0, F(x) - u)$ and the multiplicative integral of $H(T_t(x, u))$ on this interval is $e^{-iuH(x, u)} e^{iG(x)}$. Finally the multiplicative integral of $H(T_t(x, u))$ on the interval

$$\left[\sum_{k=0}^{[s+u]_x - 1} F(S^k x) - u, s\right)$$

is

$$e^{i\langle s+u\rangle_x H(S^{[s+u]_x} x)}.$$

Thus

$$\overset{\curvearrowright}{\int_0^s} e^{iH(T_t(x, u))dt} = e^{-iuH(x, u)} \bar{A}(s, (x, u)) e^{i\langle s+u\rangle_x H(T_s(x, u))}.$$

This can be seen by breaking the integral over the intervals

$$[0, F(x) - u), [F(x) - u, F(x) + F(Tx) - u), \ldots, \left[\sum_{j=0}^{[s+u]_x - 1} F(S^k x) - u, s\right).$$

Thus $\bar{A}$ and the cocycle given by multiplicative integral on the left are cohomologous (ρ) with $\rho(x, u) = e^{-iuH(x, u)}$.

Remark 1. Since every ergodic flow is isomorphic to a flow built under a function, the multiplicative integral representation for a cocycle (up to cohomogy) given above is valid for any ergodic flow.

Remark 2. The integral representation for a cocycle was introduced by Helson [7] in connection with harmonic analysis on compact groups with ordered duals. In [7, section 5], Helson showed that every cocycle (taking values in the circle group) for a certain flow $x \to x+e_t$ on a compact group K with ordered dual is cohomologous to one of the form $C\exp(\mathrm{i}\int_0^t m(x+e_s)\mathrm{d}s)$ where C is a cocycle taking values $+1$ or -1. The superfluous factor C was removed by Carlson [4], see Helson [8, Ch. 4, Section 7]. The function $m(x+e_s)$ in this representation is analytic in s and the method of proof relies on spectral theory and complex function theory. A straightforward generalization of the result by replacing real m by a hermitian operator valued function does not yield a $\mathcal{U}(\mathcal{H})$ valued cocycle unless values of m commute. Hence we have to resort to multiplicative integral. Also the use of the notion of flow built under a function permits us to avoid essentially spectral theory and complex function theory.

4. Lemma on Radon–Nikodym derivatives of measures *m* and *n*

This lemma will be used in the next section.

Lemma 2. *For any $t \in R$*

$$\frac{\mathrm{d}n_t}{\mathrm{d}n}(x,u) = \frac{\mathrm{d}m_{[t+u]_x}}{\mathrm{d}m}(x), \quad a.e.n.$$

where n_t and m_k are the measures $n_t(\cdot)=n(T_t\cdot)$ and $m_k(\cdot)=m(S^k\cdot)$.

Proof. Let t be positive and $A \subset Y$ measurable. Then

$$A = \bigcup_{k=0}^{\infty} A_k \quad \text{where } A_k = \{(x,u) \in A: [t+u]_x = k\}.$$

The sets A_k's are disjoint. Further

$$T_t A_k = \left\{ \left(S^k x, u+t-\sum_{j=0}^{k-1} F(S^j x) \right) : (x,u) \in A_k \right\}.$$

Therefore for any $x \in X$,

$$(T_t A_k)^x = (A_k)^{S^{-k}x} + t - F(S^{-1}x) - \cdots - F(S^{-k+1}x).$$

So

$$n(T_tA_k)=\int_X \lambda(T_tA_k)^x\,dm(x),\quad \lambda=\text{Lebesgue measure}$$

$$=\int_X \lambda\big((A_k)^{S^{-k}x}\big)\,dm(x)$$

$$=\int_X \lambda\big((A_k)^x\big)\frac{dm_k}{dm}(x)\,dm(x)$$

$$=\int_{A_k}\frac{dm_k}{dm}(x)\,dn(x,u)$$

$$=\int_{A_k}\frac{dm_{[t+u]_x}}{dm}(x)\,dn(x,u).$$

Since A_n's are disjoint

$$n(T_tA)=\int_A \frac{dm_{[t+u]_x}}{dm}(x)\,dn(x,u).$$

Hence for $t>0$,

$$\frac{dn_t}{dn}(x,u)=\frac{dm_{[t+u]_x}}{dm}(x)\quad \text{a.e. } n.$$

For $t<0$ the equality follows from the cocycle identity satisfied by Radon–Nikodym derivatives of quasi-invariant measures.

5. Proof of a spectral result regarding well-built flows

For any spectral measure E, we shall say that a sequence of mutually singular measures $\nu_\infty, \nu_1, \nu_2, \ldots$ determines E if the equivalence classes of $\nu_\infty, \nu_1, \nu_2, \ldots$ form a complete set of invariants for E. Let A be a (Z, X) cocycle. Define unitary groups U_k, $k\in Z$, and $\overline{U}_t$, $t\in R$, as follows:

$$(U_kf)(x)=A(k,x)\left(\frac{dm_k}{dm}(x)\right)^{1/2} f(S^kx),\quad f\in L^2(X,m,\mathcal{H}),$$

$$(\overline{U}_tg)(y)=\overline{A}(t,y)\left(\frac{dn_t}{dn}(y)\right)^{1/2} g(T_ty),\quad g\in L^2(Y,n,\mathcal{H}).$$

Let E and $\overline{E}$ be the spectral measures of U_k, $k\in Z$ and $\overline{U}_t$, $t\in R$, respectively. In general, E does not determine $\overline{E}$; however, for a well-built flow E does determine $\overline{E}$. For any measure ν on $[0,2\pi)$ we shall write $\bar{\nu}$ for the measure $(\sin\frac{1}{2}x/\frac{1}{2}x)^2 d\tilde{\nu}$ where $\tilde{\nu}$ is the σ-finite measure on R given by

$$\tilde{\nu}(A)=\sum_{k=-\infty}^{\infty}\nu[(A+2k\pi)\cap[0,2\pi)].$$

It can be shown that

$$\hat{\bar{\nu}}(t)=(1-\langle t\rangle)\hat{\nu}([t])+\langle t\rangle\hat{\nu}([t]+1)$$

where $[t]$ and $\langle t\rangle$ are the integral and fractional parts of t, respectively [3, Lemma 5.1].

Theorem 3. *Assume that T_t, $t\in R$, is a well-built flow. If $\nu_\infty, \nu_1, \nu_2, \ldots$ are mutually singular measures which determine E, then $\bar{\nu}_\infty, \bar{\nu}_1, \bar{\nu}_2, \ldots$ are the mutually singular measures which determine $\bar{E}$.*

Proof. We assume that E has uniform multiplicity, say l. The general case can be proved by breaking $L^2(X, m, \mathcal{H})$ into mutually orthogonal subspaces on each of which E has uniform multiplicity. We also assume that one is not an eigenvalue of U (this assumption will be removed later). Since E has uniform multiplicity, there are functions $f_1, f_2, \ldots, f_l$ in $L^2(X, m, \mathcal{H})$ such that

(i) $(U_k f_i, f_j)=0$ if $i\neq j$, $1\leq i, j\leq l$ for all k.

(ii) The measures $\nu(\cdot)=(E(\cdot)f_i, f_i)$, $1\leq i\leq l$ are all the same.

(iii) The closed linear span of $\{U^k f_i$: k integer, $1\leq i\leq l\}$ is $L^2(X, m, \mathcal{H})$.

Define $\bar{f}_i$ on Y by $\bar{f}_i(x,u)=f_i(x)$, $1\leq i\leq l$. Then

$$\begin{aligned}(\bar{U}_t\bar{f}_i, \bar{f}_j)&=\int_Y\left\langle \bar{A}(t,(x,u))\left(\frac{dn_t}{dn}(x,u)\right)^{1/2}\bar{f}_i(T_t(x,u)), \bar{f}_j(x,u)\right\rangle dn\\
&=\int_0^1\int_X\left\langle A([t+u],x)\left(\frac{dm_{[t+u]}}{dm}(x)\right)^{1/2}f_i(S^{[t+u]}x), f_j(x)\right\rangle dm\,du\\
&=\int_0^1\left(U^{[t+u]}f_i, f_j\right)du=(1-\langle t\rangle)\left(U^{[t]}f_i, f_j\right)+\langle t\rangle\left(U^{[t]+1}f_i, f_j\right)\\
&=\begin{cases}0 & \text{if } i\neq j,\\ \hat{\bar{\nu}}(t) & \text{if } i=j.\end{cases}\end{aligned}\tag{1}$$

We have used Lemma 2 of the last section in the second step above. Furthermore, we have used the notation $\langle ,\rangle$ for inner product in $\mathcal{H}$ and $(,)$ for inner product in both $L^2(X, m, \mathcal{H})$ and $L^2(Y, n, \mathcal{H})$. Thus we have shown that the measures $(\bar{E}(\cdot)\bar{f}_i, \bar{f}_i)$ for $i=1,2,\ldots,l$ are all the same and equal to $\bar{\nu}$. We now show that the closed linear span of $\{\bar{U}_t\bar{f}_i$: $t\in R$, $i=1,2,\ldots,l\}$ is all of $L^2(Y, n, \mathcal{H})$. Let $h\in L^2(Y, n, \mathcal{H})$ be such that $(\bar{U}_t\bar{f}_j, h)=0$ for all $t\in R$ and $i=1,2,\ldots,l$. Let k be an integer; t, s be real numbers such that $k\leq t<s<k+1$. Then

$$\begin{aligned}0&=(\bar{U}_t-\bar{U}_s\bar{f}_i, h)\\
&=\int_X\int_0^1\left\{\left\langle A([t+u],x)\left(\frac{dm_{[t+u]}}{dm}(x)\right)^{1/2}f_i(S^{[t+u]}x), h(x,u)\right\rangle\right.\\
&\qquad\left.-\left\langle A([s+u],x)\left(\frac{dm_{[s+u]}}{dm}(x)\right)^{1/2}f_i(S^{[s+u]}x), h(x,u)\right\rangle\right\}du\,dm\\
&=\int_\alpha^\beta\left\{\left(U^{k+1}f_i, h_u\right)-\left(U^k f_i, h_u\right)\right\}du,\end{aligned}$$

where $\alpha=k+1-s$, $\beta=k+1-t$ and h_u is the u-section of h. Varying t, s, this is true for all α, β, $0\leqslant\alpha<\beta<1$. Therefore, for all k and $i=1,2,\ldots,l$, $(U^{k+1}f_i, h_u)=(U^k f_i, h_u)$ or $(U^k f_i, h_u)=(U^0 f_i, h_u)$ for a.e.u. So $h_u\in E(1)(L^2(X,m,\mathcal{H}))=\{0\}$. Therefore for a.e.$u$, h_u is null function and hence, h is a null function. Hence $\bar{E}$ also has uniform multiplicity l with the associated measure $\bar{\nu}$.

If E has constant one in its point spectrum, it can be seen that A is a coboundary and that T_t, $t\in R$, and S admit equivalent finite invariant measures. We may therefore assume that $A=I$ and m and n are invariant under S and T_t, $t\in R$, respectively. Since S and T_t, $t\in R$, are ergodic the constant one appears in the spectrum of each with multiplicity one. By confining E and $\bar{E}$ to the orthogonal complements of constant functions in $L^2(X,m,\mathcal{H})$ and $L^2(Y,n,\mathcal{H})$, respectively and proceeding as above we see that spectrum of $\bar{E}$ is determined by that of E as described in the theorem.

Remark. In the above theorem the multiplicity l of E can possibly be $\aleph_0$ in which case $1\leqslant i\leqslant l$ has to be interpreted to mean $1\leqslant i\leqslant\infty$.

6. A result on the spectral measure of a unitary group associated with an ergodic flow

Let τ_t, $t\in R$, be an ergodic flow on $(\Omega,\mathcal{A},\mu)$ and let A be a $\mathcal{U}(\mathcal{H})$ valued (R,Ω) cocycle. Let U_t, $t\in R$, be the group of unitary operators on $L^2(\Omega,\mu,\mathcal{H})$ defined as follows:

$$(U_t f)(w)=A(t,w)f(T_t w)\left(\frac{\mathrm{d}\mu_t}{\mathrm{d}\mu}(w)\right)^{1/2}.$$

Theorem 4. *Given any hermitian operator H, and $K>0$ and any $\varepsilon>0$, there is a unitary operator valued function g on Ω such that*

$$\|U_t g-g\mathrm{e}^{\mathrm{i}tH}\|<\varepsilon,\quad 0\leqslant t\leqslant K.$$

To prove the above theorem we need the following result due to Ornstein [12, p. 63, Lemma 1].

Let τ_t, $t\in R$, be an ergodic non-singular flow and $\varepsilon>0$, $N>0$ be given. Then τ_t, $t\in R$, is isomorphic to a flow built under a function F with base space $(X,\mathcal{B},m)$ with $F=N$ on a set $\bar{X}\subset X$, $m(\bar{X})>(1-\varepsilon)m(X)$ and $F\leqslant N$ on the rest of X, i.e. on $X-\bar{X}$.

To be accurate Ornstein proves the above result for measure preserving flows, but essentially the same proof combined with the result of Krengel and Dani mentioned earlier yields the above result. Equipped with this we now prove Theorem 4. Let N be a positive integer so large that $1/N<\frac{1}{4}\varepsilon$. Replace the flow τ_t, $t\in R$, by an isomorphic flow T_t, $t\in R$, built under a function F on a base space $(X,\mathcal{B},m)$, such that $F=NK$ on set $\bar{X}\subset X$ of measure greater than $(1-(\varepsilon/4NK))m(X)$ and $F\leqslant NK$ on the rest of X. Since the total integral of F is one, if N is large, $m(X)$ would be less than one, and we assume this to be the case. Let $(Y, \mathcal{C}, n)$ denote the space on which T_t, $t \in R$, acts and we may assume that A is an (R, Y) cocycle. Let $Z = \bar{X} \times [0, NK)$. Then $n(Y-Z) < \frac{1}{4}\varepsilon$. We also note that

$$\frac{\mathrm{d}n_t}{\mathrm{d}n}(x, u) = 1, \quad \text{if } (x, u) \in Z \text{ and } u + t \leqslant NK.$$

Now define

$$g(x, u) = \begin{cases} A^*(u, (x, 0))\, \mathrm{e}^{\mathrm{i}Hu} & \text{if } (x, u) \in Z, \\ I, & \text{otherwise} \end{cases}$$

where A^* denotes the adjoint of A. Clearly g is a unitary operator valued function on Y and

$$(U_t g)(x, u) = g(x, u)\, \mathrm{e}^{\mathrm{i}tH}, \quad 0 \leqslant t \leqslant K$$

provided $x \in \bar{X}$ and $0 \leqslant u < NK - t$. Thus for $0 \leqslant t \leqslant K$,

$$\| U_t g - g\mathrm{e}^{\mathrm{i}tH} \| = \| (U_t g - g\mathrm{e}^{\mathrm{i}tH}) 1_B \| < 2n(B) < \varepsilon$$

where $B = Y - \bar{X} \times [0, NK - t)$.

References

[1] Ambrose, W. (1941). Representation of ergodic flows. *Ann. Math.* **42**, 723–739.

[2] Ambrose, W. and Kakutani, S. (1942). Structure and continuity of measurable flows. *Duke Math. J.* **9**, 25–42.

[3] Bagchi, S.C., Mathew, J. and Nadkarni, M.G. (1974). On systems of imprimitivity on locally compact abelian groups with dense actions. *Acta Math.* **133**, 287–304.

[4] Carlson, C.G.R. (1972). Cohomology classes in harmonic analysis. Thesis, Stanford University, Stanford, CA.

[5] Dani, S.G. (1976). Kolmogorov automorphisms on homogeneous spaces. *Am. J. Math.* **98**, 119–163.

[6] Gamelin, T.W. (1967). *Uniform Algebras.* Prentice-Hall, Englewood Cliffs, NJ.

[7] Helson, H. (1965). Compact groups with ordered duals. *Proc. London Math. Soc.* **14A**, 144–156.

[8] Helson, H. (1975). Analyticity on compact abelian groups. In: J.H. Williamson, Ed., *Algebras in Analysis.* Academic Press, New York.

[9] Krengel, U. (1969). Darstellungssätze fur Strömungen und Helbströmungen II. *Math. Ann.* **182**, 1–39.

[10] Masani, P. (1947). Multiplicative Riemann integration in normed rings. *Trans. Am. Math. Soc.* **61**, 147–192.

[11] Mathew, J. (1975). Systems of imprimitivity for ergodic action of local compact Abelian groups. Thesis, Indian Statistical Institute.

[12] Ornstein, D. (1974). *Ergodic Theory, Randomness and Dynamical Systems.* Yale Mathematical Monograph. Yale University Press, New Haven, CT.

[13] Potapov, V.P. (1960). The multiplicative structure of *J*-contractive matrix functions. *Am. Math. Soc. Transl. Ser. 2* **15**, 131–243.

G. Kallianpur, P.R. Krishnaiah, J.K. Ghosh, eds., *Statistics and Probability*: *Essays in Honor of C.R. Rao*

PROPERTIES OF THE FUNDAMENTAL BORDERED MATRIX USED IN LINEAR ESTIMATION*

Sujit Kumar MITRA
Indian Statistical Institute, New Delhi, India and Indiana University, Bloomington, U.S.A.

Consider the Gauss–Markov model $(Y, X\beta, \sigma^2 V)$ with a possibly singular covariance matrix $\sigma^2 V$. The X matrix may also be deficient in rank. The matrix,

$$F_0 = \begin{pmatrix} V & X \\ X' & O \end{pmatrix}, \tag{1}$$

has been termed the fundamental bordered matrix of linear estimation [2] on account of its role in the IPM (Inverse Partitioned Matrix) method of linear estimation proposed by Rao [5]. Let

$$G_0 = \begin{pmatrix} C_1 & C_2 \\ C_3 & -C_4 \end{pmatrix} \tag{2}$$

be a g-inverse of F_0. Khatri [3] and Rao [5, 6, and 7, pp. 294–296] established interesting interconnections between the various blocks of G_0, namely C_1, C_2, C_3 and C_4. Additional characterizations of these blocks are given in Hall and Meyer [2]. Hall [1] presents these results in a Hilbert space setting. The object of this paper is to examine the extent to which the results hitherto established depend on the specific structure of the F_0 matrix and in particular on the fact that V is a symmetric non-negative definite (n.n.d.) matrix. In what follows we shall consider matrices on the field of complex numbers. We adopt the notations of Rao and Mitra [8].

Consider the complex matrix,

$$F = \begin{pmatrix} V & X \\ Y & 0 \end{pmatrix}, \tag{3}$$

where $V \in \mathcal{C}^{n\times p}$, $X \in \mathcal{C}^{n\times m}$, $Y \in \mathcal{C}^{q\times p}$ and 0 is the null matrix of order $q \times m$. Let

$$G = \begin{pmatrix} C_1 & C_2 \\ C_3 & -C_4 \end{pmatrix} \tag{4}$$

be a g-inverse of F. We shall assume that the matrix F satisfies the following rank additivity condition:

$$\text{Rank } F = \text{Rank}(V \vdots X) + \text{Rank } Y = \text{Rank}\begin{pmatrix} V \\ Y \end{pmatrix} + \text{Rank } X. \tag{5}$$

*Research supported by National Science Foundation Grant No. MCS 76-00951 at Indiana University.

Theorem 1. *Let the matrix F satisfy the rank additivity condition* (5). *Then*
(i) $XC_3X=X, \quad YC_2Y=Y$
(ii) $YC_1X=0, \quad YC_1V=0, \quad VC_1X=0$
(iii) $XC_3V=VC_2Y=XC_4Y$
(iv) $VC_1VC_1V=VC_1V, \quad \operatorname{Tr} VC_1=\operatorname{Rank}(V \vdots X)-\operatorname{Rank} X$

$$=\operatorname{Rank}\begin{pmatrix} V \\ Y \end{pmatrix}-\operatorname{Rank} Y$$

(v) $\begin{pmatrix} C_1 \\ C_3 \end{pmatrix}$ *is a g-inverse of* $(V \vdots X)$, $(C_1 \vdots C_2)$ *is a g-inverse of* $\begin{pmatrix} V \\ Y \end{pmatrix}$.

Proof. We need the following lemma which is easily established.

Lemma 1. *Conditions* (5) *is equivalent to the following condition*

$$\mathfrak{M}\begin{pmatrix} V \\ 0 \end{pmatrix} \subset \mathfrak{M}(F), \qquad \mathfrak{M}\begin{pmatrix} V^* \\ 0 \end{pmatrix} \subset \mathfrak{M}(F^*). \tag{6}$$

(i), (ii), (iii), (v) and the first part of (iv) are simple consequences of the equations

$$FGF=F, \tag{7a}$$

$$FG\begin{pmatrix} V \\ 0 \end{pmatrix}=\begin{pmatrix} V \\ 0 \end{pmatrix}, \qquad (V \vdots 0)GF=(V \vdots 0), \tag{7b}$$

the last two equations being implied by (6).

Since $G \in \{F^-\}$, $C_3 \in \{X^-\}$, $C_2 \in \{Y^-\}$

$$\operatorname{Rank} F=\operatorname{Rank} FG=\operatorname{Tr} FG=\operatorname{Tr}(VC_1+XC_3)+\operatorname{Tr} YC_2$$
$$=\operatorname{Tr} VC_1+\operatorname{Rank} X+\operatorname{Rank} Y.$$

The second part of (iv) therefore follows from Condition (5).

Theorem 1 is strikingly similar to Theorem 1.2 of Rao [5] and most of the results in the latter are seen to follow if $Y=X^*$, provided condition (5) is satisfied. In fact here C_2 is a g-inverse of Y satisfying the additional condition

$$Y^*C_2^*VC_2Y=VC_2Y. \tag{8a}$$

Condition (8a) reminds us of a minimum V seminorm g-inverse of Y when V is hermitian n.n.d. Similarly C_3^* is a g-inverse of $Y=X^*$ satisfying the additional condition

$$Y^*C_3V^*C_3^*Y=V^*C_3^*Y. \tag{8b}$$

In fact, when V is hermitian non-negative definite, we have the Theorem 1.2 of Rao [5]. We need one part of Lemma 2 for this purpose.

Lemma 2. *If* $p=n$, $q=m$, $Y=X^*$ *and V is hermitian, the following two statements are equivalent*
(i) *V is hermitian non-negative or non-positive definite.*
(ii) *Condition* (5) *holds for all matrices X with same number of rows as in V.*

Proof. We note that condition (5) is equivalent to claiming

$$\mathfrak{M}\begin{pmatrix} V \\ X^* \end{pmatrix} \cap \mathfrak{M}\begin{pmatrix} X \\ 0 \end{pmatrix} = \{0\}, \tag{9a}$$

$$\mathfrak{M}\begin{pmatrix} V^* \\ X^* \end{pmatrix} \cap \mathfrak{M}\begin{pmatrix} X \\ 0 \end{pmatrix} = \{0\}, \tag{9b}$$

not (ii)$\Rightarrow$not (i):

Assume for example that (9a) is not true and that $\exists$ a n-tuple, a, and a m-tuple, b, such that

$$Va = Xb \neq 0, \qquad X^*a = 0.$$

This implies $a^*Va = 0 \Rightarrow$not (a). Not $(a)\Rightarrow$not (b): if V is indefinite $\exists$ a n-tuple a such that

$$a^*Va = 0, \qquad Va \neq 0.$$

Choose $X = Va$ and observe that

$$\begin{pmatrix} X \\ 0 \end{pmatrix} = \begin{pmatrix} V \\ X^* \end{pmatrix} a$$

which contradicts (9a).

The similarity in fact goes much deeper than Theorem 1. We have seen in (8a) and (8b) that when $Y = X^*$ and V is hermitian n.n.d. C_2 and C_3^* are minimum V seminorm g-inverses of Y. Hence from Theorem 2.1 of Mitra and Puri [4] it is seen that VC_2Y and VC_3^*Y are unique matrices, invariant under the choice of the g-inverse G of F. A general version in our context is given in Theorem 2.

Theorem 2. *If the matrix F satisfies the rank additivity condition* (5) *the matrices VC_1V, XC_3V, VC_2Y, XC_4Y are unique with respect to the choice of a g-inverse G of F.*

Proof. Let

$$\begin{pmatrix} C_1 & C_2 \\ C_3 & -C_4 \end{pmatrix} \quad \text{and} \quad \begin{pmatrix} \tilde{C}_1 & \tilde{C}_2 \\ \tilde{C}_3 & -\tilde{C}_4 \end{pmatrix}$$

be distinct choices of F^-. We show $VC_1V = V\tilde{C}_1V$, $XC_3V = X\tilde{C}_3V$, $VC_2Y = V\tilde{C}_2Y$, $XC_4Y = X\tilde{C}_4Y$. In fact on account of Theorem 1 (iii) it suffices to establish only the first two. The first of the two in equations (7b)$\Rightarrow$

$$V = VC_1V + XC_3V = V\tilde{C}_1V + X\tilde{C}_3V. \tag{10}$$

We now show $\mathfrak{M}(VC_1V)$ and $\mathfrak{M}(V\tilde{C}_1V)$ are both virtually disjoint with $\mathfrak{M}(X)$. If, for example

$$VC_1Va = Xb$$

is a vector in $\mathfrak{M}(VC_1V) \cap \mathfrak{M}(X)$ we have:

$$VC_1Va = VC_1VC_1Va = VC_1Xb = 0.$$

This establishes our claim. Similarly using $XC_3V = VC_2Y = XC_4Y$, $X\tilde{C}_3V = V\tilde{C}_2Y = X\tilde{C}_4Y$ as implied by Theorem 1 (iii) we observe that $\mathfrak{M}[(VC_1V)^*]$ and $\mathfrak{M}[(V\tilde{C}_1V)^*]$

are both virtually disjoint with $\mathcal{M}(Y^*)$. Hence,

$$VC_1V = V\tilde{C}_1V + (X\tilde{C}_4Y - XC_4Y), \tag{11a}$$

$$V\tilde{C}_1V = VC_1V + (XC_4Y - X\tilde{C}_4Y). \tag{11b}$$

(11a)$\Rightarrow$Rank VC_1V=Rank $V\tilde{C}_1V$+Rank$(X\tilde{C}_4Y - XC_4Y)\Rightarrow$Rank $VC_1V \geqslant$ Rank $V\tilde{C}_1V$. (11b) similarly implies Rank $V\tilde{C}_1V \geqslant$ Rank VC_1V. Hence

$$\text{Rank}(X\tilde{C}_4Y - XC_4Y) = 0$$

$$\Rightarrow X\tilde{C}_4Y = XC_4Y,$$

and Theorem 2 follows.

Remark 1. Let $\mathbb{S}$ and $\mathfrak{T}$ denote respectively the subspaces $\mathcal{M}(X)$, $\mathcal{M}(Y^*)$. Similar to the Krein–Anderson–Trapp definition of the shorted operator reproduced in Mitra and Puri [4], we could define the generalized shorted matrix $S(V|\mathbb{S},\mathfrak{T})$ as follows:

(i)
$$\mathcal{M}[S(V|\mathbb{S},\mathfrak{T})] \subset \mathbb{S}, \tag{12a}$$

$$\mathcal{M}[\{S(V|\mathbb{S},\mathfrak{T})\}^*] \subset \mathfrak{T}. \tag{12b}$$

(ii) If E is any other matrix satisfying (12a) and (12b), then

$$\text{Rank}(V-E) \geqslant \text{Rank}[V - S(V|\mathbb{S},\mathfrak{T})]. \tag{13}$$

It is not difficult to see that the generalized shorted matrix need not always be uniquely determined. In this particular case, however, when condition (5) is satisfied, from the arguments in the proof of Theorem 2 it is seen that $XC_4Y = XC_3V = VC_2Y$ is the unique generalized shorted matrix $S(V|\mathbb{S},\mathfrak{T})$ and strict inequality holds in (13) whenever E is a distinct matrix obeying (12a) and (12b). The generalized shorted matrix will be studied in greater detail in a separate publication.

Remark 2. For the limited purpose of proving Theorem 2 it suffices to appeal to Lemma 2.2.4 (iii) of ref. [8]. The invariance of VC_1V is thus a consequence of Lemma 1 of this paper. The invariance of $S = XC_3V = VC_2Y = XC_4Y$ is similarly deduced. This proof, though simple and straightforward, does not, however, reveal the other interesting properties of the S matrix we have come across in Remark 1.

Theorems 3.1 and 3.2 of ref [2] can also be presented in the more general setting as follows. For a matrix, A, let E_A and F_A denote respectively the left and right annihilators of A, namely $I - AA^-$ and $I - A^-A$.

Theorem 3. *For any choice of the g-inverse of X, Y and E_XVF_Y*

$$\begin{pmatrix} 0 & Y^- \\ X^- & -X^-VY^- \end{pmatrix} + \begin{pmatrix} I \\ -X^-V \end{pmatrix} Q \left(I \vdots -VY^- \right) \tag{14}$$

is a g-inverse of F where $Q = F_Y(E_XVF_Y)^-E_X$.

Proof. The proof is by direct computation. We omit the proof.

Theorem 4. *Let the matrix F satisfy condition* (5). *Then Q is one choice of the C_1-matrix iff Q satisfies*

$$E_X(V-VQV)=0[\text{or } (V-VQV)F_Y=0]$$

$$YQV=0$$

$$VQX=0$$

$$YQX=0$$

Proof. The "if" part does not require condition (5) and can be proved by direct computation noting that a Q matrix satisfying these four equations, when substituted in expression (14) actually leads to a g-inverse of F.

The "only if" part is virtually contained in Theorem 1, (ii). The first condition which is the only equation left out, follows from (10). One could similarly present suitable extensions of other theorems in ref [2]. We leave the details to the reader.

References

[1] Hall, F.J. (1975). Generalized inverses of a bordered matrix of operators. *SIAM J. Appl. Math.* **29**, 152–163.

[2] Hall, F.J. and Meyer, C.D. (1975). Generalized inverses of the fundamental bordered matrix used in linear estimation. *Sankhyā Ser. A.* **37**, 428–438.

[3] Khatri, C.G. (1968). Some results for the singular multivariate regression models. *Sankhyā Ser. A.* **30**, 268–280.

[4] Mitra, S.K. and Puri, M.L. (1979). Shorted operators and generalized inverses of matrices. *Linear Algebra Appl.* **25**, 45–56.

[5] Rao, C.R. (1971). Unified theory of linear estimation. *Sankhyā Ser. A.* **33**, 371–394.

[6] Rao, C.R. (1972). A note on the IPM method in the unified theory of linear estimation. *Sankhyā Ser. A.* **34**, 285–288.

[7] Rao, C.R. (1973). *Linear Statistical Inference and its Applications.* Wiley, New York.

[8] Rao, C.R. and Mitra, S.K. (1971). *Generalized Inverse of Matrices and its Applications.* Wiley, New York.

G. Kallianpur, P.R. Krishnaiah, J.K. Ghosh, eds., *Statistics and Probability: Essays in Honor of C.R. Rao*

THE SURFACE AREA OF AN ELLIPSOID

P.A.P. MORAN
The Australian National University, Canberra, Australia

Consider an ellipsoid defined by the equation,

$$x^2/A^2+y^2/B^2+z^2/C^2=1, \tag{1}$$

which is also convenient to write in the form,

$$ax^2+by^2+cz^2=1, \tag{2}$$

where $a=A^{-2}$, $b=B^{-2}$, $c=C^{-2}$. For later convenience we assume $A \geqslant B \geqslant C$ so that $a \leqslant b \leqslant c$. Then, it has been known since at least early in the nineteenth century that the surface area, S, of this ellipsoid is given by the formula,

$$S=2\pi C^2+2\pi B(A^2-C^2)^{-1/2}\{C^2F(k,\phi)+(A^2-C^2)E(k,\phi)\}. \tag{3}$$

Here $k=A(B^2-C^2)^{1/2}/B(A^2-C^2)^{1/2}$, $\phi=\cos^{-1}CA^{-1}$, and F and E are the incomplete elliptic integrals of the first and second kind, defined by

$$E(k,\phi)=\int_0^{\phi}(1-k^2\sin^2\theta)^{1/2}\,\mathrm{d}\theta,$$

$$F(k,\phi)=\int_0^{\phi}(1-k^2\sin^2\theta)^{-1/2}\,\mathrm{d}\theta.$$

These integrals (the notation varies) can be found by numerical integration provided care is used in the end corrections. However it is of some interest to reexamine the problem from first principles.

We parametrize (1) by representing a point on the surface (1) by the vector,

$$\boldsymbol{x}=(A\cos\theta,\ B\sin\theta\cos\phi,\ C\sin\theta\sin\phi), \tag{4}$$

where $0\leqslant\theta\leqslant\pi$, $0\leqslant\phi\leqslant 2\pi$. By applying elementary differential geometry, the vector differential elements of $\boldsymbol{x}$ corresponding to increments $\mathrm{d}\theta, \mathrm{d}\phi$ are then

$$\mathrm{d}\boldsymbol{s}_1=(-A\sin\theta,\ B\cos\theta\cos\phi,\ C\cos\theta\sin\phi)\,\mathrm{d}\theta,$$

$$\mathrm{d}\boldsymbol{s}_2=(0,\ -B\sin\theta\sin\phi,\ C\sin\theta\cos\phi)\,\mathrm{d}\phi.$$

The element of surface is then

$$\mathrm{d}\theta\,\mathrm{d}\phi\sqrt{(|\mathrm{d}\boldsymbol{s}_1|^2|\mathrm{d}\boldsymbol{s}_2|^2-|\mathrm{d}\boldsymbol{s}_1\cdot\mathrm{d}\boldsymbol{s}_2|^2)}=f(\theta,\phi)\,\mathrm{d}\theta\,\mathrm{d}\phi, \quad \text{say,}$$

where

$$f(\theta,\phi)^2=(B^2C^2\cos^2\theta+A^2C^2\sin^2\theta\cos^2\phi+A^2B^2\sin^2\theta\sin^2\phi)\sin^2\theta,$$

and

$$S=\int_0^{2\pi}\int_0^{\pi}|f(\theta,\phi)|\,d\theta\,d\phi \tag{5}$$

$$=8\int_0^{\pi/2}\int_0^{\pi/2}|f(\theta,\phi)|\,d\theta\,d\phi. \tag{6}$$

It is possible to approach the problem of determining S from a quite different point of view by using a well-known result of Cauchy which states that the surface area of a convex body is four times the mean projected area in all directions, the latter being uniformly distributed over all directions in space.

Let (l_1, l_2, l_3) be the direction cosines of a unit vector from the origin, and consider the area of the projection of the ellipsoid on a plane perpendicular to (l_1, l_2, l_3). This is most easily found by first considering the projection on the (x, y) plane in the direction (l_1, l_2, l_3). To do this write $x=X+ul_1$, $y=Y+ul_2$, $z=ul_3$ where u is a variable. Inserting these values in eq. (2) and finding the condition that the resulting quadratic in u has a double root we obtain the equation of an ellipse

$$X^2(abl_2^2+acl_3^2)-2XYabl_1l_2+Y^2(abl_1^2+bcl_3^2)=al_1^2+bl_2^2+cl_3^2. \tag{7}$$

By using elementary geometry, the area of this ellipse is

$$\pi(al_1^2+bl_2^2+cl_3^2)^{1/2}/l_3(abc)^{1/2}.$$

The area of the projection on a plane perpendicular to the vector (l_1, l_2, l_3) is then obtained by multiplying this by l_3 and so S is four times the average of

$$\pi(al_1^2+bl_2^2+cl_3^2)^{1/2}/(abc)^{1/2},$$

where (l_1, l_2, l_3) is uniformly distributed on the surface of the unit sphere. Thus we can write

$$S=8(abc)^{-1/2}\int_0^{\pi/2}\int_0^{\pi/2}(al_1^2+bl_2^2+cl_3^2)^{1/2}\sin\theta\,d\theta\,d\phi. \tag{8}$$

where $l_1=\cos\theta$, $l_2=\sin\theta\cos\phi$, $l_3=\sin\theta\sin\phi$. This result is, however, identical with (6). In the course of the above proof we have obtained the surprising fact that the area of the projection of the ellipsoid in the direction (l_1, l_2, l_3) is proportional to the differential element of area at the point (Al_1, Bl_2, Cl_3).

We can now represent (8) in a probabilistic form. Let X_1, X_2, X_3 be independent random variables having normal distributions with zero means and unit standard deviations. Using E to denote expectations with respect to the joint distribution of these three variates, we have

$$S=4\pi(abc)^{-1/2}E\left\{\left(\frac{aX_1^2+bX_2^2+cX_3^2}{X_1^2+X_2^2+X_3^2}\right)^{1/2}\right\}. \tag{9}$$

We now consider various methods of evaluating (9).

One approach is the following. Rescale a, b, c to a_1, b_1, c_1 so that $a_1+b_1+c_1=3$, and write

$$S=4\pi(a+b+c)^{1/2}(3abc)^{-1/2}E\left\{\left(\frac{a_1X_1^2+b_1X_2^2+c_1X_3^2}{X_1^2+X_2^2+X_3^2}\right)^{1/2}\right\}$$

$$=4\pi(a+b+c)^{1/2}(3abc)^{-1/2}E\left\{\left(1+\frac{\alpha X_1^2+\beta X_2^2+\gamma X_3^2}{X_1^2+X_2^2+X_3^2}\right)^{1/2}\right\}, \tag{10}$$

where $\alpha=a_1-1, \beta=b_1-1, \gamma=c_1-1$. Then (provided $|\alpha|+|\beta|+|\gamma|<1$, and possibly more generally) we can expand the right-hand side by the binomial theorem.

Using Pitman's theorem on the expectation of the ratio of two quadratic forms (Pitman 1937) we have, for $k=1,2,3,\ldots$

$$E\left\{\frac{(\alpha X_1^2+\beta X_2^2+\gamma X_3^2)^k}{(X_1^2+X_2^2+X_3^2)^k}\right\}=\frac{E\{(\alpha X_1^2+\beta X_2^2+\gamma X_3^2)^k\}}{E\{(X_1^2+X_2^2+X_3^2)^k\}}, \tag{11}$$

which we write as EA_k/EB_k.

By standard theory we have $EB_k=(2k+1)!/2^k k!$. To obtain EA_k we first observe that the kth cumulant of $\alpha X_1^2+\beta X_2^2+\gamma X_3^2$ is $(\alpha^k+\beta^k+\gamma^k)2^{2k-1}(k-1)!$. EA_k is the kth moment and for k not large can be easily expressed as a polynomial in the cumulants. Taking the expansion as far as the sixth moment, we have

$$S=4\pi(a+b+c)^{1/2}(3abc)^{-1/2}\{1-EA_2/120+EA_3/1680-EA_4/24192$$
$$+EA_5/380160-EA_6/6589440+\cdots\}, \tag{12}$$

where the kth term is $(-)^k EA_k/2^k k!(4k^2-1)$. This is a reasonably effective method if α, β, γ are small, and is easily programmed on a desk computer. Thus using the expansion as far as EA_6, we obtain for $(A, B, C)=(1,1,0.5), (1,0.8,0.625)$ and $(1,0.5,0.125)$ the values 8.668579, 8.151605 and 3.379205. Comparing these with the exact values (see later) the errors are -0.003304, -0.000014 and -0.045598, respectively. Although the method could be pushed a little further by using moments of higher order, the increasing complexity of the formulae for expressing moments in terms of cumulants makes the method unattractive, and furthermore it is not clear that it always converges.

A more useful expansion which is also helpful theoretically is obtained as follows. Starting from (9) and using the fact that $a\leqslant b\leqslant c$, write $\alpha=1-a/c, \beta=1-b/c$. Then using Pitman's theorem again,

$$S=4\pi(ab)^{-1/2}E\left\{\left(1-\frac{\alpha X_1^2+\beta X_2^2}{X_1^2+X_2^2+X_3^2}\right)^{1/2}\right\}$$

$$=4\pi(ab)^{-1/2}\left\{1-\sum_{n=1}^{\infty}\frac{E\{(\alpha X_1^2+\beta X_2^2)^k\}}{2^n n!(4n^2-1)}\right\}. \tag{13}$$

Since $E(X_1^{2k})=(2k)!/2^k k!$ we have

$$E\{(\alpha X_1^2+\beta X_2^2)^k\}=\sum_{m=0}^{n}\alpha^m\beta^{n-m}\frac{(2m)!(2n-2m)!n!}{2^n(m!)^2((n-m)!)^2}$$

and

$$S=4\pi(ab)^{-1/2}\left\{1-\sum_{n=1}^{\infty}\frac{1}{2^{2n}(4n^2-1)}\sum_{m=0}^{n}\alpha^m\beta^{n-m}\frac{(2m)!(2n-2m)!}{(m!)^2((n-m)!)^2}\right\}. \tag{14}$$

It is easy to prove that this series is always convergent for $0\leqslant\alpha,\beta\leqslant 1$ and it is easily programmed. (14) was tried out on a desk computer for $A=1, C=0.5(0.1)1.0$, $B=C(0.1)1$. The slowest convergence occurred when $C=0.5$. For $(A, B, C)=(1,0.5,0.5)$, forty five terms in the series were required to obtain nine decimal accuracy. For $C<0.5$, (14) is not a useful formula for calculation.

(14) can also be used to prove an inequality for S which is related to more general inequalities for the surface area of convex bodies. In Moran (1944), it was shown that for any convex body with surface area S, the mean projected area, P_1 say, on the twelve faces of a dodecahedron satisfies the inequalities $0.91759=\frac{1}{3}(1+\surd 5)(\frac{1}{2}+1/4\surd 5)\leqslant 4P_1/S\leqslant\frac{1}{3}(1+\surd 5)=1.07869$. A similar but tighter pair of inequalities hold for the mean projection of the faces of an icosahedron.

By using the same kind of argument as was used in that paper, rather cruder inequalities can be established for the mean of the projections on the faces of a cube, or on those of a regular octahedron. Consider the cube case first.

If S is the surface area of a convex body and P_1 is the mean of the areas of the projections on the faces of the cube then it is easy to prove that

$$0.6667=\tfrac{2}{3}\leqslant 4P_1/S\leqslant 2/\surd 3=1.1547\,. \tag{15}$$

Taking the convex body to be an ellipsoid with its axes parallel to the edges of the cube we have

$$\tfrac{2}{3}\leqslant 4\pi(AB+BC+CA)/3S\leqslant 2/\surd 3\,. \tag{16}$$

Similarly suppose the axes of the ellipsoid are parallel to the three diagonals of a regular octahedron. Then by using the earlier result on the projected area, we find that the mean of the areas of the projections (which are now all equal) is

$$\pi\surd\left(\tfrac{1}{3}(A^2B^2+B^2C^2+C^2A^2)\right). \tag{17}$$

It is similarly easy to show that if P_2 is the mean of the areas of the projections of any convex body on the eight faces of a regular octahedron,

$$0.8165=\surd 2/\surd 3\leqslant 4P_2/S\leqslant 2/\surd 3=1.1547\,. \tag{18}$$

Thus if S is now the area of the ellipsoid,

$$\surd 2/\surd 3\leqslant\frac{4\pi}{S}\surd\left(\tfrac{1}{3}(A^2B^2+B^2C^2+C^2A^2)\right)\leqslant 2/\surd 3\,. \tag{19}$$

These results suggest that as approximations to S we take the values

$$A_1=\tfrac{4}{3}\pi(AB+BC+CA), \tag{20}$$

and

$$A_2=4\pi\sqrt{\left(\tfrac{1}{3}(A^2B^2+B^2C^2+C^2A^2)\right)}. \tag{21}$$

Using exact values of S the accuracy of (20) and (21) was explored in the restricted range $1=A\geqslant B\geqslant C\geqslant 0.5$. In this range S/A_1 was larger than unity and its extreme value occurred at $(A, B, C)=(1,1,0.5)$ where it was equal to 1.035130. Similarly S/A_2 was less than unity and its extreme value, 0.975930, occurred also at $(A, B, C)=(1,1,0.5)$.

In this range S/A_1 appeared to be about 1.5 times as much above unity as S/A_2 was below it. This suggests that $A_3=\tfrac{1}{5}(2A_1+3A_2)$ would be a better approximation. Exploring this in the same range we find that S/A_3 lies between 0.998778 and 1.001936 so that it is a fairly good approximation. Outside this range it gets much worse and at $(A, B, C)=(1,1,0)$ the error is about 4%.

The approximations, A_1 and A_2, are probably well known but I have been unable to find any references to them, nor to the facts, which we now prove, that in fact $A_1\leqslant S\leqslant A_2$.

Consider eq. (9) and use the fact that for any random variable $U, \{E|U|\leqslant\sqrt{E(U^2)}$. Then

$$\begin{aligned} S&\leqslant 4\pi(abc)^{-1/2}\left\{E\left(\frac{aX_1^2+bX_2^2+cX_3^2}{X_1^2+X_2^2+X_3^2}\right)\right\}^{1/2}\\ &\leqslant 4\pi(abc)^{-1/2}\sqrt{\left(\tfrac{1}{3}(a+b+c)\right)}\\ &\leqslant 4\pi\sqrt{\left(\tfrac{1}{3}(A^2B^2+B^2C^2+C^2A^2)\right)}=A_2. \end{aligned}$$

The proof that $A_1\leqslant S$ is a little more awkward. Remembering that $c\geqslant b\geqslant a$, and again putting $\alpha=1-a/c, \beta=1-b/c$, we have

$$\begin{aligned} A_1&=\tfrac{4}{3}\pi(ab)^{-1/2}(1+a^{1/2}c^{-1/2}+b^{1/2}c^{-1/2})\\ &=4\pi(ab)^{-1/2}\left\{\tfrac{1}{3}+\tfrac{1}{3}(1-\alpha)^{1/2}+\tfrac{1}{3}(1-\beta)^{1/2}\right\}. \end{aligned} \tag{22}$$

By expanding the square roots, using the binomial theorem, assuming $\alpha<1, \beta<1$ (the general case following by continuity) and comparing with (13), the result will be proved if we can show that

$$\frac{E\left\{(\alpha X_1^2+\beta X_2^2)^k\right\}}{E\left\{(X_1^2+X_2^2+X_3^2)^k\right\}}\leqslant\tfrac{1}{3}(\alpha^k+\beta^k). \tag{23}$$

The kth moment of $\alpha X_1^2+\beta X_2^2$ is a polynomial, with positive coefficients, in the cumulants of orders $1,2,\ldots,k$, and the mth cumulant is

$$(\alpha^m+\beta^m)2^{m-1}(m-1)!.$$

However, for $m\leqslant k$,

$$(\alpha^m+\beta^m)/(\alpha^k+\beta^k)^{m/k}$$

attains its maximum when $\alpha=\beta$. It therefore follows that

$$E\{(\alpha X_1^2+\beta X_2^2)^k\}/(\alpha^k+\beta^k)\leqslant 2^{k-1}k!. \tag{24}$$

When inserting this in (23), the latter is equivalent to asserting that

$$\begin{aligned}2^{k-1}k!&\leqslant\tfrac{1}{3}E\{(X_1^2+X_2^2+X_3^2)^k\}\\&\leqslant\tfrac{1}{3}2^k\Gamma(\tfrac{3}{2}+k)/\Gamma(\tfrac{3}{2})\end{aligned}$$

which can be easily verified by induction. Thus $A_1\leqslant S$.

The approximations A_1 and A_2 are reminiscent of the well-known approximations $\pi(A+B)$ and $\pi\sqrt{(2A^2+2B^2)}$ for the perimeter, L say, of an ellipse with semi-axes A and B. These can be treated in the same manner as above and if L is the perimeter, it is easy to show that

$$\pi(A+B)\leqslant L\leqslant\pi\sqrt{(2A^2+2B^2)}\,. \tag{25}$$

Then

$$\tfrac{1}{2}\pi\left(A+B+\sqrt{(2A^2+2B^2)}\right)$$

is quite a good approximation. Its ratio to the true value lies between 0.999095 and unity if $A=1, 0.5\leqslant B\leqslant 1$, and between 0.948060 and unity for all A, B. Ramanujan (1914) gives two simple but better approximations which he states were obtained "empirically". The first of these is

$$\pi\{3(A+B)-\sqrt{((A+3B)(3A+B))}\}.$$

For $A=1, 0\leqslant B\leqslant 1$ the ratio of this to the true perimeter lies between 0.995845 and unity, and for $0.5\leqslant B\leqslant 1$, between 0.999997 and unity. The second approximation is

$$\pi\left\{A+B+\frac{3(A-B)^2}{10(A+B)+\sqrt{(A^2+14AB+B^2)}}\right\}.$$

For $A=1, 0\leqslant B\leqslant 1$ the ratio of this to the true value lies between 0.999597 and unity, and for $0.3\leqslant B\leqslant 1$, between 0.999999926 and unity. How Ramanujan actually obtained these formulae seems to be a mystery. They suggest that very accurate approximations of this form might be found for the surface area of an ellipsoid.

A more effective method of calculating the surface area of an ellipsoid than (14) can be obtained by returning to (5). Remembering that $A \geqslant B \geqslant C$, write $g = b\sin^2\phi + c\cos^2\phi \geqslant a$. Then (5) can be written as

$$S = ABC\int_0^{2\pi} d\phi \int_0^{\pi} d\theta\left[b\sin^2\phi + c\cos^2\phi - \cos^2\theta(b\sin^2\phi + c\cos^2\phi - a)\right]^{1/2}\sin\theta$$

$$= ABC\int_0^{2\pi} d\phi \int_0^{\pi} d\theta\{g - (g-a)\cos^2\theta\}^{1/2}\sin\theta.$$

By putting $e = 1 - a/g$, and $x = \cos\theta$, we get

$$S = ABC\int_0^{2\pi} d\phi \int_{-1}^{1} g^{1/2}(1 - ex^2)^{1/2} dx$$

$$= ABC\int_0^{2\pi} g^{1/2}\{(1-e)^{1/2} + e^{-1/2}\sin^{-1}e^{1/2}\} d\phi$$

$$= ABC\int_0^{2\pi}\{a^{1/2} + g^{1/2}\sin^{-1}\sqrt{(1-a/g)}\,/\sqrt{(1-a/g)}\,\} d\phi. \qquad (26)$$

As long as $c < \infty$, $C > 0$, the integrand is a smooth periodic function of ϕ with period π, and symmetric about $\phi = \frac{1}{2}\pi$. Numerical integration with equally spaced abscissae is therefore likely to give very accurate results [compare Hsu (1963)], and no end corrections are necessary.

The accuracy of (26) was investigated on a desk computer for $A = 1$, $B = 0.1(0.1)1$, $C = 0.1(0.1)B$. The integral was estimated by summing the ordinates in (26) for $\phi = \pi/4N, 3\pi/4N, \ldots, (2N-1)\pi/4N$, where N was taken as 5, 10, 20 or 40 until the ninth decimal place did not change. $N = 10$ gave nine decimal accuracy for $C \geqslant 0.5$, $N = 20$ gave nearly nine decimal accuracy for $C \geqslant 0.2$, and $N = 40$ was similarly sufficient for $C \geqslant 0.1$. The decrease in accuracy is dependent on the smallness of C/B rather than on C itself. When $C/B < 0.1$, this method begins to deteriorate rapidly and thus the problem of finding a good estimate of S is not here solved for $C/B \leqslant 0.1$. Note that when $b = c$ the integrand is constant and (26) becomes the known explicit formula for a prolate spheroid.

Keller (1979) discusses the calculation of S by starting from a formula in cartesian coordinates which is equivalent to (5). He observes that one can integrate explicitly with respect to the coordinate corresponding to θ in (5), and then carries out the remaining integration by using Gaussian quadrature. As examples he obtains the area when $(A, B, C) = (1, 1, 0.5)$, $(1, 0.8, 0.625)$ and $(1, 0.5, 0.125)$ and his values agree to nine decimal places with those obtained from (26). However, Gaussian quadrature is not the most suitable method here because on transposing the variable to θ, the integrand becomes a smooth periodic function for which equally spaced abscissae give accurate results.

Acknowledgements

It is a pleasure to contribute to a volume in honour of Professor C.R.Rao.

References

Hsu, L.C. (1963). Approximate integration of rapidly oscillating functions and of periodic functions. *Proc. Cam. Phil. Soc.* **59**, 81–88.

Keller, S.R. (1979). On the surface area of an ellipsoid. *Math. Comput.* **33**, 310–324.

Moran, P.A.P., (1944). Measuring the surface area of a convex body. *Ann. Math.* **45**, 793–799.

Pitman, E.J.G. (1937). The "closest" estimates of statistical parameters. *Proc. Cam. Phil. Soc.* **33**, 212–222.

Ramanujan, S. (1914). Modular equations and approximations to π. *Q. J. Math.* **45**, 350–372.

G. Kallianpur, P.R. Krishnaiah, J.K. Ghosh, eds., *Statistics and Probability: Essays in Honor of C.R. Rao*

A MODEL FOR AERIAL SURVEILLANCE OF MOVING OBJECTS WHEN ERRORS OF OBSERVATION ARE MULTIVARIATE NORMAL

Ingram OLKIN
Stanford University, Stanford, CA, U.S.A.

Sam C. SAUNDERS
Washington State University, Pullman, WA, U.S.A.

This paper presents a general theorem on the invariant behavior of a certain function of a matrix. It then shows the importance of this result principally by using it to derive properties of certain maximum likelihood estimates which arise when considering problems such as the location of a moving object being surveyed from a moving observatory when all data on location are subject to stochastic error. This problem is important in tracking objects either from an observatory satellite or from a surveillance plane bearing ground seeking radar. Some applications to this situation are made.

1. Introduction

This paper deals with statistical problems arising from the estimation of the position, heading and velocity of a moving object using data which are subject to statistical error. (For earlier work see [3,4].) These data are observations made during one overflight when the exact location of the observatory platform is not precisely known with respect to the ground. The solution of this problem was originally intended to assist in the determination of both the travel and location of ships when using data obtained from an observatory satellite. Recently, the same mathematical problems have arisen with the introduction of ground mapping radar which is being borne by airplanes and used in the observation and prediction of position of objects moving on land or sea. For a discussion of the related problem of ship surveillance by satellite using range data only, see [1].

In Section 1, a simple model with normal errors for the moving target from a moving observatory is given and then in Section 2, the maximum likelihood estimates of target position are obtained. A theorem is then utilized to yield certain invariance properties of these estimates (Section 3). The subsequent sections deal with some specializations which are of practical importance.

2. The model and its assumption

The basic components are a fixed monitor which defines coordinate axes, a moving object or target and a moving satellite or observatory. The trajectory of the target

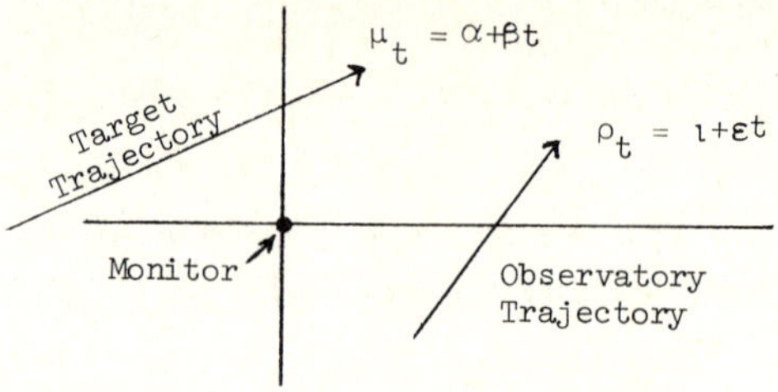

Fig. 1. A tracking model.

and observatory are linear (see Fig. 1). The linear trajectory of the target is completely unknown, and is to be estimated; the observatory has a known velocity so that the trajectory is partially known.

We now specify precisely the assumptions made:

(1) The observatory moves at constant height linearly above the plane with a known constant velocity.

(2) The object moves linearly in the plane at a constant but unknown velocity.

(3) The estimated position coordinates of the observatory over the plane, as determined from the ground at a given time, are bivariate normal random variables with known covariance.

(4) The estimated position coordinates of the object in the plane, as determined relative to the true position of the observatory, are bivariate normal random variables and successive observations of such relative positions are independent.

(5) Time between successive observations can be measured with sufficient accuracy so that errors of position due to time inaccuracy are negligible.

(6) The parameters of the covariance matrix of the observations of object position relative to the true observatory position can be determined from bearing angle and range data.

On a single overflight, an observatory may make several observations of the position of an object and we first determine the joint distribution of the observations of the target position, using the fact that the target position relative to the observatory is subject to observational error, as is the estimate of the observatory positions relative to the ground. Thus, at a given time t_i we assume the observatory is at some position, say ρ_i and the target at some position μ_i both in a plane located with respect to a given coordinate system. However, the position of the object as observed by radar from the observatory is subject to chance error and hence, the radar estimate of the object position from the observatory position is a random variable, say X_i. Now a radar measurement from the ground at time t_0 of the observatory position ρ_0 on the given coordinate system is also a random variable, call it Y.

From (1) and (2), the observatory follows a linear path in the plane, say $\rho_t = \iota + \varepsilon t$, as does the object, say, $\mu_t = \alpha + \beta t$.

From (3) and (4), each X_i $(i = 1, \ldots, n)$ is bivariate normal with mean vector $\mu_i - \rho_i$ and known covariance matrix G_i, whereas Y has a known mean vector ρ_0 and known covariance matrix ψ. That is,

$$X_i \sim \mathcal{N}(\mu_i - \rho_i, G_i) \quad \text{and} \quad Y \sim \mathcal{N}(\rho_0, \psi).$$

The data obtained from the observatory yield the observations $Z_i = X_i + Y$ for $i=1,\ldots,n$. Then $Z_1,\ldots, Z_n$ are normally distributed with $\mathrm{Var}(Z_i) = G_i + \psi$ and $\mathrm{Cov}(Z_i, Z_j) = \psi$ for $i \neq j$, $i, j = 1,\ldots, n$. Hence the covariance matrix of $Z = (Z_1,\ldots, Z_n)$ is given by

$$\Sigma \equiv \mathrm{Cov}(Z) = (\delta_{ij} G_i + \psi),$$

where δ_{ij} is the Kronecker delta. To obtain the inverse covariance with

$$\Sigma^{-1} = \Lambda = (\Lambda_{ij}),$$

we use the following lemma [2, p. 124]:

Lemma. *If $Q_1,\ldots, Q_k$ are non-singular, symmetric matrices of the same dimension and*

$$H = \mathrm{diag}(Q_1,\ldots, Q_k) + \begin{pmatrix} M_1 \\ \vdots \\ M_k \end{pmatrix} (I,\ldots, I),$$

then

$$H^{-1} = \mathrm{diag}(Q_1^{-1},\ldots, Q_k^{-1}) - \begin{pmatrix} Q_1^{-1}M_1 \\ \vdots \\ Q_k^{-1}M_k \end{pmatrix} \left(I + \sum_{i=1}^{k} Q_i^{-1}M_i \right)^{-1} (Q_1^{-1},\ldots, Q_k^{-1}).$$

It follows from the lemma with the choice $Q_i = G_i$, $M_i = \psi$ that

$$\Lambda_{ij} = \delta_{ij} G_i^{-1} - G_i^{-1} \left(\psi^{-1} + \sum_1^n G_i^{-1} \right)^{-1} G_j^{-1}. \tag{2.1}$$

The above calculations are summarized in:

Theorem 1. *The distribution of the observations of target position $Z = (Z_1,\ldots, Z_n)$, each $Z_i \in R^2$, obtained from the observatory is*

$$Z \sim \mathcal{N}(\nu, \Sigma),$$

where $\nu = (\nu_1,\ldots, \nu_n) = (\mu_1 - \rho_1 - \rho_0,\ldots, \mu_n - \rho_n - \rho_0)$, $\Lambda = \Sigma^{-1}$ is defined by (2.1).

Theorem 1 tells us that a single observation of observatory position from a bivariate normal distribution relative to the ground, combined with n observations of the object position, which are normally distributed relative to the observatory position yields a joint normal distribution of observed target positions relative to the ground. We now estimate the target course using this normal distribution of error and the assumption that the target is moving linearly at a constant velocity.

3. The maximum likelihood estimates and their distribution

To find the maximum likelihood estimate (MLE), $\hat{\mu}_t$, of future target positions μ_t, as a function of t, it is immediate that

$$\hat{\mu}_t = \hat{\alpha} + \hat{\beta} t,$$

where $\hat{\alpha}$ and $\hat{\beta}$ are the MLE's of α and β.

Recall that $\nu_i = \mu_i - \rho_i + \rho_0 = (\alpha + \varepsilon t_0) + (\beta - \varepsilon) t_i \equiv \gamma + \kappa t_i$, for $i = 1, \ldots, n$, where $\gamma = \alpha + \varepsilon t_0$, $\kappa = \beta - \varepsilon$. Since ε and t_0 are known, the MLE's of γ and κ yield the MLE's of α, β.

Write $e = (1, \ldots, 1)$ and $t = (t_1, \ldots, t_n)$, then since Σ is known, the MLE are obtained from

$$\operatorname*{Min}_{\gamma, \kappa} (z - \gamma e - \kappa t) \Lambda (z - \gamma e - \kappa t)'.$$

Differentiation with respect to γ and κ yields the equations

$$\gamma e \Lambda e' + \kappa e \Lambda t' = e \Lambda z', \qquad \gamma e \Lambda t + \kappa t \Lambda t' = t \Lambda z',$$

from which we obtain

$$(\hat{\gamma}, \hat{\kappa}) = (e \Lambda z', t \Lambda z') \begin{pmatrix} e\Lambda e' & e\Lambda t' \\ t\Lambda e' & t\Lambda t' \end{pmatrix}^{-1}. \tag{3.1}$$

We first notice that

$$E(\hat{\gamma}, \hat{\kappa}) = (\gamma, \kappa). \tag{3.2}$$

To see this, let

$$A = \begin{pmatrix} a_{11} & a_{12} \\ a_{12} & a_{22} \end{pmatrix} = \begin{pmatrix} e\Lambda e' & e\Lambda t' \\ t\Lambda e' & t\Lambda t' \end{pmatrix}.$$

Then from (3.1)

$$E(\hat{\gamma}, \hat{\kappa}) = (\gamma a_{11} + \kappa a_{12}, \gamma a_{12} + \kappa a_{22}) \begin{pmatrix} a_{22} & -a_{12} \\ -a_{12} & a_{11} \end{pmatrix} \Big/ |A| = (\gamma, \kappa).$$

It then follows that

$$E(\hat{\alpha}, \hat{\beta}) = (\alpha, \beta), \tag{3.3}$$

where

$$\hat{\alpha} = \hat{\gamma} - \varepsilon t_0, \qquad \hat{\beta} = \hat{\kappa} + \varepsilon.$$

The covariance matrix of $(\hat{\gamma}, \hat{\kappa})$ is

$$\operatorname{Cov}(\hat{\gamma}, \hat{\kappa}) = A^{-1} \begin{pmatrix} e \\ t \end{pmatrix} \Lambda (\operatorname{Cov} Z) \Lambda (e', t') A^{-1} = A^{-1} \left[\begin{pmatrix} e \\ t \end{pmatrix} \Lambda (e', t') \right] A^{-1} = A^{-1}.$$

To summarize, the distributions of the MLE are

$$(\hat{\gamma},\hat{\kappa})\sim\mathfrak{N}\big((\gamma,\kappa),A^{-1}\big),\qquad(\hat{\alpha},\hat{\beta})\sim\mathfrak{N}\big((\alpha,\beta),A^{-1}\big).\tag{3.4}$$

The object is to estimate the target trajectory μ_{t_*} at a time t_*. From (3.4),

$$\hat{\mu}_{t_*}=\hat{\alpha}+\hat{\beta}t_*\sim\mathfrak{N}\big(\alpha+\beta t_*,B_{t_*}\big),\tag{3.5}$$

where

$$\begin{aligned}B_{t_*}&=\operatorname{Var}(\hat{\alpha})+2t_*\operatorname{Cov}(\hat{\alpha},\hat{\beta})+t_*^2\operatorname{Var}(\hat{\beta})\\&=\big(a_{22}-2t_*a_{12}+t_*^2a_{11}\big)/|A|\\&=\frac{t\Lambda t'-2t_*e\Lambda t'+t_*^2t\Lambda t'}{(t\Lambda t')(e\Lambda e')-(e\Lambda t')^2}.\end{aligned}$$

We now state and prove a general result on the behavior at a particular function of positive definite matrices. In a private communication, O. Kempthorne noted that he required Theorem 2 in the analysis of a linear model.

Theorem 2. *Let c be a given real number, Σ_0, Σ_1, Σ_2 positive definite (symmetric) matrices and the matrix function $F(t)$, defined for any real t by the expression*

$$F(t)=(I,(t+c)I)\begin{pmatrix}\Sigma_0&\Sigma_1+t\Sigma_0\\\Sigma_1+t\Sigma_0&\Sigma_2+2t\Sigma_1+t^2\Sigma_0\end{pmatrix}^{-1}\begin{pmatrix}I\\(t+c)I\end{pmatrix}.\tag{3.6}$$

Then $F(t)\equiv F(0)$.

Proof. The result will hold if $f(t)=uF(t)u'$ is independent of t for all vectors u of appropriate dimension. Note

$$f(t)=(u,(t+c)u)\begin{pmatrix}\Sigma_0&\Sigma_1+t\Sigma_0\\\Sigma_1+t\Sigma_0&\Sigma_2+2t\Sigma_1+t\Sigma_0\end{pmatrix}^{-1}\begin{pmatrix}u'\\(t+c)u'\end{pmatrix}.$$

Note further that if A is non-singular, then

$$|A+w'w|=|A||I+A^{-1}w'w|=|A|(1+wA^{-1}w').$$

Consequently, $f(t)$ independent of t is equivalent to

$$\frac{\begin{vmatrix}\Sigma_0+u'u&\Sigma_1+t\Sigma_0+(t+c)u'u\\\Sigma_1+t\Sigma_0+(t+c)u'u&\Sigma_2+2t\Sigma_1+t^2\Sigma_0+(t+c)^2u'u\end{vmatrix}}{\begin{vmatrix}\Sigma_0&\Sigma_1+t\Sigma_0\\\Sigma_1+t\Sigma_0&\Sigma_2+2t\Sigma_1+t^2\Sigma_0\end{vmatrix}}\tag{3.7}$$

independent of t. If the numerator is independent of t for all u then the denominator will also be independent of t by taking $u=0$. By using the fact that if A is non-singular

$$\begin{vmatrix} A & B \\ B' & D \end{vmatrix} = |A||D - B'A^{-1}B|,$$

we obtain for the numerator of (3.7);

$$|\Sigma_0 + u'u| \cdot |\Sigma_2 + 2t\Sigma_1 + t^2\Sigma_0 + (t+c)^2 u'u$$
$$- [\Sigma_1 + cu'u + t(\Sigma_0 + u'u)](\Sigma_0 + u'u)^{-1}(\Sigma_1 + t\Sigma_0 + (t+c)u'u)|.$$

The first term is independent of t. The matrix in the second term is

$$\Sigma_2 + 2t\Sigma_1 + t^2\Sigma_0 + (t+c)^2 u'u - (\Sigma_1 + cu'u)(\Sigma_0 + u'u)^{-1}(\Sigma_1 + cu'u)$$
$$- 2t(\Sigma_1 + cu'u) - t^2(\Sigma_0 + u'u),$$

which upon expansion and simplification reduces to

$$\Sigma_2 + c^2 u'u - (\Sigma_1 + cu'u)(\Sigma_0 + u'u)^{-1}(\Sigma_1 + cu'u).$$

This expression is independent of t, which completes the proof.

We now state a result of great practical importance below in Theorem 3.

Theorem 3. *The estimate $\hat{\mu}_t$ and its covariance matrix B_t are invariant under location and scale change in time, i.e. under the transformation $t_i \to a + ct_i$, $i = 1, \ldots, n$.*

The proof is based on writing $\hat{\mu}_t$ and B_t as a function of a matrix of the form (3.6) and then invoking Theorem 2.

4. Some specializations of practical importance

Assume that at a given time t_i the observatory is at point ρ_i and traveling in a straight line at a known constant velocity. It observes, at a true bearing angle θ_i and range R_i, the object which is located at position μ_i obtaining the random variable $X_i \sim \mathfrak{N}(\mu_i - \rho_i, C_i^{-1})$. See Fig. 2.

From elementary geometrical arguments,

$$G_i = \begin{pmatrix} \sigma_{Ri}^2 \cos^2\theta_i + \sigma_{Ai}^2 \sin^2\theta_i & \dfrac{\sigma_{Ri}^2 - \sigma_{Ai}^2}{2} \sin 2\theta_i \\ \dfrac{\sigma_{Ri}^2 - \sigma_{Ai}^2}{2} \sin 2\theta_i & \sigma_{Ri}^2 \sin^2\theta_i + \sigma_{Ai}^2 \cos^2\theta_i \end{pmatrix}, \tag{4.1}$$

where at time t_i, σ_{Ri} is the standard deviation of the range error, σ_{Ai} is the standard deviation of the aximuth error and both are known functions of the range Ri, all in accord with assumption (6) (see [4]) viz., that the parameters of the covariance matrix can be determined from bearing angle and range data.

We also assume that the matrix ψ is diagonal and known.

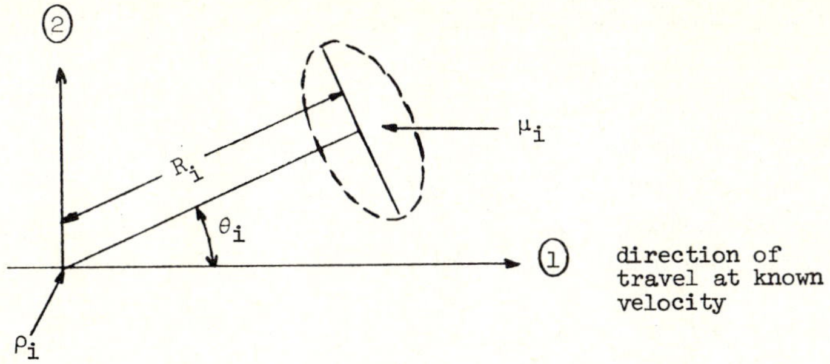

Fig. 2.

If it is true that the distribution of radar errors is constant in time, that neither the direction of travel of the observatory platform nor its position relative to the target will influence the covariance matrix of observations, then $G_i \equiv G$ for $i=1,\ldots,n$ and $\Lambda_{ij}=\Lambda_{ji}$. By a straightforward calculation, with the time chosen so that $\bar{t}=\Sigma t_i/n=0$, we obtain the MLE

$$\hat{\gamma}=\bar{z}, \qquad \hat{\kappa}=tz'/tt',$$

and hence

$$\hat{\alpha}=\bar{z}-\varepsilon t_0, \qquad \hat{\beta}=(tz'/tt')+\varepsilon.$$

The estimator $\hat{\mu}_t=\hat{\alpha}+\hat{\beta}t$ is unbiased and has covariance matrix $B_{t.}=(1/n)(1+t_*^2/tt')G+\psi$.

If the observations are symmetrically spaced in time and the origin is chosen so that

$$t_i+t_{n+1-i}=0, \qquad \theta_i+\theta_{n+1-i}=\pi, \qquad i=1,\ldots,n,$$

then $G_i=G_{n+1-i}$ is a diagonal matrix and the various computations simplify.

5. Estimation and confidence intervals for speed and heading

The parameters of interest when tracking a moving object are the speed and heading, and we consider the accuracy with which they can be estimated. The estimate of the true position $\mu_{t.}$ at any time t_*, can be written

$$\hat{\mu}(t_*)=\hat{\alpha}+\hat{\beta}t_*=(\hat{m}_1(t_*),\hat{m}_2(t_*)).$$

The true velocity s, heading angle h, are independent of time and are given by

$$s=\left\{[m_1'(t_*)]^2+[m_2'(t_*)]^2\right\}^{1/2}, \qquad h=\arctan\left(\frac{m_2(1)-m_2(0)}{m_1(1)-m_1(0)}\right).$$

Thus by analogy we have the equations

$$\hat{\beta}=(\hat{b}_1,\hat{b}_2)=(V\cos\Phi, V\sin\Phi) \tag{5.1}$$

defining the random variables V and Φ (for velocity and for heading) which are estimates of the true speed s and the true heading h.

By (3.4), $\hat{\beta}\sim\mathfrak{N}(\beta, Q_{22})$, where $A^{-1}=(Q_{ij})$, $i, j=1,2$. The joint density of Φ, V is found by simply making a transformation to polar coordinates. This yields

$$g(\varphi, v|h,s)=\frac{v}{2\pi|Q_{22}|^{1/2}}\exp\{-\tfrac{1}{2}\zeta Q_{22}^{-1}\zeta'\}, \qquad 0<v<\infty, 0<\varphi<2\pi, \tag{5.2}$$

where $\zeta=(v\cos\varphi-s\cos h,\ v\sin\varphi-s\sin h)$. This density can be used to study the distribution of velocity and heading estimates that could arise under infrequent headings and/or high velocity.

To obtain separate confidence intervals on the heading and on the velocity, we use the marginal densities of v and φ, which are unfortunately functions of both parameters h and s. The presence of the nuisance parameter prevents us from obtaining confidence intervals separately when both parameters are unknown. However, we can obtain a joint confidence region for (h,s), by using the distribution of the quadratic form:

$$P\{(\hat{\beta}-\beta)Q_{22}^{-1}(\hat{\beta}-\beta)'\leq\chi_2^2(p)\}=1-p, \tag{5.3}$$

where $\chi_2^2(p)$ is the pth percentile of the χ_2^2 distribution.

The (random) elliptical region W, so defined, determines a $100(1-p)\%$ confidence region for β. We seek the smallest area, W^*, in polar coordinates which is the Cartesian product of intervals and contains W. See Fig. 3.

If

$$W^*=\{(\theta,r):\Phi_1<\theta<\Phi_2, V_1<r<V_2\}, \tag{5.4}$$

then $(r\cos\theta, r\sin\theta)\in W$ implies $(\theta,r)\in W^*$, $P\{(h,s)\in W^*\}\geq 1-p$.

The task to which we now turn is the determination of a functional representation for the random variables (Φ_i, V_i), $i=1,2$. If we denote the elements of the symmetric

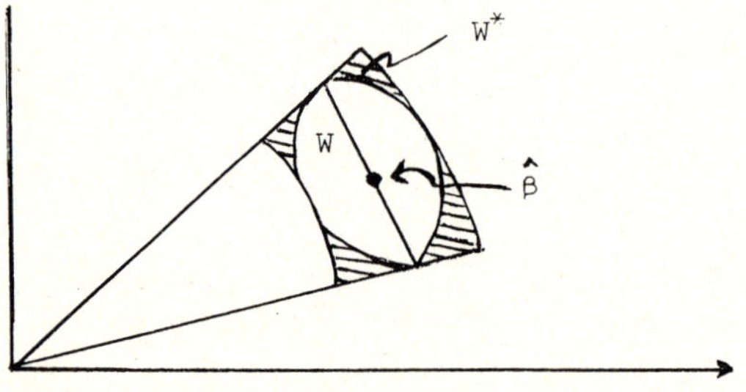

Fig. 3. Joint confidence intervals for heading angle Φ and velocity V.

matrix

$$\frac{1}{\chi_2^2(p)} Q_{22}^{-1} = \begin{pmatrix} q_{11} & q_{12} \\ q_{12} & q_{22} \end{pmatrix},$$

then we can write from (5.3)

$$q_{11}(x-\hat{b}_1)^2 + 2q_{12}(x-\hat{b}_1)(y-\hat{b}_2) + q_{22}(y-\hat{b}_2)^2 = 1. \tag{5.5}$$

We want to find the maximum and minimum of the functions

$$f(x, y) = y/x, \qquad g(x, y) = x^2 + y^2 \tag{5.6}$$

subject to the restriction (5.5).

Proceeding directly leads to the solution of a quartic equation. We first transform (5.5) to eliminate the cross product term. The angle of rotation θ_0, satisfies

$$\tan 2\theta_0 = \frac{2q_{12}}{q_{22} - q_{11}}.$$

Next we make a trigonometric substitution so that the elliptic region (5.5) becomes circular. Let

$$\delta_1 = q_{11}\cos^2\theta_0 + q_{12}\sin 2\theta_0 + q_{22}\sin^2\theta_0,$$

$$\delta_2 = q_{11}\sin^2\theta_0 - q_{12}\sin 2\theta_0 + q_{22}\cos^2\theta_0,$$

and set

$$\eta_i = \frac{\cos\theta_0}{\sqrt{\delta_i}}, \qquad \xi_i = \frac{\sin\theta_0}{\sqrt{\delta_i}}, \quad i = 1, 2.$$

The problem then becomes that of extremizing the functions of one variable corresponding to (5.6), namely,

$$f(\varphi) = \frac{\xi_1\cos\varphi + \eta_2\sin\varphi + \hat{b}_2}{\eta_1\cos\varphi - \xi_2\sin\varphi + \hat{b}_1} \tag{5.7}$$

and

$$g(\varphi) = (\eta_1\cos\varphi - \xi_2\sin\varphi + \hat{b}_1)^2 + (\xi_1\cos\varphi + \eta_2\sin\varphi + \hat{b}_2)^2 \tag{5.8}$$

for the range of values $0 < \varphi \leqslant 2\pi$. It is straightforward to show that $f'(\varphi) = 0$ if and only if

$$q_{11} + T_1\sin\varphi + T_2\cos\varphi = 0, \tag{5.9}$$

where

$$q_{11} = \xi_2\xi_1 + \eta_1\eta_2, \qquad T_1 = \hat{b}_2\eta_1 - \hat{b}_1\xi_1, \qquad T_2 = \hat{b}_1\eta_2 + \hat{b}_2\xi_2.$$

Now let $\Phi_0 \in (0, 2\pi)$ be defined by $\tan\Phi_0 = T_1/T_2$. Then (5.9) becomes $\cos(\Phi_0 - \varphi) = -\delta_1/T$.

Let $\Theta = \arccos(-\delta_1/T)$, $\Theta \in (0,\pi)$, and $\Omega_1 = \Phi_0 - \Theta, \Omega_2 = \Phi_0 + \Theta$ be the two solutions of (5.9). Then the limits Φ_1 and Φ_2 in (5.4), are given by $\Phi_i = \arctan f(\Omega_i)$ for $i = 1,2$. This yields the location of local extrema of f. Graphical examination of the function on $(0, 2\pi)$ may be necessary to determine the global extremum.

The extremum of g is obtained in a manner similar to that for f. This yields $g'(\varphi) = 0$ if and only if

$$\delta_1 \sin 2\varphi + \delta_2 \cos 2\varphi + L_1 \sin\varphi + L_2 \cos\varphi = 0, \tag{5.10}$$

where

$$2\delta_1 = \eta_1^2 - \eta_2^2 + \xi_1^2 - \xi_2^2, \qquad \delta_2 = \eta_1\xi_2 - \xi_1\eta_2,$$

$$L_1 = \hat{b}_1\eta_1 + \hat{b}_2\xi_1, \qquad L_2 = \hat{b}_1\xi_2 - \hat{b}_2\eta_2.$$

The solutions Γ_1, Γ_2 of (5.10) cannot, in general, be found explicitly, but must be found numerically. However, this is easily done.

Having found Γ_1, Γ_2, the limits on the velocity become

$$V_1 = \min\left(\sqrt{g(\Gamma_1)}, \sqrt{g(\Gamma_2)}\right), \qquad V_2 = \max\left(\sqrt{g(\Gamma_1)}, \sqrt{g(\Gamma_2)}\right).$$

A circumstance of practical interest occurs whenever the observations are drawn symmetrically on each overflight. In this case $\theta_0 = 0$, $\delta_1 = q_{11}$, $\delta_2 = q_{22}$, $\delta_{12} = 0$, and so we find

$$\eta_1 = \frac{1}{\sqrt{q_{11}}}, \qquad \xi_2 = 0 = \xi_1, \qquad \eta_2 = \frac{1}{\sqrt{q_{22}}}$$

$$2\delta_1 = \frac{1}{q_{11}} - \frac{1}{q_{22}}, \qquad \delta_2 = 0, \qquad T_1 = b_1/\sqrt{q_{11}}, \qquad T_2 = -b_2/\sqrt{q_{22}}.$$

Thus eq. (5.10) reduces to

$$\frac{1}{2}\left(\frac{1}{q_{11}} - \frac{1}{q_{22}}\right)\sin 2\varphi + \frac{\hat{b}_1}{\sqrt{q_{11}}}\sin\varphi = \frac{\hat{b}_2}{\sqrt{q_{22}}}\cos\varphi. \tag{5.11}$$

If we specialize further and assume that $q_{11} = q_{22}$ we see from (5.11) that the equation to be solved is merely $\tan\varphi = \hat{b}_2/\hat{b}_1$, which has solutions

$$\cos\varphi = \pm\frac{\hat{b}_1}{V}, \qquad \sin\varphi = \pm\frac{\hat{b}_2}{V},$$

where V is the estimate of velocity defined in (5.1), from which we obtain Φ_1 and Φ_2. In this case we attain the simple bounds on the velocity also, namely

$$V_2 = V + \frac{1}{\sqrt{q_{11}}}, \qquad V_1 = V - \frac{1}{\sqrt{q_{11}}}.$$

6. Conclusions

In this paper a mathematical framework is presented which demonstrates a surprising and important property of the estimates of position and heading for moving targets obtained by aerial surveillance with radar, namely, the invariance of both the estimate and its covariance with respect to time. This property has in fact been obtained for real systems. This feature makes it possible to survey simultaneously, and predict the future position of, a large number of moving targets when one has only minimal computational capacity.

References

[1] Gelfand, A.E. (1968). Ship tracking via satellite. Techn. Rep. No. 141, Department of Statistics, Stanford University, Stanford, CA.

[2] Householder, A.S. (1964). *The Theory of Matrices in Numerical Analysis*. Blaisdell Publishing Co., New York.

[3] Saunders, S.C. (1965). Further statistical problems of tracking a target from an observatory satellite. Boeing Document D1-82-0390. Boeing Scientific Research Laboratories, Seattle, WA.

[4] Saunders, S.C. and Johnson, D.L. (1964). On the estimation of location coordinates from data taken from a manned orbital laboratory. Boeing Document D1-82-0320, Boeing Scientific Research Laboratories, Seattle, WA.

G. Kallianpur, P.R. Krishnaiah, J.K. Ghosh, eds., *Statistics and Probability: Essays in Honor of C.R. Rao*
© North-Holland Publishing Company (1982) 531–545

AN EXTENSION OF STRASSEN'S FUNCTIONAL LAW OF THE ITERATED LOGARITHM FOR VARIABLES WITHOUT VARIANCE

R.P. PAKSHIRAJAN and R. VASUDEVA
University of Mysore, India

Necessary and sufficient conditions for the classical law of the iterated logarithm to hold for independent and identically distributed random variables are due to Feller. Under conditions assumed by Feller, the functional form of the law of the iterated logarithm is established in this paper. Strassen's results, proved under the assumption of finite variance, become immediate corrollaries.

1. Introduction

Let $(X_n, n \geqslant 1)$ be a sequence of independent and identically distributed (i.i.d.) random variables (r.v.) defined on a probability space $(\Omega, \mathcal{B}, P)$. Assume that F, the distribution function (d.f.) of X_1 has mean 0. Define $S_0 = 0$; $S_n = \Sigma_{j=1}^{n} X_j$, $n \geqslant 1$; $H(x) = \int_{-x}^{x} y^2 \mathrm{d}F(y)$, $x \geqslant 0$. Denote by a_n the largest root of $x^2 = 2nH(x)\log\log x$ and observe that, when $\sigma^2 = EX_1^2 < \infty$, $a_n \sim \sigma\sqrt{2n\log\log n}$. For any real number λ, let $[\lambda]$ stand for the greatest integer $\leqslant \lambda$. For $0 \leqslant t \leqslant 1$, define

$$a_n Z_n(t) = S_{[nt]} + (nt - [nt])X_{[nt]+1}. \tag{1.1}$$

Let $C = C[0,1]$ be the space of all continuous functions f defined on $[0,1]$ with $f(0) = 0$. For $f \in C$, let $\| f \| = \sup_{0 \leqslant t \leqslant 1} |f(t)|$. Endow C with the topology resulting from this norm. Let $K \subset C$ be the set of all absolutely continuous functions f with $\int_0^1 (\mathrm{d}f/\mathrm{d}t)^2 \mathrm{d}t \leqslant 1$. Notice that for each $w \in \Omega$, $Z_n(\cdot, w) \in C$ and that K is a compact subset of C. When $EX_1^2 = 1$, Strassen (1964) established that the sequence $(Z_n(\cdot, w))$ is relatively compact in C and has K for its limit set. We refer to this result as the functional form of the law of the iterated logarithm (FLIL). Strassen's method consists in establishing first FLIL for the normalised sequence

$$\left\{ \frac{B(nt)}{\sqrt{2n\log\log n}} \right\},$$

where $B(t)$ is a standard Brownian motion process, and then extending the result to the sequence (Z_n) via Skorokhod representation. Chover (1967) supplied an elementary proof of Strassen's FLIL, but, he needed to assume that for some $\delta > 0$, $E|X_1|^{2+\delta} < \infty$. Chover's proof makes use of Esseen's rate of convergence to the Normal law. Our object in the present paper is to prove that Strassen's results for (Z_n) hold even when $EX_1^2 = \infty$, provided that for x and t large, there exists an $M > 0$ and a μ, $0 \leqslant \mu < 1$, such that

$$H(tx) \leqslant Mt^{\mu} H(x) \tag{1.2}$$

and provided further that for some $a>0$,

$$\int_a^\infty \frac{\mathrm{d}H(x)}{H(x)\log\log x}<\infty. \tag{1.3}$$

This is done in Section 3. Our proof rests on the method of Chover (1967).

Feller (1969) imposes condition (1.2) but allows μ to take any non-negative number. Under such a condition he proves that (1.3) is both necessary and sufficient for $P\{\limsup Z_n(1)=1\}=1$. We wish to note here, that all the d.f.'s in the domain of attraction of the Normal law satisfies (1.2) for any $\mu\geqslant 0$. In Remark 1, we present an example of a d.f. which is not in the domain of attraction of the Normal law but for which both (1.2) and (1.3) hold.

In Section 2 below we introduce the notations, collect some known results and prove some needed lemmas.

2. Notations and lemmas

Define $\beta_n=a_n(\log\log a_n)^{-p}, n\geqslant 3, \beta_1=\beta_2=\beta_3$, where $p>(1-\mu)^{-1}$ and arbitrary; $X'_n=X_n$ if $|X_n|\leqslant\beta_n$ and $=0$ otherwise and $X''_n=X_n-X'_n, n\geqslant 1$. Let $\mu_n=EX'_n, \theta_n=EX''_n, Y_n=X'_n-\mu_n$ and let Y''_n be the symmetrized form of X''_n. Define $S'_n=\Sigma_{j=1}^n Y_j$, $S''_n=\Sigma_{j=1}^n(X''_j-\theta_j)$ and $T_n=\Sigma_{j=1}^n Y''_j$. Throughout the paper let R, N (integers) c and ϵ, with or without a suffix, denote positive constants which may differ in value at different positions of their occurrence. In the lemmas below, (ξ_n) denotes a sequence of independent r.v.'s defined on $(\Omega, \mathscr{B}, P)$.

Lemma 2.1. *Let $E\xi_n=0$, $E\xi_n^2=\sigma_n^2$ and $s_n^2=\Sigma_{j=1}^n\sigma_j^2$. Let $\alpha_n\geqslant\max_{1\leqslant j\leqslant n}|\xi_j|/s_n$ a.s., where (α_n) is a sequence of constants.*

(i) *if $0<\epsilon\alpha_n\leqslant 1$, then*

$$P\left(\sum_{j=1}^n \xi_j>\epsilon s_n\right)\leqslant\exp\left\{-\tfrac{1}{2}\epsilon^2\left(1-\frac{\epsilon\alpha_n}{2}\right)\right\} \tag{2.1}$$

and

(ii) *given $\gamma>0$, if $\alpha=\alpha(\gamma)$ is sufficiently small and $\epsilon=\epsilon(\gamma)$ is sufficiently large then*

$$P\left(\sum_{j=1}^n \xi_j>\epsilon s_n\right)\geqslant\exp\{-\tfrac{1}{2}\epsilon^2(1+\gamma)\}. \tag{2.2}$$

For proof, see Loeve (1968, p. 254).

Lemma 2.2. *Let $E\xi_n=0$ and $E|\xi_n|^3<\infty, n\geqslant 1$. Set $s_n^2=\Sigma_{j=1}^n E\xi_j^2$ and $\Gamma_n=\Sigma_{j=1}^n E|\xi_j|^3$. Then*

$$\sup_{-\infty<x<\infty}|P(S_n\leqslant xB_n)-\phi(x)|<c\Gamma_n/s_n^3,$$

where ϕ is the standard normal d.f.

For proof, see Loeve (1968, p. 288).

Lemma 2.3. *Let there exist a $d>0$ such that for integers i and j with $1\leqslant i, j\leqslant n$, $P(\Sigma_{k=i}^{j} Y_k>0)>d$ and $P(\Sigma_{k=i}^{j} Y_k<0)>d$. Let $m\geqslant 1$ be some integer and for $i=1,2,\ldots,m$ let λ_i and ν_i be integers with $1\leqslant\lambda_i\leqslant\nu_i\leqslant n$. Then for any $a>0$,*

$$P\left(\max_{1\leqslant i\leqslant m}\left|\sum_{k=\lambda_i}^{\nu_i} Y_k\right|\geqslant a\right)\leqslant\frac{2}{d}P\left(\left|\sum_{k=1}^{n} Y_k\right|\geqslant\tfrac{1}{2}a\right).$$

For proof, see Lemma 1 (Chover 1967).

Lemma 2.4. *Let there exist a sequence (γ_n) such that*
(i) $\lim\gamma_{n+1}/\gamma_n=1$ *and*
(ii) $P(\lim(\Sigma_{j=1}^{n}\xi_j/\gamma_n)=0)=1$.
Then

$$P\left(\lim_{n\to\infty}\frac{\xi_n}{\gamma_n}=0\right)=1.$$

Proof. By (ii) there exists a set $A\subset\Omega$ with $P(A)=1$ such that for every $w\in A$, $\Sigma_{j=1}^{n}\xi_j(w)/\gamma_n\to 0$ as $n\to\infty$. Now in view of (i) we have

$$\lim_{n\to\infty}\frac{\sum_{j=1}^{n}\xi_j(w)-\sum_{j=1}^{n-1}\xi_j(w)}{\gamma_n}=\lim_{n\to\infty}\frac{\xi_n(w)}{\gamma_n}=0,\quad w\in A, \tag{2.3}$$

This completes the proof of the lemma.

Lemma 2.5. *Let $m\geqslant 1$ be some integer. Then there exists a $d>0$ such that $P(\Sigma_{k=i}^{j} Y_k>0)>d$ and $P(\Sigma_{k=i}^{j} Y_k<0)>d$, $1<i<j\leqslant m$.*

Proof. Observe that (X_n') converges weakly to X_1 and $\mu_n\to 0$ as $n\to\infty$. Hence a N_0 exists such that

$$P(Y_n>0)\geqslant\tfrac{1}{2}P(X_1>0), \tag{2.4}$$

for all $n\geqslant N_0$.

Set $E\Sigma_{k=i}^{j} Y_k^2=s_{i,j}^2$ and $E\Sigma_{k=i}^{j}|Y_k|^3=\Gamma_{i,j}$. Then by Lemma 2.2 we have

$$\left|P\left(\sum_{k=i}^{j} Y_k>0\right)-\tfrac{1}{2}\right|\leqslant\frac{\Gamma_{i,j}}{s_{i,j}^3}.$$

Noticing that $|Y_n|\leqslant 2\beta_n$, $n\geqslant 1$, one gets $\Gamma_{i,j}/s_{i,j}^3\leqslant 2\beta_j/s_{i,j}$. By (1.2) for any $q>p$, and for any δ, $0<\delta<1$, one can find a $N_1(q)$ such that

$$\frac{(1-\delta)(j-i)H(a_j)}{(\log\log a_j)^{q\mu}}\leqslant s_{i,j}^2\leqslant(j-i)H(a_j) \tag{2.5}$$

for all $j\geqslant i+N_1(q)$.

Using the fact that $p(1-\mu)>1$, and that q is any number greater than p, one can find a q_0 such that $p-\mu q_0>1$. For the q_0 chosen this way, one obtains $\Gamma_{i,j}/s_{ij}^3 \leqslant 2\beta_j/s_{ij} \leqslant c(\log\log N_1)^{-(p-\mu q_0/2)+1/2}$ for all $j \geqslant i+N_1$, where $N_1=N_1(q_0)$. Choose N_2 such that $(\log\log N_2)^{-(p-\mu q_0/2)+1/2}<\frac{1}{4}$. Then for $N_3=\max(N_1, N_2)$, $i \geqslant N_0$ and $j \geqslant i+N_3$ one obtains:

$$P\left(\sum_{k=i}^{j} Y_k>0\right)>\tfrac{1}{4}. \tag{2.6}$$

For $i>N_0$ and $i \leqslant j \leqslant i+N_3$, we have, in view of (2.4),

$$P\left(\sum_{k=i}^{j} Y_k>0\right) \geqslant P(Y_i>0, Y_{i+1}>0 \ldots Y_j>0)$$

$$\geqslant \frac{1}{2^{N_3}}(P(X_1>0))^{N_3}.$$

$EX_1=0$ implies that $P(X_1>0)=a>0$. Hence,

$$P\left(\sum_{k=i}^{j} Y_k>0\right) \geqslant \left(\frac{a}{2}\right)^{N_3}, \tag{2.7}$$

whenever $i \geqslant N_0$ and $i \leqslant j \leqslant i+N_3$.

From the fact that $EY_k=0$, there exists a $c_k>0$, such that $P(Y_k>0)>c_k$. Consequently, for $0<i<j<N_0$,

$$P\left(\sum_{k=i}^{j} Y_k>0\right) \geqslant P(Y_i>0, Y_{i+1}>0, \ldots, Y_j>0) \geqslant \prod_{k=1}^{N_0} c_k. \tag{2.8}$$

Define $d_1=\frac{1}{4}(a/2)^{N_3}\prod_{k=1}^{N_0} c_k$. Then for $0<i<j \leqslant m$,

$$P\left(\sum_{k=i}^{j} Y_k>0\right)>d_1.$$

Similarly, one can show that for $0<i<j \leqslant m$,

$$P\left(\sum_{k=i}^{j} Y_k<0\right)>d_2>0.$$

Now the proof is complete by taking $d=\min(d_1, d_2)$.

Lemma 2.6. *The sequence $\{S_n/a_n\}$ converges to zero in probability.*

Proof. According to Theorem 4 (Gnedenko and Kolmogorov 1954, §25), it is enough if we establish that for any $x>0$,

$$\lim_{n\to\infty} nP(|X_n|>xa_n)=0, \tag{2.9}$$

and

$$\lim_{\epsilon\to 0}\ \limsup_{n\to\infty} n\left(\int_{|x|<\epsilon} x^2\,\mathrm{d}F(xa_n)\right)=0. \tag{2.10}$$

From (1.3) we know that for any $x>0$,

$$\lim_{n\to\infty}\int_{xa_n}^{\infty}\frac{\mathrm{d}H(y)}{H(y)\log\log y}=0. \tag{2.11}$$

However,

$$\int_{xa_n}^{\infty}\frac{\mathrm{d}H(y)}{H(y)\log\log y}\geqslant\frac{x^2a_n^2P(|X_1|>xa_n)}{H(xa_n)\log\log xa_n}.$$

Hence from (1.2), there exists a $X_0>0$ and a N_0 such that for all $x<X_0$ and $n>N_0$

$$\int_{xa_n}^{\infty}\frac{\mathrm{d}H(y)}{H(y)\log\log y}\geqslant x^{2-\mu}nP(|X_1|>xa_n). \tag{2.12}$$

Now (2.11) and (2.12), together, establish (2.9). For any $\epsilon>0$,

$$n\int_{|x|<\epsilon}x^2\mathrm{d}F(xa_n)=\frac{n}{a_n^2}\int_{|y|<\epsilon a_n}y^2\mathrm{d}F(y)=\frac{nH(\epsilon a_n)}{a_n^2}.$$

Consequently,

$$\lim_{n\to\infty}n\int_{|x|<\epsilon}x^2\mathrm{d}F(xa_n)=0,$$

and (2.10) is established.

Lemma 2.7. *The sequence $\{S_n''/a_n\}$ converges to zero with probability one.*

Proof. By Lemmas 3.3, 4.1 and 5.1 of Feller (1969), one has $P(\lim T_n/a_n=0)=1$. On applying strong symmetrization results, see Theorem 3.2.1, (Stout 1974, p. 117) one gets

$$P\left\{\lim_{n\to\infty}\frac{\sum_{j=1}^{n}X_j''-\tilde{m}\left(\sum_{j=1}^{n}X_j''\right)}{a_n}=0\right\}=1, \tag{2.13}$$

where $\tilde{m}(X)$ is a median of the r.v. X.

Let $\Gamma_n=E\Sigma_{k=1}^n|Y_k|^3$ and $s_n^2=E\Sigma_{k=1}^nY_k^2$. Then proceeding as in the proof of Lemma 2.5, one can find a $q_0>p$ with $p-\mu q_0>1$, and a N_0 such that for all $n\geqslant N_0$, $\Gamma_n\leqslant s_n^3(\log\log n)^{-(p-\mu q_0)+1/2}$. Hence by Lemma 2.2, the sequence $\{S_n'/s_n\}$ converges weakly to a standard normal r.v. However, $s_n^2\leqslant nH(\beta_n)\leqslant nH(a_n)$. Consequently $S_n'/a_n\xrightarrow{P}0$ ("$\xrightarrow{P}$" means converges in probability). This along with Lemma 2.6 implies $S_n''/a_n\xrightarrow{P}0$. By weak symmetrization lemma (Loeve 1968), one now gets

$$\lim_{n\to\infty}\frac{\sum_{j=1}^{n}\theta_j-\tilde{m}\left(\sum_{j=1}^{n}X_j''\right)}{a_n}=0. \tag{2.14}$$

By writing $S_n''=\{\Sigma_{j=1}^n X_j''-\tilde{m}(\Sigma_{j=1}^n X_j'')\}+\{\tilde{m}(\Sigma_{j=1}^n X_j'')-\Sigma_{j=1}^n \theta_j\}$ and by using (2.13) and (2.14), the proof of the lemma is complete.

3. Theorem. *Let* (X_n) *be a sequence of i.i.d. r.v.'s for which* (1.2) *holds. Then for a.e.* $w\in\Omega$, *the sequence* $\{Z_n(t,w), t\in[0,1]\}$ *is relatively compact with* K *as the set of all its limit functions if and only if for some* $a>0$

$$\int_a^\infty \frac{\mathrm{d}H(x)}{H(x)\log\log x}<\infty.$$

Proof. For K to be the limit set of $(Z_n(\cdot,w))$ it is necessary that

$$P(\limsup Z_n(1)=1)=1.$$

In view of Feller (1969), the convergence of the integral is hence a necessary condition. We now go to the sufficiency part.

For $t\in[0,1]$ and $n\geqslant 3$, define

$$a_nU_n(t)=S'_{[nt]}+(nt-[nt])Y_{[nt]+1}$$

and

$$a_nV_n(t)=S''_{[nt]}+(nt-[nt])(X''_{[nt]+1}-\theta_{[nt]+1})$$

and notice that $Z_n(t)=U_n(t)+V_n(t)$. We first establish that

$$P\Big(\lim_{n\to\infty}\sup_{0\leqslant t\leqslant 1}V_n(t)=0\Big)=1. \tag{3.1}$$

This is done by showing that

$$\sup_{0\leqslant t\leqslant 1} S''_{[nt]}/a_n\to 0,\quad \text{w.p. 1.} \tag{3.2}$$

and

$$\sup_{0\leqslant t\leqslant 1}\frac{X''_{[nt]+1}-\theta_{[nt]+1}}{a_n}\to 0,\quad \text{w.p. 1,} \tag{3.3}$$

where w.p. 1 stands for "with probability one".

(3.2) is immediate from Lemma 2.7; (3.3) now follows from (3.2) and Lemma 2.4. Consequently, it remains for us to show that for a.e. w, the sequence $\{U_n(t,w), t\varepsilon[0,1]\}$ is relatively compact in C and has K for its limit set. For any $\epsilon>0$, let K_ϵ be the set of all $f\in C$ such that the distance between f and K is less than ϵ and let $K_\epsilon^c=C-K_\epsilon$. The proof now follows in three stages. In the first stage, the relative compactness of (U_n) is established; in the second stage $P(U_n\in K_\epsilon^c \text{ i.o.})$ is shown to be zero, for any $\epsilon>0$, and in the third stage every element of K is shown to be an almost sure limit function of (U_n).

3.1. *To establish the relative compactness of* (U_n)

The sequence (U_n) is claimed to be relatively compact by showing that it is equicontinuous and by appealing to Ascoli's theorem.

The sequence $(U_n(\cdot, w))$ is said to be equicontinuous if for any $\epsilon>0$, there exists a $\delta_\epsilon>0$ such that for a.e. $w\in\Omega$ and for some integer N_0

$$|U_n(t,w)-U_n(t',w)|<\epsilon \tag{3.4}$$

whenever $|t-t'|<\delta_\epsilon$ and $n\geqslant N_0$.

Let $\epsilon>0$ be fixed and let q be the smallest integer such that $\epsilon^2 2^q>8$. Then we show (3.4) with $\delta_\epsilon=2^{-q}$.

For $t=i/n, t'=j/n, i<j$, the event $\{|U_n(t)-U_n(t')|\geqslant\epsilon\}$ is the same as $\{|\Sigma_{k=i+1}^{j} Y_k| \geqslant\epsilon a_n\}$. Hence for the set of all points t, t' of this type

$$P\left\{\sup_{0\leqslant|t-t'|<\delta_\epsilon}|U_n(t)-U_n(t')|\geqslant\epsilon \text{ i.o.}\right\}=P\left\{\max_{0\leqslant j-i<n2^{-q}}\left|\sum_{k=i+1}^{j} Y_k\right|\geqslant\epsilon a_n \text{ i.o.}\right\}.$$

Take $n_r=2^r$, $r=1,2,\ldots$ and define

$$A_n=\left\{\max_{0\leqslant j-i<n2^{-q}}\left|\sum_{k=i+1}^{j} Y_k\right|\geqslant\epsilon a_n\right\}$$

$$B_r=\left\{\max_{n_r\leqslant n<n_{r+1}}\ \max_{\substack{0<j-i<n2^{-q}\\ 0<i<j<n}}\left|\sum_{k=i+1}^{j} Y_k\right|\geqslant\epsilon a_{n_r}\right\}$$

and

$$C_r=\left\{\max_{0<j-i\leqslant 2^{-q}n_r}\left|\sum_{k=i+1}^{j} Y_k\right|\geqslant\epsilon a_{n_r}\right\}.$$

Observe that

$$P(A_n \text{ i.o.})\leqslant P(B_r \text{ i.o.})\leqslant P(C_r \text{ i.o.}). \tag{3.5}$$

Define

$$C_r^{(l)}=\left\{\max_{(i,j)\in I(r,l)}\left|\sum_{k=i+1}^{j} Y_k\right|\geqslant\epsilon a_{n_r}\right\}$$

and

$$D_r^{(l)}=\left\{\left|\sum_{k\in I(r,l)} Y_k\right|\geqslant\tfrac{1}{2}\epsilon a_{n_r}\right\},$$

where $I(r,l)$ is the collection of all integers, k, such that

$$(l-1)2^{(r-q+1)}\leqslant k\leqslant(l+1)2^{(r-q+1)},\quad l=1,2\ldots(2^q-1).$$

Notice that

$$C_r = \bigcup_{l=1}^{(2^q-1)} C_r^{(l)}. \tag{3.6}$$

In view of Lemmas 2.3 and 2.5 there exists a $d>0$ such that

$$P(C_r^{(l)}) \leqslant \frac{2}{d} P(D_r^{(l)}), \quad l=1,2,\ldots,(2^q-1).$$

Let $s_{r,l}^2 = \Sigma_{k\in I(r,l)} EY_k^2$. Using the facts that $E(X_k')^2 = H(\beta_k)$ and $\lim_{k\to\infty}\mu_k = 0$, one can find an integer R_1 and a positive number q_0 with $p-\mu q_0 > 1$, such that for all $r \geqslant R_1$,

$$\frac{2^{(r-q+2)}H(a_{n_r})}{(\log\log a_{n_r})^{q_0\mu}} \leqslant s_{r,l}^2 \leqslant 2^{(r-q+3)}H(a_{n_r}), \quad l=1,2,\ldots,(2^q-1). \tag{3.7}$$

Define $\alpha_{r,l} = 2\beta_{n_r}/s_{r,l}$ and $\epsilon_{r,l} = \epsilon a_{n_r}/s_{r,l}$. Then by (3.7) observe that $\alpha_{r,l} \leqslant c(\log\log n_r)^{-(p-\mu q_0)+1/2}$, for all $r \geqslant R_2$, and that $\lim_{r\to\infty}\epsilon_{r,l}\alpha_{r,l}=0$. Hence, a R_3 can be found such that for all $r \geqslant R_3, \epsilon_{r,l}\alpha_{r,l} \leqslant \epsilon_1$ where ϵ_1 satisfies the inequality $\epsilon^2 2^{(q-3)}(1-(\epsilon_1/2)) > 1$. By Lemma 2.1 one now gets, for all $r \geqslant R_3$,

$$\begin{aligned} P(D_r^{(l)}) &\leqslant \exp\left\{-\frac{\epsilon^2 a_{n_r}^2}{2s_{r,l}^2}\left(1-\frac{\epsilon_1}{2}\right)\right\} \\ &\leqslant \exp\left(-\epsilon^2 2^q (1-\epsilon_1/2)(\log\log a_{n_r})\right) \\ &\leqslant cr^{-(1+\epsilon_2)}. \end{aligned}$$

Hence $\Sigma_{r=1}^{\infty} P(c_r^{(l)}) \leqslant \Sigma_{r=1}^{R_3} P(c_r^{(l)}) + c\Sigma_{r=R_3+1}^{\infty} r^{-(1+\epsilon_2)} < \infty$. By appealing to the Borel–Cantelli lemma, one concludes that $P(c_r^{(l)}$ i.o.$)=0$. (3.5) and (3.6), together, imply that $P(A_n$ i.o.$)=0$.

Since the function U_n is piecewise continuous, the above study of (U_n), for points of the type i/n, j/n is sufficient to claim the equicontinuity of (U_n).

Note 3.1. For any $f \in C$ let $\Pi_m f$ denote the piecewise linear function $\Pi_m f(i/m) = f(i/m)$, $i=0,1,2,\ldots,m$ and $\Pi_m f$ is linear in $[(i-1)/m,(i/m)]$, $i=1,2,\ldots,m$. Then as a consequence of equicontinuity, for a.e. w and for any $\epsilon>0$, one can find an integer $N(=N(\epsilon,w))$ such that $\|\Pi_m U_n - U_n\| < \epsilon$ for all $m, n \geqslant N$.

Note 3.2. For a.e. w and for any $\epsilon > 0$, one can argue on the lines of Corollary 2 (Chover 1967), and obtain a N_0 such that $\|U_n - U_{n'}\| < \epsilon$ for all $n, n' \geqslant N_0$ with $|1-(n'/n)| < 1/N_0$. The details are omitted.

3.2 *To show that $P(U_n \in K_\epsilon^c$ i.o.$)=0$*

Define $M_r = [\beta^r]$, $\beta > 1$. In view of Notes 3.1 and 3.2, it is enough if we establish that $P(\Pi_m U_{M_r} \varepsilon K_\epsilon^c$ i.o.$)=0$, for every $\epsilon>0$ and for all arbitrary large m. Let

$$D_{m,r} = \left\{ m \sum_{v=1}^{m} \left(\Pi_m U_{M_r}\left(\frac{v}{m}\right) - \Pi_m U_{M_r}\left(\frac{v-1}{m}\right)\right)^2 \geqslant (1+\epsilon)\right\}.$$

As $\Pi_m U_n$ is piecewise linear, our problem reduces to that of proving

$$P(D_{m,r}\text{ i.o.})=0. \tag{3.8}$$

We have

$$\Pi_m U_{M_r}\left(\frac{v}{m}\right)-\Pi_m U_{M_r}\left(\frac{v-1}{m}\right)=\sum_{k=i}^{j}\frac{Y_k}{a_{M_r}}+y_{r,v}, \tag{3.9}$$

where i is the smallest integer such that $i/M_r \geqslant (v-1)/m$, j is the largest integer such that $j/M_r \leqslant v/m$ and

$$y_{r,v}=\Pi_m U_{M_r}\left(\frac{v}{m}\right)-\Pi_m U_{M_r}\left(\frac{j}{M_r}\right)+\Pi_m U_{M_r}\left(\frac{i-1}{M_r}\right)-\Pi_m U_{M_r}\left(\frac{v-1}{m}\right).$$

Observe that

$$|y_{r,v}|\leqslant\frac{|Y_{i-1}|+|Y_{j+1}|}{a_{M_r}}$$

and that $Ey_{r,v}\to 0$ as $r\to\infty$.

By Tchebychev's inequality, one has for all $r\geqslant R$ and for all $v=1,2,\dots,m$,

$$P\{|y_{r,v}|\geqslant(M_r\log\log M_r)^{-1/3}\}\leqslant c(M_r\log\log M_r)^{-1/3}. \tag{3.10}$$

Consequently, for any $\epsilon>0$, the Borel–Cantelli lemma implies that

$$P\{|y_{r,v}|\geqslant\epsilon\text{ i.o.}\}=0,\quad v=1,2,\dots,m. \tag{3.11}$$

Define $T_{r,v}=\sqrt{m}\,(|\Sigma_{k=i}^{j}Y_k|/a_{M_r})$, $v=1,2,\dots,m$, for i and j as defined in (3.9).

Let $\Delta=\{(a_1,a_2,\dots,a_m);\ a_i\geqslant 0,\ i=1,2,\dots,m;\ \Sigma_{i=1}^{m}a_i^2\geqslant 1+\epsilon\}$ and $\Delta_1=\{(a_1,a_2,\dots,a_m);\ a_i\geqslant 0,\ i=1,2,\dots,m;\ \max_{1\leqslant i\leqslant m}a_i>1+\epsilon\}$. Observe that $\Delta_1\subset\Delta$. In view of (3.11), (3.8) follows once we establish that

$$P\left\{\sum_{v=1}^{m}T_{r,v}^2\geqslant(1+\epsilon)\text{ i.o.}\right\}=0, \tag{3.12}$$

or equivalently,

$$P\{(T_{r,1},T_{r,2},\dots,T_{r,m})\in\Delta,\text{ i.o.}\}=0.$$

For any v,

$$P(T_{r,v}>(1+\epsilon))=P\left\{\left|\sqrt{m}\sum_{k=i}^{j}Y_k\right|\geqslant(1+\epsilon)a_{M_r}\right\}.$$

Let $s_{r,v}^2=E\Sigma_{k=i}^{j}Y_k^2$ and $\alpha_{r,v}=2\beta_{M_r}/s_{r,v}$. Then there exists a R and a $q_0>p$ with $p-\mu q_0>1$, such that for all $r\geqslant R$,

$$\frac{M_r H(a_{M_r})}{2m(\log\log M_r)^{\mu q_0}}\leqslant s_{r,v}^2\leqslant\frac{M_r}{m}H(a_{M_r}),$$

$$\alpha_{r,v}\leqslant c(\log\log M_r)^{-(p-\mu q_0/2)+1/2},$$

and $a_{M_r} \geqslant s_{r,v}\sqrt{2m \log\log M_r}$, $v=1,2,\dots,m$. Taking $\epsilon_{r,v}=(1+\epsilon)\sqrt{2\log\log M_r}$ and applying Lemma 2.1 we get for all $r \geqslant R$

$$P(T_{r,v}>(1+\epsilon)) \leqslant 2\exp\left\{-(1+\epsilon)^2\left(1-\frac{\epsilon_{r,v}\alpha_{r,v}}{2}\right)\log\log M_r\right\}.$$

Since $\epsilon_{r,v}\alpha_{r,v} \to 0$ as $r \to \infty$, there exists a $R_1(\geqslant R)$ such that for all $r \geqslant R_1$

$$(1+\epsilon)^2\left(1-\frac{\epsilon_{r,v}\alpha_{r,v}}{2}\right)>(1+\epsilon_2)>1,$$

and consequently,

$$P(T_{r,v}>1+\epsilon) \leqslant cr^{-(1+\epsilon_2)} \tag{3.13}$$

on applying the Borel–Cantelli lemma one gets $P(T_{r,v}>1+\epsilon \text{ i.o.})=0$, $v=1,2\dots m$. In turn we have shown that

$$P\{(T_{r,1},T_{r,2},\dots,T_{r,m}) \in \Delta_1 \text{ i.o.}\}=0. \tag{3.14}$$

In order to prove (3.11), it now remains to be established that

$$P\{(T_{r,1},T_{r,2},\dots,T_{r,m}) \in \Delta-\Delta_1 \text{ i.o.}\}=0. \tag{3.15}$$

Let $(a_1,a_2,\dots,a_m) \in \Delta-\Delta_1$ and let for any $\epsilon>0$, $N_\epsilon=(a_1-\epsilon,a_1+\epsilon)\times(a_2-\epsilon,a_2+\epsilon)\times\cdots\times(a_m-\epsilon,a_m+\epsilon)$. We will now show that for a $\epsilon>0$, which will be chosen later

$$P\{(T_{r,1},T_{r,2},\dots,T_{r,m}) \in N_\epsilon \text{ i.o.}\}=0. \tag{3.16}$$

Let $a_v>0$ and let $0<\epsilon<\epsilon_1=\{\min a_i/2, a_i>0\}$. Then from Lemma 2.1, we get for all $r \geqslant R$

$$P\{T_{r,v}>(a_v-\epsilon)\} \leqslant \exp\left\{-(a_v-\epsilon)^2\left(1-\frac{\epsilon_{r,v}\alpha_{r,v}}{2}\right)\log\log M_r\right\}$$

where

$$\epsilon_{r,v}=(a_v-\epsilon)(\log\log M_r)^{1/2}, \qquad \alpha_{r,v}=2\beta_{M_r}/s_{r,v}.$$

If $a_v=0$, then we have

$$P(T_{r,v} \geqslant a_v-\epsilon)=1.$$

Consequently, we have for all $r \geqslant R_1$,

$$\begin{aligned} P\{(T_{r,1},T_{r,2},\dots,T_{r,m}) \in N_\epsilon\} &\leqslant P\left\{\bigcap_{v=1}^{m}(T_{r,v} \geqslant a_v-\epsilon)\right\} \\ &= \prod_{v=1}^{m} P(T_{r,v}>a_v-\epsilon) \\ &\leqslant \exp\left\{-\sum_v{}^{*}(a_v-\epsilon)^2\left(1-\frac{\epsilon_{r,v}\alpha_{r,v}}{2}\right)(\log\log M_r)\right\}, \end{aligned}$$

where $*$ indicates that the sum is over terms with $a_v>0$. Let ϵ_2 be such that

$\Sigma^*(a_v-\epsilon_2)^2>1$. Noticing that $\Sigma_{v=1}^m a_v^2=\Sigma_v^* a_v^2$ and that $\epsilon_{r,v}\alpha_{r,v}\to 0$ as $r\to\infty$, one can find a R_2 such that for all $r\geqslant R_2$ and for $\epsilon<\min(\epsilon_1,\epsilon_2)$

$$\sum_v {}^*(a_v-\epsilon)^2\left(1-\frac{\epsilon_{r,v}\alpha_{r,v}}{2}\right)>(1+\epsilon_3)>1.$$

Hence for all $r\geqslant R_2$ and $\epsilon<\min(\epsilon_1,\epsilon_2)$,

$$P\{(T_{r,1},T_{r,2},\ldots,T_{r,m})\in N_\epsilon\}\leqslant cr^{-(1+\epsilon_3)}.$$

By appealing to Borel–Cantelli lemma (3.16) follows. As $\Delta-\Delta_1$ is compact (3.16) in turn implies (3.15).

3.3. *To claim that every $f\in K$ is a limit of (U_n)*

Let $f\in K$. Then it is a limit function of the sequence (U_n) if for every $\epsilon>0$

$$P\{\|U_n-f\|<\epsilon \text{ i.o.}\}=1.$$

In view of Notes 3.1 and 3.2 it is enough if we prove that for any m

$$P\{\|\Pi_m U_n-\Pi_m f\|<\epsilon \text{ i.o.}\}=1. \tag{3.17}$$

Let $N_r=m^r$, $r=1,2,\ldots$ and

$$A_r^{(v)}=\left\{\left|\Pi_m U_{N_r}\left(\frac{v}{m}\right)-\Pi_m U_{N_r}\left(\frac{v-1}{m}\right)-\overline{f\left(\frac{v}{m}\right)-f\left(\frac{v-1}{m}\right)}\right|<\frac{\epsilon}{m}\right\}. \tag{3.18}$$

Now (3.17) is claimed once we show that

$$P\left\{\limsup_{n\to\infty}\bigcap_{v=1}^m A_r^{(v)}=0\right\}=1. \tag{3.19}$$

Since $\Pi_m U_n(0)=f(0)=0$, for all $m\geqslant m_0(w)$ there exists a N such that for all $n\geqslant N$, $\sup_{0\leqslant t\leqslant 1/m}|\Pi_m U_n(t)-\Pi_m f(t)|<\epsilon/m$ a.s. Hence it is enough if we consider $\bigcap_{v=2}^m A_r^{(v)}$ in (3.19).

Let $V_r^{(v)}=\Pi_m U_{N_r}(v/m)-\Pi_m U_{N_r}((v-1)/m)$ and $a_{N_r}W_r^{(v)}=\Sigma_{k=i}^j Y_k$, where i is the least integer such that, $i/N_r\geqslant(v-1)/m$ and j is the greatest integer such that $j/N_r\leqslant v/m$, $v=2,3,\ldots,m$. Set

$$B_r^{(v)}=\left\{|V_r^{(v)}-W_r^{(v)}|\geqslant\frac{\epsilon}{2m}\right\}.$$

From Tchebychev's inequality and the Borel–Cantelli lemma, one can show that

$$P(B_r^{(v)} \text{ i.o.})=0. \tag{3.20}$$

Let $f(v/m)-f((v-1)/m))>0$. Then

$$P\left\{\left|W_r^{(v)}-\overline{f\left(\frac{v}{m}\right)-f\left(\frac{v-1}{m}\right)}\right|\leqslant\frac{\epsilon}{2m}\right\}=P\left\{W_r^{(v)}\geqslant f\left(\frac{v}{m}\right)-f\left(\frac{v-1}{m}\right)-\frac{\epsilon}{2m}\right\}$$

$$-P\left\{W_r^{(v)}\geqslant f\left(\frac{v}{m}\right)-f\left(\frac{v-1}{m}\right)+\frac{\epsilon}{2m}\right\}.$$

Using the facts that for some $q_0 > p$ with $p - \mu q_0 > 1$ and for all $r \geqslant R$

$$\frac{N_r H(a_{N_r})}{2m(\log\log N_r)^{q_0\mu}} \leqslant s_{r,v}^2 \leqslant \frac{N_r H(a_{N_r})}{m}, \quad \left(s_{r,v}^2 = E \sum_{k=i}^{j} Y_k^2 \right),$$

$$\alpha_{r,v} = \frac{2\beta_{N_r}}{s_{r,v}} \leqslant c(\log\log N_r)^{-(p-(\mu q_0/2))+1/2}$$

and setting

$$\epsilon_{r,v} = \sqrt{m}\left(f\left(\frac{v}{m}\right) - f\left(\frac{v-1}{m}\right) - \frac{\epsilon}{2m} \right)\sqrt{2\log\log N_r}, \qquad \gamma_{r,v} = (\log\log N_r)^{-1}$$

one can apply Lemma 2.1 and obtain

$$P\left\{ W_r^{(v)} \geqslant f\left(\frac{v}{m}\right) - f\left(\frac{v-1}{m}\right) - \frac{\epsilon}{2m} \right\} \geqslant \exp\left\{ -\frac{\epsilon_{r,v}^2}{2}(1+\gamma_{r,v}) \right\},$$

where $\gamma_{r,v}$ can be made as small as desired by choosing r large.

Similarly, by taking $\epsilon_{r,v}^* = \sqrt{m}\,(f(v/m) - f((v-1)/m) + \epsilon/2m)\sqrt{2\log\log N_r}$ and applying Lemma 2.1, one gets

$$P\left\{ W_r^{(v)} \geqslant f\left(\frac{v}{m}\right) - f\left(\frac{v-1}{m}\right) + \frac{\epsilon}{2m} \right\} \leqslant \exp\left\{ -\frac{(\epsilon_{r,v}^*)^2}{2}\left(1 - \frac{\alpha_{r,v}\epsilon_{r,v}^*}{2}\right) \right\}. \quad (3.21)$$

Given $\epsilon > 0$, a R_1 can be chosen such that for all $r > R_1$,

$$\frac{\epsilon_{r,v}^2}{2}(1+\gamma_{r,v}) \leqslant \frac{(\epsilon_{r,v}^*)^2}{2}\left(1 - \frac{\epsilon_{r,v}^*\alpha_{r,v}}{2}\right), \quad v = 2, 3, \ldots, m.$$

Consequently, for all $r \geqslant R_1$,

$$P\left\{ \left| W_r^{(v)} - \overline{f\left(\frac{v}{m}\right) - f\left(\frac{v-1}{m}\right)} \right| \leqslant \frac{\epsilon}{2m} \right\} \geqslant$$

$$\geqslant c_1 \exp\left\{ -\left(f\left(\frac{v}{m}\right) - f\left(\frac{v-1}{m}\right) - \frac{\epsilon}{2m} \right)^2 (1+\gamma_{r,v})\log r \right\}. \quad (3.22)$$

Let $f(v/m) - f((v-1)/m)) < 0$. Then

$$P\left\{ \left| W_r^{(v)} - \overline{f\left(\frac{v}{m}\right) - f\left(\frac{v-1}{m}\right)} \right| < \frac{\epsilon}{2m} \right\} = P\left\{ -W_r^{(v)} \geqslant f\left(\frac{v-1}{m}\right) - f\left(\frac{v}{m}\right) - \frac{\epsilon}{2m} \right\}$$

$$- P\left\{ -W_r^{(v)} \geqslant f\left(\frac{v-1}{m}\right) - f\left(\frac{v}{m}\right) + \frac{\epsilon}{2m} \right\}.$$

By noticing that $f((v-1)/m) - f(v/m) > 0$, and by proceeding as previously, one obtains for all $r \geqslant R_2$

$$P\left\{ \left| W_r^{(v)} - \overline{f\left(\frac{v}{m}\right) - f\left(\frac{v-1}{m}\right)} \right| \leqslant \frac{\epsilon}{2m} \right\} \geqslant$$

$$\geqslant c_2 \exp\left\{ -\left(f\left(\frac{v-1}{m}\right) - f\left(\frac{v}{m}\right) - \frac{\epsilon}{2m} \right)^2 (1+\gamma_{r,v})\log r \right\}. \quad (3.23)$$

If $f(v/m)-f((v-1)/m)=0$, then there exists a R_3 such that

$$P\left\{|W_r^{(v)}|\leqslant\frac{\epsilon}{2m}\right\}\geqslant\tfrac{1}{4}, \tag{3.24}$$

for all $r\geqslant R_3$.

Define $E_r^{(v)}=\{|W_r^{(v)}-\overline{f(v/m)-f((v-1)/m)}|\leqslant\epsilon/2m\}$, $v=2,3,\ldots,m$, $r=1,2,\ldots$. By (3.22), (3.23) and (3.24) one gets for all $r\geqslant R$ [$=\max(R_1, R_2, R_3)$],

$$P\left\{\bigcap_{v=2}^{m} E_r^{(v)}\right\}\geqslant cr^{-g(f,r,\epsilon)},$$

where

$$g(f,r,\epsilon)=\sum_{v=2}^{m}{}'\left(f\left(\frac{v}{m}\right)-f\left(\frac{v-1}{m}\right)-\frac{\epsilon}{2m}\right)^2(1+\gamma_{r,v})$$
$$+\sum_{v=2}^{m}{}''\left(f\left(\frac{v-1}{m}\right)-f\left(\frac{v}{m}\right)-\frac{\epsilon}{2m}\right)^2(1+\gamma_{r,v})$$

with Σ' indicating that the summation is over terms in which $f(v/m)-f((v-1)/m)>0$ and Σ'' indicating that the summation is over terms in which $f(v/m)-f((v-1)/m)<0$. Choose ϵ small such that

$$\max_{2\leqslant v\leqslant m}\left\{\frac{\epsilon}{2m\left|f\left(\frac{v}{m}\right)-f\left(\frac{v-1}{m}\right)\right|},\ \left|f\left(\frac{v}{m}\right)-f\left(\frac{v-1}{m}\right)\right|>0\right\}<1.$$

For the ϵ thus chosen, select R such that

$$(1+\gamma_{r,v})\left(1-\frac{\epsilon}{2m\left|f\left(\frac{v}{m}\right)-f\left(\frac{v-1}{m}\right)\right|}\right)^2<(1-\epsilon_1)<1,$$

for all $r\geqslant R$ and for all v with $|f(v/m)-f((v-1)/m)|>0$. Consequently, for all $r\geqslant R$ we have

$$P\left\{\bigcap_{v=2}^{m} E_r^{(v)}\right\}\geqslant cr^{-(1-\epsilon_1)\Sigma_{v=2}^{m}(f(v/m)-f((v-1)/m))^2}.$$

However, $f\in K$ implies that $\Sigma_{v=2}^{m}(f(v/m)-f((v-1)/m))^2\leqslant 1$. Hence, for all $r\geqslant R$,

$$P\left\{\bigcap_{v=2}^{m} E_r^{(v)}\right\}\geqslant cr^{-(1-\epsilon_1)}.$$

Using the fact that $E_r^{(v)}$, $v=2,3,\ldots,m$, $r=1,2,\ldots$ are mutually independent events, the Borel–Cantelli lemma helps us claim that

$$P\left\{\bigcap_{v=2}^{m} E_r^{(v)}\text{ i.o.}\right\}=1. \tag{3.25}$$

In case $f(v/m)-f((v-1)/m)=0$ for all $v=2,3,\ldots,m$, (3.25) is immediate from (3.24).

Observe that $A_r^{(v)} \supset E_r^{(v)} \cap (B_r^{(v)})^c$, $v=2,3,\ldots,m$. Hence (3.20) and (3.25) together establish that $P\{\bigcap_{v=2}^m A_r^{(v)} \text{ i.o.}\}=1$ and complete the proof of the theorem.

Remark 1. Here we present an example of a d.f. which is not in the domain of attraction of the normal law, but for which both (1.2) and (1.3) hold.

Example. Let F by a symmetric d.f. with the probability density function

$$f(x)=\begin{cases} \dfrac{c}{|x|^3\log|x|}\left(1+\tfrac{1}{12}\sin\log_3|x|+\tfrac{1}{12}\cos\log_3|x|\right), & \text{if } |x|\geqslant a,\\ f(a), & \text{if } |x|\leqslant a,\end{cases}$$

where $a=\mathrm{e}^e, \log_3 x=\log\log\log x$ and where c is a positive constant such that $\int_{-\infty}^{\infty} f(x)\mathrm{d}x=1$. Simple calculations show that

$$H(x)=\begin{cases} \tfrac{2}{3}f(a)x^3, & \text{if } x\leqslant a\\ \tfrac{2}{3}f(a)a^3+c(\log\log x)\left(1+\tfrac{1}{12}\sin\log_3 x\right)-c, & \text{if } x\geqslant a.\end{cases}$$

Consequently, $\int_a^\infty dH(x)/H(x)\log\log x<\infty$ and for any $\mu\geqslant 0$, $H(tx)<2t^\mu H(x)$ for all t and x large.

Remark 2. Let $(X_n=(X_{1,n}, X_{2,n},\ldots, X_{p,n}))$ be a sequence of i.i.d. random vectors in the Euclidean space R_p of p-dimensions. Assume that the component r.v.'s of X_n are mutually independent and that they are defined on a common probability space $(\Omega, \mathfrak{B}, P)$. Take the mean of X_n to be 0. Define

$$H_j(x)=\int_{-x}^{x} y^2 \mathrm{d}P(X_{j,n}\leqslant y), x>0, \quad j=1,2,\ldots, p.$$

Let $a_{j,n}$ be the largest root of $x^2=2nH_j(x)\log\log x$, $j=1,2,\ldots,p$. For $0\leqslant t\leqslant 1$ define $a_{j,n}Z_{j,n}(t)=\sum_{j=1}^{[nt]} X_{j,n}+(nt-[nt])X_{j,[nt]+1}$ and $Z_n(t)=(Z_{1,n}(t), Z_{2,n}(t),\ldots, Z_{p,n}(t))$. Let $C_p=C_p[0,1]$ be the space of all functions $f=(f_1, f_2,\ldots, f_p)$ from $[0,1]$ to R_p with $f(0)=0$ and with $f_j(t)$ continuous in t, $0\leqslant t\leqslant 1$, $1\leqslant j\leqslant p$. Let $K_p\subset C_p$ be the set of all absolutely continuous functions of C_p with $\int_0^1(\mathrm{d}f/\mathrm{d}t)^2\mathrm{d}t\leqslant 1$, $((\mathrm{d}f/\mathrm{d}t)^2$ stands for the inner product of $\mathrm{d}f/\mathrm{d}t)$. For $f\in C_p$ define the norm $\|f\|=\sup_{0\leqslant t\leqslant 1}|f(t)|$. Endow C_p with the topology resulting from this norm. Proceeding on the lines of proof similar to that of "our theorem", one can show that the sequence (Z_n) is relatively compact in C_p and has K_p for the set of its limit functions if and only if for some $a>0$ and for all $j=1,2,\ldots,p$

$$\int_a^\infty \frac{\mathrm{d}H_j(x)}{H_j(x)\log\log x}<\infty.$$

References

Chover, J. (1967). On Strassen's version of loglog law. *Z. Wahrscheinlichkeitstheorie Verw. Geb.* **8**, 83–90.

Feller, W. (1969). An extension of the law of the iterated logarithm to variables without variance. *J. Math. Mech.* **18**, 343–355.

Gnedenko, B.V. and Kolmogorov, A.N. (1954). *Limit Distributions for Sums of Independent Random Variables.* (English translation). Addison Wesley, New York.

Loeve, M. (1968). *Probability Theory.* East West student Edition. Van Nostrand Company, Inc.

Stout, W. (1974). *Almost Sure Convergence.* Academic Press, New York.

Strassen, V. (1964). An invariance principle for the law of the iterated logarithm, *Z. Wahrscheinlichkeitstheorie Verw. Geb.* **3**, 211—226.

G. Kallianpur, P.R. Krishnaiah, J.K. Ghosh, eds., *Statistics and Probability: Essays in Honor of C.R. Rao*

DIFFUSIONS WITH RANDOM COEFFICIENTS

George C. PAPANICOLAOU and S.R.S. VARADHAN
Courant Institute of Mathematical Sciences, New York University, NY, U.S.A.

1. Introduction

Here we will be studying diffusion processes in R^d whose coefficients are random, stationary, stochastic processes. We will be mainly concerned with the behavior of such diffusion processes for large times for which we will establish a form of the ergodic theorem and deduce a central limit theorem from it.

Our diffusion processes will always start at $t=0$ from the point $x=0$ in R^d and will be governed by the diffusion generator:

$$L_\omega = \tfrac{1}{2}\sum a_{ij}(x,\omega)\frac{\partial^2}{\partial x_i \partial x_j}.$$

Here $\omega \in \Omega$ where $(\Omega, \mathcal{F}, \mu)$ is a probability space and $\{a_{ij}(x,\omega)\}$ constitutes a stationary stochastic process with $x \in R^d$ as the parameter set. For each x, the random variables $\{a_{ij}(x,\omega)\}$ constituting the diffusion coefficients at x, can be viewed as taking values in the space S_d of symmetric positive semi-definite matrices of size d. We will assume, throughout this paper, that the coefficients $\{a_{ij}(x,\omega)]$ are all almost surely continuous in x and that there exist constants $0 < C_1 \leqslant C_2 < \infty$ which are non-random such that,

$$C_1 \sum \xi_j^2 \leqslant \sum a_{ij}(x,\omega)\xi_i\xi_j \leqslant C_2 \sum \xi_j^2, \quad \forall x \in R^d, \quad \omega \in \Omega.$$

The last assumption we need is the ergodicity of the coefficient process $\{a(x,\omega)\}$ with respect to translations in space.

For each $\omega \in \Omega$, L_ω governs a diffusion and corresponding to the initial conditions of starting from $x=0$ at $t=0$ we have a measure P_ω on the space $C[[0,\infty); R^d]$ of continuous trajectories on $[0,\infty)$ with values in R^d. We want to look at the behavior of the distribution of $x_t/\sqrt{t}$ under P_ω as $t \to \infty$. Some form of the ergodic theorem should produce an averaging effect and the limiting distribution of $x_t/\sqrt{t}$ under P_ω should be independent of ω and should be a normal distribution with mean zero and covariance $\bar{a}$. The identification of $\bar{a}$ should be made as a precise form of averaging. This is the goal of this article.

Instead of random coefficients, one can take periodic coefficients and this case has been studied extensively in refs. [2, 1]. The case of random coefficients when L_ω is in divergence form has been studied in refs. [3, 4].

When the dimension, d, is one, we can use the method of random time change to study the problem explicitly and an easy computation establishes the result and calculates the limiting variance $\bar{a}$ to be the harmonic mean:

$$\bar{a}=\left[E\frac{1}{a(0;\omega)}\right]^{-1}.$$

2. An ergodic theorem

Although we formulated the theorem as a central limit theorem it is in reality an ergodic theorem that underlies the phenomenon. To see this, from the fact that P_ω is the diffusion corresponding to L_ω it follows that,

$$E^{P_\omega}\left[\exp\left\{\mathrm{i}\langle\theta,x(t)\rangle+\tfrac{1}{2}\left\langle\theta,\int_0^t a[x(s),\omega]\,\mathrm{d}s\,\theta\right\rangle\right\}\right]=1 \quad \text{a.e. } \omega.$$

By replacing θ by $\theta/\surd t$ we can rewrite this as:

$$E^{P_\omega}\left[\exp\left\{\mathrm{i}\left\langle\theta,\frac{x(t)}{\surd t}\right\rangle+\tfrac{1}{2}\left\langle\theta,\frac{1}{t}\int_0^t a[x(s),\omega]\,\mathrm{d}s\,\theta\right\rangle\right\}\right]=1.$$

If we had an ergodic theorem asserting,

$$\lim_{t\to\infty}\frac{1}{t}\int_0^t a(x(s),\omega)\,\mathrm{d}s=\bar{a}, \quad \text{a.e. } P_\omega, \text{a.e. } \mu, \tag{2.1}$$

it then follows that,

$$\lim_{t\to\infty}E^{P_\omega}\left[\exp\left\{\mathrm{i}\left\langle\theta,\frac{x(t)}{\surd t}\right\rangle\right\}\right]=\exp\{-\tfrac{1}{2}\langle\theta,\bar{a}\theta\rangle\},$$

which is the central limit theorem that we are after.

Now we will set up the basic notation that we will be using. Let Ω denote the space of all continuous maps $\omega: R^d\to S_d$ satisfying

$$C_1\sum\xi_j^2\leqslant\sum\omega_{ij}(x)\xi_i\xi_j\leqslant C_2\sum\xi_j^2,$$

for all $\xi\in R^d$ and $x\in R^d$. We will use also the customary notation $a_{ij}(x,\omega)$ to denote $\omega_{ij}(x)$ so we may conveniently view $a(x,\omega)=\{a_{ij}(x,\omega)\}$ either as a function of x for each fixed $\omega\in\Omega$ or as a function on Ω for each fixed x. We also have the translation maps $\tau_x:\Omega\to\Omega$ defined by:

$$a(y,\tau_x\omega)=a(x+y,\omega), \quad \forall x\in R^d, \quad y\in R^d \quad \text{and} \quad \omega\in\Omega.$$

There is a canonical Markov process with Ω as state space. Let us fix $\omega\in\Omega$. Then we have the diffusion operator,

$$L_\omega=\tfrac{1}{2}\sum a_{ij}(x,\omega)\frac{\partial^2}{\partial x_i\,\partial x_j},$$

on R^d and this generates a diffusion measure P_ω corresponding to these coefficients, with initial conditions corresponding to starting from the origin at time 0. P_ω is a measure on the space $C[[0,\infty); R^d]$ of trajectories starting from 0 at time 0.

Given a trajectory $x(t)$ with $x(0)=0$ we can map it onto a trajectory in the Ω space by setting

$$\omega_t = \tau_{x(t)}\omega.$$

This induces a measure Q_ω on the space of trajectories in Ω starting from ω at time 0 and a moment's reflection tells us that this is in fact a Markov process with Ω as state space. Let us denote by $q(t,\omega,\mathrm{d}\omega')$ the transition probability of this Markov process. If we think of the topology of Ω as coming from uniform convergence on compact sets it is easy to see that $q(t,\omega,\mathrm{d}\omega')$ has the Feller property. If we denote by $A(\omega)$ the function $a(0,\omega)$ then the property (2.1) that we want is really,

$$\lim_{t\to\infty}\frac{1}{t}\int_0^t A(\omega_s)\,\mathrm{d}s=\bar{a} \quad \text{a.e. } Q_\omega \quad \text{a.e. } \mu, \tag{2.2}$$

where μ is a stationary ergodic measure on Ω. In other words, we need an ergodic theory for the canonical Markov process $\{Q_\omega\}$ on Ω. In the next section we will establish the following theorem.

Theorem 2.1. *Given any ergodic stationary measure μ on Ω, there exists a unique λ on Ω which is mutually absolutely continuous with respect to μ, and which is an ergodic invariant measure with respect to our canonical Markov process on Ω.*

As an immediate consequence we have:

Theorem 2.2. *For any ergodic stationary measure μ on Ω,*

$$\lim_{t\to\infty}\frac{1}{t}\int_0^t A(\omega_s)\,\mathrm{d}s=\int_\Omega A(\omega)\lambda(\mathrm{d}\omega), \quad a.e.\ Q_\omega, \quad a.e.\ \mu,$$

where λ is determined by μ according to Theorem 2.1.

3. Proof

In order to prove our main theorem we need two results from the theory of partial differential equations which we shall state as lemmas without proofs.

Lemma 3.1. *Let $a(x)$ be a set of coefficients on R^d which is periodic in each variable of period* 1. *We will think of*

$$L=\tfrac{1}{2}\sum a_{ij}\frac{\partial^2}{\partial x_i\,\partial x_j}$$

as a diffusion process on the d-dimensional torus, T_d. Then the resolvent,

$$R=(I-L)^{-1},$$

maps $L_d(T_d)$ into $L_\infty(T_d)$ with a bound K that depends only on the upper and lower bounds for eigenvalues of the coefficient matrix $\{a(x)\}$ and not on the smoothness of the coefficients:

Remark. This result can be found in ref. [6] in a slightly different form. The modifications needed to reduce it to Lemma 3.1 are routine.

Lemma 3.2. *Let $a(x)$ be a set of coefficients in our space Ω. Let $p(t,x,\mathrm{d}y)$ be the transition probability for the diffusion on R^d governed by a. Then for each $x \in R^d$ and $t>0$, the measure $p(t,x,\mathrm{d}y)$ is equivalent to the Lebesgue measure on R^d.*

Remark. The existence of a density for $p(t,x,\mathrm{d}y)$ can be found in ref. [7] and the fact that the density cannot vanish on a set of positive measure is essentially the strong maximum principle.

The next ingredient is the fact that any stationary process is a limit of periodic processes:

Lemma 3.3. *Let μ be a stationary process on Ω. Then we can find for a sequence $l_n \to \infty$, periodic functions $\omega_n \in \Omega$ of period l_n in each variable, such that, the measure μ_n on Ω obtained as the distribution of $\tau_x\omega_n$ where x is random and distributed uniformly on the cube of size l_n, converges weakly to μ as $n \to \infty$.*

Remark. This is a well-known fact from the theory of stationary process. See for instance reference [5] for a proof.

Proof of Theorem 2.1. Let us start with the coefficients $\omega_n \in \Omega$ given by Lemma 3.3 which is periodic of period l_n. By the map $x \to x/l_n$ we can shrink the torus of size l_n to one of unit size. If ω_n represents the coefficients $a^{(n)}(x)$ then we consider the coefficients $a^{(n)}(l_n x)$ on the unit torus. The diffusion generated by this on the torus will have an invariant density $\phi_n(x)$ representing an invariant probability distribution for this diffusion. Although the coefficients are rougher the upper and lower bounds C_1, C_2 stay the same, i.e.

$$C_1 \sum \xi_j^2 \leq a_{ij}^{(n)}(l_n x)\xi_i \xi_j \leq C_2 \sum \xi_j^2 .$$

From Lemma 3.1 we know that the resolvent R maps L_d into L_∞ boundedly with a bound K independent of n. The adjoint R^* therefore maps L_1 boundedly into $L_{d/(d-1)}$ with the same bound K. Since $R^*\phi_n = \phi_n$ and ϕ_n has L_1 norm one we obtain,

$$\|\phi_n\|_{d/(d-1)} \leq K \quad \text{for all } n .$$

If we now go back to the torus of size l_n and denote by $\tilde{\phi}_n(x)$, the function $\phi_n(x/l_n)$ then,

$$\|\tilde{\phi}_n\|_{d/(d-1)} \leq K ,$$

provided we compute the norms with respect to the normalized Lebesgue measure on the torus of size l_n. We denote the probability measure represented by $\tilde{\phi}_n$ by $\tilde{\lambda}_n$ and map it into Ω by the map $x \to \tau_x \omega_n$. The measure induced on Ω is denoted by λ_n. It is easily verified that λ_n is an invariant measure for the canonical Markov process, $\lambda_n \ll \mu_n$ and that

$$\int_\Omega \left| \frac{\mathrm{d}\lambda_n}{\mathrm{d}\mu_n} \right|^{d/(d-1)} \mathrm{d}\mu_n \leqslant K^{d/(d-1)}.$$

Since μ_n is converging weakly to μ and K is independent of n, it is routine to check that λ_n has a weakly convergent subsequence, and the limit λ so obtained is again an invariant measure for our canonical Markov process. In addition, $\lambda \ll \mu$ and:

$$\int_\Omega \left| \frac{\mathrm{d}\lambda}{\mathrm{d}\mu} \right|^{d/(d-1)} \mathrm{d}\mu \leqslant K^{d/(d-1)}.$$

We now assume that μ is ergodic and show that $\mu \ll \lambda$ and λ is ergodic for the canonical process. If μ is not absolutely continuous with respect to λ then since $\lambda \ll \mu$ we can use the Lebesgue decomposition to find a set $E \subset \Omega$ with the properties $\lambda(E)=0$, $0<\mu(E)<1$ and $\mu \ll \lambda$ on E^c. From $\lambda(E^c)=1$ we conclude,

$$q(t, \omega, E^c)=1 \quad \text{a.e. } \omega \quad \text{w.r.t. } \lambda.$$

Since $\mu \ll \lambda$ on E^c this implies

$$q(t, \omega, E^c)=1 \quad \text{a.e. } \omega \text{ on } E^c \quad \text{w.r.t. } \mu.$$

From the definition of q and Lemma 3.2, it now follows that

$$\tau_x \omega \in E^c \quad \text{a.e. } x \quad \text{a.e. } \omega \text{ on } E^c \quad \text{w.r.t. } \mu.$$

This in turn implies by Fubini's theorem that

$$\tau_x E^c = E^c \quad \text{for almost all } x,$$

as elements in the measure algebra of μ. Since $x \to \tau_x$ is continuous we conclude that E^c is an invariant set. By ergodicity of μ we must have $\mu(E^c)=0$ or 1 and we have a contradiction.

Let us now prove that as an invariant measure for our canonical process λ is ergodic; if not, we have a set $E \subset \Omega$ with $0<\lambda(E)<1$ such that

$$q(t, \omega, E)=1 \quad \text{a.e. } \omega \text{ in } E \quad \text{w.r.t. } \lambda.$$

Since $\mu \ll \lambda$, we conclude

$$q(t, \omega, E)=1 \quad \text{a.e. } \omega \text{ in } E \quad \text{w.r.t. } \mu.$$

Just as in the preceding argument we conclude now by Lemma 3.2 that E is invariant and by ergodicity of μ that $\mu(E)=0$ or 1. Since $\lambda \ll \mu$ we must also have $\lambda(E)=0$ or 1.

This completes the proof of Theorem 2.1. Theorem 2.2 of course follows immediately. Note that the ergodicity of any arbitrary λ which is absolutely continuous with respect to μ implies the uniqueness of λ.

4. Remarks

We remark that without any work we can obtain the weak convergence of the distribution of the process $(x(kt))/\sqrt{k}$ as $k \to \infty$ under Q_ω to a suitable Brownian motion.

If $d\lambda/d\mu = \psi(\omega)$ then $\psi(\tau_x \omega)$ defines a stochastic process $\psi(x, \omega)$ with the following properties: For a.e. ω $\psi(x, \omega)$ is a σ-finite non-negative invariant density for L_ω on R^d. It is a stationary stochastic process in x and is normalized to have $E\psi(x, \omega) = 1$.

References

[1] Freidlin, M.I. (1964). Dirichlet's problem for an equation with periodic coefficients depending on a small parameter. *Th. Prob. Appl.* **9**, 121–125.

[2] Bensoussan, A., Lions, J.L. and Papanicolaou, G. (1978). *Asymptotic Methods for Periodic Structures*. North-Holland, Amsterdam.

[3] Kozlov, C.M. (1978). *Dokl. Akad. Nauk.* **5**, 236–239.

[4] Papanicolaou, G. and Varadhan, S. Boundary value problems with rapidly oscillating random coefficients. *Seria Coll. Math. Soc. János Bolyai*, North-Holland, Amsterdam, to appear.

[5] Parthasarathy, K.R. (1961). On the category of ergodic measures. *Ill. J. Math.* **5**, 648–655.

[6] Pucci, C. (1966). Limitazioni per soluzioni di equazioni ellitiche. *Ann. Matematica Pura Applic.*, Series IV, 15–30.

[7] Stroock, D. and Varadhan, S.R.S. (1978). *Multidimensional Diffusion Processes*. Springer, New York.

G. Kallianpur, P.R. Krishnaiah, J.K. Ghosh, eds., *Statistics and Probability: Essays in Honor of C.R. Rao*

A RANDOM TROTTER–KATO PRODUCT FORMULA

K.R. PARTHASARATHY and Kalyan B. SINHA
Indian Statistical Institute, Delhi Centre, India

A randomised version of the Trotter–Kato product formula for semigroups of operators is obtained. This is used to construct the semigroup corresponding to the Schrödinger operator of classical quantum mechanics in the presence of a non-Abelian gauge. An abstract Feynman–Kac formula is also established.

1. Introduction

Suppose $\{A_j, j=1,2,\ldots,k\}$ are bounded operators in a Banach space, then, the Lie product formula tells us that,

$$\lim_{n\to\infty}\left(\prod_{j=1}^{k} e^{(t/n)A_j}\right)^n = e^{t(A_1+\cdots+A_k)},$$

for all $t\in\mathbb{R}$ in the uniform topology. Various generalisations of this result to the case of semigroups with unbounded generators exist and they are generally known as Trotter–Kato product formulae. One such typical result is the following: let A_j, $j=1,2,\ldots,k$ be selfadjoint operators with domains $D_j, j=1,2,\ldots,k$ respectively in a Hilbert space, $\mathcal{H}$. Suppose $A_1+A_2+\cdots+A_k$ with domain $\bigcap_{j=1}^{k} D_j$ is selfadjoint. Then,

$$\operatorname{s.lim}_{n\to\infty}\left(\prod_{j=1}^{k} e^{i(t/n)A_j}\right)^n = e^{it(A_1+\cdots+A_k)},$$

for all $t\in\mathbb{R}$, where s.lim denotes strong operator limit. (For a proof of this result see pp. 295–297, in ref. [4]; further references to such formulae may be found in the same book.) In the present article we make an attempt to construct under these conditions the contraction semigroup $\exp -\frac{1}{2}t(A_1^2+A_2^2+\cdots+A_k^2)$ from the one parameter unitary groups $\{\exp itA_j, t\in\mathbb{R}, j=1,2,\ldots,k\}$ by a suitable randomisation procedure with the help of k standard Brownian motions. As an illustration the semigroup corresponding to the Laplace operator in the presence of non-Abelian gauges is constructed. An abstract Feynman–Kac formula is obtained for the semigroup $\exp t[-\frac{1}{2}(A_1^2+\cdots+A_k^2)+B]$ when $A_1, A_2,\ldots, A_k$ satisfy the conditions mentioned earlier and B is a bounded operator. The classical Feynman–Kac formula corresponds to the situation when $A_j=(1/i)(\partial/\partial x_j), j=1,2,\ldots,k$ and B is multiplication by a bounded function V in the Hilbert space $L_2(\mathbb{R}^k)$.

2. Preliminaries

Let $\mathcal{H}$ be a complex separable Hilbert space. For any operator A on $\mathcal{H}$ we shall denote its domain by $D(A)$. Suppose $A_1, A_2, \dots, A_k$ are fixed selfadjoint operators such that $A_1^2 + A_2^2 + \cdots + A_k^2$ is selfadjoint on the domain $\bigcap_{j=1}^{k} D(A_j^2)$.

Let $w = (w_1, w_2, \dots, w_k)$ be k-independent, standard Brownian motions in $[0, \infty)$. We denote by Ω the space of trajectories w of the k-dimensional Brownian motion and $\mathcal{B}$ its complete σ-algebra. We write $\mathcal{B}(s, t)$ to denote the smallest complete sub σ-algebra with respect to which the collection of random variables $\{w(b) - w(a), s \leqslant a \leqslant b \leqslant t\}$ is measurable. We have to deal with operator valued random variables. We denote by $\mathcal{B}(\mathcal{H})$ the space of all bounded operators on $\mathcal{H}$ and endow it with the borel structure arising from the strong operator topology. Since $\mathcal{H}$ is separable this borel structure is also the one arising from the weak operator topology. By an *operator valued random variable* we mean a borel map from $(\Omega, \mathcal{B})$ into $\mathcal{B}(\mathcal{H})$.

We will use the theory of Bochner integrals of Banach space valued functions on a measure space rather extensively. In this context the reader may refer to reference [5]. Stochastic integrals of vector valued non-anticipating random functions will also be used. However, these are obvious modifications of the results in the theory of ordinary stochastic integrals for which we refer to reference [1, 3].

In order to formulate the notion of a "random Trotter–Kato product" we will frequently use the following lemma.

Lemma 2.1. *For any $t > 0$ let $[t]$ denote its integral part. The sequences $\{[2^n t]2^{-n}\}$ and $\{([2^n t] + 1)2^{-n}\}$ monotonically increase and decrease, respectively to the same limit t for every $t > 0$.*

Proof. We leave it to the reader.

For any $0 < s < t < \infty$ and any sufficiently large n, we have the following diadic picture in view of the above lemma.

$0 \quad [2^n s]2^{-n} \quad s \quad ([2^n s]+1)2^{-n} \qquad [2^n t]2^{-n} \quad t \quad ([2^n t]+1)2^{-n}$

We define the unitary operators

$$U_n(s, t, w) = \prod_{r=1}^{k} e^{i(w_r(([2^n s]+1)2^{-n}) - w_r(s))A_r}$$

$$\times \prod_{j=[2^n s]+1}^{[2^n t]-1} \prod_{r=1}^{k} e^{i(w_r(\overline{j+1}\,2^{-n}) - w_r(j2^{-n}))A_r}$$

$$\times \prod_{r=1}^{k} e^{i(w_r(t) - w_r([2^n t]2^{-n}))A_r}, \tag{2.1}$$

for $0 \leqslant s < t < \infty$ and every Brownian trajectory $w \in \Omega$. If $s = t$, we define $U_n(s, s, w) = I$, the identity operator in $\mathcal{H}$. Then $U_n(s, t, w)$ is a unitary operator valued random variable when n, s, t are fixed. We call (2.1) a *random Trotter–Kato product*.

Lemma 2.2. *For all* $0 \leqslant s < t < \infty$ *and sufficiently large* $n, U_n(s,t,w)$ *is a unitary operator valued random variable satisfying the following properties*:

(i) *If* $s_j < t_j$, $s_j \to s$, $t_j \to t$ *as* $j \to \infty$ *then* $U_n(s_j, t_j w) \to U_n(s,t,w)$ *strongly as* $j \to \infty$ *for each Brownian path* w.

(ii) $U_n(s,t,w)$ *is* $\mathfrak{B}(s,t)$*-measurable.*

Proof. This is immediate from the definition of $U_n(s,t,w)$, the continuity of the trajectories w and the strong continuity of the one parameter unitary groups $\{e^{itA_j}\}$.

Let $\mathfrak{h} = L_2([0,\infty), \mathbb{R}^k)$ be the real Hilbert space of all square integrable maps from $[0,\infty)$ into $\mathbb{R}^k$. An element $\boldsymbol{\phi} = (\phi_1, \phi_2, \ldots, \phi_k) \in \mathfrak{h}$ is said to be *simple of order m* if there exist constants $\{a_{jl}, l = 0,1,2,\ldots; j = 1,2,\ldots,k\}$ such that,

$$\phi_j(t) = a_{jl}, \quad \text{if } l2^{-m} \leqslant t < \overline{l+1}\, 2^{-m}. \tag{2.2}$$

We denote by $\mathbb{S}$ the linear manifold of all such simple elements. If $\boldsymbol{\phi} \in \mathbb{S}$ is simple of order m, then $\boldsymbol{\phi}$ is simple of order m' for every $m' \geqslant m$. We define for $0 \leqslant s < t < \infty$ and $\boldsymbol{\phi} \in \mathfrak{h}$,

$$\Gamma_n(s,t,\boldsymbol{\phi}) = EU_n(s,t,w)\exp\left\{ i\int_s^t \boldsymbol{\phi}\cdot dw \right\} \tag{2.3}$$

in the strong sense, where $\int_s^t \boldsymbol{\phi}\cdot dw$ is the Wiener integral $\int_s^t \Sigma_j \phi_j dw_j$. Then $\Gamma_n(s,t,\boldsymbol{\phi})$ is a contraction operator for fixed $n, s, t, \boldsymbol{\phi}$.

In view of Lemma 2.1, we have the following diadic picture for $n > m$ provided n is sufficiently large and m is large enough depending on s, t:

$$\begin{aligned}[2^m s]2^{-m} &< [2^n s]2^{-n} \leqslant s < ([2^n s]+1)2^{-n} < ([2^m s]+1)2^{-m} \\ &< [2^m t]2^{-m} < [2^n t]2^{-n} \leqslant t < ([2^n t]+1)2^{-n} < ([2^m t]+1)2^{-m}.\end{aligned}$$

Lemma 2.3. *Suppose* $\boldsymbol{\phi}$ *is simple of order m and defined by* (2.2). *Then for all sufficiently large n*,

$$\begin{aligned}\Gamma_n(s,t,\boldsymbol{\phi}) = &\prod_{r=1}^{k} \exp\{-\tfrac{1}{2}(([2^n s]+1)2^{-n} - s)(A_r + a_{r,[2^m s]})^2\} \\ &\times \left\{ \prod_{r=1}^{k} \exp\{-\tfrac{1}{2}2^{-n}(A_r + a_{r,[2^m s]})^2\} \right\}^{(([2^m s]+1)2^{n-m} - ([2^n s]+1))} \\ &\times \prod_{l=[2^m s]+1}^{[2^m t]-1} \left\{ \prod_{r=1}^{k} \exp\{-\tfrac{1}{2}2^{-n}(A_r + a_{r,l})^2\} \right\}^{2^{n-m}} \\ &\times \left\{ \prod_{r=1}^{k} \exp\{-\tfrac{1}{2}2^{-n}(A_r + a_{r,[2^m t]})^2\} \right\}^{([2^n t] - [2^m t]2^{n-m})} \\ &\times \prod_{r=1}^{k} \exp\{-\tfrac{1}{2}(A_r + a_{r,[2^m t]})^2\}(t - [2^n t]2^{-n}).\end{aligned}$$

Proof. This follows immediately from the definitions of U_n, Γ_n, property (ii) of Lemma 2.2, standard properties of Brownian motion and stochastic integrals and the identity,

$$(2\pi)^{-1/2}\int e^{i\xi B}e^{-\xi^2/2}d\xi = e^{-B^2/2},$$

for any selfadjoint operator B. The operator integral is in the strong sense.

Lemma 2.4. *Let $C_1, C_2, \ldots, C_k$ be positive selfadjoint operators on $\mathcal{H}$ such that $C_1+C_2+\cdots+C_k$ is selfadjoint on the domain $D=\bigcap_{j=1}^k D(C_j)$. Let $m(n)$ be a sequence of positive integers such that $[m(n)]/n \leqslant p$ and $\lim_{n\to\infty}[m(n)]/n = p$. Then*

$$\operatorname{s.lim}_{n\to\infty}\left(\prod_{j=1}^k e^{-(t/n)C_j}\right)^{m(n)} = e^{-pt(C_1+C_2+\cdots+C_k)}$$

for all $t \geqslant 0$, where s.lim *denotes a strong operator limit.*

Proof. This follows from a slight modification of the arguments in pages 295–297 of ref. [4].

3. The random Trotter–Kato product formula

Under the notations of Section 2 we have the following lemma:

Lemma 3.1. *For any simple $\phi \in \mathcal{S}$, and $0 \leqslant s < t < \infty$,*

$$\Gamma(s,t,\phi) = \operatorname{s.lim}_{n\to\infty}\Gamma_n(s,t,\phi),$$

exists. If ϕ is defined by (2.2) *then,*

$$\begin{aligned}\Gamma(s,t,\phi) = {} & \exp\left\{-\tfrac{1}{2}(([2^m s]+1)2^{-m}-s)\sum_{r=1}^k (A_r + a_{r,[2^m s]})^2\right\}\\ & \times \prod_{l=[2^m s]+1}^{[2^m t]-1} \exp\left\{-\tfrac{1}{2}2^{-m}\sum_{r=1}^k (A_r+a_{rl})^2\right\}\\ & \times \exp\{-\tfrac{1}{2}(t-[2^m t]2^{-m})\}\sum_{r=1}^k (A_r + a_{r,[2^m t]})^2. \end{aligned} \tag{3.1}$$

If $D=\bigcap_{r=1}^k D(A_r^2)$, then for all $v \in D, \phi \in \mathcal{S}$,

$$\frac{\partial}{\partial t}\Gamma(s,t,\phi)v = \Gamma(s,t,\phi)\left\{-\tfrac{1}{2}\sum_{r=1}^k (A_r+\phi_r(t))^2\right\}v, \tag{3.2}$$

in the strong sense. If $s>0$ then,

$$\frac{\partial}{\partial s}\Gamma(s,t,\phi)v=\left\{\tfrac{1}{2}\sum_{r=1}^{k}(A_r+\phi_r(s))^2\right\}\Gamma(s,t,\phi)v, \tag{3.3}$$

for all $v\in D$, $t>s$ and $\phi\in\mathbb{S}$.

Proof. We observe that $(A_r+a_{rl})^2$ is a positive selfadjoint operator with domain $D(A_r^2)$. Under the conditions of Section 2, $\Sigma_{r=1}^{k}(A_r+a_{rl})^2$ is selfadjoint with domain D in view of Kato–Rellich theorem. When m is fixed we have from Lemma 2.1:

$$\lim_{n\to\infty}\frac{[2^nt]-[2^mt]2^{n-m}}{2^n}=t-[2^mt]2^{-m}\quad\text{for all } t>0,$$

$$\lim_{n\to\infty}\frac{([2^ms]+1)2^{n-m}-([2^ns]+1)}{2^n}=([2^ms]+1)2^{-m}-s\quad\text{for all } s>0.$$

Now we apply Lemma 2.4 to the products appearing on the right-hand side of the expression for $\Gamma_n(s,t,\phi)$ in Lemma 2.3. We have

$$\operatorname*{s.lim}_{n\to\infty}\prod_{r=1}^{k}\exp\left\{-\tfrac{1}{2}(([2^ns]+1)2^{-n}-s)(A_r+a_{r,[2^ms]})^2\right\}=I;$$

$$\operatorname*{s.lim}_{n\to\infty}\left\{\prod_{r=1}^{k}\exp\left\{-\tfrac{1}{2}2^{-n}(A_r+a_{r,[2^ms]})^2\right\}\right\}^{(([2^ms]+1)2^{n-m}-([2^ns]+1))}$$

$$=\exp\{-\tfrac{1}{2}(([2^ms]+1)2^{-m}-s)\sum_{r=1}^{k}(A_r+a_{r,[2^ms]})^2\};$$

$$\operatorname*{s.lim}_{n\to\infty}\left\{\prod_{r=1}^{k}\exp\left\{-\tfrac{1}{2}2^{-n}(A_r+a_{rl})^2\right\}\right\}^{2^{n-m}}=\exp\left\{-\tfrac{1}{2}2^{-m}\sum_{r=1}^{k}(A_r+a_{rl})^2\right\};$$

$$\operatorname*{s.lim}_{n\to\infty}\{\prod_{r=1}^{k}\exp\{-\tfrac{1}{2}2^{-n}(A_r+a_{r,[2^mt]})^2\}\}^{[2^nt]-[2^mt]2^{n-m}}$$

$$=\exp\{-\tfrac{1}{2}(t-[2^mt]2^{-m})\sum_{r=1}^{k}(A_r+a_{r,[2^mt]})^2\};$$

$$\operatorname*{s.lim}_{n\to\infty}\prod_{r=1}^{k}\exp\left\{-\tfrac{1}{2}(t-[2^nt]2^{-n})(A_r+a_{r,[2^mt]})^2\right\}=I.$$

Now by taking products we conclude from Lemma 2.4 that $\operatorname{s.lim}_{n\to\infty}\Gamma_n(s,t,\phi)$ exists and that (3.1) holds. Equations (3.2) and (3.3) follow by routine differentiation and the theory of contraction semigroups. This completes the proof.

Lemma 3.2. *Let $\Gamma(s,t,\phi)$ be as in Lemma 3.1. Then for every $\phi\in\mathbb{S}$*

$$\Gamma(s,t,\phi)\Gamma(t,u,\phi)=\Gamma(s,u,\phi)\quad \textit{for all } 0\leqslant s<t<u.$$

Proof. This is immediate from the identity (3.1).

Now we consider the complex Hilbert space $L_2(\Omega, \mathcal{H})$ of all borel maps from Ω into $\mathcal{H}$ which are square integrable with respect to the Wiener measure of the stochastic process $\boldsymbol{w}$. Define the operator $V_n(s,t)$ on $L_2(\Omega, \mathcal{H})$ by,

$$(V_n(s,t)\boldsymbol{f})(\boldsymbol{w}) = U_n(s,t,\boldsymbol{w})\boldsymbol{f}(\boldsymbol{w}), \quad \boldsymbol{w} \in \Omega, \tag{3.4}$$

where U_n is defined by (2.1).

Lemma 3.3. *Let $V_n(s,t)$ be defined by* (3.4). *Then $V_n(s,t)$ is a unitary operator and*

$$V(s,t) = \operatorname{w.lim}_{n\to\infty} V_n(s,t) \tag{3.5}$$

exists for all $0 \leqslant s \leqslant t$, where w.lim *denotes weak operator limit. Further $V(s,t)$ is a contraction operator in $L_2(\Omega, \mathcal{H})$.*

Proof. Let $\boldsymbol{f}, \boldsymbol{g} \in L_2(\Omega, \mathcal{H})$ be of the form:

$$\boldsymbol{f}(\boldsymbol{w}) = \left(\exp \mathrm{i}\int_0^\infty \boldsymbol{\phi}\cdot \mathrm{d}\boldsymbol{w}\right)u$$

$$\boldsymbol{g}(\boldsymbol{w}) = \left(\exp \mathrm{i}\int_0^\infty \boldsymbol{\psi}\cdot \mathrm{d}\boldsymbol{w}\right)v \tag{3.6}$$

where $\boldsymbol{\phi}, \boldsymbol{\psi} \in \mathcal{S}$ and $u, v \in \mathcal{H}$. For all sufficiently large n we obtain from the standard properties of Wiener integrals and (2.3),

$$\langle V_n(s,t)\boldsymbol{f}, \boldsymbol{g}\rangle = \langle \Gamma_n(s,t,\boldsymbol{\phi}-\boldsymbol{\psi})u, v\rangle$$
$$\times \exp -\tfrac{1}{2}\left(\int_0^s + \int_t^\infty |\boldsymbol{\phi}(\tau)-\boldsymbol{\psi}(\tau)|^2 \mathrm{d}\tau\right).$$

By Lemma 3.1 we now get

$$\lim_{n\to\infty} \langle V_n(s,t)\boldsymbol{f}, \boldsymbol{g}\rangle = \langle \Gamma(s,t,\boldsymbol{\phi}-\boldsymbol{\psi})u, v\rangle$$
$$\times \exp -\tfrac{1}{2}\left(\int_0^s + \int_t^\infty |\boldsymbol{\phi}(\tau)-\boldsymbol{\psi}(\tau)|^2 \mathrm{d}\tau\right).$$

Since $V_n(s,t)$ is unitary and vectors of the form $\boldsymbol{f}, \boldsymbol{g}$ in (3.6) span $L_2(\Omega, \mathcal{H})$ it follows that $V_n(s,t)$ converges weakly to a contraction operator $V(s,t)$ as $n \to \infty$. This completes the proof.

Lemma 3.4. *Let the operators $V(s,t)$, $0 \leqslant s < t < \infty$ be defined by* (3.5). *Then there exists a contraction operator valued random variable $U(s,t,\boldsymbol{w})$ in $\mathcal{H}$ such that the map $(s,t,\boldsymbol{w}) \to U(s,t,\boldsymbol{w})$ is jointly measurable and,*

$$(V(s,t)\boldsymbol{f})(\boldsymbol{w}) = U(s,t,\boldsymbol{w})\boldsymbol{f}(\boldsymbol{w}), \tag{3.7}$$

for all $\boldsymbol{f} \in L_2(\Omega, \mathcal{H})$, $0 \leqslant s < t < \infty$. For fixed s, t the map $\boldsymbol{w} \to U(s,t,\boldsymbol{w})$ is $\mathcal{B}(s,t)$-measurable and,

$$E e^{\mathrm{i}\int_s^t \boldsymbol{\phi}\cdot \mathrm{d}\boldsymbol{w}} U(s,t,\boldsymbol{w}) = \Gamma(s,t,\boldsymbol{\phi}), \tag{3.8}$$

for every $\phi \in \mathcal{S}$, *where* $\Gamma(s,t,\phi)$ *is as in Lemma* 3.1 *and the expectation is in the strong sense.*

Proof. Let $P(E)$, $E \in \mathcal{B}$ denote the canonical spectral measure of multiplication by χ_E in the Hilbert space $L_2(\Omega, \mathcal{H})$. Then the operators $V_n(s,t)$ defined by (3.4) commute with $P(E)$ for all E. Hence their weak limit $V(s,t)$ also commutes with $P(E)$ for all E. Further $V(s,t)$ is a contraction and the map $(s,t) \to V(s,t)$ is measurable. Hence there exists a contraction operator valued random variable $U(s,t,\mathbf{w})$ which is jointly measurable in $s, t, \mathbf{w}$ and satisfies (3.7). Since "multiplication" by $U(s,t,\mathbf{w})$ in $L_2(\Omega, \mathcal{H})$ is the weak limit of "multiplication" by $U_n(s,t,\mathbf{w})$ property (ii) in Lemma 2.2 implies that $U(s,t,\mathbf{w})$ is $\mathcal{B}(s,t)$-measurable for fixed s, t.

Now let $\mathbf{f}, \mathbf{g}$ be as in (3.6). From the proof of Lemma 3.3 we get,

$$\begin{aligned} \langle V(s,t)\mathbf{f}, \mathbf{g}\rangle &= E\langle U(s,t,\mathbf{w}) e^{i\int_s^t (\phi-\psi)\cdot d\mathbf{w}} u, v\rangle \\ &\quad \times \exp -\tfrac{1}{2}\left(\int_0^s + \int_t^\infty |\phi-\psi|^2 \, d\tau\right) \\ &= \langle \Gamma(s,t,\phi-\psi)u, v\rangle \exp -\tfrac{1}{2}\left(\int_0^s + \int_t^\infty |\phi-\psi|^2 \, d\tau\right). \end{aligned}$$

This, in particular, implies (3.8) and completes the proof.

Lemma 3.5. *Let* $U(s,t,\mathbf{w})$ *be as in Lemma* 3.4. *Then,*

$$U(s,t,\mathbf{w})U(t,u,\mathbf{w}) = U(s,u,\mathbf{w}), \quad a.e.\ \mathbf{w},$$

for each $0 \leqslant s < t < u < \infty$.

Proof. Since $U(s,t,\mathbf{w})$ and $U(t,u,\mathbf{w})$ are respectively $\mathcal{B}(s,t)$ and $\mathcal{B}(t,u)$-measurable they are independent random variables. Hence for all simple ϕ, we have from Lemma 3.2,

$$\begin{aligned} &EU(s,t,\mathbf{w})U(t,u,\mathbf{w}) e^{i\int_s^u \phi\cdot d\mathbf{w}} = \\ &\quad = EU(s,t,\mathbf{w}) e^{i\int_s^t \phi\cdot d\mathbf{w}} \times EU(t,u,\mathbf{w}) e^{i\int_t^u \phi\cdot d\mathbf{w}} \\ &\quad = \Gamma(s,t,\phi)\Gamma(t,u,\phi) = \Gamma(s,u,\phi) = EU(s,u,\mathbf{w}) e^{i\int_s^u \phi\cdot d\mathbf{w}}. \end{aligned}$$

Since $\exp i\int_s^u \phi\cdot d\mathbf{w}$ span $L_2(\Omega)$ as ϕ runs over $\mathcal{S}$, we get the required result.

Lemma 3.6. *Let* $D = \bigcap_{j=1}^k D(A_j^2)$. *For* $v \in D$, *let*

$$\tilde{v}(s,t,\mathbf{w}) = v + \sum_j \int_s^t iU(s,\tau,\mathbf{w}) A_j v \, dw_j(\tau) - \tfrac{1}{2}\sum_j \int_s^t U(s,\tau,\mathbf{w}) A_j^2 v \, d\tau \qquad (3.9)$$

in the strong sense. Then,

$$\tilde{v}(s,t,\mathbf{w}) = U(s,t,\mathbf{w})v, \quad a.e.\ \mathbf{w},$$

for each $0 \leqslant s < t < \infty$.

Proof. Let

$$\xi(s,t,w)=\exp\left(\mathrm{i}\int_s^t \phi\cdot \mathrm{d}w+\tfrac{1}{2}\int_s^t |\phi(\tau)|^2\,\mathrm{d}\tau\right).$$

Then $\xi(s,t,w)$ is a martingale for $t\geqslant s$ and,

$$\mathrm{d}\xi=\mathrm{i}\xi\phi\cdot\mathrm{d}w. \tag{3.10}$$

We have

$$E\,\xi(s,t,w)v=v. \tag{3.11}$$

Further (3.8), (3.10) and routine properties of stochastic integrals imply

$$\begin{aligned} E\xi(s,t,w)\int_s^t U(s,\tau,w)A_j v\,\mathrm{d}w_j(\tau)= \\ =\int_s^t E(\mathrm{i}\xi(s,\tau,w)U(s,\tau,w)A_j v)\phi_j(\tau)\,\mathrm{d}\tau \\ =\mathrm{i}\int_s^t (\Gamma(s,\tau;\phi)A_j v)\phi_j(\tau)\mathrm{e}^{(1/2)\int_s^\tau |\phi(\tau')|^2\,\mathrm{d}\tau'}\,\mathrm{d}\tau. \end{aligned} \tag{3.12}$$

The martingale properties of $\xi(s,t,w)$ and (3.8) imply

$$\begin{aligned} E\,\xi(s,t,w)\int_s^t U(s,\tau,w)A_j^2 v\,\mathrm{d}\tau= \\ =\int_s^t E\,\xi(s,t,w)U(s,\tau,w)A_j^2\,\mathrm{d}\tau \\ =\int_s^t E\,\xi(s,\tau,w)U(s,\tau,w)A_j^2 v\,\mathrm{d}\tau \\ =\int_s^t \mathrm{e}^{(1/2)\int_s^\tau |\phi(\tau')|^2\,\mathrm{d}\tau'}\Gamma(s,\tau,\phi)A_j^2 v\,\mathrm{d}\tau. \end{aligned} \tag{3.13}$$

Now from (3.9) and (3.11)–(3.13) we have,

$$\begin{aligned} E\xi(s,t,w)\tilde{v}(s,t,w)= \\ =v+\int_s^t \mathrm{e}^{(1/2)\int_s^\tau |\phi(\tau')|^2\,\mathrm{d}\tau'}\Gamma(s,\tau,\phi)\{-\tfrac{1}{2}\textstyle\sum A_j^2-\sum\phi_j(\tau)A_j\}v\,\mathrm{d}\tau. \end{aligned}$$

If we put $u(s,t)$ to denote the above expression, we obtain from (3.2)

$$\begin{aligned} u(s,s)&=v, \\ \frac{\partial u(s,t)}{\partial t}&=\mathrm{e}^{(1/2)\int_s^t |\phi(\tau)|^2\,\mathrm{d}\tau}\Gamma(s,t,\phi)\{-\tfrac{1}{2}\textstyle\sum A_j^2-\sum\phi_j(t)A_j\}v \\ &=\frac{\partial}{\partial t}\left\{\mathrm{e}^{(1/2)\int_s^t |\phi(\tau)|^2\,\mathrm{d}\tau}\Gamma(s,t,\phi)v\right\}, \end{aligned}$$

for all $t>s$. Hence,

$$u(s,t)=\mathrm{e}^{1/2\int_s^t |\phi(\tau)|^2\,\mathrm{d}\tau}\Gamma(s,t,\phi)v.$$

In other words,

$$Ee^{i\int_s^t \phi \cdot dw}\tilde{v}(s,t,w) = Ee^{i\int_s^t \phi \cdot dw}U(s,t,w)v.$$

Since this holds for all simple ϕ we get,

$$\tilde{v}(s,t,w) = U(s,t,w)v, \quad \text{a.e. } w.$$

This completes the proof of the lemma.

Theorem 3.7. *Let $A_1, A_2, \dots, A_k$ be k selfadjoint operators in a complete separable Hilbert space $\mathcal{H}$ with domains $D(A_1), D(A_2), \dots, D(A_k)$, respectively and let $A_1^2 + A_2^2 + \cdots + A_k^2$ be selfadjoint on the domain $D = \bigcap_{j=1}^k D(A_j^2)$. Let $\{w(t) = (w_1(t), w_2(t), \dots, w_k(t)),\ t \geq 0\}$ be the k-dimensional standard Brownian motion with Ω as the space of trajectories. Then there exists a family $\{\tilde{U}(s,t,w), 0 \leq s < t < \infty, w \in \Omega\}$ of contraction operators on $\mathcal{H}$ satisfying the following properties*:

(i) *for each $w \in \Omega$, $\tilde{U}(s,t,w)$ is strongly continuous in s and t*;

(ii) *$\tilde{U}(s,t,w)\tilde{U}(t,u,w) = \tilde{U}(s,u,w)$ for all $s < t < u$, $w \in \Omega$*;

(iii) *for fixed $s < t$, $\tilde{U}(s,t,w)$ is an operator valued random variable which is measurable with respect to $\mathcal{B}(s,t)$ which is the smallest complete σ-algebra making the family $\{w(b) - w(a), s \leq a < b \leq t\}$ measurable on Ω. In particular, for $s_1 < t_1 \leq s_2 < t_2 \leq \cdots \leq s_n < t_n$ the operator valued random variables $\{\tilde{U}(s_j, t_j, w), j = 1, 2, \dots, k\}$ are independent*;

(iv) *$E\tilde{U}(s,t,w) = \exp[-\frac{1}{2}(t-s)\Sigma_{j=1}^k A_j^2]$ in the strong sense*;

(v) *for each $v \in D$, $\tilde{U}(s,s,w)v = v$ and $d\tilde{U}(s,t,w)v = i\tilde{U}(s,t,w)\Sigma_{j=1}^k A_j v\, dw_j - \frac{1}{2}\tilde{U}(s,t,w)\Sigma_{j=1}^k A_j^2 v\, dt$ in the sense of stochastic integrals of Hilbert space valued non-anticipating random functions*;

(vi) *if the random Trotter–Kato product $U_n(s,t,w)$ is defined by (2.1) then in the Hilbert space $L_2(\Omega, \mathcal{H})$ multiplication by $U_n(s,t,w)$ converges weakly to multiplication by $\tilde{U}(s,t,w)$ as $n \to \infty$.*

Proof. Select a complete orthonormal basis $\{v_1, v_2, \dots\}$ for $\mathcal{H}$ such that $v_j \in D$ for all j. Define,

$$\tilde{U}(s,t,w)v_j = \tilde{v}_j(s,t,w) \tag{3.14}$$

where $\tilde{v}_j(s,t,w)$ is defined by (3.9). From the theory of stochastic integrals it follows that $\tilde{v}_j(s,t,w)$ is continuous in s and t for almost all w. Hence we can select $\Omega_0 \subset \Omega$ so that $\Pr(\Omega_0) = 1$ and $\tilde{v}_j(s,t,w)$ is continuous in s and t for every j and all $w \in \Omega_0$. From Lemma 3.6, Lemma 3.1 and (3.8) we have,

$$E\tilde{U}(s,t,w)v_j = e^{-(1/2)(t-s)\Sigma_{r=1}^k A_r^2} v_j, \tag{3.15}$$

for all $j = 1, 2, \dots$. Further we can select $\Omega_1 \subset \Omega_0$ such that $\Pr(\Omega_1) = 1$ and

$$\left\| \sum_1^n c_j \tilde{U}(s,t,w) v_j \right\| \leq \left\| \sum_1^n c_j v_j \right\| \tag{3.16}$$

for all rational $s<t$, complex numbers $c_1, c_2, \ldots, c_n$ with rational real and imaginary parts, $n=1,2,\ldots$, and $w \in \Omega_1$. Since the left-hand side of (3.16) is continuous in $s, t, c_1, c_2, \ldots, c_n$ the same inequality extends for all $s<t$ and $c_1, c_2, \ldots, c_n$. Now (3.16) implies that we can extend the definition of $\tilde{U}(s,t,w)$ linearly and uniquely to the whole of $\mathcal{H}$ for each $w \in \Omega_1$ and all $s<t$ maintaining the strong continuity in s, t and the contraction property. By Lemmas 3.5 and 3.6 we can select $\Omega_2 \subset \Omega_1$ such that $\Pr(\Omega_2)=1$ and

$$\tilde{U}(s,t,w)\tilde{U}(t,u,w)=\tilde{U}(s,u,w)$$

for all rational $s<t<u$ and $w \in \Omega_2$. By strong continuity the same equation extends to all $s<t<u$. Now define

$$\tilde{U}(s,t,w)=I \quad \text{for all } s<t \text{ and } w \notin \Omega_2.$$

Then properties (i) and (ii) are fulfilled. Further, (3.15) implies property (iv). Property (iii) is immediate from Lemma 3.6. Property (v) is now immediate from the definition of $\tilde{U}$. Property (vi) is a restatement of Lemmas 3.3 and 3.4 put together. This completes the proof.

Remark 3.8. At this stage it is natural to ask whether $\tilde{U}(s,t,w)$ can be chosen to be unitary for all s, t, w. Indeed, if $A_1, A_2, \ldots, A_k$ are bounded selfadjoint operators one can show that $\tilde{U}(s,t,w)$ can be chosen to be unitary. Another case where a unitary choice is possible is when $A_1, A_2, \ldots, A_k$ generate a unitary representation of a Lie group.

When a unitary choice is possible we can strengthen property (vi) in Theorem 3.7 to

$$\lim_{n\to\infty} E\|U_n(s,t,w)v-\tilde{U}(s,t,w)v\|^2=0 \quad \text{for } v \in \mathcal{H}.$$

One can also show that if $\{\tilde{U}(s,t,w)\}$ admit an isometric choice then they admit a unitary choice.

Example 3.9. Let $\mathcal{H}=L_2(\mathbb{R}^k, \mathbb{C}^l)$ be the Hilbert space of all $\mathbb{C}^l$-valued square integrable maps with respect to the Lebesgue measure in $\mathbb{R}^k$. Let $V_j(x), j=1,2,\ldots,k$ be $l \times l$ Hermitian matrix valued C^1 functions so that

$$\sup_{x\in\mathbb{R}^k}\left(\sum_{j,r,s}|v_j^{r,s}(x)|+\sum_{i,j,r,s}\left|\frac{\partial}{\partial x_i}v_j^{r,s}(x)\right|\right)<\infty,$$

where $v_j^{r,s}$ is the r, sth element of V_j. Let A_j be the operator defined by

$$(A_j f)(x)=\frac{1}{\mathrm{i}}\frac{\partial f}{\partial x_j}(x)+V_j(x)f(x), \quad f\in\mathcal{H}, \tag{3.17}$$

whose domain is the standard domain of selfadjointness of the operator $(1/\mathrm{i})(\partial/\partial x_j)$. Then A_j is selfadjoint and $\Sigma_{j=1}^k A_j^2$ is selfadjoint on $D=\bigcap_{j=1}^k D(A_j^2)$. This follows from the standard results of perturbation theory of linear operators in a Hilbert space. (See, for example ref. [2].) We may call $\Sigma_{j=1}^k A_j^2$ as the *Laplacian in the presence of the gauge* $(V_1, V_2, \ldots, V_k)$.

We will now obtain explicitly the solution of the stochastic differential equation (v) in Theorem 3.7. To this end we consider the stochastic differential equation in matrix valued random functions in the interval $[s, \infty)$:

$$\mathrm{d}C=C\left\{ i\sum_{j=1}^{k} V_j(x+w(t)-w(s))\,\mathrm{d}w_j(t) + \tfrac{1}{2}\left[i\sum_{j=1}^{k} \frac{\partial V_j}{\partial x_j}(x+w(t)-w(s)) - \sum_{j=1}^{k} V_j^2(x+w(t)-w(s))\right]\mathrm{d}t\right\}.$$

We denote the unique solution of this equation by $C(s, t, x, w)$ with initial condition $C(s, s, x, w)=I$ and $t\geqslant s$. Then for fixed $s<t$ and $x, C(s, t, x, w)$ is a $\mathcal{B}(s, t)$-measurable random variable. A simple application of Itô's formula (see refs. [1, 3]) shows that $\mathrm{d}(CC^*)=0$ or $CC^*=I$. In other words $C(s, t, x, w)$ is a unitary matrix valued function.

Now define,

$$(\tilde{U}(s, t, w)f)(x)=C(s, t, x, w)f(x+w(t)-w(s)) \tag{3.18}$$

for all $f\in\mathcal{H}$. Then $\tilde{U}(s, t, w)$ is a unitary operator in $\mathcal{H}$ for each $s<t$, $w\in\Omega$. Once again a routine application of Itô's formula followed by elementary computations shows that,

$$\mathrm{d}\tilde{U}(s, t, w)f=\mathrm{i}\tilde{U}(s, t, w)\sum_{j=1}^{k} A_j f\,\mathrm{d}w_j-\tfrac{1}{2}\tilde{U}(s, t, w)\sum_{j=1}^{k} A_j^2 f\,\mathrm{d}t$$

for all $f\in D$. Thus Theorem 3.7 holds with a unitary choice for $\tilde{U}(s, t, w)$ and,

$$E\tilde{U}(s, t, w)f=\mathrm{e}^{\{-(1/2)(t-s)\Sigma_{j=1}^{k}A_j^2\}}f, \quad f\in\mathcal{H}.$$

In particular the random Trotter–Kato product converges in the sense of Remark 3.8.

It may be noted that if V_j are scalar valued then in (3.18),

$$C(s, t, x, w)=\exp \mathrm{i}\left[\sum\int_s^t V_j(x+w(\tau)-w(s))\,\mathrm{d}w_j(\tau) + \tfrac{1}{2}\int_s^t \sum \frac{\partial V_j}{\partial x_j}(x+w(\tau)-w(s))\,\mathrm{d}\tau\right].$$

4. An abstract Feynman–Kac formula

Suppose there exists a unitary solution $\tilde{U}(s, t, w)$ of the stochastic differential equation (v) in Theorem 3.7. Let B be any bounded operator in $\mathcal{H}$. In the Banach space $\mathcal{B}(\mathcal{H})$ the ordinary differential equation,

$$K(s,s,w)=I, \tag{4.1}$$

has a unique solution for $s<t$ and each Brownian trajectory w. The theory of ordinary differential equations implies that the unique solution K of (4.1) satisfies the identity

$$K(s,u,w)=K(s,t,w)\tilde{U}(s,t,w)K(t,u,w)\tilde{U}(s,t,w)^* \quad \text{for all } s<t<u. \tag{4.2}$$

We may call $K(s,t,w)$ as the *Feynman–Kac cocycle* corresponding to the "random evolution" $\tilde{U}(s,t,w)$. Define

$$\tilde{U}_B(s,t,w)=K(s,t,w)\tilde{U}(s,t,w). \tag{4.3}$$

Then (4.2) implies that,

$$\tilde{U}_B(s,t,w)\tilde{U}_B(t,u,w)=\tilde{U}_B(s,u,w) \quad \text{for all } s<t<u.$$

We can call $\tilde{U}_B$ the "perturbed evolution". Routine arguments of the Picard method show that there exist bounded operators $\Gamma_B(s,t)$ on $\mathcal{H}$ such that,

$$\Gamma_B(s,t)v=E\tilde{U}_B(s,t,w)v \quad \text{for all } v\in\mathcal{H}.$$

Since $\tilde{U}_B(s,t,w)$ is $\mathcal{B}(s,t)$-measurable for fixed s,t it follows that,

$$\Gamma_B(s,t)\Gamma_B(t,u)=\Gamma_B(s,u) \quad \text{for all } s<t<u.$$

For $v\in D$, we have from (4.1) and (4.3) and Theorem 3.7,

$$\lim_{h\to 0}\frac{\Gamma_B(s,t+h)v-\Gamma_B(s,t)v}{h}=$$

$$=\Gamma_B(s,t)\left[\lim_{h\to 0}E\left(\frac{K(t,t+h,w)-1}{h}\right)\tilde{U}(t,t+h,w)v+\lim_{h\to 0}E\frac{\tilde{U}(t,t+h,w)v-v}{h}\right]$$

$$=\Gamma_B(s,t)\left[Bv-\tfrac{1}{2}\sum A_j^2 v\right].$$

Thus $\Gamma_B(s,t)=\exp(t-s)(-\frac{1}{2}\Sigma A_j^2+B)v$. In other words,

$$EK(s,t,w)\tilde{U}(s,t,w)=\exp(t-s)\left[-\tfrac{1}{2}\sum A_j^2+B\right], \tag{4.4}$$

in the strong sense where K is determined by (4.1). We may call (4.4) the *abstract Feyman–Kac formula*. When $\mathcal{H}=L_2(\mathbb{R}^k)$, $A_j=(1/\mathrm{i})(\partial/\partial x_j)$ and B is multiplication by a bounded function V, (4.4) reduces to the classical Feynman–Kac formula (see ref. [1]).

Combining this argument with Example 3.9 we have the following theorem.

Theorem 4.1. *Let A_j, $j=1,2,\ldots,k$ be as in Example* 3.9. *Suppose $V(\mathbf{x})$ is a $l\times l$ bounded matrix valued borel function on $\mathbb{R}^k$. Let $L(s,t,\mathbf{x},w)$ be the solution of the* *stochastic differential equation in $l \times l$ matrix multiplication operators in $\mathcal{H}$:*

$$\mathrm{d}L = L\Bigg(\mathrm{i} \sum V_j(x + w(t) - w(s))\,\mathrm{d}w_j + \left\{ \frac{\mathrm{i}}{2} \sum \frac{\partial V_j}{\partial x_j}(x + w(t) - w(s)) - \tfrac{1}{2} \sum V_j^2(x + w(t) - w(s)) + V(x + w(t) - w(s)) \right\} \mathrm{d}t \Bigg);$$

$$L(s, s, x, w) = I, \quad t \geqslant s.$$

Let $W(s, t, w)$ be the operator defined by

$$(W(s, t, w) f)(x) = L(s, t, x, w) f(x + w(t) - w(s))$$

for all $f \in \mathcal{H}$. Then $W(s, t, w)$ is a strongly continuous family of bounded evolution operators for each $w \in \Omega$ and

$$EW(s, t, w) f = \left(\mathrm{e}^{(t-s)(-1/2 \Sigma A_j^2 + V)} f\right)(x).$$

If V_j and V are scalar valued then

$$L(s, t, x, w) = \exp\left\{ \mathrm{i} \sum_{j=1}^{k} \int_s^t V_j(x + w(\tau) - w(s))\,\mathrm{d}w_j(\tau) + \frac{\mathrm{i}}{2} \int_s^t \sum_{j=1}^{k} \frac{\partial V_j}{\partial x_j}(x + w(\tau) - w(s))\,\mathrm{d}\tau + \int_s^t V(x + w(\tau) - w(s))\,\mathrm{d}\tau \right\}.$$

References

[1] Ghiman, I.I. and Skorohod, A.V. (1972). *Stochastic Differential Equations*. Springer, Berlin.
[2] Kato, T. (1976). *Perturbation of Linear Operators*. Springer, Berlin.
[3] McKean, H.P. (1969). *Stochastic Integrals*. Academic Press, New York.
[4] Reed, M. and Simon, B. (1972). *Functional Analysis*, Vol. 1. Academic Press, New York.
[5] Yosida, K. (1974). *Functional Analysis*. Springer, International Student Edition, New Delhi.

G. Kallianpur, P.R. Krishnaiah, J.K. Ghosh, eds., *Statistics and Probability: Essays in Honor of C.R. Rao*
© North-Holland Publishing Company (1982) 567–573

ASYMPTOTIC NORMALITY OF THE AVERAGE OF DISTINCT UNITS IN SIMPLE RANDOM SAMPLING WITH REPLACEMENT

P.K. PATHAK
The University of New Mexico, Albuquerque, NM, U.S.A.

This paper develops the following result on the asymptotic normality of the mean of distinct sample units in simple random sampling with replacement. Suppose that the sample size n and the population size N approach infinity in such a way that $\lim N\exp(-n/N)[1-\exp(-n/N)]=\infty$. Then the sample mean based on the distinct sample units, $\bar{y}_\nu$ say, is asymptotically normally distributed with parameters $(E\bar{y}_\nu, V(\bar{y}_\nu))$ if and only if for each $\varepsilon>0$,

$$\lim\left[\sum_{Q(\varepsilon)}(Y_j-\bar{Y})^2\Big/\sum(Y_j-\bar{Y})^2\right]=0,$$

where $\Sigma_{Q(\varepsilon)}$ denotes the summation over those population units which satisfy the inequality:

$$(Y_j-\bar{Y})^2>\varepsilon^2\{\exp(-n/N)\}\{1-\exp(-n/N)\}\sum(Y_j-\bar{Y})^2,$$

Y_j being the Y-variate value of the jth population unit, $\bar{Y}=N^{-1}\Sigma Y_j$ and the sum, Σ, extends over all the N population units.

1. Introduction

It is well-known that in simple random sampling with replacement, the sample mean based on the distinct sample units is a more efficient estimator of the population mean than the sample mean based on the overall average of all the sample units including their repetitions (Basu, 1958). The mean of distinct units is also known to be an admissible estimator of the population mean (Joshi, 1968). It is therefore worthwhile to study the asymptotic distribution of this estimator. The main object of this paper is to furnish a necessary and sufficient condition for the asymptotic normality of this estimator. The paper makes no claims as to the relative merits and demerits of simple random sampling with replacement as such but rather makes an attempt to learn more about this well-known estimator which has attracted so much attention in the recent history of sample surveys.

For reasons of clarity in the exposition, we first introduce a sequence of populations. For each integer k, consider a finite population $\Gamma_k=(U_{k1},\ldots,U_{kj},\ldots,U_{kN_k})$ of N_k units, in which U_{kj} denotes the jth unit of Γ_k. In the sequel, unless otherwise stated the subscript j runs from 1 through N_k. Let Y_{kj} denote the value of a Y-characteristic of U_{kj}. Let $s_{n_k}=(u_{k1},\ldots,u_{ki},\ldots,u_{kn_k})$ denote a sample of size n_k selected from Γ_k by simple random sampling with replacement (SRSWR), u_{ki} being the sample unit selected from Γ_k at the ith draw. Unless otherwise stated the subscript i runs from 1 through n_k. Let y_{ki} denote the Y-variate value associated with u_{ki}. It is clear that $u_{k1},\ldots,u_{kn_k}$ are independent and identically distributed random variables with $P(u_{k1}=U_{kj})=N_k^{-1}$. Now, because of with replacement sampling, not all of the n_k units in the sample s_{n_k} will be distinct. Let ν_k ($\nu_k \leqslant n_k$) denote the number of distinct units in s_{n_k} and let $\tau_{\nu_k}=\{u_{(k1)},\ldots,u_{(kh)},\ldots,u_{(k\nu_k)}\}$ denote the set of ν_k distinct units in s_{n_k}, $u_{(kh)}$ being the hth distinct unit. Unless otherwise stated, the subscript h runs from 1 through ν_k. Similar to y_{ki}, $y_{(kh)}$ stands for the Y-variate value of $u_{(kh)}$.

For estimating the population mean $\bar{Y}_k=N_k^{-1}\Sigma_j Y_{kj}$ under SRSWR, the following two unbiased estimators of $\bar{Y}_k$ are commonly considered.

$$\bar{y}_{n_k}=(1/n_k)\sum_i y_{ki} \tag{1.1}$$

and

$$\bar{y}_{(\nu_k)}=(1/\nu_k)\sum_h y_{kh}. \tag{1.2}$$

The variances of these estimators are given by (cf., Pathak, 1962).

$$V(\bar{y}_{n_k})=(1/n_k)\sigma_k^2 \tag{1.3}$$

and

$$V(\bar{y}_{(\nu_k)})=\frac{\left[1^{n_k-1}+2^{n_k-1}+\cdots+(N_k-1)^{n_k-1}\right]}{N_k^{n_k-1}(N_k-1)}\sigma_k^2, \tag{1.4}$$

where $\sigma_k^2=N_k^{-1}\Sigma(Y_{kj}-Y_k)^2$.

An immediate consequence of the classical central limit theorem is that as n_k approaches infinity, the sequence $\{\bar{y}_{n_k}, k\geqslant 1\}$ is asymptotically normally distributed with parameters $(E\bar{y}_{n_k}, V(\bar{y}_{n_k}))$ if and only if for each $\varepsilon>0$,

$$\lim\left[\sum_{T(\varepsilon)}(Y_{kj}-\bar{Y}_k)^2\Big/\sum(Y_{kj}-\bar{Y}_k)^2\right]=0, \tag{1.5}$$

where $\Sigma_{T(\varepsilon)}$ denotes the sum over those units which satisfy the inequality:

$$(Y_{kj}-\bar{Y}_k)^2>\varepsilon^2(n_k/N_k)\sum(Y_{kj}-\bar{Y}_k)^2. \tag{1.6}$$

It is much more difficult to establish the asymptotic normality of $\bar{y}_{(\nu_k)}$. Pathak and Sethuraman (1964) established certain sufficient conditions for the asymptotic normality of $\bar{y}_{(\nu_k)}$. In this paper we establish the following much stronger result.

Theorem 1.1. *Suppose that both n_k and N_k approach infinity in such a way that*

$$\lim N_k\exp(-n_k/N_k)[1-\exp(-n_k/N_k)]=\infty. \tag{1.7}$$

Then $\bar{y}_{(\nu_k)}$ is asymptotically normally distributed with parameters $(E\bar{y}_{(\nu_k)}, V(\bar{y}_{(\nu_k)}))$ if and only if for each $\varepsilon>0$,

$$\lim\left[\sum_{Q(\varepsilon)}(Y_{kj}-\bar{Y}_k)^2\Big/\sum(Y_{kj}-\bar{Y}_k)^2\right]=0, \tag{1.8}$$

where $\Sigma_{Q(\varepsilon)}$ *denotes the summation over those population units in* Γ_k *which satisfy the inequality*:

$$\left(Y_{kj}-\bar{Y}_k\right)^2>\varepsilon^2\exp(-n_k/N_k)\{1-\exp(-n_k/N_k)\}\sum\left(Y_{kj}-\bar{Y}_k\right)^2. \tag{1.9}$$

When $\lim(n_k/N_k)$ exists, (1.7) is equivalent to one of the following two conditions:

(i) either $\lim n_k/N_k=\alpha$, $0\leqslant\alpha<\infty$,

(ii) or $\lim n_k/N_k=\infty$ and $\lim N_k\exp(-n_k/N_k)=\infty$.

It is also easily seen that when $\lim(n_k/N_k)=\alpha$, $0<\alpha<\infty$, the conditions (1.6) and (1.9) become equivalent. Thus at least in this case, the necessary and sufficient condition for asymptotic normality of the average of distinct units is no more severe than the classical Lindeberg condition for asymptotic normality of the overall average of all the units in the sample. Roughly speaking, the condition (1.8)–(1.9) simply says that in order to ensure asymptotic normality of the average of distinct units, the tail of the population must go off to zero at a certain rate.

It is worthwhile adding here that Lanke (1975) has furnished a Berry–Essen type result for Basu's estimator while Holst (1973) has provided a sufficient condition for its asymptotic normality. Convergence to asymptotic normality in both cases depends on $\max_j(Y_{kj}-\bar{Y}_k)^2/\Sigma(Y_{kj}-\bar{Y})^2$, which is a somewhat restrictive measure of the tail of the population. The results of this paper can be considered as complementing as well as strengthening these results of Lanke (1975), Holst (1973) and others.

2. Preliminaries

We need the following preliminary results for further developments.

Theorem 2.1 (Hájek, 1968). *Consider the sequence* $\{\Gamma_k\colon k\geqslant 1\}$ *of populations and suppose that for each* $k\geqslant 1$, *a sample* τ_{n_k} *of size* n_k *is drawn from* Γ_k *by simple random sampling without replacement* (*SRSWOR*). *Let*

$$\bar{y}_{(n_k)}=\frac{1}{n_k}\sum y_{ki} \tag{2.1}$$

denote the sample mean of Y-variate values based on τ_{n_k}.

Now suppose that both n_k *and* N_k *approach infinity in such a way that*

$$\lim n_k(1-n_k/N_k)=\infty. \tag{2.2}$$

Then $\bar{y}_{(n_k)}$ *is asymptotically normally distributed with parameters* $(E\bar{y}_{(n_k)}, V(\bar{y}_{(n_k)}))$ *if and only if for each* $\varepsilon>0$,

$$\lim\left[\sum_{P(\varepsilon)}\left(Y_{kj}-\bar{Y}_k\right)^2\Big/\sum\left(Y_{kj}-\bar{Y}_k\right)^2\right]=0, \tag{2.3}$$

where $P(\varepsilon)$ denotes those population units in Γ_k which satisfy the inequality:

$$\left(Y_{kj}-\bar{Y}_k\right)^2>\varepsilon^2(n_k/N_k)(1-n_k/N_k)\sum\left(Y_{kj}-\bar{Y}_k\right)^2. \tag{2.4}$$

Lemma 2.1 (Lanke, 1975). *Let ν_k denote the number of distinct units in an SRSWR sample of size n_k drawn from a population of size N_k. Then*

$$N_k[1-\exp(-n_k/N_k)]\leqslant E\nu_k\leqslant 1+N_k[1-\exp(-n_k/N_k)] \tag{2.5}$$

holds for all n_k and N_k, and

$$V(\nu_k)\leqslant 4N_k\exp(-n_k/N_k)[1-\exp(-n_k/N_k)] \tag{2.6}$$

holds for all n_k and $N_k\geqslant 3$.

3. A random experiment

The problem of asymptotic normality for estimators based on distinct units in SRSWR can be reduced to a corresponding problem for estimators in SRSWOR by performing the following random experiments:

(i) First draw an SRSWR sample of size n_k from Γ_k. Let s_{n_k} denote the sample so drawn and τ_{ν_k} the set of distinct units in s_{n_k}. Let ν_k be the observed number of distinct units and $[E\nu_k]$ denote the largest integer less than or equal to $1+N_k[1-\exp(-n_k/N_k)]$.

(ii) If $\nu_k\leqslant[E\nu_k]$, then select an SRSWOR sample of size $([E\nu_k]-\nu_k)$ from the remaining $(N_k-\nu_k)$ units in Γ_k. Let $\tau_{[E\nu_k]}$ denote the set of $[E\nu_k]$ distinct units so drawn from Γ_k.

(iii) If $\nu_k\geqslant[E\nu_k]$, then select an SRSWOR sample of size $[E\nu_k]$ from τ_{ν_k}, the set of distinct units in s_{n_k}. Let $\tau_{[E\nu_k]}$ denote the subsample of $[E\nu_k]$ distinct population units so drawn.

It is easily shown that $\tau_{[E\nu_k]}$ is an SRSWOR sample of size $[E\nu_k]$ from Γ_k, while given ν_k the set of distinct units $\tau_{\nu_k}\Delta\tau_{[E\nu_k]}$, the units in the symmetric difference of τ_{ν_k} and $\tau_{[E\nu_k]}$, is an SRSWOR sample of size $|\nu_k-[E\nu_k]|$ from Γ_k. Define

$$d_{(\nu_k)}=\sum_1\left(y_{(kh)}-\bar{Y}_k\right)=\nu_k\left(\bar{y}_{(\nu_k)}-\bar{Y}_k\right), \tag{3.1}$$

where the sum, Σ_1, is taken over all the distinct units in s_{n_k}, and

$$d^*_{[E\nu_k]}=\sum_2\left(y_{(kh)}-\bar{Y}_k\right)=[E\nu_k]\left(\bar{y}_{[E\nu_k]}-\bar{Y}_k\right), \tag{3.2}$$

where the sum, Σ_2, is taken over all the units in $\tau_{[E\nu_k]}$. Obviously,

$$\left|d_{(\nu_k)}-d^*_{[E\nu_k]}\right|=\left|\sum_3\left(y_{(kh)}-\bar{Y}_k\right)\right|, \tag{3.3}$$

in which the sum, Σ_3, is taken over the units in the symmetric difference of the sets τ_{ν_k} and $\tau_{[E\nu_k]}$.

The following lemmas play a central role in the discussion that follows.

Lemma 3.1.

$$E\left[\left(d_{(\nu_k)}-d^*_{[E\nu_k]}\right)^2\right]\leqslant\sigma_k^2\sqrt{\{1+V(\nu_k)\}}\,. \tag{3.4}$$

Proof. Since given ν_k, the conditional distribution of $\tau_{\nu_k}\Delta\tau_{[E\nu_k]}$ is that of an SRSWOR sample of size $|\nu_k-[E\nu_k]|$ from Γ_k, we have

$$E\left[\left(d_{(\nu_k)}-d^*_{[E\nu_k]}\right)^2|\nu_k\right]=V\left(\left(d_{(\nu_k)}-d^*_{[E\nu_k]}\right)|\nu_k\right)$$

$$=\frac{|\nu_k-[E\nu_k]|(N_k-|\nu_k-[E\nu_k]|)}{N_k(N_k-1)}\sum\left(Y_{kj}-\bar{Y}_k\right)^2$$

$$\leqslant|\nu_k-[E\nu_k]|\sigma_k^2. \tag{3.5}$$

So

$$E\left[\left(d_{(\nu_k)}-d^*_{[E\nu_k]}\right)^2\right]\leqslant(E|\nu_k-[E\nu_k]|)\sigma_k^2$$

$$\leqslant\sigma_k^2\sqrt{E(\nu_k-[E\nu_k])^2}\,.$$

By Lemma 2.1,

$$\leqslant\sigma_k^2\sqrt{\{1+V(\nu_k)\}}\,. \tag{3.6}$$

Lemma 3.2. *Let $n_k\to\infty$ and $N_k\to\infty$ in such a way that* (1.5) *holds. Then the limiting distributions of $d_{(\nu_k)}$ and $d^*_{[E\nu_k]}$, if they exist, are identical.*

Proof. Since $d^*_{[E\nu_k]}$ is based on an SRSWOR of size $[E\nu_k]$, it is easily seen that

$$V\left(d^*_{[E\nu_k]}\right)=\frac{[E\nu_k](N_k-E[\nu_k])}{(N_k-1)}\sigma_k^2. \tag{3.7}$$

Also, since

$$N_k[1-\exp(-n_k/N_k)]\leqslant[E\nu_k]\leqslant1+N_k[1-\exp(-n_k/N_k)],$$

it follows that

$$\left|V\left(d^*_{[E\nu_k]}\right)-N_k\exp(-n_k/N_k)[1-\exp(-n_k/N_k)]\,\sigma_k^2\right|\leqslant2\sigma_k^2. \tag{3.8}$$

Making use of (2.6), (3.8) and Lemma 3.1, we find that

$$\frac{E\left(d_{(\nu_k)}-d^*_{[E\nu_k]}\right)^2}{V\left(d^*_{[E\nu_k]}\right)}\leqslant\frac{\sqrt{(1+V(\nu_k))}}{(N_k\exp(-n_k/N_k)[1-\exp(-n_k/N_k)]-2)}$$

$$\leqslant\frac{c}{\sqrt{(N_k\exp(-n_k/N_k))[1-\exp(-n_k/N_k)]}} \tag{3.9}$$

where c is a suitable constant. So

$$\left(d_{(\nu_k)}-d^*_{[E\nu_k]}\right)\Big/\sqrt{V\left(d^*_{[E\nu_k]}\right)}$$

converges to zero in probability when (1.5) holds. Also, it follows from (3.8) that under (1.5)

$$\lim\left[V\left(d_{(\nu_k)}\right)\Big/V\left(d^*_{[E\nu_k]}\right)\right]=1. \tag{3.10}$$

Now let $\{a_k\colon k\geqslant 1\}$ and $\{b_k\colon k\geqslant 1\}$ be two preassigned sequences of numbers with the b_k's positive. Then

$$\frac{a_k+b_kd_{(\nu_k)}}{\sqrt{V\left(b_kd_{\nu(k)}\right)}}=\left[\frac{a_k+b_kd^*_{[E\nu_k]}}{\sqrt{V\left(b_kd^*_{[E\nu_k]}\right)}}+\frac{d_{(\nu_k)}-d^*_{[E\nu_k]}}{\sqrt{V\left(d^*_{[E\nu_k]}\right)}}\right]\sqrt{\frac{V\left(d^*_{[E\nu_k]}\right)}{V\left(d_{(\nu_k)}\right)}}. \tag{3.11}$$

Since, under (1.5),

$$\lim\left[V\left(d^*_{[E\nu_k]}\right)\Big/V\left(d_{(\nu_k)}\right)\right]=1$$

and

$$\left(d_{(\nu_k)}-d^*_{[E\nu_k]}\right)\Big/\sqrt{V\left(d^*_{[E\nu_k]}\right)}$$

converges to zero in probability, (3.11) shows that the limiting distributions of $a_k+b_kd_{(\nu_k)}$ and $a_k+b_kd^*_{[E\nu_k]}$, whenever either one exists, are identical. This completes the proof of the lemma.

4. The main theorem

An immediate consequence of Lemma 3.2 is that in order to study the limiting distribution of $a_k+b_kd_{(\nu_k)}$, it suffices to study the limiting distribution of $a_k+b_kd^*_{[E\nu_k]}$. Furthermore, since $d^*_{[E\nu_k]}$ is a sample total based on an SRSWOR sample of size,

$$[E\nu_k]=[1+N_k(1-\exp(-n_k/N_k))],$$

Theorem 2.1 can be applied to study the limiting distributions of $a_k+b_kd^*_{[E\nu_k]}$ with the n_k in the theorem replaced by $[E\nu_k]$.

Theorem 4.1. *Suppose that $n_k\to\infty$ and $N_k\to\infty$ in such a way that* (1.5) *holds. Then $d_{(\nu_k)}$ given in* (3.1) *is asymptotically normally distributed with parameters* $(0, V(d_{(\nu_k)}))$ *if and only if* (1.6) *holds.*

The proof of the theorem is a consequence of Theorem 2.1 and Lemma 3.2.

Corollary 1. *Suppose that $n_k\to\infty$ and $N_k\to\infty$ in such a way that* (1.5) *holds. Then $\bar{y}_{(\nu_k)}$, the average of distinct sample units given by* (1.1) *and based on an SRSWR* *sample of size* n_k *from* Γ_k, *is asymptotically normally distributed with parameters* $(\bar{Y}_k, V(\bar{y}_{(\nu_k)}))$ *if and only if* (1.6) *holds.*

Proof. To prove the claim, first note that

$$\left(\bar{y}_{(\nu_k)} - \bar{Y}_k\right) = \frac{1}{\nu_k} d_{(\nu_k)} = [(E\nu_k)/\nu_k]\left[d_{(\nu_k)}/E\nu_k\right]. \tag{4.1}$$

Now use Theorem 4.1 and the fact that under (1.5) $(\nu_k/E\nu_k)$ converges to one in probability by virtue of (2.5) and (2.6).

References

Basu, D. (1958). On sampling with and without replacement. *Sankhyā* **20**, 287–294.

Hájek, J. (1960). Limiting distributions in simple random sampling from a finite population. *Publ. Math. Inst. Hung. Acad. Sci. A* **5**, 361–374.

Holst, L. (1973). Some limit theorems with applications in sampling theory. *Ann. Statist.* **1**, 644–658.

Joshi, V.M. (1966). Admissibility of the sample mean as estimate of the mean of a finite population. *Ann. Math. Statist.* **39**, 606–620.

Lanke, J. (1975). Some contributions to the theory of survey sampling. Thesis, Department of Mathematical Statistics, University of Lund, Sweden.

Pathak, P.K. (1962). On simple random sampling with replacement. *Sankhyā A* **24**, 287–302.

Pathak, P.K. and Sethuraman, J. (1964). On the asymptotic distribution of the mean of distinct units in sampling from a finite population. *Publ. Math. Inst. Hung. Acad. Sci. A* **9**, 113–116.

G. Kallianpur, P.R. Krishnaiah, J.K. Ghosh, eds., *Statistics and Probability: Essays in Honor of C.R. Rao*

MAXIMUM PROBABILITY ESTIMATION FOR DIFFUSION PROCESSES

B.L.S. PRAKASA RAO
Indian Statistical Institute, New Dehli, India and University of Wisconsin, Madison, WI, U.S.A

The study of inference problems for stochastic processes with a continuous and discrete time parameter has attracted a large number of workers recently in view of their applications to engineering and biological problems. Among the class of continuous time processes, it has been found that the class of diffusion processes is amenable to statistical analysis. A survey of the work in this area is given in the book "Statistical Inference for Stochastic Processes" by Basawa and Prakasa Rao (1980). Further results on asymptotic theory of maximum likelihood and Bayes estimators, asymptotic theory of least squares estimators and non-parametric estimation are discussed in Prakasa Rao (1979a, b, 1981), Prakasa Rao and Rubin (1981) and Dorogovcev (1976). Here we study the asymptotic theory of maximum probability estimation for diffusion processes extending the work of Roussas (1977) for discrete time stationary Markov processes.

1. Introduction

Let us consider the stochastic differential equation

$$\mathrm{d}\xi(t)=f(\theta,\xi(t))\,\mathrm{d}t+\sigma(\xi(t))\,\mathrm{d}W(t),\quad \xi(0)=\xi_0,\quad t\geqslant 0 \tag{1.1}$$

where $f(\theta, x)$ is a real valued function defined on $\Theta\times R$, $\sigma(x)$ is a positive valued function defined on R and $W(\cdot)$ is the standard Wiener process on R. Suppose that the stochastic differential eq. (1.1) has a unique stationary ergodic solution $\{\xi(t)\}$ on $[0,T]$ for every $T>0$ and for every $\theta\in\Theta$ open interval in R. Let μ_θ^T be the measure corresponding to the solution of (1.1) on $C[0,T]$ and μ_W^T be the corresponding Wiener measure on $C[0,T]$ for fixed initial value $\xi(0)$. Suppose that $\mu_\theta^T\ll\mu_W^T$ for all $\theta\in\Theta$. It is well known that the Radon–Nikodym derivative of μ_θ^T with respect to μ_W^T is given by

$$L_T(\theta)\equiv\frac{\mathrm{d}\mu_\theta^T}{\mathrm{d}\mu_W^T}=\exp\left\{\int_0^T\frac{f(\theta,\xi(t))}{\sigma(\xi(t))}\,\mathrm{d}\xi(t)-\tfrac{1}{2}\int_0^T\left[\frac{f(\theta,\xi(t))}{\sigma(\xi(t))}\right]^2\mathrm{d}t\right\}. \tag{1.2}$$

This can be seen from Gikhman and Skorokhod (1972) or from Basawa and Prakasa Rao (1980, Thm 3.6, ch. 9). For each $\theta\in\Theta$, consider the interval $(\theta-T^{-1/2}, \theta+T^{-1/2})$ for sufficiently large T and define

$$Z_T(\theta)=\int_{\theta-T^{-1/2}}^{\theta+T^{-1/2}}L_T(t)\,\mathrm{d}t. \tag{1.3}$$

Any measurable function $\tilde{d}_T$ for which the integrated likelihood $Z_T(\theta)$ is maximized with respect to θ is called a maximum probability estimator (MPE) of θ based on the realization $\{\xi(s)\colon 0 \leqslant s \leqslant T\}$.

In order to simplify further notation, we will assume without loss of generality that $\sigma(x) \equiv 1$. Hence

$$\log L_T(\theta) = \int_0^T f(\theta, \xi(t))\,\mathrm{d}\xi(t) - \tfrac{1}{2}\int_0^T f^2(\theta, \xi(t))\,\mathrm{d}t \tag{1.4}$$

and the corresponding stochastic differential equation is

$$\mathrm{d}\xi(t) = f(\theta, \xi(t))\,\mathrm{d}t + \mathrm{d}W(t), \quad \xi(0) = \xi_0, \quad t \geqslant 0. \tag{1.5}$$

2. Assumptions and some lemmas

We assume that $L_T(\theta)$ is twice differentiable with respect to θ and differentiation twice under the integral sign is valid in (1.4). It is easy to see that

$$\frac{\mathrm{d}\log L_T(\theta)}{\mathrm{d}\theta} = \int_0^T f^{(1)}(\theta, \xi(t))\,\mathrm{d}\xi(t) - \int_0^T f(\theta, \xi(t)) f^{(1)}(\theta, \xi(t))\,\mathrm{d}t \tag{2.1}$$

and the expression on R.H.S. of (2.1) is equal to $\int_0^T f^{(1)}(\theta, \xi(t))\,\mathrm{d}W(t)$ when θ is the true parameter. Similarily

$$\frac{\mathrm{d}^2\log L_T(\theta)}{\mathrm{d}\theta^2} = \int_0^T f^{(2)}(\theta, \xi(t))\,\mathrm{d}\xi(t) - \int_0^T \left[f(\theta, \xi(t)) f^{(2)}(\theta, \xi(t)) + f^{(1)^2}(\theta, \xi(t)) \right] \mathrm{d}t \tag{2.2}$$

and the expression on the R.H.S. of (2.2) is equal to

$$\int_0^T f^{(2)}(\theta, \xi(t))\,\mathrm{d}W(t) - \int_0^T f^{(1)^2}(\theta, \xi(t))\,\mathrm{d}t$$

when θ is the true parameter. Hence $E_\theta[\mathrm{d}\log L_T(\theta)/\mathrm{d}\theta] = 0$ and

$$E_\theta\left[\frac{\mathrm{d}\log L_T(\theta)}{\mathrm{d}\theta}\right]^2 = -E_\theta\left[\frac{\mathrm{d}^2\log L_T(\theta)}{\mathrm{d}\theta^2}\right] = \int_0^T E_\theta\left[f^{(1)^2}(\theta, \xi(t))\right] \mathrm{d}t = T\sigma^2(\theta) \tag{2.3}$$

by stationarity where $\sigma^2(\theta) = E_\theta[f^{(1)^2}(\theta, \xi(0))]$. Suppose that

$$\sigma^2(\theta) > 0. \tag{2.4}$$

Let $\phi_T(\theta) = \mathrm{d}\log L_T(\theta)/\mathrm{d}\theta$, $\psi_T(\theta) = \mathrm{d}^2\log L_T(\theta)/\mathrm{d}\theta^2$ and $\Delta_T(\theta) = T^{-1/2}\phi_T(\theta)$. Further, for $h \in R$, let $\theta_T = \theta + hT^{-1/2}$. Therefore $\theta_T \in \Theta$ for large T. Define

$$\Lambda(\theta, \theta_T) = \Lambda_T(\theta) = \log \frac{\mathrm{d}\mu^T_{\theta_T}}{\mathrm{d}\mu^T_\theta} = \int_0^T [f(\theta_T, \xi(t)) - f(\theta, \xi(t))]\,\mathrm{d}\xi(t)$$
$$- \tfrac{1}{2}\int_0^T \left[f^2(\theta_T, \xi(t)) - f^2(\theta, \xi(t)) \right] \mathrm{d}t. \tag{2.5}$$

Hence

$$\Lambda_T(\theta)-h\Delta_T(\theta)=\int_0^T\left[f(\theta_T,\xi(t))-f(\theta,\xi(t))-hT^{-1/2}f^{(1)}(\theta,\xi(t))\right]\mathrm{d}W(t)$$

$$-\tfrac{1}{2}\int_0^T\left[f(\theta_T,\xi(t))-f(\theta,\xi(t))\right]^2\mathrm{d}t \tag{2.6}$$

when θ is the true parameter. This in turn is equal to

$$\tfrac{1}{2}h^2T^{-1}\int_0^T f^{(2)}(\theta_T^*,\xi(t))\,\mathrm{d}W(t)-\tfrac{1}{2}h^2T^{-1}\int_0^T f^{(1)^2}(\tilde{\theta}_T,\xi(t))\,\mathrm{d}t. \tag{2.7}$$

ote that

$$\psi_T(s)=\int_0^T f^{(2)}(s,\xi(t))\,\mathrm{d}\xi(t)-\int_0^T\left[f(s,\xi(t))f^{(2)}(s,\xi(t))+f^{(1)^2}(s,\xi(t))\right]\mathrm{d}t \tag{2.8}$$

$$=\int_0^T f^{(2)}(s,\xi(t))\,\mathrm{d}W(t)$$

$$-\int_0^T\left[\{f(s,\xi(t))-f(\theta,\xi(t))\}f^{(2)}(s,\xi(t))+f^{(1)^2}(s,\xi(t))\right]\mathrm{d}t \tag{2.9}$$

if θ is the true parameter. Let

$$W_T(s)=\frac{1}{T}\{\psi_T(s)-E_\theta(\psi_T(s))\}. \tag{2.10}$$

Lemma 2.1. *Suppose $f^{(2)}(\theta,x)$ is differentiable with respect to θ and $f^{(2)}(\theta,x)$ is Lipschitzian in θ in a closed interval $I(\theta)$ of θ with Lipschitzian function $c(x)$ such that $E_\theta[f^{(3)}(s,\xi(0))]^2<\infty$, $s\in I(\theta)$ and*

$$E_\theta\left[c(\xi(0))\right]^2<\infty.$$

Then

$$\frac{1}{T}\sup_{s\in I(\theta)}\left|\int_0^T f^{(2)}(s,\xi(t))\,\mathrm{d}W(t)\right|\to 0,\quad \textit{as } T\to\infty \tag{2.11}$$

almost surely P_θ.

This lemma follows from the uniform ergodic theorem for stochastic integrals proved in the appendix.

Lemma 2.2. *Suppose that the functions $f(s_1,x)f^{(2)}(s_2,x)$ and $f^{(1)}(s_1,x)f^{(1)}(s_2,x)$ are uniformly bounded with respect to s_1 and s_2 in a closed interval $I(\theta)$ of θ by an integrable function of x. Then*

$$\sup_{\substack{(s_1,s_2)\in\\ I(\theta)\times I(\theta)}}\left|\frac{1}{T}\int_0^T\left\{f(s_1,\xi(t))f^{(2)}(s_2,\xi(t))-E_\theta\left[f(s_1,\xi(t))f^{(2)}(s_2,\xi(t))\right]\right\}\mathrm{d}t\right|\to 0$$

$$\textit{as } T\to\infty\quad \textit{a.s.}\ [P_\theta] \tag{2.12}$$

and

$$\sup_{(s_1,s_2)\in I(\theta)\times I(\theta)}\left|\frac{1}{T}\int_0^T\{f^{(1)}(s_1,\xi(t))f^{(1)}(s_2,\xi(t))\right.$$

$$\left.-E_\theta\left[f^{(1)}(s_1,\xi(t))f^{(1)}(s_2,\xi(t))\right]\right\}\mathrm{d}t\Big|\to 0 \quad \text{as } T\to\infty \quad a.s.\,[P_\theta]. \tag{2.13}$$

Furthermore $E_\theta[f(s_1,\xi(0))f^{(2)}(s_2,\xi(0))]$, $E_\theta[f^{(1)}(s_1,\xi(0))f^{(1)}(s_2,\xi(0))]$ *are continuous in* (s_1,s_2).

This lemma follows from the uniform ergodic theorem proved in Appendix A an by the dominated convergence theorem.

Lemma 2.3. *Under the conditions stated in Lemmas* 2.1 *and* 2.2,

$$\sup_{s\in I(\theta)}|W_T(s)|\to 0 \quad \text{as } T\to\infty \quad a.s.\,[P_\theta]. \tag{2.14}$$

Proof. This lemma follows from Lemmas 2.1 and 2.2 and relations (2.9) and (2.10).

Lemma 2.4. *Under the conditions stated in Lemmas* 2.1 *and* 2.2,

$$\int_0^1|W_T(\theta+\lambda\tau_T)|\,\mathrm{d}\lambda\to 0 \quad \text{as } T\to\infty \tag{2.15}$$

$a.s.\,[P_\theta]$ *when* $\tau_T\to 0$ *as* $T\to\infty$.

Proof. This lemma follows from Lemma 2.3.

Lemma 2.5. *Under the conditions stated in Lemmas* 2.1 *and* 2.2,

$$\frac{1}{T}\int_0^1\psi_T(\theta+\lambda\tau_T)\,\mathrm{d}\lambda\to -\sigma^2(\theta) \quad \text{as } T\to\infty \tag{2.16}$$

$a.s.\,[P_\theta]$ *when* $\tau_T\to 0$ *as* $T\to\infty$.

Proof. In view of Lemma 2.4, it is sufficient to prove that

$$\frac{1}{T}\int_0^1 E_\theta(\psi_T(\theta+\lambda\tau_T))\,\mathrm{d}\lambda\to -\sigma^2(\theta) \quad \text{as } T\to\infty. \tag{2.17}$$

By stationarity of the process $\xi(t)$,

$$\frac{1}{T}E_\theta(\psi_T(s))=-E_\theta\left\{[f(s,\xi(0))-f(\theta,\xi(0))]\,f^{(2)}(s,\xi(0))+f^{(1)^2}(s,\xi(0))\right\}.$$

It is clear that

$$\frac{1}{T}E_\theta(\psi_T(\theta+\lambda\tau_T))\to -\sigma^2(\theta) \tag{2.18}$$

by the continuity proved in Lemma 2.2. Furthermore the integrand is dominated in s by an integrable function. Hence the result (2.17) follows by the dominated convergence theorem.

Hereafter we will assume the conditions stated in Lemmas 2.1 and 2.2 and the other assumptions stated earlier.

3. Existence of consistent estimators

Let us consider the equation

$$\frac{\mathrm{d}Z_T(\theta)}{\mathrm{d}\theta}=L_T(\theta+T^{-1/2})-L_T(\theta-T^{-1/2})=0 \tag{3.1}$$

where $Z_T(\theta)$ is defined by (1.3). Note that for d in a neighbourhood of θ,

$$\begin{aligned}\log L_T(d\pm T^{-1/2})&=\log L_T(\theta)+(d-\theta\pm T^{-1/2})\frac{\mathrm{d}\log L_T(\theta)}{\mathrm{d}\theta}\\&\quad+\tfrac{1}{2}(d-\theta\pm T^{-1/2})^2\int_0^1\frac{\mathrm{d}^2\log L_T(\phi)}{\mathrm{d}\phi^2}\bigg|_{\phi=\theta+\lambda(d-\theta\pm T^{-1/2})}\mathrm{d}\lambda\\&=\log L_T(\theta)+(d-\theta\pm T^{-1/2})\phi_T(\theta)\\&\quad+\tfrac{1}{2}(d-\theta\pm T^{-1/2})^2\int_0^1\psi_T(\theta+\lambda(d-\theta\pm T^{-1/2}))\,\mathrm{d}\lambda.\end{aligned} \tag{3.2}$$

Let

$$\begin{aligned}I_T^{\pm}&=\frac{1}{T}\int_0^1\psi_T(\theta+\lambda(d-\theta\pm T^{-1/2}))\,\mathrm{d}\lambda.\\&\equiv I_T((d-\theta\pm T^{-1/2})+\theta).\end{aligned} \tag{3.3}$$

Replacing θ by d in eq. (3.1) and using the relation (3.2), we obtain that

$$\begin{aligned}&(d-\theta+T^{-1/2})\phi_T(\theta)+\tfrac{1}{2}(d-\theta+T^{-1/2})^2TI_T^+=\\&=(d-\theta-T^{-1/2})\phi_T(\theta)+\tfrac{1}{2}(d-\theta-T^{-1/2})^2TI_T^-\end{aligned}$$

and hence

$$\begin{aligned}&4T^{-1/2}\phi_T(\theta)+\{T^{1/2}(d-\theta)\}^2(I_T^+-I_T^-)\\&\quad+2T^{1/2}(d-\theta)(I_T^++I_T^-)+(I_T^+-I_T^-)=0.\end{aligned} \tag{3.4}$$

Let $\gamma_T(d)$ denote the expression on the left-hand side of (3.4). Then

$$\gamma_T(\theta+MT^{-1/2})=$$

$$=4T^{-1/2}\phi_T(\theta)+M^2\{I_T(\theta+(M+1)T^{-1/2})-I_T(\theta+(M-1)T^{-1/2})\}$$
$$+2M\{I_T(\theta+(M+1)T^{-1/2})+I_T(\theta+(M-1)T^{-1/2})\}$$
$$+\{I_T(\theta+(M+1)T^{-1/2})-I_T(\theta+(M-1)T^{-1/2})\} \tag{3.5}$$

and

$$\gamma_T(\theta-MT^{-1/2})=$$

$$=4T^{-1/2}\phi_T(\theta)+M^2\{I_T(\theta-(M-1)T^{-1/2})-I_T(\theta-(M+1)T^{-1/2})\}$$
$$-2M\{I_T(\theta-(M-1)T^{-1/2})+I_T(\theta-(M+1)T^{-1/2})\}$$
$$+\{I_T(\theta-(M-1)T^{-1/2})-I_T(\theta-(M+1)T^{-1/2})\}. \tag{3.6}$$

In view of (2.1), (2.3) and (2.4), it follows that

$$T^{-1/2}\phi_T(\theta)\xrightarrow{\mathcal{L}}N(0,\sigma^2(\theta))$$

by the central limit theorem for stochastic integrals [cf. Basawa and Prakasa Rao (1980)] when θ is the true parameter. On the other hand, all the I_T's which appear in (3.5) and (3.6) converge to $-\sigma^2(\theta)$ a.s. $[P_\theta]$ by Lemma 2.5. Hence, for any fixed $M>0$,

$$\gamma_T(\theta+MT^{-1/2})\xrightarrow{\mathcal{L}}N(-4M\sigma^2(\theta),16\sigma^2(\theta))$$

and

$$\gamma_T(\theta-MT^{-1/2})\xrightarrow{\mathcal{L}}N(4M\sigma^2(\theta),16\sigma^2(\theta))$$

when θ is the true parameter. Since $\sigma^2(\theta)>0$, for $\varepsilon>0$ and for large T, $P_\theta[\gamma_T(\theta+MT^{-1/2})<0]>1-\frac{1}{2}\varepsilon$ and $P_\theta[\gamma_T(\theta-MT^{-1/2})>0]>1-\frac{1}{2}\varepsilon$ for sufficiently large M. However, $\gamma_T(d)$ is continuous in d. Hence there exists $\tilde{d}_T=\tilde{d}_T(\xi_0^T)$ such that $\tilde{d}_T\in[\theta-MT^{-1/2},\theta+MT^{-1/2}]$ with probability greater than $1-\varepsilon$ and $\gamma_T(\tilde{d}_T)=0$. Hence we have the following theorem.

Theorem 3.1. *Under the assumptions stated above, for all sufficiently large T with P_θ-probability approaching one, there exists at least one root $\tilde{d}_T=\tilde{d}_T(\xi_0^T)$ of (3.1) which is a $\sqrt{T}$-consistent estimate of θ, i.e. $\sqrt{T}(\tilde{d}_T-\theta)=O_p(1)$.*

4. Existence of maximum probability estimators

Let $\tilde{d}_T$ be as defined in the previous section. Then

$$\left.\frac{\mathrm{d}Z_T(\theta)}{\mathrm{d}\theta}\right|_{\theta=\tilde{d}_T}=0, \tag{4.1}$$

and

$$\sqrt{T}\,(\tilde{d}_T-\theta)=\mathrm{O}_p(1), \tag{4.2}$$

with P_θ-probability tending to one for large T. We will now prove that

$$\left.\frac{\mathrm{d}^2Z_T(\theta)}{\mathrm{d}\theta^2}\right|_{\theta=\tilde{d}_T}<0 \tag{4.3}$$

with P_θ-probability approaching one for large T. It is sufficient to show that

$$\left.\frac{\mathrm{d}L_T(\theta+T^{-1/2})}{\mathrm{d}\theta}\right|_{\theta=\tilde{d}_T}<\left.\frac{\mathrm{d}L_T(\theta-T^{-1/2})}{\mathrm{d}\theta}\right|_{\theta=\tilde{d}_T} \tag{4.4}$$

with P_θ-probability approaching one for large T. Since

$$\frac{\mathrm{d}L_T(\theta)}{\mathrm{d}\theta}=L_T(\theta)\frac{\mathrm{d}\log L_T(\theta)}{\mathrm{d}\theta} \tag{4.5}$$

it follows that

$$\frac{\mathrm{d}\,L_T(\theta\pm T^{-1/2})}{\mathrm{d}\theta}=L_T(\theta\pm T^{-1/2})\times\left[\int_0^T f^{(1)}\big(\theta\pm T^{-1/2},\xi(t)\big)\,\mathrm{d}\xi(t)\right.$$
$$\left.-\int_0^T f\big(\theta\pm T^{-1/2},\xi(t)\big)f^{(1)}\big(\theta\pm T^{-1/2},\xi(t)\big)\,\mathrm{d}t\right]. \tag{4.6}$$

Observe that $L_T(\tilde{d}_T-T^{-1/2})=L_T(\tilde{d}_T+T^{-1/2})$. Therefore inequality (4.4) holds with P_θ-probability approaching one for large T provided

$$\int_0^T f^{(1)}\big(\tilde{d}_T+T^{-1/2},\xi(t)\big)\,\mathrm{d}\xi(t)-\int_0^T f\big(\tilde{d}_T+T^{-1/2},\xi(t)\big)f^{(1)}\big(\tilde{d}_T+T^{-1/2},\xi(t)\big)\,\mathrm{d}t$$
$$<\int_0^T f^{(1)}\big(\tilde{d}_T-T^{-1/2},\xi(t)\big)\,\mathrm{d}\xi(t)$$
$$-\int_0^T f\big(\tilde{d}_T-T^{-1/2},\xi(t)\big)f^{(1)}\big(\tilde{d}_T-T^{-1/2},\xi(t)\big)\,\mathrm{d}t \tag{4.7}$$

with P_θ-probability approaching one as $T\to\infty$. Since

$$\mathrm{d}\xi(t)=f(\theta,\xi(t))\,\mathrm{d}t+\mathrm{d}W_t,$$

it follows that (4.7) holds provided

$$\begin{aligned}&\int_0^T f^{(1)}\big(\tilde{d}_T+T^{-1/2},\xi(t)\big)\,\mathrm{d}W(t)\\&\quad-\int_0^T\big[f\big(\tilde{d}_T+T^{-1/2},\xi(t)\big)-f(\theta,\xi(t))\big]\,f^{(1)}\big(\tilde{d}_T+T^{-1/2},\xi(t)\big)\,\mathrm{d}t\\&<\int_0^T f^{(1)}\big(\tilde{d}_T-T^{-1/2},\xi(t)\big)\,\mathrm{d}W(t)\\&\quad-\int_0^T\big[f\big(\tilde{d}_T-T^{-1/2},\xi(t)\big)-f(\theta,\xi(t))\big]\,f^{(1)}\big(\tilde{d}_T-T^{-1/2},\xi(t)\big)\,\mathrm{d}t\end{aligned}\tag{4.8}$$

with P_θ-probability approaching one as $T\to\infty$. Note that $\tilde{d}_T\xrightarrow{p}\theta$ as $T\to\infty$ with respect to P_θ-measure. Since f is differentiable with respect to θ, (4.8) holds provided

$$\begin{aligned}&\int_0^T f^{(1)}\big(\tilde{d}_T+T^{-1/2},\xi(t)\big)\,\mathrm{d}W(t)\\&\quad-\big(\tilde{d}_T+T^{-1/2}-\theta\big)\int_0^T f^{(1)}\big(d_T^*+T^{-1/2},\xi(t)\big)f^{(1)}\big(\tilde{d}_T+T^{-1/2},\xi(t)\big)\,\mathrm{d}t\\&<\int_0^T f^{(1)}\big(\tilde{d}_T-T^{-1/2},\xi(t)\big)\,\mathrm{d}W(t)\\&\quad-\big(\tilde{d}_T-T^{-1/2}-\theta\big)\int_0^T f^{(1)}\big(d_T^{**}-T^{-1/2},\xi(t)\big)f^{(1)}\big(\tilde{d}_T-T^{-1/2},\xi(t)\big)\,\mathrm{d}t\end{aligned}\tag{4.9}$$

where $|d_T^*+T^{-1/2}-\theta|<|\tilde{d}_T+T^{-1/2}-\theta|$ and $|d_T^{**}-T^{-1/2}-\theta|<|\tilde{d}_T-T^{-1/2}-\theta|$ with P_θ-probability approaching one as $T\to\infty$. Note that $d_T^*\xrightarrow{p}\theta$ and $d_T^{**}\xrightarrow{p}\theta$ as $T\to\infty$. After dividing by $T^{1/2}$, inequality (4.9) can be written in the form

$$\begin{aligned}&\frac{2}{T}\int_0^T f^{(2)}\big(\hat{d}_T,\xi(t)\big)\,\mathrm{d}W(t)\\&\quad-\Big[\big(\tilde{d}_T-\theta\big)\sqrt{T}\Big]\frac{1}{T}\int_0^T f^{(1)}\big(d_T^*+T^{-1/2},\xi(t)\big)f^{(1)}\big(\tilde{d}_T+T^{-1/2},\xi(t)\big)\,\mathrm{d}t\\&\quad-\frac{1}{T}\int_0^T f^{(1)}\big(d_T^*+T^{-1/2},\xi(t)\big)f^{(1)}\big(\tilde{d}_T+T^{-1/2},\xi(t)\big)\,\mathrm{d}t\\&<-\Big[\big(\tilde{d}_T-\theta\big)\sqrt{T}\Big]\frac{1}{T}\int_0^T f^{(1)}\big(d_T^{**}+T^{-1/2},\xi(t)\big)f^{(1)}\big(\tilde{d}_T+T^{-1/2},\xi(t)\big)\,\mathrm{d}t\\&\quad+\frac{1}{T}\int_0^T f^{(1)}\big(d_T^{**}-T^{-1/2},\xi(t)\big)f^{(1)}\big(\tilde{d}_T-T^{-1/2},\xi(t)\big)\,\mathrm{d}t\end{aligned}\tag{4.10}$$

where $\tilde{d}_T-T^{-1/2}<\hat{d}_T<\tilde{d}_T+T^{-1/2}$. Since

$$\left|\frac{1}{T}\int_0^T f^{(2)}\big(\hat{d}_T,\xi(t)\big)\,\mathrm{d}W(t)\right|<\frac{1}{T}\sup_{\phi\in I(\theta)}\left|\int_0^T f^{(2)}\big(\phi,\xi(t)\big)\,\mathrm{d}W(t)\right|$$

and the last term tends to zero almost surely P_θ as $T\to\infty$ by Lemma 2.1, it follows that

$$\frac{1}{T}\int_0^T f^{(2)}\big(\hat{d}_T,\xi(t)\big)\,\mathrm{d}W(t)\to 0 \quad \text{a.s.}\ [P_\theta] \tag{4.11}$$

as $T\to\infty$. Furthermore, by Lemma 2.2 and the dominated convergence theorem, it can be checked that

$$\left|\frac{1}{T}\int_0^T f^{(1)}\big(d_T^*+T^{-1/2},\xi(t)\big)f^{(1)}\big(\tilde{d}_T+T^{-1/2},\xi(t)\big)\,\mathrm{d}t-\sigma^2(\theta)\right|\to 0 \quad \text{a.s.}\ [P_\theta] \tag{4.12}$$

and

$$\left|\frac{1}{T}\int_0^T f^{(1)}\big(d_T^{**}-T^{-1/2},\xi(t)\big)f^{(1)}\big(\tilde{d}_T-T^{-1/2},\xi(t)\big)\,\mathrm{d}t-\sigma^2(\theta)\right|\to 0 \quad \text{a.s.}\ [P_\theta] \tag{4.13}$$

and $\sqrt{T}(\tilde{d}_T-\theta)=\mathrm{O}_p(1)$. In view of (4.11)–(4.13), inequality (4.10) holds with probability P_θ approaching one as $T\to\infty$. Hence $\tilde{d}_T$ is a $\sqrt{T}$-consistent maximum probability estimator with P_θ-probability approaching one as $T\to\infty$.

Theorem 4.1. *Under the assumptions stated above, for all sufficiently large T and with P_θ-probability approaching one, there exists a maximum probability estimator of θ which is $\sqrt{T}$-consistent.*

5. Asymptotic normality of maximum probability estimators

Let $\tilde{d}_T$ be a MPE as discussed above. Since

$$\left.\frac{\mathrm{d}Z_T(\theta)}{\mathrm{d}\theta}\right|_{\theta=\tilde{d}_T}=0, \tag{5.1}$$

it follows that

$$\begin{aligned}4T^{-1/2}\phi_T(\theta)&+\left\{\sqrt{T}\big(\tilde{d}_T-\theta\big)\right\}^2\big(\hat{I}_T^+-\hat{I}_T^-\big)\\&+2\sqrt{T}\big(\tilde{d}_T-\theta\big)\big(\hat{I}_T^++\hat{I}_T^-\big)+\big(\hat{I}_T^+-\hat{I}_T^-\big)=0\end{aligned} \tag{5.2}$$

from (3.4) where

$$\hat{I}_T^{\pm}=\frac{1}{T}\int_0^1\psi_T\big(\theta+\lambda\big(\tilde{d}_T-\theta\pm T^{-1/2}\big)\big)\,\mathrm{d}\lambda. \tag{5.3}$$

However,

$$\hat{I}_T^{\pm}\to-\sigma^2(\theta) \quad \text{a.s. } P_\theta \tag{5.4}$$

by arguments similar to those given in Lemma 2.5 since $\tilde{d}_T \xrightarrow{p} \theta$ as $T \to \infty$. eq. (5.2) can be written in the form

$$(\hat{I}_T^+ + \hat{I}_T^-)(\tilde{d}_T - \theta)T^{1/2} = -2T^{-1/2}\phi_T(\theta) - \tfrac{1}{2}\left[\left(T^{1/2}(\tilde{d}_T - \theta)\right)^2(\hat{I}_T^+ - \hat{I}_T^-)\right] - \tfrac{1}{2}(\hat{I}_T^+ - \hat{I}_T^-). \tag{5.5}$$

Since $\hat{I}_T^+ + \hat{I}_T^- \to -2\sigma^2(\theta) < 0$ a.s. $[P_\theta]$ as $T \to \infty$, dividing the eq. (5.5) throughout by $\hat{I}_T^+ + \hat{I}_T^-$, we have

$$\sqrt{T}(\tilde{d}_T - \theta) = \frac{-2T^{-1/2}\phi_T(\theta)}{\hat{I}_T^+ + \hat{I}_T^-} - \tfrac{1}{2}\left[\sqrt{T}(\tilde{d}_T - \theta)\right]^2 \frac{\hat{I}_T^+ - \hat{I}_T^-}{\hat{I}_T^+ + \hat{I}_T^-} - \frac{1}{2}\frac{\hat{I}_T^+ - \hat{I}_T^-}{\hat{I}_T^+ + \hat{I}_T^-}. \tag{5.6}$$

Since $T^{-1/2}\phi_T(\theta) \xrightarrow{\mathcal{L}} N(0, \sigma^2(\theta))$ and $\sqrt{T}(\tilde{d}_T - \theta) = O_p(1)$, it follows easily by Slutsky's theorem that

$$T^{1/2}(\tilde{d}_T - \theta) \xrightarrow{\mathcal{L}} N\left(0, \sigma^{-2}(\theta)\right) \quad \text{as } T \to \infty.$$

Theorem 5.1. *Under the assumptions stated above, the maximum probability estimator $\tilde{d}_T$ defined in the previous section is asymptotically normal. More precisely*

$$T^{1/2}(\tilde{d}_T - \theta) \xrightarrow{\mathcal{L}} N\left(0, \frac{1}{\sigma^2(\theta)}\right) \quad \textit{as } T \to \infty.$$

6. Asymptotic efficiency of maximum probability estimator

Let $\theta_T = \theta + hT^{-1/2}$ as defined in the previous section. By Lemmas 2.1 and 2.2 and stationarity and ergodicity of the process $\{\xi_t\}$

$$\Lambda_T(\theta) - h\,\Delta_T(\theta) \to -\tfrac{1}{2}h^2\sigma^2(\theta) \quad \text{a.s. } [P_\theta] \tag{6.1}$$

in view of (2.6). However,

$$\Delta_T(\theta) = T^{-1/2}\phi_T(\theta) \xrightarrow{\mathcal{L}} N\left(0, \sigma^2(\theta)\right),$$

by the central limit theorem for stochastic integrals [cf. Basawa and Prakasa Rao (1980)]. Hence

$$L(\Lambda_T(\theta)|P_\theta) \xrightarrow{\mathcal{L}} N\left(-\tfrac{1}{2}h^2\sigma^2(\theta), h^2\sigma^2(\theta)\right) \equiv L, \quad \text{(say)} \tag{6.2}$$

when θ is the true parameter. Since

$$\int \exp \lambda \, \mathrm{d}L(\lambda) = 1,$$

it follows that the measures $\{P_\theta\}$ and $\{P_{\theta_T}\}$ are contiguous by extending Theorem 6.1, of Roussas (1972, p. 33) to a family of measures. Hence by Corollary 7.2 of Roussas (1972, p. 35), it follows that

$$\Lambda_T(\theta) - h\,\Delta_T(\theta) \to -\tfrac{1}{2}h^2\sigma^2(\theta) \tag{6.3}$$

in P_{θ_T}-probability as $T \to \infty$ and

$$L(\Lambda_T(\theta)|P_{\theta_T}) \xrightarrow{\mathcal{L}} N\left(\tfrac{1}{2}h^2\sigma^2(\theta), h^2\sigma^2(\theta)\right) \tag{6.4}$$

as $T \to \infty$. Hence

$$L(\Delta_T(\theta)|P_{\theta_T}) \xrightarrow{\mathcal{L}} N\left(h\sigma^2(\theta), \sigma^2(\theta)\right). \tag{6.5}$$

By the contiguity of $\{P_\theta\}$ and $\{P_{\theta T}\}$, it follows that

$$\hat{I}_T^{\pm} \to -\sigma^2(\theta) \text{ in } P_{\theta_T}\text{-probability} \tag{6.6}$$

and

$$\sqrt{T}(\tilde{d}_T - \theta) = \mathrm{O}_p(1) \text{ in } P_{\theta_T}\text{-probability} \tag{6.7}$$

by Proposition 6.1 of Roussas (1972, p. 31). Now relation (6.5) proves that

$$L\left(\sqrt{T}(\tilde{d}_T - \theta)|P_{\theta_T}\right) \xrightarrow{\mathcal{L}} N\left(h, \frac{1}{\sigma^2(\theta)}\right) \tag{6.8}$$

as $T \to \infty$ by using eq. (5.2) and arguments in Section 5. In particular, it follows that

$$P_{\theta_T}\left(\sqrt{T}(\tilde{d}_T - \theta_T) \in (-1, 1)\right) \to 2\Phi(\sigma(\theta)) - 1 \tag{6.9}$$

as $T \to \infty$ and for any $\varepsilon > 0$, $\delta > 0$, and for large $|h|$

$$\lim_{T\to\infty} P_{\theta_T}\left(|\sqrt{T}(\tilde{d}_T - \theta_T)| < \delta|h|\right) = 2\Phi(\delta|h|\sigma(\theta)) - 1 \tag{6.10}$$

where Φ is the standard normal distribution function.

In view of (6.9) and (6.10), it can now be shown by the standard arguments [see Theorem 3.1, Weiss and Wolfowitz (1974)] that the estimator $\tilde{d}_T$ is asymptotically efficient in the sense of Weiss and Wolfowitz. We omit the details.

7. Example

Consider the class of diffusion processes satisfying the stochastic differential equation

$$d\xi(t)=[a(t,\xi(t))+\theta b(t,\xi(t))]\,dt+\sigma(t,\xi(t))\,dW(t), \quad t\geq 0, \quad \xi(0)=\xi_o. \tag{7.1}$$

Suppose the functions $a(\cdot,\cdot)$ $b(\cdot,\cdot)$ and $\sigma(\cdot,\cdot)$ are such that a unique solution exists on $[0,T]$ for every $T>0$. Let $\theta\in\Theta$ and θ_o be the true parameter. It is known that

$$\begin{aligned}\log L_T(\theta) &\equiv \log \frac{d\mu_\theta^T}{d\mu_{\theta o}^T} \\ &= (\theta-\theta_o)\int_0^T \left\{\frac{b(t,\xi(t))}{\sigma\sigma(t,\xi(t))}\right\} dW(t) \\ &\quad -\tfrac{1}{2}(\theta-\theta_o)^2 \int_0^T \left\{\frac{b(t,\xi(t))}{\sigma(t,\xi(t))}\right\}^2 dt \qquad (7.2) \\ &= (\theta-\theta_o)\alpha_T T^{1/2} - \tfrac{1}{2}(\theta-\theta_o)^2 \beta_T T \quad \text{(say)}. \qquad (7.3)\end{aligned}$$

Let

$$Z_T(\theta)=\int_{\theta-T^{-1/2}}^{\theta+T^{-1/2}} L_T(t)\,dt.$$

Then $dZ_T(\theta)/d\theta=0$ if and only if

$$\alpha_T=\beta_T T^{1/2}(\theta-\theta_o). \tag{7.4}$$

Let $\tilde{\theta}_T$ denote the solution of (7.4). Observe that

$$\begin{aligned}\left.\frac{d^2 Z_T(\theta)}{d\theta^2}\right|_{\theta=\tilde{\theta}_T} &= \\ &= L_T(\tilde{\theta}_T+T^{-1/2})\left[\alpha_T T^{1/2}-(\tilde{\theta}_T+T^{-1/2}-\theta_o)\beta_T T\right] \\ &\quad - L_T(\tilde{\theta}_T-T^{-1/2})\left[\alpha_T T^{1/2}-(\tilde{\theta}_T-T^{-1/2}-\theta_o)\beta_T T\right]. \qquad (7.5)\end{aligned}$$

In view of the fact that $L_T(\tilde{\theta}_T+T^{-1/2})=L_T(\tilde{\theta}_T-T^{-1/2})$ and $\alpha_T=\beta_T T^{1/2}(\tilde{\theta}_T-\theta_o)$, it is easily seen that

$$\left.\frac{d^2 Z_T(\theta)}{d\theta^2}\right|_{\theta=\tilde{\theta}_T} < 0.$$

Hence $\tilde{\theta}_T$ is the maximum probability estimator. This is the same as the maximum likelihood estimator here since the likelihood equation is given by $d\log L_T(\theta)/d\theta=0$, i.e. $\alpha_T=\beta_T T^{1/2}(\theta-\theta_o)$ which is the same as eq. (7.4).

Akritas (1978) studied asymptotic properties of maximum probability estimators for stochastic differential equations of the type (7.1).

Appendix A. Uniform ergodic theorem

Let $\{\xi(t),\ t\geqslant 0\}$ be a stationary ergodic process. Suppose that $g(s,x)$ is continuous in s a.s. $[P_\theta]$ and $|g(s,x)|\leqslant h(x)$ for all $s\in I(\theta)$ where $I(\theta)$ is a closed interval containing θ. Furthermore, suppose that $E_\theta[g(s,\xi(0))]=0$ for all $s\in I(\theta)$ and $E_\theta[h(\xi(0))]<\infty$. Then, as $T\to\infty$,

$$\sup_{s\in I(\theta)}\left|\frac{1}{T}\int_0^T g(s,\xi(t))\,\mathrm{d}t\right|\to 0 \quad a.s.\ [P_\theta]. \tag{A.1}$$

Proof. Without loss of generality assume that $g(s,x)$ is continuous in s for all $x\in R$. Let $I(s,\rho)$ be an open interval of length 2ρ centered at s and define

$$U(x;s,\rho)=\inf\{g(s;x): s\in I(\theta)\cap\bar{I}(s,\rho)\}. \tag{A.2}$$

Since $g(s,x)$ is continuous and $I(\theta)\cap\bar{I}(s,\rho)$ is closed, it follows that $U(x;s,\rho)\uparrow g(s,x)$ as $\rho\downarrow 0$ a.s. $[P_\theta]$. Note that $|E_\theta\{U(\xi(0);s,\rho)\}|<\infty$ since $E_\theta[h(\xi(0))]<\infty$. Hence, by the monotone convergence theorem, it follows that

$$E_\theta[U(\xi(0);s,\rho)]\to 0 \quad \text{as } \rho\downarrow 0. \tag{A.3}$$

Therefore, given $\varepsilon>0$, there exists $\rho_1=\rho_1(s,\varepsilon)>0$ such that

$$E_\theta[U(\xi(0);s,\rho_1)]>-\varepsilon. \tag{A.4}$$

Let $\rho_2=\frac{1}{2}\rho_1(s,\varepsilon)$ and consider the closed interval $\bar{I}(s,\rho_2)$ centered at s and length $2\rho_2$. Note that for any $s'\in I(\theta)\cap\bar{I}(s,\rho_2)$,

$$U(x;s',\rho_2)\geqslant U(x;s,\rho_1), \quad x\in R$$

since $I(s',\rho_2)\subset I(s,\rho_1)$. Therefore, given $s\in I(\theta)$,

$$E_\theta[U(\xi(0);s',\rho_2)]\geqslant -\varepsilon \tag{A.5}$$

for all $s'\in I(\theta)\cap\bar{I}(s,\rho_2)$.

Cover $I(\theta)$ by $\{I(s,\rho_2): s\in I(\theta)\}$. Since $I(\theta)$ is compact, there exists a finite number of open intervals $I(s_i,\rho_{2i})$, $1\leqslant i\leqslant k$ covering $I(\theta)$. It is clear that

$$\min_{1\leqslant i\leqslant k}\{E_\theta(U(\xi(0);s_i,\rho_{2i}))\}\geqslant -\varepsilon. \tag{A.6}$$

Given $s\in I(\theta)$, there exists s_i such that $s\in I(s_i,\rho_{2i})$ for some $1\leqslant i\leqslant k$ and

$$g(s,x)\geqslant U(x;s_i,\rho_{2i}), \quad x\in R \tag{A.7}$$

and hence

$$\frac{1}{T}\int_0^T g(s,\xi(t))\,\mathrm{d}t\geqslant\frac{1}{T}\int_0^T U(\xi(t);s_i,\rho_{2i})\,\mathrm{d}t \tag{A.8}$$

for all $T\geqslant 0$. Therefore

$$\inf_{s\in I(\theta)}\frac{1}{T}\int_0^T g(s,\xi(t))\,\mathrm{d}t\geqslant\min_{1\leqslant i\leqslant k}\left\{\frac{1}{T}\int_0^T U(\xi(t);s_i,\rho_{2i})\,\mathrm{d}t\right\}. \tag{A.9}$$

However, by the ergodic theorem,

$$\frac{1}{T}\int_0^T U(\xi(t); s_i, \rho_{2i})\,\mathrm{d}t \to E_\theta[U(\xi(0); s_i, \rho_{2i})] \tag{A.10}$$

as $T\to\infty$ a.s. $[P_\theta]$ for $1\leqslant i\leqslant k$. Hence, by (A.6), we have that

$$\liminf_{T\to\infty}\ \inf_{s\in I(\theta)}\left\{\frac{1}{T}\int_0^T g(s,\xi(t))\,\mathrm{d}t\right\}\geqslant -\varepsilon \quad \text{a.s. } [P_\theta]. \tag{A.11}$$

Similarly we obtain that

$$\limsup_{T\to\infty}\ \sup_{s\in I(\theta)}\left\{\frac{1}{T}\int_0^T g(s,\xi(t))\,\mathrm{d}t\right\}\leqslant \varepsilon \quad \text{a.s. } [P_\theta]. \tag{A.12}$$

Relations (A.11) and A.12) together prove the theorem.

Remarks. The proof given above is analogous to that of Roussas (1975) in the i.i.d. case. The theorem can be easily extended to higher dimensions.

Appendix B. Uniform ergodic theorem for stochastic integrals

Let $\{W_t,\ t\geqslant 0\}$ be the standard Wiener process and $\{\mathcal{F}_t\}$ be an increasing family of σ-algebras such that $\sigma\{W_s: 0\leqslant s\leqslant t\}\subset\mathcal{F}_t$ and $\mathcal{F}_t$ is independent of $\sigma\{W_s - W_t: s\geqslant t\}$. Let $\{\xi(t)\}$ be a stationary stochastic process such that $\{\xi(t)\}$ is adapted to $\{\mathcal{F}_t\}$. Let $h(\theta, x)$ be differentiable with respect to θ for $\theta\varepsilon[-1,1]$ and further suppose that $h^{(1)}(\theta, x)$ is Lipschitz in θ with Lipschitzian function $C(x)$. Let us assume further that

$$E[C(\xi(0))]^2<\infty, \tag{B.1}$$

$$h(-1,x)=h(1,x)=0, \tag{B.2}$$

and

$$h^{(1)}(-1,x)=h^{(1)}(1,x)=0. \tag{B.3}$$

It is easy to see that $\int_0^T h(\theta,\xi(t))\,\mathrm{d}W(t)$ exists as an Ito-integral for every $\theta\in[-1,1]$. Furthermore, by Lemma 4.3 of Prakasa Rao and Rubin (1981), given any $\gamma>\frac{1}{2}$, there exists $H>0$ such that

$$\limsup_{T\to\infty}\frac{\sup_\theta\left|\int_0^T h(\theta,\xi(t))\,\mathrm{d}W(t)\right|}{T^{1/2}(\log T)^\gamma}\leqslant H \quad \text{a.s.}$$

Therefore

$$\sup_\theta\left|\frac{1}{T}\int_0^T h(\theta,\xi(t))\,\mathrm{d}W(t)\right|\leqslant\frac{HT^{1/2}(\log T)^\gamma}{T}$$

with probability one for large T. By letting $T\to\infty$ we obtain that

$$\sup_\theta\left|\frac{1}{T}\int_0^T h(\theta,\xi(t))\,\mathrm{d}W(t)\right|\to 0\quad\text{as } T\to\infty$$

almost surely.

In general, let $g(\theta,x)$ be differentiable with respect to θ on $[-1,1]$. Then there exists a cubic polynomial $P(\theta,x)$ in θ such that

$$g(-1,x)=P(-1,x);\qquad g(1,x)=P(1,x), \tag{B.4}$$

$$g^{(1)}(-1,x)=P^{(1)}(-1,x);\qquad g^{(1)}(1,x)=P^{(1)}(1,x), \tag{B.5}$$

and the coefficients of $P(\theta,x)$ are linear functions of $g(-1,x)$, $g(1,x)$, $g^{(1)}(-1,x)$ and $g^{(1)}(1,x)$. Define

$$h(\theta,x)=g(\theta,x)-P(\theta,x). \tag{B.6}$$

Then $h(\theta,x)$ satisfies (B.2) and (B.3). Suppose that $g^{(1)}(\theta,x)$ is Lipschitz in θ with Lipschitzian function $C(x)$, satisfying

$$E\left[g^{(1)}(1,\xi(0))\right]^2<\infty \tag{B.7}$$

and

$$E[C(\xi(0))]^2<\infty. \tag{B.8}$$

It is easy to see that conditions (B.4), (B.5), (B.7) and (B.8) imply that $h(\theta,x)$ defined by (B.6) satisfies the conditions (B.1) to (B.3) and hence

$$\sup_\theta\left|\frac{1}{T}\int_0^T h(\theta,\xi(t))\,\mathrm{d}W(t)\right|\to 0\quad\text{as } T\to\infty\quad\text{a.s.}$$

However, it is obvious that

$$\sup_\theta\left|\frac{1}{T}\int_0^T P(\theta,\xi(t))\,\mathrm{d}W(t)\right|\to 0\quad\text{as } T\to\infty\quad\text{a.s}$$

since $\theta\in[-1,1]$ and $P(\theta,x)$ is a polynomial in θ. Hence

$$\sup_\theta\left|\frac{1}{T}\int_0^T g(\theta,\xi(t))\,\mathrm{d}W(t)\right|\to 0\quad\text{as } T\to\infty\quad\text{a.s.}$$

Remarks. The restriction that $\theta\in[-1,1]$ can be changed to θ ranging over a compact set by suitable reparametrization.

References

[1] Akritas, M.G. (1978). Contiguity of probability measures associated with continuous time stochastic processes. Thesis. The University of Wisconsin, Madison, WI.

[2] Basawa, I.V. and Prakasa Rao, B.L.S. (1980). *Statistical Inference for Stochastic Processes*. Academic Press, London.

[3] Dorogovcev, A.Ja. (1976). The consistency of an estimate of a parameter of a stochastic differential equation. *Theory Prob. Math. Statist.* **10**, 73–82.
[4] Gikhman, I.I. and Skorokhod, A.V. (1972). *Stochastic Differential Equations*. Springer-Verlag, Berlin.
[5] Prakasa Rao, B.L.S. (1979a). Asymptotic theory for nonlinear least squares estimators for diffusion processes. (Preprint). Indian Statistical Institute, New Delhi.
[6] Prakasa Rao, B.L.S. (1979b). Nonparametric estimation for continuous time Markov processes via delta-families. *Publ. Inst. Stat. Univ. Paris* **24**, 79–97.
[7] Prakasa Rao, B.L.S. (1981). The Bernstein–von Mises theorem for a class of diffusion processes (in Russian). *Theory Random Process.* **9**, to appear.
[8] Prakasa Rao, B.L.S. and Rubin, H. (1981). Asymptotic theory of estimation in non-linear stochastic differential equations. *Sankhyā Ser. A*, to appear.
[9] Roussas, G.G. (1972). *Contiguity of Probability Measures*. Cambridge Univ. Press.
[10] Roussas, G.G. (1975). Asymptotic properties of the maximum probability estimates in the i.i.d. case. In. M.L. Puri, Ed., *Statistical Inference and Related Topics*. Academic Press, New York.
[11] Roussas, G.G. (1977). Asymptotic properties of the maximum probability estimates in Markov processes. *Ann. Inst. Statist. Math.* **29**, 203–219.
[12] Weiss, L. and Wolfowitz, J. (1974). *Maximum Probability Estimators and Related Topics*. Springer-Verlag, Berlin.

G. Kallianpur, P.R. Krishnaiah, J.K. Ghosh, eds., *Statistics and Probability: Essays in Honor of C.R. Rao*

ON THE DEGENERATION OF THE VARIANCE IN THE ASYMPTOTIC NORMALITY OF SIGNED RANK STATISTICS

Madan L. PURI and Stefan S. RALESCU
Indiana University, Bloomington, IN, and Brown University, Providence, RI, U.S.A.

The purpose of this paper is to establish the asymptotic normality of simple linear signed rank statistics S_N^+ considered by Hušková (1970), Koul and Staudte (1972), and Puri and Ralescu (1980) for the case when the score-generating function is discontinuous and $\mathrm{Var}(S_N^+)$ compared with the variance of S_N^+ under the hypothesis of symmetry is allowed some degree of degeneracy.

The results obtained are extensions of those by Hájek (1968), Dupač and Hájek (1969), Dupač (1970), Koul and Staudte (1972) and Puri and Ralescu (1980).

1. Preliminaries

Let $X_{N1},\dots, X_{NN}$, $N\geqslant 1$ be independent random variables, with continuous distribution function $F_{N1},\dots, F_{NN}$ respectively, and let R_{Ni}^+ be the rank of $|X_{Ni}|$ among $|X_{N1}|,\dots,|X_{NN}|$. Consider the statistic

$$S_N^+ = \sum_{i=1}^{N} c_{Ni} a_N(R_{Ni}^+) \operatorname{sgn} X_{Ni} \tag{1.1}$$

where $c_{N1},\dots, c_{NN}$ are known regression constants, $a_N(1),\dots, a_N(N)$ are scores and $\operatorname{sgn} x=1$ if $x\geqslant 0$, $\operatorname{sgn} x=-1$ if $x<0$.

For simplicity of notation, we shall drop the subscript N in X_{Ni}, c_{Ni} and R_{Ni}^+ in the sequel.

In order to study the asymptotic behavior of S_N^+, the ratio $\mathrm{Var}(S_N^+)/\sum_{i=1}^N c_i^2$ plays an important role [see Hájek (1968), Dupač and Hájek (1969), Dupač (1970) and Koul and Staudte (1972)]. For the case of the unit step score-generating function $\psi(t)=1$ for $t\geqslant v$, $\psi(t)=0$ for $t<v$ $(0<v<1)$, under suitable conditions on the distribution functions and regression constants, we shall prove that if the ratio $\mathrm{Var}(S_N^+)/\sum_{i=1}^N c_i^2$ goes to zero at most at the rate $N^{-\alpha}$ for some $0<\alpha<\frac{1}{2}$, then S_N^+ is asymptotically normal with natural parameters $(E(S_N^+), \mathrm{Var}(S_N^+))$ as well as with some other simpler parameters (μ_N^+, σ_N^2).

We assume that the c_i's satisfy the condition

$$\max_{1\leqslant i\leqslant N} c_i^2 / \sum_{i=1}^{N} c_i^2 = \mathrm{O}(N^{-1/2}). \tag{1.2}$$

Let $F_i^*(x)$ be the distribution function of $|X_i|$ and define

$$H_N^*(x) = \frac{1}{N} \sum_{i=1}^{N} F_i^*(x),$$

$$H_N^{*-1}(t) = \inf\{x\colon H_N^*(x) \geqslant t\}, \quad 0<t<1,$$

$$L_i(t) = F_i(H^{*-1}(t)), \quad 0<t<1,$$

$$M_i(t) = -F_i(-H^{*-1}(t)), \quad 0<t<1,$$

$$G_i(t) = F_i^*(H^{*-1}(t)) = L_i(t) + M_i(t), \quad 0<t<1. \tag{1.3}$$

Assume that the scores are generated by a function $\psi(t)$, $0<t<1$, either by interpolation

$$a_N(i) = \psi(i/(N+1)), \quad 1 \leqslant i \leqslant N \tag{1.4}$$

or by a procedure satisfying

$$\sum_{i=1}^{N} |a_N(i) - \psi(i/(N+1))| = O(1). \tag{1.5}$$

If $v \in (0,1)$ represents a jump point of the score-generating function ψ, then for every $K>0$ we assume the existence of the derivatives $L_i'(t)$ and $M_i'(t)$ in the interval $|t-v| \leqslant KN^{-1/2}Lg^{1/2}N$ and the satisfaction of the following conditions.

$$\max_{1 \leqslant i \leqslant N} |L_i'(t)| = O(1), \tag{1.6}$$

$$\max_{1 \leqslant i \leqslant N} |M_i'(t)| = O(1), \tag{1.7}$$

$$\max_{1 \leqslant i \leqslant N} \sup_{|t-v| \leqslant KN^{-1/2}Lg^{1/2}N} |L_i'(t) - L_i'(v)| = O(N^{-1/2}Lg^{1/2}N), \tag{1.8}$$

$$\max_{1 \leqslant i \leqslant N} \sup_{|t-v| \leqslant KN^{-1/2}Lg^{1/2}N} |M_i'(t) - M_i'(v)| = O(N^{-1/2}Lg^{1/2}N). \tag{1.9}$$

Another condition concerning the G_i's that we use is:

$$\liminf_{N\to\infty} N^{-1} \sum_{i=1}^{N} G_i(v)(1-G_i(v)) > 0. \tag{1.10}$$

Sometimes, mainly for purposes of applications, we replace (1.6)–(1.10) by the following condition which is easier to verify:

Suppose that each F_i has a density f_i. For each $\varepsilon>0$ denote $I_\varepsilon = (H^{*-1}(v) - \varepsilon, H^{*-1}(v) + \varepsilon)$.

Suppose that there exist $\varepsilon_1, \varepsilon_2, \varepsilon_3 > 0$ such that:

(a)

$$\liminf_{N\to\infty} H_N^{*-1}(v) > 0, \tag{1.11}$$

(b) $f_i'(x)$ are uniformly bonded (in x, i, N) on $I_{\varepsilon_1} \cup (-I_{\varepsilon_1})$,
(c) for all $N \geqslant 1$,

$$\frac{1}{N} \operatorname{Card}\left\{1 \leqslant i \leqslant N: \inf_{x \in I_{\varepsilon_1}} f_i^*(x) > \varepsilon_2\right\} > \varepsilon_3$$

where f_i^* is the density of F_i^*.

The last condition that we require concerns some possible degeneration of $\operatorname{Var}(S_N^+)$ in the form

$$\liminf_{N \to \infty} \operatorname{Var}(S_N^+) \Big/ \left(N^{-\alpha} \sum_{i=1}^{N} c_i^2 \right) > 0 \tag{1.12}$$

for some $0 < \alpha < \frac{1}{2}$.

Alternatively we shall assume that (1.12) holds with $\operatorname{Var}(S_N^+)$ replaced by some approximate variance σ_N^2:

$$\liminf_{N \to \infty} \sigma_N^2 \Big/ \left(N^{-\alpha} \sum_{i=1}^{N} c_i^2 \right) > 0 \quad (0 < \alpha < \tfrac{1}{2}). \tag{1.13}$$

2. Main theorems

Let $u(t)$ be 1 or 0 according to $t \geqslant 0$ or $t < 0$.

The main result of this paper is the following theorem:

Theorem 2.1. *Let S_N^+ given by* (1.1) *have scores given by* (1.5) *where* $\psi(t) = u(t - v)$, $0 < v < 1$.

Then S_N^+ is asymptotically normal with natural parameters $(E(S_N^+), \operatorname{Var}(S_N^+))$ *if any of the following sets of conditions is satisfied*:

$$(\tilde{C}_1^+): (1.2), (1.6), (1.7), (1.8), (1.9), (1.10), (1.12),$$

$$(\tilde{C}_2^+): (1.2), (1.11), (1.12).$$

Proof. We show that S_N^+ is asymptotically equivalent to its projection $\hat{S}_N^+$ onto the space of linear statistics and then that $\hat{S}_N^+$ is asymptotically equivalent to a sum of independent random variables to which the Lindeberg central limit theorem applies.

Let us begin by assuming that scores are given by (1.4) and that $(\tilde{C}_1^+)$ holds.

First we would like to derive an upper bound for the residual variance $E(S_N^+ - \hat{S}_N^+)^2$, where:

$$\hat{S}_N^+ = \sum_{i=1}^{N} E(S_N^+ | X_i) - (N-1) E(S_N^+).$$

This will be accomplished by using the Residual variance inequality [see Hájek (1968) and Koul and Staudte (1972)]:

$$E(S_N^+ - \hat{S}_N^+)^2 \leqslant \sum_{i=1}^{N} c_i^2 E(a(R_i^+) - E(a(R_i^+)|X_i))^2$$

$$+ \sum_{i \neq j} c_i c_j \Big\{ E(\text{sgn } X_i \text{sgn } X_j \text{Cov}(a(R_i^+), a(R_j^+)|X_i, X_j))$$

$$+ E\{\text{sgn } X_i \text{sgn } X_j [E(a(R_i^+)|X_i, X_j) - E(a(R_i^+)|X_i)]$$

$$\times [E(a(R_j^+)|X_i, X_j) - E(a(R_j^+)|X_j)]\}$$

$$- \sum_{k \neq i, j} \text{Cov}\{E(\text{sgn } X_i a(R_i^+)|X_k), E(\text{sgn } X_j a(R_j^+)|X_k)\} \Big\}.$$

We investigate each term in the above inequality. The proof is divided in several steps:

Lemma 2.1. *Let $x, y \in \mathbb{R}$. Then for each $K_1 > \sqrt{6}$ there exists a $K_2 \geqslant \frac{3}{2}$ such that for all $N > N_0(K_1)$ we have*

(i) $v - H^*(|x|) > K_1 N^{-1/2} Lg^{1/2} N \Rightarrow P(R_i^+ \geqslant V | X_i = x, X_j = y) < N^{-K_2}$,

(ii) $v - H^*(|x|) < -K_1 N^{-1/2} Lg^{1/2} N \Rightarrow P(R_i^+ \leqslant V | X_i = x, X_j = y) < N^{-K_2}$ *where* $V = [(N+1)v]$. ($[\cdot]$ = *integer part*).

Furthermore, (i) *and* (ii) *remain true even when the condition $X_j = y$ is omitted.*

Let

$$D^2 = N^{-1} \sum_{i=1}^{N} G_i(v)(1 - G_i(v)).$$

Lemma 2.2. *Suppose that $|v - H^*(|x|)| \leqslant K_3 N^{-1/2} Lg^{1/2} N$. Then for sufficiently large N, we have*

(i) $$\left| \sum_{i=1}^{N} F_i^*(|x|)(1 - F_i^*(|x|)) - ND^2 \right| \leqslant K_4 N^{1/2} Lg^{1/2} N,$$

(ii) $$\left| \phi\left(V; \sum_{i=1}^{N} F_i^*(|x|), \sum_{i=1}^{N} F_i^*(|x|)(1 - F_i^*(|x|)) \right) - \phi\left(Nv; \sum_{i=1}^{N} F_i^*(|x|), ND^2 \right) \right| \leqslant K_5 N^{-1} Lg^{1/2} N,$$

(iii)
$$\left|\Phi\left(V;\ \sum_{i=1}^{N} F_i^*(|x|),\ \sum_{i=1}^{N} F_i^*(|x|)(1-F_i^*(|x|))\right)\right.$$
$$\left. -\Phi\left(Nv;\ \sum_{i=1}^{N} F_i^*(|x|),\ ND^2\right)\right| \leqslant K_6 N^{-1/2} Lg^{1/2} N$$

where $\phi(x;\mu,\sigma^2)$, $\Phi(x;\mu,\sigma^2)$ *denote the normal density, respectively the normal distribution function with parameters* (μ,σ^2).

The proofs of Lemmas 2.1 and 2.2 are analogous to those of Lemmas 5 and 6 of Dupač and Hájek (1969) and are therefore omitted.

Lemma 2.3. *For* $N\to\infty$, *we have*
$$E\big(a(R_i^+)-E(a(R_i^+)|X_i)\big)^2=\mathrm{o}(N^{-\alpha})$$
uniformly in $1\leqslant i\leqslant N$.

Proof. Let $\nabla^+(X_i)=E(a(R_i^+)|X_i)-[E(a(R_i^+)|X_i)]^2$. Then, by conditioning, we obtain
$$E\big[a(R_i^+)-E(a(R_i^+)|X_i)\big]^2=E(\nabla^+(X_i)).$$

Now, by definition
$$E(a(R_i^+)|X_i=x)=P(R_i^+>V|X_i=x).$$

Thus
$$\nabla^+(X_i=x)=P(R_i^+>V|X_i=x)\cdot P(R_i^+\leqslant V|X_i=x).$$

Let $I=\{x\colon |H^*(|x|)-v|\leqslant K_1N^{-1/2}Lg^{1/2}N\}$ with $K_1>\sqrt{6}$. By Lemma 2.1, if $x\notin I$ we have:
$$\nabla^+(X_i=x)<N^{-K_2}\quad \text{for every } N>N_0(K_1),\ K_2\geqslant\tfrac{3}{2}.$$

On the other hand, if $x\in I$, then since $P(R_i^+=k\,|\,X_i=x)=B^i(k, F_1^*(|x|),\ldots,F_N^*(|x|))$ (in the notation used by Dupač and Hájek (1969)) we obtain, using Lemmas 2.1 and 2.2, that
$$\nabla^+(X_i=x)=\Big\{\sum_{k>V} B^i(k, F_1^*(|x|),\ldots,F_N^*(|x|))\Big\}$$
$$\times\Big\{\sum_{l\leqslant V} B^i(l, F_1^*(|x|),\ldots,F_N^*(|x|))\Big\}$$
$$=\cdots=\Phi\left(\frac{H^*(|x|)-v}{DN^{-1/2}}\right)\left\{1-\Phi\left(\frac{H^*(|x|)-v}{DN^{-1/2}}\right)\right\}+\theta_1N^{-1/2}Lg^{1/2}N$$

for sufficiently large N, $|\theta_1|\leqslant K_7$. Here Φ denotes the standard normal distribution function. We use ϕ for its density function in the sequel.

We observe that the last equality remains true even if we enlarge I to

$$I' = \{x: |H^*(|x|) - v| \leqslant K_9 D N^{-1/2} L g^{1/2} N\},$$

where K_9 is such that $K_9 = {}^K 1/2 K_8$ with $K_8 \leqslant D \leqslant \frac{1}{2}$.

Now, using (1.6)–(1.9) it is easy to show that

$$N^{-1/2} L g^{1/2} N \int_{I'} \theta_1 \, \mathrm{d} F_i(x) = \mathrm{o}(N^{-\alpha})$$

and

$$\int_{I'} \Phi\left(\frac{H^*(|x|) - v}{D N^{-1/2}}\right)\left\{1 - \Phi\left(\frac{H^*(|x|) - v}{D N^{-1/2}}\right)\right\} \mathrm{d} F_i(x) = \mathrm{o}(N^{-\alpha})$$

uniformly in $1 \leqslant i \leqslant N$. Hence

$$E(\nabla^+(x_i)) = \mathrm{o}(N^{-\alpha}) \text{ uniformly in } 1 \leqslant i \leqslant N$$

and the proof follows.

Lemma 2.4. *For $N \to \infty$ we have*

$$E\{\operatorname{sgn} X_i \operatorname{sgn} X_j [E(a(R_i^+)|X_i, X_j) - E(a(R_i^+)|X_i)]$$
$$\times [E(a(R_j^+)|X_i, X_j) - E(a(R_j^+)|X_j)]\} = \mathrm{o}(N^{-1-\alpha}),$$

uniformly in $1 \leqslant i, j \leqslant N$.

The proof of this lemma is similar to that of Lemma 2.3 and is therefore omitted.

Lemma 2.5. *For $N \to \infty$ we have*

$$E[\operatorname{sgn} X_i \operatorname{sgn} X_j \operatorname{Cov}(a(R_i^+), a(R_j^+)|X_i, X_j)]$$
$$= N^{-1} D^2 (L_i'(v) - M_i'(v))(L_j'(v) - M_j'(v)) + \mathrm{o}(N^{-1-\alpha})$$

uniformly in $1 \leqslant i, j \leqslant N$.

Proof. We have

$$\Delta^+ = \operatorname{Cov}(a(R_i^+), a(R_j^+)|X_i = x, X_j = y)$$
$$= \begin{cases} P(R_i^+ > V|X_i = x, X_j = y) P(R_j^+ \leqslant V|X_i = x, X_j = y), \\ \quad \text{if } |x| < |y|, \\ P(R_j^+ > V|X_i = x, X_j = y) P(R_i^+ \leqslant V|X_i = x, X_j = y), \\ \quad \text{if } |x| \geqslant |y|. \end{cases}$$

Let $K_1 > \sqrt{6}$. Denote

$$I = \{(x, y): \max(|H^*(|x|) - v|, |H^*(|y|) - v|) \leqslant K_1 N^{-1/2} L g^{1/2} N\}.$$

By considerations as used in the derivation of (4.11) and (4.12) in Dupač and Hájek (1969) we obtain:

$$\Delta^+(x,y)\begin{cases} <N^{-K_2} \quad \text{for } (x,y)\notin I,\ N>N_0(K_2) \\ =\Phi\left(\dfrac{H^*(|x|)-v}{DN^{-1/2}}\right)\left\{1-\Phi\left(\dfrac{H^*(|y|)-v}{DN^{-1/2}}\right)\right\}+\theta_2 N^{-1/2}Lg^{1/2}N \\ \text{for } N \text{ sufficiently large, } (x,y)\in I,\ |x|<|y| \\ \text{and } |\theta_2|\leqslant K_{10} \\ =\Phi\left(\dfrac{H^*(|y|)-v}{DN^{-1/2}}\right)\left\{1-\Phi\left(\dfrac{H^*(|x|)-v}{DN^{-1/2}}\right)\right\}+\theta_3 N^{-1/2}Lg^{1/2}N \\ \text{for } N \text{ sufficiently large, } (x,y)\in I,\ |x|\geqslant|y| \\ \text{and } |\theta_3|\leqslant K_{11} \end{cases} \tag{2.1}$$

where $K_2 \geqslant \frac{3}{2}$. We note that the *equality* in (2.1) remains true even if we enlarge I to I' where

$$I'=\{(x,y):\ \max(|H^*(|x|)-v|,|H^*(|y|)-v|)\leqslant K_9 DN^{-1/2}Lg^{1/2}N\}$$

where $K_9={}^K1/2K_8$, $K_8\leqslant D\leqslant\frac{1}{2}$.

We have, using (2.1) that

$$\begin{aligned} &E(\operatorname{sgn} X_i \operatorname{sgn} X_j \operatorname{Cov}(a(R_i^+),a(R_j^+))|X_i,X_j)= \\ &=\iint_{I'\cap\{|x|<|y|\}} \operatorname{sgn} x \operatorname{sgn} y\,\Phi\left(\frac{H^*(|x|)-v}{DN^{-1/2}}\right)\left\{1-\Phi\left(\frac{H^*(|y|)-v}{DN^{-1/2}}\right)\right\}\mathrm{d}F_i(x)\,\mathrm{d}F_j(y) \\ &+\iint_{I'\cap\{|x|\geqslant|y|\}} \operatorname{sgn} x \operatorname{sgn} y\,\Phi\left(\frac{H^*(|y|)-v}{DN^{-1/2}}\right)\left\{1-\Phi\left(\frac{H^*(|x|)-v}{DN^{-1/2}}\right)\right\}\mathrm{d}F_i(x)\,\mathrm{d}F_j(y) \\ &\qquad +N^{-1/2}Lg^{1/2}N\iint_{I'} \operatorname{sgn} x \operatorname{sgn} y\,\theta_4(x,y)\,\mathrm{d}F_i(x)\,\mathrm{d}F_j(y)+\theta_5 N^{-K_2} \end{aligned} \tag{2.2}$$

with $|\theta_4|\leqslant K_{12}$, $|\theta_5|\leqslant 1$.

The last two terms are $o(N^{-1-\alpha})$ uniformly in i, j as follows by using (1.6)–(1.9) and $K_2\geqslant\frac{3}{2}$. It remains to estimate the first two terms.

Denote the first term by $\mathfrak{T}$. Consider

$$\mathfrak{T}_1=\iint_{A_{xy}} \Phi\left(\frac{H^*(x)-v}{DN^{-1/2}}\right)\left\{1-\Phi\left(\frac{H^*(y)-v}{DN^{-1/2}}\right)\right\}\mathrm{d}F_i(x)\,\mathrm{d}F_j(y)$$

where

$$A_{xy}=\left\{(x,y):\ \begin{matrix} x>0 \\ y>0 \\ x<y \end{matrix},\ \max(|H^*(x)-v|,|H^*(y)-v|)\leqslant K_9 DN^{-1/2}Lg^{1/2}N\right\}.$$

Set

$$p=\frac{H^*(x)-v}{DN^{-1/2}}, \qquad q=\frac{H^*(y)-v}{DN^{-1/2}},$$

and

$$I''=\{(p,q)\colon \max(|p|,|q|)\leqslant K_9Lg^{1/2}N\}.$$

Then

$$\mathfrak{T}_1=\iint_{I''\cap\{p<q\}}\Phi(p)(1-\Phi(q))\,\mathrm{d}L_i(v+DN^{-1/2}p)\,\mathrm{d}L_j(v+DN^{-1/2}q).$$

Let $\Omega=I''\cap\{p<q\}$ and $\Omega^*=\{p<q\}\setminus\Omega$. Then using (1.6)–(1.9), one can easily show that

$$\mathfrak{T}_1=D^2N^{-1}L_i'(v)L_j'(v)\iint_{\Omega}\Phi(p)(1-\Phi(q))\,\mathrm{d}p\,\mathrm{d}q+o(N^{-1-\alpha}) \tag{2.3}$$

uniformly in $1\leqslant i,\ j\leqslant N$.

Let $\Omega_1^*=\Omega^*\cap\{p>-q\}$. Then, by Fubini's theorem we have:

$$\begin{aligned}N^{-1}\iint_{\Omega_1^*}\Phi(p)(1-\Phi(q))\,\mathrm{d}p\,\mathrm{d}q&=N^{-1}\int_{K_9Lg^{1/2}N}^{\infty}(1-\Phi(q))\left(\int_{-q}^{q}\Phi(p)\,\mathrm{d}p\right)\mathrm{d}q\\&\leqslant 2N^{-1}\int_{K_9Lg^{1/2}N}^{\infty}q\Phi(-q)\,\mathrm{d}q.\end{aligned}$$

Using integration by parts it follows that:

$$N^{-1}\iint_{\Omega_1^*}\Phi(p)(1-\Phi(q))\,\mathrm{d}p\,\mathrm{d}q=o(N^{-1-\alpha}). \tag{2.4}$$

Similarly we can show that:

$$N^{-1}\iint_{\Omega_2^*}\Phi(p)(1-\Phi(q))\,\mathrm{d}p\,\mathrm{d}q=o(N^{-1-\alpha}) \tag{2.5}$$

where $\Omega_2^*=\Omega^*\setminus\Omega_1^*$.

By (2.4) and (2.5)

$$D^2N^{-1}L_i'(v)L_j'(v)\iint_{\Omega^*}\Phi(p)(1-\Phi(q))\,\mathrm{d}p\,\mathrm{d}q=o(N^{-1-\alpha}). \tag{2.6}$$

From (2.3), (2.6) and the fact that

$$\iint_{\{p<q\}}\Phi(p)(1-\Phi(q))\,\mathrm{d}p\,\mathrm{d}q=\tfrac{1}{2}$$

we obtain:

$$\mathfrak{T}_1=\tfrac{1}{2}D^2N^{-1}L_i'(v)L_j'(v)+o(N^{-1-\alpha}) \tag{2.7}$$

uniformly in $1\leqslant i,\ j\leqslant N$.

Let

$$\mathfrak{T}_2 = \iint_{B_{xy}} -\Phi\left(\frac{H^*(-x)-v}{DN^{-1/2}}\right)\left\{1-\Phi\left(\frac{H^*(y)-v}{DN^{-1/2}}\right)\right\} \mathrm{d}F_i(x)\,\mathrm{d}F_j(y)$$

$$= \iint_{I''\cap\{p<q\}} -\Phi(p)(1-\Phi(q))\,\mathrm{d}M_i(v+DN^{-1/2}p)\,\mathrm{d}L_j(v+DN^{-1/2}q)$$

where

$$B_{xy} = \left\{(x,y):\ \begin{matrix} x<0 \\ y>0 \\ -x<y \end{matrix},\max(|H^*(-x)-v|,|H^*(y)-v|)\leqslant K_9 DN^{-1/2}Lg^{1/2}N\right\},$$

$$\mathfrak{T}_3 = \iint_{C_{xy}} \Phi\left(\frac{H^*(-x)-y}{DN^{-1/2}}\right)\left\{1-\Phi\left(\frac{H^*(-y)-v}{DN^{-1/2}}\right)\right\} \mathrm{d}F_i(x)\,\mathrm{d}F_j(y)$$

and

$$\mathfrak{T}_4 = \iint_{D_{xy}} -\Phi\left(\frac{H^*(x)-y}{DN^{-1/2}}\right)\left\{1-\Phi\left(\frac{H^*(-y)-v}{DN^{-1/2}}\right)\right\} \mathrm{d}F_i(x)\,\mathrm{d}F_j(y)$$

where

$$C_{xy} = \left\{(x,y):\ \begin{matrix} x<0 \\ y<0 \\ -x<-y \end{matrix},\max(|H^*(-x)-v|,|H^*(-y)-v|)\leqslant K_9 DN^{-1/2}Lg^{1/2}N\right\}$$

and

$$D_{xy} = \left\{(x,y):\begin{matrix} x>0 \\ y<0 \\ x<-y \end{matrix},\max(|H^*(x)-v|,|H^*(-y)-v|)\leqslant K_9 DN^{-1/2}Lg^{1/2}N\right\}.$$

We now repeat the steps used in the derivation of (2.7), this time applying them to $\mathfrak{T}_2$, $\mathfrak{T}_3$ and $\mathfrak{T}_4$ to obtain

$$\mathfrak{T}_2 = -\tfrac{1}{2}D^2N^{-1}M_i'(v)L_j'(v)+\mathrm{o}(N^{-1-\alpha}) \quad \text{uniformly in } 1\leqslant i,\ j\leqslant N,$$

$$\mathfrak{T}_3 = \tfrac{1}{2}D^2N^{-1}M_i'(v)M_j'(v)+\mathrm{o}(N^{-1-\alpha}) \quad \text{uniformly in } 1\leqslant i,\ j\leqslant N$$

and

$$\mathfrak{T}_4 = -\tfrac{1}{2}D^2N^{-1}L_i'(v)M_j'(v)+\mathrm{o}(N^{-1-\alpha}) \quad \text{uniformly in } 1\leqslant i,\ j\leqslant N.$$

Thus

$$\mathfrak{T} = \mathfrak{T}_1+\mathfrak{T}_2+\mathfrak{T}_3+\mathfrak{T}_4 = \tfrac{1}{2}D^2N^{-1}(L_i'(v)-M_i'(v))(L_j'(v)-M_j'(v))+\mathrm{o}(N^{-1-\alpha})$$

uniformly in $1\leqslant i,\ j\leqslant N$. (2.8)

Proceeding as above, it can be shown that the second term of (2.2) is the same as (2.8). The proof follows.

Lemma 2.6. *For $N \to \infty$ we have (for $i \neq j$)*

$$\sum_{k \neq i, j} \operatorname{Cov}\{E(\operatorname{sgn} X_i a(R_i^+)|X_k), E(\operatorname{sgn} X_j a(R_j^+)|X_k)\} =$$

$$= D^2 N^{-1}(L_i'(v) - M_i'(v))(L_j'(v) - M_j'(v)) + o(N^{-1-\alpha})$$

uniformly in $1 \leqslant i, j \leqslant N$.

Proof. By Lemma 3.2 in Hájek (1968) we have

$$E(a(R_i^+)\operatorname{sgn} X_i|X_i = x, X_k = z) - E(a(R_i^+)\operatorname{sgn} X_i|X_i = x) =$$

$$= \operatorname{sgn} x[u(|x| - |z|) - F_k^*(|x|)] \cdot P(R_i^+ = V + 1|X_i = x, |X_k| = |x| - 1).$$

From Lemma 2.1, we have

$$P(R_i^+ = V + 1|X_i = x, |X_k| = |x| - 1) < N^{-K_2} \tag{2.9}$$

for some $K_2 \geqslant \frac{3}{2}$ and all $|H^*(|x|) - v| \geqslant K_1 N^{-1/2} Lg^{1/2} N$.

Furthermore Lemmas 2.1 and 2.2 imply:

$$P(R_i^+ = V + 1|X_i = x, |X_k| = |x| - 1) = \phi\left(Nv;\ \sum_{j=1}^{N} F_j^*(|x|), ND^2\right) + \theta_6 N^{-1} Lg^{1/2} N$$

for some $|\theta_6| \leqslant K_{13}$ and all $|H^*(|x|) - v| \leqslant K_1 N^{-1/2} Lg^{1/2} N$.

As before, the last equality remains true even if $|H^*(|x|) - v| \leqslant K_9 D N^{-1/2} Lg^{1/2} N$. Let

$$I' = \{x\colon |H^*(|x|) - v| \leqslant K_9 D N^{-1/2} Lg^{1/2} N\}.$$

Then

$$E(a(R_i^+)\operatorname{sgn} X_i|X_k = z) - E(a(R_i^+)\operatorname{sgn} X_i) =$$

$$= \int \operatorname{sgn} x[u(|x| - |z|) - F_k^*(|x|)]$$

$$\times P(R_i^+ = V + 1|X_i = x, |X_k| = |x| - 1)\,\mathrm{d}F_i(x)$$

$$= \int_{I'} (\cdots)\,\mathrm{d}F_i(x) + \int_{\mathbb{R}\setminus I'} (\cdots)\,\mathrm{d}F_i(x).$$

The second integral is $o(N^{-1-\alpha})$ by (2.9), while the first is equal to

$$\int_{I'} \operatorname{sgn} x[u(|x| - |z|) - F_k^*(|x|)]\,\phi\left(Nv;\ \sum_{j=1}^{N} F_j^*(|x|), ND^2\right)\mathrm{d}F_i(x)$$

$$+ \int_{I'} \operatorname{sgn} x[u(|x| - |z|) - F_k^*(|x|)]\,\theta_6 N^{-1} Lg^{1/2} N\,\mathrm{d}F_i(x).$$

In the last expression, let us denote by $\mathfrak{T}_5$ the first term and $\mathfrak{T}_6$ the second. From (1.6)–(1.9) and the Mean Value Theorem, it follows that:

$$\max_{1 \leqslant k \leqslant N} \ \operatorname*{Sup}_{|p| \leqslant K_9 Lg^{1/2} N} |G_k(v + D N^{-1/2} p) - G_k(v)| = O(N^{-1/2} Lg^{1/2} N). \tag{2.10}$$

Then it is easy to show that

$$D^{-1}N^{-1/2}\int_{\{|p|\leqslant K_9Lg^{1/2}N\}} G_k(v+DN^{-1/2}p)\phi(p)\,\mathrm{d}L_i(v+DN^{-1/2}p)=$$

$$=D^{-1}N^{-1/2}\int_{\{|p|\leqslant K_9Lg^{1/2}N\}} G_k(v)\phi(p)\,\mathrm{d}L_i(v+DN^{-1/2}p)+\mathrm{o}(N^{-1-\alpha})$$

uniformly in i and k.

We write

$$\mathfrak{T}_5=D^{-1}N^{-1/2}\int_{\{x>0:|H^*(x)-v|\leqslant K_9DN^{-1/2}Lg^{1/2}N\}}(\cdots)\,\mathrm{d}F_i(x)$$

$$+D^{-1}N^{-1/2}\int_{\{x<0:\,|H^*(-x)-v|\leqslant K_9DN^{-1/2}Lg^{1/2}N\}}(\cdots)\,\mathrm{d}F_i(x)$$

$$=\mathfrak{T}_5'+\mathfrak{T}_5''.$$

We have:

$$\mathfrak{T}_5'=N^{-1}L_i'(v)\int_{\{|p|\leqslant K_9Lg^{1/2}N\}}[u(p-q)-G_k(v)]\phi(p)\,\mathrm{d}p$$

$$+N^{-1}\mathrm{O}(N^{-1/2}Lg^{1/2}N)\int_{\{|p|\leqslant K_9Lg^{1/2}N\}}[u(p-q)-G_k(v)]\phi(p)\,\mathrm{d}p$$

$$+\mathrm{o}(N^{-1-\alpha}).$$

In the last expression, the second term is $\mathrm{o}(N^{-1-\alpha})$ while the first is equal to

$$N^{-1}L_i'(v)[1-\Phi(q)-G_k(v)]$$

$$-N^{-1}L_i'(v)\int_{\{|p|>K_9Lg^{1/2}N\}}[u(p-q)-G_k(v)]\phi(p)\,\mathrm{d}p.$$

But

$$\left|N^{\alpha}\int_{K_9Lg^{1/2}N}^{\infty}[u(p-q)-G_k(v)]\phi(p)\,\mathrm{d}p\right|\leqslant$$

$$\leqslant N^{\alpha}\int_{K_9Lg^{1/2}N}^{\infty}\phi(p)\,\mathrm{d}p=N^{\alpha}\Phi(-K_9Lg^{1/2}N)\to 0$$

and we obtain:

$$\mathfrak{T}_5'=N^{-1}L_i'(v)[1-\Phi(q)-G_k(v)]+\mathrm{o}(N^{-1-\alpha})$$

uniformly in z, $1\leqslant i\leqslant N$, where $q=(H^*(|z|)-v)/DN^{-1/2}$.

Similarly

$$\mathfrak{T}_5''=-N^{-1}M_i'(v)[1-\Phi(q)-G_k(v)]+\mathrm{o}(N^{-1-\alpha}).$$

Also, it is easy to check that $\mathcal{T}_6=o(N^{-1-\alpha})$. Hence

$$\begin{aligned}E(a(R_i^+)\operatorname{sgn} X_i|X_k=z)-E(a(R_i^+)\operatorname{sgn} X_i)\\ =\mathcal{T}_5'+\mathcal{T}_5''+o(N^{-1-\alpha})\\ =N^{-1}(L_i'(v)-M_i'(v))[1-\Phi(q)-G_k(v)]+o(N^{-1-\alpha})\end{aligned} \tag{2.11}$$

uniformly in $-\infty<z<\infty$.

We now show that under (1.6)–(1.9)

$$\int_{-v/DN^{-1/2}}^{(1-v)/DN^{-1/2}}(1-\Phi(q))\,dG_k(v+DN^{-1/2}q)=G_k(v)+o(N^{-\alpha}). \tag{2.12}$$

Indeed, using integration by parts

$$\int_{-v/DN^{-1/2}}^{(1-v)/DN^{-1/2}}(1-\Phi(q))\,dG_k(v+DN^{-1/2}q)-G_k(v)=$$

$$=\left[1-\Phi\left(\frac{1-v}{DN^{-1/2}}\right)\right]$$

$$+\int_{-\infty}^{\infty}\left[I_{(-(v/DN^{-1/2}),((1-v)/DN^{-1/2}))}(q)G_k(v+DN^{-1/2}q)-G_k(v)\right]\phi(q)\,dq.$$

Let $\mathcal{Q}$ denote the last integral in the above relation. Then:

$$\mathcal{Q}=\int_{\{|q|\geqslant Lg^{1/2}N\}}(\cdots)\phi(q)\,dq+\int_{\{|q|\leqslant Lg^{1/2}N\}}(\cdots)\phi(q)\,dq=\mathcal{Q}_1+\mathcal{Q}_2.$$

Since

$$N^{\alpha}\int_{Lg^{1/2}N}^{\infty}\phi(q)\,dq=N^{\alpha}\Phi(-Lg^{1/2}N)\to 0$$

and

$$|\mathcal{Q}_1|\leqslant\int_{\{|q|\geqslant Lg^{1/2}N\}}\phi(q)\,dq$$

it follows that $\mathcal{Q}_1=o(N^{-\alpha})$. On the other hand, (2.10) entails for sufficiently large N that:

$$\begin{aligned}|N^{\alpha}\mathcal{Q}_2|&\leqslant N^{\alpha}\int_{\{|q|\leqslant Lg^{1/2}N\}}|G_k(v+DN^{-1/2}q)-G_k(v)|\phi(q)\,dq\\ &=O(N^{\alpha-1/2}Lg^{1/2}N).\end{aligned}$$

Thus $\mathcal{Q}_2=o(N^{-\alpha})$ and we get

$$\mathcal{Q}=\mathcal{Q}_1+\mathcal{Q}_2=o(N^{-\alpha}).$$

This, together with the fact that

$$1-\Phi\left(\frac{1-v}{DN^{-1/2}}\right)=o(N^{-\alpha})$$

proves (2.12).

Similarly we can show that

$$\int_{-v/DN^{-1/2}}^{(1-v)/DN^{-1/2}} (1-\Phi(q))^2 \mathrm{d}G_k(v+DN^{-1/2}q) = G_k(v) + \mathrm{o}(N^{-\alpha}). \tag{2.13}$$

Finally, using (2.11), (2.12) and (2.13) and proceeding as in Dupač and Hájek (1969), the proof follows.

By Lemmas 2.3–3.6 and the Residual variance inequality

$$E(S_N^+ - \hat{S}_N^+)^2 = \mathrm{o}\left(N^{-\alpha} \sum_{i=1}^{N} c_i^2 \right). \tag{2.14}$$

Let us show now that

$$E\left\{ \Phi\left(\frac{v - H^*(|X_i|)}{DN^{-1/2}} \right) - u(v - H^*(|X_i|)) \right\}^2 = \mathrm{o}(N^{-\alpha}). \tag{2.15}$$

The left-hand side of (2.15) equals $I_1 + I_2$ where

$$\begin{aligned} I_1 &= \int_{-v/DN^{-1/2}}^{0} \Phi^2(p)\,\mathrm{d}G_i(v+DN^{-1/2}p) \\ &= -\int_{-\infty}^{0} \left[I_{(-v/DN^{-1/2},0)}(p) G_i(v+DN^{-1/2}p) - G_i(v) \right] 2\Phi(p)\phi(p)\,\mathrm{d}p \end{aligned}$$

and

$$\begin{aligned} I_2 &= \int_0^{(1-v)/DN^{-1/2}} [1-\Phi(p)]^2 \mathrm{d}G_i(v+DN^{-1/2}p) \\ &= \Phi^2\left(-\frac{(1-v)}{DN^{-1/2}} \right) + \int_0^{\infty} \left[I_{(0,((1-v')/DN^{-1/2}))}(p) G_i(v+DN^{-1/2}p) - G_i(v) \right] \\ &\qquad \times 2\Phi(-p)\phi(p)\,\mathrm{d}p. \end{aligned}$$

Then, proceeding as in the derivation of (2.12) and (2.13) it follows that $I_1 = \mathrm{o}(N^{-\alpha})$ and $I_2 = \mathrm{o}(N^{-\alpha})$ and hence (2.15) holds true.

Now, using (1.6)–(1.9) and (2.15) we obtain

$$E(Y_i - Z_i)^2 = \mathrm{o}\left(N^{-1-\alpha} \sum_{i=1}^{N} c_i^2 \right) \tag{2.16}$$

where

$$Y_i = \sum_{j=1}^{N} c_j \{ E(a(R_j^+)\,\mathrm{sgn}\, X_j | X_i) - E(a(R_j^+)\,\mathrm{sgn}\, X_j) \}, \quad 1 \leq i \leq N \tag{2.17}$$

and

$$\begin{aligned} Z_i = N^{-1} &\left[\sum_{\substack{j=1 \\ j \neq i}}^{N} c_j (L_j'(v) - M_j'(v)) \right] [u(v - H^*(|X_i|)) - G_i(v)] \\ &+ c_i [E(\mathrm{sgn}\, X_i a(R_i^+) | X_i) - E(\mathrm{sgn}\, X_i a(R_i^+))], \quad 1 \leq i \leq N. \end{aligned} \tag{2.18}$$

Then, since $\hat{S}_N^+ - E(\hat{S}_N^+) = \sum_{i=1}^N Y_i$, (2.14) and (2.16) entail

$$\operatorname{Var}\left(S_N^+ - \sum_{i=1}^N Z_i \right) = \mathrm{o}\left(N^{-\alpha} \sum_{i=1}^N c_i^2 \right). \tag{2.19}$$

Proceeding as in Lemma 13 of Dupač and Hájek (1969), it is easy to show that (1.12) holds if and only if (1.13) holds with $\sigma_N^2 = \sum_{i=1}^N \operatorname{Var}(Z_i)$ and in this case

$$\lim_{N\to\infty} \operatorname{Var}(S_N^+)/\sigma_N^2 = 1.$$

Finally, the asymptotic normality of $\sum_{i=1}^N Z_i$ with parameters $(0, \sigma_N^2)$ follows as in Lemma 14 of Dupač and Hájek (1969) with the help of (1.2), (1.6), (1.7), (1.13) and the Lindeberg central limit theorem.

Hence, since we have proved that

$$\sum_{i=1}^N Z_i/\sigma_N \xrightarrow{\mathfrak{D}} N(0,1), \qquad \left(S_N^+ - E(S_N^+) - \sum_{i=1}^N Z_i \right) \Big/ \sigma_N \xrightarrow{\mathfrak{L}^2} 0$$

and

$$\operatorname{Var}(S_N^+)/\sigma_N^2 \to 1,$$

we obtain

$$(S_N^+ - E(S_N^+))/(\operatorname{Var} S_N^+)^{1/2} \xrightarrow{\mathfrak{D}} N(0,1).$$

Suppose we want to relax condition (1.4) to (1.5). Let us denote the statistic corresponding to (1.4) by S_N^+ and the statistic corresponding to (1.5) by S_N^{+*}. Then, using (1.2) and (1.5)

$$\operatorname{Var}(S_N^+ - S_N^{+*}) = \mathrm{o}\left(N^{-\alpha} \sum_{i=1}^N c_i^2 \right).$$

Consequently, the asymptotic normality of S_N^{+*} easily follows from the last relation and the asymptotic normality of S_N^+.

We have proved Theorem 2.1 under condition $(\tilde{C}_1^+)$. It remains to show that this set of conditions is implied by the conditions $(\tilde{C}_2^+)$. The proof of this fact is similar to the implications $(C_3) \Rightarrow (C_1)$ and $(C_2) \Rightarrow (C_1)$ in Dupač and Hájek (1969, Section 5) and is therefore omitted.

The following theorem shows that under the same conditions $(\tilde{C}_1^+)$ or $(\tilde{C}_2^+)$, S_N^+ is asymptotically normal with (simpler) parameters (μ_N^+, σ_N^2). This problem is of practical interest since μ_N^+ and σ_N^2 are easier to evaluate:

Let us define:

$$\mu_N^+ = \sum_{i=1}^N c_i E[\operatorname{sgn} X_i \psi(H^*(|X_i|))] \tag{2.20}$$

and

$$\sigma_N^2 = \sum_{i=1}^N \operatorname{Var} Z_i' \tag{2.21}$$

where

$$Z_i' = N^{-1}\left[\sum_{\substack{j=1\\ j\neq i}}^{N} c_j\big(L_j'(v)-M_j'(v)\big)\right]\big[u(v-H^*(|X_i|))-G_i(v)\big]$$

$$+c_i\big[\operatorname{sgn} X_i\psi(H^*(|X_i|))-E(\operatorname{sgn} X_i\psi(H^*(|X_i|)))\big]. \tag{2.22}$$

Theorem 2.1. *Let S_N^+ be given by* (1.1) *with scores satisfying* (1.5) *where $\psi(t)=u(t-v)$.*

Assume that ($\tilde{C}_1^+$) *or* ($\tilde{C}_2^+$) *holds, with* (1.12) *replaced by* (1.13), *where σ_N^2 is given by* (2.21).

Then S_N^+ is asymptotically normal with parameters (μ_N^+, σ_N^2) defined in (2.20) *and* (2.21).

Proof. We shall follow the proof of Theorem 2.1 (where it is first assumed that $a(i)=\psi(i/(N+1))$ and that ($\tilde{C}_i^+$) holds).

With Y_i and Z_i defined by (2.17) and (2.18) respectively, we have:

$$E(Y_i-Z_i)^2=o\left(N^{-1-\alpha}\sum_{j=1}^{N} c_j^2\right) \tag{2.23}$$

uniformly in $1\leqslant i\leqslant N$, as follows from (2.11) and (2.15).

Define

$$\Delta_i(X_i)=\{E(\operatorname{sgn} X_i a(R_i^+)|X_i)-E(\operatorname{sgn} X_i a(R_i^+))\}$$

$$-\{\operatorname{sgn} X_i u(H^*(|X_i|)-v)-E(\operatorname{sgn} X_i u(H^*(|X_i|)-v))\}.$$

Proceeding as in Dupač (1970) it can be shown (omitting the details of computation) that

$$\operatorname{Var}(\Delta_i)=E(\Delta_i^2)=O(N^{-1/2})$$

where $\Delta_i=\Delta_i(X_i)$.

Then, since $Z_i=Z_i'+c_i\Delta_i$, we have

$$E(Z_i-Z_i')^2=c_i^2O(N^{-1/2}) \tag{2.24}$$

uniformly in $1\leqslant i\leqslant N$.

But $\hat{S}_N^+-E(\hat{S}_N^+)=\sum_{i=1}^N Y_i$ and from (2.23) and (2.24) we obtain:

$$\operatorname{Var}\left(\hat{S}_N^+-\sum_{i=1}^{N} Z_i'\right)=o\left(N^{-\alpha}\sum_{i=1}^{N} c_i^2\right).$$

This together with (2.14) entails

$$\operatorname{Var}\left(S_N^+-\sum_{i=1}^{N} Z_i'\right)=o\left(N^{-\alpha}\sum_{i=1}^{N} c_i^2\right).$$

Then, proceeding precisely as in the proof of Theorem 2.1, it follows that:

$$\sum_{i=1}^{N} Z_i'/\sigma_N \xrightarrow{\mathfrak{D}} N(0,1),$$

$$\left(S_N^+ - E(S_N^+) - \sum_{i=1}^{N} Z_i'\right)/\sigma_N \xrightarrow{\mathcal{L}^2} 0$$

and

$$\operatorname{Var}(S_N^+)/\sigma_N^2 \to 1. \tag{2.25}$$

Further set

$$\rho_i = E(\operatorname{sgn} X_i a(R_i^+) - E(\operatorname{sgn} X_i u(H^*(|X_i|) - v)), \quad 1 \leqslant i \leqslant N.$$

It can be shown (again omitting the details of computation) that:

$$\rho_i = o(N^{-\alpha/2-1/2}) \quad \text{uniformly in } 1 \leqslant i \leqslant N. \tag{2.26}$$

Now, using the inequality:

$$(E(S_N^+) - \mu_N^+)^2 \leqslant \left(\sum_{i=1}^{N} c_i^2\right) \cdot \left(\sum_{i=1}^{N} \rho_i^2\right)$$

together with (2.26) we get

$$(E(S_N^+) - \mu_N^+)^2 = o\left(N^{-\alpha} \sum_{i=1}^{N} c_i^2\right). \tag{2.27}$$

Finally, making use of (1.13), (2.25) and (2.27), the proof follows.

3. An example

Assume that $X_1, \ldots, X_N$ are i.i.d. with common density function f, and consider the problem of testing the hypothesis of symmetry with normal underlying density H_0: $f(x) = \phi(x)$ against the sequence of shift alternatives

$$H_1: f(x) = \phi(x - \Delta) \quad (\Delta = \Delta_N > 0).$$

Assume that $\Delta \to \infty$ sufficiently slow such that:

$$\limsup_{N \to \infty} \Delta \cdot Lg^{-1/2} N < \tfrac{1}{2}. \tag{3.1}$$

We shall prove that under H_1, (3.1) implies the asymptotic normality of S_N^+ where

$$S_N^+ = \sum_{i=1}^{N} u\left(\frac{R_i^+}{N+1} - v\right) \operatorname{sgn} X_i.$$

First we note that since $\Delta \to \infty$, $H^{*-1}(v) \to \infty$, in such a way that:

$$\lim_{N \to \infty} (H^{*-1}(v) - \Delta) = 0. \tag{3.2}$$

Using (3.2) it is easy to see that condition (1.11) is satisfied. Also, since $c_1 = \cdots = c_N = 1$, it is easy to check that $\sigma_N^2/\sum_{i=1}^N c_i^2 \to 0$, where σ_N^2 is defined by (2.21).

Now, (3.1) implies the existence of a constant $0 < C < \frac{1}{2}$ such that, for sufficiently large N

$$\Delta^2 \leqslant C^2 LgN. \tag{3.3}$$

It can be shown (omitting the details of computation) that for sufficiently large N,

$$\sigma_N^2 / \sum_{i=1}^{N} c_i^2 \geqslant C'\Phi(-2\Delta - 1) \tag{3.4}$$

for some constant $C' > 0$.

Let α and C'' satisfy $\frac{1}{2} < C'' < (1/8C^2)$ and $4C''C^2 < \alpha < \frac{1}{2}$. Thus from (3.3) and (3.4) we have:

$$\sigma_N^2 \Big/ \left(N^{-\alpha} \sum_{i=1}^{N} c_i^2 \right) > C' N^{\alpha - 4C''C^2}.$$

The last relation clearly implies the satisfaction of (1.13). The result follows by an application of Theorem 2.2.

References

Dupač, V. (1970). *A Contribution to the Asymptotic Normality of Simple Linear Rank Statistics. Nonparametric Techniques in Statistical Inference*. Cambridge Univ. Press, London, pp. 75–88.

Dupač, V. and Hájek, J. (1969). Asymptotic normality of simple linear rank statistics under alternatives II. *Ann. Math. Statist.* **40**, 1992, 2017.

Hájek, J. (1968). Asymptotic normality of simple linear rank statistics under alternatives. *Ann. Math. Statist.* **39**, 325–346.

Husková, M. (1970). Asymptotic distribution of simple linear rank statistics for testing symmetry. *Z. Wahrscheinlichkeitstheorie verw. Geb.* **14**, 308–322.

Koul, H.L. and Staudte, R.G. (1972). Asymptotic normality of signed rank statistics. *Z. Wahrscheinlichkeitstheorie verw. Geb.* **22**, 295–300.

Puri, M.L. and Ralescu, S.S. (1980). Asymptotic normality of signed rank statistic with discontinuous score generating function. Preprint.

G. Kallianpur, P.R. Krishnaiah, J.K. Ghosh, eds., *Statistics and Probability: Essays in Honor of C.R. Rao*
© North-Holland Publishing Company (1982) 609–616

AN INTEGRAL EQUATION IN PROBABILITY THEORY AND ITS IMPLICATIONS

B. RAMACHANDRAN
Indian Statistical Institute, India

Let $X_1, X_2, \ldots, X_n$ be $\boldsymbol{n}$ independent and identically distributed random variables (i.i.d. r.v.'s), with F as their common probability distribution function (d.f.). [$F(x)$ will be defined as $P[X_1 \leqslant x]$, so that F is (and all d.f.'s below will be taken to be) right continuous.] Let $X_{1,n} \leqslant \cdots \leqslant X_{n,n}$ be the order-statistics generated by $X_1, \ldots, X_n$. Rossberg (1972) considered the property, which obtains if F is an exponential d.f., that, for any fixed k, $1 \leqslant k < n$, $X_{k+1,n} - X_{k,n}$ has the same distribution as $X_{1,n-k}$, and proved that if this property holds for some $\boldsymbol{n}$ and some $k<n$, and if F is not a lattice d.f. (in particular non-degenerate), then F is necessarily an exponential d.f. The above property, namely,

$$P[X_{k+1,n} - X_{k,n} \geqslant x] = [1-F(x-)]^{n-k} \quad \text{for all real } x \tag{1}$$

is equivalent to

$$F(0-)=0, \tag{2}$$

$$\binom{n}{k} \int \{1-F[(t+x)-]\}^{n-k} \mathrm{d}F^k(t) = [1-F(x-)]^{n-k} \quad \text{for all } x>0.$$

Setting $K=\binom{n}{k}F^k$, we have from (2) that, for all $x>0$,

$$\int_{(0,\infty)} \{1-F[(t+x)-]\}^{n-k} \mathrm{d}K(t) = [1-K(0)][1-F(x-)]^{n-k}$$

so that, if $K(0)>1$, we must have $F=\delta_0$, where δ_a denotes the d.f. degenerate at $\boldsymbol{a}$. If $K(0)=1$, then it follows from the above relation that

$$\int_{(x_1-\varepsilon,\, x_1+\varepsilon)} \{1-F[(t+x_0)-]\}^{n-k} \mathrm{d}K(t) = 0$$

for (in particular) any point of increase $x_1>0$ of K, any $x_0>0$ and any ε with $0<\varepsilon<x_1$. It easily follows that $F(x_1+x_0)=1$ for such x_1 and x_0 and thence again that $F(x_1)=1$ for such x_1, i.e. for all $0<x_1 \in S(F)$, the support of F (equivalently of K). If $x^*=\inf\{(0,\infty)\cap S(F)\}$, then $x^*>0$ ($x^*=0$ would imply that $F(0)=F(0+)=1$ or $F=\delta_0$, but then $(0,\infty)\cap S(F)=\emptyset$ in such a case) and $x^* \in S(F)$, so $F(x^*)=1$ by what has been proved above, i.e. $S(F)=\{0, x^*\}$ or F is a "two-point distribution". Such F *can* be solutions of (2) as easily checked. In what follows, we will therefore be concerned with the case:

$$K(0)<1 \tag{3}$$

(in the present context, this is equivalent to: $\binom{n}{k}F^k(0)<1$). Setting $\tilde{F}(y)=[1-F(-y-)]^{n-k}$ for all real y, (2) is equivalent to

$$\tilde{F}(0)=1, \qquad \int \tilde{F}(x-t)\,\mathrm{d}K(t)=\tilde{F}(x) \quad \text{for all } x<0, \tag{4}$$

$\tilde{F}$ and $K/\binom{n}{k}$ are d.f.'s, with support $\subset(-\infty,0]$ and $[0,\infty)$, respectively and by the commutativity of the convolution operation for d.f.'s, (4) is in turn equivalent to:

$$\tilde{F}(0)=1, \qquad \int K(x-t)\,\mathrm{d}\tilde{F}(t)=\tilde{F}(x) \quad \text{for all } x<0 \tag{5}$$

where

$$\operatorname{Var} K>1 \tag{6}$$

being $=\binom{n}{k}$ here. We shall consider solutions F of (5) subject to the conditions (3) and (6) — the fact that K itself happens to be defined in terms of F in our present context plays no distinguished role in the analysis below, except towards the end — in deriving the final form of F.

Passing to a different context, let X be a r.v. distributed exponentially, and Y a non-negative r.v. independent of X. Then the "strong memorylessness property" of the exponential law is given by the relation

$$P[X\geqslant Y+x]=P[X\geqslant Y]\cdot P[X\geqslant x] \quad \text{for all } x\geqslant 0 \tag{7}$$

whatever be the distribution of Y. If F and G are the d.f.'s of X and Y, respectively, then (7) is equivalent to

$$\tilde{F}(0)=1, \qquad \int G(x-t)\,\mathrm{d}\tilde{F}(t)=c\tilde{F}(x) \quad \text{for all } x<0 \tag{8}$$

where $\tilde{F}$ is the d.f. conjugate to F, i.e., $\tilde{F}(t)=1-F(-t-)$ for all real t, and

$$c=\int G\,\mathrm{d}F=\int \tilde{F}(-y)\,\mathrm{d}G(y). \tag{9}$$

The cases $c=0$ and $c=1$ are of no interest — for their disposal, see Ramachandran (1977, 1979), abbreviated from here on as (R-1977) and (R-1979). Assuming therefore that $0<c<1$, we have from (8):

$$\tilde{F}(0)=1, \qquad \int K(x-t)\,\mathrm{d}\tilde{F}(t)=\tilde{F}(x), \qquad \operatorname{Var} K(=c^{-1})>1 \tag{10}$$

where $K=c^{-1}G$ and G is a d.f. with support on $[0,\infty)$, while $\tilde{F}$ is a d.f. with support on $(-\infty,0]$. Corresponding to (3): $K(0)<1$, we have

$$(0\leqslant)G(0)<c(<1) \tag{11}$$

which we shall assume satisfied: if $G(0)=c$ for instance, F need not have a structure of the kinds described in Theorem 2 below, as seen from the following example due to Huang: take $F(x)=x$ for $0\leqslant x\leqslant 1$ and $G(x)=1-\mathrm{e}^{-1}$ for $0\leqslant x\leqslant 1$ and $=1-\mathrm{e}^{-x}$ for $x\geqslant 1$: these F, G satisfy (8).

Thus, in both the contexts discussed above, we come up with the same integral eq. (5), which we will solve in the presence of conditions (3) and (6). Under the assumption that the F there was non-lattice (in particular, non-degenerate), eq. (2) was solved in Rossberg (1972), invoking the Wiener–Hopf technique. Eq. (8) was solved under assumption (11) in refs. (R-1977) and (R-1979), also by (different) complex analysis methods. In the present paper, we show how the basic idea of employing the Wiener–Hopf technique, together with some of the arguments of refs. (R-1977) and (R-1979), provides an optimal complex analysis method of solving (5) subject to (3) and (6), incidentally yielding a more comprehensive solution (covering the case of lattice F as well) for eq. (2) than in Rossberg (1972).

Then, let H be defined according to

$$\int K(x-t)\,\mathrm{d}\tilde{F}(t)=\tilde{F}(x)+H(x) \tag{12}$$

so that (i) $H(x)=0$ for $x<0$, by (5), and (ii) H is a function of bounded variation on $[0,\infty)$. We may therefore speak of the Laplace–Stieltjes transforms (LST's) $\tilde{f}$, k and h of $\tilde{F}$, K and H, respectively: thus, let:

$$\tilde{f}(s)=\int_{(-\infty,0]}\mathrm{e}^{-sx}\,\mathrm{d}\tilde{F}(x),\qquad k(s)=\int_{[0,\infty)}\mathrm{e}^{-sx}\,\mathrm{d}K(x),$$

$$h(s)=\int_{[0,\infty)}\mathrm{e}^{-sx}\,\mathrm{d}H(x); \tag{13}$$

$\tilde{f}$ is analytic in $C_L=\{s\colon \operatorname{Re}s<0\}$ and continuous in $\bar{C}_L$ while k and h are analytic in $C_R=\{s\colon \operatorname{Re}s>0\}$ and continuous in $\bar{C}_R$. On taking the Fourier–Stieltjes transforms of both sides in (12), we have

$$k(s)\tilde{f}(s)=\tilde{f}(s)+h(s),\quad\text{or,}\quad \tilde{f}(s)[k(s)-1]=h(s) \tag{14}$$

for all purely imaginary values of s: $s=\mathrm{i}t$, t real. Now $k(0)=\operatorname{Var}K>1$ by (6); hence there is an interval of the imaginary axis, say $I=(-\mathrm{i}\alpha,\mathrm{i}\alpha)$ for some $\alpha>0$, where $k(s)\neq 1$, by continuity. Hence, setting $k^*=k-1$, we have that h/k^* is continuous on $I\cup C_R$ and meromorphic (analytic except for possible poles – at the zeros of k^* there) in C_R, and by (14), coincides on I with the function $\tilde{f}$, which is analytic in C_L and continuous on $C_L\cup I$. Hence, by a familiar "reflection principle", there exists a function f^* defined and analytic on $C_L\cup I\cup C_R$ except possibly for poles at the zeros of k^*, such that $\tilde{f}$ and h/k^* are restrictions of f^* to their respective domains of definition. In particular, f^* is analytic at the origin, and since $\tilde{f}$ is the LS-transform of a pr. measure, it follows that $\tilde{f}$ is either (a) analytic everywhere and the integral representation (13) holds for $\tilde{f}$ for all complex s, or (b) analytic in a (maximal) strip of the form $\{s\colon \operatorname{Re}s<\lambda\}$ for some $\lambda>0$, where, further λ itself is a singularity for f^*. Case (a) is ruled out for the following reason: if $s=\sigma+\mathrm{i}\tau$, $f^*(\sigma)=\tilde{f}(\sigma)\to+\infty$ as $\sigma\to+\infty$ in view of the integral representation in (13) for $\tilde{f}$ being valid for all s, while, on the other hand, $f^*(\sigma)=h(\sigma)/k^*(\sigma)$ tends to a finite limit as $\sigma\to+\infty$.

Turning, therefore, to case (b), we note that λ must be a zero for k^* on the positive real axis, but, by assumption, $k(0)=\operatorname{Var} K>1$, while $k(\sigma)\to K(0)<1$ as $\sigma\to+\infty$, and $k'(\sigma)<0$ for $\sigma>0$, so $k^*(\lambda)=0$ for a *unique* $\lambda>0$. We also note that for $\operatorname{Re} s=\sigma>\lambda$, $|k(s)|\leqslant k(\sigma)<1$, so k^* does not vanish in $\{s\colon \operatorname{Re} s>\lambda\}$. Also, $f^*=\tilde{f}$ is analytic in $\{s\colon \operatorname{Re} s<\lambda\}$, hence the only possible singularities for f^* are poles at those zeros of k^* which lie on the line $\operatorname{Re} s=\lambda$. Regarding these zeros, we have the following easily verified (cf. (R-1977))

Lemma 1. *Let $\mathfrak{L}_\rho^+$ denote the class of all positive multiples of d.f.'s with support $\subset\{0,\rho,2\rho,\dots\}$, and let $\mathfrak{L}^+=\bigcup_{\rho>0}\mathfrak{L}_\rho^+$. Then*:

(i) *if $K\notin\mathfrak{L}^+$, the only zero of k^* on $\{s\colon \operatorname{Re} s=\lambda\}$ is λ itself*;

(ii) *if $K\in\mathfrak{L}_\rho^+$ for some $\rho>0$, then the zeros of k^* on $\{s\colon \operatorname{Re} s=\lambda\}$ form the set $\{\lambda+(2n\pi i/\rho)\colon n$ integer$\}$ and*

(iii) *the zeros of k^* on $\operatorname{Re} s=\lambda$ are all simple (in case* (i) *or in case* (ii)*: the derivative k' does not vanish at these points).*

Then, as in ref. (R-1977), we may invoke the following two results concerning the zeros of k^* (these are not needed for the more limited purposes of Huang (1978) or Rossberg (1972)).

Lemma 2. *The number of zeros of k^* in any closed rectangle of the form $0<a\leqslant\operatorname{Re} s\leqslant b$, $y\leqslant\operatorname{Im} s\leqslant y+1$ is bounded by a number $N(a,b)$ depending only on a and b and not on y.*

Lemma 3. *Given $\gamma,\varepsilon>0$, there exists an $m(\gamma,\varepsilon)>0$ such that $|k^*(s)|>m(\gamma,\varepsilon)$ for all s lying in the strip, $\varepsilon<\operatorname{Re} s<\gamma-\varepsilon$, but outside of discs of radius, ε, with centres at the zeros of k^*.*

Taking for instance $a=\frac{1}{2}\lambda$ and $b=\frac{3}{2}\lambda$ in the statement of Lemma 2, we easily see that there exists a $\delta>0$ and a sequence $\{T_n\}$, $n=\cdots,-2,-1,0,1,2,\dots$ of real numbers with $T_{-n}=-T_n$ for $n\geqslant1$ such that the zeros of k^* in the set $\{\frac{1}{2}\lambda\leqslant\operatorname{Re} s\leqslant\frac{3}{2}\lambda, n\leqslant\operatorname{Im} s<n+1\}$ lie at a distance $\geqslant\delta$ from the line $\operatorname{Im} s=T_n$. On the strength of Lemma 3, we then see that $|k^*(s)|>$ some $m>0$, where m depends only on λ and δ, for all points s such that $\operatorname{Im} s=T_n$, $\frac{1}{2}\lambda\leqslant\operatorname{Re} s\leqslant\frac{3}{2}\lambda$.

It easily follows that the meromorphic function f^* is bounded, uniformly with respect to n, on each of the rectangles Γ_n whose sides are formed by the lines $\operatorname{Re} z=\pm n$, $\operatorname{Im} z=T_{\pm n}$. It then follows from a well-known representation theorem for meromorphic functions [cf. Titchmarsh (1939, p. 110)] that f^*, which has only simple poles at most, admits the representation,

$$f^*(s)=f^*(0)+\sum a_n\left(\frac{1}{s-s_n}+\frac{1}{s_n}\right) \tag{15}$$

where $\{s_n\}$ is some enumeration of the simple poles of f^* in non-decreasing order of moduli, and the $\{a_n\}$ are suitable complex constants (the residues of f^* at the s_n).

In particular, if $K \notin \mathfrak{L}^+$, λ is the only possible pole for f^* and we have easily,

$$f^*(s) = c + \frac{a}{(s-\lambda)}$$

where c and a are some constants.

It follows that F is a mixture of δ_0 and E_μ, the exponential law with a (suitable) parameter μ: $F = a\delta_0 + bE_\mu$, where $0 \leqslant a, b \leqslant 1$, $a+b=1$ and $b>0$ in fact in view of $F \neq \delta_0$. If F satisfies (2), then $\mu = \lambda/(n-k)$, and if F satisfies (8), $\mu = \lambda$, itself. We show below that in either case $b=1$, so that $F = E_\mu$, i.e. F is exponential (if $K \notin \mathfrak{L}^+$).

Taking solutions of (2) first, we have from the definition of λ in this case:

$$\binom{n}{k}\int_{[0,\infty)} e^{-\lambda x} dF^k(x) = 1, \quad \text{or,} \quad \lambda\binom{n}{k}\int_0^\infty F^k(x) e^{-\lambda x} dx = 1.$$

Substituting $F = a\delta_0 + bE_\mu$ and $y = e^{-\mu x}$, we have:

$$(n-k)\binom{n}{k}\int_0^1 (1-by)^k y^{n-k-1} dy = 1.$$

Since the expression on the LHS is a strictly decreasing function of b for $0 \leqslant b \leqslant 1$ and takes the value 1 for $b=1$, it is clear that the above relation can be satisfied iff $b=1$, i.e. iff $F = E_\mu$.

Taking solutions of (8) next, we have

$$\int_{[0,\infty)} e^{-\lambda y} dG(y) = (c=) \int_{[0,\infty)} [1 - F(y-)] dG(y) (\neq 0);$$

it is possible when $F = a\delta_0 + bE_\lambda$ iff $b=1$, i.e. $F = E_\lambda$.

If, however, $K \in \mathfrak{L}_\rho^+$ for some $\rho > 0$, then f^* is analytic everywhere except possibly for simple poles at (some or all of) the points $b_n = \lambda + (2n\pi i)/\rho$, n integer; and hence, as pointed out earlier, admits the representation, valid for all $s \neq$ any of the b_n:

$$f^*(s) = f^*(0) + s \sum_{-\infty}^{\infty} \frac{a_n}{b_n(s-b_n)} \tag{16}$$

where a_n is the residue of f^* at b_n: $a_n = h(b_n)/k'(b_n)$. Now, since $K \in \mathfrak{L}_\rho^+$, we have for *all* n:

$$|k'(b_n)| = \int_{[0,\infty)} x e^{-\lambda x} dK(x) > 0$$

(since K is not degenerate at 0), while

$$|h(b_n)| \leqslant \int_{[0,\infty)} e^{-\lambda x} d|H|(x);$$

hence $|a_n| \leqslant A$ for all n for some $A > 0$.

In order to enable us to appeal to the familiar form of a complex inversion formula for Laplace transforms, we now switch over to f, the LST of F: f has $\{s: \operatorname{Re} s > -\lambda\}$ as its half plane of convergence and in view of (16) admits the representation:

$$f(s)=f(0)+s\cdot\sum_{-\infty}^{\infty}\frac{a_n}{b_n(s+b_n)} \tag{17}$$

for all $s\neq -b_n$ for any n. An integration by parts shows that ϕ given by $\phi(s)=[f(s)-f(0)]/s$ is the (ordinary) Laplace transform of $(1-F)$, and it follows that ϕ also has the same half-plane of convergence as f, and from (17) that ϕ admits the representation

$$\phi(s)=\sum_{-\infty}^{\infty}\frac{a_n}{b_n(s+b_n)} \tag{18}$$

for all $s\neq -b_n$ for any n.

We may now appeal to the complex inversion formula for Laplace transforms, according to which:

$$-\int_t^{\infty}[1-F(x)]\mathrm{d}x=\lim_{T\to\infty}\frac{1}{2\pi\mathrm{i}}\int_{x_0-\mathrm{i}T}^{x_0+\mathrm{i}T}\frac{\mathrm{e}^{ts}}{s}\phi(s)\mathrm{d}s\quad\text{for all } t>0, \tag{19}$$

valid for any $x_0\in(-\lambda,0)$: cf. Doetsch (1974, Theorem 27.2, p. 181).

Now consider the lines $\operatorname{Im} s=T_N=(2N+1)(\pi/\rho)$, N integer. Considering the case $\rho=2\pi$ for simplicity of writing, we have for such s and all $N\geq 1$,

$$|\phi(s)|\leq A\sum_{-\infty}^{\infty}\frac{1}{|(\lambda+in)|(N+n+\frac{1}{2})}=\sum_{-\infty}^{-(N+1)}+\sum_{-N}^{0}+\sum_{1}^{\infty}.$$

Each of the three sums on the right above is bounded by a constant not depending on $N\geq 1$; similarly for $N\leq -1$ as well, and for the case of arbitrary $\rho>0$; thus,

$$|\phi(s)|\leq \text{some } B>0\quad\text{for all } s \text{ with } \operatorname{Im} s=T_N=\frac{(2N+1)\pi}{\rho},\quad N\text{ integer}.$$

Now consider the rectangular contour with vertices $-R\pm\mathrm{i}T_n$ (R arbitrary), $x_0\pm\mathrm{i}T_n$; as $R\to\infty$, $\phi(-R+\mathrm{i}y)\to 0$ whatever be y real, $|y|\leq T_n$, so that:

$$\frac{1}{2\pi\mathrm{i}}\int_{x_0-\mathrm{i}T_N}^{x_0+\mathrm{i}T_N}\frac{\mathrm{e}^{ts}}{s}\phi(s)\mathrm{d}s=S_1(N,t)+S_2(N,t)$$

where

$$2\pi\mathrm{i}S_1(N,t)=\int_{-\infty+\mathrm{i}T_N}^{x_0+\mathrm{i}T_N}-\int_{-\infty-\mathrm{i}T_N}^{x_0-\mathrm{i}T_N}\frac{\mathrm{e}^{ts}}{s}\phi(s)\mathrm{d}s$$

and

$$S_2(N,t)=\text{sum of the residues of }\frac{\mathrm{e}^{ts}}{s}\phi(s)\text{ at the points } b_n \text{ with } -N\leq n\leq N.$$

$$=\sum_{-N}^{N}c_n\exp\left\{\left(-\lambda+\frac{2\pi n\mathrm{i}}{\rho}\right)t\right\}$$

$$=\eta_N(t)\mathrm{e}^{-\lambda t}$$

where η_N is periodic with period ρ. In view of (20), it follows that $S_1(N,t)\to 0$ as $N\to\infty$, for any fixed $t>0$, so that we have from (19):

$$-\int_t^\infty [1-F(t)]\,\mathrm{d}t=\eta(t)\mathrm{e}^{-\lambda t}$$

where $\eta(t)=\lim_{N\to\infty}\eta_N(t)$ exists for all $t>0$ and is periodic with period ρ.

Thus, if F satisfies (4), then, for all $t>0$, $1-F(t)$ is either $=\exp(-\lambda t)$ or $=\exp(-\lambda t)\cdot\xi(t)$ where ξ is periodic with period ρ, according to whether $K\notin\mathfrak{L}^+$ or $K\in\mathfrak{L}_\rho^+$ for some $\rho>0$, provided of course conditions (3) and (6) are satisfied. In particular, we have the following special cases:

Theorem 1. *Let $X_1,\ldots,X_n$ be i.i.d.r.v.'s with $F\neq\delta_0$ as their common d.f., and let $X_{j,n}$ be the jth order statistic generated by them. If $X_{k+1,n}-X_{k,n}$ has the same distribution as $X_{1,n-k}$, for some $k<n$, then:*

$$\binom{n}{k}F^k(0)\leqslant 1$$

and either:

(i) *$F=p\delta_0+q\delta_a$ for some $a>0$, with $\binom{n}{k}p^k=1$, $q=1-p$; or*

(ii) *F is a mixture of δ_0 and a geometric type d.f.:*

$$F=a\delta_0+b\sum_{N=1}^{\infty}pq^{N-1}\delta_{N\rho}$$

for some $\rho>0$, $0\leqslant a\leqslant 1$, $0<p<1$, $b=1-a$, $q=1-p$, where, further, $\binom{n}{k}a^k<1$ and the parameters are related according to,

$$\binom{n}{k}\int_{[0,\infty)}\mathrm{e}^{-\lambda x}\,\mathrm{d}F^k(x)=1\quad \text{with } \lambda=-(n-k)\log q,$$

or equivalently,

$$\binom{n}{k}\left[Q_0+\sum_{N=1}^{\infty}(Q_N-Q_{N-1})q^{(n-k)N\rho}\right]=1,\quad \text{where } Q_N=(1-bq^N)^k$$

or

(iii) *F is an exponential law.*

In contrast to the exponential laws, not every geometric law satisfies (2) for all n and k: for example, take $F=\sum_{N=0}^{\infty}pq^N\delta_N$ for some $p=1-q\in(0,\frac{1}{3})$ and $n=3$, $k=1$; then (2) is not satisfied (for the value $x=1$ for instance). Thus assertion (ii) of the theorem above cannot be improved upon.

We can obtain similar results which include Theorem 2 (relating to the Pareto law) and Theorem 3 of Rossberg (1972); statements are omitted.

Theorem 2. *If $G\neq\delta_0$ be a given d.f., and F a d.f. on R^1, both with support $\subset[0,\infty)$, and satisfying (8), and if further,*

$$G(0)<c<1, \tag{20}$$

then $\lambda>0$ *being the unique solution of the equation,*

$$\int_{[0,\infty)} e^{-\lambda y}\,dG(y)=c,$$

and $\mathfrak{L}_{\rho}^{+}$, $\mathfrak{L}^{+}$ *being defined as in the statement of Lemma* 1, *we have either*:

(i) $1-F(x)=e^{-\lambda x}$ *for all* $x>0$, *if* $G\notin\mathfrak{L}^{+}$ *or*

(ii) $1-F(x)=\xi(x)e^{-\lambda x}$ *if* $G\in\mathfrak{L}_{\rho}^{+}$, *for some* $\rho>0$, *where* ξ *is periodic with period* ρ.

Note: Again, as already pointed out, if (20) is violated (for instance if $c=G(0)$), then F need not have a structure of either of the forms stated above.

References

[1] Doetsch, G. (1974). *Introduction to the Theory and Application of the Laplace Transformation*. English tr., Springer-Verlag, Berlin.

[2] Huang, J.S. (1978). On a "lack of memory" property. Univ. of Guelph Statistical Series No. 84 of 1978.

[3] Ramachandran, B. (1977). On the strong Markov property of the exponential laws. *Colloquia Mathematica Societatis Janos Bolyai* Vol. 21. *Analytic Function Methods in Probability Theory*. pp. 277–292, Debrecen.

[4] Ramachandran, B. (1979). On the "strong memorylessness property" of the exponential and geometric probability laws. *Sankhyā Ser. A* **41**, 244–251.

[5] Rossberg, H.-J. (1972). Characterization of the exponential and Pareto distributions by means of some properties of the distributions which the differences and quotients of order statistics are subject to. *Math Operationsforsch. Statist.* **3**, Pt.-3, 207–216.

[6] Titchmarsh, E.C. (1939). *The Theory of Functions*. 2nd ed., Oxford.

G. Kallianpur, P.R. Krishnaiah, J.K. Ghosh, eds., *Statistics and Probability: Essays in Honor of C.R. Rao*

APPLICATION AND EXTENSION OF CRAMÉR'S THEOREM ON DISTRIBUTIONS OF RATIOS*

M.M. RAO
University of California, Riverside, CA, USA

1. Introduction

In the asymptotic theory of the distributions of estimators of parameters in stochastic models, one often has to find the distributions of ratios of random variables which are not necessarily independent. In several cases, this is to be obtained from the knowledge of the joint characteristic functions (ch.f.) of such limit random variables, when the small sample (exact) distribution is complicated. Thus if the limit ch.f. is available, then a technique introduced by Cramér [3, p. 47] in 1937, and virtually ignored by later authors of textbooks, will be exemplified and extended here. In many cases, it obviates the need to invert the limit ch.f., and in some cases it is the only technique available. Applications of ch.f's in the inference theory have been brought out forcefully in the important book by Kagan et al. [7]. The present paper can be regarded as a contribution in this direction in that the method of ch.f is the central theme.

To motivate the study, it will be convenient to present the classical result of Cramér in the following form (cf. [4, p. 317]):

Theorem 1. *Let X and Y have a joint distribution of continuous type with $\varphi(\cdot,\cdot)$ as their joint ch.f. Suppose that $P(Y>0)=1$ and $E(|X|)<\infty$, $E(|Y|)<\infty$, and let $G(x)=P(X/Y<x)$. Then*

$$g(x)=G'(x)=\frac{1}{2\pi \mathrm{i}}\int_{-\infty}^{\infty}\left(\frac{\partial\varphi}{\partial u}\right)(t,u)\bigg|_{u=-tx}\,\mathrm{d}t,$$

whenever the integral is uniformly convergent for x in compact sets of $\mathbb{R}$. (Here and throughout, E stands for "expectation".)

This result was used by Cramér in [3], to derive the distributions of (non-central) F- and t-random variables and certain generalizations of these. In each case, X and Y are independent. Here, by using Theorem 1, for the case of *dependent* X and Y, the main results of White [11, 12] will be deduced in a somewhat more direct manner. This theorem also enabled a solution of an unsolved problem of several years [10].

It must be noted that Theorem 1 itself is a consequence of another and more powerful result of Cramér's [3, p. 33]. The latter will be extended in Section 3 below.

*Partially supported under the NSF Grant MCS76-15544-A01.

As a consequence a generalization of Theorem 1 will be obtained in Section 4. Two related papers which have considered similar problems are due to Curtiss [5] and Gurland [6]. Of these, [5] is particularly inspired by [3], but a multivariate extension of the crucial result of [3, Thm. 12] has not been found in the literature. Thus, there is practically no overlap between the following work and these earlier papers. First the above-stated application will be treated in the next section, as it gives a better understanding of the type of results considered.

2. An application

Consider the simple autoregressive model:

$$X_t = \alpha X_{t-1} + \varepsilon_t, \quad t \geqslant 1, \alpha \in \mathbb{R}, \tag{1}$$

where $X_0 = 0$, α is an unknown parameter and ε_t's are independent and identically distributed random variables with means zero and unit variances. If $X_1, X_2, \ldots, X_T$ are T random variables which are governed by (1), the least squares estimator $\hat{\alpha}_T$ of α is given by:

$$\hat{\alpha}_T = \sum_{t=1}^{T} X_t X_{t-1} \Big/ \sum_{t=1}^{T} X_{t-1}^2. \tag{2}$$

Even if the ε_t are normal, the exact distribution of $(\hat{\alpha}_T - \alpha)$ is complicated, but its asymptotic distribution can be obtained with relative ease.

First suppose that the ε_t are also normal, and note that by using (1)

$$\hat{\alpha}_T - \alpha = \sum_{t=1}^{T} \varepsilon_t X_{t-1} \Big/ \sum_{t=1}^{T} X_{t-1}^2 = U_T / V_T \quad \text{(say)}. \tag{3}$$

If $I_T(\cdot)$ is the "Fisher information" function, then

$$\begin{aligned} I_T(\alpha) &= -E\left(\frac{\partial^2 \log f(X_1, \ldots, X_T)}{\partial \alpha^2} \right) = E(V_T) \\ &= \frac{1}{1-\alpha^2}\left(T - \frac{1-\alpha^{2T}}{1-\alpha^2} \right), \quad \text{if } |\alpha| \neq 1 \\ &= \tfrac{1}{2} T(T-1), \qquad\qquad \text{if } |\alpha| = 1. \end{aligned} \tag{4}$$

It is known that $(\hat{\alpha}_T - \alpha) \to 0$ in probability (i.e., $\hat{\alpha}_T$'s are consistent estimators of α) and this also follows from the next result in that $\sqrt{I_T(\alpha)}\,(\hat{\alpha}_T - \alpha)$ has a (non-degenerate) limit distribution for any $\alpha \in \mathbb{R}$. Equivalently, if $\beta^2(T) = I_T(\alpha)$ or omitting the factors converging to zero so that taking

$$\beta^2(T) = \begin{cases} T(1-\alpha^2)^{-1} & \text{if } |\alpha| < 1, \\ \alpha^{2T}/\alpha^2 - 1 & \text{if } |\alpha| > 1, \\ \tfrac{1}{2} T^2 & \text{if } |\alpha| = 1, \end{cases}$$

then $U_T/\beta(T)$ tends in distribution to U and that $V_T/\beta^2(T)$ tends likewise to V. In fact, if

$$\varphi_T(u,v)=E\left(\exp\left\{\mathrm{i}\,u\frac{U_T}{\beta(T)}+\mathrm{i}\,v\frac{V_T}{\beta^2(T)}\right\}\right),$$

then it is not hard to show by an explicit computation that, as $T\to\infty$,

$$\varphi_T(u,v)\to\varphi(u,v)=\begin{cases}\mathrm{e}^{\mathrm{i}v-(u^2/2)} & \text{if } |\alpha|<1,\\ (1-2\mathrm{i}v+u^2)^{-1/2} & \text{if } |\alpha|>1,\\ \left\{\mathrm{e}^{\sqrt{2}\alpha\mathrm{i}u}\left(\cos 2(\mathrm{i}v)^{1/2}-\dfrac{\mathrm{i}\alpha u}{(2\mathrm{i}v)^{1/2}}\sin 2(\mathrm{i}v)^{1/2}\right)\right\}^{-1/2} & \text{if } \alpha=1.\end{cases} \tag{5}$$

Actually the work of [1] and [9] shows that $U_T/\beta(T)\to U$ and $V_T/\beta^2(T)\to V$ in probability also, even without the additional assumption of normality of ε_t's. The normality assumption gives the explicit formulas (5), as shown by White [11].

The limit distribution of $\beta(T)(\hat{\alpha}_T-\alpha)$ can be given as follows. It was established by White, in [11] and [12], in the case of $|\alpha|>1$, and by Mann and Wald [8] when $|\alpha|<1$ and in [10] when $|\alpha|=1$. A more general version of the first and second cases in the context of higher-order models is contained in the papers [1, 9]. However, the present work, based on Theorem 1, gives a better insight into the structure of the problem.

Theorem 2. *Let* $\{X_t, t\geqslant 1\}$ *be random variables given by the model* (1) *with* ε_t's *as independent and identically distributed with zero means and unit variances. If* $\hat{\alpha}_T$ *is given by* (2) *and* $\beta(T)$ *is as defined above, then*

$$\lim_{T\to\infty} P\{\beta(T)(\hat{\alpha}_T-\alpha)<x\}=\int_{-\infty}^{x} f(t)\,\mathrm{d}t \tag{6}$$

exists where (i) $f(t)=(1/\sqrt{2\pi})\mathrm{e}^{-(t^2/2)}$ *for* $|\alpha|<1$, (ii) $f(t)=1/[\pi(1+t^2)]$ *for* $|\alpha|>1$ *when the* ε_t's *are also normal and* (iii) *for* $|\alpha|=1$ (ε_t's *need not be normal*),

$$f(x)=\frac{1}{\sqrt{8\pi^2}}\int_{-\infty}^{\infty}\frac{\rho(x,t)}{[r(x,t)]^{3/2}}\cos\!\left(\delta(x,t)-\tfrac{3}{2}\theta(x,t)\right)$$

$$\times\left(\chi_{\mathbb{R}^2_+}(x,t)+\chi_{\mathbb{R}^2_+}(-x,t)\right)\frac{\mathrm{d}t}{(tx)^{1/2}}$$

where

$$r(x,t)^2 = \sinh^2(2tx)^{1/2} + \cos^2(2tx)^{1/2} + \frac{t}{2x}\left(\sinh^2(2tx)^{1/2} + \sin^2(2tx)^{1/2}\right)$$
$$+ \frac{\alpha}{2}\left(\frac{t}{x}\right)^{1/2}\left(\sin(8tx)^{1/2} - \sinh(8tx)^{1/2}\right),$$

$$\theta(x,t) =$$
$$\arctan\left\{\frac{1-(\alpha/2)(t/x)^{1/2}\left(\coth(2tx)^{1/2}+\cot(2tx)^{1/2}\right)}{1-(\alpha/2)(t/x)^{1/2}\left(\tanh(2tx)^{1/2}-\tan(2tx)^{1/2}\right)}\tan(2tx)^{1/2}\tanh(2tx)^{1/2}\right\},$$

$$\rho(x,t)^2 = 2\left(1-\frac{\alpha}{(8x^2)^{1/2}}\right)^2\left(\sinh^2(2tx)^{1/2}+\sin^2(2tx)^{1/2}\right)$$
$$+\frac{t}{x}\left(\sinh^2(2tx)^{1/2}+\cos^2(2tx)^{1/2}\right)$$
$$-\alpha\left(\frac{t}{x}\right)^{1/2}\left(1-\frac{\alpha}{(8x^2)^{1/2}}\right)\left(\sin(8tx)^{1/2}+\sinh(8tx)^{1/2}\right),$$

$$\delta(x,t) = \arctan\left(\frac{C\cos(\alpha t/\surd 2)+D\sin(\alpha t/\surd 2)}{D\cos(\alpha t/\surd 2)-C\sin(\alpha t/\surd 2)}\right),$$

with

$$C = \left(1-\frac{\alpha}{(8x^2)^{1/2}}\right)\left(\sinh(2tx)^{1/2}\cos(2tx)^{1/2}-\cosh(2tx)^{1/2}\sin(2tx)^{1/2}\right)$$
$$+\alpha\left(\frac{t}{x}\right)^{1/2}\sinh(2tx)^{1/2}\sin(2tx)^{1/2},$$

$$D = \left(1-\frac{\alpha}{(8x^2)^{1/2}}\right)\left(\sinh(2tx)^{1/2}\cos(2tx)^{1/2}+\cosh(2tx)^{1/2}\sin(2tx)^{1/2}\right)$$
$$-\alpha\left(\frac{t}{x}\right)^{1/2}\cosh(2tx)^{1/2}\cos(2tx)^{1/2}.$$

Furthermore, if $|\alpha|>1$ *and* ε_t*'s are also normal, or* $|\alpha|<1$ *and* ε_t*'s satisfy conditions of case* (i), *then*

$$\lim_{T\to\infty} P\left(\sqrt{\sum_{t=1}^{T} X_{t-1}^2}\,(\hat{\alpha}_T-\alpha)<x\right) = \frac{1}{\sqrt{2\pi}}\int_{-\infty}^{x} e^{-t^2/2}\,dt. \tag{7}$$

Proof. The result for $|\alpha|=1$ uses Theorem 1 and is given in [10]. So only the cases that $|\alpha|\neq 1$ of the theorem need be proved now, and the case that $|\alpha|=1$ is stated here for completeness. Note that in (7), since for $|\alpha|<1$, $V_T^2/\beta(T)\to 1$ in probability, (7) and (6) yield the same result in that case and thus only the case $|\alpha|>1$ is new and (7) is thus different in this case. To prove (6), consider the two cases separately. First let the ε_t be normal.

Case 1. $|\alpha|<1$. Then by (5) and Theorem 1,

$$f(x)=\frac{1}{2\pi \mathrm{i}}\int_{-\infty}^{\infty}\frac{\partial\varphi}{\partial u}(v,-xv)\,\mathrm{d}v=\frac{1}{2\pi\mathrm{i}}\int_{-\infty}^{\infty}\mathrm{i}\mathrm{e}^{-\mathrm{i}vx-(v^2/2)}\,\mathrm{d}v.$$

$$=\frac{1}{\sqrt{2\pi}}\int_{-\infty}^{\infty}\mathrm{e}^{-\mathrm{i}vx-(v^2/2)}\frac{\mathrm{d}v}{\sqrt{2\pi}}=\frac{1}{\sqrt{2\pi}}\mathrm{e}^{-x^2/2}. \tag{8}$$

The general result follows since if $|\alpha|\leqslant 1$, the invariance principle applies (cf. [10] on this point).

Case 2. $|\alpha|>1$. Then (5) and Theorem 1 yield, since $(\partial\varphi/\partial u)(v,-xv)=\mathrm{i}(1+v^2+2\mathrm{i}vx)^{-3/2}$,

$$f(x)=\frac{1}{2\pi\mathrm{i}}\int_{-\infty}^{\infty}\frac{\mathrm{i}\,\mathrm{d}v}{[1+v^2+2\mathrm{i}vx]^{3/2}}$$

$$=\frac{1}{2\pi}\int_{-\infty}^{\infty}\mathrm{d}\left(\frac{v+\mathrm{i}x}{(1+x^2)\sqrt{(1+v^2+2\mathrm{i}vx)}}\right)^{*} \quad \text{where one takes the positive square root for definiteness,} \tag{9}$$

$$=\frac{1}{2\pi}\left[\lim_{T\to\infty}\frac{1+(\mathrm{i}x/T)}{(1+x^2)\sqrt{\dfrac{1+T^2+2\mathrm{i}Tx}{T^2}}}-\lim_{T\to\infty}\frac{1+(\mathrm{i}x/T)}{(1+x^2)(-1)\sqrt{\dfrac{1+T^2+2\mathrm{i}Tx}{T^2}}}\right]$$

$$=\frac{1}{2\pi}\frac{2}{1+x^2}=\frac{1}{\pi(1+x^2)}.$$

Note that $f(\cdot)$ is the density function of the ratio of *dependent non-normal* random variables, which is Cauchy! Also, as discussed in [10], in the case that $|\alpha|>1$, an invariance principle does not apply and hence the limit distribution of $\hat{\alpha}_T$ depends on that of ε_t's.

It remains to prove (7) when $|\alpha|>1$. In this case, Theorem 1 applies after a computation. An inversion (at an intermediate step) is necessary before it can be applied. This will be detailed here.

If $|\alpha|>1$, (5) gives $\varphi(u,v)=E(\mathrm{e}^{\mathrm{i}uU+\mathrm{i}vV})$. However, $\sum_{t=1}^{T}X_{t-1}^2/\beta^2(T)\to V$ in probability as $T\to\infty$ and, as seen above, $\beta(T)(\hat{\alpha}_T-\alpha)\to U/V$ in distribution where $P(V>0)=1$. Hence it follows by Slutsky's Theorem [4, p. 255] that $\sqrt{\sum_{t=1}^{T}X_{t-1}^2}\,(\hat{\alpha}_T-\alpha)\to\left(U/\sqrt{V}\right)$ in distribution as $T\to\infty$. Thus the desired limit distribution is obtained from Theorem 1, if one has in hand $\psi(u,v)=E(\mathrm{e}^{\mathrm{i}uU+\mathrm{i}v\sqrt{V}})$. By (5) $\varphi(u,v)$

*My colleague, T. Barth, has pointed this simpler evaluation which obviated my attempt through complex integration!

is known. So if $f_{U,V}$ is the density function with φ as its ch.f., then

$$\psi(u,v)=\int_{-\infty}^{\infty}\int_{-\infty}^{\infty} e^{iux+iv\sqrt{y}} f_{U,V}(x,y)\,dx\,dy. \tag{10}$$

That $f_{U,V}$ exists is a simple consequence of the Riemann-Lebesgue lemma and the fact that φ is differentiable. To simplify (10), one has [with (5)],

$$\psi(u,v)=\int_{-\infty}^{\infty}\int_{-\infty}^{\infty} e^{iux+iv\sqrt{y}}\,dx\,dy\frac{1}{(2\pi)^2}\int_{-\infty}^{\infty}\int_{-\infty}^{\infty} e^{-irx-isy}\varphi(r,s)\,dr\,ds$$

$$=\int_{-\infty}^{\infty}\int_{-\infty}^{\infty} e^{iux+iv\sqrt{y}}\,dx\,dy\frac{1}{2\pi}\int_{-\infty}^{\infty} e^{-irx}(1+r)^{-1/2}\,dr\frac{1}{2\pi}\int_{-\infty}^{\infty} e^{-isy}\frac{ds}{\left(1-\dfrac{2is}{1+r^2}\right)^{1/2}}$$

$$=\int_{x=-\infty}^{\infty}\int_{y=0}^{\infty} e^{iux+iv\sqrt{y}}\,dx\,dy\frac{1}{2\pi}\int_{-\infty}^{\infty} e^{-irx}\frac{y^{-1/2}e^{-(1/2)y(1+r^2)}}{\Gamma(1/2)\sqrt{2}}\,dr,$$

using the form of the ch.f. of a gamma distribution,

$$=\int_{y=0}^{\infty}\frac{1}{\sqrt{2\pi y}}e^{iv\sqrt{y}-(y/2)}\,dy\frac{1}{\sqrt{2\pi y}}\int_{-\infty}^{\infty} e^{iux-x^2/2y}\,dx$$

$$=\frac{1}{\sqrt{2\pi}}\int_0^{\infty}\frac{e^{iv\sqrt{y}-(y/2)}}{\sqrt{y}}\cdot e^{-(u^2y)/2}\,dy$$

$$=\frac{2}{\sqrt{2\pi}}\int_0^{\infty} e^{ivt-(t^2/2)(1+u^2)}\,dt. \tag{11}$$

Thus

$$\frac{\partial\psi}{\partial v}(u,v)=\sqrt{\frac{2}{\pi}}\int_0^{\infty} it e^{ivt-(u^2+1)t^2/2}\,dt.$$

Hence the required density is given by Theorem 1 as:

$$g(x)=\frac{1}{2\pi i}\int_{-\infty}^{\infty}\frac{\partial\psi}{\partial v}(u,-ux)\,du$$

$$=\frac{1}{2\pi}\sqrt{\frac{2}{\pi}}\int_{-\infty}^{\infty}du\int_0^{\infty} te^{-iutx-((u^2t^2)/2)-(t^2/2)}\,dt$$

$$=\frac{1}{\sqrt{2\pi}\cdot\pi}\int_0^{\infty} te^{-t^2/2}\,dt\int_{-\infty}^{\infty} e^{-ix(ut)-(ut)^2/2}\,du$$

$$=\frac{1}{\pi}\int_0^{\infty} e^{-t^2/2}\,dt\cdot e^{-x^2/2}=\frac{e^{-x^2/2}}{\sqrt{2\pi}}. \tag{12}$$

The interchange of integrals and the differentiation under the integral sign used above are easily justified. Thus (12) is (7), and the result follows.

Remark 1. Note that in computing ψ for the last part, an inversion of φ was needed. Thus in these cases, the advantage of the ch.f. method is lessened. Indeed, one can use a more elementary technique to compute g when the density $f_{U,V}$ is known. That was done in [12]. However, the unified method used above has certain advantages.

Remark 2. For $|\alpha|<1$, since $\Sigma_{t=1}^{T} X_t^2/T \to 1$ in probability (cf. (5)), part (i) and (7) agree, and the result is new only for $|\alpha|>1$, ε_t's being normal. In the nonnormal case, the theory of [1] and [9] imply the existence of the limiting distribution, but it depends on that of ε_t's. If $|\alpha|=1$, then again $\Sigma_{t=1}^{T} X_t^2/T^2 \to V$ in distribution and $E(\mathrm{e}^{\mathrm{i}tV})=(\sec(4\mathrm{i}t)^{1/2})^{1/2}$. In this case also $\sqrt{\Sigma_{t=1}^{T} X_t^2}\,(\hat{\alpha}_T-\alpha)\to Y$ in distribution, $P(Y=(U/V^{1/2}))=1$, where U and V are random variables whose ch.f. φ is given by (5). This assertion is a consequence of the fact that if $Z_n=(Z_{1n},\ldots,Z_{kn})$ is a random vector and $Z_n \to Z=(Z_1,\ldots,Z_k)$ in distribution, and if $g:\mathbb{R}^k\to\mathbb{R}^k$ is a Borel measurable mapping whose discontinuity points form a set of F_Z-measure zero (F_Z is the distribution function of Z), then $g(Z_n)\to g(Z)$ in distribution – a known (classical) result due to Mann and Wald. Thus if $\psi(u,v)=E(\mathrm{e}^{\mathrm{i}uV+\mathrm{i}v\sqrt{V}})$, $P(V>0)=1$, and $f_{U,V}$ is the joint density with ch.f. φ of (5) for $|\alpha|=1$ (which exists), then Theorem 1 and the proof for (7) yields the density for Y as:

$$h(y)=\frac{1}{(2\pi)^3}\int_{t=0}^{\infty}\sqrt{t}\,\mathrm{d}t\int_{u=-\infty}^{\infty}\mathrm{d}u\int_{x=-\infty}^{\infty}\mathrm{d}x\int_{r=-\infty}^{\infty}\mathrm{e}^{-\alpha\mathrm{i}r/\sqrt{2}}\,\mathrm{d}r$$

$$\times\int_{s=-\infty}^{\infty}\frac{\mathrm{e}^{\mathrm{i}u(x-y\sqrt{t})-\mathrm{i}rx-\mathrm{i}st}\,\mathrm{d}s}{\left(\cos 2(\mathrm{i}s)^{1/2}-\dfrac{\mathrm{i}\alpha r}{(2\mathrm{i}s)^{1/2}}\sin 2(\mathrm{i}s)^{1/2}\right)^{1/2}}. \tag{13}$$

This is not a normal density, and an explicit evaluation needs a detailed and involved computation as in [10]. This will not be attempted here.

3. An extension of Cramér's theorem

It was remarked in Section 1 that Theorem 1 was derived from a more powerful result [3, Thm. 12]. In extending the univariate results to the multivariate case, Cramér states [3, p. 107] about this as follows: "We shall not enter here upon the question of a k-dimensional generalization of Theorem 12". There he introduced a useful method (the Cramér-Wold technique) to reduce the multivariate results to the univariate case. This, however, does not apply to the above noted theorem. A bivariate extension of it will be given in this section, and the corresponding specialization, extending Theorem 1, will be sketched in the next section. Of course, N-variate extensions are possible for both results.

The desired extension can be presented as:

Theorem 3. *Let* $R:\mathbb{R}^2\to\mathbb{R}$ *be a normalized function of bounded variation in the plane in the sense that*

$$\Sigma|\Delta R(x,y)|\leq K_0<\infty \tag{14}$$

where $(\Delta R)(x,y)=R(y_1,y_2)-R(y_1,x_2)-R(x_1,y_2)+R(x_1,x_2)$, $(x,y)=(x_1,y_1)\times(x_2,y_2)$ *and the sum is taken over these finite sets of disjoint rectangles ranging over the plane. Suppose that* (i) $R(x_1,x_2)\to 0$ *as* $x_i\to-\infty$, $i=1,2$, (ii) $R(x_1,\infty)=R_1(x_1)$, $R(\infty,x_2)=R_2(x_2)$ *for* $(x_1,x_2)\in\mathbb{R}^2$, *where* $R_i(u)\to 0$ *as* $u\to\pm\infty$, $i=1,2$, *and* (iii)

$$\int_{-\infty}^{\infty}\int_{-\infty}^{\infty}(|x_1|+|x_2|)|R(\mathrm{d}x_1,\mathrm{d}x_2)|<\infty. \tag{15}$$

Let r *be the Fourier-Stieltjes transform of* R *on* $\mathbb{R}^2$. *Then for each* $h_i>0$, $x_i\in\mathbb{R}$, $0<\alpha_i<1$, $i=1,2$ *one has*

$$\int_{x_1}^{x_1+h_1}\int_{x_2}^{x_2+h_2}(y_1-x_1)^{\alpha_1-1}R(y_1,y_2)\,\mathrm{d}y_1\,\mathrm{d}y_2-\frac{h_1^{\alpha_1}}{2\alpha_1}\int_{x_2}^{x_2+h_2}(y_2-x_2)^{\alpha_2-1}R_2(y_2)\,\mathrm{d}y_2$$

$$-\frac{h_2^{\alpha_2}}{2\alpha_2}\int_{x_1}^{x_1+h_1}(y_1-x_1)^{\alpha_1-1}R_1(y_1)\,\mathrm{d}y_1$$

$$=\left(\frac{1}{2\pi\mathrm{i}}\right)^2\int_{-\infty}^{\infty}\int_{-\infty}^{\infty}\frac{r(t_1,t_2)}{t_1t_2}\mathrm{e}^{-\mathrm{i}t_1x_1-\mathrm{i}t_2x_2}\,\mathrm{d}t_1\,\mathrm{d}t_2$$

$$\times\int_0^{h_1}\int_0^{h_2}u_1^{\alpha_1-1}u_2^{\alpha_2-1}\mathrm{e}^{-\mathrm{i}t_1u_1-\mathrm{i}t_2u_2}\,\mathrm{d}u_1\,\mathrm{d}u_2. \tag{16}$$

Moreover, if $r(t_1,t_2)/t_1t_2)$ *is Lebesgue integrable on* $\mathbb{R}^2$, *then*

$$R(x,y)=\left(\frac{1}{2\pi\mathrm{i}}\right)^2\int_{-\infty}^{\infty}\int_{-\infty}^{\infty}\frac{r(t_1,t_2)}{t_1t_2}\mathrm{e}^{-\mathrm{i}t_1x-\mathrm{i}t_2y}\,\mathrm{d}t_1\,\mathrm{d}t_2+\tfrac{1}{2}(R_1(x)+R_2(y)), \tag{17}$$

at all (x,y) *which are continuity points of* R, R_1 *and* R_2.

Remark. The conditions (i)–(iii) are automatic if R is the difference of two distributions in $\mathbb{R}^2$ having the first order moments. (Cf., [2, p. 254 and p. 241] on normalized — also called mean continuous — functions.) The result is a two-dimensional version of Cramér's theorem. However, the proof is more involved than the univariate case.

Proof. First let $x_i\in\mathbb{R}$, $h_i>0$ such that x_i+h_i are continuity points of R and R_j, $j=1,2$, $i=1,2$. Let us start with a simplification of the right-hand side (R.H.S.) of

(16). Thus,

$$\text{R.H.S.} = \lim_{M\to\infty}\left(\frac{1}{2\pi i}\right)^2 \int_{-M}^{M}\int_{-M}^{M}\int_{-\infty}^{\infty}\int_{-\infty}^{\infty} e^{it_1y_1+it_2y_2}R(dy_1,dy_2)e^{-it_1x_1-it_2x_2}\frac{dt_1\,dt_2}{t_1t_2}$$

$$\times\int_0^{h_1}\int_0^{h_2} u_1^{\alpha_1-1}u_2^{\alpha_2-1}e^{-iu_1t_1-iu_2t_2}du_1\,du_2$$

$$= \lim_{M\to\infty}\left(\frac{1}{2\pi i}\right)^2(2i)^2\int_{-\infty}^{\infty}\int_{-\infty}^{\infty}R(dy_1,dy_2)\int_0^{h_1}\int_0^{h_2}u_1^{\alpha_1-1}u_2^{\alpha_2-1}du_1\,du_2$$

$$\times\int_0^{M}\int_0^{M}\frac{\sin t_1(y_1-x_1-u_1)}{t_1}\cdot\frac{\sin t_2(y_2-x_2-u_2)}{t_2}dt_1\,dt_2$$

by Fubini's theorem,

$$=\tfrac{1}{4}\int_{-\infty}^{\infty}\int_{-\infty}^{\infty}R(dy_1,dy_2)\int_0^{h_1}\int_0^{h_2}u_1^{\alpha_1-1}u_2^{\alpha_2-1}\operatorname{sgn}(y_1-x_1-u_1)$$

$$\operatorname{sgn}(y_2-x_2-u_2)\,du_1\,du_2, \quad (18)$$

since

$$\lim_{M\to\infty}\frac{2}{\pi}\int_0^M\frac{\sin ht}{t}=\operatorname{sgn} h,$$

and the Dominated Convergence Theorem is applicable [by condition (iii)],

$$=\tfrac{1}{4}\int_{y_1=-\infty}^{\infty}\int_0^{h_1}u_1^{\alpha_1-1}\operatorname{sgn}(y_1-x_1-u_1)\,du_1\left(-\int_{y_2=-\infty}^{x_2}\frac{h_2^{\alpha_2}}{\alpha_2}R(dy_1,dy_2)\right.$$

$$+\int_{y=x_2}^{x_2+h_2}\left[\int_0^{y_2-x_2}u_2^{\alpha_2-1}du_2-\int_{y_2-x_2}^{h_2}u_2^{\alpha_2-1}du_2\right]R(dy_1,dy_2)$$

$$\left.+\int_{y_2=x_2+h_2}^{\infty}\frac{h_2^{\alpha_2}}{\alpha_2}R(dy_1,dy_2)\right)$$

$$=\frac{1}{4\alpha_2}\int_{y_1=-\infty}^{\infty}\int_0^{h_1}u_1^{\alpha_1-1}\operatorname{sgn}(y_1-x_1-u_1)\,du_1$$

$$\times\left[-h_2^{\alpha_2}R(dy_1,x_2)+h_2^{\alpha_2}R(dy_1,-\infty)\right.$$

$$+2\int_{x_2}^{x_2+h_2}(y_2-x_2)^{\alpha_2}R(dy_1,dy_2)-h_2^{\alpha_2}R(dy_1,x_2+h_2)+h_2^{\alpha_2}R(dy_1,x_2)$$

$$\left.+h_2^{\alpha_2}R(dy_1,\infty)-h_2^{\alpha_2}R(dy_1,x_2+h_2)\right],$$

by Condition (i) and [2, Thm. 2 on p. 267],

$$=\frac{1}{4\alpha_2}\int_{y_1=-\infty}^{\infty}\int_0^{h_1}u_1^{\alpha_1-1}\operatorname{sgn}(y_1-x_1-u_1)\,du_1\,a(dy_1;x_2,h_2) \quad (19)$$

where

$$a(\mathrm{d}y_1;x_2,h_2)=2\left(\int_{y_2=x_2}^{x_2+h_2}(y_2-x_2)^{\alpha_2}R(\mathrm{d}y_1,\mathrm{d}y_2)-h_2^{\alpha_2}R(\mathrm{d}y_1,x_2+h_2)\right)$$

$$+h_2^{\alpha_2}R_1(\mathrm{d}y_1),$$

$$=\frac{1}{4\alpha_2}\left(-\int_{y_1=-\infty}^{x_1}\frac{h_1^{\alpha_1}}{\alpha_1}a(\mathrm{d}y_1;x_2,h_2)+\int_{x_1}^{x_1+h_1}\left[\frac{(y_1-x_1)^{\alpha_1}}{\alpha_1}-\frac{h_1^{\alpha_1}-(y_1-x_1)^{\alpha_1}}{\alpha_1}\right]\right.$$

$$\left.\times a(\mathrm{d}y_1;x_2,h_2)+\frac{h_1^{\alpha_1}}{\alpha_1}(a(\infty,x_2,h_2)-a(x_1+h_1;x_2,h_2))\right)$$

$$=\frac{1}{2\alpha_1\alpha_2}\left\{\int_{x_1}^{x_1+h_1}(y_1-x_1)^{\alpha_1}a(\mathrm{d}y_1;x_2,h_2)+h_1^{\alpha_1}\int_{x_2}^{x_2+h_2}(y_2-x_2)^{\alpha_2}R_2(\mathrm{d}y_2)\right.$$

$$\left.-h_1^{\alpha_1}(a(x_1+h_1;x_2,h_2)+h_2^{\alpha_2}R_2(x_2+h_2))\right\}, \tag{20}$$

since $R_1(+\infty)=0$, and the value of $a(+\infty,x_2,h_2)$ is substituted. Now using the expression of $a(\cdot;\cdot,\cdot)$, (20) becomes

$$=\frac{1}{\alpha_1\alpha_2}\left(\int_{x_1}^{x_1+h_1}\int_{x_2}^{x_2+h_2}(y_1-x_1)^{\alpha_1}(y_2-x_2)^{\alpha_2}R(\mathrm{d}y_1,\mathrm{d}y_2)\right.$$

$$-h_2^{\alpha_2}\int_{x_1}^{x_1+h_1}(y_1-x_1)^{\alpha_1}R(\mathrm{d}y_1,x_2+h_2)$$

$$\left.-h_1^{\alpha_1}\int_{x_2}^{x_2+h_2}(y_2-x_2)^{\alpha_2}R(x_1+h_1,\mathrm{d}y_2)+h_1^{\alpha_1}h_2^{\alpha_2}R(x_1+h_1,x_2+h_2)\right)$$

$$+\frac{1}{2\alpha_1\alpha_2}\left(\int_{x_1}^{x_1+h_1}(y_1-x_1)^{\alpha_1}R_1(\mathrm{d}y_1)+h_1^{\alpha_1}\int_{x_2}^{x_2+h_2}(y_1-x_1)^{\alpha_2}R_2(\mathrm{d}y_2)\right.$$

$$\left.-h_1^{\alpha_1}h_2^{\alpha_2}(R_1(x_1+h_1)+R_2(x_2+h_2))\right)$$

$$=\frac{1}{\alpha_1\alpha_2}I_1+\frac{1}{2\alpha_1\alpha_2}I_2\quad\text{(say)}. \tag{21}$$

To simplify (21) consider each term separately, and use integration by parts componentwise. Then using the fact that x_i+h_i are continuity points of R and R_j in (21), one has (cf. the Stieltjes integration theory in [2, Thm. 2, p. 267]):

$$\frac{1}{\alpha_1\alpha_2}I_2=\frac{1}{\alpha_1\alpha_2}\int_{x_1}^{x_1+h_1}\int_{x_2}^{x_2+h_2}(y_1-x_1)^{\alpha_1}(y_2-x_2)^{\alpha_2}R(\mathrm{d}y_1,\mathrm{d}y_2)$$

$$+\frac{h_1^{\alpha_1}}{\alpha_1}\int_{x_2}^{x_2+h_2}R(x_1+h_1,y_2)(y_2-x_2)^{\alpha_2-1}\mathrm{d}y_2$$

$$+\frac{h_2^{\alpha_2}}{\alpha_2}\int_{x_1}^{x_1+h_1}R(y_1,x_2+h_2)(y_1-x_1)^{\alpha_1-1}\mathrm{d}y_1-\frac{h_1^{\alpha_1}h_2^{\alpha_2}}{\alpha_1\alpha_2}R(x_1+h_1,x_2+h_2). \tag{22}$$

By a similar reasoning applied to the first integral of (22), one gets:

$$\int_{x_1}^{x_1+h_1}\int_{x_2}^{x_2+h_2}(y_1-x_1)^{\alpha_1}(y_2-x_2^{\alpha_2}R(\mathrm{d}y_1,\mathrm{d}y_2))$$

$$=h_1^{\alpha_1}h_2^{\alpha_2}R(x_1+h_1,x_2+h_2)-\alpha_1h_2^{\alpha_2}\int_{x_1}^{x_1+h_1}(y_1-x_1)^{\alpha_1-1}R(y_1,x_2+h_2)\,\mathrm{d}y_1$$

$$-\alpha_2h_1^{\alpha_1}\int_{x_2}^{x_2+h_2}(y_2-x_2)^{\alpha_2-1}R(x_1+h_1,y_2)\,\mathrm{d}y_2$$

$$+\alpha_1\alpha_2\int_{x_1}^{x_1+h_1}\int_{x_2}^{x_2+h_2}(y_1-x_1)^{\alpha_1-1}(y_2-x_2)^{\alpha_2-1}R(y_1,y_2)\,\mathrm{d}y_1\,\mathrm{d}y_2, \tag{23}$$

using condition (iii) again. Substitution of (23) in (22) gives, after cancellations:

$$\frac{1}{\alpha_1\alpha_2}I_1=\int_{x_1}^{x_1+h_1}\int_{x_2}^{x_2+h_2}(y_1-x_1)^{\alpha_1-1}(y_2-x_2)^{\alpha_2-1}R(y_1,y_2)\,\mathrm{d}y_1\,\mathrm{d}y_2. \tag{24}$$

The second term of (21) can be simplified by a similar reasoning to get:

$$\frac{1}{2\alpha_1\alpha_2}I_2=-\frac{h_2^{\alpha_2}}{2\alpha_2}\int_{x_1}^{x_1+h_1}(y_1-x_1)^{\alpha_1-1}R_1(y_1)\,\mathrm{d}y_1$$

$$-\frac{h_1^{\alpha_1}}{2\alpha_1}\int_{x_2}^{x_2+h_2}(y_2-x_2)^{\alpha_2-1}R_2(y_2)\,\mathrm{d}y_2. \tag{25}$$

When substituting (24) and (25) in (21), it reduces to the left-hand side of (16).

Thus far x_i+h_i were assumed to be continuity points of R and R_j, $j=1,2$, $i=1,2$. However both sides of (16) are actually continuous in x_i and h_i and since the continuity points of R, R_j, $j=1,2$ are dense in $\mathbb{R}^2$, eq. (16) holds for all $x_i\in\mathbb{R}$ and $h_i>0$. In fact, to verify the continuity assertion, let $v_i=(y_i-x_i)/h_i$, $i=1,2$. Then the left side of (16) becomes:

$$h_1^{\alpha_1}h_2^{\alpha_2}\int_0^{h_1}\int_0^{h_2}v_1^{\alpha_1-1}v_2^{\alpha_2-1}R(x_1+v_1h_1,x_2+v_2h_2)\,\mathrm{d}v_1\,\mathrm{d}v_2$$

$$-\frac{h_1^{\alpha_1}h_2^{\alpha_2}}{2\alpha_1}\int_0^{h_2}v_2^{\alpha_2-1}R_2(x_2+v_2h_2)\,\mathrm{d}v_2$$

$$-\frac{h_1^{\alpha_1}h_2^{\alpha_2}}{2\alpha_2}\int_0^{h_1}v_1^{\alpha_1-1}R_1(x_1+v_1h_1)\,\mathrm{d}v_1\to 0, \tag{26}$$

as $h_i\to 0$, $i=1,2$, for each fixed $0<\alpha_i<1$. Also the Dominated Convergence implies that the quantity in (26) is continuous in x_i, because of the "strong uniform continuity" of R and R_j in the sense of [2, p. 255]. Thus the left side of (16) is continuous. Regarding the right side, since r is bounded and continuous [even differentiable by condition (iii)] for $t_i\neq 0$, the second integral satisfies:

$$\left|\int_0^{h_1}\int_0^{h_2}u_1^{\alpha_1-1}u_2^{\alpha_2-1}\mathrm{e}^{-it_1u_1-it_2u_2}\,\mathrm{d}u_1\,\mathrm{d}u_2\right|=$$

$$=\left|\int_0^{t_1h_1}\mathrm{e}^{-iv_1}\left(\frac{v_1}{t_1}\right)^{\alpha_1-1}\frac{\mathrm{d}v_1}{t_1}\int_0^{t_2h_2}\mathrm{e}^{-iv_2}\left(\frac{v_2}{t_2}\right)^{\alpha_2-1}\frac{\mathrm{d}v_2}{t_2}\right|<\frac{h_1^{\alpha_1}h_2^{\alpha_2}}{\alpha_1\alpha_2}.$$

Thus the right side is also continuous in x_i, h_i, $i=1,2$ as asserted.

To establish (17), with the additional hypothesis of integrability on r, one can let $\alpha_i \to 1$, $i=1,2$ in (16) to obtain (Dominated Convergence again)

$$\left(\frac{1}{2\pi\mathrm{i}}\right)^2 \int_{-\infty}^{\infty}\int_{-\infty}^{\infty} \frac{r(t_1,t_2)}{t_1 t_2} \mathrm{e}^{-\mathrm{i}t_1x_1-\mathrm{i}t_2x_2}\mathrm{d}t_1\,\mathrm{d}t_2 \int_0^{h_1}\int_0^{h_2} \mathrm{e}^{-\mathrm{i}t_1u_1-\mathrm{i}t_2u_2}\mathrm{d}u_1\,\mathrm{d}u_2 =$$

$$= \int_{x_1}^{x_1+h_1}\int_{x_2}^{x_2+h_2} R(y_1,y_2)\,\mathrm{d}y_1\,\mathrm{d}y_2 - \tfrac{1}{2}h_1\int_{x_2}^{x_2+h_2} R_2(y)\,\mathrm{d}y - \tfrac{1}{2}h_2\int_{x_1}^{x_1+h_1} R_1(y)\,\mathrm{d}y. \tag{27}$$

By differentiating (27) relative to h_1, h_2 in succession, by the Fundamental Theorem of Calculus, one has at all continuity points x_i+h_i of R and R_j,

$$\left(\frac{1}{2\pi\mathrm{i}}\right)^2 \int_{-\infty}^{\infty}\int_{-\infty}^{\infty} \frac{r(t_1,t_2)}{t_1 t_2} \mathrm{e}^{-\mathrm{i}t_1(x_1+h_1)-\mathrm{i}t_2(x_2+h_2)}\mathrm{d}t_1\,\mathrm{d}t_2 =$$

$$= R(x_1+h_1, x_2+h_2) - \tfrac{1}{2}(R_1(x_1+h_1) + R_2(x_2+h_2)). \tag{28}$$

By letting $x=x_1+h_1$ and $y=x_2+h_2$ in (28), it reduces to (17). This completes the proof.

Remark. Since R_1, R_2 satisfy the hypothesis of [3, Thm. 12], if r_j are the Fourier-Stieltjes transformations of R_j, $j=1,2$, then using Cramér's theorem, one can present (16) in the following form:

$$\int_{x_1}^{x_1+h_1}\int_{x_2}^{x_2+h_2} (y_1-x_1)^{\alpha_1-1}(y_2-x_2)^{\alpha_2-1} R(y_1,y_2)\,\mathrm{d}y_1\,\mathrm{d}y_2 =$$

$$= \left(\frac{1}{2\pi\mathrm{i}}\right)^2 \int_{-\infty}^{\infty}\int_{-\infty}^{\infty} \frac{r(t_1,t_2)}{t_1 t_2} \mathrm{e}^{-\mathrm{i}t_1x_1-\mathrm{i}t_2x_2}\mathrm{d}t_1\,\mathrm{d}t_2 \int_0^{h_1}\int_0^{h_2} u_1^{\alpha_1-1}u_2^{\alpha_2-1}\mathrm{e}^{-\mathrm{i}t_1u_1-\mathrm{i}t_2u_2}\mathrm{d}u_1\,\mathrm{d}u_2$$

$$-\frac{1}{4\pi\mathrm{i}}\left(\frac{h_1^{\alpha_1}}{\alpha_1}\int_{-\infty}^{\infty}\frac{r_2(t)}{t}\mathrm{e}^{-\mathrm{i}tx_2}\mathrm{d}t\int_0^{h_2} u_2^{\alpha_2-1}\mathrm{e}^{-\mathrm{i}tu_2}\mathrm{d}u_2\right.$$

$$\left.+\frac{h_2^{\alpha_2}}{\alpha_2}\int_{-\infty}^{\infty}\frac{r_1(t)}{t}\mathrm{e}^{-\mathrm{i}tx_1}\mathrm{d}t\int_0^{h_1} u_1^{\alpha_1-1}\mathrm{e}^{-\mathrm{i}tu_1}\mathrm{d}u_1\right).$$

In some cases, $R(x_1,x_2)\to 0$ as $x_i\to+\infty$, $i=1,2$ can also hold. In this event $R_j\equiv 0$, $j=1,2$, and (16) simplifies further. This may be stated as:

Corollary 1. *Let $R:\mathbb{R}^2\to\mathbb{R}$ be a normalized function of bounded variation as in the theorem such that*

(i) $R(x_1,x_2)\to 0$ *as* $x_i\to\pm\infty$, $i=1,2$, *and*

(ii) $\int_{-\infty}^{\infty}\int_{-\infty}^{\infty}(|x_1|+|x_2|)|R(\mathrm{d}x_1,\mathrm{d}x_2)|<\infty$.

Then for each $0<\alpha_i<1$, *and for all* $x_i\in\mathbb{R}$, $h_i>0$, $i=1,2$, *one has*

$$\int_{x_1}^{x_1+h_1}\int_{x_2}^{x_2+h_2} (y_1-x_1)^{\alpha_1-1}(y_2-x_2)^{\alpha_2-1} R(y_1,y_2)\,\mathrm{d}y_1\,\mathrm{d}y_2 =$$

$$= \left(\frac{1}{2\pi i}\right)^2 \int_{-\infty}^{\infty}\int_{-\infty}^{\infty} \frac{r(t_1,t_2)}{t_1 t_2} \mathrm{e}^{-\mathrm{i}t_1x_1-\mathrm{i}t_2x_2}\mathrm{d}t_1\,\mathrm{d}t_2$$

$$\times \int_0^{h_1}\int_0^{h_2} u_1^{\alpha_1-1}u_2^{\alpha_2-1}\mathrm{e}^{-\mathrm{i}t_1u_1-\mathrm{i}t_2u_2}\mathrm{d}u_1\,\mathrm{d}u_2. \tag{29}$$

Also if $[r(t_1, t_2]/(t_1t_2)$ *is Lebesgue integrable, then for* (x, y) *continuity point of* R,

$$R(x, y)=\left(\frac{1}{2\pi i}\right)^2 \int_{-\infty}^{\infty} \int_{-\infty}^{\infty} \frac{r(t_1, t_2)}{t_1t_2} e^{-it_1x-it_2y} dt_1 dt_2. \tag{30}$$

4. Specialization to distributions of ratios

With the above theorem, let us obtain the distribution of the ratios of random variables by taking R as the difference of a pair of bivariate distribution functions. Thus let (X_i, Y_i), $i=1,2$ be random variables such that $P(Y_i=0)=0$, $i=1,2$. For $z_i \in \mathbb{R}$, consider

$$H(z_1, z_2)=P\left(\frac{X_1}{Y_1}<z_1, \frac{X_2}{Y_2}<z_2\right)=P(C_z) \quad \text{(say)}. \tag{31}$$

Given the joint distribution of X_1, X_2, Y_1, Y_2, it is desired to obtain the distribution of the ratios so that H of (31) is determined. This will be deduced from Theorem 3, after introducing some necessary notation.

Let $A_1=[Y_1>0]$, $A_2=[Y_1<0]$, $B_1=[Y_2>0]$, $B_2=[Y_2<0]$ and $D=[Y_1=0]\cup[Y_2=0]$ so that $P(D)=0$. If $C_{ij}=A_i \cap B_j$, $\Omega_0=\cup_{i,j}C_{ij}$, then C_{ij} being disjoint, $P(\Omega_0)=1$. Let $W_j=X_j-z_jY_j$, $\tilde{W}_j=(-1)^{j-1}W_1$, $\overline{W}_j=(-1)^{j-1}W_2$, $j=1,2$. Also define the non-decreasing left continuous bounded functions G_{ij}^z by the equations

$$\begin{aligned} G_{jk}^z(u_1, u_2)&=P_{C_{jk}}(\tilde{W}_j<u_1, \overline{W}_k<u_2) \\ &=P([\tilde{W}_j<u_1]\cap[\overline{W}_j<u_2]\cap C_{jk}), \quad j, k=1,2. \end{aligned} \tag{32}$$

Since $P(C_z \cap \Omega_0)=P(C_z)$ (because $P(\Omega_0)=1$) one has, if $G^z: \mathbb{R}^2 \to \mathbb{R}^+$ is given by

$$G^z(u_1, u_2)= \sum_{j,k=1}^{2} G_{jk}^z(u_1, u_2), \tag{33}$$

then $G^z(0,0)=P(C_z)=H(z_1, z_2)$. Clearly G^z is a bivariate probability distribution function for each fixed $z=(z_1, z_2)\in\mathbb{R}^2$. Using an idea due to Curtiss [5], consider the Fourier–Stieltjes transformation r_{jk}^z of G_{jk}^z and the ch.f. g^z of G^z, so that

$$g^z(t_1, t_2)= \sum_{j,k=1}^{2} r_{jk}^z(t_1, t_2).$$

Let $F: \mathbb{R}^2 \to \mathbb{R}^+$ be a distribution function and set $R^z=G^z-F$. Let r^z, $r_1^{z_1}$, $r_2^{z_2}$ and s be the Fourier–Stieltjes transformations of R^z, $R_1^{z_1}$, $R_2^{z_2}$ and F, so that $r^z=g^z-s$. Here $R_1^{z_1}(u)=R^z(u, \infty)$, $R_2^{z_2}(u)=R^z(\infty, u)$. That $R_1^{z_1}$ and $R_2^{z_2}$ depend only on z_1 and z_2 (but not on both) is seen as follows. By definition $R_1^z(u_1)=R^z(u_1, \infty)=G^z(u_1, \infty)-F(u_1, \infty)$, where

$$G^z(u_1, \infty)= \sum_{i,j=1}^{2} G_{ij}^z(u_1, \infty).$$

However, by writing $P_{C_{ij}}(A)$ as $P(A; C_{ij})$ one has:

$$\sum_{i,j=1}^{2} G_{ij}^{z}(u_1,\infty)=P(\tilde{W}_1<u_1,\overline{W}_1<\infty;C_{11})+P(\tilde{W}_1<u_1,\overline{W}_2<\infty;C_{12})$$
$$+P(\tilde{W}_2<u_1,\overline{W}_1<\infty;C_{21})+P(\tilde{W}_2<u_1,\overline{W}_2<\infty;C_{22})$$
$$=P(\tilde{W}_1<u_1;C_{11}\cup C_{12})+P(\tilde{W}_2<u_1;C_{21}\cup C_{22})$$
$$=P(W_1<u_1;A_1)+P(-W_1<u_2;A_1)$$
$$=P(X_1-z_1Y_1<u_1,Y_1>0)+P(-X_1+z_1Y_1<u_1,Y_1<0).$$

Hence

$$G^z(0,\infty)=\sum_{i,j=1}^{\infty} G_{ij}^{z}(0,\infty)=P\left(\frac{X_1}{Y_1}<z_1,Y_1>0\right)+P\left(\frac{X_1}{Y_1}<z_1,Y_1<0\right)$$
$$=P\left(\frac{X_1}{Y_1}<z_1\right)=G_1^{z_1}(0). \tag{34}$$

Similarly

$$G^z(\infty,u_2)=P(X_2-z_2Y_2<u_2,Y_2>0)+P(-X_2+z_2Y_2<u_2,Y_1<0)$$
$$G^z(\infty,0)=P\left(\frac{X_2}{Y_2}<z_2\right)=G_2^{z_2}(0). \tag{35}$$

If $F_1(u)=F(u,\infty)$ and $F_2(u)=F(\infty,u)$ define the marginal distributions of F, then $R_1^z(u_1)$ depends only on z_1 and similarly $R_2^z(u_1)$ depends on z_2. Further

$$R_1^{z_1}(0)=P\left(\frac{X_1}{Y_1}<z_1\right)-F_1(0),\qquad R_2^{z_2}(0)=P\left(\frac{X_2}{Y_2}<z_2\right)-F_2(0).$$

Consequently for $R_1^{z_1}$ and $R_2^{z_2}$ their Fourier–Stieltjes transforms depend only on z_1 and z_2, and Cramér's result [3, Thm. 12] applies to them. With this the desired extension can be presented as follows:

Theorem 4. *Let (X_i, Y_i), $i=1,2$ be integrable random variables on a probability space (Ω,Σ,P) and let F be a distribution function on $\mathbb{R}^2$ with first order moments. If $R^z=G^z-F$, $R_1^{z_1}$, $R_2^{z_2}$, F_1, F_2, are the corresponding functions introduced above, with r^z, $r_1^{z_1}, r_2^{z_2}$, s, s_1, s_2 as their Fourier–Stieltjes transforms, suppose $r^z(t_1,t_2)/t_1t_2$ is Lebesgue integrable on $\mathbb{R}^2$. If $P(Y_i=0)=0$, $i=1,2$, and $r_j^{z_j}(t)/t$ is also Lebesgue integrable, $j=1,2$, then*

$$G^z(0,0)=H(z_1,z_2)=F(0,0)+\tfrac{1}{2}(F_1(0)+F_2(0))$$
$$-\frac{1}{4\pi^2}\int_{-\infty}^{\infty}\int_{-\infty}^{\infty}\frac{r^z(t_1,t_2)-s(t_1,t_2)}{t_1t_2}\,\mathrm{d}t_1\,\mathrm{d}t_2$$
$$+\frac{1}{4\pi\mathrm{i}}\left[\int_{-\infty}^{\infty}\frac{s_1(t_1)-r_1^{z_1}(t_1)}{t_1}\,\mathrm{d}t_1+\int_{-\infty}^{\infty}\frac{s_2(t_2)-r_2^{z_2}(t_2)}{t_2}\,\mathrm{d}t_2\right]. \tag{36}$$

In particular, if $F(y_1, y_2)=P(Y_1<y_1, Y_2<y_2)$ *and* $P_{jk}(\cdot)=P(\cdot\cap C_{jk})$, $P_j(\cdot)=P(\cdot\cap A_j)$, $Q_j(\cdot)=P(\cdot\cap B_j)$ *are the finite measures on* (Ω,Σ) *thus defines, let*

$$\varphi_{jk}(t_1,t_2,t_3,t_4)=\int_\Omega e^{it_1X_1+it_2Y_1+it_3X_2+it_4Y_2}\,dP_{jk},$$

$$\varphi_j(t_1,t_2)=\int_\Omega e^{it_1X_1+it_2Y_1}\,dP_j,\qquad \psi_j(t_1,t_2)=\int_\Omega e^{it_1X_2+it_2Y_2}\,dQ_j.$$

Then (36) *reduces to*:

$$\begin{aligned}H(z_1z_2)={}&F(0,0)+\tfrac{1}{2}(F_1(0)+F_2(0))-\frac{1}{16\pi^2}\sum_{j,k=1}^{2}\int_{-\infty}^{\infty}\int_{-\infty}^{\infty}\\&\frac{4\varphi_{jk}\big((-1)^{j-1}t_1,(-1)^jz_1t_1,(-1)^{k-1}t_2,(-1)^kz_2t_2\big)-s(t_1,t_2)}{t_1t_2}\,dt_1\,dt_2\\&-\frac{1}{4\pi i}\left(\int_{-\infty}^{\infty}\frac{\varphi_1(t_1,-t_1z_1)+\varphi_2(-t_1,t_1z_1)-s_1(t_1)}{t_1}\,dt_1\right.\\&\left.+\int_{-\infty}^{\infty}\frac{\psi_1(t_1,-t_1z_2)+\psi_2(-t_1,t_1z_2)-s_2(t_1)}{t_1}\,dt_1\right).\end{aligned}\tag{37}$$

If the integrand in (37) *obtained by formally differentiating relative to* z_1 *and* z_2 *is uniformly convergent in an open interval of* $\mathbb{R}^2$ *containing* z_1, z_2, *then the density of* H *exists there and is given by*

$$\begin{aligned}h(z_1,z_2)&=\frac{\partial^2 H}{\partial z_1\partial z_2}(z_1,z_2)\\&=\frac{1}{4\pi^2}\sum_{j,k=1}^{2}(-1)^{j+k+1}\int_{-\infty}^{\infty}\int_{-\infty}^{\infty}\frac{\partial^2\varphi_{jk}}{\partial t_1\partial t_4}\Bigg|_{\substack{t_1=(-1)^{j-1}\tau_1,\,t_2=(-1)^jz_1\tau_1\\ t_3=(-1)^{k-1}\tau_2,\,t_4=(-1)^kz_2\tau_2}}d\tau_1\,d\tau_2.\end{aligned}\tag{38}$$

If further $P(Y_i>0)=1$, $i=1,2$, so that $P(C_{11})=1$, then (38) becomes

$$h(z_1,z_2)=\left(\frac{1}{2\pi i}\right)^2\int_{-\infty}^{\infty}\int_{-\infty}^{\infty}\frac{\partial^2\varphi}{\partial t_2\partial t_4}\Bigg|_{\substack{t_1=\tau_1,\,t_2=-z_1\tau_1\\ t_3=\tau_2,\,t_4=-z_2\tau_2}}d\tau_1\,d\tau_2\tag{39}$$

where φ *replaces* φ_{11} *now*.

Remark. If X_1, X_2, Y_1, Y_2 are mutually independent and satisfy the integrability (and vanishing) hypotheses for (39), then the joint density of X_1/Y_1 and X_2/Y_2 factors and one gets

$$h(z_1,z_2)=\frac{1}{2\pi i}\int_{-\infty}^{\infty}\varphi_1(t_1)\varphi_2'(-t_1z_1)\,dt_1\cdot\frac{1}{2\pi i}\int_{-\infty}^{\infty}\varphi_3(t_2)\varphi_4'(-t_2z_2)\,dt_2,\tag{40}$$

where the φ_i's are the ch.f.'s of X_i's and Y_i's. The result therefore reduces to Cramér's theorem [3, Thm. 16] as it should.

Proof. If $R^z = G^z - F$, then the assumed integrability of r^z and $r_j^{z_j}$ imply the hypothesis of Theorem 3 and [3, Thm. 12]. So (36) follows from a substitution of the relevant quantities. For (37) note that

$$r_{jk}^z(t_1, t_2) = \int_{-\infty}^{\infty}\int_{-\infty}^{\infty} e^{it_1u_1 + it_2u_2} G_{jk}^z(du_1, du_2) = \int_\Omega e^{it_1\tilde{W}_j + it_2\overline{W}_k} dP_{jk}$$

$$= \int_\Omega e^{i(-1)^{j-1}t_1W_1 + i(-1)^{k-1}t_2W_2} dP_{jk}$$

$$= \int_\Omega e^{i(-1)^{j-1}t_1X_1 + it_1(-1)^j z_1Y_1 + i(-1)^{k-1}t_2X_2 + i(-1)^k t_2z_2Y_2} dP_{jk}$$

$$= \varphi_{jk}((-1)^{j-1}t_1, (-1)^j t_1z_1, (-1)^{j-1}t_2, (-1)^k t_2z_2).$$

Similarly

$$r_1^{z_1}(t) = \varphi_1(t, -tz_1) + \varphi_2(-t, tz_1),$$
$$r_2^{z_2}(t) = \psi_1(t, -tz_2) + \psi_2(-t, tz_2).$$

Substitution of these in (36) yields (37). Differentiating (37) relative to z_1 and z_2 and using the additional hypothesis implies (38) immediately. Then (39) is a simple consequence since $P_{11} = P$ and $P(C_{jk}) = 0$ for $j > 1$ or $k > 1$. This completes the proof of the theorem.

Remark. (i) Analogous results for related problems have been considered by Gurland (cf. [6, Thm. 6]). The above one extends Cramér's theorem [4, p. 317] as well as Curtiss' theorem [5, Thm. 5.1] to the bivariate case.

(ii) The result of Theorem 4 is useful in applications such as the problems of analysis of variance — e.g., the two-way layout, randomized blocks, or lattin squares (cf. [4, pp.543–547]) — where it is of general interest to compare two (or more) F-ratios F_1, F_2 simultaneously (related to the row and column effects in a randomized block design, for instance). Thus one needs to know $H(z_1, z_2) = P(F_1 < z_1, F_2 < z_2)$, and this is given by a suitable specialization of (38). The same formula is also of use in obtaining the joint limiting distribution of the estimators of parameters of a second order autoregressive model when the roots of its characteristic equation may be on the boundary of the unit circle. It will extend the work of [10] which is the motivation for this paper. The details, however, are involved and the analysis needs additional work.

References

[1] Anderson, T.W. (1959). On the asymptotic distributions of estimates of parameters of stochastic difference equations. *Ann. Math. Statist.* **30**, 676–687.

[2] Bergström, H. (1963). *Limit Theorems for Convolutions*. Wiley, New York.

[3] Cramér, H. (1962). *Random Variables and Probability Distributions*. 2nd ed. Cambridge Univ. Press, Cambridge.

[4] Cramér, H. (1946). *Mathematical Methods of Statistics*. Princeton Univ. Press, Princeton, NJ.

[5] Curtiss, J.H. (1941). On the distribution of the quotient of two chance variables. *Ann. Math. Statist.* **12**, 409–421.

[6] Gurland, J. (1948). Inversion formulae for the distribution of ratios. *Ann. Math. Statist.* **19**, 228–237.

[7] Kagan, A.M., Linnik, Yu.V. and Rao, C.R. (1973). *Characterization Problems in Mathematical Statistics*. Wiley, New York.

[8] Mann, H.B. and Wald, A. (1943). On the statistical treatment of linear stochastic difference equations. *Econometrica* **11**, 173–200.

[9] Rao, M.M. (1961). Consistency and limit distributions of estimators of parameters in explosive stochastic difference equations. *Ann. Math. Statist.* **32**, 195–218.

[10] Rao, M.M. (1978). Asymptotic distribution of an estimator of the boundary parameter of an unstable process. *Ann. Statist.* **6**, 185–190. [Erratum, *ibid.* **8**(1980) 1403].

[11] White, J.S. (1958). The limiting distribution of the serial correlation coefficient in the explosive case. *Ann. Math. Statist.* **29**, 1188–1197.

[12] White, J.S. (1959). The limiting distribution of the serial correlation coefficient in the explosive case II. *Ann. Math. Statist.* **30**, 831–834.

G. Kallianpur, P.R. Krishnaiah, J.K. Ghosh, eds., *Statistics and Probability*: *Essays in Honor of C.R. Rao*

BOUNDARY EFFECTS ON THE BEHAVIOR OF SMOOTHING SPLINES*

J. RICE and M. ROSENBLATT
University of California, San Diego, La Jolla, CA 92093, U.S.A.

1. Introduction

Assume that we observe

$$x_i = g(t_i) + \varepsilon_i, \tag{1}$$

at the points,

$$t_i = i/n, \quad i = 0, 1, \ldots, n-1,$$

with $g(t)$ a smooth unknown function on [0, 1] and the ε_i are orthogonal random variables with

$$E(\varepsilon_i) = 0, \qquad E(\varepsilon_i \varepsilon_j) = \delta_{i,j}\sigma^2, \quad \sigma^2 > 0. \tag{2}$$

A smoothing spline has been proposed as an estimate of g, that is, a function $f(t; \lambda, n)$ minimizing

$$\frac{1}{n}\sum_{i=0}^{n-1}[f(t_i) - x_i]^2 + \frac{\lambda}{(2\pi)^2}\int_0^1 [f''(t)]^2 \, dt. \tag{3}$$

(See refs. [9, 5, 7].) Analyses of the asymptotic behavior of the smoothing spline as $\lambda = \lambda(n) \downarrow 0$ $n \to \infty$ have been carried out [2, 8]. Some of these analyses have been in terms of the expected integrated square error

$$E\int_0^1 |f(t; \lambda(n), n) - g(t)|^2 \, dt. \tag{4}$$

However, these analyses do not take into account a dependence of the asymptotic behavior of the integrated mean square error on the behavior of g at 0 and 1. The results we report on here show some aspect of this dependence but the detailed derivation of these results will appear elsewhere. Some computations in specific cases are carried out to illustrate how the boundary effect can make the major contribution to the integrated square error.

*The research was supported in part by NSF Grant MCS-7901800 and ONR Contract N00014-75-C-0428.

2. Statement of results

It is convenient to write the expected integrated square error as

$$E\int_0^1 |f(t)-g(t)|^2\,dt = \int_0^1 |Ef(t)-g(t)|^2\,dt + \int_0^1 \sigma^2[f(t)]\,dt, \tag{5}$$

where

$$\sigma^2[f(t)] = E[f(t)-Ef(t)]^2.$$

In our results we assume that $f, g \in C^2[0,1]$ with

$$f(0)=f(1), \qquad f'(0)=f'(1), \qquad g(0)=g(1), \qquad g'(0)=g'(1). \tag{6}$$

Theorem 1. *Let $g \in C^2[0,1]$ with boundary conditions (6). A necessary and sufficient condition that*

$$\int_0^1 \sigma^2[f(t)]\,dt \to 0 \tag{7}$$

as $n\to\infty$ and $\lambda(n)\to 0$ is that

$$\lambda(n)n^4 \to \infty. \tag{8}$$

If (8) holds then

$$\int_0^1 \sigma^2[f(t)]\,dt \cong \frac{\sigma^2}{n\lambda(n)^{1/4}} \int_{-\infty}^{\infty} \frac{dx}{(1+x^4)} + o\left(n^{-1}\lambda(n)^{-1/4}\right). \tag{9}$$

The analyses use Fourier methods and so it is convenient to use the Fourier series of g

$$g(t) = \sum_{k=-\infty}^{\infty} a_k \exp(2\pi i k t).$$

Theorem 2. *Assume $g\in C^2[0,1]$ and that the conditions (6) hold. If $\lambda(n)\downarrow 0$ and $n^4\lambda(n)\to\infty$ as $n\to\infty$ and*

$$|a_k|^2 = O\left(|k|^{-5-\varepsilon}\right) \tag{10}$$

for some $\varepsilon>0$, then

$$\int_0^1 |Ef(t)-g(t)|^2\,dt = \sum_{|j|\le n/2} |a_j|^2 \frac{\lambda^2 j^8}{(1+\lambda j^4)^2} + O(n^{-5-\varepsilon}\lambda^{-1/4} + n^{-4}). \tag{11}$$

If

$$|a_k|^2 \cong |k|^{-5-\varepsilon}, \quad 0<\varepsilon<4, \tag{12}$$

expression (11) is asymptotically proportional to $\lambda^{1+\varepsilon/4}$. If

$$\sum |a_k|^2 k^8 < \infty \tag{13}$$

expression (11) is asymptotically proportional to λ^2.

Corollary. *Given the conditions of Theorem 2, if*

$$|a_k|^2 \cong |k|^{-5-\varepsilon}, \quad 0<\varepsilon<4,$$

the optimal rate of decrease of the expected integrated square deviation to zero is given by $n^{-(4+\varepsilon)/(5+\varepsilon)}$. *If*

$$\sum |a_k|^2 k^8 < \infty, \tag{14}$$

the optimal rate of decrease is $n^{-8/9}$.

Notice that if (14) holds, we must have

$$g''(0)=g''(1)$$

together with (6). It is this case that was analyzed in ref. [8]. Of course, it is unnatural to assume that (6) holds. However, if one just assumes $g \in C^2[0,1]$, the rate of decrease of integrated mean square error to zero is even slower than the rate indicated in the corollary.

The effect of the boundary on the asymptotic behavior is perhaps not so surprising since a similar effect occurs in the Gibbs phenomenon when using a Fourier representation (see Hall [3] relative to Fourier estimates of a probability density function). Gasser and Muller [4] have also pointed out boundary effects when making kernel estimates of a probability density function. This type of behavior is also suggested by results in Rosenblatt [6]. There the integrated square error of a "natural spline" interpolator is shown to converge to zero at a slower rate than the squared error away from the boundary. This is due to the influence of the boundary which dominates the integrated squared error asymptotically. In the paper it is noted that this can be compensated for by properly estimating the derivative of the function at the boundary and incorporating the estimate in the interpolator. This also suggests that the smoothing spline estimator might be "saved", that is, one could improve the asymptotic behavior of the integrated mean square deviation of the smoothing spline by incorporating appropriate estimates of the boundary values and their derivatives. This boundary effect leads one to have possible doubts about the efficacy of cross-validation used in conjunction with smoothing splines. A least squares procedure (see ref. [1]) using splines does not appear to have these difficulties with the boundary.

3. Numerical examples

To assess the relevance of the asymptotic theory, we have computed some numerical examples. For the first example, we consider the function $g(x)=\cos(\pi x)$ estimated by a smoothing spline. We have also computed an estimate by a modified method in which a cubic polynomial is added to the data and then subtracted from a smoothing spline which is fitted to the modified data. The polynomial $p(x)$ is chosen so that $\tilde{g}(x)=g(x)+p(x)$ satisfies $\tilde{g}^{(k)}(1)-\tilde{g}^{(k)}(0)=0$, $k=0,1,2$. Note that this is not the case for g.

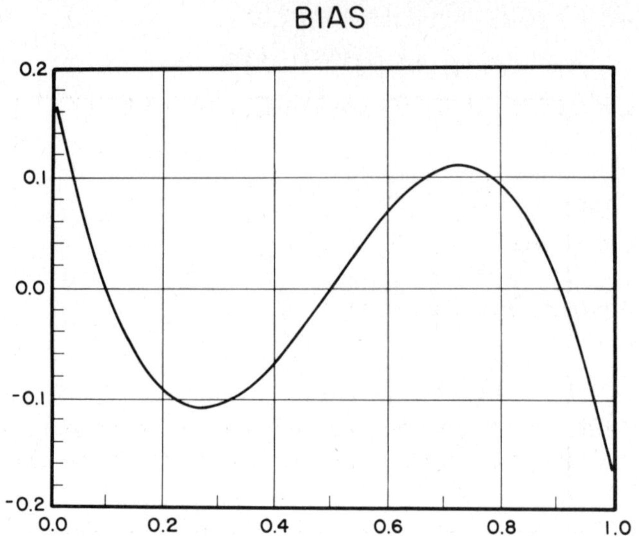

Fig. 1. Bias of smoothing spline for the function $g(x)=\cos(\pi x)$, $\lambda/(2\pi)^2=5\times10^{-3}$, $n=21$.

For $n=21$ points the following table shows the integrated squared bias B^2 and the integrated variance divided by $\sigma^2, V/\sigma^2$, for several values of λ. V/σ^2 is the same for both methods since they only differ by a deterministic factor. B^2 and V/σ^2 have been computed by noting that $f(t)$ is of the form $f(t)=\boldsymbol{l}_t^{\mathrm{T}}\boldsymbol{x}$ and thus $Ef(t)=\boldsymbol{l}_t^{\mathrm{T}}E(\boldsymbol{x})$ and $\sigma^2[f(t)]=\sigma^2\boldsymbol{l}_t^{\mathrm{T}}\boldsymbol{l}_t$.

$\lambda/(2\pi)^2$	B^2	B^2 (mod.)	V/σ^2
5×10^{-3}	6.6×10^{-3}	4.8×10^{-4}	9.0×10^{-2}
10^{-4}	4.8×10^{-4}	3.3×10^{-5}	1.6×10^{-1}
10^{-5}	1.7×10^{-5}	6.0×10^{-7}	2.5×10^{-1}

It is seen that the deterministic error (B^2) is roughly a factor of 10 larger for the unmodified estimate; the magnitude of this difference, predicted on asymptotic grounds, is surprising for such a small number of observations.

Figures 1 and 2 show the biases of the two estimates for $\lambda/(2\pi)^2=5\times10^{-3}$ and Fig. 3 shows the variance. It is interesting that the reduction in bias takes place throughout the interval [0, 1] and not merely at the ends. The bias appears roughly proportional to the fourth derivative of g.

We next took the same function with 100 points. The results are summarized below:

$\lambda/(2\pi)^2$	B^2	B^2 (mod.)	V/σ^2
10^{-3}	4.7×10^{-3}	3.3×10^{-4}	1.8×10^{-2}
10^{-4}	6.6×10^{-4}	4.1×10^{-5}	3.3×10^{-2}
10^{-5}	3.5×10^{-5}	1.0×10^{-6}	5.9×10^{-2}

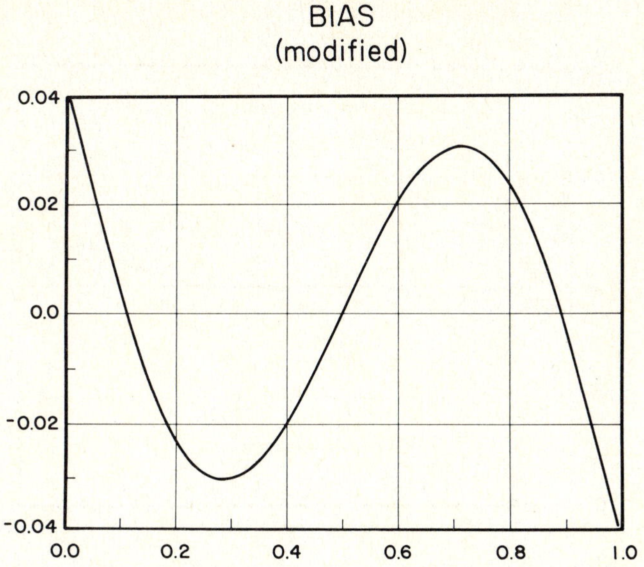

Fig. 2. Bias of modified smoothing spline. $g(x)=\cos(\pi x)$, $\lambda/(2\pi)^2=5\times10^{-3}$, $n=21$.

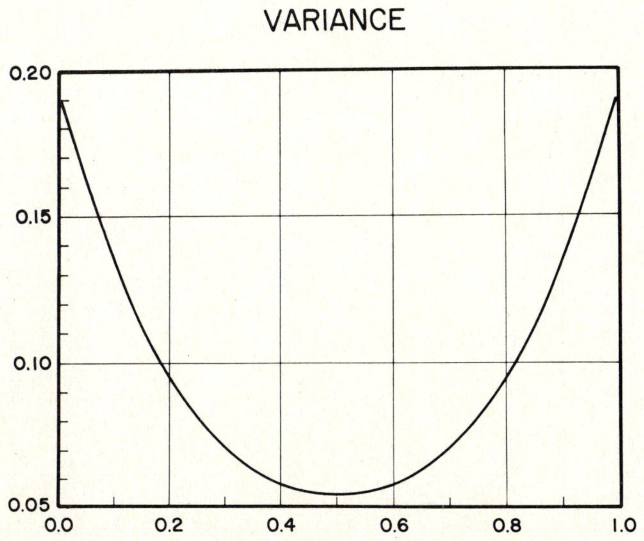

Fig. 3. Variance of smoothing spline. $g(x)=\cos(\pi x)$, $\lambda/(2\pi)^2=5\times10^{-3}$, $n=21$.

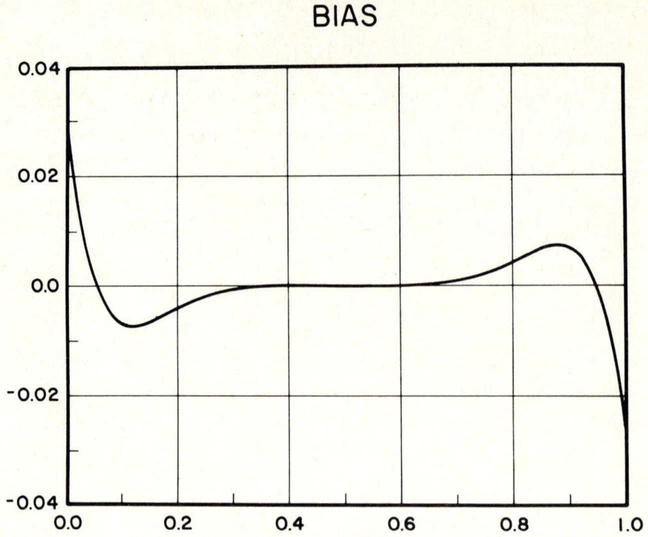

Fig. 4. Bias of smoothing spline. $g(x)=\cos(\pi x)$, $\lambda/(2\pi)^2=10^{-5}$, $n=100$.

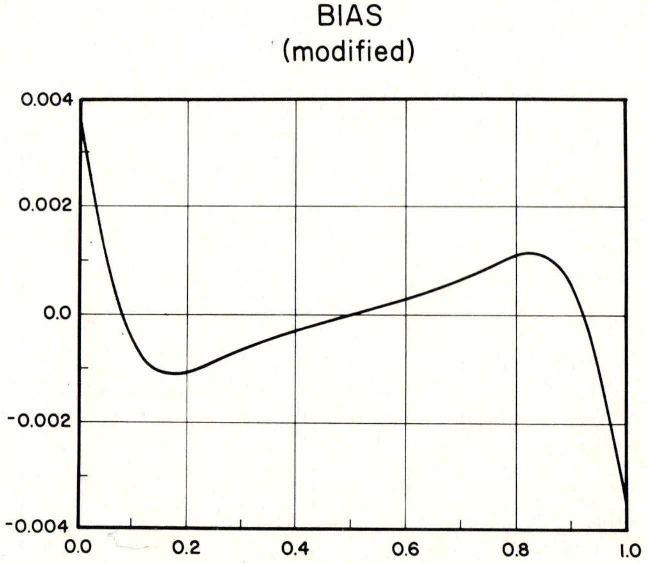

Fig. 5. Bias of modified smoothing spline. $g(x)=\cos(\pi x)$, $\lambda/(2\pi)^2=10^{-5}$, $n=100$.

Again the reduction in the integrated squared bias is about a factor of 10. Examination of the bias locally shows that for the larger values of λ, the decrease again takes place throughout the interval. For $\lambda/(2\pi)^2=10^{-5}$, Figs. 4 and 5 show that the major contribution to B^2 for the unmodified estimate comes from the boundary region. It appears that this region is of width approximately $2\lambda^{1/4}=0.28$, which may correspond to something like a bandwidth for the smoothing procedure.

In interpreting these results, it is useful to recall that what value of λ is reasonable depends on the value of σ^2; λ is optimal if $B^2=V$. Thus in the previous example $\lambda/(2\pi)^2=10^{-5}$ would be optimal for $\sigma=2.4\times10^{-2}$, which is a fairly large signal-to-noise ratio.

We next consider a function g which satisfies the assumptions of the theorems above and also modify g so that $\tilde{g}^{(2)}(1)-\tilde{g}^{(2)}(0)=0$; $g(x)=\sin(\pi x)\cdot\sin(2\pi x)$. The results are as follows ($n=100$):

$\lambda/(2\pi)^2$	B^2	B^2 (mod.)	V/σ^2
10^{-3}	9.6×10^{-2}	2.4×10^{-1}	1.8×10^{-2}
10^{-4}	1.5×10^{-2}	2.9×10^{-2}	3.3×10^{-2}
10^{-5}	8.6×10^{-3}	5.7×10^{-3}	5.9×10^{-2}
10^{-6}	3.4×10^{-5}	6.8×10^{-6}	1.1×10^{-1}

It is curious that in this case B^2 is smaller for the unmodified case for larger values of λ. $\lambda/(2\pi)^2=10^{-4}$ would be optimal for $\sigma=0.67$ which would be a very small signal-to-noise ratio. For smaller values of λ, the modified estimate does better. Figures 6 and 7 show the biases for $\lambda/(2\pi)^2=10^{-6}$.

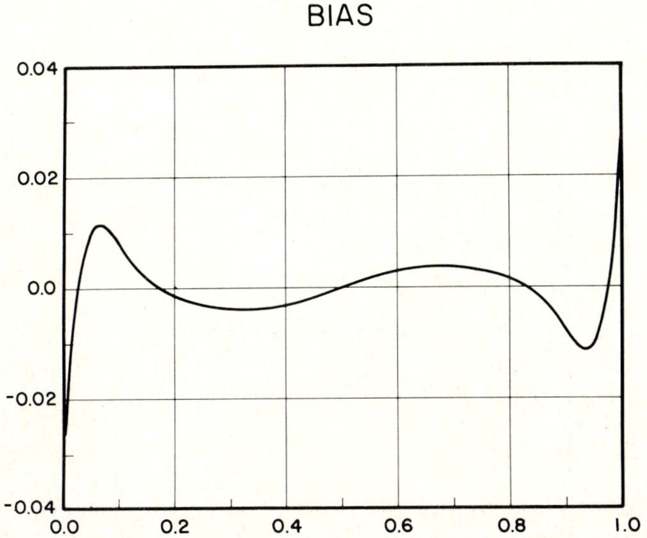

Fig. 6. Bias of smoothing spline. $g(x)=\sin(\pi x)\cdot\sin(2\pi x)$, $\lambda/(2\pi)^2=10^{-6}$, $n=100$.

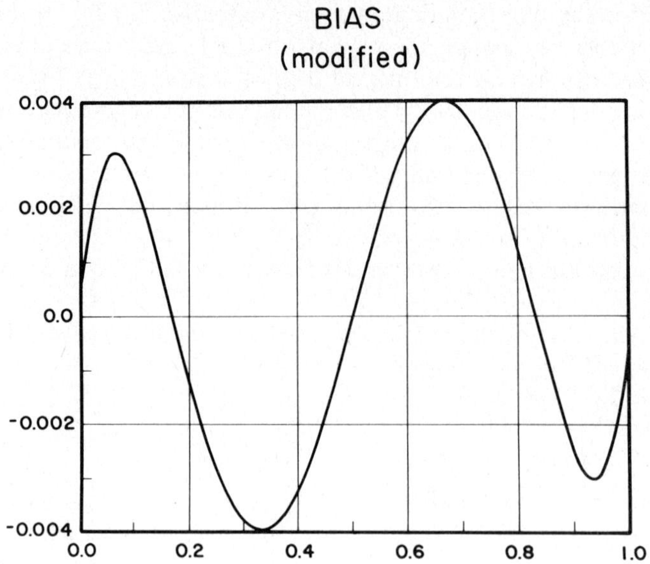

Fig. 7. Bias of modified smoothing spline. $g(x)=\sin(\pi x)\cdot\sin(2\pi x)$. $\lambda/(2\pi)^2=10^{-6}$, $n=100$.

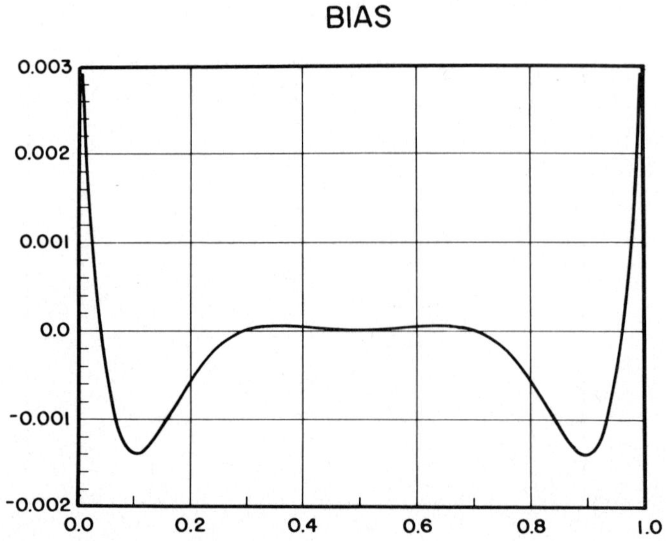

Fig. 8. Bias of smoothing spline. $g(x)=x(1-x)$. $\lambda/(2\pi)^2=10^{-3}$, $n=21$.

One reason for the deficient performance of the smoothing splines is that even though it is in a sense a third order method, it is biased for cubic polynomials. Figure 8 shows the bias for the function $g(t)=t(1-t), n=21, \lambda/(2\pi)^2=10^{-3}$. $(B^2=4.9\times 10^{-4}, V/\sigma^2=1.1\times 10^{-1}.)$ This value of λ is optimal if $\sigma=6.7\times 10^{-2}$.

4. Final remarks

It should be noted that the conditions

$$f(0)=f(1), \qquad f'(0)=f'(1)$$

are not usually imposed on smoothing splines. Since this information is incorporated in the approximation, one would expect the asymptotic behavior of this approximation (in terms of expected mean square error) to be at least as good as that of the standard smoothing spline.

References

[1] G. Agarwal and W. Studden (1980). Asymptotic integrated mean square error using least squares and bias minimizing splines. Manuscript.

[2] P. Craven and G. Wahba (1979). Smoothing noisy data with spline functions. *Numer. Math.* **31**, 377–403.

[3] P. Hall (1980). On trigonometric series estimates of densities. To appear in *Ann. Stat.*

[4] T. Gasser and H.-G. Muller (1979). Kernel estimation of regression functions. p. 23–68 In *Smoothing Techniques for Curve Estimation*, Gasser and Rosenblatt, Eds. Springer Lecture Notes in Mathematics 757, pp. 23–68. Springer Verlag, Berlin.

[5] C. Reinsch (1967). Smoothing by spline functions. *Numer. Math.* **10**, 177–183.

[6] M. Rosenblatt (1976). Asymptotics and representation of cubic splines. *J. Approx. Th.* **17**, 332–343.

[7] I.J. Schoenberg (1964). Spline functions and the problem of graduation. *Proc. Nat. Acad. Sci. (USA)* **52**, 947–950.

[8] G. Wahba (1975). Smoothing noisy data with spline functions. *Numer. Math.* **24**, 383–393.

[9] E. Whittaker (1923). On a new method of graduation. *Proc. Edinburgh Math. Soc.* **41**, 63–75.

G. Kallianpur, P.R. Krishnaiah, J.K. Ghosh, eds., *Statistics and Probability: Essays in Honor of C.R. Rao*
© North-Holland Publishing Company (1982) 645–647

ON MULTIDIRECTIONAL MARKOV PROCESSES

Yu.A. ROZANOV
Steklov Mathematical Institute, Academy of Sciences of the USSR, Moscow, U.S.S.R.

Let $\xi(t), t=(t_1,\ldots,t_n)\in G$, be a family of variables with a parameter, t, in some domain $G\subseteq R^n$; one can treat

$$\bar{\xi}_k(u)=\{\xi(t), t\in G: t_k=u\}, \quad u\in R^1,$$

as a process of ξ-evolution in the corresponding kth direction ($k=1,\ldots,n$). Suppose all the processes $\xi_k(u), u\in R^1$, are $\forall\varepsilon$-Markovian with respect to the usual "past" and "future". It is interesting to know whether the multiparameter family $\xi(t), t\in G$, is $\forall\varepsilon$-Markovian with respect to any complementary domains in G, namely for any bounded domain $S\subseteq G$ with a boundary Γ the families

$$\{\xi(t), t\in S\} \quad \text{and} \quad \{\xi(t), t\in G\setminus(S\cup\Gamma)\}$$

are conditionally independent under given variables $\{\xi(t), t\in\Gamma^\varepsilon\}$ from ε-neighborhood of Γ, $\varepsilon>0$.

Generally it is not so. For example, if

$$\xi(t)=\eta(t_1)+\eta(t_2); \quad t=(t_1,t_2)\in R^2,$$

where $\eta(u), u\in R^1$, is an ergodic stationary process with zero mean then

$$\eta(t_1)=\lim\frac{1}{2T}\int_{-T}^{T}\xi(t)\,\mathrm{d}t_2,$$

$$\eta(t_2)=\xi(t)-\eta(t_1), \quad t_2\in R^1,$$

are determined with $\xi(t), t_2\in R^1$, and

$$\xi_1(u)=\{\eta(t_1)+\eta(t_2), t_2\in R^1\}$$

as well as $\xi_2(u), u\in R^1$, is the singular Markov process but the 2-parameter function $\xi(t_1,t_2); t_1,t_2\in R^1$ is generally not of Markov type.

Suppose the function $\xi(t), t\in G$, is Gaussian. Let us denote $H(S), S\subseteq G$, the linear closure of all variables $\xi(t), t\in S$, and set

$$H_+(S)=\bigcap H(S^\varepsilon).$$

Suppose the following fairly wide condition of regularity holds true:

$$H_+(S_1)\cap H_+(S_2)=H_+(S_1\cap S_2)$$

for any closed domains S_1, S_2 in G (see ref. [1]).

Theorem. *The multiparameter function $\xi(t_1,\dots,t_n)$, $(t_1,\dots,t_n)\in G$, is $\forall\varepsilon$-Markovian with respect to all complementary domains in G.*

One can apply this general result to the well-known Brownian sheet $\xi(t_1,t_2)$; $t_1,t_2\geqslant 0$, with zero mean and correlation

$$E[\xi(s_1,s_2)\xi(t_1,t_2)]=\min(s_1,t_1)\min(s_2,t_2).$$

Obviously the increments, $\xi(u,t_2)-\xi(u_0,t_2)$, are independent of $\xi(t_1,t_2)$; $t_1\leqslant u_0$, $t_2\in R^1$, and

$$\xi_1(u)=\{\xi(u,t_2);\ t_2\in R^1\},$$

as well as $\xi_2(u)$, $u\in R^1$, is a pure Markov process (with independent increments) and therefore the Brownian sheet is $\forall\varepsilon$-Markovian with respect to all complementary domains.

Note it is not so if one means the pure Markov property namely $H_+(\Gamma)$ can be richer then the linear closure of the boundary variables $\xi(t_1,t_2)$, $(t_1,t_2)\in\Gamma$. As known (see ref. [1]), $H_+(\Gamma)$ consists of all variables

$$\eta=\int\int v(t_1,t_2)\,\mathrm{d}\xi(t_1,t_2)$$

where $v\in\mathfrak{L}^2(G)$ and

$$\frac{\partial^2}{\partial t_1\partial t_2}v(t_1,t_2)=0,\quad (t_1,t_2)\notin\Gamma.$$

Suppose a domain, S, is under a smooth monotone curve $\Gamma: t_2=\varphi(t_1)$, see fig. 1. Let us take any nonconstant smooth function $v_1(t_1)$ and set

$$f(t_1)=\frac{1}{t_1}\int_0^{t_1}v_1(t_1)\,\mathrm{d}t_1,\quad v_2(t_2)=[t_2 f(\psi(t_2))]',$$

where ψ is the inverse function to φ. Then

$$\frac{1}{t_2}\int_0^{t_2}v_2(t_2)\,\mathrm{d}t_2=\frac{1}{t_1}\int_0^{t_1}v_1(t_1)\,\mathrm{d}t_1,\quad (t_1,t_2)\in\Gamma,$$

so the non-zero function

$$v(t)=\begin{cases}v_1(t_1)-v_2(t_2), & t\in S,\\ 0, & t\in\Gamma\end{cases}$$

satisfies our conditions and the corresponding $\eta\in H_+(\Gamma)$ is independent of variables $\{\xi(t), t\in\Gamma\}$ because

$$E\eta\xi(t)=\int_0^{t_1}\int_0^{t_2}v(s)\,\mathrm{d}s=0,\quad t\in\Gamma.$$

If we show that $\eta(v)\in H(S)\cap H(T)$, where T denotes a complementary domain to $S\cup\Gamma$ then one can see that the Markov property: $H(S)$ and $H(T)$ are conditionally independent under the given $H(\Gamma)$, does not hold true because the variable $\eta(v)\neq 0$ which is independent on $H(\Gamma)$ can not be independent on itself.

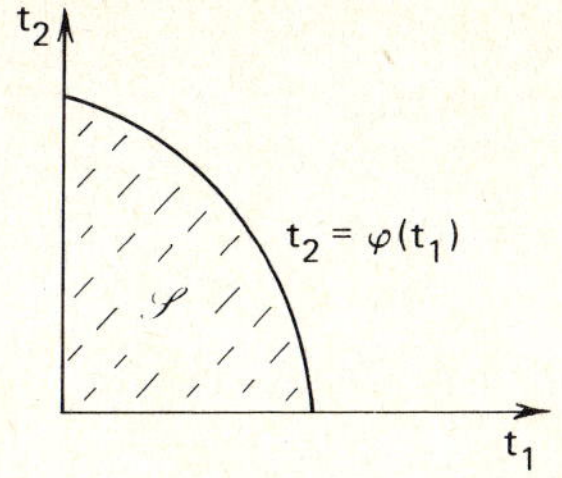

Fig. 1

Let us set $u_\alpha(t)=v(\alpha t)$; we have

$$\int |v(t)-u_\alpha(t)|^2 \mathrm{d}t \to 0, \quad \alpha \to 1,$$

and $\eta(u_\alpha) \in H_+((1/\alpha)\Gamma)$ because

$$\frac{\partial^2}{\partial t_1 \partial t_2} u_\alpha(t)=0, \quad t \notin (1/\alpha)\Gamma,$$

(note that $u_\alpha(t)=0$ above the corresponding curve $(1/\alpha)\Gamma$). Obviously $(1/\alpha)\Gamma \subset S$, $\alpha>1$ and $(1/\alpha)\Gamma \subset T, \alpha<1$, so

$$\eta(v)=\lim_{\alpha\to 1} \eta(u_\alpha) \subset H(S) \cap H(T).$$

Reference

[1] Rozanov, Yu.A. (1979). Markov random fields. In: *Developments in Statistics*, Vol. 2. Krishnaiah, P.R. Ed., Academic Press, New York.

G. Kallianpur, P.R. Krishnaiah, J.K. Ghosh, eds., *Statistics and Probability: Essays in Honor of C.R. Rao*

ASYMPTOTIC THEORY OF SOME TIME-SEQUENTIAL TESTS BASED ON PROGRESSIVELY CENSORED QUANTILE PROCESSES* **

Pranab Kumar SEN
Department of Biostatistics, University of North Carolina, Chapel Hill, NC 27514, U.S.A.

In the context of time-sequential tests, a general class of (parametric as well as non-parametric) testing procedures rests on progressively censored linear combinations of functions of order statistics with stochastic coefficient-vectors. Invariance principles for such quantile processes, developed in Sen (1979), are extended here to more general models, and these are incorporated in the study of the asymptotic properties of the allied time-sequential tests.

1. Introduction

In *clinical trials* and *life testing problems*, usually, one draws statistical inference from *censored* or *truncated* data. In this context, a *progressive censoring scheme* (PCS) allows a monitoring of the experimentation continuously from the beginning with the objective of an *early termination* whenever the accumulating outcome indicates so; hence, a PCS involves a *time-sequential procedure*. For a *simple regression model* (containing the classical two-sample problem as a special case), Chatterjee and Sen (1973) have studied the general theory of PCS testing procedures based on *linear rank statistics*. For some parametric models, Sen (1976b) and Gardiner and Sen (1978) have employed the *likelihood ratio principle* in a PCS and the resulting test statistics are based on certain *linear combinations of functions of order statistics*. Sen (1979) has also shown that for a general class of location, scale and regression models, *progressively censored likelihood ratio statistics* (PCLRS) are based on some *quantile processes* which involve partial sequences of linear combinations of functions of order statistics with stochastic coefficients depending progressively on the censoring stages; along with some martingale characterizations of these PC quantile processes, some invariance principles for them have been derived there and these have been incorporated in the study of the asymptotic properties of some proposed tests.

Majumdar and Sen (1978) have extended the results of Chatterjee and Sen (1973) for the *multiple regression model* (containing the several sample problem as a particular case); see also Sinha and Sen (1979b) in this context. A multiparameter extension of the results of Sen (1976b) has been considered by Sen and Tsong (1981). The object of the present investigation is to formulate suitable multiparameter generalizations of the model considered in Sen (1979) and to develop the

*Work supported by the National Heart, Lung and Blood Institute, Contract NIH-NHLBI-71-2243 from the National Institutes of Health.
**This research is dedicated to Professor C.R. Rao on the occasion of his 60th birthday.

asymptotic theory of PCS tests relating to these models. Thus, the results in the current paper are the multiparameter extensions of those in Sen (1979). In view of this, in the sequel, whenever possible, we shall attempt to minimize the technical manipulations by suitable cross reference to the earlier paper.

Along with the preliminary notions, the proposed progressively censored quantile processes (PCQP) are introduced in Section 2. Invariance principles for these PCQP's are considered in Section 3. Section 4 is devoted to the proposal and study of some PCS tests based on these PCQP's with special emphasis on their asymptotic properties. In particular, the multiple linear regression model and the location-scale model are treated elaborately in this context.

2. The proposed PCQP's

Let $X_1, \ldots, X_n$ be the *survival times* (with distribution functions (d.f.) $F_1, \ldots, F_n$, respectively) of $n(\geqslant 1)$ items under life testing and let $\boldsymbol{Z}_n = (Z_{n1}, \ldots, Z_{nn})$ be the vector of *order statistics* corresponding to $\boldsymbol{X}_n = (X_1, \ldots, X_n)$. The F_i are all assumed to be continuous [and defined on the real line $(-\infty, \infty)$], so that ties among them are neglected, in probability. Let $\boldsymbol{Q}_n = (Q_{n1}, \ldots, Q_{nn})$, the vector of *anti-ranks* be defined by:

$$Z_{nj} = X_{Q_{nj}} \quad \text{for } j = 1, \ldots, n. \tag{2.1}$$

From time and cost considerations, often, the experimentation is terminated at the *rth failure* Z_{nr}, where

$$r = [np] + 1 \quad \text{for some } p \in (0, 1); \tag{2.2}$$

($[s]$ being the largest integer contained in s), so that for such a *censored case*, the observable random vectors (r.v.) are

$$\boldsymbol{Z}^{(r)} = (Z_{n1}, \ldots, Z_{nr}) \quad \text{and} \quad \boldsymbol{Q}^{(r)} = (Q_{n1}, \ldots, Q_{nr}), \tag{2.3}$$

and a test statistic $\mathfrak{L}_{nr} = \mathfrak{L}_n(\boldsymbol{Z}_n^{(r)}, \boldsymbol{Q}_n^{(r)})$ which depends on $\boldsymbol{Z}_n^{(r)}$ and $\boldsymbol{Q}_n^{(r)}$.

In a PCS, the experiment is monitored from the beginning , so that at each failure Z_{nk}, $\mathfrak{L}_n(Z_n^{(k)}, Q_n^{(k)})$ is constructed: if $\mathfrak{L}_{nk}$ favors a clear cut terminal decision, experimentation is curtailed at that time-point and if no such $k(\leqslant r)$ exists then experimentation is stopped at the preplanned rth failure Z_{nr}; here, r may even be taken as n. Thus, in a PCS, one confronts a *time-sequential* testing procedure based on the partial sequence $\{\boldsymbol{Z}_n^{(k)}, \boldsymbol{Q}_n^{(k)}; k \leqslant r\}$, where neither the Z_{ni} nor the Q_{ni} are mutually independent, and hence, the $\mathfrak{L}_{nk}$ may not have independent increments.

To introduce the PCQP's we consider a multiparameter extension of the model considered in Sen (1979) and assume (for the time being) that the d.f. F_i has a continuous probability density function (p.d.f.) f_i (almost everywhere), where

$$f_i(x) = f(x; \boldsymbol{\Delta}(\boldsymbol{c}_i - \bar{\boldsymbol{c}}_n)'), \quad i = 1, \ldots, n, \quad x \in R = (-\infty, \infty); \tag{2.4}$$

$\boldsymbol{\Delta}$ is an $m \times q$ matrix of unknown parameters, the $\boldsymbol{c}_i$ are specified q-vectors ($m \geqslant 1, q \geqslant 1$) and $\bar{\boldsymbol{c}}_n = n^{-1} \Sigma_{i=1}^n \boldsymbol{c}_i$. The two (as well as several) samples location-scale models and the multiple regression models are special cases of (2.4). Suppose now that we intend to test for:

$$H_0: \boldsymbol{\Delta}=\mathbf{0} \quad \text{against} \quad H_1: \boldsymbol{\Delta}\neq\mathbf{0}. \tag{2.5}$$

The likelihood function of $(\boldsymbol{Z}_n^{(k)}, \boldsymbol{Q}_n^{(k)})$ is given by:

$$L_{nk}(\boldsymbol{Z}_n^{(k)}, \boldsymbol{Q}_n^{(k)})=\prod_{i=1}^{k} f(Z_{ni}; \boldsymbol{\Delta}(\boldsymbol{c}_{Q_{ni}}-\bar{\boldsymbol{c}}_n)')\prod_{i=k+1}^{n}[1-F(Z_{nk}; \boldsymbol{\Delta}(\boldsymbol{c}_{Q_{ni}}-\bar{\boldsymbol{c}}_n)')]. \tag{2.6}$$

Let Θ be an mq-dimensional open rectangle containing $\mathbf{0}$ as an inner point $\boldsymbol{\theta}=((\theta_{jl}))_{j=1,\dots,m;\, l=1,\dots,q}$ and let $\{f(x; \boldsymbol{\theta}), \boldsymbol{\theta}\in\Theta\}$ be a family of absolutely continuous p.d.f.'s. For every $x\in R$ and $j(=1,\dots,m)$, we let

$$g_j(x)=-(\partial/\partial\theta_{j1})\log f(x; \boldsymbol{\theta})|_{\mathbf{0}}, \tag{2.7}$$

$$\bar{G}_j(x)=[1-F(x; \mathbf{0})]^{-1}\int_x^{\infty} g_j(z)\,\mathrm{d}F(z; \mathbf{0}), \tag{2.8}$$

$$\boldsymbol{g}(x)=(g_1(x),\dots, g_m(x))' \quad \text{and} \quad \bar{\boldsymbol{G}}(x)=(\bar{G}_1(x),\dots,\bar{G}_m(x))'. \tag{2.9}$$

Then, by (2.4) through (2.9), we obtain

$$\begin{aligned}\boldsymbol{T}_{nk} &= -(\partial/\partial\boldsymbol{\Delta})\log L_{nk}(\boldsymbol{Z}_n^{(k)}, \boldsymbol{Q}_n^{(k)})|_{\boldsymbol{\Delta}=\mathbf{0}} \\ &= \sum_{i=1}^{k}[\boldsymbol{g}(Z_{ni})-\bar{\boldsymbol{G}}(Z_{nk})](\boldsymbol{c}_{Q_{ni}}-\bar{\boldsymbol{c}}_n)', \quad 1\leqslant k\leqslant n,\end{aligned} \tag{2.10}$$

and $T_{n0}=0$. Proceeding as in (2.8) through (2.12) of Sen (1979) we may consider the related sequence

$$\boldsymbol{T}_{nk}^{*}=\begin{cases}0, \quad k=0\\ \displaystyle\sum_{i=1}^{k}\boldsymbol{g}(Z_{ni})\left[(\boldsymbol{c}_{Q_{ni}}-\bar{\boldsymbol{c}}_n)+\frac{1}{n-k}\sum_{s=1}^{k}(\boldsymbol{c}_{Q_{ns}}-\bar{\boldsymbol{c}}_n)\right]', & 1\leqslant k\leqslant n-1,\\ \boldsymbol{T}_{nn-1}^{*}, \quad k=n.\end{cases} \tag{2.11}$$

We will find it more convenient to study the asymptotic properties of $\{\boldsymbol{T}_{nk}^{*}\}$.

Let $\{\boldsymbol{d}_i,\ 1\leqslant i\leqslant n\}$ be q-vectors for which $\Sigma_{i=1}^{n}\boldsymbol{d}_i=\mathbf{0}$ and

$$\boldsymbol{D}_n=\sum_{i=1}^{n}\boldsymbol{d}_i\boldsymbol{d}_i' \quad \text{is positive definite (p.d.).} \tag{2.12}$$

Also, let $\boldsymbol{Q}=(Q_1,\dots,Q_n)$ assume each permutation of $(1,\dots,n)$ with the common probability $(n!)^{-1}$. Note that for every k: $1\leqslant k\leqslant n$,

$$\sup_{\boldsymbol{l}\neq\mathbf{0}}(\boldsymbol{l}'\boldsymbol{D}_n\boldsymbol{l})^{-1/2}\left|\sum_{i=1}^{k}\boldsymbol{l}'\boldsymbol{d}_{Q_i}\right| =\left[\left(\sum_{i=1}^{k}\boldsymbol{d}_{Q_i}\right)'\boldsymbol{D}_n^{-1}\left(\sum_{i=1}^{k}\boldsymbol{d}_{Q_i}\right)\right]^{1/2}=U_{nk}^{*}, \tag{2.13}$$

say, where

$$EU^{*2}_{nk}=E\left(\text{Trace}\left\{D_n^{-1}\left[\left(\sum_{i=1}^{k} d_{Q_i}\right)\left(\sum_{i=1}^{k} d_{Q_i}\right)'\right]\right\}\right)$$

$$=\text{Trace}\left\{D_n^{-1}[D_n k(n-k)/n(n-1)]\right\}=qk(n-k)/n(n-1), \quad 1\leqslant k\leqslant n. \tag{2.14}$$

Furthermore, proceeding as in the proof of Lemma 2.1 of Sen (1979), we claim that $\{U_{nk}=(n-k)^{-1}U^*_{nk},\ 1\leqslant k<n\}$ is a non-negative submartingale. Hence, by a direct vector-extension of the proof of Lemma 2.1 of Sen (1979), we arrive at the following:

Let $\omega=\{\omega(t), 0<t<1\}$ be a continuous, non-negative, U-shaped and square integrable function [inside (0, 1)] and $\{d_i\}$, Q be defined as in above. Then,

$$P\left\{\sup_{l\neq 0}\max_{1\leqslant k<n}\omega\left(\frac{k}{n}\right)(l'D_n l)^{-1/2}\left|\sum_{i=1}^{k} l'd_{Q_i}\right|\geqslant 1\right\}\leqslant q\int_0^1\omega^2(t)\,\mathrm{d}t. \tag{2.15}$$

Note that Lemma 2.2 of Sen (1979) also directly extends to the case where both g and $\bar{G}$ are q-vectors (as is the case here), and hence, by using (2.15) and some standard inequalities, we obtain that under (2.2) and H_0 in (2.5), as $n\to\infty$,

$$\max_{k\leqslant r}\ \sup_{a:\,a'a=1}\ \sup_{l\neq 0}|a'(T_{nk}-T^*_{nk})l|(l'C_n l)^{-1/2}\xrightarrow[p]{}\mathbf{0}, \tag{2.16}$$

where

$$C_n=\sum_{i=1}^{n}(c_i-\bar{c}_n)(c_i-\bar{c}_n)' \quad \text{is assumed to be p.d.} \tag{2.17}$$

(2.16) provides the justification for replacing $\{T_{nk}\}$ by $\{T^*_{nk}\}$. Also, to be more general, we consider a class of $h(x)=(h_1(x),\ldots,h_m(x))'$, $x\in R$, [quite arbitrary and need not be g, defined by (2.7)], where (i) each $h_j(x)$ is assumed to be expressible as a difference of two non-decreasing functions and (ii)

$$\nu=\int_{-\infty}^{\infty}[h(x)][h(x)]'\,\mathrm{d}F(x) \quad \text{is assumed to be p.d. and finite;} \tag{2.18}$$

here $F(x)=F(x;\,0)$. Furthermore, let ξ_α: $F(\xi_\alpha)=\alpha$, $0<\alpha<1$ and let

$$\nu_\alpha=\int_{-\infty}^{\xi_\alpha}[h(x)][h(x)]'\,\mathrm{d}F(x)+(1-\alpha)^{-1}\left(\int_{-\infty}^{\xi_\alpha}h(x)\,\mathrm{d}F(x)\right)\left(\int_{-\infty}^{\xi_\alpha}h(x)\,\mathrm{d}F(x)\right)'. \tag{2.19}$$

Now, in (2.11), we replace the $g(Z_{ni})$ by $h(Z_{ni})$ and post-multiply the matrix by $C_n^{-1/2}$. The resulting matrix is denoted by T^{**}_{nk} i.e.

$$T^{**}_{nk}=(T^*_{nk}|_{g=h})C_n^{-1/2} \quad \text{for } k=0,1,\ldots,n. \tag{2.20}$$

We roll out T^{**}_{nk} into an mq-vector and denote it by $\tilde{T}^{**}_{nk}$, $0\leqslant k\leqslant n$.

3. Invariance principles for $\{T_{nk}^{**};\ 0 \leqslant k \leqslant r\}$

First, we consider the case of the null hypothesis H_0: $X_1, \ldots, X_n$ are independent and identically distributed (i.i.d.) random variables with an absolutely continuous d.f. F (and p.d.f.f.). Then, on denoting by E_0 the expectation under H_0, we have

$$E_0(\boldsymbol{T}_{nk}^{**}) = \boldsymbol{0}, \quad \forall 0 \leqslant k \leqslant 1, \tag{3.1}$$

and proceeding as in the proof of Lemma 3.1 of Sen (1979), we obtain that

$$[k/n \to \alpha] \Rightarrow E_0\{(\tilde{\boldsymbol{T}}_{nk}^{**})(\tilde{\boldsymbol{T}}_{nk}^{**})'\} \to \boldsymbol{\nu}_\alpha \times \boldsymbol{I}_q, \tag{3.2}$$

for every $0 < \alpha < 1$, where $\boldsymbol{\nu}_\alpha$ is defined by (2.19).

For every $n(\geqslant 1)$ and r, satisfying (2.2), we introduce an mq-variate stochastic process $\boldsymbol{W}_n = \{\boldsymbol{W}_n(t), t \in [0,1]\}$ by letting

$$\boldsymbol{W}_n(t) = \tilde{\boldsymbol{T}}_{nn(t)}^{**}, \quad n(t) = \max\{k : k/n \leqslant pt\}, \quad t \in [0,1]. \tag{3.3}$$

Thus, $\boldsymbol{W}_n$ belongs to the $D^{mq}[0,1]$ space, endowed with the (extended) Skorokhod J_1-topology. Also, let $\boldsymbol{W} = [(\boldsymbol{W}(t), t \in [0,1]\}$ be an mq-variate Gaussian function on $[0,1]$, such that $E\boldsymbol{W}(t) = \boldsymbol{0}, \forall t \in [0,1]$ and

$$E[(\boldsymbol{W}(t))(\boldsymbol{W}(s))'] = \boldsymbol{\nu}_{p(s \wedge t)} \times \boldsymbol{I}_q, \quad \forall s, t \in [0,1]. \tag{3.4}$$

Then, the main theorem of this section is the following, where we impose the following condition: as $n \to \infty$,

$$\max_{1 \leqslant k \leqslant n} (\boldsymbol{c}_k - \bar{\boldsymbol{c}}_n)' \boldsymbol{C}_n^{-1} (\boldsymbol{c}_k - \bar{\boldsymbol{c}}_n) \to 0. \tag{3.5}$$

[(3.5) may be regarded as the vector-version of the classical Noether condition.]

Theorem 3.1. *Under H_0, (2.2) and (3.5), as $n \to \infty$,*

$$\boldsymbol{W}_n \underset{\mathfrak{D}}{\to} \boldsymbol{W}, \textit{ in the extended } J_1\textit{-topology on } D^{mq}[0,1]. \tag{3.6}$$

Proof. Since this theorem is multiparameter generalization of Theorem 1 of Sen (1979), we shall only provide a sketch of the proof by repeated cross reference to the proof in Sen (1979). We need to show that (i) the finite-dimensional distributions (f.d.d.) of $\{\boldsymbol{W}_n\}$ converge to those of W and (ii) $\{\boldsymbol{W}_n\}$ is tight. Let $\mathfrak{B}_{nk} = \mathfrak{B}(\boldsymbol{Z}_n^{(k)}, \boldsymbol{Q}_n^{(k)})$ and $\mathfrak{B}_{nk}^* = \mathfrak{B}(\boldsymbol{Z}_n^{(n)}, \boldsymbol{Q}_n^{(k)})$ be, respectively, the σ-fields generated by $(\boldsymbol{Z}_n^{(k)}, \boldsymbol{Q}_n^{(k)})$ and $(\boldsymbol{Z}_n^{(n)}, \boldsymbol{Q}_n^{(k)})$, so that both are non-decreasing in k $(0 \leqslant k \leqslant n)$. Then, repeating the proof of Lemma 3.2 of Sen (1979) for each coordinate of T_{nk}^{**}, we obtain that:

$$E_0\{\tilde{\boldsymbol{T}}_{nk+1}^{**} | \mathfrak{B}_{nk}^*\} = \tilde{\boldsymbol{T}}_{nk}^{**} \quad \text{a.e.} \quad \forall 0 \leqslant k \leqslant n-1. \tag{3.7}$$

Thus, if we let

$$\hat{\boldsymbol{T}}_{nk}^{**} = \tilde{\boldsymbol{T}}_{nk}^{**} - \tilde{\boldsymbol{T}}_{nk-1}, \quad 1 \leqslant k \leqslant n; \qquad \hat{\boldsymbol{T}}_{n0} = \tilde{\boldsymbol{T}}_{n0} = \boldsymbol{0}, \tag{3.8}$$

and if $\boldsymbol{\lambda}_{nk}$, $0 \leqslant k \leqslant n$ are arbitrary (non-stochastic) mq-vectors, then:

$$E_0\{\boldsymbol{\lambda}_{nk} \hat{\boldsymbol{T}}_{nk}^{**} | \mathfrak{B}_{nk-1}^*\} = E_0\{\boldsymbol{\lambda}_{nk} \hat{\boldsymbol{T}}_{nk}^{**} | \mathfrak{B}_{nk-1}\} = 0 \quad \text{a.e.} \ \forall 1 \leqslant k \leqslant n. \tag{3.9}$$

Now, to prove (i), for any (fixed) $d(\geqslant 1),(0\leqslant)t_1(<\cdots<)t_d\leqslant 1$ and arbitrary $\boldsymbol{\lambda}_i$, $1\leqslant i\leqslant d$ (all mq-vectors), consider the linear compound

$$\begin{aligned} W_n^* &= \sum_{i=1}^{d} \boldsymbol{\lambda}_i' \boldsymbol{W}_n(t_i) = \sum_{i=1}^{d} \boldsymbol{\lambda}_i' \tilde{T}^{**}_{nn(t_i)} \quad [\text{by } (3.3)] \\ &= \sum_{k\leqslant n} \left\{ \sum_{i=1}^{d} I(k\leqslant n(t_i))\boldsymbol{\lambda}_i' \right\} \hat{\boldsymbol{T}}^{**}_{nk} \quad [\text{by } (3.8)] \\ &= \sum_{k\leqslant n} \boldsymbol{\omega}_{nk} \hat{\boldsymbol{T}}^{**}_{nk} = \sum_{k\leqslant n} W^*_{nk}, \end{aligned} \tag{3.10}$$

say, where the $\boldsymbol{\omega}_{nk}$ depend on $\lambda_1,\ldots,\lambda_d$ and $n(t_1),\ldots,n(t_d)$, and by (3.9),

$$E_0\{W^*_{nk}|\mathfrak{B}^*_{nk-1}\}=0 \quad \text{a.e.} \quad \forall\, 1\leqslant k\leqslant n. \tag{3.11}$$

For the W^*_{nk}, we may readily extend the proof of the Lemma 3.3 of Sen (1979) and show that under (2.18) and (3.5),

$$\sum_{k\leqslant n} E_0\{W^{*2}_{nk}|\mathfrak{B}^*_{nk-1}\} \underset{p}{\rightarrow} \sigma_0^2, \quad \text{as } n\rightarrow\infty, \tag{3.12}$$

where σ_0^2 [is the variance of $\Sigma_{i=1}^d \lambda_i' W(t_i)$, see (3.4), and] is given by

$$\sigma_0^2 = \sum_{i=1}^{d} \sum_{j=1}^{d} \boldsymbol{\lambda}_i' \left(\boldsymbol{\nu}_{p(t_i\wedge t_j)} \times \boldsymbol{I}_q\right) \boldsymbol{\lambda}_j. \tag{3.13}$$

Further, by a direct extension, using the C_r-inequality, of the proof of Lemma 3.4 of Sen (1979), we obtain that for every $\varepsilon>0$,

$$\sum_{k\leqslant n} E_0\{W^{*2}_{nk} I(|W^*_{nk}|>\varepsilon|\mathfrak{B}^*_{nk-1}\} \underset{p}{\rightarrow} 0, \quad \text{as } n\rightarrow\infty. \tag{3.14}$$

Hence, the asymptotic normality of W_n^* follows from (3.10) through (3.14) and Dvoretzky's (1972) dependent central limit theorem. Finally, to prove (ii), we note that the *tightness* of each (of the mq) component(s) of $\{W_n\}$ follows from Theorem 1 of Sen (1979) and since mq is fixed,the tightness of $\{\boldsymbol{W}_n\}$ is ensured by the tightness of its mq-components.

In the above theorem, by (2.2), we have limited $r=[np]+1$ for some $p<1$. The remarks following Theorem 1 of Sen (1979) also apply to this more general case (coordinatewise) when we like to take $p=1$.

Let us now consider the non-null case and conceive of a model similar to (2.4); but, we confine ourselves to local alternatives for which the limiting results are non-degenerate. We conceive of a triangular array $\{\boldsymbol{d}_{ni},\ 1\leqslant i\leqslant n;\ n\geqslant 1\}$ of q-vectors, define $\boldsymbol{D}_n=\Sigma_{i=1}^n d_{ni} d_{ni}'$ and assume that:

$$\sum_{i=1}^{n} \boldsymbol{d}_{ni}=\boldsymbol{0}, \quad \forall n, \quad \sup_n \operatorname{Tr}(\boldsymbol{D}_n)<\infty, \tag{3.15}$$

$$\boldsymbol{D}_n \text{ is p.d. for every } n(\geqslant n_0), \tag{3.16}$$

$$\lim_{n\rightarrow\infty} \left\{ \max_{1\leqslant k\leqslant n} \boldsymbol{d}_{nk}' \boldsymbol{D}_n^{-1} \boldsymbol{d}_{nk} \right\}=0. \tag{3.17}$$

Then let $\{K_n\}$ be a sequence of alternative hypotheses, where

$$K_n: f_i(x)=f(x;\boldsymbol{\beta}\boldsymbol{d}_{ni}), \quad 1\leqslant i\leqslant n, \quad x\in R, \tag{3.18}$$

$\boldsymbol{\beta}$ is an $m\times q$ matrix of (fixed) unknown parameters (all finite) and f satisfies all the regularity conditions, stated after (2.6). We also define $\boldsymbol{g}$, $\overline{\boldsymbol{G}}$, etc., as (2.7)–(2.9). Let us also define $\xi_\alpha(0<\alpha<1)$ as after (2.18) and define

$$\zeta(\alpha)=\int_{-\infty}^{\xi_\alpha}(\boldsymbol{h}(x))(\boldsymbol{g}(x))'\,\mathrm{d}F(X)+(1-\alpha)\left(\int_{-\infty}^{\xi_\alpha}\boldsymbol{h}(x)\,\mathrm{d}F(x)\right)\left(\int_{-\infty}^{\xi_\alpha}\boldsymbol{g}(x)\,\mathrm{d}F(x)\right)'. \tag{3.19}$$

Further, we assume that the $\boldsymbol{d}_{ni}$ and $\boldsymbol{c}_i$ satisfy the condition that

$$\lim_{n\to\infty}\left(\sum_{i=1}^{n}\boldsymbol{d}_{ni}(\boldsymbol{c}_i-\bar{\boldsymbol{c}}_n)'\right)\boldsymbol{C}_n^{-1/2}=\boldsymbol{P}\text{ exists.} \tag{3.20}$$

Let then

$$\boldsymbol{M}(\alpha)=[\zeta(\alpha)]\boldsymbol{\beta}[\boldsymbol{P}], \quad 0<\alpha<1. \tag{3.21}$$

We roll out $\boldsymbol{M}(\alpha)$ into an mq-vector and denote it by $\boldsymbol{\mu}(\alpha)$, $0<\alpha<1$. Finally, we define r as in (2.2) with $0<p<1$ and denote by:

$$\boldsymbol{\mu}=\{\boldsymbol{\mu}_t=\boldsymbol{\mu}(pt), 0\leqslant t\leqslant 1\}. \tag{3.22}$$

Then, we have the following:

Theorem 3.2. *Let* $\{\boldsymbol{W}_n\}$, $\boldsymbol{W}$ *and* $\boldsymbol{\mu}$ *be defined as in* (3.3), (3.4), *and* (3.22), *then, under* (2.2), (3.5), (3.15), (3.16), (3.17) *and* (3.20), *for* f *having a finite Fisher information* (*matrix*) $\mathcal{I}=E[\boldsymbol{g}(x)][\boldsymbol{g}(x)]'$,

$$\boldsymbol{W}_n-\boldsymbol{\mu}\xrightarrow[\mathcal{D}]{}\boldsymbol{W}, \text{ in the extended } J_1\text{-topology on } D^{mq}[0,1]. \tag{3.23}$$

Proof. It follows from Hájek and Šidak (1967, Ch. VI, pp. 239-240) that under the assumed regularity conditions, P_n^*, the probability measure (for $\boldsymbol{Z}_n^{(n)}, \boldsymbol{Q}_n^{(n)}$) under K_n, is *contiguous* to P_n^0, the same under H_0. Hence, as in the proof of Theorem 2 of Sen (1979), the tightness of $\{\boldsymbol{W}_n\}$ under H_0, proved in Theorem 3.1, and the contiguity of $\{P_n^*\}$ to $\{P_n^0\}$ ensure that tightness of $\{\boldsymbol{W}_n\}$ under $\{K_n\}$ as well. Hence, to prove the theorem, it suffices to show that the f.d.d.'s of $\{\boldsymbol{W}_n-\boldsymbol{\mu}\}$ converge to those of $\boldsymbol{W}$. Suppose that in (2.8), we replace the g_j by h_j, denote the resulting quantities by $\overline{H}_j$, $1\leqslant j\leqslant m$ and let $\overline{\boldsymbol{H}}(x)=(H_1(x),\ldots,\overline{H}_m(x))'$, $x\in R$. Let then

$$\check{\boldsymbol{T}}_{nk}^*=\sum_{i=1}^{n}I(x_i\leqslant\xi_{k/n})\{\boldsymbol{h}(x_i)+\overline{\boldsymbol{H}}(x_i)\}(\boldsymbol{c}_i-\bar{\boldsymbol{c}}_n)'\boldsymbol{C}_n^{-1/2}, \quad 0\leqslant k\leqslant n, \tag{3.24}$$

where the $\xi_{k/n}$ are defined by (2.19) for $\alpha=k/n$. Then, by repeating the proof of Theorem 2 of Sen (1979) for each coordinate of $\check{\boldsymbol{T}}_{nk}^*-\boldsymbol{T}_{nk}^{**}$, we obtain that for any $k: k/n\to\alpha$: $0<\alpha<1$, as $n\to\infty$,

$$\boldsymbol{T}_{nk}^{**}-\check{\boldsymbol{T}}_{nk}^*\xrightarrow[p]{}\boldsymbol{0}, \quad \text{under } \{K_n\}. \tag{3.25}$$

On the other hand, $\check{T}_{nk}^{*}$ involves independent summands (random matrices) and the classical multivariate central limit theorem along with a theorem of Behnen and Neuhaus (1975) yields that under $\{K_n\}$, for any (fixed) $a(\geqslant 1)$ and $0 \leqslant t_1 \leqslant \cdots \leqslant t_a \leqslant 1$, letting $k_j = [npt_j]+1$, $1 \leqslant j \leqslant a$, $\check{T}_{nk_1}^{*}, \ldots, \check{T}_{nk_a}^{*}$ have (jointly) asymptotically a mqa-variate normal distribution with means $\boldsymbol{M}(ptj)$, $1 \leqslant j \leqslant a$, defined by (3.21) and covariance functions conforming to that of $\boldsymbol{W}$. Hence, (3.25) and (3.3) along with the above lead us to the desired result.

We may remark that both Theorems 3.1 and 3.2 remain true if the T_{nk}^{**} are replaced by $\boldsymbol{\tau}_{nk} = E_0(\boldsymbol{T}_{nk}^{**} | \boldsymbol{Q}^{(k)})$ which amounts to replacing the $\boldsymbol{h}(Z_{ni})$ by $E_0 \boldsymbol{h}(Z_{ni}) = \boldsymbol{h}_n(i)$, say, $1 \leqslant i \leqslant n$. In this case, $\boldsymbol{\nu}_\alpha$ in (2.19) may also be replaced by

$$\boldsymbol{\nu}_{n\alpha} = n^{-1} \sum_{i \leqslant n\alpha} [\boldsymbol{h}_n(i)][\boldsymbol{h}_n(i)]' + n^{-2}(1-\alpha)^{-1}\Big(\sum_{i \leqslant n} \boldsymbol{h}_n(i)\Big)\Big(\sum_{i \leqslant n} \boldsymbol{h}_n(i)\Big)', \quad (3.26)$$

for $0 < \alpha < 1$; $\boldsymbol{\nu}_{n0} = \boldsymbol{0}$. Note that, by definition, the τ_{nk} depends only on the $\boldsymbol{Q}_n^{(k)}$ and therefore are PCS rank statistics. The necessary modifications in the proofs are quite straightforward, and hence, the details are omitted.

4. Asymptotic time-sequential tests

The quantile processes, studied in the previous sections, depend, in general, on both $\boldsymbol{Z}_n$ and $\boldsymbol{Q}_n$. For the simple regression model [i.e., in (2.4), $m=q=1$), Hájek (1963) considered an invariance principle related to the T_{nk}, where $h(t) \equiv 1$, wherein a *tied-down Wiener process* approximation was developed and the same was incorporated in the study of the asymptotic properties of some Kolmogorov–Smirnov, Cramér–von Mises' and Rényi type statistics (based on $\boldsymbol{Q}_n$ alone) for testing the hypothesis of no regression. Sinha and Sen (1979a) have developed a reverse-martingale approach to this problem, extended the Hájek results in a PCS setup and also considered the use of some weighted test statistics. Sinha and Sen (1979b) have also extended their theory to the multiple regression model, which is again a special case of (2.4) with $m=1$, $q \geqslant 1$. Their procedure is based on an invariance principle related to the tied-down Wiener process in the vector case. We shall see later on that an alternative formulation of this problem involving the *Bessel processes* can be made by using our results in Sections 2 and 3.

We will find convenient to classify the PCQP's into two-types:

(i) *Type A PCQP's*: Here, as in the end of Section 3, we let $\boldsymbol{h}(\boldsymbol{Z}_n^{(k)})$ depend only on (k, n) (but not on $\boldsymbol{Z}_n^{(k)}$), so that the $\boldsymbol{T}_{nk}^{*}$ becomes censored rank statistics. These statistics have been studied in detail by Chatterjee and Sen (1973), Majumdar and Sen (1978) and others. At the same time, the needed asymptotic theory also follows from our Theorems 3.1 and 3.2. These are generally distribution-free procedures (under H_0).

(ii) *Type B PCQP's*: The Type A PCQP's do not utilize the information contained in the set of (ordered) failures and therefore may not be fully efficient in many cases. For Type B PCQP's, the information in these failures is incorporated in some convenient way. Here, the $\boldsymbol{T}_{nk}^{*}$ depend on both the $\boldsymbol{Z}_n^{(k)}$ and $\boldsymbol{Q}_n^{(k)}$. The dependence on $\boldsymbol{Z}_n^{(k)}$, though, on the one hand, may enhance the power of the tests, on the other hand, fails to make the tests genuinely distribution-free, and to construct asymptotically distribution-free tests, one needs to estimate $\boldsymbol{\nu}_\alpha$ in (2.18)–(2.19). However, asymptotically, these tests can be worked out nicely, and such procedures will be mainly discussed here.

Note that by (3.2) and the martingale property in (3.7), for every $0<\alpha<\alpha'<1$, $\boldsymbol{\nu}_{\alpha'}-\boldsymbol{\nu}_\alpha$ is positive semi-definite (p.s.d.). We consider the case of $m=1$ first. For a Type A PCQP, for every $\alpha\in(0,1)$, ν_α is known and we may choose:

$$\mathfrak{L}_{nk}=(\boldsymbol{T}_{nk}^{**})'(\nu_p\times\boldsymbol{I}_q)^{-1}(\boldsymbol{T}_{nk}^{**})=\nu_p^{-1}[\boldsymbol{T}_{nk}^{**\prime}\boldsymbol{T}_{nk}^{**}]. \tag{4.1}$$

Let then $\mathfrak{L}_{nr}^{*}=\max_{1\leqslant k\leqslant r}\mathfrak{L}_{nk}$. It follows from Theorem 3.1 that under H_0,

$$\mathfrak{L}_{nr}^{*}\underset{\mathfrak{D}}{\rightarrow}\sup\{[\boldsymbol{W}(t)]'[\boldsymbol{W}(t)]:0\leqslant t\leqslant 1\}=W^{**}, \tag{4.2}$$

say where $W^{*}=\{W^{*}(t)=(\boldsymbol{W}(t))'(\boldsymbol{W}(t),t\in[0,1]\}$ is a q-variate Bessel process. Thus, if W_α^{**} be the upper $100\alpha\%$ point of the d.f. of W^{**}, then we have the following PCS testing procedure:

At each failure, (k), compute $\mathfrak{L}_{nk}$; if, for the first time, for some $k=N(\leqslant r)$, $\mathfrak{L}_{nN}$ exceeds W_α^{**}, stop experimentation along with the rejection of H_{01} and, if no such N exists, stop experimentation at the preplanned rth failure along with the acceptance of H_0.

It follows from (4.2) that the Type A error for this PCS procedure is asymptotically equal to α $(0<\alpha<1)$. Also, from Theorem 3.2, we conclude that under the hypothesis of Theorem 3.2, the asymptotic power of the test is given by:

$$P\{[\boldsymbol{W}(t)+\boldsymbol{\mu}_t]'[\boldsymbol{W}(t)+\boldsymbol{\mu}_t]>W_\alpha^{**},\quad\text{for some } t\in[0,1]\}. \tag{4.3}$$

Further, it follows from Theorems 3.1 and 3.2 that for $k/n\rightarrow\gamma$, $0<\gamma\leqslant p$,

$$P\{N>k|K_n\}\rightarrow P\{[\boldsymbol{W}(t)+\boldsymbol{\mu}_t]'[\boldsymbol{W}(t)+\boldsymbol{\mu}_t)]\leqslant W_\alpha^{**},\ \ \forall\, 0\leqslant t\leqslant p^{-1}\gamma\}. \tag{4.4}$$

Thus, noting that $Z_{nk}\rightarrow\xi_\alpha$ (a.s. as well as in the tth mean, $r\geqslant 1$) and that $n(\xi_{k/n}-\xi_{(k-1)/n})\rightarrow[f(\xi_{k/n})]^{-1}$ for $k/n\rightarrow\gamma$, $0<\gamma\leqslant p$, we obtain from (4.4) that the *average stopping time* is asymptotically equal to

$$\lim_{n\rightarrow\infty}E(Z_{nN}|K_n)=\int_0^1\frac{p}{f(\xi_{p\theta})}P\{[W(t)+\mu_t]'[W(t)+\mu_t]\leqslant W_\alpha^{**},0\leqslant t\leqslant\theta\}\,\mathrm{d}\theta. \tag{4.5}$$

A second type of test statistics (more common in RST procedures) may also be considered where we take:

$$\tilde{\mathfrak{L}}_{nk}=\nu_{k/n}^{-1}[\boldsymbol{T}_{nk}^{**\prime}\boldsymbol{T}_{nk}];\quad k\leqslant r. \tag{4.6}$$

Note that by (4.1) and (4.6), $\tilde{\mathfrak{L}}_{nk}=\mathfrak{L}_{nk}(\nu_p/\nu_{k/n})$. Let then $\tilde{\mathfrak{L}}_{nr}^{*}=\max_{n_1\leqslant k\leqslant r}\tilde{\mathfrak{L}}_{nk}$ where $r_1=[r\varepsilon]$, for some $\varepsilon>0$. Parallel to (4.2), we have for every $0<\varepsilon<1$ and under H_0,

$$\tilde{\mathfrak{L}}_{nr}^{*}\underset{\mathfrak{D}}{\rightarrow}\sup\{t^{-1}W^{*}(t):\varepsilon'\leqslant t\leqslant 1\}=\tilde{W}_\varepsilon^{**}, \tag{4.7}$$

say, where $\varepsilon'=\nu_{p\varepsilon}/\nu_p(>0)$. Hence, if $\tilde{W}^{**}_{\varepsilon,\alpha}$ be the upper $100\alpha\%$ point of the d.f. of $\tilde{W}^{**}_{\varepsilon}$, then, we may proceed as in the case of $\mathfrak{L}^*_{nr}$ but replacing the $\mathfrak{L}_{nk}$ and W^{**}_{α} by $\tilde{\mathfrak{L}}_{nk}$ and $\tilde{W}_{\varepsilon,\alpha}$, respectively and starting the RST only when Z_{nr_1} is observed. The reason for choosing an $\varepsilon>0$ is quite clear: as $t\downarrow 0$, $t^{-1}W^*(t)$ (or $t^{-1}[W_n(t)]'[W_n(t)]$) does not behave regularly; in fact, $t^{-1}W^*(t)\to\infty$ a.s., as $t\downarrow 0$. However, a small exclusion $[0,\varepsilon)$ eliminates this problem. Sinha and Sen (1979a) have discussed (for $q=1$) the choice of ε (>0) in this context and a very similar picture holds for general $q(\geqslant 1)$. The asymptotic power of the test based on $\tilde{\mathfrak{L}}^*_{nr}$ (for the local alternative treated in Theorem 3.2) is given by

$$P\{t^{-1}[W(t)+\mu_t]'[W(t)+\mu_t]>\tilde{W}^{**}_{\varepsilon,\alpha},\text{ for some } t\in[\varepsilon',1]\},\qquad(4.8)$$

where $\mu(t)$ and ε' are defined as before. Finally, the asymptotic value of the average stopping time is:

$$\xi_{p\varepsilon}+p\int_{\varepsilon}^{1}\frac{1}{f(\xi_{p\theta})}P\{t^{-1}[W(t)+\mu_t]'[W(t)+\mu_t]\leqslant\tilde{W}^{**}_{\varepsilon,\alpha},\ \forall\,\varepsilon'\leqslant t\leqslant\theta\}\,\mathrm{d}\theta.\qquad(4.9)$$

Let us now consider RST based on Type B PCQP's. Since ν_α is not known in advance, we shall find it convenient to use the second type of tests, described above, with the $\nu_{k/n}$ being estimated by

$$V_{nk}=n^{-1}\sum_{i=1}^{k}h^2(Z_{ni})+\frac{1}{n(n-k)}\left(\sum_{i=1}^{k}h(Z_{ni})\right)^2,\quad 1\leqslant k<n,\qquad(4.10)$$

where the Z_{ni} are defined by (2.1). Thus, in (4.6), we need to replace $\nu_{k/n}$ by V_{nk}, while the rest of the discussions in (4.7) through (4.9) remains true for this case too.

As has been noted earlier, the classical multiple regression model is a special case of (2.4) where $m=1$ and $q\geqslant 1$. In this case, we have

$$f_i(x)=f(x-\Delta'(c_i-\bar{c}_n)),\quad x\in R,\quad i=1,\ldots,n,\qquad(4.11)$$

where Δ is a q-vector of regression coefficients and the c_i are the known regression constants (vectors). Both the Type A and Type B procedures described before are applicable here to test for $H_0:\Delta=0$ against $H_1:\Delta\neq 0$ (under progressive censoring). We may further add that this multiple regression model includes, as a special case, the several sample location model: Suppose there are k $(=q+1)$ samples of sizes $n_1,\ldots,n_k$, respectively (where $n=\Sigma_{i=1}^k n_i$) drawn from distributions $F_1,\ldots,F_k$ where

$$F_i(x)=F(x-\theta_i),\quad i=1,\ldots,k,\ x\in R,\qquad(4.12)$$

and let $\Delta_i=\theta_i-\theta_1$, for $i=2,\ldots,k$. Then, if we let $c_1=\cdots=c_{n_1}=0$, $c_{n_1+1}=\cdots=c_{n_1+n_2}=(1,0,\ldots,0)',\ldots,c_{n-n_k+1}=\cdots=c_n=(0,\ldots,0,1)'$, (4.12) corresponds to (2.4). Thus, the proposed Type A and Type B tests are applicable for testing the identity (under PCS) of $F_1,\ldots,F_k$ (against shift alternatives); PCS rank tests for this problem are due to Majumdar and Sen (1978). It may be remarked that instead of the location/regression model in (4.11)–(4.12), one could have considered the regression model in the scale parameter which would have the several sample scale model as a special case. The proposed procedures remain applicable for this scale model too.

We now proceed to the case of $m>1$. The most notable case pertaining to this model [in (2.4)] is the so-called location-scale model where we allow both the location and scale parameters to vary (under the alternative hypothesis), retaining the identity of the d.f.'s under the null hypothesis. Though the theory developed earlier extends readily to the case of $m \geqslant 1, q \geqslant 1$, for the Gaussian process $\boldsymbol{W}$ in Theorem 3.1 [see (3.4)], $\{[\boldsymbol{W}(t)]'[\boldsymbol{\nu}_p \otimes \boldsymbol{I}_q]^{-1}[\boldsymbol{W}(t)],\ 0 \leqslant t \leqslant 1\}$ is, in general, not a Bessel process, and for this process, the distribution of the supremum may depend on the sequence $\{\boldsymbol{\nu}_\alpha:\ 0 \leqslant \alpha \leqslant p\}$ in a rather involved way, so that the computation of the critical values may pose a serious problem. The situation, however, becomes quite manageable, if we assume that

$$\boldsymbol{\nu}_\alpha = \gamma_p(\alpha)\boldsymbol{\nu}_p, \quad \forall 0 \leqslant \alpha \leqslant p, \tag{4.13}$$

where $\gamma_p(\alpha)$ is a non-decreasing function of $\alpha (0 \leqslant \alpha \leqslant p)$, for every fixed p $(0 < p < 1)$. In such a case, in (4.1), we may replace ν_p by $\boldsymbol{\nu}_p$, while, in (4.6), we need to take $\tilde{\mathfrak{L}}_{nk} = \boldsymbol{T}_{nk}^{**\prime}[\boldsymbol{\nu}_{k/n} \times \mathcal{G}_q]^{-1}\boldsymbol{T}_{nk}^{**}$; for Type B PCQP's, $\nu_{k/n}$ need to be replaced by

$$\boldsymbol{V}_{nk} = \frac{1}{n}\sum_{i=1}^{k}[\boldsymbol{h}(Z_{ni})][\boldsymbol{h}(Z_{ni})]' + \frac{1}{n(n-k)}\left[\sum_{i=1}^{k}\boldsymbol{h}(Z_{ni})\right]\left[\sum_{i=1}^{k}\boldsymbol{h}(Z_{ni})\right]', \tag{4.14}$$

for $1 \leqslant k \leqslant n$. (The Bessel process appearing in (4.2) [or (4.7)] will involve mq processes, instead of the q ones in the earlier cases.) The rest of the discussion will be the same. We conclude this section with the remark that Majumdar (1977), Majumdar and Sen (1978) and DeLong (1980, 1981) have considered the evaluation of the critical points W_α^{**} and $\tilde{W}_{\varepsilon,\alpha}^{**}$, for the k-parameter Bessel process for some typical k and these may be used for the actual RST based on the proposed PCQP's.

Acknowledgements

Thanks are due to Professor J.K. Ghosh for his critical reading of the manuscript which resulted in the elimination of numerous errors.

References

[1] Chatterjee, S.K. and Sen, P.K. (1973). Nonparametric testing under progressive censoring. *Calcutta Statist. Assoc. Bull.* **22**, 13–50.

[2] DeLong, D. M. (1980). Some asymptotic properties of a progressively censored nonparametric test for multiple regression. *J. Multivar. Anal.* **10**, 363–370.

[3] DeLong, D.M. (1981). Crossing probabilities for a square root boundary by a Bessel process. *Comm. Statist. Theor. Meth.* **A10**, in press.

[4] Dvoretzky, A. (1972). Central limit theorems for dependent random variables. *Proc. 6th Berkeley Symp. Math. Statist. Prob.* **2**, 513–535.

[5] Gardiner, J.C. and Sen, P.K. (1978). Asymptotic normality of a class of time-sequential statistics and applications. *Comm. Statist. A* **7**, 373–388.

[6] Hájek, J. (1963). Extension of the Kolmogorov-Smirnov test to regression alternatives. In: *Proc. Bernoulli-Bayes-Laplace Seminar*, LeCam, L. Ed. Berkeley, CA, pp.45–60.

[7] Hájek, J. and Šidák, Z. (1967). *Theory of Rank Tests*. Academic Press, New York.

[8] Majumdar, H. (1977). Rank order tests for multiple regression for grouped data under progressive censoring. *Calcutta Statist. Assoc. Bull.* **26**, 1–16.

[9] Majumdar, H. and Sen, P.K. (1978). Nonparametric tests for multiple regression under progressive censoring. *J. Multivar. Anal.* **8**, 93–95.

[10] Sen, P.K. (1976a). Asymptotically optimal rank order tests for progressive censoring. *Calcutta Statist. Assoc. Bull.* **25**, 65–78.

[11] Sen, P.K. (1976b). Weak convergence of progressively censored likelihood ratio statistics and its role in asymptotic theory of life testing. *Ann. Statist.* **4**, 1247–1257.

[12] Sen, P.K. (1979). Weak convergence of some quantile processes arising in progressively censored tests. *Ann. Statist.* **7**, 414–431.

[13] Sen, P.K. (1981). The Cox regression model, invariance principles for some induced quantile processes and some repeated significance tests. *Ann. Statist.* **9**, 109–121.

[14] Sen, P.K. and Tsong, Y. (1981). An invariance principle for progressively truncated likelihood ratio statistics. *Metrika* (to appear).

[15] Sinha, A.N. and Sen, P.K. (1979a). Progressively censored tests for clinical experiments and life testing problems based on weighted empirical distributions. *Comm. Statist. A* **8**, 871–898.

[16] Sinha, A.N. and Sen, P.K. (1979b). Progressively censored tests for multiple regression based on weighted empirical distributions. *Calcutta Statist. Assoc. Bull.* **28**, 57–82.

G. Kallianpur, P.R. Krishnaiah, J.K. Ghosh, eds., *Statistics and Probability: Essays in Honor of C.R. Rao*

ON THE STABILITY OF CHARACTERIZATIONS OF THE NORMAL DISTRIBUTION

Ryoichi SHIMIZU
The Institute of Statistical Mathematics, Tokyo, Japan

Let $X_1, X_2, \ldots, X_n$ be i.i.d. random variables with common distribution, F, and let $a_1, a_2, \ldots, a_n$ be either real constants or random variables independent of the X's such that $\Sigma a_j^2 = 1$ (with probability one). It is proved that F is close to the normal distribution in some sense whenever the distribution of the linear statistic $\Sigma a_j X_j$ is close to F. Multivariate extensions are also given.

1. Introduction

Linnik (1962) obtained a necessary and sufficient condition under which two linear statistics $\Sigma a_j X_j$ and $\Sigma b_j X_j$ based on i.i.d. random variables $X_1, X_2, \ldots, X_n$ have the same distribution if and only if the distribution F of X_1 is normal. The result contains as a special case the following characterization theorem for the normal distribution: if $\Sigma a_j^2 = 1$ and $\Sigma a_j \neq 1$, then $\Sigma a_j X_j$ has the same distribution as X_1 if and only if X_1 is normally distributed with mean zero. Laha and Lukacs (1965) obtained a similar characterization theorem for an infinite sum, while Eaton (1966) extended it to the multivariate case. Shimizu (1968), Ramachandran and Rao (1968, 1970), Davies and Shimizu (1976) and Shimizu (1978) proved similar characterization theorems for the non-normal stable distributions. Quite recently, Shimizu and Davies (1979) obtained a general result, which implies that most of the characterization theorems mentioned above can be extended to the case when the coefficients a's are random variables independent of the X's.

The purpose of the present article is to prove stability of these characterizations. Assuming $\Sigma a_j^2 = 1$, we will prove in Section 2 that under some conditions on the function $S(x) \equiv \Pr\{\Sigma a_j X_j \leqslant x\} - \Pr\{X_1 \leqslant x\}$, F has finite mean μ and variance σ^2 and that $\sup_x |F(x) - \Phi(x; \mu, \sigma^2)|$ is small, where $\Phi(x; \mu, \sigma^2)$ is the distribution function of $N(\mu, \sigma^2)$. In Sections 3 and 4 we prove similar stability for the cases when the a's are random variables and when the X's are p-variate random vectors and the a's are $p \times p$ symmetric matrices, respectively.

2. Constant coefficients

Let $X_1, X_2, \ldots, X_n$ ($n \geqslant 2$) be independently and identically distributed random variables with common distribution, F, and let $a_1, a_2, \ldots, a_n$ be non-zero real numbers such that:

$$a_1^2 + a_2^2 + \cdots + a_n^2 = 1. \tag{1}$$

Put, for real x,

$$S(x) \equiv \Pr\{a_1 X_1 + a_2 X_2 + \cdots + a_n X_n \leqslant x\} - \Pr\{x_1 \leqslant x\},$$

and let $|S|(x)$ be the total variation of the function of bounded variation $S(x)$. We assume that the following conditions are satisfied:

$$\varepsilon \equiv \int_{-\infty}^{\infty} |x|^3 \mathrm{d}|S|(x) < \infty, \tag{2}$$

and

$$\int_{-\infty}^{\infty} x \mathrm{d}S(x) = \int_{-\infty}^{\infty} x^2 \mathrm{d}S(x) = 0. \tag{3}$$

We shall prove:

Theorem 1. *Under the assumptions* (1)–(3), *the distribution F has finite mean μ and variance σ^2 and the following inequality holds*:

$$\Delta \equiv \sup_x |F(x) - \Phi(x; \mu, \sigma^2)| \leqslant 1.8(1-a)^{-1/4} \sigma^{-3/4} \varepsilon^{1/4}, \tag{4}$$

where $a = \max\{|a_1|, |a_2|, \ldots, |a_n|\}$.

Proof. Throughout the present article the symbols η and B denote quantities bounded, respectively, by 1 and by a constant independent of t, θ and x. Let $\phi(t)$ be the characteristic function of F and let $\psi(t)$ be the Fourier–Stieltjes transform of $S(x)$. By the definition of S, $\phi(t)$ and $\psi(t)$ satisfy the relation:

$$\phi(t) = \prod_{j=1}^{n} \phi(a_j t) + \psi(t). \tag{5}$$

The assumptions (2) and (3) imply

$$\psi(t) = \varepsilon_1 \eta t^3, \quad \text{where } \varepsilon_1 = \varepsilon/6. \tag{6}$$

We can find a positive number t_0 such that

$$|\phi(t)| \geqslant \tfrac{1}{2} \geqslant |\psi(t)/\phi(t)| \geqslant 0, \quad |t| \leqslant t_0.$$

Then the symmetric functions,

$$g(t) \equiv -\log|\phi(t)|/t^2 \quad \text{and} \quad \varepsilon(t) \equiv \log|1 - \psi(t)/\phi(t)|,$$

are well defined for $0 < |t| \leqslant t_0$ and $\varepsilon(t) = Bt^3$ by (6). The equation (5) implies

$$g(t) = \sum_{j=1}^{n} a_j^2 g(a_j t) + Bt \geqslant \sum_{j=1}^{n} a_j^2 g(a_j t) - C|t|, \quad 0 < |t| \leqslant t_0, \tag{7}$$

where C is a positive constant. Then there exists an index k, which may depend on t, such that:

$$g(t) \geqslant g(a_k t) - C|t|. \tag{8}$$

On iterating (8) we obtain

$$g(t) \geqslant g(a_{k_1}a_{k_2}\cdots a_{k_m}t) - C(1+a+a^2+\cdots+a^{m-1})|t|, \quad m=1,2,\ldots.$$

It follows that:

$$0 \leqslant c \equiv \varliminf_{t\to 0} g(t) \leqslant \varliminf_{m\to\infty} \inf_{k_1,\ldots,k_m} g(a_{k_1}a_{k_2}\cdots a_{k_m}t_0) \leqslant g(t_0)$$
$$+C(1-a)^{-1}t_0 < \infty. \tag{9}$$

Let $\tilde{F}(x)$ be the symmetric distribution function corresponding to the characteristic function $|\phi(t)|^2$. In view of (9) we have,

$$\infty > c = \varliminf_{t\to 0} g(t) \geqslant \tfrac{1}{2} \varliminf_{t\to 0} \left(1-|\phi(t)|^2\right)/t^2$$
$$= \tfrac{1}{2} \varliminf_{t\to 0} \int_{-\infty}^{\infty} \frac{1-\cos tx}{(tx)^2} x^2 \mathrm{d}\tilde{F}(x) \geqslant \tfrac{1}{6} \varliminf_{t\to 0} \int_{|tx|\leqslant 1} x^2 \mathrm{d}\tilde{F}(x)$$
$$= \tfrac{1}{6} \int_{-\infty}^{\infty} x^2 \mathrm{d}\tilde{F}(x),$$

which means that $\mu \equiv E(X_1)$ and $\sigma^2 \equiv \mathrm{Var}(X_1)$ exist finitely. If t_1 is a sufficiently small positive number, then $\phi(t)$ can be put in the form

$$\phi(t) = \exp\{\mathrm{i}\mu t - \tfrac{1}{2}\sigma^2 t^2 + \gamma(t)t^2\}, \quad |t| \leqslant t_1, \tag{10}$$

where $\gamma(t)$ is a complex-valued function such that $\lim_{t\to 0}|\gamma(t)| = 0$. In particular, we can assume $|\gamma(t)| \leqslant \sigma^2/2$ for $|t| \leqslant t_1$. The condition (3) implies

$$\mu \sum_{j=1}^{n} a_j = \mu. \tag{11}$$

For $m=2,3,\ldots$ put $t_m = t_1/a^{m-1}$ and let $\{\gamma_m(t)\}_{m=1}^{\infty}$ be the sequence of positive functions defined recursively by $\gamma_1(t) = |\gamma(t)|$ and

$$\gamma_m(t) = \max\{\gamma_{m-1}(a_1 t), \gamma_{m-1}(a_2 t), \ldots, \gamma_{m-1}(a_n t)\}, \quad |t| \leqslant t_m.$$

Note that the γ's satisfy the inequalities

$$\gamma_m(t) \leqslant \tfrac{1}{2}\sigma^2, \quad |t| \leqslant t_m. \tag{12}$$

We shall show by mathematical induction on m that $\phi(t)$ can be put in the form

$$\phi(t) = \exp\{\mathrm{i}\mu t - \tfrac{1}{2}\sigma^2 t^2 + \eta\gamma_m(t)t^2\} + \varepsilon_1\eta(1+a+\cdots+a^{m-1})t^3, \quad |t| \leqslant t_m. \tag{13}$$

This is a direct consequence of (10) for $m=1$. Suppose (13) is true for some m and let $|t| \leqslant t_{m+1}$. As $|a_j t| \leqslant t_m$, we can write

$$\phi(a_j t) = Y_j(t) + \varepsilon_1\eta Z_j(t), \quad j=1,2,\ldots,n, \tag{14}$$

where

$$Y_j(t) \equiv \exp\{\mathrm{i}\mu a_j t - \tfrac{1}{2}\sigma^2 a_j^2 t^2 + \eta\gamma_{m+1}(t)a_j^2 t^2\}$$

and

$$Z_j(t) \equiv (1+a+a^2+\cdots+a^{m-1})aa_j^2t^3.$$

Note that $|Y_j(t)+\varepsilon_1\eta Z_j(t)|=|\phi(a_jt)|\leqslant 1$ and that:

$$|Y_j(t)|\leqslant \exp\left\{\operatorname{Re}\left(-\tfrac{1}{2}\sigma^2a_j^2t^2+\eta\gamma_{m+1}(t)a_j^2t^2\right)\right\}$$
$$\leqslant \exp\left\{a_j^2t^2\left(-\tfrac{1}{2}\sigma^2+\gamma_{m+1}(t)\right)\right\}\leqslant 1.$$

Substituting (14) into (5) and remembering the relations (1) and (11) we obtain

$$\begin{aligned}\phi(t)&=\prod_{j=1}^{n}\left(Y_j(t)+\varepsilon_1\eta Z_j(t)\right)+\varepsilon_1\eta t^3\\ &=Y_1(t)\prod_{j=2}^{n}\left(Y_j(t)+\varepsilon_1\eta Z_j(t)\right)+\varepsilon_1\eta\left(t^3+Z_1(t)\right)\\ &=\cdots\\ &=\prod_{j=1}^{n}Y_j(t)+\varepsilon_1\eta\left(t^3+\sum_{j=1}^{n}Z_j(t)\right)\\ &=\exp\left\{\mathrm{i}\mu t-\tfrac{1}{2}\sigma^2t^2+\eta\gamma_{m+1}(t)t^2\right\}+\varepsilon_1\eta(1+a+\cdots+a^m)t^3\end{aligned}$$

as was to be proved.

Now let T be an arbitrary positive number and let m_0 be a sufficiently large positive integer such that $t_{m_0}\geqslant T$. Then the expression (13) is valid for $|t|\leqslant T$ and $m\geqslant m_0$. As $\lim_{t\to 0}\gamma_1(t)=0$, it follows that $\lim_{m\to\infty}\gamma_m(t)=0$ uniformly in the interval $[-T,T]$. As T is arbitrary we conclude that $\phi(t)$ can be written in the form

$$\phi(t)=\mathrm{e}^{\mathrm{i}\mu t-\sigma^2t^2/2}+\varepsilon_1\eta(1-a)^{-1}t^3, \tag{15}$$

for all real t, and hence we have for any positive T:

$$\begin{aligned}\delta(T)&\equiv\int_{-T}^{T}\left|\frac{\phi(t)-e^{\mathrm{i}\mu t-\sigma^2t^2/2}}{t}\right|\mathrm{d}t\leqslant 2\varepsilon_1(1-a)^{-1}\int_0^T t^2\,\mathrm{d}t\\ &=\tfrac{1}{9}\varepsilon(1-a)^{-1}T^3.\end{aligned}$$

By Esseen's inequality [see, e.g. Feller (1966, p. 512)] we obtain

$$\begin{aligned}\Delta&\equiv\sup_x|F(x)-\Phi(x;\mu,\sigma^2)|\leqslant\frac{1}{\pi}\delta(T)+\frac{24}{\pi\sqrt{2\pi}\,\sigma}\frac{1}{T}\\ &\leqslant\frac{1}{9\pi}\varepsilon(1-a)^{-1}T^3+\frac{24}{\pi\sqrt{2\pi}\,\sigma}\frac{1}{T}.\end{aligned}$$

The desired result follows by taking:

$$T=\left(72(1-a)/\sqrt{2\pi}\,\sigma\varepsilon\right)^{1/4}.$$

3. Variable coefficients

In this section, we will generalize Theorem 1 to the case where the coefficients, a, are random variables independent of the X's. Let $(\Theta, \mathcal{Q}, P)$ be the probability space on which the a's are defined. Corresponding to the condition (1) we assume in this section

$$\Pr\left\{\sum_{j=1}^{n} a_j^2(\theta)=1\right\}=1 \quad \text{and} \quad a \equiv \max_j \operatorname{ess.sup}_\theta \{|a_j(\theta)|\}<1, \tag{16}$$

and also

$$\sum_{j=1}^{n} E(a_j(\theta)) \neq 1. \tag{17}$$

Assuming the conditions (2)–(4), the relation (5) now becomes

$$\phi(t)=\int_\Theta \prod_{j=1}^{n} \phi(a_j(\theta)t)\,\mathrm{d}P(\theta)+\psi(t), \tag{18}$$

where $\psi(t)$ is of the form (6). We will first prove that (18) implies the existence of the second order moment $E(X_1^2)$. To this end we introduce $g(t)$ and $\varepsilon(t)$ as in the preceding section. By Jensen's inequality we have for $0<|t|\leqslant \min\{t_0, 1\}$

$$\begin{aligned}|\phi(t)-\psi(t)| &\leqslant \int_\Theta \prod_{j=1}^{n} |\phi(a_j(\theta)t)|\,\mathrm{d}P(\theta)\\ &= \int_\Theta \exp\left\{-t^2 \sum_{j=1}^{n} a_j^2(\theta) g(a_j(\theta)t)\right\}\mathrm{d}P(\theta)\\ &\leqslant \left[\int_\Theta \exp\left\{-\sum_{j=1}^{n} a_j^2(\theta) g(a_j(\theta)t)\right\}\mathrm{d}P(\theta)\right]^{t^2}\\ &\leqslant \left[\int_\Theta \sum_{j=1}^{n} a_j^2(\theta)\exp\{-g(a_j(\theta)t)\}\,\mathrm{d}P(\theta)\right]^{t^2}.\end{aligned}$$

On the other hand, if $0<|t|\leqslant t_0$ and if $C(\geqslant 1)$ is sufficiently large

$$\begin{aligned}|\phi(t)-\psi(t)| &= |\phi(t)|\cdot|1-\psi(t)/\phi(t)| = \mathrm{e}^{-t^2 g(t)}\mathrm{e}^{\varepsilon(t)}\\ &\geqslant \mathrm{e}^{-t^2(g(t)+Bt)} \geqslant \left[\mathrm{e}^{-g(t)}(1-C|t|)\right]^{t^2}.\end{aligned}$$

It follows that:

$$\mathrm{e}^{-g(t)}(1-C|t|) \leqslant \int_\Theta \sum_{j=1}^{n} a_j^2(\theta)\exp\{-g(a_j(\theta)t)\}\,\mathrm{d}P(\theta), \quad 0<|t|\leqslant t_0. \tag{19}$$

Put $t^0=\min\{t_0, (1-a)/2C\}$ and let:

$$\xi(t) \equiv \mathrm{e}^{-g(t)}\left(1-C(1-a)^{-1}|t|\right), \quad 0<|t|\leqslant t_0.$$

Let $Q(x)$ be the distribution function on the finite interval $[0, a]$ defined by:

$$Q(x)\equiv\int_{\Theta}\sum_{j=1}^{n} a_j^2(\theta)\mathrm{e}(x-|a_j(\theta)|)\,\mathrm{d}P(\theta).$$

where $\mathrm{e}(x)$ is the distribution function degenerate at $x=0$. We have from (19):

$$0<\xi(t)\leqslant\int_{[0,\,a]}\xi(ut)\,\mathrm{d}Q(x),\quad 0<|t|\leqslant t_0. \tag{20}$$

Then for each t $(0<t\leqslant t_0)$ there exists a u_1 such that $0<u_1\leqslant a$ and that:

$$0<\xi(t)\leqslant\xi(u_1 t). \tag{21}$$

On iterating (21) we can find a sequence $\{u_m\}$ of positive numbers such that $0<u_m\leqslant a$ and that $0<\xi(t)\leqslant\xi(u_1u_2\cdots u_m t)$. In particular, we have $c_1\equiv\overline{\lim}_{t\to 0}\xi(t)>0$, and hence $c\equiv\underline{\lim}_{t\to 0}g(t)<\infty$, which in turn implies, as in the preceding section, the existence of $\mu=E(X_1)$ and $\sigma^2=\mathrm{Var}(X_1)$. However, by conditions (3) and (17), we have $\mu=0$. Therefore, $\phi(t)$ can be put in the form (10) with $\mu=0$. Almost the same argument as in the preceding section can apply to conclude that $\phi(t)$ is of the form (15) with $\mu=0$ and we obtain:

Theorem 2. *The result of Theorem* 1 *holds true with* $\mu=0$ *if the coefficients*, a, *are random variables, provided they are independent of the* X*'s and satisfy the conditions* (16) *and* (17).

Both Theorems 1 and 2 can readily be generalized to the cases of infinite sums. In particular, we can prove the following stability theorems for the characterization given by Shimizu and Davies (1979, Theorem 7.1).

Theorem 3. *Let* $\{X_n\}_{n=1}^{\infty}$ *be a sequence of i.i.d. random variables with common distribution* F *and let* N $(\geqslant 2)$ *be an integer valued random variable independent of the* X*'s. Put for real* x:

$$S(x)\equiv\Pr\left\{X_1+X_2+\cdots+X_N\leqslant\sqrt{N}\,x\right\}-\Pr\{X_1\leqslant x\}.$$

If the function $S(x)$ *satisfies conditions* (2)–(4), *then* F *has zero mean and finite variance* σ^2 *and the inequality* (4) *holds with* $\mu=0$ *and* $a=E(N^{-1/2})$.

Proof. Writing $p(n)=\Pr\{N=n\}$, we have

$$\phi(t)=\sum_{n=2}^{\infty}p(n)\phi^n\left(t/\sqrt{n}\right)+\psi(t), \tag{22}$$

where $\psi(t)=\varepsilon_1\eta t^3$. Let the continuous functions $g(t)$ and $\varepsilon(t)$ be defined as in

Section 2. Then for sufficiently small $|t|$, Jensen's inequality leads to:

$$\left[e^{-g(t)}(1-C|t|)\right]^{t^2} \leqslant |\phi(t)-\psi(t)| \leqslant \sum_{n=2}^{\infty} p(n)|\phi^n(t/\sqrt{n})|$$

$$\leqslant \sum_{n=2}^{\infty} p(n) e^{-t^2 g(t/\sqrt{n})} \leqslant \left(\sum_{n=2}^{\infty} p(n) e^{-g(t/\sqrt{n})} \right)^{t^2}. \tag{23}$$

Let $\xi(t)$ be as in Section 2 and let $Q(x)$ be the distribution function on the interval $[0,1/\sqrt{2}]$ defined by

$$Q(x) \equiv \sum_{n=2}^{\infty} p(n) e\left(x-1/\sqrt{n}\right).$$

Then $\xi(t)$ satisfies the inequality (20), and we conclude that $\mu=E(X_1)$ and $\sigma^2=\operatorname{Var}(X_1)$ exist finitely. Then the condition (3) implies the relation $\mu=\mu\sum_{n=2}^{\infty} p(n)\sqrt{n}$. However, as $1<\sum_{n=2}^{\infty} p(n)\sqrt{n} \leqslant \infty$, we have $\mu=0$. The rest of the proof is quite similar to that of Theorem 1 and we omit the details.

4. Multivariate case

We now extend the results of the preceding sections to the p-variate case. Let $\boldsymbol{X}_1, \boldsymbol{X}_2, \ldots, \boldsymbol{X}_n$ $(n \geqslant 2)$ be i.i.d. p-variate random vectors with common distribution $F(x)$. We assume that for each constant non-zero vector $\boldsymbol{c}$, the random variable $\boldsymbol{c}'\boldsymbol{X}_1$ is non-degenerate. Let $A_1, A_2, \ldots, A_n$ be $p \times p$ symmetric non-singular matrices such that

$$A_1^2 + A_2^2 + \cdots + A_n^2 = I_p, \tag{24}$$

where I_p denotes the $p \times p$ identity matrix. For p-dimensional real vectors $\boldsymbol{x}'=(x_1, x_2, \ldots, x_p)$ and $\boldsymbol{y}'=(y_1, y_2, \ldots, y_p)$, we write $\boldsymbol{x} \leqslant \boldsymbol{y}$ to mean $x_i \leqslant y_i$, for $i=i,2,\ldots,p$. Set

$$S(\boldsymbol{x}) \equiv \Pr\{A_1\boldsymbol{X}_1 + A_2\boldsymbol{X}_2 + \cdots + A_n\boldsymbol{X}_n \leqslant \boldsymbol{x}\} - \Pr\{\boldsymbol{X}_1 \leqslant \boldsymbol{x}\}$$

and

$$\psi(\boldsymbol{t}) \equiv \int_{\boldsymbol{R}^p} e^{i\boldsymbol{t}'\boldsymbol{x}} \, dS(\boldsymbol{x}).$$

We assume

$$\int_{\boldsymbol{R}^p} x_j \, dS(\boldsymbol{x}) = 0, \quad j=1,2,\ldots,p, \tag{25}$$

$$\int_{\boldsymbol{R}^p} x_j x_k \, dS(\boldsymbol{x}) = 0, \quad j,k=1,2,\ldots,p \tag{26}$$

and

$$0 \leqslant \varepsilon \equiv \int_{\boldsymbol{R}^p} |\boldsymbol{x}|^3 \, d|S|(\boldsymbol{x}) < \infty. \tag{27}$$

In view of the relations

$$e^{it'x}=1+it'x-\tfrac{1}{2}(t'x)^2+\frac{\eta}{6}(t'x)^3 \quad \text{and} \quad |t'x|\leqslant|t|\cdot|x|,$$

and of the conditions (25)–(27), $\psi(t)$ can be expressed as

$$\psi(t)=\frac{\eta}{6}\int_{R^p}(t'x)^3\,dS(x)=\varepsilon_1\eta|t|^3, \quad \text{where } \varepsilon_1=\varepsilon/6. \tag{28}$$

The characteristic function $\phi(t)$ of F satisfies the equation:

$$\phi(t)=\prod_{j=1}^{n}\phi(A_jt)+\psi(t). \tag{29}$$

If $t_0>0$ is taken sufficiently small, the functions

$$g(t)\equiv-\log|\phi(t)|/|t|^2 \quad \text{and} \quad \varepsilon(t)\equiv\log|1-\psi(t)/\phi(t)|$$

are defined for $0<|t|\leqslant t_0$ and $\varepsilon(t)=\varepsilon_1\eta|t|^3$. Moreover $g(t)$ satisfies the inequality

$$g(t)\geqslant\sum_{j=1}^{n}p_j(t)g(A_jt)-C|t|,$$

where C is a positive constant and

$$p_j(t)\equiv t'A_j^2t/|t|^2, \quad j=1,2,\ldots,n.$$

As $\Sigma_{j=1}^{n}p_j(t)=t'(\Sigma_{j=1}^{n}A_j^2)t=1$ for any t, there exists an index k, which may depend on t, such that

$$g(t)\geqslant g(A_kt)-C|t|.$$

On iterating the inequality we obtain

$$g(t)\geqslant g(A_{k_1}A_{k_2}\cdots A_{k_m}t)-C(1+a+\cdots+a^{m-1})|t|, \tag{30}$$

where a is the maximum of the absolute values of eigenvalues of the symmetric matrices $A_1, A_2,\ldots, A_n$. In view of the condition (24), a is less than 1.

For a fixed t_0 $(0<|t_0|\leqslant t_0)$, put

$$t_m=A_{k_1}A_{k_2}\cdots A_{k_m}t_0 \quad \text{and} \quad u_m=t_m/|t_m|.$$

Note that

$$t_m\equiv|t_m|\leqslant a^m|t_0|\leqslant a^m t_0\to0, \quad \text{as } m\to\infty,$$

while u's are unit vectors. Let $\tilde{F}(x)$ be the distribution function corresponding to the characteristic function $|\phi(t)|^2$. Then (30) implies

$$\begin{aligned}\infty>c_0&\equiv g(t_0)+C(1-a)^{-1}|t_0|\geqslant g(t_m)\\ &\geqslant\tfrac{1}{2}(1-|\phi(t_m)|^2)/|t_m|^2=\tfrac{1}{2}\int_{R^p}(1-\cos t_m'x)/|t_m|^2\,d\tilde{F}(x)\\ &\geqslant\tfrac{1}{6}\int_{|t_m'x|\leqslant1}\left|\frac{t_m'x}{t_m}\right|^2 d\tilde{F}(x)=\tfrac{1}{6}\int_{|u_m'x|\leqslant t_m^{-1}}(u_m x)^2\,d\tilde{F}(x).\end{aligned}$$

If T is an arbitrary positive number, we can find an m_0 such that $t_m^{-1} \geq T$ for all $m \geq m_0$. As $|x| \leq T$ implies $|u_m' x| \leq t_m^{-1}$, we have

$$\int_{|x| \leq T} (u_m' x)^2 \, \mathrm{d}\tilde{F}(x) \leq 6c_0 < \infty. \tag{31}$$

It follows that if u_0 is the limit of any convergent subsequence of $\{u_m\}$ then (31) holds true also for u_0. As T is arbitrary we conclude that the non-degenerate random variable $u_0' X_1$ has the finite second moment. We can assume without loss of generality that the first q $(1 < q \leq p)$ components of the random vector $X_1' = (X_1, X_2, \ldots, X_q, X_{q+1}, \ldots, X_p)$ have finite second moments. We shall show that we can take $q = p$. To this end suppose the contrary and let $q < p$ and let $E(X_k^2) = \infty$, for $k = q+1, \ldots, p$. In view of the condition (26) the first q components of the vector $A_1 X_1 + A_2 X_2 + \cdots + A_n X_n$ have finite second moments. As $c'X$ is assumed to be non-degenerate for all non-zero c, this is possible if and only if the A's are of the form

$$A_j = \begin{pmatrix} A_{1j} & \mathbf{0} \\ \mathbf{0} & A_{2j} \end{pmatrix},$$

where A_{1j} and A_{2j} are $q \times q$ and $(p-q) \times (p-q)$ symmetric matrices satisfying $A_{11}^2 + A_{12}^2 + \cdots + A_{1n}^2 = I_q$ and $A_{21}^2 + A_{22}^2 + \cdots + A_{2n}^2 = I_{q-p}$, respectively. Therefore, for $p-q$ dimensional vector $v' = (t_{q+1}, \ldots, t_p)$, the characteristic function $\phi_0(v)$ of the random vector $(X_{q+1}, \ldots, X_p)$, which is equal to $\phi(0, 0, \ldots, 0, t_{q+1}, \ldots, t_p)$, satisfies the equation

$$\phi_0(v) = \prod_{j=1}^{n} \phi_0(A_{2j} v) + B|v|^3.$$

We have thus arrived at a contradiction: at least one of the variables $X_{q+1}, \ldots, X_p$ has finite second moment.

Thus, we conclude that the characteristic function $\phi(t)$ of the random vector X_1 can be put in the form

$$\phi(t) = \exp\{\mathrm{i}\mu' t - \tfrac{1}{2} t' \Sigma t + \gamma(t)|t|^2\}, \quad |t| \leq t_0,$$

where μ and Σ are the mean vector and the variance-covariance matrix of X_1, respectively, and where $\gamma(t)$ is a complex valued function such that the matrix $\Sigma - 2|\gamma(t)| \cdot I_p$ is non-negative definite for all $|t| \leq t_0$. The conditions (25) and (26) imply

$$(A_1 + A_2 + \cdots + A_n)\mu = \mu \tag{32}$$

and

$$A_1 \Sigma A_1 + A_2 \Sigma A_2 + \cdots + A_n \Sigma A_n = \Sigma, \tag{33}$$

and a similar argument as in Section 2 leads to

$$\phi(t) = \exp\{\mathrm{i}\mu' t - \tfrac{1}{2} t' \Sigma t\} + \varepsilon_1 \eta (1-a)^{-1} |t|^3 \quad \text{for all } t.$$

In particular, if a is a non-zero vector, then the characteristic function of the random variable $\boldsymbol{a}'\boldsymbol{X}_1$ is given by

$$\phi(t;\boldsymbol{a})=\phi(t\boldsymbol{a})=e^{i\mu t-\sigma^2t^2/2}+\varepsilon_2\eta(1-a)^{-1}t^3,\quad -\infty<t<\infty,$$

where $\mu=\boldsymbol{\mu}'\boldsymbol{a}$, $\sigma^2=\boldsymbol{a}'\boldsymbol{\Sigma}\boldsymbol{a}$ and $\varepsilon_2=\varepsilon_1|\boldsymbol{a}|^3$.

Using Esseen's inequality we obtain:

Theorem 4. *Let* $\boldsymbol{a}$ *be a p-dimensional non-zero vector and let* $F(x;\boldsymbol{a})$ *be the distribution function of the random variable* $\boldsymbol{a}'\boldsymbol{X}_1$. *Under the conditions stated at the beginning of this section the random vector* $\boldsymbol{X}_1$ *has finite second moments and on writing* $\boldsymbol{\mu}=E(\boldsymbol{X}_1)$ *and* $\boldsymbol{\Sigma}=Var(\boldsymbol{X}_1)$ *we have the following inequality*

$$\sup_x|F(x;\boldsymbol{a})-\Phi(x;\boldsymbol{a}'\boldsymbol{\mu},\boldsymbol{a}'\boldsymbol{\Sigma}\boldsymbol{a})|\leqslant 1.8(1-a)^{-1/4}(\boldsymbol{a}'\boldsymbol{a}/\boldsymbol{a}'\boldsymbol{\Sigma}\boldsymbol{a})^{3/8}\varepsilon^{1/4}.$$

Finally we remark that Theorem 4 can be extended to the case when the A's are random matrices independent of the $\boldsymbol{X}$'s and the equality (24) holds with probability 1, and also to the case of infinite sums. We omit the details.

References

Davies, P.L. and Shimizu, R. (1976). On identically distributed linear statistics. *Ann. Inst. Stat. Math.* **28**, 469–489.

Eaton, M.L. (1966). Characterization of distributions by the identical distribution of linear forms. *J. Appl. Probability* **3**, 481–494.

Feller, W. (1966). *An Introduction to Probability Theory and its Applications*, Vol. II, Wiley, New York.

Laha, R.C. and Lukacs, E. (1965). On a linear form whose distribution is identical with that of a monomial. *Pacific J. Math.* **15**, 207–214.

Linnik, Yu.V. (1953). Linear forms and statistical criteria. *Ukrain. Math. Zurnal* (in Russian) **5**. Eng. translation (1962). *Selected Transl. Math. Stat. Probability* **3**, 1–90.

Ramachandran, B. and Rao, C.R. (1968). Some results on characteristic functions and characterizations of the normal and generalized stable laws. *Sankhyā Ser. A* **30**, 125–140.

Ramachandran, B. and Rao, C.R. (1970). Solutions of functional equations arising in some regression problems and a characterization of the Cauchy law. *Sankhyā Ser A* **32**, 1–30.

Shimizu, R. (1968). Characteristic functions satisfying a functional equation (I). *Ann. Inst. Stat. Math.* **20**, 187–209.

Shimizu, R. (1978). Solution to a functional equation and its application to some characterization problems. *Sankhyā Ser. A.* **40**, 319–332.

Shimizu, R. And Davies, P.L. (1979). General characterization theorems for the Weibull and the stable distributions. Research Memorandum no. 173, The Institute of Statistical Mathematics, Tokyo.

G. Kallianpur, P.R. Krishnaiah, J.K. Ghosh, eds., *Statistics and Probability*: *Essays in Honor of C.R. Rao*

INVERSION OF INFORMATION MATRICES OF BALANCED 3^m FACTORIAL DESIGNS OF RESOLUTION V, AND OPTIMAL DESIGNS*

J.N. SRIVASTAVA and W.M. ARIYARATNA
Department of Statistics, Colorado State University, Fort Collins, CO, U.S.A.

Using a simple, intuitively appealing, and statistically meaningful technique, we obtain the inverse of the information matrix of balanced 3^m fractional factorial designs with N runs. [This approach avoids complicated (multidimensional) association schemes, linear associative algebras, ideal theory, group fields, etc. which have been used in similar problems by Srivastava and/or other authors.] It is shown that this inversion is reduced to that of three interesting matrices, of sizes 6×6, 6×6, and 4×4. By using this, a new formula for the trace of the inverse (which is an important step in obtaining trace-optimal designs) is obtained.

1. Introduction

Consider 3^m factorials. As usual, the effects will be denoted as follows. The symbol F_0 will stand for the general mean, F_i $(i=1,\dots,m)$ for the linear main effects, F_i^2 $(i=1,\dots,m)$ for quadratic main effects, F_iF_j, $F_i^2F_j^2$ $(i<j;\ i,j=1,2,\dots,m)$ and $F_iF_j^2$ $(i,j=1,2,\dots,m)$ for various two factor interactions, and similar other symbols for the higher order interactions. The information matrix (M), from Bose and Srivastava (1964a), for resolution V designs for 3^m factorials is given by

$$M=\begin{bmatrix} M_1 & M_2 & M_3 & M_4 & M_5 & M_6 \\ & M_7 & M_8 & M_9 & M_{10} & M_{11} \\ & & M_{12} & M_{13} & M_{14} & M_{15} \\ & & & M_{16} & M_{17} & M_{18} \\ & & & & M_{19} & M_{20} \\ & & & & & M_{21} \end{bmatrix}. \tag{1.1}$$

In the above, the matrix M has been partitioned into 6 row and 6 column blocks thus giving 36 submatrices. Since M is symmetric, only the distinct submatrices in the upper triangular part have been numbered. The six row (and also column) blocks, respectively, correspond to $\{F_0\}$, $\{F_i^2\}$, $\{F_iF_j\}$, $\{F_i^2F_j^2\}$, $\{F_i\}$ and $\{F_iF_j^2\}$, where $\{F_0\}$ is the set consisting of the single element F_0, $\{F_i\}$ is the set of (linear) main effects $F_1, F_2,\dots, F_m$, $\{F_i^2\}$ is the set of effects $F_1^2, F_2^2,\dots, F_m^2$, $\{F_iF_j\}$ is the set $F_1F_2, F_1F_3,\dots, F_{m-1}F_m$, $\{F_i^2F_j^2\}$ is the set $F_1^2F_2^2, F_1^2F_3^2,\dots, F_{m-1}^2F_m^2$, $\{F_iF_j^2\}$ is the

*This research was supported by AFOSR Grant No. 77-3127.

set $F_1F_2^2, F_1^2F_2, F_1F_3^2, F_1^2F_3, \ldots, F_{m-1}F_m^2, F_{m-1}^2F_m$. Thus the matrix, M, is of the size $\nu_m \times \nu_m$, where

$$\nu_m = 1 + m + \binom{m}{2} + \binom{m}{2} + m + m(m-1) = 1 + 2m^2. \tag{1.2}$$

In this paper, we shall restrict ourselves to the case where the design under consideration is balanced. This was studied in Srivastava (1961), and Bose and Srivastava (1964b). A design is said to be balanced, if the variance–covariance matrix of the estimates for the parameters (Assuming that the parameters not of interest are negligible) is invariant under the permutations of factor symbols. In other words, the variance–covariance matrix of the estimators can have at most 49 distinct elements as indicated below:

$$\begin{gathered}
\mathrm{v}(\hat F_0, \hat F_0),\ \mathrm{c}(\hat F_0, \hat F_i^2),\ \mathrm{c}(F_0, \widehat{F_iF_j}),\ \mathrm{c}(\hat F_0, \widehat{F_i^2F_j^2}),\ \mathrm{c}(\hat F_0, \hat F_i),\ \mathrm{c}(\hat F_0, \widehat{F_iF_j^2}),\ \mathrm{v}(\hat F_i^2, \hat F_i^2),\\
\mathrm{c}(\hat F_i^2, \hat F_j^2),\ \mathrm{c}(\hat F_i^2, \widehat{F_iF_j}),\ \mathrm{c}(\hat F_i^2, \widehat{F_jF_k}),\ \mathrm{c}(\hat F_i^2, \widehat{F_i^2F_j^2}),\ \mathrm{c}(\hat F_i^2, \widehat{F_j^2F_k^2}),\ \mathrm{c}(\hat F_i^2, \hat F_i),\\
\mathrm{c}(\hat F_i^2, \hat F_j),\ \mathrm{c}(\hat F_i^2, \widehat{F_iF_j^2}),\ \mathrm{c}(\hat F_i^2, \widehat{F_i^2F_j}),\ \mathrm{c}(\hat F_i^2, \widehat{F_jF_k^2}),\ \mathrm{v}(\widehat{F_iF_j}, \widehat{F_iF_j}),\ \mathrm{c}(\widehat{F_iF_j}, \widehat{F_iF_k}),\\
\mathrm{c}(\widehat{F_iF_j}, \widehat{F_kF_l}),\ \mathrm{c}(\widehat{F_iF_j}, \widehat{F_i^2F_j^2}),\ \mathrm{c}(\widehat{F_iF_j}, \widehat{F_i^2F_k^2}),\ \mathrm{c}(\widehat{F_iF_j}, \widehat{F_k^2F_l^2}),\ \mathrm{c}(\widehat{F_iF_j}, \hat F_i),\\
\mathrm{c}(\widehat{F_iF_j}, \hat F_k),\ \mathrm{c}(\widehat{F_iF_j}, \widehat{F_iF_j^2}),\ \mathrm{c}(\widehat{F_iF_j}, \widehat{F_iF_k^2}),\ \mathrm{c}(\widehat{F_iF_j}, \widehat{F_i^2F_k}),\ \mathrm{c}(\widehat{F_iF_j}, \widehat{F_kF_l^2}),\\
\mathrm{v}(\widehat{F_i^2F_j^2}, \widehat{F_i^2F_j^2}),\ \mathrm{c}(\widehat{F_i^2F_j^2}, \widehat{F_i^2F_k^2}),\ \mathrm{c}(\widehat{F_i^2F_j^2}, \widehat{F_k^2F_l^2}),\ \mathrm{c}(\widehat{F_i^2F_j^2}, \hat F_i),\ \mathrm{c}(\widehat{F_i^2F_j^2}, \hat F_k),\\
\mathrm{c}(\widehat{F_i^2F_j^2}, \widehat{F_iF_j^2}),\ \mathrm{c}(\widehat{F_i^2F_j^2}, \widehat{F_iF_k^2}),\ \mathrm{c}(\widehat{F_i^2F_j^2}, \widehat{F_i^2F_k}),\ \mathrm{c}(\widehat{F_i^2F_j^2}, \widehat{F_kF_l^2}),\ \mathrm{v}(\hat F_i, \hat F_i),\ \mathrm{c}(\hat F_i, \hat F_j),\\
\mathrm{c}(\hat F_i, \widehat{F_iF_j^2}),\ \mathrm{c}(\hat F_i, \widehat{F_i^2F_j}),\ \mathrm{c}(\hat F_i, \widehat{F_jF_k^2}),\ \mathrm{v}(\widehat{F_iF_j^2}, \widehat{F_iF_j^2}),\ \mathrm{c}(\widehat{F_iF_j^2}, \widehat{F_i^2F_j}),\\
\mathrm{c}(\widehat{F_iF_j^2}, \widehat{F_iF_k^2}),\ \mathrm{c}(\widehat{F_iF_j^2}, \widehat{F_i^2F_k}),\ \mathrm{c}(\widehat{F_iF_j^2}, \widehat{F_j^2F_k}),\ \mathrm{c}(\widehat{F_iF_j^2}, \widehat{F_kF_l^2}).
\end{gathered}$$

Here, the symbol v stands for the variance, and c stands for the covariance and the symbol "^" above the different parameters indicates the estimates of the corresponding parameters. Using the method of Bose and Srivastava (1964b), it can be easily checked that if the design, T, is balanced, then the information matrix, M, will have the same structure as the variance covariance matrix, V, of the estimates.

Following Bose–Srivastava (1964a, b) and Srivastava and Chopra (1973), the following notations will be used. The vector $\boldsymbol{F}(\nu_m \times 1)$ will stand for the set consisting of the general mean, main effects and the two factor interactions taken in the same order in which they correspond to the rows and columns of M. Now, let θ, θ' denote any two elements of $\boldsymbol{F}$, which may or may not be distinct. Then, the element of M corresponding to θ and θ' will be denoted by $\varepsilon(\theta, \theta')$. Then, it is well known that T is balanced if and only if the following conditions are satisfied for some fixed (known) ε's:

$$\begin{gathered}
\varepsilon(F_0, F_0) = N = \varepsilon_1, \quad \varepsilon(F_0, F_i^2) = \varepsilon_2, \quad \varepsilon(F_0, F_iF_j) = \varepsilon(F_i^2, F_iF_j) = \\
= \varepsilon(F_i, F_j) = \varepsilon(F_i, F_i^2F_j) = \varepsilon(F_iF_j, F_i^2F_j^2) = \varepsilon(F_iF_j^2, F_jF_i^2) = \varepsilon_3,\\
\varepsilon(F_0, F_i^2F_j^2) = \varepsilon(F_i^2, F_j^2) = \varepsilon_4, \quad \varepsilon(F_0, F_i) = \varepsilon(F_i^2, F_i) = \varepsilon_5,\\
\varepsilon(F_0, F_i^2F_j) = \varepsilon(F_i^2, F_j) = \varepsilon(F_i^2F_j^2, F_i) = \varepsilon(F_i^2, F_iF_j^2) = \varepsilon_6, \quad \varepsilon(F_i^2, F_i^2) = \varepsilon_7,\\
\varepsilon(F_i^2, F_i^2F_k^2) = \varepsilon_8, \quad \varepsilon(F_i^2, F_j^2F_k^2) = \varepsilon_9, \quad \varepsilon(F_i^2, F_i^2F_j) = \varepsilon(F_i^2F_j^2, F_i^2F_j) = \varepsilon_{10},
\end{gathered}$$

$$\varepsilon(F_i^2, F_j^2F_k)=\varepsilon(F_i^2F_k^2, F_j^2F_k)=\varepsilon(F_i, F_j^2F_k^2)=\varepsilon_{11}, \quad \varepsilon(F_iF_j, F_iF_j)=\varepsilon_{12},$$
$$\varepsilon(F_iF_j, F_iF_k)=\varepsilon_{13}, \quad \varepsilon(F_iF_j, F_kF_l)=\varepsilon_{14},$$
$$\varepsilon(F_i, F_jF_k^2)=\varepsilon(F_iF_j, F_i^2F_k^2)=\varepsilon(F_iF_j^2, F_jF_k^2)=\varepsilon_{15},$$
$$\varepsilon(F_iF_j, F_k^2F_l^2)=\varepsilon(F_iF_j^2, F_kF_l^2)=\varepsilon_{16}, \quad \varepsilon(F_i, F_iF_j)=\varepsilon(F_iF_j^2, F_iF_j)=\varepsilon_{17}, \tag{1.3}$$
$$\varepsilon(F_i, F_jF_k)=\varepsilon(F_iF_j^2, F_jF_k)=\varepsilon_{18}, \quad \varepsilon(F_iF_j, F_iF_k^2)=\varepsilon_{19},$$
$$\varepsilon(F_iF_j, F_kF_l^2)=\varepsilon_{20}, \quad \varepsilon(F_i^2F_j^2, F_i^2F_j^2)=\varepsilon_{21}, \quad \varepsilon(F_i^2F_j^2, F_i^2F_k^2)=\varepsilon_{22},$$
$$\varepsilon(F_i^2F_j^2, F_k^2F_l^2)=\varepsilon_{23}, \quad \varepsilon(F_i^2F_j^2, F_i^2F_k)=\varepsilon_{24}, \quad \varepsilon(F_i^2F_j^2, F_k^2F_l)=\varepsilon_{25},$$
$$\varepsilon(F_i, F_i)=\varepsilon_{26}, \quad \varepsilon(F_i, F_iF_j^2)=\varepsilon_{27}, \quad \varepsilon(F_iF_j^2, F_iF_j^2)=\varepsilon_{28},$$
$$\varepsilon(F_iF_j^2, F_iF_k^2)=\varepsilon_{29}, \quad \varepsilon(F_iF_j^2, F_j^2F_k)=\varepsilon_{30}.$$

In other words, M is invariant under permutation of factor symbols. It is well known that the normal equations for estimating $\boldsymbol{F}$ are given by

$$M\hat{\boldsymbol{F}}=\boldsymbol{Z}, \tag{1.4}$$

where $\boldsymbol{Z}$ is the stochastic vector which equals to a known matrix multiplied by a vector of observations $\boldsymbol{Y}$ $(N\times 1)$. The method of computation of $\boldsymbol{Z}$ from $\boldsymbol{Y}$ is quite simple and is explained in Bose and Srivastava (1964a). Also, in the above, $\hat{\boldsymbol{F}}$ denotes the best linear unbiased estimate (BLUE) of $\boldsymbol{F}$. If the matrix M is non-singular, we may call the design to be non-singular. In this case, the equation (1.4) can be solved to get

$$\hat{\boldsymbol{F}}=M^{-1}\boldsymbol{Z}. \tag{1.5}$$

Thus, knowing M^{-1}, the estimate vector $\hat{\boldsymbol{F}}$ can be easily computed. Also, it is well known that,

$$\operatorname{Var}(\hat{\boldsymbol{F}})=\sigma^2M^{-1}=\sigma^2V, \tag{1.6}$$

say, where σ^2 is the common variance of each element of the vector $\boldsymbol{Y}$.

In Section 2, we obtain M^{-1} by a simple method originally developed by Srivastava (1961). Next, we make comments on optimal designs. Finally, for the convenience of the reader, some of the important works related to this field are listed at the end.

2. Inverse of M

Consider the normal equations (1.4) again. We introduce the following notation:

$$\sum_{i=1}^{m} F_i^2=mP_0, \quad \sum_{i<j}\sum F_iF_j=\tfrac{1}{2}m(m-1)Q_0, \quad \sum_{i<j}\sum F_i^2F_j^2=\tfrac{1}{2}m(m-1)R_0,$$

$$\sum_{i=1}^{m} F_i = mS_0, \qquad \sum\sum_{i\neq j} F_i^2 F_j = m(m-1)T_0, \qquad \sum_{j\neq k} F_k F_j = (m-1)Q_{k0},$$

$$\sum_{j\neq k} F_k^2 F_j^2 = (m-1)R_{k0}, \qquad \sum_{k\neq k} F_k^2 F_j = (m-1)T_{k0}, \qquad \sum_{j\neq k} F_j^2 F_k = (m-1)T_{0k},$$

$$\sum_k Q_{k0} = mQ_0, \qquad \sum_k R_{k0} = mR_0, \qquad \sum_k T_{k0} = \sum_k T_{0k} = mT_0. \tag{2.1}$$

Without loss of generality, we shall assume that the vector $\boldsymbol{Z}$ has elements of exactly the same form as the corresponding elements of $\boldsymbol{F}$, except that instead of $\boldsymbol{F}$ we shall be using $\boldsymbol{Z}$. Also, corresponding to element F_0 of $\boldsymbol{F}$, we shall assume that $\boldsymbol{Z}$ has the element Z_0. Thus, in order, the elements of $\boldsymbol{Z}$ are Z_0; $Z_1^2,\ldots,Z_m^2$; $Z_1Z_2,\ldots,Z_{m-1}Z_m$; etc. Corresponding to the sums of the elements of $\boldsymbol{F}$ defined in (2.1), we can define the sums of the elements of $\boldsymbol{Z}$. The sum of a set of elements of $\boldsymbol{Z}$ will be denoted by the same symbols (with a prime) as the ones used for the corresponding sums of the elements of $\boldsymbol{F}$. Thus, we have $\Sigma Z_i^2 = mP_0'$; $\Sigma\Sigma_{i<j} Z_i Z_j = \frac{1}{2}m(m-1)Q_0'$, $\Sigma\Sigma_{i<j} Z_i^2 Z_j^2 = \frac{1}{2}m(m-1)R_0'$, etc. Now, let

$$d = \tfrac{1}{2}m(m-1), \tag{2.2}$$

$$\boldsymbol{\alpha}_0' = [F_0, P_0, Q_0, R_0, S_0, T_0], \qquad \boldsymbol{\alpha}_l' = [F_l^2, Q_{l0}, R_{l0}, F_l, T_{l0}, T_{0l}],$$

$$\boldsymbol{\alpha}_k' = [F_k^2, Q_{k0}, R_{k0}, F_k, T_{k0}, T_{0k}], \qquad \boldsymbol{\alpha}_{kl}' = [F_kF_l, F_k^2F_l^2, F_kF_l^2, F_k^2F_l],$$

$$\boldsymbol{\beta}_0' = [Z_0, P_0', Q_0', R_0', S_0', T_0'], \qquad \boldsymbol{\beta}_l' = [Z_l^2, Q_{l0}', R_{l0}', Z_l, T_{l0}', T_{0l}'],$$

$$\boldsymbol{\beta}_k' = [Z_k^2, Q_{k0}', R_{k0}', Z_k, T_{k0}', T_{0k}'], \qquad \boldsymbol{\beta}_{kl}' = [Z_kZ_l, Z_k^2Z_l^2, Z_kZ_l^2, Z_k^2Z_l],$$

$$\underset{(6\times 6)}{D_1} = \mathrm{diag}(1, m, d, d, m, 2d),$$

$$\underset{(6\times 6)}{D_2} = \mathrm{diag}[1,(m-1),(m-1),1,(m-1),(m-1)],$$

$$\underset{(6\times 6)}{D_3} = \begin{bmatrix} (m-2) & 0 & 0 & 0 & 0 & 0 \\ 0 & (m-1) & 0 & 0 & 0 & 0 \\ 0 & 0 & (m-1) & 0 & 0 & 0 \\ 0 & 0 & 0 & (m-2) & 0 & 0 \\ 0 & 0 & 0 & 0 & \dfrac{(m-1)^2}{m} & \dfrac{(m-1)}{m} \\ 0 & 0 & 0 & 0 & \dfrac{(m-1)}{m} & \dfrac{(m-1)^2}{m} \end{bmatrix}.$$

Notice that the normal equations (1.4), written in the form of ordinary equations would give rise to one equation corresponding to each element of vector $\boldsymbol{Z}$. These equations are divisible into six sets, these sets being exactly the same as the six sets used for partitioning the matrix, M, in equation (1.1). Now add all the equations in each set. Doing this operation and writing the resulting six equations in matrix form, we obtain,

$$X_A' M\boldsymbol{F} = X_A'\boldsymbol{Z}, \tag{2.3}$$

where

$$\underset{(6\times v_m)}{X'_A} = \begin{bmatrix} J_{11} & 0_{1m} & 0_{1d} & 0_{1d} & 0_{1m} & 0_{1,2d} \\ 0_{11} & J_{1m} & 0_{1d} & 0_{1d} & 0_{1m} & 0_{1,2d} \\ 0_{11} & 0_{1m} & J_{1d} & 0_{1d} & 0_{1m} & 0_{1,2d} \\ 0_{11} & 0_{1m} & 0_{1d} & J_{1d} & 0_{1m} & 0_{1,2d} \\ 0_{11} & 0_{1m} & 0_{1d} & 0_{1d} & J_{1m} & 0_{1,2d} \\ 0_{11} & 0_{1m} & 0_{1d} & 0_{1d} & 0_{1m} & J_{1,2d} \end{bmatrix}, \tag{2.4}$$

and $J_{mn}, 0_{mn}$ denote $(m\times n)$ matrices with elements 1 and 0, respectively. Now, it is easy to check that,

$$X'_A \boldsymbol{F} = D_1 \boldsymbol{\alpha}_0, \qquad X'_A \boldsymbol{Z} = D_1 \boldsymbol{\beta}_0, \tag{2.5}$$

$$X'_A M = A D_1^{-1} X'_A, \tag{2.6}$$

where A is a 6×6 symmetric matrix given below:

$$A = ((a_{ij})), \tag{2.7}$$

$$a_{11} = \varepsilon_1, \qquad a_{12} = m\varepsilon_2, \qquad a_{13} = \tfrac{1}{2}m(m-1)\varepsilon_3,$$

$$a_{14} = \tfrac{1}{2}m(m-1)\varepsilon_4, \qquad a_{15} = m\varepsilon_5, \qquad a_{16} = m(m-1)\varepsilon_6,$$

$$a_{22} = m[\varepsilon_7 + (m-1)\varepsilon_4], \qquad a_{23} = \tfrac{1}{2}m(m-1)[2\varepsilon_3 + (m-2)\varepsilon_{15}],$$

$$a_{24} = \tfrac{1}{2}m(m-1)[2\varepsilon_8 + (m-2)\varepsilon_9], \qquad a_{25} = m[\varepsilon_5 + (m-1)\varepsilon_6],$$

$$a_{26} = m(m-1)[\varepsilon_{10} + \varepsilon_6 + (m-2)\varepsilon_{11}],$$

$$a_{33} = \tfrac{1}{2}m(m-1)\left[\varepsilon_{12} + (2m-4)\varepsilon_{13} + \tfrac{1}{2}(m^2 - 5m + 6)\varepsilon_{14}\right],$$

$$a_{34} = \tfrac{1}{2}m(m-1)\left[\varepsilon_3 + (2m-4)\varepsilon_{15} + \tfrac{1}{2}(m^2 - 5m + 6)\varepsilon_{16}\right],$$

$$a_{35} = \tfrac{1}{2}m(m-1)[2\varepsilon_{17} + (m-2)\varepsilon_{18}],$$

$$a_{36} = \tfrac{1}{2}m(m-1)\left[(m-2)\{2\varepsilon_{18} + 2\varepsilon_{19} + (m-3)\varepsilon_{20}\} + 2\varepsilon_{17}\right],$$

$$a_{44} = \tfrac{1}{2}m(m-1)\left[\varepsilon_{21} + (2m-4)\varepsilon_{22} + \tfrac{1}{2}(m^2 - 5m + 6)\varepsilon_{23}\right],$$

$$a_{45} = \tfrac{1}{2}m(m-1)[2\varepsilon_6 + (m-2)\varepsilon_{11}],$$

$$a_{46} = \tfrac{1}{2}m(m-1)\left[2\varepsilon_{10} + (m-2)\{2\varepsilon_{11} + 2\varepsilon_{24} + (m-3)\varepsilon_{25}\}\right],$$

$$a_{55} = m[\varepsilon_{26} + (m-1)\varepsilon_3], \qquad a_{56} = m(m-1)[\varepsilon_3 + \varepsilon_{27} + (m-2)\varepsilon_{15}],$$

$$a_{66} = m(m-1)\left[\varepsilon_{28} + \varepsilon_3 + (m-2)\{\varepsilon_{30} + \varepsilon_{29}(m-3)\varepsilon_{16} + 2\varepsilon_{15}\}\right]. \tag{2.8}$$

Substituting (2.5)–(2.8) in (2.3) gives:

$$A\boldsymbol{\alpha}_0 = D_1 \boldsymbol{\beta}_0. \tag{2.9}$$

We now obtain another set of six equations by considering other linear combinations of equation (1.4). In order to express this in a compact form, we first define a vector $\boldsymbol{\delta}'_l$ corresponding to the various sets of effects contained in $\boldsymbol{Z}$ and $\boldsymbol{F}$. Thus, let

$\boldsymbol{\beta}_l'\{Z_i^2\}$ be a $(1\times m)$ vector whose elements correspond, respectively, to the elements of the set $\{Z_i^2\}$, and which contains zero everywhere except corresponding to the element Z_l^2. Thus, this vector equals $(0,\ldots,0,1,0,\ldots,0)$ where the 1 occurs in the lth position. Clearly, these vectors are defined for $l=1,\ldots,m$. Next, the vector, $\boldsymbol{\delta}_l'\{Z_iZ_j\}$, is a vector of size $(1\times d)$, and contains zeros and 1's, where the 1's correspond to pairs (i, j), such that either $i=l$ or $j=l$. Note that in the set $\{Z_iZ_j\}$, we have $i<j$, $i, j=1,2,\ldots,m$. Thus, for example, when $m=5$ the vector $\boldsymbol{\delta}_2'\{Z_iZ_j\}$ equals $(1,0,0,0,1,1,1,0,0,0)$. Finally, corresponding to the set $\{Z_iZ_j^2\}$, which contains $2d$ elements, we define two vectors $\boldsymbol{\delta}_l'\{Z_iZ_j^2\}$, and $\boldsymbol{\delta}_l'\{Z_i^2Z_j\}$, as follows. Each of these vectors is of size $(1\times 2d)$. The elements in these two vectors are 0's and 1's. The vectors $\boldsymbol{\delta}_l'\{Z_iZ_j^2\}$ and $\boldsymbol{\delta}_l'\{Z_i^2Z_j\}$ have the element 1 in places where $j=l$. Thus, for $m=5$ the vectors $\boldsymbol{\delta}_2'(Z_iZ_j^2)$ and $\boldsymbol{\delta}_2'(Z_i^2Z_j)$ are, respectively, $(1, 0, 0, 0, 0, 0, 0, 0, 0, 1, 0, 1, 0, 1, 0, 0, 0, 0, 0, 0)$ and $(0, 1, 0, 0, 0, 0, 0, 0, 1,0,1,0,1,0,0,0,0,0,0,0)$. Now, let

$$\underset{(6\times\nu_m)}{X_{lB}'} =$$

$$\begin{bmatrix} 0_{11} & (m-2)\boldsymbol{\delta}_l'\{Z_i^2\} & 0_{1d} & 0_{1d} & 0_{1m} & 0_{1,2d} \\ 0_{11} & 0_{1m} & (m-1)\boldsymbol{\delta}_l'\{Z_iZ_j\} & 0_{1d} & 0_{1m} & 0_{1,2d} \\ 0_{11} & 0_{1m} & 0_{1d} & (m-1)\boldsymbol{\delta}_l'\{Z_i^2Z_j^2\} & 0_{1m} & 0_{1,2d} \\ 0_{11} & 0_{1m} & 0_{1d} & 0_{1d} & (m-1)\boldsymbol{\delta}_l'\{Z_i\} & 0_{1,2d} \\ 0_{11} & 0_{1m} & 0_{1d} & 0_{1d} & 0_{1m} & \frac{(m-1)^2}{m}\boldsymbol{\delta}_l'\{Z_iZ_j^2\} + \frac{(m-1)}{m}\boldsymbol{\delta}_l'\{Z_i^2Z_j\} \\ 0_{11} & 0_{1m} & 0_{1d} & 0_{1d} & 0_{1m} & \frac{(m-1)^2}{m}\boldsymbol{\delta}_l'\{Z_i^2Z_j\} + \frac{(m-1)}{m}\boldsymbol{\delta}_l'\{Z_iZ_j^2\} \end{bmatrix}. \tag{2.10}$$

Similarly, let X_{kB}' be the $(6\times\nu_m)$ matrix obtained from X_{lB}' by replacing l with k. Then the above set of 6 equations can be expressed as:

$$X_{lB}'\boldsymbol{MF}=X_{lB}'\boldsymbol{Z}. \tag{2.11}$$

It can be easily checked that,

$$X_{lB}'\boldsymbol{F}=D_2D_3\boldsymbol{\alpha}_l,\qquad X_{lB}'\boldsymbol{Z}=D_2D_3\boldsymbol{\beta}_l,\qquad X_{kB}'\boldsymbol{F}=D_2D_3\boldsymbol{\alpha}_k,\qquad X_{kB}'\boldsymbol{Z}=D_2D_3\boldsymbol{\beta}_k, \tag{2.12}$$

$$X_{lB}'\boldsymbol{M}=\boldsymbol{B}D_2^{-1}D_3^{-1}X_{lB}'+\boldsymbol{G}D_1^{-1}X_A',\qquad X_{kB}'\boldsymbol{M}=\boldsymbol{B}D_2^{-1}D_3^{-1}X_{kB}'+\boldsymbol{G}D_1^{-1}X_A', \tag{2.13}$$

where B is a (6×6) symmetric matrix and G is a (6×6) matrix given below:

$$B=((b_{ij})), \tag{2.14}$$

$$b_{11}=(m-2)(\varepsilon_7-\varepsilon_4), \quad b_{12}=(m-1)(m-2)(\varepsilon_3-\varepsilon_{15}),$$
$$b_{13}=(m-1)(m-2)(\varepsilon_8-\varepsilon_9), \quad b_{14}=(m-2)(\varepsilon_5-\varepsilon_6),$$
$$b_{15}=(m-1)(m-2)(\varepsilon_{10}-\varepsilon_{11}), \quad b_{16}=(m-1)(m-2)(\varepsilon_6-\varepsilon_{11}),$$
$$b_{22}=(m-1)^2[\varepsilon_{12}+(m-4)\varepsilon_{13}-(m-3)\varepsilon_{14}],$$
$$b_{23}=(m-1)^2[\varepsilon_3+(m-4)\varepsilon_{15}-(m-3)\varepsilon_{16}], \quad b_{24}=(m-1)(m-2)(\varepsilon_{17}-\varepsilon_{18}),$$
$$b_{25}=(m-1)^2[\varepsilon_{17}+(m-3)\varepsilon_{18}-\varepsilon_{19}-(m-3)\varepsilon_{20}],$$
$$b_{26}=(m-1)^2[\varepsilon_{17}-\varepsilon_{18}+(m-3)(\varepsilon_{19}-\varepsilon_{20})],$$
$$b_{33}=(m-1)^2[\varepsilon_{21}+(m-4)\varepsilon_{22}-(m-3)\varepsilon_{23}], \quad b_{34}=(m-1)(m-2)(\varepsilon_6-\varepsilon_{11}),$$
$$b_{35}=(m-1)^2[(m-3)\varepsilon_{24}-(m-3)\varepsilon_{25}+\varepsilon_{10}-\varepsilon_{11}],$$
$$b_{36}=(m-2)^2[\varepsilon_{10}+(m-3)\varepsilon_{11}-\varepsilon_{24}-(m-3)\varepsilon_{25}],$$
$$b_{44}=(m-2)(\varepsilon_{26}-\varepsilon_3), \quad b_{45}=(m-1)(m-2)(\varepsilon_3-\varepsilon_{15}),$$
$$b_{46}=(m-1)(m-2)(\varepsilon_{27}-\varepsilon_{15}),$$
$$b_{55}=\frac{(m-1)^2}{m}\big[\varepsilon_3-2\varepsilon_{15}-m(m-3)\varepsilon_{16}+(m-1)\varepsilon_{28}-(m-1)\varepsilon_{29}$$
$$+(m^2-3m+1)\varepsilon_{30}\big],$$
$$b_{56}=\frac{(m-1)^2}{m}\big[(m-1)\varepsilon_3+(m^2-4m+2)\varepsilon_{15}-m(m-3)\varepsilon_{16}+\varepsilon_{28}-\varepsilon_{29}-\varepsilon_{30}\big],$$
$$b_{66}=\frac{(m-1)^2}{m}\big[\varepsilon_3-2\varepsilon_{15}-m(m-3)\varepsilon_{16}$$
$$+(m-1)\varepsilon_{28}+(m^2-3m+1)\varepsilon_{29}-(m-1)\varepsilon_{30}\big], \tag{2.15}$$

$$G=((g_{ij})), \tag{2.16}$$

$$g_{11}=(m-2)\varepsilon_2, \quad g_{12}=m(m-2)\varepsilon_4, \quad g_{13}=\tfrac{1}{2}m(m-1)(m-2)\varepsilon_{15},$$
$$g_{14}=\tfrac{1}{2}m(m-1)(m-2)\varepsilon_9, \quad g_{15}=m(m-2)\varepsilon_6, \quad g_{16}=m(m-1)(m-2)\varepsilon_{11},$$
$$g_{21}=(m-1)^2\varepsilon_3, \quad g_{22}=m(m-1)[\varepsilon_3+(m-2)\varepsilon_{15}],$$
$$g_{23}=\tfrac{1}{2}m(m-1)^2[2\varepsilon_{13}+(m-3)\varepsilon_{14}],$$
$$g_{24}=\tfrac{1}{2}m(m-1)^2[2\varepsilon_{15}+(m-3)\varepsilon_{16}], \quad g_{25}=m(m-1)[\varepsilon_{17}+(m-2)\varepsilon_{18}],$$
$$g_{26}=m(m-1)^2[\varepsilon_{18}+\varepsilon_{19}+(m-3)\varepsilon_{20}],$$
$$g_{31}=(m-1)^2\varepsilon_4, \quad g_{32}=m(m-1)[\varepsilon_8+(m-2)\varepsilon_9],$$

$$g_{33}=\tfrac{1}{2}m(m-1)^2[2\varepsilon_{15}+(m-3)\varepsilon_{16}], \quad g_{34}=\tfrac{1}{2}m(m-1)^2[2\varepsilon_{22}+(m-3)\varepsilon_{23}],$$
$$g_{35}=m(m-1)[\varepsilon_6+(m-2)\varepsilon_{11}], \quad g_{36}=m(m-1)^2[\varepsilon_{11}+\varepsilon_{24}+(m-3)\varepsilon_{25}],$$
$$g_{41}=(m-2)\varepsilon_5, \quad g_{42}=m(m-2)\varepsilon_6, \quad g_{43}=\tfrac{1}{2}m(m-1)(m-2)\varepsilon_{18},$$
$$g_{44}=\tfrac{1}{2}m(m-1)(m-2)\varepsilon_{11}, \quad g_{45}=m(m-2)\varepsilon_3, \quad g_{46}=m(m-1)(m-2)\varepsilon_{15},$$
$$g_{51}=(m-1)^2\varepsilon_6, \quad g_{52}=(m-1)[(m-1)\varepsilon_6+\varepsilon_{10}+m(m-2)\varepsilon_{11}],$$
$$g_{53}=\tfrac{1}{2}(m-1)^2[2\varepsilon_{18}+2(m-1)\varepsilon_{19}+m(m-3)\varepsilon_{20}],$$
$$g_{54}=\tfrac{1}{2}(m-1)^2[2(m-1)\varepsilon_{11}+2\varepsilon_{24}+m(m-3)\varepsilon_{25}],$$
$$g_{55}=(m-1)[\varepsilon_3+m(m-2)\varepsilon_{15}+(m-1)\varepsilon_{27}],$$
$$g_{56}=(m-1)^2[m\varepsilon_{15}+m(m-3)\varepsilon_{16}+(m-1)\varepsilon_{29}+\varepsilon_{30}], \quad g_{61}=(m-1)^2\varepsilon_6,$$
$$g_{62}=(m-1)[\varepsilon_6+(m-1)\varepsilon_{10}+m(m-2)\varepsilon_{11}],$$
$$g_{63}=\tfrac{1}{2}(m-1)^2[2(m-1)\varepsilon_{18}+2\varepsilon_{19}+m(m-3)\varepsilon_{20}],$$
$$g_{64}=\tfrac{1}{2}(m-1)^2[2\varepsilon_{11}+2(m-1)\varepsilon_{24}+m(m-3)\varepsilon_{25}],$$
$$g_{65}=(m-1)[(m-1)\varepsilon_3+m(m-2)\varepsilon_{15}+\varepsilon_{27}],$$
$$g_{66}=(m-1)^2[m\varepsilon_{15}+m(m-3)\varepsilon_{16}+\varepsilon_{29}+(m-1)\varepsilon]. \tag{2.17}$$

Substituting (2.12)–(2.17) in (2.11) gives

$$B\boldsymbol{\alpha}_l+G\boldsymbol{\alpha}_0=D_3(D_2\boldsymbol{\beta}_l), \qquad B\boldsymbol{\alpha}_k+G\boldsymbol{\alpha}_0=D_3(D_2\boldsymbol{\beta}_k). \tag{2.18}$$

Finally, we obtain a set of 4 equations by considering the 4 equations in (1.4) corresponding to the elements Z_kZ_l, $Z_k^2Z_l^2$, $Z_kZ_l^2$ and $Z_k^2Z_l$. Such sets of 4 equations are defined for all permissible values of k and l. We shall express these equations in a matrix notation. Let

$$\underset{(4\times\nu_m)}{X'_{klC}}=\begin{bmatrix} 0_{11} & 0_{1m} & \boldsymbol{\delta}'_{kl}\{Z_iZ_j\} & 0_{1d} & 0_{1m} & 0_{1,2d} \\ 0_{11} & 0_{1m} & 0_{1d} & \boldsymbol{\delta}'_{kl}\{Z_i^2Z_j^2\} & 0_{1m} & 0_{1,2d} \\ 0_{11} & 0_{1m} & 0_{1d} & 0_{1d} & 0_{1m} & \boldsymbol{\delta}'_{kl}\{Z_iZ_j^2\} \\ 0_{11} & 0_{1m} & 0_{1d} & 0_{1d} & 0_{1m} & \boldsymbol{\delta}'_{kl}\{Z_i^2Z_j\} \end{bmatrix}, \tag{2.19}$$

where $\boldsymbol{\delta}'_{kl}\{Z_iZ_j\}$ and $\boldsymbol{\delta}'_{kl}\{Z_i^2Z_j^2\}$ are $(1\times d)$ vectors with zeros everywhere except at the place where $i=k$, $j=l$, and $\boldsymbol{\delta}'_{kl}\{Z_iZ_j^2\}$ and $\boldsymbol{\delta}'_{kl}(Z_i^2Z_j)$ are $(1\times 2d)$ vectors with zeros everywhere except at the place where $i=k, j=l$. For example, when $m=5$ the vectors $\boldsymbol{\delta}'_{24}\{Z_iZ_j\}=\boldsymbol{\delta}'_{24}(Z_i^2Z_j^2)$, $\boldsymbol{\delta}'_{24}\{Z_iZ_j^2\}$ and $\boldsymbol{\delta}'_{24}\{Z_i^2Z_j\}$ are (0,0,0,0,0,1,0,0,0,0), (0, 0, 0, 0, 0, 0, 0, 0, 0, 1, 0, 0, 0, 0, 0, 0, 0, 0, 0, 0) and (0, 0, 0, 0, 0, 0, 0, 0, 0, 0,0,1,0,0,0,0,0,0,0,0), respectively. The above set of 4 equations can be expressed as

$$X'_{klC}\boldsymbol{MF}=X'_{klC}\boldsymbol{Z}. \tag{2.20}$$

It can be easily checked that

$$X'_{klC}\boldsymbol{F}=\boldsymbol{\alpha}_{kl}, \qquad X'_{klC}\boldsymbol{Z}=\boldsymbol{\beta}_{kl}, \tag{2.21}$$

$$X'_{klC}M=CX'_{klC}+LD_3^{-1}D_2^{-1}X'_{lB}+KD_3^{-1}D_2^{-1}X'_{kB}+HD_1^{-1}X'_A, \tag{2.22}$$

where C is symmetric (4×4), and H, L and K are (4×6), and as given below:

$$C=((c_{ij})), \tag{2.23}$$

$$c_{11}=\varepsilon_{12}-2\varepsilon_{13}+\varepsilon_{14}, \qquad c_{12}=\varepsilon_3-2\varepsilon_{15}+\varepsilon_{16},$$

$$c_{13}=c_{14}=\varepsilon_{17}-\varepsilon_{18}-\varepsilon_{19}+\varepsilon_{20},$$

$$c_{22}=\varepsilon_{21}-2\varepsilon_{22}+\varepsilon_{23}, \qquad c_{23}=c_{24}=\varepsilon_{10}-\varepsilon_{11}-\varepsilon_{24}+\varepsilon_{25},$$

$$c_{33}=c_{44}=\varepsilon_{16}+\varepsilon_{28}-\varepsilon_{29}-\varepsilon_{30}, \qquad c_{34}=\varepsilon_3-2\varepsilon_{15}+\varepsilon_{16}, \tag{2.24}$$

$$L=(m-1)(L_1-L_2), \tag{2.25}$$

where

$$\underset{(4\times 6)}{L_1}=\begin{bmatrix} \frac{\varepsilon_3}{(m-1)} & \varepsilon_{13} & \varepsilon_{15} & \frac{\varepsilon_{17}}{(m-1)} & \varepsilon_{18} & \varepsilon_{19} \\ \frac{\varepsilon_8}{(m-1)} & \varepsilon_{15} & \varepsilon_{22} & \frac{\varepsilon_6}{(m-1)} & \varepsilon_{24} & \varepsilon_{11} \\ \frac{\varepsilon_{10}}{(m-1)} & \varepsilon_{18} & \varepsilon_{24} & \frac{\varepsilon_3}{(m-1)} & \varepsilon_{30} & \varepsilon_{15} \\ \frac{\varepsilon_6}{(m-1)} & \varepsilon_{19} & \varepsilon_{11} & \frac{\varepsilon_{27}}{(m-1)} & \varepsilon_{15} & \varepsilon_{29} \end{bmatrix},$$

$$\underset{(4\times 6)}{L_2}=\begin{bmatrix} \frac{\varepsilon_{15}}{(m-1)} & \varepsilon_{14} & \varepsilon_{16} & \frac{\varepsilon_{18}}{(m-1)} & \varepsilon_{20} & \varepsilon_{20} \\ \frac{\varepsilon_9}{(m-1)} & \varepsilon_{16} & \varepsilon_{23} & \frac{\varepsilon_{11}}{(m-1)} & \varepsilon_{25} & \varepsilon_{25} \\ \frac{\varepsilon_{11}}{(m-1)} & \varepsilon_{20} & \varepsilon_{25} & \frac{\varepsilon_{15}}{(m-1)} & \varepsilon_{16} & \varepsilon_{16} \\ \frac{\varepsilon_{11}}{(m-1)} & \varepsilon_{20} & \varepsilon_{25} & \frac{\varepsilon_{15}}{(m-1)} & \varepsilon_{16} & \varepsilon_{16} \end{bmatrix},$$

$$K=(m-1)(K_1-K_2), \tag{2.26}$$

where

$$\underset{(4\times 6)}{K_1} = \begin{bmatrix} \frac{\varepsilon_3}{(m-1)} & \varepsilon_{13} & \varepsilon_{15} & \frac{\varepsilon_{17}}{(m-1)} & \varepsilon_{18} & \varepsilon_{19} \\ \frac{\varepsilon_8}{(m-1)} & \varepsilon_{15} & \varepsilon_{22} & \frac{\varepsilon_6}{(m-1)} & \varepsilon_{24} & \varepsilon_{11} \\ \frac{\varepsilon_6}{(m-1)} & \varepsilon_{19} & \varepsilon_{11} & \frac{\varepsilon_{27}}{(m-1)} & \varepsilon_{15} & \varepsilon_{29} \\ \frac{\varepsilon_{10}}{(m-1)} & \varepsilon_{18} & \varepsilon_{24} & \frac{\varepsilon_3}{(m-1)} & \varepsilon_{30} & \varepsilon_{15} \end{bmatrix},$$

$$\underset{(4\times 6)}{K_2} = \begin{bmatrix} \frac{\varepsilon_{15}}{(m-1)} & \varepsilon_{14} & \varepsilon_{16} & \frac{\varepsilon_{18}}{(m-1)} & \varepsilon_{20} & \varepsilon_{20} \\ \frac{\varepsilon_9}{(m-1)} & \varepsilon_{16} & \varepsilon_{23} & \frac{\varepsilon_{11}}{(m-1)} & \varepsilon_{25} & \varepsilon_{25} \\ \frac{\varepsilon_{11}}{(m-1)} & \varepsilon_{20} & \varepsilon_{25} & \frac{\varepsilon_{15}}{(m-1)} & \varepsilon_{16} & \varepsilon_{16} \\ \frac{\varepsilon_{11}}{(m-1)} & \varepsilon_{20} & \varepsilon_{25} & \frac{\varepsilon_{15}}{(m-1)} & \varepsilon_{16} & \varepsilon_{16} \end{bmatrix},$$

$$\underset{(4\times 6)}{H} = \begin{bmatrix} \varepsilon_3 & m\varepsilon_{15} & \frac{1}{2}m(m-1)\varepsilon_{14} & \frac{1}{2}m(m-1)\varepsilon_{16} & m\varepsilon_{18} & m(m-1)\varepsilon_{20} \\ \varepsilon_4 & m\varepsilon_9 & \frac{1}{2}m(m-1)\varepsilon_{16} & \frac{1}{2}m(m-1)\varepsilon_{23} & m\varepsilon_{11} & m(m-1)\varepsilon_{25} \\ \varepsilon_6 & m\varepsilon_{11} & \frac{1}{2}m(m-1)\varepsilon_{20} & \frac{1}{2}m(m-1)\varepsilon_{25} & m\varepsilon_{15} & m(m-1)\varepsilon_{16} \\ \varepsilon_6 & m\varepsilon_{11} & \frac{1}{2}m(m-1)\varepsilon_{20} & \frac{1}{2}m(m-1)\varepsilon_{25} & m\varepsilon_{15} & m(m-1)\varepsilon_{16} \end{bmatrix}. \tag{2.27}$$

Substituting (2.21)–(2.27) in (2.20) gives

$$C\boldsymbol{\alpha}_{kl} + L\boldsymbol{\alpha}_l + K\boldsymbol{\alpha}_k + H\boldsymbol{\alpha}_0 = \boldsymbol{\beta}_{kl}. \tag{2.28}$$

As before, these equations are defined for all permissible values of k, l.

We now establish the relation between M and A, B, C.

Theorem 2.1. *The matrix M is non-singular if A, B and C are so.*

Proof. Suppose the matrices A, B and C are non-singular. Then, from the above discussion, it follows that the normal equations can be uniquely solved for $\hat{\boldsymbol{F}}$. Hence, M is non-singular.

We next establish a stronger result.

Theorem 2.2. *The matrix M is non-singular if and only if each of the matrices A, B and C are so.*

Proof. We proceed to show that if M is non-singular, then so are A, B and C. Let

$$\underset{(6\times 6)}{\phi_1} = X'_{lB}X_A = \begin{bmatrix} 0 & (m-2) & 0 & 0 & 0 & 0 \\ 0 & 0 & (m-1)^2 & 0 & 0 & 0 \\ 0 & 0 & 0 & (m-1)^2 & 0 & 0 \\ 0 & 0 & 0 & 0 & (m-2) & 0 \\ 0 & 0 & 0 & 0 & 0 & (m-1)^2 \\ 0 & 0 & 0 & 0 & 0 & (m-1)^2 \end{bmatrix}, \tag{2.29}$$

$$\underset{(6\times 6)}{D_4} = \frac{\sqrt{m/2}}{(m-2)} \left[\begin{array}{cccc|c} \frac{1}{\sqrt{(m-2)}} & 0 & 0 & 0 & \\ 0 & \frac{1}{(m-1)} & 0 & 0 & \\ 0 & 0 & \frac{1}{(m-1)} & 0 & 0_{4,2} \\ 0 & 0 & 0 & \frac{1}{\sqrt{(m-2)}} & \\ \hline & & 0_{2,4} & & p_1 I + q_1 J \end{array}\right], \tag{2.30}$$

where,

$$p_1 = \frac{\sqrt{m}}{(m-1)\sqrt{(m-2)}} \quad \text{and} \quad q_1 = \frac{1}{2(m-1)} - \frac{\sqrt{m}}{2(m-1)\sqrt{(m-2)}},$$

$$\underset{(6\times 6)}{D_5} = -2D_4\phi_1 D_1^{-1/2}, \tag{2.31}$$

$$P'_1 = D_1^{-1/2}X'_A, \tag{2.32}$$

$$P'_2 = D_4(X'_{lB} + X'_{kB}) + D_5 P'_1. \tag{2.33}$$

Then by using (2.4), (2.10) and (2.29) we get:

$$X'_A X_A = D_1, \qquad X'_{kB}X_A = \phi_1. \tag{2.34}$$

Also, by direct computation, it can be shown that:

$$X'_{lB}X_{lB} = X'_{kB}X_{kB} = D_2 D_3^2, \tag{2.35}$$

$$X'_{lB}X_{kB}=\left[\begin{array}{cccc|c} 0 & 0 & 0 & 0 & \\ 0 & (m-1)^2 & 0 & 0 & \\ & & & & 0_{4,2} \\ 0 & 0 & (m-1)^2 & 0 & \\ 0 & 0 & 0 & 0 & \\ \hline & & 0_{2,4} & & a_1I+b_1J \end{array}\right], \tag{2.36}$$

where $a_1=[2(m-1)^3/m^2]-[(m-1)^2/m^2][m^2-2m+2]$, $b_1=[(m-1)^2/m^2][m^2-2m+2]$,

$$D_2D_3^2+X'_{lB}X_{kB}-2\phi_1D_1^{-1}\phi'_1=\tfrac{1}{2}D_4^{-1}D_4^{-1'}, \tag{2.37}$$

$$D_2^{-1}D_3^{-1}=\frac{2(m-2)^2}{m}D'_4D_4. \tag{2.38}$$

Then by using equations (2.31), (2.32) and (2.34), we get:

$$P'_1P_1=D_1^{-1/2}X'_AX_AD_1^{-1/2}=I_6, \tag{2.39}$$

$$\begin{aligned} P'_1P_2&=D_1^{-1/2}X'_A[(X_{lB}+X_{kB})D'_4+P_1D'_5] \\ &=D_1^{-1/2}X'_AX_{lB}D'_4+D_1^{-1/2}X'_AX_{kB}D'_4+D_1^{-1/2}X'_AX_AD_1^{-1/2}D'_5=0_{66}, \end{aligned} \tag{2.40}$$

$$\begin{aligned} P'_2P_2&=[D_4(X'_{lB}+X'_{kB})+D_5P'_1][(X_{lB}+X_{kB})D'_4+P_1D'_5] \\ &=D_4X'_{lB}X_{lB}D'_4+D_4X'_{kB}X_{lB}D'_4+D_4X'_{lB}X_{kB}D'_4 \\ &\quad+D_4X'_{kB}X_{kB}D'_4+D_4X'_{lB}X_AD_1^{-1/2}D'_5+D_4X'_{kB}X_AD_1^{-1/2}D'_5 \\ &=2D_4(D_2D_3^2+X'_{lB}X_{kB}-2\phi_1D_1^{-1}\phi'_1)D'_4 \\ &=I_6. \end{aligned} \tag{2.41}$$

Next consider equation (2.32). Then, we can write:

$$P'_1M=D_1^{-1/2}X'_AM. \tag{2.42}$$

Using equation (2.6) gives:

$$P'_1M=A^*P'_1, \tag{2.43}$$

where $A^*=D_1^{-1/2}AD_1^{-1/2}$. We next show that:

$$P'_2M=B^*P'_2, \tag{2.44}$$

where $B^*=\{2(m-2)^2/m\}D_4BD'_4$. Using (2.6), (2.13) and (2.33), it follows that:

$$\begin{aligned} P'_2M&=D_4X'_{lB}M+D_4X'_{kB}M+D_5D_1^{-1/2}X'_AM \\ &=D_4BD_2^{-1}D_3^{-1}X'_{lB}+D_4BD_2^{-1}D_3^{-1}X'_{kB} \\ &\quad+2D_4GD_1^{-1}X'_A-2D_4\phi_1D_1^{-1}AD_1^{-1}X'_A. \end{aligned} \tag{2.45}$$

Also, by using equations (2.37) and (2.43), we obtain:

$$B^*P_2' = \frac{2(m-2)^2}{m} D_4 B D_4' [D_4(X_{lB}' + X_{kB}') + D_5 P_1']$$
$$= D_4 B D_2^{-1} D_3^{-1} X_{lB}' + D_4 B D_2^{-1} D_3^{-1} X_{kB}'$$
$$-2 D_4 B D_2^{-1} D_3^{-1} \phi_1 D_1^{-1} X_A'. \tag{2.46}$$

Thus, in order to show (2.44), we need to prove that the coefficients of X_A' on the right-hand side of (2.45) and (2.46) are equal. However, it is easily seen that this is equivalent to proving that

$$G = \phi_1 D_1^{-1} A - B D_2^{-1} D_3^{-1} \phi_1, \tag{2.47}$$

which can be easily checked by direct computation. Now, from (2.43) and (2.44) we get

$$P_1' M P_1 = A^*, \qquad P_2' M P_2 = B^*. \tag{2.48}$$

The equation (2.48) implies that

$$A = D_1^{1/2} P_1' M P_1 D_1^{1/2'}, \qquad B = \{m/2(m-2)^2\} D_4^{-1} P_2' M P_2 D_4^{-1'},$$

from which it follows that A and B are non-singular (and, indeed, positive definite) if M is so.

Next consider (2.28). From (2.28), we obtain:

$$C\boldsymbol{\alpha}_{kl} = \boldsymbol{\beta}_{kl} - (L\boldsymbol{\alpha}_l + K\boldsymbol{\alpha}_k + H\boldsymbol{\alpha}_0). \tag{2.49}$$

Notice that the matrices L, K and H are made up of pure constants. Also, the vectors $\boldsymbol{\alpha}_l$, $\boldsymbol{\alpha}_k$, $\boldsymbol{\alpha}_0$ and $\boldsymbol{\beta}_{kl}$ are each equal to a matrix of pure constants multiplied by the vector $\boldsymbol{Z}$. Thus, as soon as $\boldsymbol{Z}$ is given, the right-hand side of the last equation gets completely fixed. This is also true in our present situation where we are taking $\boldsymbol{Z}$ to be a vector of indeterminates. Now, if C is singular, obviously the last equation will imply that $\boldsymbol{\alpha}_{kl}$ equals $\psi\boldsymbol{Z}$, where ψ can have many distinct values. Clearly, this would imply that M would have many inverses, which is impossible if M is non-singular.

The matrices A, B and C are interesting and important because these matrices arise in a natural way in an effort to solve the equation (1.4). Notice also that their elements are simple linear functions of the ε's which can be easily computed.

Next, we use the above results to obtain an expression for the distinct elements of M^{-1}. Consider (2.9), (2.18) and (2.28). If A, B and C are non-singular, then the solution to these three equations are:

$$\boldsymbol{\alpha}_0 = A^{-1}(D_1\boldsymbol{\beta}_0), \tag{2.50}$$

$$\boldsymbol{\alpha}_l = -B^{-1}GA^{-1}(D_1\boldsymbol{\beta}_0) + B^{-1}D_3(D_2\boldsymbol{\beta}_l), \tag{2.51}$$

$$\boldsymbol{\alpha}_{kl} = C^{-1}\Big[\{(K+L)B^{-1}GA^{-1} - HA^{-1}\}D_1\boldsymbol{\beta}_0$$
$$-(LB^{-1}D_3)D_2\boldsymbol{\beta}_l - (KB^{-1}D_3)D_2\boldsymbol{\beta}_k + \boldsymbol{\beta}_{kl}\Big]. \tag{2.52}$$

It is clear from (1.5) that the elements of the vector $\hat{\boldsymbol{F}}$ can be expressed as a linear combination of the elements of $\boldsymbol{Z}$.

We now proceed to explicitly obtain the diagonal elements of M^{-1}. First consider equation (2.50). Notice that the element in the first row and first column of A^{-1} is the element corresponding to F_0 and Z_0 in M^{-1}. Let the element of M^{-1} corresponding to f in $\boldsymbol{F}$ and z in $\boldsymbol{Z}$ be denoted by $M^{-1}(f, z)$. Then,

$$M^{-1}(F_0, Z_0)=a^{11}, \tag{2.53}$$

where

$$A^{-1}=((a^{ij})), \qquad B^{-1}=((b^{ij})), \qquad C^{-1}=((c^{ij})). \tag{2.54}$$

Next we use equation (2.51) to obtain $M^{-1}(F_i, Z_i)$ and $M^{-1}(F_i^2, Z_i^2)$. Now, define:

$$R=((r_{ij}))=GA^{-1}. \tag{2.55}$$

Then, it is easy to check that:

$$M^{-1}(F_i^2, Z_i^2)=(m-2)b^{11}-\sum_{n=1}^{6} b^{1n}r_{n2}, \tag{2.56}$$

$$M^{-1}(F_i, Z_i)=(m-2)b^{44}-\sum_{n=1}^{6} b^{4n}r_{n5}. \tag{2.57}$$

Finally, we consider equation (2.52). We define,

$$S=((s_{ij}))=(K+L)B^{-1}R-HA^{-1}, \tag{2.58}$$

$$W=((w_{ij}))=LB^{-1}D_3, \tag{2.59}$$

$$U=((u_{ij}))=KB^{-1}D_3. \tag{2.60}$$

Then, (2.52) leads to

$$M^{-1}(F_iF_j, Z_iZ_j)=c^{11}-\sum_{n=1}^{4} c^{1n}(w_{n2}+u_{n2}-s_{n3}), \tag{2.61}$$

$$M^{-1}(F_i^2F_j^2, Z_i^2Z_j^2)=c^{22}-\sum_{n=1}^{4} c^{2n}(w_{n3}+u_{n3}-s_{n4}), \tag{2.62}$$

$$M^{-1}(F_i^2F_j, Z_i^2Z_j)=c^{44}-\sum_{n=1}^{4} c^{4n}(w_{n6}+u_{n5}-s_{n6}). \tag{2.63}$$

Thus, from the above

$$\begin{aligned}\text{Trace}\, M^{-1}=M^{-1}(F_0, Z_0)&+m\{M^{-1}(F_i, Z_i)+M^{-1}(F_i^2, Z_i^2)\}\\ &+\tfrac{1}{2}m(m-1)\{M^{-1}(F_iF_j, Z_iZ_j)+M^{-1}(F_i^2F_j^2, Z_i^2Z_j^2)\}\\ &+m(m-1)\{M^{-1}(F_i^2F_j, Z_i^2Z_j)\}.\end{aligned} \tag{2.64}$$

Finally, proceeding in an analogous manner, one may find the remaining 43 possibly distinct elements of M^{-1}. Because of their potential usefulness in future work, these are being presented below:

$$M^{-1}=\left(F_0, Z_i^2\right)=a^{12}, \quad M^{-1}\left(F_0, Z_iZ_j\right)=a^{13}, \quad M^{-1}\left(F_0, Z_i^2Z_j^2\right)=a^{14},$$

$$M^{-1}(F_0, Z_i)=a^{15}, \quad M^{-1}\left(F_0, Z_i^2Z_j\right)=a^{16}, \quad M^{-1}\left(F_i^2, Z_j^2\right)=-\sum_{n=1}^{6} b^{1n}r_{n2},$$

$$M^{-1}\left(F_i^2, Z_iZ_j\right)=(m-1)b^{12}-\sum_{n=1}^{6} b^{1n}r_{n3}, \quad M^{-1}\left(F_i^2, Z_jZ_k\right)=-\sum_{n=1}^{6} b^{1n}r_{n3},$$

$$M^{-1}\left(F_i^2, Z_i^2Z_j^2\right)=(m-1)b^{13}-\sum_{n=1}^{6} b^{1n}r_{n4}, \quad M^{-1}\left(F_i^2, Z_j^2Z_k^2\right)=-\sum_{n=1}^{6} b^{1n}r_{n4},$$

$$M^{-1}\left(F_i^2, Z_i\right)=(m-2)b^{14}-\sum_{n=1}^{6} b^{1n}r_{n5}, \quad M^{-1}\left(F_i^2, Z_j^2\right)=-\sum_{n=1}^{6} b^{1n}r_{n5},$$

$$M^{-1}\left(F^2, Z_iZ_j^2\right)=\frac{(m-1)}{m}b^{15}+\frac{(m-1)^2}{m}b^{16}-\sum_{n=1}^{6} b^{1n}r_{n6},$$

$$M^{-1}\left(F_i^2, Z_i^2Z_j\right)=\frac{(m-1)^2}{m}b^{15}+\frac{(m-1)}{m}b^{16}-\sum_{n=1}^{6} b^{1n}r_{n6},$$

$$M^{-1}\left(F_i^2, Z_j^2Z_k\right)=-\sum_{n=1}^{6} b^{1n}r_{n6}, \quad M^{-1}\left(F_iF_j, Z_iZ_k\right)=\sum_{n=1}^{4} c^{1n}(s_{n3}-u_{n2}),$$

$$M^{-1}\left(F_iF_j, Z_kZ_l\right)=\sum_{n=1}^{4} c^{1n}s_{n3}, \quad M^{-1}\left(F_iF_j, Z_i^2Z_j^2\right)=c^{12}+\sum_{n=1}^{4} c^{1n}(s_{n4}-w_{n3}-u_{n3}),$$

$$M^{-1}\left(F_iF_j, Z_i^2Z_k^2\right)=\sum_{n=1}^{4} c^{1n}(s_{n4}-u_{n3}), \quad M^{-1}\left(F_iF_j, Z_k^2Z_l^2\right)=\sum_{n=1}^{4} c^{1n}s_{n4},$$

$$M^{-1}\left(F_iF_j, Z_i\right)=\sum_{n=1}^{4} c^{1n}(s_{n5}-u_{n4}), \quad M^{-1}\left(F_iF_j, Z_k\right)=\sum_{n=1}^{4} c^{1n}s_{n5},$$

$$M^{-1}\left(F_iF_j, Z_iZ_j^2\right)=c^{13}+\sum_{n=1}^{4} c^{1n}(s_{n6}-w_{n5}-u_{n6}),$$

$$M^{-1}\left(F_iF_j, Z_jZ_k^2\right)=\sum_{n=1}^{4} c^{1n}(s_{n6}-w_{n6}), \quad M^{-1}\left(F_iF_j, Z_iZ_k^2\right)=\sum_{n=1}^{4} c^{1n}(s_{n6}-u_{n5}),$$

$$M^{-1}\left(F_iF_j, Z_k^2Z_l\right)=\sum_{n=1}^{4} c^{1n}s_{n6}, \quad M^{-1}\left(F_i^2F_j^2, Z_i^2Z_k^2\right)=\sum_{n=1}^{4} c^{2n}(s_{n4}-u_{n3}),$$

$$M^{-1}\left(F_i^2F_j^2, Z_k^2Z_l^2\right)=\sum_{n=1}^{4} c^{2n}s_{n4}, \quad M^{-1}\left(F_i^2F_j^2, Z_i\right)=\sum_{n=1}^{4} c^{2n}(s_{n5}-u_{n4}),$$

$$M^{-1}(F_i^2F_j^2, Z_k) = \sum_{n=1}^{4} c^{2n} s_{n5}, \qquad M^{-1}(F_i^2F_j^2, Z_i^2Z_j) = c^{24} + \sum_{n=1}^{4} c^{2n}(s_{n6} - w_{n6} - u_{n5}),$$

$$M^{-1}(F_i^2F_j^2, Z_i^2Z_k) = \sum_{n=1}^{4} c^{2n}(s_{n6} - u_{n5}), \qquad M^{-1}(F_i^2F_j^2, Z_jZ_k^2) = \sum_{n=1}^{4} c^{2n}(s_{n6} - w_{n6}),$$

$$M^{-1}(F_i^2F_j^2, Z_k^2Z_l) = \sum_{n=1}^{4} c^{2n} s_{n6}, \qquad M^{-1}(F_i, Z_i) = -\sum_{n=1}^{6} b^{4n} r_{n5},$$

$$M^{-1}(F_i, Z_i^2Z_j) = \frac{(m-1)^2}{m} b^{45} + \frac{(m-1)}{m} b^{46} - \sum_{n=1}^{6} b^{4n} r_{n6},$$

$$M^{-1}(F_i, Z_iZ_j^2) = \frac{(m-1)}{m} b^{45} + \frac{(m-1)^2}{m} b^{46} - \sum_{n=1}^{6} b^{4n} r_{n6},$$

$$M^{-1}(F_i, Z_j^2Z_k) = \sum_{n=1}^{6} b^{4n} r_{n6}, \qquad M^{-1}(F_i^2F_j, Z_i^2Z_k) = \sum_{n=1}^{4} c^{4n}(s_{n6} - u_{n5}),$$

$$M^{-1}(F_i^2F_j, Z_iZ_j^2) = c^{43} + \sum_{n=1}^{4} c^{4n}(s_{n6} - w_{n5} - u_{n6}),$$

$$M^{-1}(F_i^2F_j, Z_iZ_k^2) = \sum_{n=1}^{4} c^{4n}(s_{n6} - u_{n6}), \qquad M^{-1}(F_i^2F_j, Z_jZ_k^2) = \sum_{n=1}^{4} c^{4n}(s_{n6} - w_{n6}),$$

$$M^{-1}(F_i^2F_j, Z_k^2Z_l) = \sum_{n=1}^{4} c^{4n} s_{n6}, \tag{2.65}$$

where i, j, k, l are distinct numbers from the set $\{1, 2, \ldots, m\}$. Notice that the problem of inverting the matrix $M(\nu_m \times \nu_m)$ is reduced to that of inverting the (6×6) matrices in (2.50) and (2.51) and the (4×4) matrix in (2.52).

Notice that the matrices A, B, C arise in an intuitive manner, and have statistical interpretations. Thus, in a sense, A corresponds to the sum of the effects in each of the six sets under consideration, these being $\{\mu\}$, $\{F_i\}$, $\{F_i^2\}$, $\{F_iF_j\}$, $\{F_i^2F_j^2\}$ and $\{F_iF_j^2\}$. Similarly, B corresponds to the six sets that emerge from the above, when we restrict attention to effects which do contain a particular fixed factor; notice that this means that $\{\mu\}$ should drop out, and $\{F_iF_j^2\}$ gives rise to 2 sets such that the fixed factor is linear in one of them, and quadratic in the other. Finally, C corresponds to four sets (each set containing only one element), such that these sets arise out of the above six sets by the fixation of two factors.

3. Optimal balanced designs

We now make some remarks on obtaining designs which are trace-optimal in the class of balanced resolution V 3^m factorial designs with N runs. The set of N treatment combinations of a balanced design of this type form a balanced array (with m rows, N columns) of strength 4 and 3 symbols. (The converse is also true provided the array is such that the information matrix M is non-singular.) A balanced array of the said type is characterized by 15 parameters ($\mu_1, \ldots, \mu_{15}$, say), which are non-negative integers. The 15 μ's correspond to the 15 ways in which we can make unordered 4-plets using three objects. The value of μ_i (for a given i) equals the number of distinct ordered 4-plets which can be made from the ith unordered 4-plet. Also, we have $N=\sum_{i=1}^{15} N_i \mu_i$, where the N_i are non-negative integers depending upon the array. Given a set of admissible parameters $\{\mu_i\}$, an array (say T) having these parameters does not necessarily exist, unless $m \leqslant 4$. (When $m \leqslant 4$, the combinatorial problem of the construction of such an array is trivial; the reverse is true for $m \geqslant 5$.)

The element ε_j of the information matrix M are each linear functions of the μ's. Thus, we have essentially obtained $\operatorname{tr} M^{-1}$ in terms of the μ's.

The optimal design problem has two major aspects: analytical and combinatorial. In the present context, the first one involves expressing $\operatorname{tr} M^{-1}$ in terms of the μ's in some "convenient" form, and the second one is concerned with constructing (or proving the nonexistence of) an array corresponding to a given set $\{\mu_i\}$. For a given N, $\operatorname{tr} M^{-1}$ is computed for "promising" candidates $\{\mu_i\}$. Suppose, somehow, we can verify or prove that a given $\{\mu_i^*\}$ minimizes $\operatorname{tr} M^{-1}$. If $m \leqslant 4$, our optimal-design problem is solved at this stage; otherwise, we study the combinatorial existence of the array. For $m \geqslant 5$, sometimes, it greatly helps to narrow down the choice of the competing sets $\{\mu_i\}$ by showing that arrays corresponding to certain (hopefully) large classes of sets $\{\mu_i\}$ are nonexistent. Generally, this work is tedious; so is the study of $\operatorname{tr} M^{-1}$ as a function of the 15 μ_i's. Some breakthroughs are needed in these two directions.

For $m=5$, trace-optimal balanced designs are obtained in Ariyaratna (1979), and designs that are trace-optimal in the subclass of balanced designs in which the underlying balanced array is of strength 5 are published in Kuwada (1979); for $m=4$, both of these present trace-optimal designs.

Acknowledgements

The designs for $m=4$ and the associated dispersion matrix M^{-1} were included in the original version of this paper, which was submitted to *J. Stat. Plann. Inf.* (JSPI) in March 1978. In October 1978, the editor concerned (Dr. Yamamoto) recommended publication subject to some relatively minor revision. However, because of individual personal difficulties, neither of the two authors could work on this paper until April 1980. However, since one of the authors had to meet the deadline for submitting a paper to this Volume, this paper was withdrawn from JSPI, and submitted here. The authors are thankful to Dr. Yamamoto and his referees for their comments on the earlier version of this paper.

References

[1] Ariyaratna, W.M. (1979). Optimal balanced resolution V designs of the 3^m series. Ph.D. Dissertation, Colorado State University, CO.

[2] Bose, R.C. and J.N. Srivastava (1964a). Analysis of irregular factorial fractions. *Sankhya Ser. A* **26**, 117–144.

[3] Bose, R.C. and J.N. Srivastava (1964b). Multidimensional partially balanced designs and their analysis with applications to partially balanced factorial fractions. *Sankhya Ser. A* **26**, 145–168.

[4] Hoke, A.T. (1974). Economical second order designs based on irregular fractions of the 3^n factorial. *Technometrics* **16**, 375–384.

[5] Hoke, A.T. (1975). The characteristic polynomial of the information matrix for second order models. *Ann. Statist.* **3**, 780–786.

[6] Kuwada, M. (1979). Optimal balanced fractional 3^m factorial designs of resolution V and balanced third-order designs, *Hiroshima Math. J.*, 347–450.

[7] Kuwada, M. (1980). Balanced arrays of strength 4 and balanced fractional 3^m factorial designs. *J. Stat. Planning Inf.* **4**(1).

[8] Kuwada, M. (1980). Characteristic polynomials of the information matrices of balanced fractional 3^m factorial designs of resolution V. *J. Stat. Planning Inf.* **4**(3).

[9] Srivastava, J.N. (1961). *Contributions to the Construction and Analysis of Designs*, University of North Carolina, Institute of Statistics, CA. Memio Series, No. 301.

[10] Srivastava, J.N. and W.M. Ariyaratna (1977). On the inverse of the information matrix of a balanced 3^m fractional factorial design of resolution V. *I.M.S. Bull.* **6** (Abstract) 237.

[11] Srivastava, J.N. and D.V. Chopra (1971c). Balanced optimal 2^m fractional factorial designs of resolution V, $m \leqslant 6$, *Technometrics* **13**, 357–369.

[12] Srivastava, J.N. and D.V. Chopra (1973). Balanced fractional factorial designs of resolution V for 3^m series. *Proc. (Vienna Session) Int. Stat. Inst.* 271–276.

G. Kallianpur, P.R. Krishnaiah, J.K. Ghosh, eds., *Statistics and Probability: Essays in Honor of C.R. Rao*

SOME REMARKS ON SPHERICAL FUNCTIONS ON REAL SEMISIMPLE LIE GROUPS

V.S. VARADARAJAN*
University of California, Los Angeles, CA, U.S.A.

1. Introduction

Let G be a connected semisimple Lie group (over $\mathbb{R}$) with finite center and $K \subset G$ a maximal compact subgroup. A function (or distribution) on G is said to be *spherical* if it is invariant under both left and right translations by elements of K. A C^∞ spherical function, f, is *elementary* if $f(1)=1$ and f is an eigenfunction for the algebra of all G-invariant differential operators on G/K. This algebra is commutative; in fact, if $r=\mathrm{rk}(G/K)$ is the rank of the symmetric Riemannian space G/K, it is isomorphic to $\mathbb{C}[\Delta_1, \ldots, \Delta_r]$, where the Δ_j are self-adjoint, and Δ_1 is the Laplace–Beltrami operator with respect to the G-invariant metric on G/K induced by the Cartan–Killing form. The central problem in the theory of spherical functions is then to obtain explicitly the spectral decomposition of this algebra of differential operators in $L^2(G//K)$ where we write $G//K$ for $K \setminus G/K$. This problem was completely solved by Harish Chandra [8] [9] who introduced what is now known as the Harish Chandra transform on the space $\mathcal{C}^2(G//K)$ of rapidly decreasing spherical functions in $L^2(G//K)$, and established explicit Plancherel and inversion formulae for it. Subsequently, the transforms of the elements of $C_c^\infty(G//K)$ as well as those of the $\mathcal{C}^p(G//K)$ $(1 \leq p < 2)$ which are the L^p-analogues of the $\mathcal{C}^2(G//K)$ were completely characterized, by Helgason [13] in the C_c^∞ case when G is complex or $r=1$, by Gangolli [5] in the C_c^∞ case for all G, and by Trombi and Varadarajan [16] in the $\mathcal{C}^p$ case; the $\mathcal{C}^p$ case for $G=SL(2,\mathbb{R})$ was worked out earlier by Ehrenpreis and Mautner [1–3]. Of course, as is well known, the Harish Chandra eigenfunction expansion in $L^2(G//K)$ leads, among other things, to the explicit decomposition of the natural representation of G in $L^2(G/K)$ as a direct integral of irreducible unitary representations.

The main problem in the theory of harmonic analysis is the study of $L^2(G)$, and $L^2(G/K)$ is only a small part of it. Nevertheless, the spherical theory has played an important role mainly because the *qualitative* features of the problem in the general case are to a large extent, similar to the ones arising in the spherical case (cf. Harish Chandra [12]). In view of this, it will seem worthwhile to review the spherical theory

*Supported in part by NSF MCS 78-03184. In addition, I would like to thank the authorities of the Tata Institute of Fundamental Research in Bombay, India, for their hospitality during the Fall of 1978, when some of the material in this note was presented in an informal seminar.

and try to understand its main results in as many ways as possible. This article is written with this viewpoint in mind. Its aim is to give a brief overview of the harmonic analysis of spherical functions. While mainly expository, it contains some simplifications and alternative arguments in a few places, and it is my hope that these remarks add a little more perspective and focus to the general theory.

Since I wish to keep this a brief note, I shall confine myself only to the main lines of arguments. However a detailed presentation of the entire theory including the variations sketched here will appear in a forthcoming book on spherical functions being written jointly with Gangolli (cf. Gangolli and Varadarajan [6]). I wish to acknowledge my gratitude to Gangolli for many discussions on this subject over the past couple of years.

2. Notation

We will mostly follow standard notation which will now be reviewed briefly. For general facts concerning symmetric spaces the reader may consult Helgason's book [14]; and for the semisimple theory, in addition to Helgason's book mentioned above, one may wish to refer to Varadarajan [17, 18].

Throughout, we work with a connected real semisimple Lie group, G, with finite center. $K\subset G$ will be a fixed maximal compact subgroup of G; θ, the corresponding Cartan involution of G. $\mathfrak{g}$ (resp. $\mathfrak{k}$) is the Lie algebra of G. K (resp. $\mathfrak{k}$) is the fixed point set of θ in G (resp. $\mathfrak{g}$), and $\mathfrak{g}=\mathfrak{k}\oplus\mathfrak{s}$ where $\mathfrak{s}$ is the subset of all $X\in\mathfrak{g}$ with $\theta X=-X$. $\mathfrak{s}$ is the orthogonal complement of $\mathfrak{k}$ in $\mathfrak{g}$ with respect to the Killing form. Let $\mathfrak{a}$ be a maximal abelian subspace of $\mathfrak{s}$, and Δ, the set of roots of $(\mathfrak{g},\mathfrak{a})$. We choose a positive system Δ^+ and write $\mathfrak{g}=\mathfrak{k}\oplus\mathfrak{a}\oplus\mathfrak{n}$, $G=KAN$, for the corresponding Iwasawa decompositions; here, as usual, $A=\exp\mathfrak{a}$ $(\mathfrak{a}=\log A)$, $N=\exp\mathfrak{n}$, $\mathfrak{n}=\Sigma_{\alpha\in\Delta^+}\mathfrak{g}_\alpha$, the $\mathfrak{g}_\alpha$ being the root spaces. If $x\in G$, we write $x=\kappa(x)a(x)n(x)$ where $\kappa(x)\in K$, $a(x)\in A$, $n(x)\in N$ and $H(x)=\log a(x)\in\mathfrak{a}$. We denote by ρ the element of $\mathfrak{a}^*$ given by $\rho=\frac{1}{2}\Sigma_{\alpha\in\Delta^+}\dim(\mathfrak{g}_\alpha)\alpha$ so that $\rho(H)=\frac{1}{2}\mathrm{tr}(\mathrm{Ad}\,H|_{\mathfrak{n}})$ $(H\in\mathfrak{a})$. For each $\alpha\in\Delta$, s_α is the Weyl reflexion corresponding to α and $\mathfrak{w}$ is the (Weyl group) group generated by the s_α. We put $r=\dim(\mathfrak{a})=\mathrm{rk}(G/K)$ and denote by $\{\alpha_1,\dots,\alpha_r\}$ the simple roots in Δ^+. $\mathfrak{a}'$ is the subset of $\mathfrak{a}$ where no root vanishes; $\mathfrak{a}^+$, the subset where all roots are >0; $A'=\exp\mathfrak{a}'$, $A^+=\exp\mathfrak{a}^+$.

Let $\mathfrak{g}_c$ be the complexification of $\mathfrak{g}$. We write $U(\mathfrak{g}_c)$ for the universal enveloping algebra of $\mathfrak{g}_c$. For any subalgebra $\mathfrak{l}$ of $\mathfrak{g}$, $U(\mathfrak{l}_c)$ is the subalgebra of $U(\mathfrak{g}_c)$ generated by 1 and $\mathfrak{l}_c$. The elements of $U(\mathfrak{g}_c)$ are interpreted as left invariant differential operators on G. As a general rule we use Harish Chandra's notation for the action of differential operators; if M is a smooth manifold and D a differential operator on M, then for any $f\in C^\infty(M)$ and $m\in M$, $(Df)(m)$ will often be written as $f(m;D)$. Thus, if $X_1,\dots,X_p\in\mathfrak{g}$ and $u=X_1X_2\dots X_p\in U(\mathfrak{g}_c)$, then for any $f\in C^\infty(G)$ and $x\in G$,

$$f(x;u)=(\partial^p/\partial t_1\dots\partial t_p)_0 f(x\exp t_1X_1\dots\exp t_pX_p)$$

(the suffix 0 means that the derivative is computed at $t_1=\cdots=t_p=0$).

The Cartan–Killing form $\langle\cdot,\cdot\rangle$ is positive definite on $\mathfrak{s}\times\mathfrak{s}$. For any $\lambda\in\mathfrak{F}$ we write H_λ for the image of λ in $\mathfrak{a}_c$ under the isomorphism $\mathfrak{F}\simeq\mathfrak{a}_c$ induced by $\langle\cdot,\cdot\rangle$; $H_\lambda\in\mathfrak{a}$ for $\lambda\in\mathfrak{F}_R$.

3. The elementary spherical functions and their estimates

Let $\mathfrak{F}=\mathfrak{a}_c^*$ be the complex vector space of linear functions on $\mathfrak{a}_c$, and $\mathfrak{F}_I$ (resp. $\mathfrak{F}_R$) the $\mathbb{R}$-linear subspace of those that take only pure imaginary (resp. real) values on $\mathfrak{a}$. The elementary spherical functions are parametrized by $\mathfrak{F}$: for $\lambda\in\mathfrak{F}$, $\varphi_\lambda(\cdot)=\varphi(\lambda:\cdot)$ is given by,

$$\varphi(\lambda:x)=\int_K e^{(\lambda-\rho)(H(xk))}\,dk,\quad (x\in G), \tag{1}$$

where dk is the normalized Haar measure on K. We have $\varphi_\lambda=\varphi_{\lambda'}$ if and only if $\lambda'=s\cdot\lambda$ for some $s\in\mathfrak{w}$; and $\varphi_\lambda(1)=1$ for all λ. If $\lambda\in\mathfrak{F}_I$, φ_λ is the matrix coefficient of an irreducible unitary representation of G defined by a K-invariant unit vector, and is hence positive definite. Although there are $\lambda\notin\mathfrak{F}_I$ for which φ_λ is positive definite, only the φ_λ for $\lambda\in\mathfrak{F}_I$ enter the eigenfunction expansion in $L^2(G//K)$.

Let $\mathfrak{D}$ be the centralizer of K in $U(\mathfrak{g}_c)$. Elements of $\mathfrak{D}$ act naturally on $C^\infty(G/K)$ and this action gives a homomorphism of $\mathfrak{D}$ onto the algebra of all G-invariant differential operators on $C^\infty(G/K)$, the kernel of this homomorphism being $\mathfrak{D}\cap(\mathfrak{k}U(\mathfrak{g}_c))$. The φ_λ are eigenfunctions for $\mathfrak{D}$; and there is a homomorphism (the Harish Chandra homomorphism) $\gamma:\mathfrak{D}\to U(\mathfrak{a}_c)$ such that γ maps $\mathfrak{D}$ onto the subalgebra $U(\mathfrak{a}_c)^{\mathfrak{w}}$ of $\mathfrak{w}$-invariant elements in $U(\mathfrak{a}_c)$, kernel of $\gamma=\mathfrak{D}\cap(\mathfrak{k}U(\mathfrak{g}_c))$, and,

$$q\varphi_\lambda=\gamma(q)(\lambda)\varphi_\lambda,\quad (q\in\mathfrak{D},\lambda\in\mathfrak{F}). \tag{2}$$

Note that $U(\mathfrak{a}_c)$ is canonically $S(\mathfrak{a}_c)$, the symmetric algebra of $\mathfrak{a}_c$, and so we have a natural isomorphism of $U(\mathfrak{a}_c)$ with the algebra of polynomial functions on $\mathfrak{F}$. One knows also how to compute $\gamma(q)$ for any q; we note first that $U(\mathfrak{g}_c)$ is the direct sum of $U(\mathfrak{a}_c)$ and $\mathfrak{k}U(\mathfrak{g}_c)+U(\mathfrak{g}_c)\mathfrak{n}$, so that it makes sense to speak of the projection $\gamma'(b\mapsto\gamma'(b))$ of $U(\mathfrak{g}_c)$ onto $U(\mathfrak{a}_c)\bmod(\mathfrak{k}U(\mathfrak{g}_c)+U(\mathfrak{g}_c)\mathfrak{n})$; we have $\deg(\gamma'(b))\leqslant\deg(b)$ and (with e^ρ denoting the character $\exp H\mapsto e^{\rho(H)}$ of A),

$$\gamma(q)=e^\rho\cdot\gamma'(q)e^{-\rho},\quad (q\in\mathfrak{D}). \tag{3}$$

The differential equations (2) are the fundamental ones of the theory. Harish Chandra's method of studying these is through polar coordinates. Since $G=K\mathrm{Cl}(A^+)K$, we consider the map τ of $K\times A\times K$ into G given by $(k_1,h_1k_2)\mapsto k_1hk_2$ and use it to transfer (2) to a system of differential equations on A. This is not possible everywhere since τ is not always submersive; but it is submersive on $K\times A'\times K$. So there is a unique homomorphism δ' of $\mathfrak{D}$ into the algebra of $\mathfrak{w}$-invariant differential operators on A' with analytic coefficients such that,

$$f(h;q)=f(h;\delta'(q)),\quad (q\in\mathfrak{D},h\in A',f\in C^\infty(G//K)). \tag{4}$$

It is natural to call $\delta'(q)$ *the radial component* of q on A'. It can be shown that,

$$\delta'(q)=\gamma'(q)+\sum_{1\leqslant i\leqslant p} g_i v_i, \tag{5}$$

where the $v_i \in U(\mathfrak{a}_c)$ are of degree $<\deg(q)$, and the g_i are in the ring $\mathfrak{R}$ (without unit element) generated by the functions $(e^{2\alpha}-1)^{-1}$ $(\alpha\in\Delta^+)$. In particular, for instance, if $q=\omega$ is the Casimir element of $\mathfrak{g}$,

$$\delta'(\omega)=\omega_{\mathfrak{a}}+2H_\rho+2\sum_{\alpha\in\Delta^+} n(\alpha)(e^{2\alpha}-1)^{-1}H_\alpha, \tag{6}$$

where $\omega_{\mathfrak{a}}=H_1^2+\cdots+H_r^2$ for any orthonormal basis (H_i), and $n(\alpha)=\dim(\mathfrak{g}_\alpha)$.

Let $\mathfrak{L}$ be the set of elements $m_1\alpha_1+\cdots+m_r\alpha_r$ where the m_j are integers $\geqslant 0$ and $m_1+\cdots+m_r>0$. The elements of the ring $\mathfrak{R}$ have expansions of the form $\Sigma_{q\in\mathfrak{L}} c_q e^{-q}$, $c_q\in\mathbb{C}$, on A^+ and so it is reasonable to expect that the elementary spherical functions φ_λ, which satisfy

$$\varphi_\lambda(h;\delta'(q))=\gamma(q)(\lambda)\varphi_\lambda(h),\quad (q\in\mathfrak{D}, h\in A'), \tag{7}$$

also have such expansions. Such a formal expansion of φ_λ is *already uniquely determined by the single equation in (7) corresponding to $q=\omega$*; Harish Chandra then gave a striking argument that this formal solution (which is actually convergent) *is a solution to the entire system* (7). This gives an expansion of each φ_λ (at least when λ is not in the union of a locally finite set of hyperplanes of $\mathfrak{F}$), known as the Harish Chandra series for φ_λ. Among other things, this series leads to the basic a priori estimate of the φ_λ:

$$|\varphi(\lambda:h)|\leqslant \text{const. } e^{-\rho(\log h)}(1+\|\log h\|)^d,\quad (\lambda\in\mathfrak{F}_I, h\in U(A^+)). \tag{8}$$

Note that although the $\varphi(\lambda:\cdot)$ are unitary matrix coefficients, the estimate (8) shows that $\varphi(\lambda:h)\to 0$ rather rapidly when $h\in A^+$ and goes to infinity. This is natural; for, the Haar measure dx in the (k_1,h,k_2)-coordinates is given by:

$$dx=J(h)\,dk_1\,dh\,dk_2,\qquad J(h)=\prod_{\alpha\in\Delta^+}(e^{\alpha(\log h)}-e^{-\alpha(\log h)})^{n(\alpha)}, \tag{9}$$

so that, on A^+, $J=0(e^{2\rho})$, showing that the $\varphi(\lambda:\cdot)$ are in $L^{2+\varepsilon}$ for every $\varepsilon>0$ and so might be expected to describe the continuous spectrum.

It is natural to ask whether one can devise a more straightforward argument for the estimate (8) as well as the existence of asymptotic expansions for the $\varphi(\lambda:\cdot)$. This can be done and the method is the same as that which is used by Harish Chandra [9, 12], and Trombi–Varadarajan [16]. It consists in formulating the system (2) as a perturbation of a first order linear system.

4. View of the system (2) as a perturbation of a first order linear system

The functions in $\mathfrak{R}$ go to zero exponentially fast on conical regions of the form,

$$A^+[L]=\{\exp tH \mid t\geqslant 1, H\in L\},\quad L\subset\mathfrak{a}^+ \text{ and is compact.} \tag{10}$$

On such regions $\delta'(q)$ is a nice perturbation of $\gamma'(q)$. Since the $\gamma(q)$ exhaust only $U(\mathfrak{a}_c)^{\mathfrak{w}}$ as q varies in $\mathfrak{D}$, the unperturbed system is not of the first order. As in the classical cases, where we go from nth order equations to first order systems, we can get a first order system as follows. We observe that $U(\mathfrak{a}_c)$ is a free module over $U(\mathfrak{a}_c)^{\mathfrak{w}}$ of rank $w=|\mathfrak{w}|$ having a basis $u_1=1, u_2, \ldots, u_w$ of homogeneous elements. We then use an induction on degree to conclude that if $u \in U(\mathfrak{a}_c)$ and has a representation,

$$u = \sum_{1\leq i\leq w} u_i \gamma(q_i^u), \quad (q_i^u \in \mathfrak{D}),$$

then for suitable $g_{ij}^u \in \mathfrak{R}$ and $q_{ij}^u \in \mathfrak{D}$ we can write,

$$u = \sum_{1\leq i\leq w} u_i \circ \delta(q_i^u) + \sum_{1\leq i\leq w} \sum_{1\leq j\leq s} g_{ij}^u \circ \delta(q_{ij}^u), \tag{11}$$

where $\delta(\cdot) = e^{\rho} \circ \delta'(\cdot) \circ e^{-\rho}$. Now, for any $u \in U(\mathfrak{a}_c)$, the operator of multiplication by u can be represented by a matrix:

$$uu_j = \sum_{1\leq i\leq w} u_i \gamma(q_{ij}^u) \quad (q_{ij}^u \in \mathfrak{D}). \tag{12}$$

So, for suitable $g_{lij}^u \in \mathfrak{R}$, $q_{lij}^u \in \mathfrak{D}$,

$$uu_j = \sum_{1\leq i\leq w} u_i \circ \delta(q_{ij}^u) + \sum_{1\leq i\leq w} \sum_{1\leq l\leq s} g_{lij}^u u_i \circ \delta(q_{lij}^u). \tag{13}$$

Let us now write Φ for the $w\times 1$ vector function on $\mathfrak{F}\times A$ whose jth component is $(\lambda, h) \mapsto \varphi(\lambda : h; u_j \circ e^{\rho})$, and let $\Gamma(u:\lambda)$ be the $w\times w$ matrix whose ijth component is $\gamma(q_{ji}^u)(\lambda)$. Then (13) leads to the differential equations on A', for any $u \in U(\mathfrak{a}_c)$:

$$u\Phi(\lambda:\cdot) = \Gamma(u:\lambda)\Phi(\lambda:\cdot) + \Omega(u:\lambda:\cdot)\Phi(\lambda:\cdot). \tag{14}$$

Here $\lambda \in \mathfrak{F}$ and $\Omega(u:\cdot:\cdot)$ is a $w\times w$ matrix whose elements are in $\mathfrak{R}[X]$. The system (14) is a perturbation of a linear system which is of order 1 because we can take $u = H_i$ where H_i $(1\leq i\leq r)$ form a basis of $\mathfrak{a}$.

For estimates on $\mathrm{Cl}(A^+)$ this is not enough; we must consider conical regions $A^+[L]$ where L is a compact neighborhood of an arbitrary element $H_0 \neq 0$ in $\mathrm{Cl}(\mathfrak{a}^+)$. It is therefore necessary to generalize suitably the preceding considerations.

Fix $H_0 \neq 0$ in $\mathrm{Cl}(\mathfrak{a}^+)$. Let M_0 (resp. $\mathfrak{m}_0$) be the centralizer of H_0 in G (resp. $\mathfrak{g}$); $K_0 = K\cap M_0$, $\mathfrak{k}_0 = \mathfrak{k}\cap\mathfrak{m}_0$. $M_0 = \theta(M_0)$ and K_0 is the maximal compact subgroup of M_0 fixed by θ. Let $\mathfrak{n}_0$ be the span of the root spaces $\mathfrak{g}_\alpha$ for $\alpha \in \Delta^+$ with $\alpha(H_0) > 0$. If M_0' is the open subset of M_0 of all m for which $\lambda(m) = (1 - \mathrm{Ad}(\theta(m^{-1})m))_{\mathfrak{n}_0}$ is invertible (suffix means restriction to $\mathfrak{n}_0$), the map $(k_1, m, k_2) \mapsto k_1 m k_2$ of $K_0 \times M_0' \times K_0$ into G is submersive and one can transcribe the system (2) to M_0'. Let $\mathfrak{R}$ be the ring (without unit element) generated by all the functions $\mu c_{\alpha\beta}$ where $\mu \in U(\mathfrak{m}_{0c})$ and the $c_{\alpha\beta}(\cdot)$ are the matrix entries with respect to a basis of $\mathfrak{n}_0$ of $c(\cdot) = \lambda(\cdot)^{-1}$. Now one can define a linear map γ_0 from $\mathfrak{D}$ to $\mathfrak{D}_0 = U(\mathfrak{m}_{0c})^{K_0}$ with the property that it vanishes on $\mathfrak{D}\cap U(\mathfrak{g}_c)\mathfrak{k} = \mathfrak{D}\cap\mathfrak{k}U(\mathfrak{g}_c)$ and that for any $q \in \mathfrak{D}$,

$$q \equiv d_0^{-1} \circ \gamma_0(d) \circ d_0 \begin{cases} \mathrm{mod}(\mathfrak{k}U(\mathfrak{g}_c) + U(\mathfrak{g}_c)\mathfrak{n}_0) \\ \mathrm{mod}(U(\mathfrak{g}_c)\mathfrak{k} + \theta(\mathfrak{n}_0)\cup(\mathfrak{g}_c)), \end{cases} \tag{15}$$

where d_0 is the homomorphism $m \mapsto |\det(\mathrm{Ad}(m))_{\mathfrak{n}_0}|^{1/2}$ of M_0 into the positive reals (d_0 is the analogue of e^{ρ}); $\gamma_0(q)$ is even uniquely defined mod $\mathfrak{D}_0 \cap U(\mathfrak{m}_{0c})\mathfrak{k}_0$. It is then possible to prove that given any $q \in \mathfrak{D}$, there are elements $g_i \in \mathfrak{R}$ and $\mu_i \in U(\mathfrak{m}_{0c})$ with $\deg(\mu_i) < \deg(q)$ $(1 \leqslant i \leqslant s)$ such that the differential operator,

$$\delta_0'(q) = d_0^{-1} \circ \gamma_0(q) \circ d_0 + \sum_{1 \leqslant i \leqslant s} g_i \mu_i, \tag{16}$$

is a radial component of q on M_0', i.e. for all $f \in C^{\infty}(G//K)$,

$$f(m; q) = f(m; \delta_0'(q)), \quad (m \in M_0'). \tag{17}$$

The map γ_0 induces an injective homomorphism of $\mathfrak{D}/\mathfrak{D} \cap U(\mathfrak{g}_c)\mathfrak{k}$ into $\mathfrak{D}_0/\mathfrak{D}_0 \cap U(\mathfrak{m}_{0c})\mathfrak{k}_0$. Moreover, if we write γ_G for γ and γ_{M_0} for its counterpart in the group M_0, and denote for convenience

$$I(\mathfrak{w}_G) = U(\mathfrak{a}_c)^{\mathfrak{w}_G}, \quad I(\mathfrak{w}_{M_0}) = U(\mathfrak{a}_c)^{\mathfrak{w}_{M_0}}, \tag{18}$$

then the following diagram is commutative:

$$\begin{array}{ccc} \mathfrak{D}/\mathfrak{D} \cap U(\mathfrak{g}_c)\mathfrak{k} & \overset{\gamma_0}{\hookrightarrow} & \mathfrak{D}_0/\mathfrak{D}_0 \cap U(\mathfrak{w}_{0c})\mathfrak{k}_0 \\ \gamma_G \downarrow & \hookrightarrow & \downarrow \gamma_{M_0} \\ I(\mathfrak{w}_G) & \underset{\text{inclusion}}{\text{natural}} & I(\mathfrak{w}_{m_0}). \end{array} \tag{19}$$

This diagram shows that γ_0 is canonical.

The situation is now very similar to the earlier case where we worked with A' instead of M_0'. $I(\mathfrak{w}_{M_0})$ is a free $I(\mathfrak{w}_G)$-module of rank$=[\mathfrak{w}_G : \mathfrak{w}_{M_0}]$, and one can select a basis,

$$u_1 = 1, u_2, \ldots, u_d, \quad d = [\mathfrak{w}_G : \mathfrak{w}_{M_0}],$$

of homogeneous elements for this module. Select $v_j \in \mathfrak{D}_0$ such that:

$$u_j = \gamma_0(v_j), \quad (1 \leqslant j \leqslant d). \tag{20}$$

It is then obvious that, with respect to this basis, the regular representation of the algebra $I(\mathfrak{w}_{M_0})$ may be identified with a representation by $d \times d$ matrices of elements from $I(\mathfrak{w}_G)$. As $I(\mathfrak{w}_{M_0})$ is abelian, the transpose of this matrix representation, denoted by Γ_0, is again a representation. We may regard Γ_0 as a representation of $\mathfrak{D}_0$ and write $\Gamma_0(v : \cdot)$ for $\Gamma_0(v)$ (or $\Gamma_0(\gamma_{M_0}(v))$ to be precise) regarded as a polynomial on $\mathfrak{F}$. The formula (17) then leads to the following result which may be regarded as the formulation of the system of differential equations (2) as a perturbation of a linear system. Let Φ_0 be the $d \times 1$ vector function on $\mathfrak{F} \times M_0$ whose jth component is $(\lambda, m) \mapsto \varphi(\lambda : m; u_j \circ d_0)$. Then, for any $v \in \mathfrak{D}_0$, there is a $d \times d$ matrix differential operator B_v on M_0' whose entries are of the form $\sum g_i \mu_i$, $g_i \in \mathfrak{R}$, $\mu_i \in U(m_{0c})$, such that for all $\lambda \in \mathfrak{F}$, $m \in M_0'$,

$$\Phi_0(\lambda : m; v) = \Gamma_0(v : \lambda)\Phi_0(\lambda : m) + \Phi_0(\lambda : m; B_v). \tag{21}$$

Let $\mathfrak{a}_0 = \mathfrak{a} \cap \mathrm{center}(\mathfrak{m}_0)$. Let us choose the enumeration of the simple roots in Δ^+ so that $\alpha_j(H_0) > 0$ if and only if $1 \leqslant j \leqslant l$. Let $\{H_1, \ldots, H_l\}$ be the basis of $\mathfrak{a}_0$ dual to

$\{\alpha_1,\ldots,\alpha_l\}$ and let $\mathfrak{a}_0^+$ be the positive chamber in $\mathfrak{a}_0$ of all $t_1H_1+\cdots+t_lH_l$ where $t_1,\ldots,t_l$ are >0. We now take $M_0^+=K_0{}^+AK_0$ where ${}^+A$ is the subset of $\mathrm{Cl}(A^+)$ of all elements $\exp H$ with $H\in\mathrm{Cl}(\mathfrak{a}^+)$ and $\alpha_j(H_0)>0$ for $1\leqslant j\leqslant l$. Then M_0^+ is open in M_0', and for any $m\in M_0$, $H\in\mathfrak{a}_0^+$, $m\exp tH\in M_0^+$ if $t\gg 1$; moreover, if g is any function in the ring $\mathfrak{R}$, $g(m\exp tH)\to 0$ exponentially fast when $t\to+\infty$.

Actually the functions in $\mathfrak{R}$ behave much more nicely under right translations by $\exp H$, $H\in\mathfrak{a}_0^+$. To formulate this we begin by recalling that $X, X'\mapsto -\langle X,\theta X'\rangle$ is a positive definite scalar product on $\mathfrak{g}\times\mathfrak{g}$. Let $\delta(m)=\|\mathrm{Ad}(m^{-1})_{\mathfrak{n}_0}\|$ for $m\in M_0$. Then $M_0^+=\{m: m\in M_0, \delta(m)<1\}$, $\delta(\exp H)=\mathrm{e}^{-\beta(H)}$ for $H\in\mathfrak{a}_0^+$ where $\beta(H)=\min_{1\leqslant j\leqslant l}\alpha_j(H)$; and for any $g\in\mathfrak{R}$ we have, for all $m\in M_0^+$, $H\in\mathfrak{a}_0^+$,

$$|g(m\exp H)|\leqslant c\delta(m)^2\left(1-\delta(m)^2\right)^{-k}\mathrm{e}^{-2\beta(H)}, \tag{22}$$

where $c=c(g)$ and $k=k(g)$ are constants >0. Furthermore, let $\mathfrak{L}_0$ be the set of elements $\lambda=m_1\alpha_1+\cdots+m_l\alpha_l$ where the m_j are all integers $\geqslant 0$ with $o(\lambda)=m_1+\cdots+m_l>0$. Then, each $g\in\mathfrak{R}$ has an expansion:

$$g(m\exp H)=\sum_{\lambda\in\mathfrak{L}_0} g_\lambda(m)\mathrm{e}^{-\lambda(H)}, \quad (m\in M_0^+, H\in\mathfrak{a}_0^+). \tag{23}$$

Here the g_λ are functions defined on all of M_0 and in fact are elements of the algebra of finite dimensional matrix coefficients of the group M_0; $g_\lambda(m\exp H)=g_\lambda(m)\mathrm{e}^{-\lambda(H)}$ for all $m\in M_0$, $H\in\mathfrak{a}_0$; and for constants $c=c(g)>0$, $k=k(g)\geqslant 0$,

$$|g_\lambda(m)|\leqslant c o(\lambda)^k\delta(m)^2, \quad (\lambda\in\mathfrak{L}_0, m\in M_0^+). \tag{24}$$

Obviously the location of the eigenvalues of $\Gamma_0(v:\lambda)$ is important for the stability of the perturbation problem (21). These are known to be the numbers $\gamma_{M_0}(v)(s^{-1}\lambda)$, $s\in\mathfrak{w}_G$. In particular, they are the numbers $\lambda(sH_i)$ if we take $v=H_i$, $1\leqslant i\leqslant l$; the system (21) is then of order 1. If $\lambda\in\mathfrak{F}_I$, these eigenvalues are all purely imaginary and so (21) is stable. *This is the central analytical fact of the entire theory.*

5. The fundamental estimates for φ_λ

It is clear from (1) that:

$$|\varphi(\lambda:x)|\leqslant\Xi(x)=\varphi(0:x), \quad (x\in G).$$

The function Ξ is the matrix coefficient $x\mapsto(\pi(x)1,1)$ of the unitary representation, π, acting on $L^2(K/M)$ by the formula,

$$(\pi(x)f)(k)=\mathrm{e}^{-\rho(H(x^{-1}k))}f\left(\kappa(x^{-1}k)\right), \quad (k\in K).$$

If $u,v\in U(\mathfrak{g}_c)$, $\Xi(u;x;v)=(\pi(x)\pi(u)1,\pi(v^\dagger)1)$ so that,

$$|\Xi(u;x;v)|\leqslant\text{const.}\,\Xi(x), \quad (x\in G). \tag{25}$$

In particular, $0\leqslant\Xi(x)=\Xi(x^{-1})\leqslant 1$ and each derivative $\Xi(u;\cdot;v)$ is bounded on G.

We now take $v=H_i$ $(1\leqslant i\leqslant l)$ in the equations (21). Writing $\boldsymbol{t}=(t_1,\dots,t_l)$, $h(\boldsymbol{t})=\exp(t_1H_1+\cdots+t_lH_l)$, we have:

$$\frac{\partial}{\partial t_j}\Phi_0(\lambda:mh(\boldsymbol{t}))=\Gamma(H_j:\lambda)\Phi_0(\lambda:mh(\boldsymbol{t}))+\Phi_0(\lambda:mh(\boldsymbol{t});B_{H_j}), \tag{26}$$

for all $m\in M_0^+, t_1,\dots,t_l>0$, $\lambda\in\mathfrak{F}$. We are in the case $\lambda=0$. Then the matrices $\Gamma_j=\Gamma(H_j:0)$ are commuting and *nilpotent*, so that $\exp(t_1\Gamma_1+\cdots+t_l\Gamma_l)$ is a *polynomial* in $t_1,\dots,t_l$. Write $F(m:\boldsymbol{t})$ (resp. $G_j(m:\boldsymbol{t})$) for $\exp(-t_1\Gamma_1-\cdots-t_l\Gamma_l)\Phi_0(0:mh(\boldsymbol{t}))$ (resp. $\exp(-t_1\Gamma_1-\cdots-\gamma_l\Gamma_l)\Phi_0(0:mh(\boldsymbol{t});B_{H_j})$). Then,

$$\frac{\partial}{\partial t_j}F=G_j,\quad (1\leqslant j\leqslant l), \tag{27}$$

where the F and G_j satisfy the estimates below which follow from (22) and (25):

$$|F(m:\boldsymbol{t})|\leqslant d_0(m)e^{a(t_1+\cdots+t_l)},$$

$$|G_j(m:\boldsymbol{t})|\leqslant C\big(1-\delta(m)^2\big)^{-k}(1+|\boldsymbol{t}|)^s e^{-\min(\boldsymbol{t})}|F(m:\boldsymbol{t})|.$$

The key point is that the $|\partial F/\partial t_j|$ are majorized *in terms of* $|F|$, so that any estimate of F can be improved by integration. An iterative argument leads to the following: fix $b>0$; then there are $s'=s'(b)\geqslant 0$, $k'=k'(b)\geqslant 0$ such that if $\min(t_1,\dots,t_l)>b(t_1+\cdots+t_l)$ and $m\in M_0^+$, we have:

$$\Xi(mh(\boldsymbol{t}))\leqslant C'(1+|\boldsymbol{t}|)^{s'}\big(1-\delta(m)^2\big)^{-k'}e^{-\rho_0(\log h(\boldsymbol{t}))}.$$

From this it follows that for any $\varepsilon>0$ we can find a compact neighborhood L of H_0 in $\mathrm{Cl}(\mathfrak{a}^+)$ such that $\Xi(h)=0(e^{-(1-\varepsilon)\rho(\log h)})$ for h in the conical region $A^+[L]$. This implies that $\Xi(h)=0(e^{-(1-\varepsilon)\rho(\log h)})$ for $h\in\mathrm{Cl}(A^+)$, for each $\varepsilon>0$.

This is a much stronger estimate than the boundedness of Ξ. Using them in (27) and (28) we can show that Φ_0 can be approximated by a solution of the unperturbed system. More precisely, we can find a spherical function θ on M_0 satisfying:

$$\gamma_0(q)\theta=\gamma(q)(0)\theta,\quad (q\in\mathfrak{D}), \tag{29}$$

and having the following property: for a suitable compact neighborhood L of H_0 in $\mathrm{Cl}(\mathfrak{a}^+)$ and $\varepsilon_0>0$, we have the estimate below, valid for all $h\in A^+[L]$:

$$|\Xi(h)-d_0(h)^{-1}\theta(h)|\leqslant Ce^{-(1+\varepsilon_0)\rho(\log h)}. \tag{30}$$

Now, the diagram (19) shows that $\gamma(q)(0)=\gamma_{M_0}(\gamma_0(q))(0)$. So, (29) suggests that θ is closely related to the function Ξ_{M_0} on M_0 that is the counterpart of Ξ. In fact, one can show that

$$\theta(m)=\int_{K_0}u(H_{M_0}(mk))e^{-\rho_{M_0}(H_{M_0}(mk))}dk,\quad (m\in M_0) \tag{31}$$

where u is a polynomial. An induction on $\dim(G)$ allows us to conclude from (31) and (30) that for a suitable $p\geqslant 0$ $\Xi(h)=0(e^{-\rho(\log h)}(1+\|\log h\|)^p)$ for all $h\in A^+[L]$. This leads to (8).

For full details, see Varadarajan [18] (Part II, pp. 156–163). It is not unreasonable that this argument (and many others) uses induction on dim(G). The question studied is the behaviour of a function at infinity on G/K, and it is to be expected that the nature of the compactification of G/K would enter in its study. The boundary of G/K consists of many parts and is, roughly speaking, composed of symmetric spaces of the type M_0/K_0. The estimate (30) may then be regarded as an approximation of Ξ by eigenfunctions on M_0/K_0 in a neighborhood of suitable boundary points. For a detailed geometric study of the compactification of G/K and its probabilistic interpretation, see Karpelevič [15], Furstenberg [4].

6. The Harish Chandra series for φ_λ

We now go back to the system (14) on A^+. For $\lambda \in \mathfrak{F}$, $H \in \mathfrak{a}^+$, $t > 0$,

$$\frac{\mathrm{d}}{\mathrm{d}t} \mathrm{e}^{-t\Gamma(H:\lambda)}\Phi(\lambda : \exp tH) = \mathrm{e}^{-t\Gamma(H:\lambda)}\Omega(H:\lambda:\exp tH)\Phi(\lambda:\exp th). \tag{32}$$

In view of the fundamental estimates (8) it is clear that the right side of this equation goes to 0 exponentially fast as $t \to +\infty$, as long as $\lambda \in \mathfrak{F}_I$ or even in a "strip" $\mathfrak{F}(\delta)$ around $\mathfrak{F}_I$, $\mathfrak{F}(\delta)$ being the set of $\lambda \in \mathfrak{F}$ with $\|\lambda_R\| < \delta$. So,

$$\Theta(\lambda) = \lim_{t \to +\infty} \mathrm{e}^{-t\Gamma(H:\lambda)}\Phi(\lambda:\exp tH), \quad (\lambda \in \mathfrak{F}(\delta)), \tag{33}$$

exists and can be shown to be independent of H and holomorphic in $\mathfrak{F}(\delta)$. We now rewrite (32) as an integral equation,

$$\Phi(\lambda:\exp tH) = \mathrm{e}^{t\Gamma(H:\lambda)}\Theta(\lambda) - \int_0^\infty \mathrm{e}^{-\tau\Gamma(H:\lambda)}\Omega(H:\lambda:\exp(\tau+t)H)\Phi(\lambda:\exp(\tau+t)H)\,\mathrm{d}\tau, \tag{34}$$

and construct the standard sequence of approximations by the Cauchy–Picard method. Using the fact that the matrix, Ω, can be expanded as a series in the exponentials e^{-q} we then obtain an asymptotic expansion for $\Phi(\lambda: h)$, and hence, for $\varphi(\lambda: h)\mathrm{e}^{\rho(\log h)}$. For the full details of the approximation method see Trombi–Varadarajan [16]. Denoting asymptotic expansions by $\sim$ we get the following: there is a function, $\boldsymbol{c}$, meromorphic on $\mathfrak{F}(\delta)$, such that for all $\lambda \in \mathfrak{F}(\delta)'$ (prime means we consider only regular points), $h \in A^+$,

$$\mathrm{e}^{\rho(\log h)}\varphi(\lambda:h) \sim \sum_{s \in \mathfrak{m}} \boldsymbol{c}(s\lambda) \sum_{q \in \mathfrak{L} \cup \{0\}} \Gamma_q(s\lambda)\mathrm{e}^{(s\lambda - q)(\log h)}, \tag{35}$$

where the Γ_q are rational functions having no singularities in $\mathfrak{F}(\delta)$. If we define $\Psi(\lambda: h)$ to be the *formal series*,

$$\Psi(\lambda:h) = \sum_{q \in \mathfrak{L} \cup \{0\}} \Gamma_q(\lambda)\mathrm{e}^{(\lambda - q)(\log h)}, \tag{36}$$

we may then regard (35) as the statement that,

$$\sum_s \boldsymbol{c}(s\lambda)\Psi(s\lambda:\cdot), \tag{37}$$

is a formal solution to the system of differential equations:

$$\delta(v)\psi=\gamma(v)(\lambda)\psi, \quad (v\in\mathfrak{D}). \tag{38}$$

From this it follows easily that $\Psi(\lambda:\cdot)$ is a formal solution to the same system, at least for λ in a dense open subset of $\mathfrak{F}(\delta)$. As we have already remarked, a formal solution of (38) with v as the Casimir operator is already uniquely determined and is, by definition, the Harish Chandra series. Hence (37) is the Harish Chandra series for $e^{\rho}\varphi(\lambda:\cdot)$.

7. The constant term. The L^2 theory of Harish Chandra. Characterizations of the Harish Chandra transforms of $\mathcal{C}^p(G//K)$ and $C_c^\infty(G//K)$

The perturbative method explained above allows us to associate with any elementary spherical function, φ, satisfying the estimate (8) a spherical function θ on M_0 with the following properties:

(1) θ satisfies an estimate of the form (8).

(2) If χ is the homomorphism $\mathfrak{D}\to\mathbb{C}$ to which φ belongs, then:

$$\gamma_0(v)\theta=\chi(v)\theta, \quad (v\in\mathfrak{D}). \tag{39}$$

(3) $\exists\ \varepsilon_0>0$ and a compact neighborhood L of H_0 in $\mathrm{Cl}(\mathfrak{a}^+)$ such that on $A^+[L]$ we have the estimate,

$$|\varphi(h)-d_0(h)^{-1}\theta(h)|\leqslant Ce^{-(1+\varepsilon_0)\rho(\log h)}.$$

We may regard θ as the *constant term* of φ relative to the perturbation defined by the system (26). It is convenient to extend θ to all of G by $(G=KM_0N_0)$:

$$\theta(kmn)=\theta(m). \tag{40}$$

It is then possible to supplement the estimate (3) of (39) by,

$$\varphi(xa)d_0(xa)-\theta(xa)\to 0, \tag{41}$$

exponentially fast in the $C^\infty(G)$ topology when $a=\exp H$ and $H\in\mathfrak{a}_0^+$ goes to infinity in such a way that $\alpha(H)\to+\infty$ for every root α taking a positive value at H_0. If $\varphi=\varphi_\lambda$ and H_0 is (regular) in $\mathfrak{a}^+$, $\theta=\theta(\lambda:\cdot)$ is given by:

$$\theta(\lambda:h)=\sum_{s\in\mathfrak{m}} \boldsymbol{c}(s\lambda)e^{s\lambda(\log h)}, \quad (\lambda\in\mathfrak{F}_I'). \tag{42}$$

The function $\boldsymbol{c}$ is the famous c-function of Harish Chandra, and can be explicitly computed (cf. Gindikin–Karpelevič [7]); $\boldsymbol{b}=\pi\boldsymbol{c}$ (π=product of the short roots >0) is holomorphic and invertible on $\mathfrak{F}(\delta)$, and $\boldsymbol{b}$ and $\boldsymbol{b}^{-1}$ have both at most polynomial growth on $\mathfrak{F}(\delta)$.

The transition from φ to θ allows us to use induction on $\dim(G)$. Using this technique Harish Chandra carried out his harmonic analysis of $\mathcal{C}^2(G//K)$ and even for $\mathcal{C}(G)$ (cf. [11]). For $f\in\mathcal{C}^2(G//K)$ its Harish Chandra transform $\mathcal{H}f$ is the function on $\mathfrak{F}_I$ defined by:

$$(\mathcal{H}f)(\lambda)=\int_G f(x)\overline{\varphi(\lambda:x)}\,\mathrm{d}x, \quad (\lambda\in\mathfrak{F}_I). \tag{43}$$

Harish Chandra's main results may now be described as follows:

I I. $\mathcal{H}$ is an isomorphism of $\mathcal{C}^2(G//K)$ with $\mathcal{S}(\mathfrak{F}_I)^{\mathfrak{w}}$, as topological algebras. Here $\mathcal{C}^2(G//K)$ is regarded as a Frechet algebra under convolution; $\mathcal{S}(\mathfrak{F}_I)$ is the usual Schwartz space of the real vector space $\mathfrak{F}_I$, regarded as a multiplication algebra; and $\mathfrak{w}$ means the subalgebra of $\mathfrak{w}$-invariant elements.

II. We have the inversion and Plancherel formulae given by:

$$f(x)=|\mathfrak{w}|^{-1}\int_{\mathfrak{F}_I}(\mathcal{H}f)(\lambda)|\boldsymbol{c}(\lambda)|^{-2}\mathrm{d}\lambda,$$

$$\int_G|f(x)|^2\mathrm{d}x=|\mathfrak{w}|^{-1}\int_{\mathfrak{F}_I}|\mathcal{H}f(\lambda)|^2|\boldsymbol{c}(\lambda)|^{-2}\mathrm{d}\lambda. \tag{44}$$

In (44) $\mathrm{d}x$ and $\mathrm{d}\lambda$ have to be normalized as follows. First we normalize the Haar measure $\mathrm{d}n$ on N so that the corresponding $\mathrm{d}\bar{n}$ on $\bar{N}=\theta(N)$ induced by θ satisfies $\int_{\bar{N}}\mathrm{e}^{-2\rho(H(\bar{n}))}\mathrm{d}\bar{n}=1$. Then, for any Haar measure $\mathrm{d}a$ on A, we take $\mathrm{d}x=\mathrm{e}^{2\rho(\log a)}\mathrm{d}k\,\mathrm{d}a\,\mathrm{d}n$ where $\int_K\mathrm{d}k=1$; $\mathrm{d}\lambda$ is the Lebesgue measure on $\mathfrak{F}_I$ *dual* to $\mathrm{d}a$ in the sense of the usual Fourier transform theory on A.

In Trombi–Varadarajan [16] the above idea of associating θ to φ was extended to associating with φ its entire asymptotic expansion. The inversion formulae in (44) may now be taken over suitable "cycles" in $\mathfrak{F}$ (for f in $\mathcal{C}^p(G//K)$, $p<2$) and we are led to the following result. Let $\varepsilon>0$ and let $\mathfrak{F}^\varepsilon$ be the subset of $\mathfrak{F}$ defined by:

$$\mathfrak{F}^\varepsilon=\{\lambda\in\mathfrak{F}\,|\,|\mathrm{Re}(s\lambda)(H)|\leqslant\varepsilon\rho(H)\forall H\in\mathfrak{a}^+,s\in\mathfrak{w}\}. \tag{45}$$

Let $Z(\mathfrak{F}^\varepsilon)$ be the multiplication algebra of all functions F defined and holomorphic on the interior of $\mathfrak{F}^\varepsilon$ and such that for any polynomial differential operator D, DF is bounded on it; let $Z(\mathfrak{F}^\varepsilon)^{\mathfrak{w}}$ be the sub-algebra of $\mathfrak{w}$-invariant elements of $Z(\mathfrak{F}^\varepsilon)$. Then $\mathcal{H}$ establishes a topological algebra isomorphism of $\mathcal{C}^p(G//K)$ with $Z(\mathfrak{F}^\varepsilon)^{\mathfrak{w}}$, if we take $\varepsilon=(2/p)-1$ $(0<p<2)$.

Finally, the work of Gangolli [5] establishes that $\mathcal{H}$ is a topological algebra isomorphism of $C_c^\infty(G//K)$ with $\mathcal{P}(\mathfrak{F})^{\mathfrak{w}}$ where $\mathcal{P}(\mathfrak{F})$ is the Paley–Wiener space of $\mathfrak{F}$.

Let us now go back to the fundamental results I and II stated above. There are three main steps in it. First one forms the "wave packet":

$$\psi_a=\int_{\mathfrak{F}_I}a(\lambda)\pi(\lambda)\varphi(\lambda:\cdot)\mathrm{d}\lambda, \quad (a\in\mathcal{S}(\mathfrak{F}_I)). \tag{46}$$

The wave packet ψ_a is proved to lie in $\mathcal{C}^2(G//K)$ using the method of approximating $\varphi(\lambda:\cdot)$ by $\theta(\lambda:\cdot)$; the factor, π, takes care of the poles of the functions $c(s\lambda)$. Then one determines the Harish Chandra transform of ψ_a in terms of a, as follows:

$$\mathcal{H}\psi_a(\lambda)=\frac{\Sigma_s\,\varepsilon(s)a(s\lambda)}{\pi(\lambda)}|\boldsymbol{b}(\lambda)|^2. \tag{47}$$

Finally one proves the completeness: the wave packet map,

$$u \mapsto \varphi_u(\cdot)=|\mathfrak{w}|^{-1}\int_{\mathfrak{F}_I} u(\lambda)\varphi(\lambda:\cdot)|c(\lambda)|^{-2}\mathrm{d}\lambda, \tag{48}$$

maps onto all of $\mathcal{C}^2(G//K)$.

Harish Chandra's argument for completeness was given in ref. [5] and uses the theory of the discrete series. The point is that if $f\in\mathcal{C}^2(G//K)$ is nonzero and $\mathcal{H}f=0$, $\Theta(f)\neq 0$ for some discrete series character Θ. This is impossible since one knows from Harish Chandra's discrete series theory [10] that no discrete series representation is of class 1, i.e. has a K-invariant vector. Subsequently, Gangolli [5] showed that his Paley–Wiener theory gave a direct proof of this completeness (cf. also Helgason [13]).

We now remark that this can be done using the Trombi–Varadarajan theory also. The idea (as in Gangolli [5] is to show that for $f\in C_c^\infty(G//K)$, $f=\varphi_{\mathcal{H}f}$. Actually one knows that $\mathcal{H}f=\mathcal{H}\varphi_{\mathcal{H}f}$ and so $\mathcal{H}g=0$ where $g=f-\varphi_{\mathcal{H}f}$. But now, as $f\in C_c^\infty(G//K)$, $\mathcal{H}f\in Z(\mathfrak{F}^\varepsilon)^{\mathfrak{w}}$ for every $\varepsilon>0$ and hence $\varphi_{\mathcal{H}f}\in\mathcal{C}^p(G//K)$ for all $p>0$. So g belongs to $\mathcal{C}^p(G//K)$ for all $p>0$. But then, the assumption $\mathcal{H}g=0$ on $\mathfrak{F}_I$ together with the fact that $\mathcal{H}g$ is an entire function on $\mathfrak{F}$ implies that $\int_G g\psi\,\mathrm{d}x=0$ for every elementary spherical function ψ, and hence that $\pi(g)=0$ for *all* irreducible unitary representations π. This means that g must be zero, i.e. $f=\varphi_{\mathcal{H}f}$.

8. Computation of the Plancherel measure

It is not difficult to show that the Plancherel measure for the decomposition of $L^2(G//K)$ is a continuous $\mathfrak{w}$-invariant density on $\mathfrak{F}_I$, and hence we need to compute it only on a dense open set, say $\mathfrak{F}_I'$. We consider $a\in C_c^\infty(\mathfrak{F}_I')$ and form the wave packet ψ_a as in (46). It is then a question of proving the formula (47) where $\boldsymbol{b}=\pi\boldsymbol{c}$.

Let $\bar{N}=\theta(N)$. The starting point is the observation that the Harish Chandra transform of any $f\in\mathcal{C}^2(G//K)$ is the Fourier transform of the function $\mathcal{A}f$ on A defined by:

$$(\mathcal{A}f)(h)=\mathrm{e}^{\rho(\log h)}\int_{\bar{N}} f(\bar{n}h)\,\mathrm{d}\bar{n}. \tag{49}$$

At the moment $\mathrm{d}\bar{n}$ is not normalized but $\mathrm{d}x=\mathrm{e}^{-2\rho(\log a)}\mathrm{d}k\,\mathrm{d}a\,\mathrm{d}\bar{n}$. So (49) allows us to reduce (47) to the proof of the following relation:

$$\mathrm{e}^{\rho(\log h)}\int_{\bar{N}}\psi_a(\bar{n}h)\,\mathrm{d}\bar{n}=\int_{\mathfrak{F}_I}|\boldsymbol{b}(\lambda)|^2\Big(\pi(\lambda)^{-1}\sum_s\varepsilon(s)a(s\lambda)\Big)\mathrm{e}^{\lambda(\log h)}\mathrm{d}\lambda, \tag{50}$$

where $\mathrm{d}\bar{n}$ has the normalization $\int_{\bar{N}}\mathrm{e}^{-2\rho(H(\bar{n}))}\mathrm{d}\bar{n}=1$ and $\mathrm{d}\lambda$ is dual to $\mathrm{d}h$.

The function $\mathrm{e}^{\rho(H(x)+\log a)}\varphi(\lambda:xa)$ is asymptotic to $\theta(\lambda:xa)=\Sigma_s\, c(s\lambda)\times\mathrm{e}^{s\lambda(H(x)+\log a)}$ when $a\in A^+$ and goes to infinity [cf. (40)–(42)]. We now replace, in the left side of (50), ψ_a by the "truncated wave packet" θ_a obtained by replacing

$\varphi(\lambda:\cdot)$ by $\theta(\lambda:\cdot)$. A computation that uses the integral representation,

$$c(\lambda)=\int_{\bar{N}} e^{-(\lambda+\rho)(H(\bar{n}))} d\bar{n}, \quad (\mathrm{Re}(\lambda,\alpha)>0 \forall \alpha \in \Delta^+),$$

then shows that (50) is true with θ_a in place of ψ_a. So one must show that the integrals over $\bar{N}$ of $\theta_a - \psi_a$ are all zero. This is true because, roughly speaking, these are functions which are *derivatives with respect to* $\bar{\mathfrak{n}}$ *of rapidly decreasing functions on* $\bar{N}$, and hence have vanishing integrals on $\bar{N}$. However, the actual proofs of these remarks are rather delicate (cf. Harish Chandra [9, pp. 592–610]).

References

[1] Ehrenpreis, L. and Mautner, F.I. (1955). Some properties of the Fourier transform on semisimple Lie groups, I. *Ann. of Math.* **61**, 406–439.

[2] Ehrenpreis, L. and Mautner, F.I. (1957). Some properties of the Fourier transform on semisimple Lie groups, II. *Trans. Amer. Math. Soc.* **84**, 1–55.

[3] Ehrenpreis, L. and Mautner, F.I. (1959). Some properties of the Fourier transform on semisimple Lie groups, III. *Trans. Amer. Math. Soc.* **90**, 431–484.

[4] Furstenberg, H. (1963). A Poisson formula for semisimple Lie Groups. *Ann. of Math.* **77**, 335–386.

[5] Gangolli, R.A. (1971). On the Plancherel formula and the Paley–Wiener theorem for spherical functions on semisimple Lie groups. *Ann. of Math.* **93**, 150–165.

[6] Gangolli, R.A. and Varadarajan, V.S. Spherical functions on reductive groups. (to appear).

[7] Gindikin, S.G. and Karpelevič, F.I. (1962). Plancherel measure for symmetric Riemannian spaces of non-positive curvature. *Dokl. Akad. Nauk SSSR.* **145**, 252–255.

[8] Harish Chandra (1958). Spherical functions on a semisimple Lie group, I. *Amer. J. Math.* **80**, 241–310.

[9] Harish Chandra (1958). Spherical functions on a semisimple Lie group, II. *Amer. J. Math.* **80**, 553–613.

[10] Harish Chandra (1966). Discrete series for semisimple Lie groups, II. *Acta Math.* **116**, 2–111.

[11] Harish Chandra (1975). Harmonic Analysis on real reductive groups, I. The theory of the constant term. *J. Functional Analysis.* **19**, 104–204.

[12] Harish Chandra (1976). Harmonic Analysis on real reductive groups, II. Wave packets in the Schwartz space. *Invent. Math.* **36**, 1–55.

[13] Helgason, S. (1966). An analogue of the Paley–Wiener theorem for the Fourier transform on certain symmetric spaces. *Math. Ann.* **165**, 297–308.

[14] Helgason, S. (1978). *Differential Geometry, Lie Groups, and Symmetric Spaces.* Academic Press, New York.

[15] Karpelevič, F.I. (1965). The geometry of geodesics and the eigenfunctions of the Laplace–Beltrami operator on symmetric spaces. *Trans. Moscow Math. Soc.* **14**, 48–185.

[16] Trombi, P.C. and Varadarajan, V.S. (1971). Spherical transforms on semisimple Lie groups. *Ann. of Math.* **94**, 246–303.

[17] Varadarajan, V.S. (1974). *Lie Groups, Lie Algebras, and Their Representations.* Prentice-Hall.

[18] Varadarajan, V.S. (1977). *Harmonic Analysis on Real Reductive Groups.* Springer Verlag Lecture Notes in Mathematics. Vol. 576. Springer Verlag, Berlin.

G. Kallianpur, P.R. Krishnaiah, J.K. Ghosh, eds., *Statistics and Probability: Essays in Honor of C.R. Rao*

THE ESTIMATION OF PALAEOMAGNETIC POLE POSITIONS*

G.S. WATSON
Princeton University, Princeton, NJ, U.S.A.

The use of the Fisher distribution to estimate the mean direction of magnetization of a rock at a sampling site is now standard. Sampling sites are chosen to cover 10^4 to 10^5 years to average out the effect of secular variation. The controversy about how to combine these site means has never been satisfactorily resolved. By using statistical models for secular variation, this paper suggests how methods should be derived. A number of interesting statistical distributions and estimation problems are shown.

1. Introduction

When a lava containing iron cools below its Curie point, it becomes magnetized in the direction of the earth's field at that place and time. Sediments containing magnetic particles also acquire (a much weaker) local magnetization as they are formed. The study of palaeomagnetism led to the current revolution in geology—plate tectonics. For assuming that the earth's magnetic field has always been much the same as it is now—except for reversals of polarity—a magnetic measurement gives a good estimate of the latitude of the site when the rock was magnetized. These latitudes make no sense unless one is willing to admit that continental drift has taken place, in fact, they allow continental reconstructions to be made. A full account may be found in the book by McElhinny (1973).

The strength of the magnetization of a specimen may have altered over time and so it is not used in this work—only the *direction* of natural remanent magnetism, i.e. the original magnetization. Later changes can be "cleaned" out by methods not relevant here. (If the rocks are unstably magnetized, they cannot be used.) If N specimens are taken from one site with directions, the N unit vectors, $r_1, \ldots, r_N$, after cleaning, may be assumed to be a sample from Fisher's distribution

$$\frac{\kappa}{4\kappa \sinh \kappa} \exp \kappa r'\mu \quad (\kappa \geqslant 0), \tag{1}$$

where μ is the (unit) mean direction. Fisher (1953) showed that the maximum likelihood estimator of μ is the direction of $R = \Sigma r_i$, the vector resultant of the sample and, further, that if $|R|$ is the length of R, $\hat{\kappa}$ is the solution of

$$\coth \kappa - (1/\kappa) = |R|/N. \tag{2}$$

*Research supported in part by a contract with the Office of Naval Research, No. N00014-79-C-0322, awarded to the Department of Statistics, Princeton University, Princeton, New Jersey.

An approximation to $\hat{\kappa}$ is κ given by

$$\kappa = \frac{N-1}{N-|R|}. \tag{2'}$$

κ measures the precision of the Fisher distribution. Further statistical methods appropriate for palaeomagneticists were given by Watson (1956a, b) and Watson and Irving (1957).

The latter paper gives an approximate within and between sites method of analyzing data from several sites. It is assumed that the within sites distribution is Fisher with $\kappa = \kappa_W$ about the site mean and that the site means are independently Fisher with $\kappa = \kappa_B$ about the true palaeomagnetic direction. Unfortunately, this compound of Fisher distributions is only approximately Fisher. The estimate of κ_B was meant to measure the secular variation since it is supposed that the sites sample the full cycle of secular variations. This is some 10^4 to 10^5 years when the pole moves around its mean position, a short period on continental drift time scales.

In designing this analysis, no real thought was given to modelling secular variation. Furthermore, the immense amount of experimental work done since then has shown that site means often do *not* vary symmetrically around some mean direction as implied by the Fisher distribution (1), as we supposed.

When the earth's field is averaged over the time period of secular variation, it is found to be approximated, with surprising accuracy, by an earthcentered magnetic dipole currently inclined at about 11 degrees to the rotational axis. From elementary physics, the magnetic force, H, at a point r from a dipole M (H, M, r are all vectors) is given by

$$H = \left(3\frac{rr'}{|r|^2} - I\right)\frac{M}{|r|^3} \tag{3}$$

where the prime denotes a transpose, I is the 3×3 identity matrix, and $|r|$ is the length of r. Of course, the center of the earth is too hot to support a dipole! The field is thought to be due to a self-exciting dynamo made of asymmetric flows of conducting rock (mainly iron and nickel) through stray permanent fields of the rotating earth. Minor changes in the flows and the eastward movement of the crust relative to the interior (core) of the earth cause the secular variation and even the reversals.

Irving and Ward (1964) supposed that the secular variation was due to a smaller geocentric dipole m whose orientation varied uniformly at random over the period but whose magnitude $|m|$ was constant. Then the field at site, r, at different times in the period is independent and has the representation

$$H = \left(3\frac{rr'}{|r|^2} - I\right)(M+m)/|r|^3. \tag{4}$$

Here and below we will regard M as fixed and define $\mu = M/|M|$. Creer et al. (1959) assume that the main dipole M has fixed strength $|M|$ but "wobbles" — that the direction of the main dipole is independent from time to time and has a Fisher distribution. Creer et al.'s model says

$$H=\left(3\frac{rr'}{|r|^2}-I\right)M^*/|r|^3, \tag{5}$$

where $M^*=|M|d$ and d has a Fisher distribution about μ, and some large κ. Cox (1970) supposes that the main dipole oscillates in strength and wobbles in direction and that there is, in addition and independently, a randomly oriented smaller dipole, m, as in the Irving and Ward model. In his case

$$H=\left(3\frac{rr'}{|r|^2}-I\right)(M^+ +m)/|r|^3 \tag{6}$$

where the average value of M^+ is to be M about which M^+ has a rotationally symmetric distribution.

In the last five years some more complex models have been suggested which generate secular variations with dipoles on the surface of the core, see, e.g. Harrison and Watkins (1979). Here we will be content to explore models expressed in (4), (5) and (6), or closely related to them. We will assume that large samples have been taken at each site so that site mean directions are known exactly. Let these directions be $L_1, L_2, \ldots, L_N$. The problem is to estimate μ, the direction or mean direction of M, M^*, M^+.

We will choose, as a unit of length, the earth's radius (assuming it to be a sphere) and set $u=r/|r|$. Then the common factor in (4), (5) and (6) is $3uu'-I$, where u is a fixed but arbitrary unit vector. We will write

$$U=3uu'-I. \tag{7}$$

Note that $U=U'$, $U^2=3uu'+I$ and that

$$U=2uu'-u_2u_2'-u_3u_3',$$

where u, u_2, u_3 are orthonormal. Hence the eigenvalues and eigenvectors of U are trivially known. Each of the models has the form

$$H=UX,$$

where X is a random vector, U is fixed and only the direction of H is observed. In the next section we will explore the distributional and estimation problems this raises. They have some intrinsic statistical interest and may have other applications. In the third section, we return explicitly to our main problem.

2. Statistical preliminaries

(a) Fisher observed that his distribution may be derived from the Gaussian as follows. If X has a trivariate Gaussian distribution with mean vector μ and covariance matrix $\sigma^2 I$, set $R=|X|$, $L=X/R$, $\lambda=\mu/|\mu|$. Then the joint density of R

and L is

$$\frac{R^2}{(2\pi)^{3/2}\sigma^3}\exp-\frac{1}{2\sigma^2}\left(R^2+|\mu|^2\right)\exp\frac{R|\mu|}{\sigma^2}L'\lambda, \tag{8}$$

so that the density of L on the unit sphere, conditional on a fixed R, is proportional to

$$\exp\frac{R|\mu|}{\sigma^2}L'\lambda. \tag{9}$$

From (1) we see that L, then, has a Fisher distribution about the mean direction λ with a κ of $R|\mu|/\sigma^2$. If (9) were appropriate, we would know how to estimate λ and $R|\mu|/\sigma^2$; use Fisher's estimations for (1).

If, however, (8) were appropriate but we were only given the directions $L_1,\ldots, L_N$ of $X_1,\ldots, X_N$, we would have to use the density, f, of L,

$$\int_0^\infty \frac{R^2}{(2\pi)^{3/2}\sigma^3}\exp-\frac{1}{2\sigma^2}\left(R^2+|\mu|^2-2R|\mu|L'\lambda\right)\mathrm{d}R. \tag{10}$$

This is simpler if we set $S=R/\sigma$, $\nu=|\mu|/\sigma$, since then

$$f(L,\lambda,\nu)=\int_0^\infty \frac{S^2}{(2\pi)^{3/2}}\exp\left\{-\tfrac{1}{2}\left(S^2+\nu^2-2S\nu L'\lambda\right)\right\}\mathrm{d}S, \tag{11}$$

with the vector derivative

$$\frac{\partial f}{\partial\lambda}=\int_0^\infty \frac{S^3\nu}{(2\pi)^{3/2}}\exp\left\{-\tfrac{1}{2}\left(S^2+\nu^2-2S\nu L'\lambda\right)\right\}L\,\mathrm{d}S, \tag{12}$$

$$=\nu J(L,\lambda,\nu)L, \tag{12'}$$

say. Since $\lambda'\lambda=1$, the maximum likelihood (m.l.) estimate of λ requires us to solve

$$\frac{\partial}{\partial\lambda}\left(\sum_{j=1}^{N}\log f(L_j)+\theta\lambda'\lambda\right)=0,$$

i.e.

$$\sum_{j=1}^{N}\frac{J(L_j,\lambda,\nu)}{f(L_j,\lambda,\nu)}L_j+\frac{2\theta\lambda}{\nu}=0. \tag{13}$$

If ν is known, (13) may be solved numerically by an iteration. An initial estimate of λ, $\lambda^{(1)}$ would be the direction of $\Sigma_1^N L_j$. With computer programs to compute $J(L_j,\lambda^{(1)})$ and $f(L_j,\lambda^{(1)})$, (13) would be used to find $\lambda^{(2)}$, etc. The two integrals have well-behaved integrands so that numerical integration will not be difficult.

To estimate the non-negative parameter, ν, we observe that

$$\frac{\partial f}{\partial\nu}=\int_0^\infty \frac{S^2}{(2\pi)^{3/2}}(-\nu+SL'\lambda)\exp\left\{-\tfrac{1}{2}\left(S^2+\nu^2-2S\nu L'\lambda\right)\right\}\mathrm{d}S$$

$$=-\nu f+L'\lambda J. \tag{14}$$

Hence the m.l. equation for $\hat{\nu}$ is

$$\sum_{j=1}^{N} -\frac{\nu f(L_j,\lambda,\nu)+L_j'\lambda J(L_j,\lambda,\nu)}{f(L_j,\lambda,\nu)}=0$$

or

$$\frac{1}{N}\sum_{j=1}^{N}\frac{L_j'\lambda J(L_j,\lambda,\nu)}{f(L_j,\lambda,\nu)}=\nu, \tag{15}$$

a form ideal for iteration. It is convenient that no more integrals need be evaluated. To start off the ν iteration we again use the analogy between (8) and (9) and suggest

$$\nu^{(1)}=\frac{N-1}{N-|\Sigma L_j|}. \tag{16}$$

Thus to solve jointly (13) and (15) for $\hat{\lambda}$ and $\hat{\nu}$, one seesaws between them using the suggested $\lambda^{(1)}$ and $\nu^{(1)}$. Hence given only the directions $L_1,\ldots, L_N$ of a sample of N from $\mathcal{G}(\mu,\sigma^2 I)$, we may estimate the direction of μ and $\nu=|\mu|/\sigma$. To find the covariance matrix, the second derivatives of the log-likelihood will be evaluated in the usual way. We will leave this calculation for an applied paper to appear elsewhere. This paper will also show the shape of distributions defined by (10) or (11) and compare it with others. This distribution has of course appeared before (see e.g. Kendall (1974)) under the names "off-set" or "displaced" normal — I prefer "angular" normal.

(b) If we observed the directions of n copies of $Y=TX$, where X is Gaussian mean μ with known covariance matrix $\sigma^2 I$, and T is a known non-singular 3×3 matrix, we see that we may set $\sigma^2=1$ without loss of generality when we wish to estimate the direction of μ. The density of Y is

$$\frac{1}{(2\pi)^{3/2}\|T\|}\exp-\tfrac{1}{2}\left(Y'(TT')^{-1}Y-2\mu'T^{-1}Y+\mu'\mu\right). \tag{17}$$

Setting $Y=RL$ where $R>0$ and $|L|=1$, the joint density of R and L is

$$\frac{R^2}{(2\pi)^{3/2}\|T\|}\exp-\tfrac{1}{2}\left(R^2L'(TT')^{-1}L-2R\mu'T^{-1}L+\mu'\mu\right).$$

Thus the density of L is

$$h(L,\mu)=\int_0^\infty\frac{R^2}{(2\pi)^{3/2}\|T\|}\exp-\tfrac{1}{2}\left(R^2L'(TT')^{-1}L-2R\mu'T^{-1}L+\mu'\mu\right)\mathrm{d}R,$$

so

$$\frac{\partial h}{\partial\mu}=\int_0^\infty\frac{R^2}{(2\pi)^{3/2}\|T\|}\exp-\tfrac{1}{2}\left(R^2L'(TT')^{-1}L-2R\mu'T^{-1}L+\mu'\mu\right)$$
$$\times\left(RT^{-1}L-\mu\right)\mathrm{d}R.$$

Define

$$\kappa(L,\mu)=\int_0^\infty \frac{R^3}{(2\pi)^{3/2}\|T\|}\exp-\tfrac{1}{2}\left(R^2L'(TT')^{-1}L-2R\mu'T^{-1}L+\mu'\mu\right)\mathrm{d}R.$$

We see that the m.l. equation for μ is

$$\frac{1}{N}\sum_{j=1}^{N}\frac{\kappa(L_j,\mu)}{h(L_j,\mu)}T^{-1}L=\mu. \tag{18}$$

A natural initial value is $\mu^{(1)}=N^{-1}T^{-1}\Sigma L_j$ for the iteration to find $\hat{\mu}$. Then we will take the direction of $\hat{\mu}$. To get the accuracy of $\hat{\mu}$, we may use second derivatives in the usual way.

(c) To go one step further, if $Y=UZ$ where Z is Gaussian with mean vector μ and a known covariance matrix Σ and we wish to estimate the direction of μ given the direction of N Y's, we may write

$$\begin{aligned} Y=UZ&=U\Sigma^{+1/2}\Sigma^{-1/2}Z\\ &=TX, \end{aligned}$$

where

$$T=U\Sigma^{1/2},\qquad X=\Sigma^{-1/2}Z$$

and X is Gaussian $(\Sigma^{-1/2}\mu, I)$. Thus we may use the method of Subsection 2(b) to estimate $\Sigma^{-1/2}\mu$. Given an estimate of the latter, we have an estimate of μ, since $\Sigma^{1/2}$ is known, and hence of its direction.

Subsection 2(c) leads to an algorithm to solve the problem of Section 1 for a model of secular variation where the geocentric dipole has a multivariate Gaussian distribution. This is not a model included in (4) and (5); it is a limiting case in (6).

(d) To deal with model (6) we are led to study the distribution of $Y=X+v$ where X is Gaussian — mean μ, covariance Σ — and v is uniformly distributed over a sphere of radius δ. This leads to a plethora of apparently new multivariate densities.

Because of their intrinsic interest, we consider some *very special cases* not directly relevant to our main problem. In one dimension, the density of y clearly is

$$\begin{aligned} &\frac{1}{\sqrt{2\pi}}\tfrac{1}{2}\left(\exp-\tfrac{1}{2}(y-\mu-\delta)^2+\exp-\tfrac{1}{2}(y-\mu+\delta)^2\right)=\\ &=\frac{1}{\sqrt{2\pi}}\exp\left\{-\tfrac{1}{2}\left[(y-\mu)^2+\delta^2\right]\right\}\cosh\delta(y-\mu). \end{aligned}$$

In two dimensions the density of $y=(y_1,y_2)'$ is

$$\begin{aligned} &\frac{1}{(2\pi)^2}\int_0^{2\pi}\exp-\tfrac{1}{2}(y_1-\mu_1-\delta\cos\theta)^2-\tfrac{1}{2}(y_2-\mu_2-\delta\sin\theta)^2\,\mathrm{d}\theta\\ &=\frac{1}{2\pi}\exp-\tfrac{1}{2}\left\{(y_1-\mu_1)^2+(y_2-\mu_2)^2+\delta^2\right\}\\ &\quad\times\frac{1}{2\pi}\int_0^{2\pi}\exp\{\delta(y_1-\mu_1)\cos\theta+\delta(y_2-\mu_2)\sin\theta\}\,\mathrm{d}\theta\\ &=\frac{1}{2\pi}\exp\left\{-\tfrac{1}{2}\left\{(y_1-\mu_1)^2+(y_2-\mu_2)^2+\delta^2\right\}\right\}I_0(\delta|y-\mu|), \end{aligned}$$

where $I_0(z)$ is the imaginary Bessel function of zero order.

In three dimensions, a similar averaging of the Gaussian with mean μ and covariance matrix, I, leads to

$$\frac{1}{(2\pi)^{3/2}}\exp\{-\tfrac{1}{2}(y-\mu)'(y-\mu)-\tfrac{1}{2}\delta^2\}\frac{\sinh\delta|y-\mu|}{\delta|y-\mu|}.$$

In each case, as $\delta\to 0$ the modifying factor tends to unity. Let us now return to our real problem.

Model (6) requires a general form of the interesting new distributions just given. To put it in neutral notation, let $Y=X+d$ where d is uniformly distributed over a sphere of radius δ, X is Gaussian mean μ and covariance Σ and the density of Y at y is given by

$$\operatorname*{Ave}_{|d|=\delta}\frac{1}{(2\pi)^{3/2}|\Sigma|}\exp-\tfrac{1}{2}(y-d-\mu)'\Sigma^{-1}(y-d-\mu)=$$

$$=\frac{1}{(2\pi)^{3/2}}\frac{1}{|\Sigma|}\exp-\tfrac{1}{2}(y-\mu)\Sigma^{-1}(y-\mu)$$

$$\times\operatorname*{Ave}_{|d|=\delta}\exp\{d'\Sigma^{-1}(y-\mu)-\tfrac{1}{2}d'\Sigma^{-1}d\}. \tag{19}$$

If $\Sigma=H'DH$, where H is orthogonal and D diagonal, $z=\delta D^{-1}H(y-\mu)$ and $E^{-1}=\delta^2 D^{-1}$, the averaging term is

$$\operatorname*{Ave}_{|e|=1}\exp\{e'z-\tfrac{1}{2}e'E^{-1}e\}$$

which cannot be further evaluated. In fact, with $z=0$, it is the normalizing constant of the Bingham distribution [see, e.g. Mardia (1972)] and, when $z\neq 0$, of a generalization recently studied by Beran (1979). To get to the working form of the Cox model, we must find the density of the direction of the vector, y, from (19). Thus the Cox model (6) is hard to deal with mathematically, so estimation schemes based on it do not seem very practical. However, the Gaussian model without the random smaller dipole seems an adequate approximation since, typically, $|m|/|M|$ is about one-tenth.

(e) The model (4) raises interesting problems. Here M is fixed and m is uniformly distributed over a sphere of radius $|m|$, also fixed. Thus $Em=0$, $Emm'=|m|I/3$.

First, we consider several related academic problems similar to familiar textbook estimation examples. Suppose that $N\geqslant 2$ points y are uniformly distributed along a line segment in space, i.e. $y=M+vd$ where v is uniformly distributed on $(0,|m|)$ with M, $|m|$ and d unknown and to be estimated. If, when the points are arranged on the segment, they fall in the order $y_1, y_2,\ldots, y_N$, then $\pm d$ is known exactly, and we have $|y_2-y_1|,|y_3-y_2|,\ldots,|y_N-y_{N_1}|$ as the interior gaps when N points are randomly distributed on $(0,|m|)$. The sufficient statistic is (y_1, y_N). Simple estimators are (with $m=d|m|$) $y_1=M$, $y_N=M+m$ or $y_1=M-m$, $y_N=M$. These will be biased, as are the analogues on the real line. To give another interesting example, suppose we could observe N copies of $y=M+m$ where M and m are as in model (4). For $N\geqslant 3$, $|m|$ and M can be found exactly from the perpendicular bisectors of chords which must intersect at M. As is so often the case, things are simpler on the circle and sphere than on the line.

In practice, we can only observe the directions L of $U(M+m)$. Assuming that $M=|m|\mu/c$, $m=|m|d$, with $c=|m|/|M|$ known and approximately 0.1 and d uniformly distributed over the surface of the unit sphere, it suffices to consider the case of observing L where

$$RL=U(\mu+cd)$$

with known c. $M.l.$ here is very awkward. Since

$$\mu+cd_i=R_iU^{-1}L_i, \quad (i=1,\ldots,N),$$

one might use (since $Ed=0$) the estimator

$$\tilde{\mu}\propto\sum_1^N U^{-1}L_i. \tag{20}$$

A cone of confidence can be obtained from the approximate multivariate distribution of $\Sigma_1^N U^{-1}L_i$. To this end we note that

$$R^2=\mu'U^2\mu+2c\mu'U^2d+c^2d'U^2d,$$

so

$$ER^2=3(\mu'u)^2+1+2c^2,$$

where $2c^2$ may be ignored. If $\bar{\mu}$ is an approximate pole position, define

$$\bar{R}^2=3(\bar{\mu}'u)^2+1,$$

and write

$$\mu+cd=\bar{R}U^{-1}L.$$

Then

$$\mu\doteqdot\bar{R}EU^{-1}L$$

$$\bar{R}U^{-1}L-\mu=cd$$

so

$$E(\bar{R}U^{-1}L-\mu)(\bar{R}U^{-1}L-\mu)'=(c^2/3)I \tag{21}$$

which leads easily to the procedure.

(f) The model (5) of Creer et al. has $M=|M|d$ where $|M|$ is fixed and known, and d has a Fisher distribution about μ with accuracy κ. If the direction of $H=UM$ is known, i.e. the direction L of Ud is known, how to estimate μ? Since $R_iL_i=Ud_i$, $d_i=R_iU^{-1}L_i$, we may compute each $U^{-1}L_i$, reduce to unit length (so not knowing R does not matter) to obtain the d_i. Since these have a Fisher distribution, Fisher's estimator and method of getting a confidence case for μ may be used. This is, in fact, a standard method of estimating the position of the pole. It is important to observe that the Fisherian method is not applied here to the site means but their transforms. Note that it is related to but different from estimators (20),(18),(15),(13), which are very similar.

3. Estimation of pole positions

We have seen that a rational method of estimating pole positions should follow, by the application of maximum likelihood, from a statistical model for secular variation. In cases where the model makes strict m.l. estimation too difficult, one must try to improve and check the approximations and simplifications made. Since, however, there is unlikely to be any agreement on the model, the chosen method should not be sensitive to model changes within the range of disagreement, i.e., the method should be robust.

The multivariate normal method of Subsection 2(c) allows a wide variety of models to be tried. These will be tried out in an applied paper to see what model changes are most crucial. The Fisher estimation method has been shown to be robust against deviations from the Fisher distribution but the cone of confidence is not so reliable, e.g. Watson (1967).

A counsel of greater perfection is to make a more detailed statistical study of secular variation in the hope of finding a more convincing model. Certainly the newer, more complex models mentioned earlier should be explored. This will be attempted elsewhere. Further, we have here assumed that the site means are known, whereas they will only be estimated, often with different accuracies. This must be taken into account in the final practical method.

References

Beran, R.J.W. (1979). Exponential models for directional data. *Ann. of Statist.* **7**, (Nov.) 1162–1178.

Cox, A. (1970). Latitude dependence of the angular dispersion of the geomagnetic field. *Geophys. J. Roy. Astron. Soc.* **20**, 253–269.

Creer, K.M., Irving, E. and Nairn, A.E.M. (1959). Palaeomagnetism of the Great Whin Sill. *Geophys. J. Roy. Astron. Soc.* **2**, 306–323.

Fisher, R.A. (1953). Dispersion on a sphere. *Proc. R. Soc. London A* **217**, 295–305.

Harrison, C.G.A. and Watkins, N.D. (1959). Comparison of the offset dipole and zonal nondipole geomagnetic field models using Icelandic palaeomagnetic data. *J. Geophys. Res. B2*, **84**, 627–635.

Irving, E. and Ward, M.A. (1964). A statistical model of the geomagnetic field. *Pure and Applied Geophys.* **57**, 47–52.

Kendall, D.G. (1974). Pole-seeking Brownian motion and bird navigation. *J. R. Stat. Soc. Series B* **36**, 365–417.

Mardia, K.V. (1972). *Statistics of Directional Data*. Academic Press, London and New York.

McElhinny, M.W. (1973). *Palaeomagnetism and Plate Tectonics*. Cambridge University Press, London and New York.

Watson, G.S. (1956a). Analysis of dispersion on a sphere. Monthly Notices, *Roy. Astro. Soc. Geophys. Suppl.* **7**(4) 153–159.

Watson, G.S. (1956b). A test for randomness of directions. Monthly Notices, *Roy. Astro. Soc. Geophys. Suppl.* **7**(4) 160–161.

Watson, G.S. and Irving, E. (1957). Statistical methods in rock magnetism. Monthly Notices. *Roy. Astro. Soc. Geophys. Suppl.* **7**(6), 289–300.

Watson, G.S. (1967). Some problems in the statistics of directions. *Bull. of Int'l. Stat'l. Inst.* (*36th Session of I.S.I., Australia*)

G. Kallianpur, P.R. Krishnaiah, J.K. Ghosh, eds., *Statistics and Probability: Essays in Honor of C.R. Rao*
© North-Holland Publishing Company (1982) 713–722

PUBLICATIONS OF C.R. RAO

Books

1. *Linear Statistical Inference and Its Applications*

First Edition	1965	John Wiley, New York
Second Edition	1973	John Wiley, New York
Taiwan (English) Edition	1976	Taiwan
Interscience (English) Edition	1975	Wiley Eastern, India
Russian Translation	1969	USSR Academy of Sciences, Moscow
German Translation	1972	Springer Verlag
Japanese Translation	1977	Tokyo Inc., Tokyo
Czech Translation	1978	Academia Publishers, Prague
Polish Translation	1980	In press

2. *Advanced Statistical Methods in Biometric Research*

First Edition	1952	John Wiley, New York
Reprint	1971	Haffner, New York
Reprint	1974	Haffner, New York

3. (with S. K. Mitra) *Generalized Inverse of Matrices and Its Applications*

First Edition	1971	John Wiley, New York
Japanese Translation	1973	Tokyo

4. (with A. Kagan and Yu. V. Linnik) *Characterization Problems of Mathematical Statistics*

Russian Edition	1972	USSR Academy of Science, Moscow
English Translation	1973	John Wiley, New York

5. *Computers and the Future of Human Society*. Andhra University, Waltair and Statistical Publishing Society, Calcutta, 1970.

6. (with R. K. Mukherji and J. C. Trevor) *Ancient Inhabitants of Jebel Moya*, Cambridge University Press, 1955.

7. (with P. C. Mahalanobis and D. N. Majumdar) *Anthropometric Survey of the United Provinces, 1941. A Statistical Study. Sankhya* **9**, 90–324.

8. (with D. N. Majumdar) *Race Elements of Bengal, a Quantitative Study*. Asia Publishing House, 1958.

9. (with A. Matthai, S. K. Mitra and B. Ramamurthy) *Formulae and Tables for Statistical Work.*

First Edition	1966	Statistical Publishing Society, Calcutta
Second Edition	1975	Statistical Publishing Society, Calcutta

Research Papers

1941

a. (with K. R. Nair) Confounded designs for asymmetrical factorial experiments. *Science and Culture* **7**, 361.

1942

a. On the volume of a prismoid in N-space and some problems in continuous probability. *Math Student* **10**, 68–74.
b. On the sum of observations from different Gamma type populations. *Science and Culture* **7**, 614.
c. (with K. R. Nair) Confounded designs for $k \times p^m \times q^n \times \ldots$ type of factorial experiments. *Science and Culture* **7**, 361.
d. (with K. R. Nair) A general class of quasi-factorial designs leading to confounded designs for factorial experiments. *Science and Culture* **7**, 457.
e. (with K. R. Nair) A note on partially balanced incomplete block designs. *Science and Culture* **7**, 568.
f. (with K. R. Nair) Incomplete block designs involving several groups of varieties. *Science and Culture* **7**, 625.

1943

a. Certain experimental arrangements in quasi-latin squares. *Current Science* **12**, 322.
b. On bivariate correlation surfaces. *Science and Culture* **8**, 236.

1944

a. (with R. C. Bose and S. Chowla) On the integral order mod p of quadratics x^2+ax+b with applications to the construction of minimum functions for $GF(p^2)$ and to some number theory results. *Bull. Cal. Math. Soc.* **36**, 153–174.
b. On balancing parameters. *Science and Culture* **9**, 554.
c. On linear estimation and testing of hypotheses. *Current Science* **13**, 154.

1945

a. Familial correlations or the multivariate generalization of the intraclass correlation. *Current Science* **14**, 66.
b. Generalization of Markoff's theorem and tests of linear hypotheses. *Sankhya* **7**, 9–16.
c. Markoff's theorem with linear restrictions on parameters. *Sankhya* **7**, 16–19.
d. Information and accuracy attainable in the estimation of statistical parameters. *Bull. Cal. Math. Soc.* **37**, 81–91.
e. Finite geometries and certain derived results in theory of numbers. *Proc. Nat. Inst. Sc.* **11**, 136–149.
f. (with R. C. Bose and S. Chowla) On the roots of a well-known congruence. *Proc. Nat. Acad. Sc.* **14**, 193.
g. (with R. C. Bose and S. Chowla) Minimum functions in Galois fields. *Proc. Nat . Acad. Sc.* **14**, 191.
h. On a linear set up leading to intra and inter block informations. *Science and Culture* **10**, 259.

1946

a. Difference sets and combinatorial arrangements derivable from finite geometries. *Proc. Nat. Inst. Sc.* **12**, 123–135.
b. On the linear combination of observations and the general theory of least squares. *Sankhya* **7**, 237–256.

c. Confounded factorial designs in quasi-latin squares. *Sankhya* **7**, 295–304.
d. Hypercubes of strength '*d*' leading to confounded designs in factorial experiments. *Bull. Cal. Math. Soc.* **38**, 67–78.
e. On the mean conserving property. *Proc. Ind. Acad. Sc.* **23**, 165–173.
f. Tests with discriminant functions in multivariate analysis. *Sankhya* **7**, 407–414.
g. On the most efficient designs in weighing. *Sankhya* **7**, 440.
h. Minimum variance and the estimation of several parameters. *Proc. Camb. Phil. Soc.* **43**, 280–283.
i. (with S. J. Poti) On locally most powerful tests when alternatives are one sided. *Sankhya* **7**, 441.
j. Studentised tests of linear hypotheses. *Science and Culture* **11**, 202.

1947

a. The problem of classification and distance between two populations. *Nature* **159**, 30.
b. Note on a problem of Ragnar Frisch. *Econometrica* **15**, 245–249. A correction to Note on a problem of Ragnar Frisch. *Econometrica* **17**, 212.
c. General methods of analysis for incomplete block designs. *J. Am. Statist. Assoc.* **42**, 541–561.
d. Factorial experiments derivable from combinatorial arrangements of arrays. *J. Roy. Statist. Soc. B* **9**, 128–140.
e. Large sample tests of statistical hypotheses concerning several parameters with applications to problems of estimation. *Proc. Camb. Phil. Soc.* **44**, 50–57.
f. On the significance of additional information obtained by the inclusion of some extra variables in the discrimination of populations. *Current Science* **16**, 216–217.

1948

a. A statistical criterion to determine the group to which an individual belongs. *Nature* **160**, 835.
b. Tests of significance in multivariate analysis. *Biometrika* **35**, 58–79.
c. The utilization of multiple measurements in problems of biological classification. *J. Roy. Statist. Soc.* **B10**, 159–203.
d. Sufficient statistics and minimum variance estimates. *Proc. Camb. Phil. Soc.* **45**, 215–218.
e. (with D. C. Shaw) On a formula for the prediction of cranial capacity. *Biometrics* **4**, 247–253.
f. (with K. R. Nair) Confounding in asymmetrical factorial experiments. *J. Roy. Statist. Soc. B* **10**, 109–131.

1949

a. On a class of arrangements. *Edin. Math. Proc.* **8**, 119–125.
b. On some problems arising out of discrimination with multiple characters. *Sankhya* **9**, 343–364.
c. A note on unbiased and minimum variance estimates. *Cal. Statist. Bull.* **3**, 36.
d. (with P. Slater) Multivariate analysis applied to differences between neurotic groups. *British J. Psychology, Statist. Sec.* **2**, 17–29.

1950

a. Methods of scoring linkage data giving the simultaneous segregation of three factors. *Heredity* **4**, 37–59.
b. The theory of fractional replication in factorial experiments. *Sankhya* **10**, 84–86.
c. Statistical inference applied to classificatory problems. *Sankhya* **10**, 229–256.
d. A note on the distribution of $D^2_{p+q} - D^2_p$ and some computational aspects of D^2 statistic and discriminant function. *Sankhya* **10**, 257–268. (50)
e. Sequential tests of null hypotheses. *Sankhya* **10**, 361–370.

1951

a. A theorem in least squares. *Sankhya* **11**, 9–12.
b. Statistical inference applied to classificatory problems. Part II. Problems of selecting individuals for various duties in a specified ratio. *Sankhya* **11**, 107–116.
c. A simplified approach to factorial experiments and the punched card technique in the construction and analysis of designs. *Bull. Inst. Inter. Statist.* XXXIII(2), 1–28.
d. An asymptotic expansion of the distribution of Wilks' criterion. *Bull. Inst. Inter. Statist.* XXXIII(2), 177–180.
e. The applicability of large sample tests for moving average and auto regressive schemes to series of short length—an experimental study. Part 2: The discriminant function approach in the classification of time series. *Sankhya* **11**, 257–272.
f. Progress of statistics in India. *Progress of Science in India* (1939–1950), 68–94, National Institute of Sciences, India.

1952

a. Some theorems on minimum variance estimation. *Sankhya* **12**, 27–42.
b. Minimum variance estimation in distributions admitting ancillary statistics. *Sankhya* **12**, 53–56.
c. On statistics with uniformly minimum variance. *Science and Culture* **17**, 483.

1953

a. Discriminant function for genetic differentiation and selection. *Sankhya* **12**, 229–246.
b. On transformations useful in the distribution problems of least squares. *Sankhya* **12**, 339–346.

1954

a. A general theory of discrimination when the information about alternative population distributions is based on samples. *Ann. Math. Statist.* **25**, 651–670.
b. On the use and interpretation of distance functions in statistics. *Bull. Inst. Inter. Statist.* **34**, 90–100.
c. Estimation of relative potency from multiple response data. *Biometrics* **10**, 208–220.

1955

a. Analysis of dispersion for multiply classified data with unequal numbers in cells. *Sankhya* **15**, 253–280.
b. Estimation and tests of significance in factor analysis. *Psychometrika* **20**, 92–111.
c. Theory of the method of estimation by minimum chi-square. *Bull. Inst. Inter. Statist.* **35**, 25–32.
d. (with G. Kallianpur) On Fisher's lower bound to asymptotic variance of a consistent estimate. *Sankhya* **16**, 331–342.

1956

a. On the recovery of interblock information in varietal trials. *Sankhya* **17**, 105–114.
b. A general class of quasi-factorial and related designs. *Sankhya* **17**, 165–174.
c. Analysis of dispersion with missing observations. *J. Roy. Statist. Soc. B* **18**, 259–264.
d. (with I. M. Chakravarthy) Some small sample tests of significance for a Poisson distribution. *Biometrics* **12**, 264–282.

1957

a. Maximum likelihood estimation for multinomial distribution. *Sankhya* **18**, 139–148.

1958

a. Quantitative studies in Sociology, need for increased use in India. *Sociology, Social Research, and Social Problems in India.* Asia Publishing House, 1961, 53–74.
b. Some statistical methods for comparison of growth curves. *Biometrics* **14**, 1–17.
c. Maximum likelihood estimation for the multinomial distribution with an infinite number of cells. *Sankhya* **20**, 211–218.

1959

a. Some problems involving linear hypotheses in multivariate analysis. *Biometrika* **46**, 49–58.
b. Expected values of mean squares in the analysis of incomplete block experiments and some comments based on them. *Sankhya* **21**, 327–336.
c. Sur une Characterization de la Distribution Normal Etablie d'apres une Proprieté Optimum des Estimations Lineares. *Coll. Inter. due C.N.R.S.* France, LXXXVII, 165.
d. (with I. M. Chakravarthy) Tables for some small sample tests of significance for Poisson distributions and 2×3 contingency tables. *Sankhya* **21**, 315–326.

1960

a. Multivariate analysis: An indispensable statistical aid in applied research. *Sankhya* **22**, 317–338.
b. Experimental designs with restricted randomization. *Bull. Inst. Inter. Statist.* XXXVII, 397–404.

1961

a. A study of large sample test criteria through properties of efficient estimates. *Sankhya A* **23**, 25–40.
b. Asymptotic efficiency and limiting information. *Proc. Fourth Berkeley Symposium on Mathematical Statistics and Probability*, University of California, **1**, 531–546.
c. A study of BIB designs with replications 11 to 16. *Sankhya A* **23**, 117–127.
d. A combinatorial assignment problem. *Nature* **191**, 100.
e. Generation of random permutations of given number of elements using random sampling numbers. *Sankhya A* **23**, 305–307.
f. Combinatorial arrangements analogous to orthogonal arrays. *Sankhya A* **23**, 283–286.
g. Some observations on multivariate statistical methods in anthropological research. *Bull. Inst. Inter. Statist.* XXXVII(4), 99–109.

1962

a. Ronald Aylmer Fisher, F. R. S. (an obituary) *Science and Culture* **29**, 80–81.
b. Some observations on anthropometric surveys: *Anthropology Today. Essays in memory of D. N. Majumdar*, 135–149, Asia Publishing House.
c. Use of discriminant and allied functions in multivariate analysis. *Sankhya A* **24**, 149–154.
d. Efficient estimates and optimum inference procedures in large samples (with discussion). *J. Roy. Statist. Soc. B* **24**, 46–72.
e. Problems of selection with restriction. *J. Roy. Statist. Soc. B* **24**, 401–405.
f. A note on a generalized inverse of a matrix with applications to problems in mathematical statistics. *J. Roy. Statist. Soc. B* **24**, 152–158.
g. Apparent anomalies and irregularities in maximum likelihood estimation (with discussion), *Sankhya B* **24**, 73–102.

1963

a. Criteria of estimation in large samples. *Sankhya A* **25**, 189–206.
b. (with V. S. Varadarajan) Discrimination of Gaussian processes. *Sankhya A* **25**, 303–330.

1964

a. Problems of selection involving programming techniques. *Proceedings of the IBM Scientific Computing Symposium on Statistics*, 29–51. (100)
b. The use and interpretation of principal components analysis in applied research. *Sankhya A* **26**, 329–358.
c. Sir Ronald Fisher—The architect of multivariate analysis, *Biometrics* **20**, 286–300.
d. (with H. Rubin) On a characterization of the Poisson distribution. *Sankhya A* **26**, 295–298.

1965

a. Efficiency of an estimator and Fisher's lower-bound to asymptotic variance. *Bull. Inst. Inter. Statist.* XLI(1), 55–63.
b. The theory of least squares when the parameters are stochastic and its applications to the analysis of growth curves. *Biometrika* **52**, 447–458.
c. On discrete distributions arising out of methods of ascertainment. *Sankhya A* **27**, 311–324.
d. Covariance adjustment and related problems in multivariate analysis. (Dayton Symposium on Multivariate Analysis) *Multivariate Analysis* **1**, 87–103. Academic Press, Inc., New York, Ed. P. R. Krishnaiah.
e. (with A. M. Kagan and Yu. V. Linnik) On a characterization of the normal law based on a property of the sample average. *Sankhya A* **27**, 405–406.

1966

a. Characterization of the distribution of random variables in linear structural relations. *Sankhya A* **28**, 251–260.
b. Generalized inverse for matrices and its applications in mathematical statistics—*Festschrift for J. Neyman. Research Papers in Statistics*, 263–299.
c. Discriminant function between composite hypotheses and related problems. *Biometrika* **53**, 315–321.
d. Discrimination among groups and assigning new individuals. *The Role of Methodology of Classification in Psychiatry and Psychopathology*, 229–240, U.S. Department of Health, Education and Welfare, Public Health Service.
e. On some characterizations of the normal law. *Sankhya A* **29**, 1–14.
f. (with M. N. Rao) Linked cross sectional study for determining norms and growth curves: A pilot survey on Indian school-going boys. *Sankhya B* **28**, 231–252 (Partly published under the title, Methods for determining norms and growth rates, A study amongst Indian school-going boys.) *Gerentologia* **12**, 200–216.

1967

a. Calculus of generalized inverses of matrices: Part I—general theory. *Sankhya A* **29**, 317–350.
b. Cyclical generation of linear subspaces in finite geometries, 515–535, Chapter 28, *Combinatorial Mathematics and Its Applications*. The University of North Carolina Monograph Series No. 4.
c. Least squares theory using an estimated dispersion matrix and its application to measurement of signals. *Proc. Fifth Berkeley Symposium on Mathematical Statistics and Probability* **1**, 355–372, University of California Press.
d. On vector variables with a linear structure and a characterization of the multivariate normal distribution. *Bull. Inst. Inter. Statist.* XLII(2), 1207–1213.

1968

a. A note on a previous lemma in the theory of least squares and some further results. *Sankhya A* **30**, 259–266.

b. (with C. G. Khatri) Some characterizations of the gamma distribution. *Sankhya A* **30**, 157–166.
c. (with C. G. Khatri) Solutions to some functional equations and their applications to characterization of probability distributions. *Sankhya A* **30**, 167–180.
d. (with B. Ramachandran) Some results on characteristic functions and characterizations of the normal and generalized stable laws. *Sankhya A* **30**, 125–140.
e. (with S. K. Mitra) Simultaneous reduction of a pair of quadratic forms. *Sankhya A* **30**, 313–322.
f. (with S. K. Mitra) Some results in estimation and tests of linear hypotheses under the Gauss–Markoff model. *Sankhya A* **30**, 281–290.

1969

a. A decomposition theorem for vector variables with a linear structure. *Ann. Math. Statist.* **40**, 1845–1849.
b. Some characterizations of the multivariate normal distribution. *Multivariate Analysis* **2**, 322–328. Academic Press, Inc., New York.
c. Recent advances in discriminatory analysis. *J. Ind. Soc. Agri. Statist.* XXI, 3–15.
d. A multidisciplinary approach for teaching statistics and probability. *Sankhya B* **31**, 321–340.
e. (with S. K. Mitra) Conditions for optimality and validity of least squares theory. *Ann. Math. Statist.* **40**, 1716–1724.

1970

a. Estimation of heteroscedastic variances in a linear model. *J. Am. Statist. Assoc.* **65**, 161–172.
b. Computers: A great revolution in scientific research. *Proc. Indian National Science Academy* **36**, 123–139.
c. Inference on discriminant function coefficients. *Essays in Probability and Statistics*, Chapter 30, 587–602. Edited by R. C. Bose and others. University of North Carolina and Statistical Publishing Society.
d. (with B. Ramachandran) Solutions of functional equations arising in some regression problems and a characterization of the Cauchy law. *Sankhya A* **32**, 1–30.

1971

a. Characterization of probability laws through linear functions. *Sankhya A* **33**, 265–270.
b. Some aspects of statistical inference in problems of sampling from finite populations. *Foundations of Statistical Inference*, Holt, Rinehart, and Winston of Canada, 177–202.
c. Estimation of variance and covariance components—Minque Theory. *J. Multivariate Analysis* **1**, 257–275.
d. Minimum variance quadratic unbiased estimation of variance components. *J. Multivariate Analysis* **1**, 445–456.
e. Unified theory of linear estimation. *Sankhya A* **33**, 370–396 and, correction, *Sankhya A* **34**, 477.
f. Data, analysis, and statistical thinking. *Economic and Social Development, Essays in Honour of C. D. Deshmukh*, Vora and Co. Bombay, 383–392.
g. (with J. K. Ghosh) A note on some translation parameter families of densities for which the median is an m.l.e. *Sankhya A* **33**, 91–93.
h. (with S. K. Mitra) Further contributions to the theory of generalized inverse of matrices and its applications. *Sankhya A* **33**, 289–300.
i. (with S. K. Mitra) Generalized inverse of a matrix and applications. *Sixth Berkeley Symposium of Mathematical Statistics and Probability* **1**, 601–620, University of California Press.
j. Some comments on the logarithmic series distribution. *Statistical Ecology* **1**, 131–142. Pennsylvania State University Press. Ed. G. P. Patil.
k. Taxonomy in Anthropology. *Mathematics in Archaeological and Historical Sciences*, 19–29, Edin. Univ. Press.

1972

a. Recent trends of research work in multivariate analysis. *Biometrics* **28**, 3–22.
b. Estimation of variance and covariance components in linear models. *J. Am. Statist. Assoc.* **67**, 112–115.
c. A note on IPM method in the unified theory of linear estimation. *Sankhya A* **34**, 285–288.
d. Some recent results in linear estimation. *Sankhya B* **34**, 369–377.
e. (with P. Bhimasankaram and S. K. Mitra) Determination of a matrix by its subclasses of *g*-inverses. *Sankhya A* **24**, 5–8.
f. (with C. G. Khatri) Functional equations and characterization of probability laws through linear functions of random variables. *J. Multivariate Analysis* **2**, 162–173. (150)

1973

a. Unified theory of least squares. *Communications in Statistics*, **1**, 1–8.
b. Some combinatorial problems of arrays and applications to design of experiments. *A Survey of Combinatorial Theory*, North Holland, Chapter 29, 349–360.
c. Prasantha Chandra Mahalanobis. *Biographical Memoirs of the Fellows of the Royal Society*, Vol. 19, 485–492.
d. Representation of best linear unbiased estimators in the Gauss–Markoff model with a singular dispersion matrix. *J. Multivariate Analysis* **3**, 276–292.
e. (with S. K. Mitra) Theory and application of constrained inverse of matrices. *SIAM J. Appl. Math.* **24**, 473–488.
f. (with A. Kagan and Yu. V. Linnik) Extension of Darmois–Skitovic theorem to functions of random variables satisfying an addition theorem. *Communications in Statistics* **1**, 471–474.
g. (with D. C. Rao and R. Chakraborthy) The generalized Wright's model. *Genetic Structure of Populations*, 55–59. Ed. by N. Morton, University Press, Hawaii.
h. (with C. G. Khatri) Solutions to some functional equations and their application to characterization of probability distributions. Contributions to Statistics and Agricultural Sciences (Presented to Dr. V. G. Panse), 147–160.

1974

a. Functional equations and characterization of probability distributions. *Proc. Int. Cong. of Mathematicians*, **2**, 163–168.
b. Statistical analysis and prediction of growth. *Proc. 8th Int. Biom. Conference*, 15–21.
c. Projectors, generalized inverses and the BLUE's. *J. Roy. Statist. Soc.* **36**, 442–448.
d. Teaching of statistics at the secondary level—an interdisciplinary approach. *Statistics at the School Level*, 121–140. Almquist and Wiksell Int., Amsterdam.
e. (with S. K. Mitra) Projections under semi-norms and generalized inverse of matrices. *Linear Algebra and Appl.* **9**, 155–167.

1975

a. Simultaneous estimation of parameters in different linear models and applications to biometric problems. *Biometrics* **31**, 545–554.
b. Growing responsibilities of government statisticians. Occasional Papers No. 4, Asian Institute of Statistics, Tokyo.
c. Some thoughts on regression and prediction. *Sankhya C* **37**, 102–120.
d. Theory of estimation of parameters in the general Gauss–Markoff model. *A Survey of Statistical Design and Linear Models*, 475–487. North-Holland, Ed. J. N. Srivastava.
e. Inaugural Linnik Memorial Lecture—Some problems in the characterization of the multivariate normal distribution. *Statistical Distributions in Scientific Work* **3**, 1–13, D. Reidel Publishing Co., Ed. G. P. Patil.

f. (with M. L. Puri) Augmenting Wilk–Shapiro test for normality. Volume dedicated to A. Linder, 129–139.
g. (with S. K. Mitra) Extension of a duality theorem concerning *g*-inverse of matrices. *Sankhya A* **37**, 439–445.
h. Some problems of sample surveys. *Adv. Appl. Prob.* **7**, 50–61.

1976

a. Statistics, statisticians and public policy making. Silver Jubilee Publication. 25th Anniversary of the Central Statistical Organization, Souvenir Volume.
b. Estimation of parameters in a linear model—Wald Lecture 1. *Ann. Statist.* **4**, 1023–1037.
c. Characterization of prior distributions and solution to a compound decision problem. *Ann. Statist.* **4**, 823–835.
d. Prediction of future observations with special reference to linear models. *Multivariate Analysis*-IV. Ed: P. R. Krishnaiah, 193–208
e. On a unified theory of linear estimation in linear model—A review of recent results. *Perspectives in Probability*, 89–104 (Papers in honour of M. S. Bartlett) Ed. J. Gani, Academic Press.
f. (with C. G. Khatri) Characterization of multivariate normality through independence of some statistics. *J. Multivariate Analysis* **3**, 81–94.

1977

a. A natural example of weighted distributions—a classroom exercise. *American Statistician* **31**, 24–26.
b. Simultaneous estimation of parameters—a compound decision problem. Statistical Decision Theory and Related Topics. Eds. S. S. Gupta and D. S. Moore, 327–350.
c. Cluster analysis applied to a study of race mixture in human populations, Michigan University Symposium, 175–197.
d. Statistics for accelerating economic and social development. *Anniversary address*, 25th Anniversary of the Central Statistical Organization.
e. (with G. P. Patil) The weighted distributions—a survey of their applications. *Applications of Statistics.* Ed. P. R. Krishnaiah, North-Holland, 383–405.

1978

a. Least squares theory for possibly singular models. *The Canadian J. Statist.* **6**, 19–23.
b. (with G. P. Patil) Weighted distributions and size biased sampling with applicatons to wildlife populations and human families. *Biometrics* **34**, 179–189.
c. (with N. Shinozaki) Precision of individual estimators in simultaneous estimation of parameters. *Biometrika*, **65**, 23–30.
d. Choice of best linear estimators in the Gauss–Markoff model with a singular dispersion matrix. *Comm. Statist. Theor. Math.*, **A7(13)**, 1199–1208.

1979

a. (with H. Yanoi) General definition and decomposition of projectors and some applications to statistical problems. *J. Statist. Planning and Inference*, **3**, 1–17.
b. Estimation of parameters in the singular Gauss–Markoff model. *Comm. Statist. Theor. Math.*, **A8(14)**. 1353–1358.
c. A note on the unified theory of least squares. *Comm. Statist. Theor. Math.*, **A7(5)**, 409–411.
d. Minque theory and its relation to ML and MML estimation of variance components. *Sankhya B* **41**, 138–153.

e. (with J. Kleffe) Estimation of variance components. *Handbook of Statistics*, Vol. **1**, 1–40 (Ed. P. R. Krishnaiah). North-Holland.
f. Separation theorems for singular values of matrices and their applications in multivariate analysis. *J. Multivariate Analysis* **9**, 362–377.
g. Perspectives in statistics. *Sankhya B* **41**, 129–137.

1980

a. Matrix approximations and reduction of dimensionality in multivariate analysis. *Multivariate Analysis V*, 3–22. Ed. P. R. Krishnaiah.
b. (with R. C. Srivastava, Sheela Talwalkar & A. Edgar) Characterization of probability distributions based on a generalized Rao–Rubin condition. *Sankhya A* **42**, 161–169.
c. (with R. C. Srivastave) On a characterization of the poisson distribution based on a multinomial splitting model. *Sankhya* **41**, 124–128.
d. Discussion on a paper by Berkson. *Ann. Statist.* **8**, 482–485.
e. Some comments on the minimum mean square error as a criterion of estimation. Univ. of Pittsburgh. Tech. Rept. 80–21.
f. Diversity and dissimilarity coefficients: a unified approach. Univ. of Pittsburgh. Tech. Rept. 80–10 (1980). (Theoretical Population Biology, in press.)
g. (with J. Burbea) On the convexity of divergence measures based on entropy functions. Univ. of Pittsburgh. Tech Rept. 80-13 (1980). (I.E.E.E. Trans. Inf. Theory, in press.) (200)
h. (with J. Burbea) Entropy differential metric, distance and divergence measures in probability spaces. Univ. of Pittsburgh. Tech. Rept. 80–18 (1980).
i. (with C. G. Khatri) Some extensions of Kantorovich inequality and statistical applications. (*J. Multivariate Analysis*, in press).

1981

a. A lemma on *g*-inverse of a matrix and computation of correlation coefficient in the singular case. *Communications in Statistics A* **10**, 1–10.
b. (with C. G. Khatri) Some generalizations of Kantorovich inequality. (*Sankhya*, in press.)
c. (with Ka-Sing Lau) Integrated Cauchy functional equation and characterizations of the expotential law. (*Sankhya*, in press.)
d. (with Ka-Sing Lau) Solution to the integrated Cauchy functional equation on the whole line. University of Pittsburgh, *Tech. Rept.* 81–18. (*Sankhya*, in press.)
e. Gini-Simpson index of diversity: a characterization, generalization and applications. University of Pittsburgh, *Tech. Rept.* 81–22 (Utilitas Mathematica, in press.)
f. Analysis of diversity: a unified approach. University of Pittsburgh, *Tech. Rept.* 81–26.
g. (with J. Burbea) On the convexity of higher order Jensen differences based on entropy functions. University of Pittsburgh, *Tech. Rept.* 81–20.
h. (with B. K. Sinha and K. Subramanyam) Third order efficiency of the maximum likelihood estimator in the multinomial distribution. University of Pittsburgh, *Tech. Rept.* 81–21.
i. (with J. Burbea) Differential matrics in probability spaces. University of Pittsburgh, *Tech. Rept.* 91–29.
j. Optimization of functions of matrices with applications to statistical problems. University of Pittsburgh, *Tech. Rept.* 81–31.